Intelligent Transport Systems Standards

For a listing of recent titles in the *Artech House Intelligent Transportation Systems Series,* turn to the back of this book.

Intelligent Transport Systems Standards

Bob Williams

ARTECH HOUSE
BOSTON | LONDON
artechhouse.com

Library of Congress Cataloging-in-Publication Data
A catalog record for this book is available from the U.S. Library of Congress.

British Library Cataloguing in Publication Data
A catalogue record for this book is available from the British Library.

ISBN-13: 978-1-59693-438-2

Cover design by

685 Canton Street
Norwood, MA 02062

10 9 8 7 6 5 4 3 2 1

Contents

Preface

The development of this book arose from a general frustration of one who, despite being involved in Intelligent Transport Systems (ITS) and the development of standards for the sector since 1991, finds information about these standards scattered across the libraries of many different standards organizations.

Where does one find out if a standard even exists? Do we spend man years developing a standard only to find out that another group has already done the work, or even worse, done something similar but slightly different and incompatible?

Additionally, and most importantly, ITS does not live in a world of isolation. ITS exists in a world of information technology. But those developing ITS standards are not necessarily IT/ICT experts, nor RFID experts, nor biometrics experts; they are more likely to be automotive engineers, traffic control managers, electronics design companies, information service providers, and representatives of governments and user groups.

Yet these technologies are an essential part of ITS. It does not make sense to start from scratch every time an ITS standard is needed when there are many IT/ICT standards already developed that can provide much of what we need; we can simply specify by reference, if only we could find them! In any case, as much of ITS is about transferring data from system to system in order achieve the provision of the desired service, ITS has to use and feed information through and to other IT/ICT systems. It has to be compatible with these standards.

What IT, ICT, RFID, and biometrics standards are available for us to use by reference? Which standards committees develop these standards? Where do we look to find out? Which standards are available for free download and which ones require payment? Does the title of the standard, which seemed so appropriate, really specify what we need? How can we find a summary of the standard and its contents without buying the document only to find out it was not what we needed? And if we find a relevant standard, from which Web site can it be obtained?

Of course, standards development organizations (SDOs) codify their standards in a systematic way, but in general they are not well organized to service the lay visitor. If you do not know which SDO or which committee developed a standard, or its reference number, or its exact title, it may take hours to wade through often slow and cumbersome Web sites.

If you wonder what those in standards committees do apart from develop the published papers, the answer in part is that we moan about how these issues make our life unnecessarily difficult. In one of these "feeling sorry for ourselves" sessions, Chris Skinner and I came to the conclusion that someone ought to put all these things together in one book. It would save everyone so much time and help us work more efficiently. And so the concept of this book was conceived. Originally Chris and I were going to develop the book together, but his career path meant that he had to bow out at an early stage.

Although, in the end, as the author and collator of this work, it is not just my work. I thank the Secretariats of the principal SDOs (ISO, CEN, ETSI, SAE) and others for their help and encouragement, and for giving permission to quote directly from the abstracts, introductions, scopes, and table of contents of various standards. Most particularly, I thank the Secretariats of ISO/TC 204, CEN/TC 278 (Tyler Messa of TIA and Jelte Dijkstra of NEN), and the convenors of the subcommittees and working groups referred to in this work, for their patience, help, and support in my searches to obtain the details which I hope are faithfully summarized in this work.

But most importantly, I thank my partner, Isabelle, for her patience and tolerance in sharing my limited time at home with the development of this book. Without her help and tolerance, it would not have been possible.

To all of you who have assisted me develop this work, you have my deepest thanks. To you the reader, I hope you find this work, whether in print or soft copy, useful in making your work in the ITS sector easier and more efficient. Keep up the good work; it is vital if we are to achieve the reductions in death and injury on our roads, reduce emissions by making transport more efficient, and improve the traveling experience, that ITS can make possible.

PART I

Introduction to Intelligent Transport Systems

CHAPTER 1
Introduction

1.1 Objectives of the Book

Intelligent Transport Systems (ITS) is a name used to describe complex and very often state-of-the-art systems to improve the driving experience. These systems involve vehicles, drivers, passengers, road operators, and managers all interacting with each other and the environment, and linking with often complex backbone infrastructure systems. Reducing the number of deaths and injuries in road traffic accidents is a key objective of many ITS services, as is improving the efficiency of vehicles and traffic networks. Increasingly, these systems are also being used to lessen adverse effects of transport systems on the environment. These systems are by their nature, interactive. To operate successfully, data must be sent accurately and in a timely manner; and also, data must find the correct recipient and be understood by that recipient, who may be in a completely different system. Such systems can only interoperate successfully if they are designed to comply with international standards.

Although ITS is often seen in the context of road traffic, because travelers and freight share modes of transport, ITS also includes interface with rail, water, and air transportation systems.

At first it may appear simple to identify ITS standards because there is a dedicated ITS committee within the International Organization for Standardization (ISO). However, the situation is more complex than this. There are also regional standards to take into account, such as those of the European Telecommunication Standards Institute (ETSI) and the European Committee for Standardization (CEN), as well as national standards, such as those of the Association of Radio Industries and Businesses (ARIB) in Japan.

Standards, be they international, national, or regional, have to work within a framework of regulations. These are largely national regulations, such as those of the Federal Communications Commission (FCC) in the United States (U.S.) and the Australian Communications and Media Authority (ACMA) in Australia. Regulations are also the responsibility of European nations, but, in order to develop a common market, the European Conference of Postal and Telecommunications Administrations (CEPT) developed "Recommendations" for common use of frequency bands among European Union (EU) countries and other CEPT members who see the benefit of a common system. On an international level, the International Telecommunication Union (ITU) (through its ITU-Telecommunications and ITU-Radio divisions) similarly makes "Recommendations" to encourage global interoperability.

When we think of ITS services, our minds may focus on delivering a service to a vehicle and its occupants, but it must not be forgotten that many ITS services are based solely within the transport network, for example, to reduce congestion, and also that most services provided to vehicles and their occupants have to be supported via the infrastructure at some point in their path.

It is most important to remember that ITS systems do not work in isolation. They are frequently dependent on backbone infrastructure systems for telecommunications, infrastructure networks, and the Internet. Over the air, they frequently have to share spectrum with other users, or may use general wireless communications systems to provide or support ITS-specific applications. In order to be complete, this book also identifies many of these communications support standards.

Therefore, in addition to an awareness of direct standards developed for ITS, there is a whole range of generic radio and telecommunications recommendations and standards that can be, will be, and are already used to support the provision of ITS services. In addition to describing standards developed specifically to support ITS services, this book identifies many of the generic communications and service support related standards that may be used to enable, provide, or support ITS services.

While we may read much about potential ITS systems, many of the systems remain concepts, and neither the technologies nor the standards are yet fully developed. ITS is a developing sector, and standards are being developed as the technology evolves. Indeed, the technology can *only* evolve to its potential if the technical developments are undertaken cooperatively, often in precompetitive phases; and the development of standards is an intrinsic part of this process.

So where does this leave the transport planner, the transport operator, the student of the subject, or indeed standards developers themselves? He (or she) is presented with a plethora of organizations, standards, and "black holes."

This book has two primary objectives:

- To list, summarize, and categorize ITS standards (and other international standards on which ITS systems may rely) that currently exist, or are under development, at the time of this writing, or in subsequent updates;
- To provide direction and guidance about where to look in the future to find relevant standards for ITS systems.

If you are looking for an overall description of what can be achieved by ITS systems, or a detailed look at some of the specific ITS technologies, there are other books in the Artech House series that may better suit your purpose; in particular, for an understanding of the concepts and technologies behind ITS, I recommend the companion book from Artech House: *Intelligent Vehicle Technology and Trends*, by Richard Bishop. However, if you are familiar with the concepts of ITS and want to know where the relevant standards are being developed, what they encompass, and what work is still to be done, this book is for you.

As the author, I acknowledge that the source of much of the content describing ITS services and ITS communications has been provided by the CALM Forum, Ltd.; I further acknowledge CALM's copyright and their kind permission to freely make use of and reproduce any of their material in this book.

1.2 How to Use This Book

1.2.1 Structure of the Book

Intelligent Transport Systems Standards is structured to make it easy for you to quickly work your way to the standard(s) that you seek. It is designed to be a reference handbook, rather than a "must-read-cover-to-cover" monograph.

Each section of this book has a similar methodology. Each section starts with a layman's overview description, moving further into the details as the section evolves.

Thus, if you want a quick "manager's" appraisal, you will find it in the first part of each section, and you only have to work your way through the whole section if you require the fuller details of the scope of the standards and other relevant information as to what the standard(s) do or will encompass.

Where specific international standards—or work items to create an international standard—are cited, the following information is provided: information concerning the name, reference number, standards development organization (SDO), and stage of development (at the time of this writing), as well as a summary description of the scope and a summary of the subject content of the standard's requirements clauses.

Where national standards are cited, the following information is provided: the reference information, the SDO identification (note that Web site links to all SDOs referenced in this book are also provided in a specific section of this book), and a summary description (if available in English; however, some countries understandably do not provide English language versions of their standards, or do not make them easily available). We have cited those national standards that are referred to in international standards, or for which the SDO is particularly active in international ITS standardization activities. This book does not, however, purport to be a comprehensive compendium of all national ITS standards around the world (many of which are simply a national instantiation adopting an international standard).

The organization responsible for the development of the standard is always stated, and in the case of regional and international standards, the committee responsible for the development of these types of standards is also cited. Readers will also find the structure of these committees explained where they are ITS committees, and for more generic committees a summary of the scope of the committee. This is important for the book's usefulness to the reader in the future, as it provides guidance as to where to look for any new standards subsequent to the publication of the latest edition or volume of this book.

One of the problems of ITS standards is that different aspects are developed by different SDOs, and so, in order to completely cover a subject the standards from more than one SDO may need to be grouped together. Therefore, except for the tabular lists of standards in the appendixes, we have deliberately avoided structuring the book around the activities of each of the SDOs in turn. Frankly, that would give you little more than going to the Websites of the SDOs, and would still leave you in the dark as to where else you should be looking.

Rather, we have organized the book around *the technologies and the service domains of ITS that enable the provision of ITS services*, and *the provision of ITS services to stakeholders*. Other issues, such as procedures and development processes, national architectures, and so on, are separated from this focused approach so that they do not clutter these focused sections with details relating to processing requirements; these details are available in a separate section of this book for those interested readers.

The book is grouped into parts, and each part considers a different aspect.

Part I provides a general overview and context of the book and of ITS standardization, as well as a general overview and background about ITS services and their supporting technologies.

Part II considers the technical means to enable the provision of ITS services and the technology standards behind them. Some of these are standards designed especially for ITS systems, and some are generic standards that can be used to enable ITS service provision. This part of the book is focused on the technologies that enable the provision of ITS services, rather than the services themselves.

Part III is focused on the provision of ITS services to stakeholders. It is here that you will find standards related to specific ITS services, a description of the ongoing work developing such standards, as well as gaps where such work has still to be undertaken.

Part IV considers standards related to ITS architecture. ITS services do not exist in a vacuum, and tend to be part of a growingly complex interactive environment. Understanding, and in some cases standardizing, the way that systems interact, depends on understanding the overall systems architecture in which they fit today and will evolve tomorrow. However, the architecture is not suitable for traditional standards as it will vary both from country to country and will evolve and change over time. Fossilizing an architecture in a standard is not appropriate. However, while not "standards" in the traditional sense, this part of the book describes national and regional ITS architectures and provides pointers for further study.

Part V considers strategies to use standards in ITS and details standards that are available to support such strategies, their evaluation, and their implementation.

Part VI considers the process of standards development. It addresses the questions: How do you start off on a course to develop an ITS standard? Which committees are most appropriate for what type of standard? What are various committees' development and approval procedures? This part also considers national standards and details some of the most relevant ones.

Part VII provides some conclusions about ITS standards and their past, current, and future development.

Finally, there are annexes which contain numerical lists of standards for reference, a glossary of terms, a list of acronyms and abbreviations, and a bibliography.

We hope that you will find this format easy to use and functionally efficient, and that it will provide an easy reference guide to specific standards, yet provide easy to read summaries and explanations. Most of the explanations describe the scope and general information regarding the standard, however, where appropriate and copyright-free, additional information is provided for important generic aspects, particularly where they involve general schemes of identification and classification that are in widespread usage.

1.2.2 Structure of Standards References

All standards referred to in this book will be presented initially in the following format (referred to as the "standard title box"):

A B C D E F

Field A: SDO Acronym. For example: ISO, CEN, ITU, ETSI.

Field B: Status of Deliverable. The following acronyms shall apply to this field:

- IS = International Standard;
- FDIS = Final Draft International Standard;
- DIS = Draft International Standard;
- FCD = Final Committee Draft;
- TS = Technical Specification;
- PAS = Publicly Available Specification (not standardized);
- TR = Technical Report;
- ITR = Internal Technical Report;
- CD = Committee Draft;
- WD = Working Draft;
- NP = New Work Item;
- PWI = Preliminary Work Item;
- EN = European Norm (European Standard);
- PrENV = Provisional European Standard;
- Prf = Final Proof version.

Details of the exact meaning of these types of deliverables vary slightly according to the SDO, and the reader is encouraged to visit the respective SDO Web site for more precise descriptions of their types of deliverable. For example:

- www.iso.org;
- www.itu.org;
- www.etsi.org;
- www.cenorm.org;
- www.cenelec.org;
- www.sae.org;
- www.ieee.org.

Field C: Standard Reference Number. This is the reference number of the standard as allocated by the SDO.

Field D: Name of Standard. For most SDOs the title of the standard normally contains multiple (usually three) parts: Subject—Group—Name. For example: Intelligent transport systems—Reference model architecture(s) for the ITS sector—Part 1: ITS service domains, service groups and services.

Field E: Current Actions. This field identifies any current actions or observations. For example: Published, Awaiting Publication, Under FDIS Ballot.

Field F: Technical Committee. This field identifies the reference number of the committee within the SDO that has developed the deliverable.

Thus the final reference takes the following form:

ISO	FDIS	14813.1	Intelligent transport systems – Reference model architecture(s) for the ITS sector – Part 1: ITS service domains, service groups and services	Published	TC204

1.3 Obtaining Copies of Standards

This book provides summaries of ITS standards. Normally, this comprises the name of the SDO, the formal name and reference number of the standard, a description of the subject area, and objectives and scope of the standard. In some cases some additional information is provided (e.g., some public domain information about a coding scheme). In order to assure accuracy, I have used or précised the actual content of Abstracts, Introduction, Scope, and in some cases Table of Contents of standards as provided by the relevant SDO or made freely available on their Web sites. I acknowledge the SDOs to be the source and copyright holders of such material and am grateful for the permission to use this material in this book.

To get the precise content of a standard, and in particular, its requirements, you need to obtain the full Standard. Some SDOs (e.g., ETSI) provide them free of charge for download. Others (e.g., CEN, ISO) recover some of their costs by charging for the standards. Some SDOs charge for most standards, but some standards and technical reports are made available without charge, because they are strategic to some objective of the SDO, or largely used for information and training. Where the author has established that a standard is available for download, the link to the appropriate Web site has been provided. If the deliverable is downloadable free of charge (FOC), it is identified as such. Only links to official SDO download sites are provided, so anything you download from these sites is legal. However, please read the specific download conditions where provided by the SDO.

NOTE: The reference number contains a hotlink that will take you directly to the download site.

If you need to purchase a standard, it is best practice to contact your National Standards Body. Normally, members of the national standards body will be given discounts on the cost of the products. CEN/CENELEC standards are usually made available via the national member body organizations—where you can also buy ISO standards—but you can also get them from the ISO Web site.

If a standard or reference document is available free of charge, then a second box will appear following the standard title box; for example:

Downloadable FOC ISO/IEC 7498-1:1994
http://standards.iso.org/ittf/PubliclyAvailableStandards/s020269_ISO_IEC_7498-1_1994(E).zip

Click on the name next to "Downloadable FOC" and, you, the reader, are online, you will be automatically directed to the site where it can be downloaded. Otherwise, the link can be manually entered into a Web browser in order for the reader to be taken to the download site.

If the standard is not available free of charge, but is available to purchase online, the Web link to the Web site where it can be purchased is shown in the standard title box, below the name of the standard. Again, if you are online, you can click on the link and be directed automatically to the site where it can be purchased. Otherwise, the link can be manually entered into a Web browser. For example:

ISO	TR	14813-3	Transport Information and Control Systems- System Architecture- Example Elaboration	Published TR	ISO TC204

http://www.iso.org/iso/en/CatalogueDetailPage.CatalogueDetail?CSNUMBER=29534

The SDO Web sites where standards can be accessed or purchased are as shown in Table 1.1.

It is generally not too difficult to navigate these sites to find where and how to purchase standards, but we have made it easier by providing specific links in the references below. The exception to this is IEEE, which, at the time of this writing, was redesigning their standards sales site, and you need to open an online account with them to sign in and be able to buy standards.

1.4 Use of the Terms "Infrastructure" and "Roadside"

Throughout ITS standards, and therefore also throughout this book, the word "infrastructure" is taken in a very wide sense, normally to incorporate both the means of delivering services to the roadside (excluding the air interface to the vehicle) and the nonmedium-specific service provision.

Thus, the reader will see expressions such as, "Data transferred from the infrastructure via an ITS air interface." This should be taken to imply that there is a (usually nonmedium-specific) service provider who is connecting to the roadside (the point at which the infrastructure side of an ITS air interface is operating) to provide services to a vehicle, and possibly receive data from a vehicle across the ITS air interface.

The term "roadside" is used to describe the last node of the infrastructure, where the service provider uses the "roadside" node to transmit across the ITS air interface. In 5.9-GHz, 63-GHz, and infrared dedicated systems, the roadside will normally mean beacons physically located at or near the roadside. However, where the ITS air interface is a mobile wireless broadband, cellular, or satellite communications link, "roadside" simply means the last point at which the signal is transmitted from the "infrastructure" to the vehicle. This may physically be kilometers away

Table 1.1 SDO Web Sites

ISO	International Organization for Standardization	Usually for purchase	www.iso.org
CEN	European Committee for Standardization (Comité Européen de Normalisation)	Usually for purchase	www.cenorm.be
CENELEC	European Committee for Electrotechnical Standardization (Comité Européen de Normalisation Electrotechnique)	Usually for purchase	www.cenelec.org
ETSI	European Telecommunication Standards Institute	Free download	www.etsi.org
ITU	International Telecommunication Union	Usually for purchase but some available FOC	www.itu.int http://www.itu.int/publications/bookshop/how-to-buy.html http://www.itu.int/rec/T-REC-G/e http://www.itu.int/ITUR/index.asp?category=publications&link=rec-search&lang=en
IEEE	Institute of Electrical and Electronics Engineers	Usually for purchase	www.ieee.org
SAE	Society of Automotive Engineers	Usually for purchase	www.sae.org
ARIB	Association of Radio Industries and Businesses		www.arib.or.jp/english
TIA	Telecommunications Industry Association		www.tia.org
ATIS	Alliance for Telecommunications Industry Solutions		www.atis.org

from the physical roadside, or, in the case of satellite communications, more than a hundred kilometers up in space.

1.5 ITS Standards and Generic Standards Used for ITS

There is one final observation about the structure and level of detail in this book. We have already noted that the provision of ITS services is dependent not only on standards developed specifically for ITS, but also on the use of more generic standards that can enable, support, or otherwise assist the provision of ITS services and standards. We live in a world of converging technology interdependence, and reference to such generic standards is essential. There would be little point in ITS services reinventing the wheel where adequate communication systems already exist and have capacity to be used.

However, this book, while providing these essential links, is focused on ITS standards. Detail will be provided of the ITS-specific aspects of such generic standards where applicable, and a short summary will be provided. However, the level of detail concerning generic standards that can be used to enable, support, or assist ITS service provision will be less detailed than that for ITS specific standards. Many of these areas (e.g., interconnectivity, the Internet, networking standards, RFID, and smart card standards, and so forth) are subject areas in their own right. References to further reading in these areas will be provided, but such detail is not the content of this book.

1.6 Introduction to Intelligent Transport Systems

Intelligent Transport Systems is a name that we use to describe transport systems where vehicles interact with the environment, and with each other, to provide an enhanced driving experience, and where intelligent infrastructure improves the safety and capacity of road systems.

There is a joke in the ITS standards development sector: "we have given up trying to make drivers intelligent, so now we are making the vehicles and roads intelligent!" This is somewhat deprecating, but it captures the essence of what ITS is trying to achieve. ITS is about reducing the risks in transportation and improving the driving experience.

Intelligent Transport Systems do not have to be only about vehicles and roads. Air transport, marine transport, and rail transport systems can, and frequently are, increasingly "intelligent." Indeed, air and rail transport systems have used advanced system and electronics design as part of their operation and infrastructure for decades. Marine navigation systems for all but small vessels have for a long time also used electronics and radio for location finding, obstacle avoidance, and collision avoidance.

However, for reasons that we will attempt to explain below, the emphasis of what is generally known as ITS is focused on land transport systems and, in particular, roads—and railways and ferries where they interact with road systems.

But first, allow me to introduce you to Mr. Joe Soap. Mr. Soap is a car driver, bus driver, truck driver, and motorcyclist. For just over a century he has been able to jump into or onto a mechanical device with wheels and steerage, and propel himself around with little assistance from outside (other than someone conveniently laying tarmac, concrete, or gravel strips between and around cities) and with little external control (other than rules and regulations which are enforced occasionally by a passing police officer). This may satisfy Joe's sense of freedom, but is often not very efficient, and is sometimes downright dangerous. Once upon a time there were not many Joe Soaps around, but now there are many, and they are growing in number, and the chances of them "bumping in to each other" are increasing significantly.

ITS services are about enhancing the driving experience. This may indeed make it easier for Joe Soap to endure or enjoy his journey. It may help pass the time of passengers in his vehicle as well. It may provide Joe with some connectivity en-route, some infotainment services, some services to ease the task of driving. It may extract money from him to pay road tolls, and provide information about the next leg of his journey. However, the most important focus of ITS service provision is safety, and environmental aspects (minimizing pollution and minimizing emissions) are becoming increasingly important.

Two reports published in the 1990s, *Global Burden of Disease* (World Health Organization, World Bank, and Harvard University; 1996) and *World Health Report—Making a Difference* (World Health Organization; 1999), showed that in 1990 road accidents as a cause of death or disability were the ninth most significant cause of human death and injury (out of a total of more than 100 separately identified causes) and predicted that by 2020 will move up to sixth place in terms of years of life lost and "disability-adjusted life years."

The actual figures are even more shocking. Jones and Aeron-Thomas ("A Review of Global Road Accident Fatalities," TRL) estimate that global road deaths were between 750,000 and 880,000 for 1999. Later estimates have put this figure closer to 1.25 million deaths per year, and the toll is increasing. The study also estimates that global road injuries (of whatever severity) amounted to between 23 and 34 million road accident injuries per annum in the late 1990s. Later estimates (Commission for Global Road Safety, June 2006, "Make Roads Safe: A New Priority for Sustainable Development") have estimated 1.25 million deaths per year, and other estimates calculate over 30 million injuries per year in addition to the death rate.

The rapid development of emerging economies, and particularly Asia, means that despite the natural evolutional improvement of safety provisions in vehicles and transport systems, without some significant changes to the way that road transport is conducted, these dreadful figures are likely to increase significantly.

Even among developed countries with organized transport networks and well-disciplined drivers, the death and injury toll remains unacceptably high. Although there have been significant improvements in the last half century, the improvement curve has flattened off, so that the death rate in the 15 long-standing countries of the EU (excluding recent expansions) has remained at around 40,000 deaths per year, and over 1 million injuries per year, for several years. The figures for North America are similar.

These attrition rates are unacceptable to civilized society, and most leading nations have committed themselves to measures to attempt to halve the death and injury rates within a decade.

Traditional safety improvements, infrastructure improvements, and driver training will contribute, but will by no means achieve, this target. Intelligent Transport Systems are now seen as the only way that these goals can be achieved. Safety is therefore one of the principal driving forces behind the evolution, development, standardization, and implementation of ITS systems.

But what is an "Intelligent Transport System"?

Surprisingly, there is a shortage of good generally applicable definitions.

The Chairman of the Australian National Road Transport Commission explained in his keynote speech to ITS Australia National Conference in 1999: "It is possible to read a great quantity of the literature, and yet still not quite understand the plot. What is really going on in this great technological revolution, and why? Computers and lots of benefits, yes, but that's (frankly) . . . ho-hum. Why might *Intelligent Transport Systems* be more exciting than *Intelligent Outdoor Barbeques*?" He went on to pose the question, "What is the difference between an intelligent transport system and a washing machine with a Ph.D.?" "I don't know," was the answer, "but at least I'll be able to recognize the washing machine."

The U.S. Department of Transport (DOT) describes ITS as follows: "ITS improves transportation safety and mobility and enhances global connectivity by means of productivity improvements achieved through the integration of advanced communications technologies into the transportation infrastructure and in vehicles. Intelligent transportation systems encompass a broad range of wireless and wire line communications-based information and electronics technologies."

This says something about what ITS is seeking to achieve, but it is a less than adequate definition of ITS.

Doug Morgan, Department of Transport, Calgary, Alberta, provides a more down-to-earth definition: "Intelligent Transport Systems or ITS is the application of technology to better manage traffic and maximize the utilization of our existing transportation infrastructure. The technologies are potential tools that will help to manage congestion, improve emergency vehicle response, optimize the operational effectiveness of transit systems and provide travelers with real-time information."

This is somewhat better, but the important aspects of safety are not adequately covered.

ISO TC204—the International Standards Organization Committee for ITS Standards—defines ITS as "information, communication and control systems in the field of urban and rural surface transportation, including intermodal and multimodal aspects, traveler information, traffic management, public transport, commercial transport, emergency services and commercial services, generally referred to as 'Intelligent Transport Systems (ITS)'."

This is an improvement, but still not adequate. A good definition is important in order to gain the involvement and commitment of stakeholders. People (whether investors, governments, or citizens) do not often buy things which they don't understand or don't know what they do, and can't readily see the benefit in their function. So let us be specific about what we mean by ITS within this book.

Here is the Williams and Skinner definition of ITS, and this is the definition that we use within this book:

> Services to support travelers of all classes—drivers, passengers and pedestrians—and to assist road network management and performance by using systems for information, communication, and control in the field of urban and rural surface transportation, to provide improved safety and an enhanced traveling experience, including intermodal and multimodal aspects. Such services include accident prevention and mitigation, emergency services response and support, driver assistance, traveler information, traffic management, infotainment en-route, public transport, commercial transport and services, theft prevention and after theft recovery, and public safety and security.

Further description of ITS services is given at the beginning of Part III of this book, ITS Services to Stakeholders.

Why do we not usually include principal railway systems, marine systems, and aircraft systems in ITS? The answer is in part due to practicality, and in part institutional.

With respect to the practicality, railway, waterway, airway, and roadway systems are, to a great extent, mutually exclusive systems. Each has its domain of expertise and system solutions, very little of which may be of benefit to the other.

It is true that there are points of overlap. Traveler information systems are good examples—it is useful, for example, to get arrival and departure information in advance of arriving at an airport, bus station, or railway station. Other overlaps exist; for example, so-called "grade" or "level" crossings where roads cross railway lines. Light rail and tram systems that use roadways are another potential crossover.

By and large, however, trains travel on tracks, and these require different systems than those for boats which travel on water or cars that travel on roads, and planes that use airstrips and space management and (with the exception of Cambridge airport in Cambridgeshire, England) do not use or cross roads. Hopefully, with a few specialized exceptions, the road vehicle does not find its way into water, and trains can only operate on tracks.

Institutionally, the separation is historic. We have already noted that air, rail, and marine transport systems have developed technology-based "standards" over a considerable period of time. These have required the development of more formal standards. Within the main SDOs, particularly ISO and CEN, committees were created many decades ago to manage the standardization programs for air, sea, and rail, or the existence of mode-specific standards organizations were recognized.

ITS standardization is more recent—starting in CEN in 1991 and ISO in 1993. So when the remit for these committees was formed, it was, of course, a condition of their creation that they did not stray into areas already handled by other committees. Therefore, the role for these ITS committees was limited (in the case of ISO TC204 by the following text):

> The following aspects of intercity rail are included in the work of ISO/TC 204: intermodal movement of passengers and freight, information systems relating to passenger and freight rail transport, and the use of ITS technology at the intersection of roads and rails ("grade crossings" or "level crossings"). Other aspects of intercity rail are not included in the work of ISO/TC 204.
>
> ISO/TC 204's work does not include ITS systems which are completely self-contained in the vehicle and which do not interact with other vehicles or the infrastructure (responsibility of ISO/TC 22).
>
> ISO/TC 204 is responsible for the overall system and infrastructure aspects of ITS as well as the coordination of the overall ISO work program in this field including the schedule for standards development, taking into account the work of existing international standardization bodies.
>
> So-called light rail and tramway systems that use roads are within the scope of ISO TC204 (although it must be noted that, apart from public transport ticketing, at the time of writing, there is little activity in these areas).

The limitation to exclude systems that are completely self-contained in the vehicle is again institutional because a committee for "Road Vehicles" had existed for many decades. Clearly, however, while this book accepts the limitation of ITS as accepted by the SDOs, systems that are completely self-contained in the vehicle that can be considered ITS exist, and the relevant standards are included in this book.

1.7 Standardization for Intelligent Transport Systems

Intelligent Transport Systems can exist without interaction with an intelligent infrastructure or other vehicles. Systems such as *lane departure warning systems* and *adaptive cruise control* use technologies such as video pattern recognition and radar to provide assistance to the driver. Similarly, forward or backward obstacle warning systems assist the driver in parking and other low-speed maneuvers. In the case of lane departure warning systems, a computer monitors video images of white lines or roadside indicators and indicates either visually, verbally, or by vibration when the line is crossed. Adaptive cruise control systems monitor the distance from the vehicle in front and changes to that distance. If the vehicle in front slows, the speed of the following vehicle is reduced in order to maintain a safe distance. Backward and forward obstacle warning systems are becoming common as well, even in lower cost vehicles. These systems are clearly ITS, but while they sense obstacles and vehicles and may characterize them, there is no two-way communication with them.

Road tolling systems, by comparison, normally establish a two-way communication with the vehicle, both identifying its presence and establishing at the least its identity. They may well lodge information on the on-board unit, to be retrieved at a later stage of the journey. Road charging using satellite navigation systems and cellular telephony to conduct more complex bidirectional transactions.

These are all early examples of ITS systems, and they are in production models or in commercial use today.

However, the bulk of the safety and service provision examples of ITS are yet to be commercially implemented and are still in trial or research and development phase (at least outside of the military), and most of these services require wireless communications between vehicles and the infrastructure, between vehicles and other vehicles, and infrastructure to infrastructure, sometimes using wireless technology.

Whereas the early electronic tolling systems use a "point" communication, which can identify a vehicle at a specific and precise point, many ITS systems will require continuous (or quasi-continuous) communications with and between vehicles.

The classes of ITS services have been defined and are to be found in Part III of this book, but it is important at this point to consider the nature of ITS at a high level.

In order to perform most ITS services there are two elements:

- The means to communicate;
- The performance of the service.

ITS standards will fall into one of these two types.

ITS standards may therefore enable the means to communicate, or they may provide the performance of the service. There is a deliberate strategy to separate the provision of the communications link from the provision of the service. This may be achieved in several ways. For infrastructure–infrastructure communications, the communications link may be wired or wireless and will generally use an existing generic communications standard.

Links between the infrastructure and vehicles, or vehicle–vehicle links, will, of course, need to use wireless communications. These communications may use an existing generic wireless medium such as cellular telephony or mobile wireless broadband, or may use a wireless medium designed especially for the provision of ITS services.

Standards supporting the provision of specific ITS services will, of course, be ITS specific. However, a link to a vehicle designed for ITS may be used for the provision of non-ITS-specific services—for example, the provision of Internet to the vehicle.

These issues are discussed in greater detail in Chapter 2.

1.8 Benefits of ITS Standardization

Some ITS services can be provided without international standards. Adaptive cruise control, lane departure warning systems, parking assistance systems, forward and backward obstacle warning systems, and noninteractive route guidance are good examples of such systems, and these can be found in widespread use in vehicles for sale today. Such systems can be designed within an automotive manufacturer's research and development department and implemented without any standardization. If Citroen achieves obstacle warning by using infrared and Mercedes by radar or some other technology, it does not require a standard, and has little consequence other than one technology may be more efficient than the other. Wherever possible, it is generally the preference of automotive manufacturers to follow this path, both for reasons of product feature separation, but also because there is an antipathy to the length of time and cost of standards development. However, most vehicle manufacturers, and also regulators, do support the development of standards to monitor the comparative performance of such systems, in order that claims made are comparable.

Automotive manufacturers participate in international standards where they can see clear commercial advantage to do so (or disadvantage if they do not participate). Nearly 500 standards developed by ISO TC22 (Road Vehicles) attest to this willingness to participate.

While some ITS can be provided without standards, this is not the case for the majority of ITS services. A more careful look at the services in the first paragraph of this section will quickly identify that these services can be characterized as being of an insular type. That is, they do not need to interact with other vehicles or the infrastructure in any interactive way. An adaptive cruise control system needs to sense the presence and speed of the preceding vehicle, but it does not have to communicate with that vehicle. A lane departure warning system identifies the positioning of white lines and/or road pavement but does not require it to communicate; similarly with forward and backward obstacle sensing systems.

The bulk of ITS services do, however, require communications with other vehicles or the infrastructure to perform their services. Indeed it is the bidirectional communication that provides the new capabilities that will enable the safety and service promises of ITS to be fulfilled.

Clearly, a collision avoidance system where a Ford vehicle speaks a different language than a Mercedes is of limited use (unless you are traveling in a space full of only one marque). An ice alert or skid alert system that only works with one marque of cars may not be quite as useless, and may offer an apparent product advantage (e.g., buy a BMW and other BMWs in the area will alert you to skid hazards), but would only work in densely populated areas, and in reality does not offer much. A roadside beacon providing variable message sign data to the vehicle is not going to provide different "languages" to different marques. For the vast majority of ITS service provision, standards are not only desirable, they are an essential component if the ITS system is to function.

For ITS standards developers, the process of ITS standards development also presents some additional significant business benefits.

Traditionally, standards have been developed once a technology has been developed or a system operational environment is determined. All too often, in these situations there may be a winner and a loser, or all parties may have to redesign their product. In these circumstances it is not surprising that manufacturers and systems designers prefer a de facto standard where their technology dominates the market before a de jure standard can be developed by multiple companies acting together.

ITS presents a different situation. Many ITS services can *only* be defined in collaboration. This is a relatively new situation. ITS is not unique in this respect: the connected home, visible supply chain, and other examples which require connectivity are in a similar situation. However these situations have arisen as the opportunities presented by connectivity have increased. They represent a relatively new scenario.

This changes the business model with both a carrot and a stick. The biggest immediate effect on the business model is the stick. Manufacturers who do not participate will be disadvantaged and may take several years longer to bring their products to the market, with the consequent risks that this incurs, or may indeed be forced to participate in order to meet new legislative requirements for safety, in order to retain their market. The stick may be effective, but it is rarely popular.

The carrot is, in this case, however, also very significant. Because there is a requirement for the backbone of ITS systems to be developed cooperatively, largely in precompetitive stages, the standardization process, instead of being an additional cost burden, offers very significant financial benefit by sharing the costs of research and development, and by reducing the risk of failure. This clearly has a positive effect on the bottom line. Section 1.10 will discuss these issues further.

1.9 Principles of Enterprise Architecture

The general approach to organizing the collaboration of disparate standards and organizations that develop and use those standards is best managed through a process called Enterprise Architecture. Architecture continues to deal with the framework in which components and subsystems are made to work with each other, based on the achievement of high levels of interoperability. However, what is also included in Enterprise Architecture is the recognition and inclusion of the business processes on which the organizations depend for their continuing survival and success.

There are several widely used methodologies for application of Enterprise Architecture, including the widely used Zachmann framework, the IEEE Std 1471-2000 "IEEE Recommended Practice for Architectural Description of Software-Intensive Systems—Description," and the architecture frameworks described elsewhere in this book. All of them are valid and useful; however, there has been a trend towards the use of architecture described in the Unified Modeling Language (UML) due to its popularity in general ICT software development.

There are practical ways of translating or converting between differing notations, but in practice it is probably best in the future to use a tool-based approach that is based on UML and provides for the export of architectural models between differing tools using a standardized interchange format.

1.10 The Business Case for ITS Standards

What, then, is the true business case for ITS standards? The fundamental argument is greater safety and efficiency at lower cost.

There is the obvious advantage of being able to share work between practitioners and between organizations for comparison and verification, which could not be readily achieved without standardization even if only bilateral. Secondly, there is less likelihood of mistakes and ambiguities when the artifacts in question are standardized. Thirdly, there is greater opportunity for reuse of previous work when it is described or defined in a standardized manner. Finally, there is greater reassurance for such use if the same component has been used successfully before.

There is some overhead attached to creating standards, but if the resulting standards are used worldwide then the cost of development is shared among a very large user base. There is also an overhead in dealing with new requirements if many current users need to be consulted to ensure compatibility.

Nevertheless, the savings are enormous and the assurance of correctness is much greater, and the risks by using standard solutions are significantly reduced. This makes the business case compelling, but it is often difficult to provide a quantitative justification.

CHAPTER 2

Introduction to ITS Services and Supporting Technologies

2.1 What Are ITS Services?

So far in this book we have referred rather loosely to "ITS services" or "ITS service provision." In this chapter we will summarize current thinking about ITS services and the characterization of them in the ITS world and in ITS standards in particular.

When we address the fundamental problem of what are Intelligent Transport Systems, we need to consider first who the customers are and how we characterize the benefit to them.

First we have to classify what the benefit to any user is. An ITS system has no point if it provides no benefit. Although the benefits are varied, they can be typified as the receipt of a "service." Thus we can say that the world of ITS is made up of the provision and receipt of *services*.

If we next consider the "customers" (the beneficiaries of the service provided), we find that they are a broad group; they comprise a very large range of individuals and organizations that can be categorized into more specific groups. These groups may be car drivers, truck drivers, emergency services, public transport, police, road managers, and so on. Indeed, any user or manager of a transport system could potentially be a customer for one or more ITS services.

But, as there are potentially thousands of ITS services, how can we classify them into manageable groups?

We cannot use customer groups as a means of classification, because different client groups may use the same service(s). So we need to turn our attention, therefore, to the types of service provided.

In making these considerations, we soon find ourselves in the world of system architecture design. Part IV of this book deals with architecture in greater detail. At this stage it is sufficient to consider that an intelligent transport system will comprise services, some of which are interdependent and related, some of which are unrelated to others. However, we can classify these services into groups.

Now we come to the first international standard referenced in this book.

ISO	IS	14813-1	Intelligent transport systems- Reference model architecture(s) for the ITS sector-Part 1: ITS service domains, service groups and services	Published	TC204

http://www.iso.org/iso/en/CatalogueDetailPage.CatalogueDetail?CSNUMBER=43664

As you can see from the title box, this standard is available to purchase online, but it is not available free of charge.

This international standard addresses the issues of classifying ITS services into groups. This deliverable was first developed in the mid-1990s and was issued as an ISO Published Technical Report. In recent years it has been substantially revised to bring it up to date to include developments in the sector over the last decade. This revision has recently been approved as a full international standard. However, this will not be the end of the story. As the ITS sector evolves and develops, the standard will again be reviewed and revised at regular intervals.

ISO 14813-1:2006 provides a definition of the primary services and application areas that can be provided to ITS users. Those with a common purpose can be collected together in ITS service domains, and within these there can be a number of ITS service groups for particular parts of the domain. ISO 14813-1:2006 identifies 11 service domains, within which numerous groups are then defined. Within this framework, there are varying levels of detail related to definition of different services. These details differ from nation to nation, depending on whether the specific national architecture building blocks are based directly upon services or on groups of functions. Thus, the intent is to address groups of services and the respective domains within which they fit. As these domains and service groups evolve over time, it is intended that this international standard be revised to include them.

ISO 14813-1:2006 is applicable to the working groups of ISO/TC 204 and other TCs which are developing international standards for the ITS sector and associated sectors whose boundaries cross into the ITS sector (such as some aspects of urban light railways, intermodal freight, and fleet). It is designed to provide information and explanation to those developing ITS international standards and to those developing specifications, implementations, and deployments for ITS.

ISO 14813-1:2006 is advisory and informative. It is designed to assist the integration of services into a cohesive reference architecture, assist interoperability and with common data definition. Specifically, services defined within the service groups will be the basis for definition of use cases and the resultant reference architecture functionality, along with definition of applicable data within data dictionaries, as well as applicable communications and data exchange standards.

Since this international standard is at the core of how we consider and classify ITS—and all of the standards and groups of standards that comprise ITS as we know it now and as it will evolve in the future—we will consider it in greater detail than most.

"Intelligent transport systems—Reference model architecture(s) for ITS, Part 1—ITS service domains, service groups and services" groups the ITS sector into 11 service domains (general areas), and within each area, it identifies types of service provided to that domain.

Thus the previous version of this standard was a published Technical Report and has been referred to as "Fundamental Services." The new version, "ITS service domains, service groups and services," reflects the evolution of technology-oriented transportation practices and applications.

The introduction to the current version identifies that ITS is now also expected to address the following:

- Transport network operations and maintenance activities;
- Freight mobility and intermodal connectivity;
- Multimodal travel including both pretrip and on-trip information and journey planning;
- Variable road pricing strategies for freight and personal travel;
- Emergency and natural disaster-related response activities and coordination;
- National security needs related to transportation infrastructure.

The international standard recognizes that ITS activities will interface with more generalized activities and environments outside the transportation sector. For example, road pricing and revenue systems activities may interface with electronic commerce (e-commerce) activities, and may thus utilize standards and principles associated with the banking industry along with generally accepted accounting principles. The addressing of national security and coordination issues also requires addressing specific national standards related to civil defense, emergency communications, and other procedures. These interfaces, while largely outside the scope of TC204, are nevertheless critical external influences on the functionality of the various services supported by "ITS service domains, service groups and services."

Figure 1 of ISO 14813-1 (reproduced here as Figure 2.1) shows a hierarchy upon which the domains and services are grouped:

- Service domains (A, B, C, . . . , *n*) = defines the nature of activities provided;
- Service groups (*n*1, *n*2, . . . , *nx*) = more specific activities carried out in service domain, but does not define the actors;
- Services = further defines activity in terms of the actors involved (e.g., users, travel modes); also serves as basis for most elemental "use cases" (user view of architecture).

2.2 ITS Users

We must next consider who the users are of ITS services. ISO 14813-1 defines an ITS user as: "one who directly receives and can act on ITS data or control products. An ITS user is one who receives, directly or indirectly, or provides to, the transaction of an ITS service; these users of ITS services may be human, systems or environment monitoring."

At the end of the chain, the final user is the driver and/or other occupants of a vehicle, a pedestrian, or user of public transit, public transit operator, commercial vehicle operator, emergency assistance provider, or road operator.

Behind these end users are those that enable the transport to function. The road manager, control center, road maintenance provider, and so on. These too are users of ITS services. But at the same time, in many cases they are also providers of components of ITS to other ITS service providers.

And to complicate matters further, when used as a provider of probe data or enquiry response data, ad hoc network link, the "end user" may also be a provider of data components to ITS service providers.

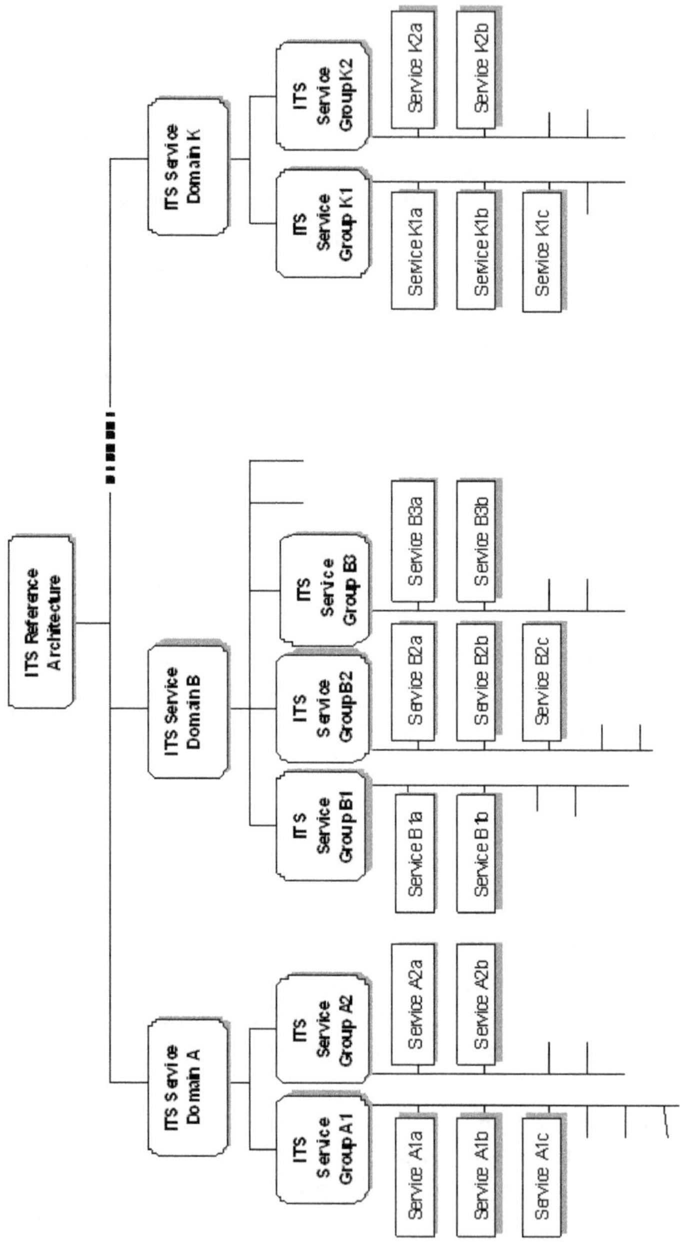

Figure 2.1 ITS services: hierarchy of definitions for "ITS reference architecture." (*Source:* CSI submission to ISO 14813-1.)

In general, we can broadly say that services are provided to groups known as: drivers, vehicle occupants, public transit users, public transit service providers, commercial vehicle operators, emergency services, pedestrians, road managers, regulators, and enforcers.

However, it is not possible to distribute ITS services by user group as different user groups will often use the same or similar ITS service. ISO 14813-1 therefore divides ITS services into service types.

2.3 Types of ITS Services

2.3.1 Introduction to ITS Service Types

ITS service domains and groups are built upon existing U.S., European Union, Japanese, and other international and national taxonomies (or classification systems) and provide a common descriptive basis for comparing these taxonomies, as well as others being developed throughout the world.

These architectures are based on national overviews of what the ITS sector comprises in their countries; and of course there are national differences. However, they are all developed from the perspective of national implementation and management and focus on the types of services that ITS can and will provide.

ISO 14813-1 identifies the following *service domains*:

- Traveler information;
- Traffic management and operations;
- Vehicle services;
- Freight transport;
- Public transport;
- Emergency;
- Transport-related electronic payment;
- Road transport related personal safety;
- Weather and environmental conditions monitoring;
- Disaster response management and coordination;
- National security.

The international standard makes it clear that categorization of the services into 11 domains does not imply that all ITS architectures should be required to follow this construction. The construction that they use should be that which is best suited to their ultimate use and should be independent of the services that they support. It should also be made clear that the standard is focused on ITS services, and not on supporting enabling technologies (e.g., media provision).

It also notes that services are often interdependent on, or providers to, other services within a service group or are key enablers for the provision of services in other service groups. It further observes that in architecture elaborations based on these services it is important that the proposed classification schema identify *who* is responsible for the provision of the service.

For each service domain, ISO 14813-1 goes on to elaborate the *service groups* within the domain and in some cases identifies specific *services* that comprise the groups. A summary description and identification of the service groups are provided in the following pages. Further detail can be obtained by reading ISO 14813-1.

2.3.2 Traveler Information

The ISO international standard 14813-1 describes the traveler information domain as, "Provision of both static and dynamic information about the transport network to users, including modal options and transfers."

The traveler information domain includes the following service groups:

- Pretrip information;
- On-trip information;
- Route guidance and navigation pretrip;
- Route guidance and navigation on-trip;
- Trip planning support;
- Travel services information.

2.3.3 Traffic Management and Operations

The ISO International Standard 14813-1 describes this domain as, "The management of the movement of vehicles, travelers and pedestrians throughout the road transport network."

The traffic management and operations domain includes the following service groups:

- Traffic management and control;
- Transport related incident management;
- Demand management;
- Transport infrastructure maintenance management;
- Policing/enforcing traffic regulations.

2.3.4 Vehicle Services

The ISO international standard 14813-1 describes this domain as, "Enhancement of safety, security and efficiency in vehicle operations, by warnings and assistances to users or control vehicle operations."

The vehicle services domain includes the following service groups:

- Transport-related vision enhancement;
- Automated vehicle operation;
- Collision avoidance;
- Safety readiness;
- Precrash restraint deployment.

2.3.5 Freight Transport and Logistics

The ISO international standard 14813-1 describes this domain as:

> The management of commercial vehicle operations; freight and fleet management; activities that expedite the authorization process for cargo at national and jurisdictional boundaries and expedite cross-modal transfers for authorized cargo.

The freight transport and logistics domain includes the following service groups:

- Administrative functions:
  - Commercial vehicle preclearance;
  - Commercial vehicle administrative processes;
  - Automated roadside safety inspection;
  - Commercial vehicle on-board safety monitoring;
- Commercial functions:
  - Freight transport fleet management;
  - Intermodal information management;
  - Management and control of intermodal centers;
  - Management of dangerous freight.

2.3.6 Public Transport

The ISO international standard 14813-1 describes this domain as, "Operation of public transport services and the provision of operational information to the operator and user, including multimodal aspects."

The public transport domain includes the following service groups:

- Public transport management;
- Demand responsive and shared transport.

2.3.7 Emergency

The ISO international standard 14813-1 describes this domain as, "Services delivered in response to incidents that are categorized as emergencies."

The emergency domain includes the following service groups:

- Transport-related emergency notification and personal security;
- After theft vehicle recovery;
- Emergency vehicle management;
- Emergency vehicle preemption;
- Emergency vehicle data;
- Hazardous materials and incident notification.

2.3.8 Transport-Related Electronic Payment

The ISO international standard 14813-1 describes this domain as, "Transactions and reservations for transport related services."

The transport-related electronic payment domain includes the following service groups:

- Transport-related electronic financial transactions;
- Integration of transport-related electronic payment services.

2.3.9 Road Transport-Related Personal Safety

The ISO international standard 14813-1 describes this domain as, "Protection of transport users including pedestrians and vulnerable users."

The road transport related personal safety domain includes the following service groups:

- Public travel security;
- Safety enhancements for vulnerable road users;
- Safety enhancements for disabled road users;
- Safety provisions for pedestrians using intelligent junctions and links.

2.3.10 Weather and Environmental Conditions Monitoring

The ISO international standard 14813-1 describes this domain as, "Activities that monitor and notify weather and environmental conditions."

The weather and environmental conditions monitoring domain includes the following service group:

- Environmental conditions monitoring.

2.3.11 Disaster Response Management and Coordination

The ISO international standard 14813-1 describes this domain as, "Road transport based activities in response to natural disasters, civil disturbances, or terror attacks."

The disaster response management and coordination domain includes the following service groups:

- Disaster data management;
- Disaster response management;
- Coordination with emergency agencies.

2.3.12 National Security

The ISO international standard 14813-1 describes this domain as, "Activities that directly protect or mitigate physical or operational harm to persons and facilities due to natural disasters, civil disturbances, or terror attacks."

The national security domain includes the following service groups:

- Monitoring and control of suspicious vehicles;
- Utility or pipeline monitoring.

2.3.13 ITS Data Management

The ISO international standard 14813-1 describes this domain as, "The collation, management, and supply of ITS data to legitimate interested parties."

The ITS data management domain includes the following service groups:

- Data registries;
- Data dictionaries;
- Emergency messages;
- Control center data;
- Enforcement;
- Traffic management data.

2.4 Other "Views" of ITS

2.4.1 Services to Drivers

In addition to the analysis of services into service domains, as analyzed in ISO 14813-1, there are also other views that should be considered in order to understand ITS. Each of these views considers an aspect of ITS that can be used by interested parties to group some of the services in different ways. These are complementary, not competitive means of analysis, and when considering standards that support ITS they can provide a very useful view.

We can see from the list of domains—and especially if we were to look more deeply into the services which comprise these domains—that many of the services are provided to drivers. Services to drivers can be categorized into five types:

- Driver/user information services;
- Driver assistance services;
- Collaborative driver assistance services;
- Collaborative driving services;
- Subconscious services to the driver.

An understanding of the generic characteristics of these services to drivers can further assist our understanding of some aspects of ITS services.

2.4.1.1 Driver/User Information Services

Driver/user information services provide relevant information to the driver/user. These may comprise, for example, satellite navigation information (excluding route guidance) and congestion and incident information.

The characterizing nature of this group of services is that they are passive or semi-passive with respect to driving or vehicle control—passive in that they provide general information, but no specific parameters are entered and no direct driving assistance is offered or suggested.

Often loosely incorporated into this group are indirect services made possible by an ITS link to the vehicle, such as in-vehicle Internet for passengers, and the ability of passengers to directly, or via the Internet, book restaurant and hotel reservations, for example.

2.4.1.2 Driver Assistance Services

The next group of services is those that provide direct driving support and assistance to drivers by proposing modification to driving behavior, but do not enact such behavior. These systems are further characterized in being stand-alone. That is, they do not require the communication or cooperation of other vehicles.

An example of this type of service is a lane departure warning system, where the driver receives an audible, visual, or sensory (usually vibration) warning when he is about to stray from the lane. Other examples include forward and backward obstacle warning systems, where the driver is alerted that he is getting close to an obstacle; and round blind corner assistance systems, which provide a CCTV image from the front of the vehicle. Route guidance, where the driver programs his destination and the system advises him of route directions, also falls into this group.

With these services, the information to the driver is driving specific, and advises the driver to modify driver behavior.

Many of the early instances of ITS, which are already appearing in production models, provide services of this nature as they are significantly easier to design and install and, with the exception of congestion sensitive route guidance systems, do not need a communication link to third parties outside of the vehicle.

2.4.1.3 Collaborative Driver Assistance Services

Collaborative driving services also provide driver assistance services, but require a communication link to other vehicles and/or the infrastructure to provide the service.

Early examples of these services were electronic road toll collection and vehicle access control systems. However, the characterization of these systems more typically requires information from others in order to provide the service.

Collision warning advice systems, where a vehicle collects location, movement, and danger information from other vehicles, is a more typical example of where this type of service is headed.

An example would be where a vehicle detects ice, or an otherwise slippery surface, and sends that information to other vehicles nearby, advising them of the danger. Once received, the driver receives advice of the distance location and nature of the warning. When he is approaching the dangerous area, he receives a second warning.

The primary characteristic of this type of service is that it can only be performed where there is quasi-continuous communication with other vehicles and/or the infrastructure.

2.4.1.4 Collaborative Driving Services

Collaborative driving services are of a similar communications nature to collaborative driver assistance services, except that these systems directly effect control, rather than advise, the driver.

Examples of these systems will be collision avoidance systems, grade (level crossing) collision avoidance systems, and platooning.

The nature of these services require that most, if not all, vehicles are equipped, and are therefore, at the time of writing this book, *future systems* that may not appear for another decade or more. However, in order for them to be possible, the communications architectures have to at least start to be implemented in vehicles in the near term.

2.4.1.5 Subconscious Services to the Driver

Subconscious services to the driver also require a communications link to the vehicle, but not to directly affect driver action, nor, usually, to advise the driver.

Automatic software updates to the vehicle's engine or system management software is one example of these types of services. They can be implemented on a much shorter timescale than collaborative driving services, or collaborative driver assistance services, but require a communications link to the vehicle.

2.5 Means of Achieving ITS Services

Another important view of ITS service provision is the means of achieving the ITS service. This is generally a communications-centric view. For a while in the national architectures and in ISO 14813-1, the perspective was to divide the sector into different service groups and domains; but when it comes to equipping vehicles with ITS interfaces and providing a network of beacons to communicate with them, the architectural separation of service types and characteristics becomes less important and, indeed, is of a low order of importance. Here we are putting one communication facility in a vehicle to talk to a limited number of roadside beacons. It is true that there may be multiple types of communication media links, but the emphasis is to be able to network different media, so that there is one facility in the vehicle, and one or more communications media through which it conducts its ITS communications. That one communications facility will carry whatever ITS communications services are required.

From this perspective we need to categorize by the type of communication; and a list of services can be provided using each type of communication.

NOTE: A single type of communication is not the same as a single communications medium, but more the nature of the communication (i.e., infrastructure to vehicle, vehicle to infrastructure, vehicle to vehicle, and infrastructure to infrastructure).

While ISO 14813-1 is focused on service provision in an architectural context of service domains and groups, in order to provide an ITS service there usually (but not always) needs to be one or more communications media in place.

For infrastructure-to-infrastructure ITS, such as many aspects of traffic control systems, a general purpose IT/ICT medium can be used. This may well be a wired medium, but increasingly, wireless media are being used within the infrastructure. However, these media are generic standardized media used for ITS, rather than specifically designed for ITS. Generic wireless media standards can also be used to provide infrastructure-to-infrastructure ITS services.

Of course, when considering vehicle-to-infrastructure, infrastructure-to-vehicle, or vehicle-to-vehicle ITS communications, the medium has to be wireless.

ITS services can in some circumstances be provided by generic standardized media, such as cellular telephony and mobile wireless broadband. However, the demands of a wireless system with moving transient participants mean that the same performance for bandwidth cannot be achieved (compared with performance of similar nonmobile links).

Where the service, therefore, becomes time critical (as, for example, with many safety related driver assistance services), it becomes desirable—even essential—to use media that are specifically designed and optimized for the provision of ITS services. Where safety related vehicle behavior modification services are involved, it becomes absolutely necessary to use dedicated, and preferably protected, media specifically designed for ITS.

There are other aspects of ITS service provision that can, in some cases, be generic standardized techniques used for ITS, generic standardized techniques adapted for ITS, or techniques specifically developed for ITS. Examples include: (1) data definition techniques, such as Abstract Syntax Notation One (ASN.1)—a data notation technique that can be used for ITS data definition, and in some cases encoding and transfer; (2) XML, Web services, IPv6 (NEMO)—generic standardized Internet services, where options can be optimized for ITS use and interoperability; and (3) specific services such as generic fare and toll transactions, which are often best specifically designed for ITS, but can be interoperable within the ITS sector, and in multiple services in the sector.

2.6 ITS Services for Vehicles and Their Occupants

The previous sections of this chapter have considered the service domains and groups (i.e., the group headings to describe functional areas of service provision). In considering the standards required to support ITS communications, it is worthwhile describing ITS services for vehicles and their occupants. The priority, and focus of work, as explained above, is to define the communications media standards that support services to vehicles and their occupants. However, in order to achieve this objective successfully, it is necessary to understand the requirements of these services.

Some services, particularly in-vehicle services, are already defined, implemented, and currently operating. Others are in advanced design, others are in early research and development stages, and some others are still in conceptual design and architecture definition.

Without comparative testing, it is not yet clear which mode of service provision is best for which service. Some services are best provided strictly in-vehicle, without the need for ITS communications with the infrastructure or other vehicles; other services are best provided via an installed infrastructure; and other services are best performed between vehicles communicating directly with each other or by ad hoc mesh networks that evolve between traffic using a particular stretch of road at a particular time.

Each mode has its proponents. Each mode is in a different stage of development. Then there is the complexity of the choice of which air interface medium is most appropriate to consider as well.

Each method has its advantages and disadvantages. Some of the disadvantages are already known to be (or trials and deployment will subsequently prove them to be) fundamentally unsuitable for the provision of a particular service. Actual deployment will depend on a mix of technical suitability, local regulations, and market forces.

This chapter provides a list of actual and envisaged services. It does not comprise a complete list, and the list evolves over time. Some of the services described in this chapter are already implemented and available, or at least have been considered conceptually and architecture options evaluated. Most are in some form of test or development. The services are briefly described, and it should be noted that these definitions are not necessarily the final service definitions, and rather are example summary definitions.

Many of the definitions have been obtained from *The CALM Handbook*, and from several ITS architecture descriptions around the world. As such, the short descriptions in this book are usually a representative amalgam of slightly varying definitions in different architectures, rather than direct quotes from any one. However, the sources for these amalgamated descriptions are acknowledged in the bibliography at the end of this book. In-vehicle ITS service definitions have been collated from standards and standards working papers and from automotive manufacturer descriptions of current and future products.

The services are briefly summarized. There will be variations from implementation to implementation, and will vary over time. These descriptions should be viewed as general descriptions and not precise specifications.

2.6.1 In-Vehicle ITS Services

Within the vehicle, and not relating to anything outside the sphere of the vehicle, it is possible to provide services that may be considered a part of ITS. Perhaps the most well known and most widespread of these services are airbags, which inflate immediately prior to collision. A lesser known, but becoming widespread application are tire pressure sensors. These are radio frequency identification (RFID) devices, embedded in the tires or wheels, which report the tire pressures to the vehicle management system; this system can alert the driver in case of abnormal tire pressure, or lack thereof. First generation forward and rear collision systems, first generation adaptive cruise control systems, and first generation lane departure warning systems are other examples of in-vehicle ITS services. The provision of an in-vehicle video screen and the human-machine interface to react with it also provides intelligent driver assistance and is clearly an in-vehicle service.

Drawing the line between what is an in-vehicle system and what is not an in-vehicle system is a fine line, and in the standards world this can be an important issue as to whether a standard is developed within ISO/TC 22 (Road Vehicles) or ISO/TC 204 (Intelligent Transport Systems). Clearly, the example of a video screen system to support the driver is an in-vehicle application, as are tire pressure sensors. In the latter case there is a wireless communication, but its functionality lies entirely within the vehicle. However, an in-vehicle video screen system may well be used to convey data from an ITS system, such as satellite navigation, collision warnings, and variable message sign data. The two principal technical committees (ISO/TC 204 and ISO/TC 22) have a formal liaison in place to deal with these issues of overlapping services.

Systems that use sensors to the environment around them lie on the dividing line between TC22 and TC204. In general, where there is sensing of the outside world, but no communication with it (for example, first generation adaptive cruise control), it is considered within the purview of an in-vehicle system and standards, where appropriate, and is (or will be) developed by ISO/TC 22. Where there is a two-way communication with another vehicle, person, or infrastructure, it is generally considered within the purview of ISO/TC 204. This delineation is, of course, important to understand if you are looking for the most likely source of a standard.

Within this book, this type of service will be found within Section 8.4.

One important characteristic of this type of service in relation to international standardization is that there is often little need for international standards as the application is entirely in-vehicle, and therefore can be adequately provided by the vehicle manufacturer. Indeed, any intellectual property may be protected to give commercial advantage over competitor automotive manufacturers. We discussed some of these issues in Section 1.5. Thus, if Citroen develops one system for forward or reverse obstacle warning, and BMW uses a different technology to provide a similar service, they can argue about the merits of their systems in the market place, not in a standardization committee, because standards are not required unless some means of assessing performance of the service, or minimum service standards requirements, become desirable or required by law.

However, when it comes to issues that, while in-vehicle, involve consumable or third party equipment that may be supplied to multiple vehicle manufacturers, then standards are required. A good example of this is for tire pressure sensor systems. Tires are manufactured by different companies than vehicles, and one model of tire is available to multiple vehicle manufacturers. Tires will be replaced several times during the lifetime of the vehicle, and in a competitive world may be replaced by tires from a different manufacturer. Standards for tire sensors have therefore been developed. Further, the exact means of achieving airbag or seatbelt performance does not need to be standardized. But performance and conformance tests to measure that a system achieves minimum requirements required by regulations are required, and so standards have been developed.

2.6.1.1 Adaptive Cruise Control[1]

Adaptive cruise control is similar to conventional cruise control in that it maintains the vehicle's preset speed. Unlike conventional cruise control, however, this newer system can automatically adjust speed in order to maintain a proper distance

1. The author acknowledges that the source of much of the material in this section is from The CACM Forum, Ltd., and thanks them for their kind permission to reproduce this material.

between vehicles in the same lane. This is achieved through a radar headway sensor, a digital signal processor, and a longitudinal controller. If the lead vehicle slows down, or if another object is detected, the system sends a signal to the engine or braking system to decelerate. Then, when the road is clear, the system will reaccelerate the vehicle back to the set speed.

2.6.1.2 Antilock Brake System

An antilock brake system (ABS) prevents the wheels from locking while braking, giving the driver more control and shortening braking distance.

2.6.1.3 Active Head Restraint

When vehicle sensors detect an imminent front/rear collision, the seat head restraints are designed to move up and forward to catch the occupant's head before the violent whiplash movement can start.

2.6.1.4 Adaptive Drivetrain Management

The development of electronic vehicle systems already enables the extension of the driver's visual horizon to an electronic horizon. By anticipating certain driving situations, an on-board system can recommend handling strategies to the driver. By means of the grouping, complex analysis, and selection of all information, specific influences can be exerted on drivetrain management and driving strategies aimed at reducing fuel consumption can be derived. New driving strategies can be developed to utilize such potential. The results calculated from simulations indicate substantial cuts in fuel consumption.

2.6.1.5 Airbag Control

Airbags inflate at the time of a crash to protect the occupants. The timing of airbag inflation is decisive to achieve maximum protection. The airbag must be opened in the right millisecond. If it opens too late, occupants could be injured. If it opens too early, they are not protected adequately, since the airbag no longer has its ideal form upon impact.

Airbag control systems process information from a multiplicity of sensors in the car to adjust the activated protection measures precisely, depending on the analyzed situation. This ensures that the appropriate protective measures are taken at precisely the right moment. The airbag control device also instantly triggers additional measures like unlocking doors or switching off the fuel pump.

2.6.1.6 Airbag Control—Intelligent

Airbag control can utilize additional sensors and information processing. Accident scenarios are defined by type and heaviness of impacts combined with special driver and passenger situations. Variable gas output accommodates the requirements for specific crash situations, occupancy and the driver/passenger position. This enables airbags to inflate moderately or aggressively depending on the crash evaluation. In-seat weight sensors can identify if the occupant is a large person or a small persons (such as child), or an empty seat, and inflate accordingly. Out-of-position sensors identify a passenger's or driver's position (usually by CCD and/or CMOS cameras). The electronic control unit analyzes the crash situation, evaluating the sensor signals, and performs sophisticated risk management to initiate appropriate gas output of the two stages, igniting only one stage or both in variable time intervals.

2.6.1.7 Automatic Light/Headlamp

Sensors detect available light in the vehicle's surroundings and automatically switch on or off headlamps and tail lamps.

2.6.1.8 Adaptive Headlight Aiming

Adaptive headlight aiming provides information to vehicles and adjusts headlamp settings to local conditions. Utilizing sensors that read such factors as the speed of the vehicle, steering, GPS information, and weather, headlight manufacturers are developing automatic systems that can deliver the best nighttime vision for any given road scenario. These intelligent lighting technologies (adaptive headlight aiming, or adaptive front lighting systems, and similar) include side lights that are activated as needed as well as front headlights that are smarter than their predecessors. Each headlight can be individually aimed so that light can be trained on the driver's lane and not into the oncoming lane.

In low speed situations, like residential driving, the system lowers the headlamp aim and increases the horizontal spread of the light to illuminate sidewalks, intersections, pedestrians and curbs. This feature is automatically activated at certain vehicle speeds (usually below 37 mph). The lighting system reveals pedestrians or bicyclists far earlier than standard lighting systems.

On the highway, sensors react to the speedometer and, as speed increases, raise the headlamp beam to illuminate a longer view of the road ahead and narrow the beam so it does not intrude into the path of oncoming traffic. Sensors automatically activate this feature at speeds above 50 mph. These systems can also move the headlamp beam to follow a curve, relying on sensors in the steering wheel and aiming the light in the direction that the vehicle is traveling. This feature also keeps the light in the driver's lane and prevents blinding oncoming motorists.

In a left-hand turn, the left headlight will pivot up to 15 degrees (the right-hand headlight remains pointed straight ahead), and in a right-hand turn, the right headlight will pivot up to 5 degrees (the left-hand headlight remains pointed straight ahead). In bad weather such as fog or snow, the system lowers the headlamp aim to improve side lighting and narrows the forward beam to a sharp pencil beam decreasing the glare. This feature can be activated with a manual control as well as through vehicle sensors.

Early instantiations are in-vehicle devices controlled solely by in-vehicle sensors and are already in the market place.

2.6.1.9 Backwards Obstacle Warning

These systems detect obstacles (by ultrasound, radar, or infrared) when the vehicle is moving backward and sounds alarm when an obstacle is sensed.

2.6.1.10 Blind Spot Information

These systems are designed to detect vehicles, pedestrians, or obstacles in a driver's blind spot by using cameras or sensing devices that are fitted to monitor blind spots (such as by the side-mirrors, at right angles to the front of the vehicle, or at a low level at the back of the vehicle). An indicator lamp is illuminated, or the driver's video display is activated and/or an audible warning is provided when activity is detected in the blind spot and recognized as a potential problem. The system is aware of the vehicle direction and speed so that it can make appropriate judgments as to what cameras to activate and warnings/images to supply. For example, when the car is put into reverse gear, the rearwards camera is activated; when the vehicle starts forward from stop, the sideways looking cameras at the front of the vehicles are activated and shown split screen for a few seconds.

2.6.1.11 Collision Warning

Collision warning systems assist the driver in evaluating surrounding traffic by warning of the unreasonable approach of an adjacent vehicle. In-vehicle systems achieve this by the use of distance sensing radars and a video screen and audio alert when a vehicle is approaching too rapidly, indicating the direction of the intrusion, and possibly the time to impact and speed of oncoming vehicle. The objective is to predict threatening collisions early enough to prevent them from occurring, and by warning the driver to take evasive action. When collision avoidance is not possible, the damages can be limited to a minimum by reducing the speed and stopping distance.

2.6.1.12 Driver Alert

This is a system that sounds a warning signal if the driving pattern changes erratically within the lane. The system alerts the driver, for example, if he becomes drowsy and starts to veer into another lane. This results in better safety margins and greater chances of avoiding accidents caused by fatigue.

2.6.1.13 Emergency Lane Assist

This system monitors oncoming vehicles in addition to lane keeping aid assistance, using both camera and radar. Should the driver not react to a warning buzzer, for example, the system adds steering force to help turn the car back into the original lane.

2.6.1.14 Electronic Stability Program (ESP)

This is an electronic device that corrects the path of a vehicle. The main function of this ECU is to detect loss of adherence of the vehicle and to influence its path by applying braking to one or more wheels. A set of sensors installed on the wheels of the vehicle will detect any loss of adherence. The ESP ECU will then determine the correction required to maintain the vehicle's path according to the position of the steering wheel. ESP is particularly effective in situations of oversteering and understeering. By affecting the vehicle's behavior, ESP ensures adherence and puts the vehicle back on the right path automatically. ESP technology helps drivers in case of loss of adherence through curves.

2.6.1.15 Forward Collision Warning

Forward collision warning systems (FCWS) are in-vehicle electronic systems that monitor the roadway in front of the vehicle and warn the driver when a potential collision risk exists. For example, currently available radar-based systems use algorithms to interpret transmitted and received radar signals with the FCWS to determine distance, azimuth, and relative speed between the host vehicle and the vehicle or object ahead of it in the lane. When the host vehicle is traveling along the roadway, FCWS can warn the driver when a vehicle or object is in its lane within a predefined closing time threshold. Currently, FCWS do not take any automatic action to avoid a collision or to control the vehicle; therefore, drivers remain responsible for the safe operation of their vehicles using both steering and braking to avoid a crash.

2.6.1.16 Forward Obstacle Warning

These systems detect obstacles (by ultrasound, radar, or infrared) when the vehicle is moving forward and sounds an alarm when an obstacle is sensed.

2.6.1.17 Full Automatic Brake Power

This is a second generation collision warning system designed to be activated at a very early stage, and if the driver does not hit the brakes when needed, the system brakes automatically to slow the vehicle down. By avoiding rear-end collisions, people in the car that would otherwise have been hit avoid suffering material damage or personal injuries such as whip-lash.

2.6.1.18 Intelligent Driver Information System

Any activity that the system interprets as overtaking or braking will delay the provision of other information or services that are not critical to safety (such as incoming phone calls and text messages) until the situation is less stressful. The idea is to help the driver concentrate on the right things, for more relaxed and controlled driving.

2.6.1.19 Lane Departure Warning System

Lane departure warning system (LDWS) detects unintentional lane departures on motorways and dual carriageways (usually at speeds of 80 km/h and above).

The system is triggered automatically when the driver mistakenly allows the vehicle to stray out of lane. This is usually achieved by infrared sensors behind the front bumper detecting abnormal lateral movement when the vehicle moves across road markings (white line lane markers) without the indicator being used. The driver is advised by audible, visual, or tactile means (such as a vibrating signal on the left or right side of the driver's seat, depending which way the vehicle is drifting), or a combination of these, allowing the driver to take immediate action to get back in lane.

To detect unintentional lane departures, LDWS use multiple (usually six) infrared sensors which are fitted under the front bumper, and along each side. Each sensor is equipped with an infrared light-emitting diode and a detection cell. Lane departures are detected by variations in the reflections from the infrared beams emitted by the diode onto the road.

2.6.1.20 Lane Keeping Aid

This is a second generation LDWS which not only provides a warning if the car drifts out of its lane, but if the driver fails to steer the vehicle back on its course, the system momentarily takes over to guide the car back into the lane.

2.6.1.21 Optimal Speed Advisory

This system provides advice from in-vehicle data to maximize fuel efficiency, while considering driver-provided parameters such as destination and time-of-arrival targets. This usually combines satellite navigation with information from the engine management system.

2.6.1.22 Parking Assistance/Automatic Parking

The driver identifies the type of parking (e.g., street parking, bay parking–head in, bay parking–reverse in, or similar). The driver drives past the parking space at low speed. Sensors at the front and rear detect the gap and measure its size. The system will identify if the gap is adequate. The driver puts the car into reverse which activates wide angle back-up cameras. The driver activates the system. With the vehicle in reverse creep mode and the driver's foot lightly on the brake (in some systems) and hands off the wheel, the sonar detection systems will measure position and steer the vehicle into the space. The vehicle advises the driver when parking is complete. Some manufacturers are developing fully automatic systems.

2.6.1.23 Precrash Sensing

Precrash sensing functionality is defined in functional steps that require an increasing situation analysis performance and a growing amount of application effort. Each functional step makes it necessary to define the appropriate range of view—the virtual barrier. It is subject to various constraints and the configurations possible for precrash sensing. Precrash sensing technology uses platform radar sensors that are designed for the functional integration of possible functions that rely on sensor information from the close surroundings of the vehicle.

The main focus of precrash sensing is to help passive safety devices in protecting the passenger in all crash situations.

2.6.1.24 Rain-Sensing Wipers

Infrared sensors automatically detect rainfall intensity and adjust wiper speed accordingly, improving driving safety and driver convenience.

2.6.1.25 Speed Control—Overspeed Warning

This system triggers a sound signal when the vehicle exceeds the speed setting programmed by the driver. The driver is free to turn the system on or off, using a manual switch. The system does not affect the real speed of the vehicle and provides only a warning.

2.6.1.26 Speed Control—Speed Limiter

This system is similar to the overspeed warning mechanism. The speed limiter enables the driver to program the desired maximum speed. But, unlike the overspeed warning, if the driver tries to go faster than the programmed speed, the accelerator does not respond. The driver can usually override the system if necessary by pressing down hard on the pedal.

2.6.1.27 Tire Pressure Sensors

Real-time sensing of the exact pressure inside the tire is normally achieved by locating the sensing device in the tire (although systems with the sensor in the rim or on the valve have been developed). This pressure measurement information is then carried to the driver and displayed in the cabin of the car. The remote sensing module is comprised of a pressure sensor, a signal processor, and an RFID transponder, which, when interrogated responds with its identity, tire pressure, and temperature. A temperature sensor is also required because the system has to compensate for pressure variations due to temperature.

2.6.1.28 Vehicle Safety Inspection

This system provides a warning to drivers, generated by in-car equipment, that a vehicle safety inspection is required. Such warnings may be generated on an elapsed time basis, a miles driven basis, or be caused by data collected by the vehicle management system when a potential safety problem is detected.

2.6.1.29 Visibility Enhancer

This system uses infrared or low light camera techniques to provide enhanced visibility of dark objects, persons, and animals in the roadway at night via a head-up display or enhanced display on the driver's video screen. This system may be supported by audible warnings.

2.6.2 Infrastructure Based Wireless ITS

Infrastructure based ITS generally describes ITS services that are provided to the driver without two-way communication. Many of these services are transparent to the driver, although he will be the beneficiary. Traffic signal management, control center–control center communications, and information systems are typical examples. Such systems may manage traffic flows and even redirect traffic, and in general these systems have been implemented using copper and fiber optic cables and general communications technology generic standards. Some of these standards are referred to in this book, but the reader who is interested in generic IT/telecommunications standards is directed towards books on these subjects. This area is too complex and too generic to be covered here. However, aspects of such standards that are specifically adapted for ITS are covered in several chapters of this book.

The traveler becomes aware of such systems most usually when there is a visual interface. Examples here include: (1) variable message signs, which provide relevant information to the driver, but are performed by a computer to a remote display screen using fiber optic or copper cables; (2) ramp management, where traffic lights control entry to highways; and (3) "green wave" sequenced traffic light operation to both control traffic flow and encourage compliance with speed limits. Arrival and departure boards at stations and airports provide other examples.

Two trends in this area are worth noting. The first is a trend to use wireless communications for infrastructure↔infrastructure communication. General wireless, usually microwave, standards are used for such systems which are often carried over public networks. Traffic managers are increasingly using dedicated local systems to collect environmental data, and in some cases communicate with remote traffic signals. Such communications are also commonly carried out using GPRS cellular communications.

There is also a dedicated microwave band allocated within Europe specifically for infrastructure↔infrastructure communications at 64 to 66 GHz, and this is foreseen as particularly appropriate for ITS communications to end infrastructure nodes or between nodes. This band is adjacent to a band at 63 GHz that is dedicated for ITS. Here it is envisioned that 64 to 66 GHz will be used to get the communication to the roadside, and 63 GHz will be used to communicate to/from and between vehicles.

The second trend is to communicate to mobile equipment using wireless communications. Currently, this is done using GPRS or 3G cellular communications, but it is envisaged that once infrastructures that enable quasi-continuous communications to/from vehicles are in place, these will also be able to be used for wireless infrastructure–mobile infrastructure communications.

Examples of infrastructure based ITS follow.

2.6.2.1 Adaptive Traffic Signal Control

This is a traffic control system that continuously senses and monitors traffic conditions and adjusts the timing of traffic lights according to the actual traffic load.

2.6.2.2 After Theft Vehicle Recovery

After theft vehicle recovery services include:

- User-initiated distress calls;
- Automated theft warning;
- Automated vehicle intrusion and stolen vehicle monitoring;
- Stolen vehicle tracking;
- Remote vehicle immobilization.

2.6.2.3 Commercial Vehicle Preclearance

The commercial vehicle preclearance service group provides services that enable commercial vehicles, including trucks and buses, to have credentials and other documents, safety status, and weights checked automatically at normal road speeds. A principal objective of this service is to establish preclearances with minimal disruption to the vehicle journey and traffic flow. Examples services include:

- Weigh-in-motion;
- Nonstop preclearance;
- Vehicle safety records monitoring.

2.6.2.4 Commercial Vehicle Administrative Processes

This service group is complementary to the commercial vehicle preclearance service group described above. It enables haulers and shippers to purchase annual and ad hoc credentials, using communications and computer technologies. Example services include:

- Automated credential filing;
- Automated commercial vehicle administration;
- Automated border crossings.

2.6.2.5 Control Center Information Sharing

This system provides for the automated sharing of traffic management, traffic load, and incident data between adjacent traffic management centers, in order to improve planning, load management, and incident management.

2.6.2.6 Corridor Traffic Management—Surface Street (Local Road) and Freeway/Highway

This is a traffic management system designed to coordinate usage of surface street and freeway/highway traffic management. When traffic congestion occurs on one roadway, travelers typically respond by shifting to another route, selecting a different roadway (freeway versus surface street), adjusting their trip to another time of day, or remaining on their current route and encountering significant delays. These disruptions range in scale, frequency, predictability, and duration. Depending on the cause, they have the potential to affect a number of transportation facilities or modes. Corridor traffic management is a management control technique to encourage or discourage certain driver decisions.

Proactively managing and coordinating the control of traffic is a viable and effective strategy to improve the safety, efficiency, and reliability of traffic on and between freeways and surface streets (minor roads) within urban corridors. Corridor traffic management can reduce travel times, improve travel reliability, increase traffic throughput, decrease crashes, and reduce the number of stops and delays at traffic signals.

Achieving these results requires an aggressive traffic operations program, strong partnerships between agencies, commitment of necessary resources and support, deployment of technology and traffic control systems, development of operational strategies and control plans, and proactive management and control of traffic within freeway corridors in metropolitan areas. Using managed lane strategies, alternate routing of traffic, operational strategies, coordinated control plans, proactive management and control of traffic, and coordinated response to changing conditions offers the potential to achieve significantly greater use of the existing roadway capacity. Within an ITS context, this refers to the use of data collected through cameras, sensors, probe vehicles, and the transmission of information to VMS signage and ITS equipped vehicles.

2.6.2.7 Data Archiving

An archived data user service provides an ITS historical data archive for all relevant ITS data and incorporates the planning, safety, operations, and research communities into ITS. It provides the data collection, manipulation, and dissemination functions of these groups, as they relate to data generated by ITS.

2.6.2.8 Data Warehouse

A data warehouse service integrates the planning, safety, operations, and research communities into ITS and processes data products for a regional ITS community. ITS data warehouse management supports the archiving and retrieval of data generated by other ITS applications and enables ITS applications that use archived information. Decision support systems, predictive information, and performance monitoring are some ITS applications enabled by ITS information management. In addition, ITS data warehouse systems can assist in transportation planning, research, and safety management activities.

2.6.2.9 Detection and Confirmation of Incident Presence

This service uses data gathered from ITS sources, camera monitoring, and so on, to support confirmation of incidents and issue advisories to relevant parties; instigate on-site motorist assistance response and traveler assistance; and provide incident coordination and clearance.

2.6.2.10 Electronic Payment Systems—Infrastructure Only

Electronic payment systems employ various communication and electronic technologies to facilitate commerce between drivers and transportation agencies, typically for the purpose of paying tolls and transit fares. While most involve direct transactions between a vehicle and the infrastructure, or a payment card in the hands of a driver and the infrastructure (see appropriate subsection below for details), some schemes, including the world's largest congestion payment scheme (London), utilize only infrastructure means (video cameras) to collect payment.

2.6.2.11 Emergency Management Systems

ITS applications in emergency management include hazardous materials management, the deployment of emergency medical services, and large and small-scale emergency response and evacuation operations. (See appropriate subsections below for emergency management systems that communicate directly with emergency vehicles and vehicles involved in the incident.)

2.6.2.12 Enforcement

This provides centralized management of violation enforcement using ITS data captured from (usually a combination of) sensors and cameras (and in the future potentially also from other ITS-equipped vehicles).

Examples include:

- Access control;
- High occupancy vehicle facility usage;
- Parking regulation enforcement;
- Speed limit enforcement;
- Signal enforcement (e.g., red light violation);
- Emissions monitoring.

2.6.2.13 Freight Transport Fleet Management

At a multimodal level, commercial fleet management includes logistics and freight management systems. (It also covers the use of automatic vehicle location (AVL) to achieve automatic freight carrier and container location, as well as vehicle-to-control center communications in order to provide vehicle location and other status information to the fleet operators dispatched. For services that require interaction with the vehicle, see the appropriate subsection below.) With respect to infrastructure based ITS service provision, this includes the use of dynamic dispatching systems to improve the efficiency of the fleet management process. These services are implemented in conjunction with the traffic management service group services. This service group includes:

- Pretrip information;
- Intermodal terminal conditions.

Example services include:

- Commercial vehicle fleet tracking;

- Commercial vehicle fleet dispatching;
- Freight container tracking.

This service group includes transactions to maintain the ITS information about a shipment from the time of the order by the consignor to the reception of goods by the consignee. The key ITS transactions provide registers of service providers and enable the goods to be tracked throughout intermodal journeys.

This service group covers the exchange of information about transport of goods across modes. This includes knowledge of where the units transporting the goods are located, plus their condition and status, as well as similar information about the vehicle transporting the unit. It is also possible to locate subunits and provide customers with information about progress with the movement of the goods.

Example services include:

- Vehicle and container arrival information exchange (users are fleet and intermodal carriers and nodes)
- Customer freight information access (users are customers and shippers);
- Intermodal center facility management;
- Intermodal vehicle and container control;
- Dangerous goods movement data sharing;
- Dangerous goods movement data registry;
- Dangerous goods movement fleet coordination;
- Dangerous goods movement police/safety coordination.

2.6.2.14 Freeway/Arterial/Highway Traffic Management

Arterial management systems manage traffic along arterial roadways, employing technologies such as traffic detectors, traffic signals, and various means of communicating information to travelers. These systems make use of information collected by traffic surveillance devices to smooth the flow of traffic along travel corridors. They also disseminate important information about travel conditions to travelers via technologies such as dynamic variable message signs and highway advisory radio.

ITS functions that provide freeway/arterial/highway management systems include:

- Traffic surveillance systems using detectors and video equipment;
- Traffic control measures on entrance ramps using sensor data to optimize freeway travel speeds and ramp meter wait times;
- Lane management to optimize the effective capacity of freeways and promote the use of high-occupancy commuting modes;
- Special event transportation management systems to control the impact of congestion at stadiums or convention centers;
- Dynamic VMS management.

2.6.2.15 Hazardous Materials Monitoring and Management

The central control for the movement of hazardous materials on the road network. includes pretrip driver authentication, route planning, and resource allocation; planning and management of difficult movements; notification of slow-moving/wide transports to traveler information advisory systems; and liaison with departments for maintenance of national security. Hazardous materials monitoring and management services include:

- Hazardous vehicle preclearance;
- Hazardous vehicle routing data:
  - Route guidance;
  - Route enforcement;
- Hazardous vehicle incident data:
  - Issuing post-incident instructions to drivers;
  - Location of vehicle;
  - Nature of incident;
  - Nature of cargo.

2.6.2.16 High Occupancy Vehicle Facility Management

This involves the organization and management of high occupancy vehicle (HOV) provisions, including remote enforcement measures, dynamic lane assignment, and traveler advisory support.

2.6.2.17 Highway Maintenance Management

ITS can be used to support highway maintenance management, including the deployment of portable VMS and dynamic updating of signage, speed limit management, on-site worker protection, and also the use of data from ITS sources to predict and manage maintenance activities and to predict and avoid incidents.

2.6.2.18 Incident Management Systems

Incident management systems can reduce the effects of incident-related congestion by decreasing the time to detect incidents, the time for responding vehicles to arrive, and the time required for traffic to return to normal conditions. Incident management systems make use of a variety of surveillance technologies, often shared with freeway and arterial management systems, as well as enhanced communications and other technologies that facilitate coordinated response to incidents. (See appropriate subsections below for incident management systems that communicate directly with emergency vehicles.)

2.6.2.19 Intermodal Highway Junction Management

This involves freight intermodal change connection points (typically road/rail but also road/air). Intermodal connectors are often not formally part of a state's highway system. With respect to infrastructure ITS, this involves the management of traffic flows and interchange to/at/from intermodal change points.

2.6.2.20 Parking Management

Parking management includes a variety of strategies that encourage more efficient use of existing parking facilities, improve the quality of service provided to parking facility users, and improve parking facility design. Parking management can help address a wide range of transportation problems and help achieve a variety of transportation, land use development, economic, and environmental objectives. ITS can be used to assist parking management by the dynamic interchange of data. ITS can assist parking management, operation and deployment by direct interaction with ITS equipped vehicles. (Examples of ITS assisted parking management which involve direct interaction with ITS equipped vehicles are given in the appropriate sections of this chapter, below.) However, ITS assistance for parking management does not have to wait until vehicles are ITS equipped; infrastructure based ITS is already assisting these operations. Examples of parking management strategies include:

- VMS parking assistance: Providing automatically updated identification of available parking places, together with road signage to the parking facilities.
- Dynamic share parking: The allocation of parking facilities, by day or time of day, to different organizations based on recent identified usage. This can increase effective car park capacity by 20% to 40% (Victoria Policy Transport Institute).
- Dynamic parking demand management: Making the most convenient parking spaces available to certain higher-value uses, like:
  - Rideshare vehicles;
  - Dynamic allocation of spaces reserved for disabled people;
  - Dynamic time related occupancy (e.g., short-term shoppers in day, residents at night);
  - Time related demand pricing;
  - More flexible pricing methods which allow motorists to pay for only the amount of time they park (which makes shorter parking periods relatively attractive).
- Dynamic park and ride management;
- Dynamic Internet based parking pricing and availability (pretrip).

2.6.2.21 Reversible Lane Management/Counterflow/Tidal Flow

These are terms used in different parts of the world to describe a traffic control system where the direction of traffic is changed according to prevailing or expected traffic conditions. This system is typically used to improve traffic flow during rush hours. Traffic signals are controlled from a traffic control center operating lights external VMS and, in the future, messages to ITS equipped vehicles.

2.6.2.22 Response to On-Site Incident Information

This ITS function uses available data to formulate appropriate response actions to each identified incident and revise those actions when necessary to minimize incident impacts and post-event effects. This function includes the proposal and facilitation of the appropriate scheduling of predicted incidents in order to minimize incident potential, incident impacts, and/or the resources required for incident management.

2.6.2.23 Specific Vehicle Types Priority and Preemption

This service allows central management system control to provide traffic signal priority and preemption for specific classes of vehicles (e.g., emergency service vehicles, public transit). This service can be affected by a number of technologies, but the dominant technologies are ITS communications equipped vehicles or vehicles equipped with an RFID tag, together with ITS beacons or RFID readers at traffic control signals. The data of an approaching priority vehicle can also be relayed to VMS signs both outside the vehicle and on the screens of ITS equipped vehicles, as well as to traffic control officers by means of audible or visual messages.

2.6.2.24 Surface Street Traffic Management

This provides real-time accurate information to the general public on individual arterial performance. The provision of real-time data from sensors, cameras, probe data, and a communications backbone enables agencies to more aggressively and accurately respond to unscheduled traffic events. This system will typically include an incident detection system, arterial surveillance system, traffic control system, and infrastructure to link each of the traffic management centers in the region. It is often linked to VMS devices or other ITS components located along the selected corridors to include modification of driver behavior to improve traffic flow.

2.6.2.25 Traffic Information Dissemination

This service provides driver information using roadway equipment including dynamic variable message signs or highway advisory radio broadcasts. A wide range of information can be disseminated, including traffic and road conditions, closure and detour information, incident information, emergency alerts, and driver advisories. This service provides information to drivers at specific equipped locations on the road network. Careful placement of the roadway equipment provides the information at points in the network where drivers have recourse and can tailor their routes to account for the new information. This service also includes the equipment and interfaces that provide traffic information from a traffic management center to the media (for example, via a direct information provision from a traffic management center or police station to radio or television computer systems). Transit management, emergency management, and information service providers. A link to the maintenance and construction management subsystem will also allow the dissemination of real-time information on road/bridge closures due to maintenance and construction activities.

2.6.2.26 Traffic Monitoring

This involves the use of sensor technology, data recording and transfer, and data sampling and analysis to measure the passage of vehicles on roadways. Traffic monitoring is an essential component for traffic management.

2.6.2.27 Transit Management Systems

Infrastructure based transit ITS services include surveillance and communications such as automated vehicle location (AVL) systems, computer-aided dispatch (CAD) systems, and remote vehicle and facility surveillance cameras, which enable transit agencies to improve the operational efficiency, safety, and security of the public transportation systems. (For transit management systems that involve communications with transit vehicles, see the appropriate subsection below.) Infrastructure based transit systems usually also interact with infrastructure based traveler information systems to provide arrival and departure information.

2.6.2.28 Traveler Information—Infrastructure Based

Infrastructure oriented traveler information applications use a variety of technologies, including Internet Web sites and telephone hotlines, as well as television, radio, and VMS, to allow users to make more informed decisions regarding trip departures, routes, and modes of travel.

2.6.2.29 Weather and Environmental Conditions Monitoring

This involves the collection of information related to weather and other environmental conditions in order to support traveler information systems (e.g., VMS, radio alerts). This function includes:

- Automated weather sensors;
- Water level/tidal monitoring and prediction;
- Seismic monitoring;
- Pollution monitoring;
- Avalanche, mudslide, fallen rock monitoring.

2.6.2.30 Work Zone Traffic Management

There are a number of ways that ITS can assist work zone traffic management. ITS can help secure the safety of workers and travelers in a work zone while facilitating traffic flow through and around the construction area. This is often achieved through the temporary deployment of other ITS services, such as elements of traffic management and incident management programs. Examples where interactions with vehicles are involved are described in the appropriate section below. Other infrastructure based ITS assistance to work zone traffic management includes:

- Portable changeable message signs (PCMS);
- Dynamic lane merge systems;
- Variable speed limits (according to traffic load or work zone activity).

2.6.3 Vehicle/Infrastructure ITS

Infrastructure based ITS systems require a communication link between the infrastructure and the vehicle. Early examples of such systems are toll roads—electronic road pricing—and vehicular access control systems.

It is worth noting that many vehicle access control systems are not the subject of any ITS or vehicle specific standardization committee, but rather fall within generic standards for RFID. These communications standards are generally found within the purview of ISO/IEC JTC1 SC31 WG4 (Information Technology, Automatic Identification, Radio Frequency Identification). Within this book, standards for such systems are identified in Section 7.4.

General aspects of automatic vehicle identification and automatic (transport related) equipment identification, as well as issues of electronic registration identification are identified in Section 7.3.

Standards concerned with electronic tolling (often known as ETC, AFC, and ATC) are developed by ISO/TC 204 WG5/CEN TC278 WG1, and such systems are identified in Section 8.8.

It is worth noting that in Europe (with the exception of Italy) there is a coalescence around the ETSI/CEN standards for EFC, and these standards are widely used in Australia and parts of Asia; in North America, however, there are no properly developed EFC/ETC standards and consequently the market in those countries has become dominated by three commercially proprietary systems that are not interoperable. China currently uses several different systems and is developing its own national standard and Japan and Korea have national standards.

For early generation systems, standards were developed as complete, whole closed systems, including the communications aspects—the air interface protocols. Later work has separated the applications from the carrier communications, and it is envisioned that, in the future, such fee collection transactions, whether for simple toll collection or more complex road pricing and demand management systems, will be transacted using a generic vehicle communications system, perhaps linked to a smart card or other device within the vehicle.

In order to perform many ITS services, as currently envisioned, it is expected that vehicles will be equipped with one or more communication system that uses one or more wireless air interface to interact between the vehicle and the infrastructure.

Examples (but not a complete list) of vehicle–infrastructure services where WIP is known to be underway are described in this section. It should be noted, however, that there is no absolute consensus as to the extent of many of the services, nor of the means of achieving them. For this reason, the descriptions are generalized and not specific, and services might exist in multiple categories.

2.6.3.1 Vehicle/Infrastructure

2.6.3.1.1 Infrastructure to On-Board Equipment

2.6.3.1.1.1 Accident Site Advisory. After a crash is reported, either electronically by an emergency call message transmission, or by a first responder or PSAP, a warning is transmitted from the nearest infrastructure beacon, from a portable beacon carried by emergency services, or, if it is capable post-crash, from one or more of the affected vehicles, to all other vehicles in the area, warning of the proximity of the accident. The warning is transmitted into the vehicle audio and/or visual display via an ITS communications interface.

2.6.3.1.1.2 Animal Crossing Zone Information. This is the provision of warnings to drivers at farm crossings, horse trail crossings/bridleways, and so on. The information is transmitted into the vehicle audio and/or visual display via an ITS communications interface.

2.6.3.1.1.3 Adaptive Drivetrain Management—Infrastructure Assisted. (See also Section 2.6.1.4.) By anticipating certain driving situations, an on-board system can recommend handling strategies to the driver for such purposes as improving fuel efficiency, and achieving safety and performance improvements. By interacting with the infrastructure, information about upcoming driving conditions, collated by sensor or probe vehicles and reported to the infrastructure, can be introduced to the calculation models to further improve the efficiency of such systems. The information is transmitted into the vehicle via an ITS communications interface.

2.6.3.1.1.4 Adaptive Headlight Aiming. (See also Section 2.6.1.8.) This is the provision of information to vehicles for the purpose of changing headlamp settings to adjust to local conditions. Early instantiations are in-vehicle devices controlled solely by in-vehicle sensors, and these are already in the marketplace. Devices that are under development interact with the infrastructure and/or other vehicles, and in these devices the information is transmitted to the vehicle via an ITS communications interface.

2.6.3.1.1.5 Blind Merge Warning. This is the provision of warnings from the infrastructure to drivers where highways or lanes merge. These in-vehicle electronic systems monitor the position of a vehicle within a roadway lane and warn a driver if it is unsafe to change lanes or merge into a line of traffic. These systems are rearward looking, radar-based systems. They assist drivers who are intentionally changing lanes by detecting vehicles in the driver's blind spot. The warning is transmitted to the vehicle via an ITS communications interface.

2.6.3.1.1.6 Curve Speed Warning—Infrastructure Based. This is the provision of warnings from the infrastructure to drivers to adapt one's speed to curve (bend) conditions. These warnings may be adaptive to current weather and road conditions. Curve speed warning systems use roadside detectors and electronic warning signs to warn drivers, typically those in commercial trucks and other heavy vehicles, of potentially dangerous speeds when approaching curves on highways. The warning is transmitted into the vehicle audio and/or visual display via an ITS communications interface.

2.6.3.1.1.7 Emergency Vehicle Signal Preemption. This is the control of traffic signals to give priority to emergency vehicles (obviating the need for emergency vehicles to cross signals set at stop). Emergency vehicle preemption allows fire trucks and ambulances to intervene in the normal operation of traffic control systems using wireless communications installed on traffic intersections and emergency vehicles. As the emergency vehicle approaches a traffic signal, it is recognized by the traffic signal controller through light, radio waves, or sound. The normal green/yellow/red cycle can then be interrupted to change the light to green.

2.6.3.1.1.8 Emergency Vehicle Video Replay. This is the provision of video data streaming to/from emergency vehicles to enable remote assistance to be provided en route. This also involves the replay of vehicle "black box" data and video to assist emergency response teams with understanding cause, actual event, and probable injury consequences of an accident.

2.6.3.1.1.9 Emergency Vehicle Warning—From Infrastructure. Infrastructure based systems, having knowledge of the site of an incident and having been notified by an emergency response vehicle moving towards the incident, either broadcast or notify vehicles registered by the beacon as being within its range of the approach of an emergency vehicle. The warning is transmitted into the vehicle audio and/or visual display via an ITS communications interface.

2.6.3.1.1.10 External Speed Limitation. This is the limitation of speed either to local road regulations or at high-risk points such as near schools or parks. The limitation is achieved by the receipt of a message over an air interface which instructs the engine management system to control speed to a certain maximum.

2.6.3.1.1.11 Fog Warning. This is the provision of warnings from the infrastructure to drivers of the presence of fog at or ahead of the vehicle. Such systems may be provided from a central reference to the location of the vehicle and known ground conditions at that point or ahead of that point, or may comprise a self-contained ITS broadcast transmitter, or vehicle presence sensing transmitter installed throughout the road and/or in known fog zone areas where during foggy conditions motorists are alerted of the fog zone condition ahead. The warning is transmitted into the vehicle audio and/or visual display via an ITS communications interface.

2.6.3.1.1.12 Freezing/Icy Bridge Warning. This is the provision of warnings from the infrastructure to drivers of the presence of ice on a bridge ahead. Such systems may be provided from a central reference to the location of the vehicle matched to information collected from sensors in the bridge, or may comprise a self-contained ITS broadcast transmitter, or vehicle presence sensing transmitter, connected to sensors in the bridge, and attached to (or in place of) an appropriately placed warning sign. The warning is transmitted into the vehicle audio and/or visual display via an ITS communications interface.

2.6.3.1.1.13 Freezing/Icy Road Surface Warning. This is the provision of warnings from the infrastructure to drivers of the presence of ice on the road. Such systems may be provided from a central reference to the location of the vehicle matched to information collected from sensors in the road, or may comprise a self-contained ITS broadcast transmitter, or vehicle presence sensing transmitter, connected to sensors in the road, and attached to (or in place of) an appropriately placed warning sign. The warning is transmitted into the vehicle audio and/or visual display via an ITS communications interface.

2.6.3.1.1.14 GNSS Corrections. This is the provision of services to provide greater accuracy to GNSS (GPS/Galileo/GLONASS) positioning. Such accuracy is essential for collision avoidance and other safety critical services.

2.6.3.1.1.15 Hazardous Warnings Restricted Area. There are multiple approaches to such situations. One approach is the provision of warnings from the infrastructure to drivers of HAZMAT vehicles when approaching restricted areas. Such systems may be provided from a central reference to the location of the vehicle matched to known HAZMAT restricted zones, or may comprise a self-contained ITS broadcast transmitter, or vehicle presence sensing transmitter, installed on a HAZMAT restriction sign (or in place of) to provide warning to the driver. The warning is transmitted into the vehicle audio and/or visual display via an ITS communications interface.

Another approach is for HAZMAT vehicles to intermittently, continuously, or on command, transmit (with or without driver consent) the nature of the HAZMAT cargo, driver authentication and identification, and so on. On receipt of a signal approaching or in a HAZMAT restricted area for that type of material, the driver is warned by audio and/or visual display via an ITS communications interface, or possibly the vehicle is immobilized and emergency/police summoned.

2.6.3.1.1.16 Highway/Rail Collision Warning. This is the provision of warnings to both road and rail of potential collisions at grade (level) crossings, or the control of road/rail vehicles to prevent collisions at grade (level) crossings. The warning is transmitted into the vehicle audio and/or visual display via an ITS communications interface.

2.6.3.1.1.17 Homeland Security Identification and Management. This involves a range of vehicle and potentially occupant identification measures to support national security and border crossing identification and management.

2.6.3.1.1.18 Intelligent On-Ramp Metering. This involves the development of ramp metering and monitoring systems to adapt more precisely to actual traffic flows; bringing ramp signals on board to video/audio presentation; and the potential extension to prevent vehicles from entering the highway when the signal indicates stop.

2.6.3.1.1.19 Intelligent Traffic Lights. This involves the collection of traffic information from vehicles to enable more intelligent operation of traffic lights to adapt to traffic flows, average speeds, and incidents. In such systems, all cars communicate to the traffic light using an ITS communications interface both their specific place in the queue and their destination address. The traffic light control system decides which option (i.e., which lanes are to be put on green) is optimal to minimize the long-term average waiting time until all cars have arrived at their destination address. The traffic light controllers solve this problem by estimating how long it would take for a car to arrive at its destination address (for which the car may need to pass many different traffic lights) when currently the light would be put on green, and how long it would take if the light would be put on red. The difference between the waiting time for red and the waiting time for green is the gain for the car. The traffic light controllers set the lights in such a way to maximize the average gain of all cars standing before the crossing. Waiting times of individual cars are used to compute the long term average waiting times using dynamic programming algorithms.

2.6.3.1.1.20 Intersection Collision—Infrastructure Based Warning. This is the provision of warnings initiated by the infrastructure to vehicles approaching or at intersections of potential collisions (especially at blind intersections).

Infrastructure based intersection collision avoidance systems use roadside sensors, processors, and warning devices; roadside-vehicle communication devices; other roadside informational or warning devices; and traffic signals to provide driving assistance to motorists. The intersection collision avoidance systems can be classified as either infrastructure-only or as infrastructure vehicle cooperative. Infrastructure-only systems rely solely on roadside warning devices to communicate with drivers. The warning is transmitted into the vehicle audio and/or visual display via an ITS communications interface.

2.6.3.1.1.21 Keep Clear Warning. This is the provision of warnings to drivers to avoid specific locations, which may be because of obstacles in the road, damage to the road, congestion, road work, standing water, high winds, accident, spillage, and so on, or because of local regulations. The warning is transmitted into the vehicle audio and/or visual display via an ITS communications interface.

2.6.3.1.1.22 Left Turn Assistant—Infrastructure Assisted. This is the provision of assistance to drivers who are turning left. This countermeasure involves warning motorists making a left turn at a traffic signal of a potential conflict with vehicles approaching from the opposite direction (which causes a total of 192,000 crashes per year in the United States [Barr, 2001]). The basic sensing requirements are to identify potential conflicts by determining the speed and the acceleration or deceleration rate of each vehicle approaching the intersection from the opposite direction, including vehicles executing through and right-turn movements. Simple point measurements will not be sufficient, since vehicles can assume various trajectories and acceleration/deceleration/stopping movements, particularly when other vehicles are present. The assistance is transmitted into the vehicle audio and/or visual display via an ITS communications interface.

2.6.3.1.1.23 Low Bridge Warning. This is the provision of warnings to drivers at low bridges; this is probably matched to vehicle characteristics. The warning is transmitted into the vehicle audio and/or visual display via an ITS communications interface.

2.6.3.1.1.24 Low Parking Structure Warning. This is the provision of warnings to drivers of height restrictions in vehicle parks; this is probably matched to vehicle characteristics. The warning is transmitted into the vehicle audio and/or visual display via an ITS communications interface.

2.6.3.1.1.25 Merge Assistant. This is the provision of assistance to drivers merging into traffic lanes. The warning is transmitted into the vehicle audio and/or visual display via an ITS communications interface.

2.6.3.1.1.26 On-Board VMS Signage. This is the provision to on-board video screens of messages transmitted to external variable message signs, or the replacement of such signage by location related traffic information to vehicles from infrastructure communication points. The warning is transmitted into the vehicle audio and/or visual display via an ITS communications interface.

2.6.3.1.1.27 Pedestrian Crossing Information. This is the provision of information to drivers approaching pedestrian crossings, as well as the provision of information to pedestrians at crossings (especially visually, aurally, and mobility impaired pedestrians). The warning is transmitted into the vehicle audio and/or visual display via an ITS communications interface.

2.6.3.1.1.28 Pedestrian Crossing Control. This is vehicle control to prevent vehicles from entering pedestrian crossings while set to pedestrian priority/stop traffic, or alternatively while physically in use by pedestrians. Control is achieved via an ITS communications interface.

2.6.3.1.1.29 Pedestrian/Children Warning. This is the provision of warnings to drivers at points where pedestrians, and particularly children, are likely to appear in the roadway (e.g., near schools and playgrounds). The warning is transmitted into the vehicle audio and/or visual display via an ITS communications interface.

2.6.3.1.1.30 Post-Crash Warning. This is the provision of warnings to drivers approaching the site of a recent accident. The warning is transmitted into the vehicle audio and/or visual display via an ITS communications interface.

2.6.3.1.1.31 Rail Road Crossing Warning. This is the provision of warnings from the infrastructure to drivers of rail crossings (grade crossings/level crossings) ahead. Such systems may be provided from a central reference to the location of the vehicle matched to known crossings, or may comprise a self-contained ITS broadcast transmitter, or vehicle presence sensing transmitter, installed on a rail crossing warning sign (or in place of) to provide warning of the crossing ahead. Systems may also detect the presence of approaching trains to raise the warning level, and may provide data to trains where it appears a car is either stationary on the crossing, or likely to ignore the warnings. Later systems may force vehicles to stop when trains are approaching. The warning is transmitted into the vehicle audio and/or visual display via an ITS communications interface, and in the latter case would directly link to vehicle control systems.

2.6.3.1.1.32 Rest Area Ahead Advisory. A self-contained intelligent beacon may be installed on a sign or the surface or side of the road to provide an in-vehicle reminder of the facilities provided at the rest area ahead. Such information may also be broadcast from an infrastructure beacon, transmitted once a vehicle is registered, and may be linked with the provision of commercial publicity information where the rest area offers commercial services. The notice is transmitted into the vehicle audio and/or visual display via an ITS communications interface.

2.6.3.1.1.33 Right Turn Assistant—Infrastructure Assisted. This is the provision of assistance to drivers turning right. See "Left Turn Assistant" above in this section. The assistance data is transmitted into the vehicle audio and/or visual display via an ITS communications interface.

2.6.3.1.1.34 Road Condition Warning—Infrastructure Assisted. This is the provision of warnings to drivers where there are adverse road conditions (e.g., ice, flood, obstacle in road, pothole, spillage, surface deformation). The warning is transmitted into the vehicle audio and/or visual display via an ITS communications interface.

2.6.3.1.1.35 Rollover Warning. This is the provision of warnings to drivers of risk of vehicle rollover (due to adverse camber, crosswinds, etc.). Rollover warning systems notify drivers when they are traveling too fast for an approaching curve, given their vehicle's operating characteristics. This function is primarily targeted to heavy goods and high sided vehicles. The warning is transmitted into the vehicle audio and/or visual display via an ITS communications interface.

2.6.3.1.1.36 School Bus Warning. An ITS communications device on-board the bus alerts approaching drivers (in both directions) of the impending danger surrounding the vehicles. Transmission signals may be intelligently programmed and transmitted in the forward (parallel) and/or reverse direction of the bus, or surrounding (up to 360° radius) the bus, as needed, to alert approaching vehicles in the vicinity of the potential danger.

2.6.3.1.1.37 School Zone Warning. This is the provision of warnings to drivers in the proximity of school entrances. The warning is transmitted into the vehicle audio and/or visual display via an ITS communications interface.

2.6.3.1.1.38 Sign Information (Warning Assistance). This is the transmission of road signs to in-vehicle video/audio display. The warning is transmitted into the vehicle audio and/or visual display via an ITS communications interface.

2.6.3.1.1.39 SOS Services—Infrastructure Assisted. This is the provision of connections to vehicles from police and emergency services for the transmission of voice and/or data communications. The warning is transmitted into the vehicle audio and/or visual display via an ITS communications interface.

2.6.3.1.1.40 Speed Limit Advisory. This is the provision of warnings from the infrastructure to drivers of local speed limits and specifically when exceeding those speed limits. Such systems may be provided from a central reference to the location of the vehicle matched to known speed limits, or may comprise a self-contained ITS broadcast transmitter, or vehicle presence sensing transmitter, installed on a speed limit sign (or in place of) to provide regulatory or recommended speed limit (in-vehicle signing) based on weather conditions, ice, rain, or other potential hazards. The warning is transmitted into the vehicle audio and/or visual display via an ITS communications interface.

2.6.3.1.1.41 Speed Limit Control. This is the control of vehicle speed to prevent the vehicle from exceeding the local speed limit. The control instruction is effected via an ITS communications interface.

2.6.3.1.1.42 Stop Sign Movement Assistance—Infrastructure Assisted. This countermeasure involves warning motorists leaving from a stop sign that their movement may conflict with another vehicle. The vehicle movement could be left turn, right turn, or through. (These incidents cause a total of 362,000 crashes per year in the United States [Barr, 2001].) The basic sensing requirements are to identify potential conflicts by determining the speed, acceleration, or deceleration rate of each vehicle approaching the intersection and the discharge from the stop line for vehicles at stop-controlled approaches. The data and warnings are transmitted into the vehicle via an ITS communications interface and warnings are displayed using audio and/or visual display.

2.6.3.1.1.43 Stop Sign Warning. This is the provision of in-vehicle warnings to drivers approaching stop signs. It involves the use of roadside equipment that detects approaching vehicles and illuminates or otherwise highlights roadside signs to warn of a stop sign being approached. The warning is transmitted into the vehicle audio and/or visual display via an ITS communications interface.

2.6.3.1.1.44 Traffic Signal Warning. This is the provision of warnings to drivers approaching a traffic signal. At its simplest, an ITS communications link associated with the traffic signal advises approaching vehicles (regardless of direction) of its status, and probably the time to change of status. The warning is transmitted into the vehicle audio and/or visual display via an ITS communications interface.

In more advanced systems, sensors may identify that the speed of an approaching vehicle to the intersection is outside the normal range, and ITS communications link associated with the traffic signal sends a distinct warning signal to the potentially violating vehicle and activates traffic signal enforcement cameras.

2.6.3.1.1.45 Traffic Signal Violation Warning. This countermeasure involves: (1) warning potential violators of a traffic signal to recognize the control device, and (2) warning motorists on adjacent approaches of the potential conflict. The target crashes reflect causal factors of "did not see," "tried to beat the light," or "deliberate violation." (These incidents cause 288,000 crashes per year in the United States [Barr, 2001].)

The basic sensing requirements are to identify potential (very highly likely) violators by determining the speed and possibly also the deceleration rate of each vehicle at a fixed location. The processing system identifies vehicles at an upstream control point that are unlikely to stop at the intersection. Preliminary calculations suggest, for example, that a vehicle traveling 30 mph at a point upstream from the line will very likely be incapable of stopping in time (at least without a severe braking event), and hence can be identified as a potential violator.

Once a violator is identified, warnings will be conveyed to the violator and also to other drivers on adjacent approaches to the intersection. The violator could be warned by: (1) warning signs and lights activated once the potential violation is detected (e.g., "Stop Ahead" warning signs could be used with a flashing amber light to draw attention to the signs and could be located on both sides of the roadway to increase the likelihood that the driver would readily see the warning); (2) a warning light could be incorporated directly in the traffic signal display itself, again to draw attention to the traffic control device (e.g., strobe lights have been used to heighten the conspicuity of traffic signal displays for rural intersections where motorists may not expect a signal); (3) an intelligent rumble strip could be activated to warn the violator to slow down, and possibly heighten awareness of the need to stop at the intersection; and (4) variable message signs could convey the warning to the driver.

Motorists on adjacent approaches also need to be warned of the potential violation and conflict, and could be warned by: (1) warning lights activated to indicate a need for caution and possibly to indicate the source of the conflict; (2) an intelligent rumble strip activated to warn the other motorists to slow down and proceed cautiously at the intersection; and (3) a VMS or graphic display sign used to warn drivers of the potential conflict with the signal violator (but these might only be effective only where there is sufficient time for the motorists to comprehend the message and respond). The warning is transmitted into the vehicle audio and/or visual display via an ITS communications interface.

2.6.3.1.1.46 Transit Vehicle Data Transfer—Safety. Transit (i.e., public transport) vehicles are increasingly monitored for both safety and commercial reasons. With respect to safety, transit vehicles are commonly monitored using on-board video cameras, and the ITS link can be used to manage and transfer video data and automatically raise alerts when problems occur. Transit vehicle priority at crossings can also be expedited using ITS technologies. Vehicle performance and maintenance issues can also obviate vehicle breakdowns (which can be the cause of accidents either directly related to the breakdown or indirectly related as passengers stand around in the road space following a breakdown).

2.6.3.1.1.47 Work Zone Warning. This is the provision of information to drivers of approaching roadwork zones, including the identification of upcoming changes to lanes available, speed limits, and other restrictions. Multiple self-contained intelligent beacons may be installed in critical areas as deemed appropriate to provide for continued warnings of workers present in the immediate work zones. The warning is transmitted into the vehicle audio and/or visual display via an ITS communications interface. In the future, in-vehicle alerts may include self-controlled speed limitation of vehicle and autonomous deceleration in such areas.

2.6.3.1.1.48 Wrong-Way Driver Warning—Infrastructure Assisted. This is the provision of warnings to the road operator, road user, and other road users of vehicles traveling in the wrong direction on a controlled direction roadway. Typically, a digital camera would constantly observe the motorway exit. The movements of the vehicles driving past are analyzed from the images. If a vehicle drives opposite the permitted direction, the camera would trigger an alarm message to the traffic management and information center, to the violating vehicle itself (if equipped with ITS communications), and to other traffic participants and response personnel. The warning is transmitted into the vehicle audio and/or visual display via an ITS communications interface.

2.6.3.1.2 On-Board Equipment to Infrastructure

2.6.3.1.2.1 Automatic Crash Notification. This is an automatic system that provides data notification to public safety answering points by means of any available wireless communications media that a vehicle has crashed, and that provides coordinates and other relevant information in a message of standardized data concepts, not limited in length. Information is transmitted via an ITS communications interface.

2.6.3.1.2.2 Blind Merge Warning. This is the provision of warnings from vehicles using the road network to other vehicles using the road network to where highways or lanes merge. Such systems monitor the position of other vehicles approaching a roadway lane in front of the monitoring vehicle and communicate to warn the driver of that vehicle whether it is safe or unsafe to change lanes or merge into a line of traffic. These systems are forward-looking, usually radar-based systems. Information is transmitted via an ITS communications interface.

2.6.3.1.2.3 eCall. This is a post-event user-instigated or automatic system that provides notification to public safety answering points, by means of wireless communications, that a vehicle has crashed, and that provides coordinates and a defined minimum set of data. Some eCall systems provide a voice channel communication with data attached; others are envisaged to use data communications with the possibility of voice over Internet (VOIP). eCall systems use cellular communications to transmit their message/enable voice communications.

2.6.3.1.2.4 Incident Mapping and Warning. This is the collection and collation of data from vehicles to assist in the mapping and management of incidents.

2.6.3.1.2.5 Intelligent Traffic Light Preemption for Priority Vehicles. A variant of emergency vehicle signal preemption (see above) where the preemption is not limited to emergency vehicles, but may provide priority to other defined classes of vehicles, such as public transport and high occupancy vehicles (HOV).

2.6.3.1.2.6 Intersection Collision Avoidance. Intersection safety has begun to receive new attention from traffic engineers, human factors specialists, and others who see that emerging intelligent systems offer significant potential for improvements [Ferlis, 1999]. Crossing path crashes at intersections, as defined by Volpe

[Barr, 2001], involve one vehicle cutting across the path of another, both initially traveling from either perpendicular or opposite directions, in such a way that they collide (see "Intersection Collision: Infrastructure-Based Warning" above, and "Intersection Collision: Vehicle-Based Warning"). Information is transmitted via an ITS communications interface.

2.6.3.1.2.7 Intersection Collision—Vehicle-Based Warning. This is the provision of warnings of potential collisions from vehicles to the infrastructure when approaching or at intersections (especially at blind intersections). Such systems are said to be "cooperative." Cooperative systems communicate information directly to vehicles and drivers. The major advantages of cooperative systems lie in their capabilities to improve the interface to the driver, and hence to virtually ensure that a warning is received. This could also take advantage of the potential to exert control over the vehicle, at least in situations where the system can be confirmed as reliable and the driver cannot reasonably be expected to take appropriate actions given the imminent hazard and response time available. Information is transmitted via an ITS communications interface.

2.6.3.1.2.8 Probe Data. Probe data is data collected by vehicles and shared with the infrastructure for probe data management (usually anonymously) using an ITS communications interface. Probe data can also be gathered by the external monitoring of the movement of vehicles (usually anonymously), for example, to calculate average speed and predicted travel times to particular points en route. Traffic information from probe vehicles has great potential for improving the estimation accuracy of traffic situations, especially where no traffic detectors are installed. Probe data is fed into state equations in macroscopic traffic-flow models, and can be combined with stationary detectors.

2.6.3.1.2.9 SOS Services. This is the provision of connections from vehicles to emergency services for the transmission of voice and/or data communications. This may use an ITS communications interface or cellular communications.

2.6.3.1.2.10 Vehicle-Based Warning. This is a generic term covering a number of services where a vehicle is the originator of the warning. Information is transmitted via an ITS communications interface.

2.6.3.2 Infrastructure–Vehicle Commercial Services

2.6.3.2.1 Infrastructure to On-Board Equipment

2.6.3.2.1.1 Border Clearance. In-vehicle equipment can communicate with customs check points to prescreen trucks for safety records, border clearance, and proper credentials. This can be initiated on the highway, before the truck reaches the border control point to obviate the need for real-time clearance at the control point, thus reducing processing delays and queues at the border. The information is transmitted via an ITS communications interface.

2.6.3.2.1.2 Commercial Service Payments. Commercial vehicles are often subject to payment demands in addition to toll and road pricing. These may be payments at the border, parking, and so on. The advantage of using ITS systems for border clearance is lost if the vehicle has to stop for the driver to make payments. ITS systems, linked to payment cards or accounts, can automatically extract payment in the same manner as road toll payment. Such payments do not all have to be real-time transactions (as with tolling) and may be part of preclearance.

Other commercial services for both commercial and private traffic can be provided using ITS communications systems: for example, fuel purchase, drive through food purchase, amusement park entry, and parking charges.

2.6.3.2.1.3 Drivers Daily Log. ITS can be used to collate and transmit drivers daily log and tachograph information. The information is transmitted via an ITS communications interface.

2.6.3.2.1.4 Driver Validation. ITS can be used for driver validation without which the vehicle would be prevented from operating. This may simply work using an access smartcard, or may compare data on a smartcard or central database (using ITS communications link) to verify driver biometrics such as fingerprint. The information is transmitted via an ITS communications interface.

2.6.3.2.1.5 Enhanced Route Guidance and Navigation. This is increased positioning capability due an increased number of satellites, as envisaged in the Galileo program. To assure continuity of localization in covered or dense urban environments, links to additional equipment such as odometers or gyroscopes are used. Higher power levels emitted from satellites will allow for better penetration of the signal inside the vehicle.

Dynamic integration of current road traffic information can be combined with the static geographical map data to provide best routes for the specific journey, which, combined with the technical improvements to accuracy, provide enhanced route guidance and navigation. The information is transmitted via an ITS communications interface.

2.6.3.2.1.6 Freight and Fleet Operations. This function covers three elements:

- Road freight transport demand management;
- Overall road freight transport supply management;
- Permanent management of individual supply components in order to satisfy demand.

This last element can be split into four high-level functions:

- Organize goods movement;
- Manage carriage operations;
- Manage resources;
- Manage hubs.

These functions can be supported using ITS technologies to provide the means of managing commercial and operational freight transport operations. If necessary, goods may also be stored at certain points along a route in order to optimize operations or to change to a different mode of transport. The function includes using ITS communications to manage operations relating to the management of a vehicle fleet and human means required for goods transportation. This includes the planning and specification of driver tasks and vehicle maintenance.

National security requirements and the management of safety and emergency services may also be included so as to enable the provision of information linked to the transportation of hazardous goods. Travel planning, route guidance, and performance against plan can also be monitored and managed, and techniques such as such as weigh-in-motion or driver alertness monitoring can be incorporated. In some circumstances regulations can be electronically enforced.

2.6.3.2.1.7 Infotainment. Infotainment is used to describe information and entertainment services, in this scenario, from a service provider, via the vehicle's ITS communication(s) links, to the driver and passengers of the vehicle. Infotainment may include Internet services, video, online games, and television and video to passenger screens, as well as VOIP, digital radio, and music downloads to all occupants of the vehicle.

2.6.3.2.1.8 Internet In-Vehicle. Part of the generic infotainment, the provision of Internet services on-board is not only seen as desirable in a connected world, but also obviates the need to develop services dependent on specific ITS technologies. This is a key part of separating the services from the medium carrying the services (as far as possible). The Internet is enabled via an ITS communications interface.

2.6.3.2.1.9 Instant Messaging. An early use of this function is seen as the provision of messages using the in-vehicle screen, converted to synthesized audio to prevent driver distraction. There is debate about this service, however, because of the possibility of distracting the driver. The information is transmitted via an ITS communications interface or cellular communications.

2.6.3.2.1.10 Just-in-Time Repair Notification—Safety. Just-in-time repair notification is described in the next section as a commercial service between vehicle user and vehicle maintenance service provider/manufacturer. However, as more and more of vehicle operation is computer controlled, the possibility of the incident of safety critical software defects increases. Traditionally this has led to the commercially embarrassing, and expensive, recall of whole model ranges. Automotive manufacturers can use the ITS communications link to provide safety critical software updates, performance and efficiency updates, and to provide notification where service visits are required and arrange these on a just-in-time mode.

2.6.3.2.1.11 Optimal Speed Advisory. This is the provision of advice from data held within the vehicle and additional data and congestion information collected from the infrastructure to maximize fuel efficiency set against driver provided parameters and desired destination arrival targets. This usually combines satellite navigation and congestion information for a determined route with information from the engine management system. The information is transmitted via an ITS communications interface.

2.6.3.2.1.12 Parking Space Identification/Navigation. This is the provision of guidance (provided via the infrastructure) to identify the most suitable available parking, to navigate to the vehicle park, and sometimes navigate around the vehicle park to free or reserved spaces. The information is transmitted via an ITS communications interface.

2.6.3.2.1.13 Open Road (No Barrier) Tolling. This is the collection of road tolls without barriers on open roads. (CEN standards already exist to support this service. See Sections 4.6.1 and 8.8.)

The information is transmitted via an ITS communications interface.

2.6.3.2.1.14 Rental Car Processing. The checking in and out of rental cars requires verification of vehicle ID, mileage, and fuel level. ITS communications media can provide this data shortly before the vehicle reaches the return area, thus speeding the return process, and the processing of data within the rental company. ITS communications can also be used to verify check and wash procedures and advise automatically when maximum target mileage is achieved and the vehicle is due for replacement.

2.6.3.2.1.15 Route Guidance. Route guidance is achieved by an on-board processor and display, combined by triangulation from three or more GNSS satellites (such as GPS (NAVSTAR), GLONASS, and Galileo). These constantly transmitting signals are triangulated by the on-board equipment to provide location. The more satellite signals that are visible, the higher the accuracy. The new generation Galileo system claims to be more precise than other systems used for nonmilitary purposes. On-board equipment is therefore a passive radio receiver and processor with maps contained in memory.

2.6.3.2.1.16 Transit Vehicle Refueling Management. In order to prevent theft, transit (i.e., public transport) vehicle refueling is an administratively heavily supervised and documented process. The ITS communications link can provide regular updates on consumption/mileage driven and either direct ITS communications, or RFID linked to ITS can be used to authorize, release, and measure the amount of fuel released and feed this information into a fuel management and control system.

2.6.3.2.1.17 Transit Vehicle Data Transfer—Commercial. Transit vehicles are increasingly monitored for both safety and commercial reasons. Safety issues have been described above. Vehicle performance and driver behavior, adherence to timetables, passenger information systems updates, and priority at crossings can also be provided using an ITS communications link (RFID in earlier generation systems). Maintenance optimization and maintenance request and software downloads can also be provided using an ITS communications link.

2.6.3.2.1.18 Vehicle Emissions Monitoring. Early generation on-road vehicle emissions monitoring initially used vans with sensing equipment to record and photograph offenders. Current generations are tending to use roadside or over-road monitoring combined with an RFID tag in the vehicle both to identify the vehicle accurately and also report the date of its last safety and emissions check and certification.

2.6.3.2.1.19 Video Downloads. Part of infotainment, the provision of video downloads to passengers in the rear seats, who are often subject to "are we there yet" syndrome, is seen by some as a major driver for the provision of ITS services because vehicle owners may be more likely to invest in ITS communications equipment in their vehicles if this perennial problem can be eased. High data rate ITS communications links can be used to download videos. Toll highway providers, many of whom provide specific radio services accessible only to those who use the toll road as part of their value proposition to encourage usage of the toll road, appear interested to provide video's as part of their value proposition to road users.

2.6.3.2.1.20 Yellow Page Services (Via In-Vehicle Internet). Part of infotainment, the provision of yellow page services was once seen as being a specific ITS service sector. The provision of such services via the Internet means that service providers can be separated from media provision and thus reduce the commercial risks of service provision. Generic yellow page services, commonly available on the Internet, can be brought into the vehicle via an ITS communications interface.

2.6.3.2.2 On-Board Equipment to Infrastructure

2.6.3.2.2.1 Just-in-Time Repair Notification—Commercial. In-vehicle diagnostics can identify operating malfunctions and poor performance and can use the ITS communications link for software updates, downloads, and repair notification (and booking) on a just-in-time basis. (See also "Just-in-Time Repair Notification—Safety" above).

2.6.4 Vehicle-to-Vehicle ITS

The provision of vehicle-to-vehicle ITS services is very much in its infancy, and it tends to center specifically on two communications technologies: communications at 5.9 GHz and communications at 63 to 64 GHz, both of which have been allocated. The provision of vehicle-to-vehicle ITS services is particularly challenging, but some services, particularly in the field of collision avoidance, cannot be practically provided through the infrastructure.

Proponents of vehicle-to-vehicle communications also believe that many of the services envisaged to be provided via the infrastructure can equally be provided directly vehicle-to-vehicle. Some even feel that mesh networks of vehicles can obviate the need to lay much of the infrastructure otherwise envisaged.

Until larger volume trials in real traffic conditions have taken place, the next stage from technical feasibility trials, the capacity limits of 5.9-GHz technology for vehicle–vehicle communications, and the transmission limitations of 63-GHz technology for vehicle↔vehicle communications cannot be fully established. It is probable that both technologies, and a mix of infrastructure–vehicle,

vehicle↔vehicle, and vehicle↔infrastructure–vehicle modes of operation will be used in the end. However, the provision of standards at an early stage, as soon as the most probable technical solutions can be agreed upon, is essential, and work is already well advanced.

With respect to early generation vehicle↔vehicle services, these will mostly be driver advisory systems. Indeed, vehicle control systems using vehicle↔vehicle communications can only be effective when a high proportion of the vehicle population is equipped with compatible ITS communications equipment.

2.6.4.1 Vehicle↔Vehicle Safety Services

2.6.4.1.1 Approaching Emergency Vehicle Warning—Vehicle-to-Vehicle. Emergency vehicles traveling to incident sites use their ITS communications system, probably in broadcast mode, to warn all vehicles on single carriageway roads and when approaching junctions, or vehicles ahead of them on separated carriageways, of their approach as a prioritized vehicle. The warning is transmitted into the vehicle audio and/or visual display via an ITS communications interface.

2.6.4.1.2 Blind Merge Warning. This is the provision of warnings (from preceding/adjacent vehicles) to drivers where highways or lanes merge. These in-vehicle electronic systems monitor the position of a vehicle within a roadway lane and warn a driver if it is unsafe to change lanes or merge into a line of traffic. These systems are rearward-looking, radar-based systems. They assist drivers who are intentionally changing lanes by detecting vehicles in the driver's blind spot and passing the information to following vehicles. The warning is transmitted into the vehicle audio and/or visual display via an ITS communications interface.

2.6.4.1.3 Blind Spot Warning. Cooperative driving systems warn vehicles ahead, via an ITS communications interface, of their approach to overtake. The warning is transmitted into the vehicle audio and/or visual display via an ITS communications interface and in some cases is highlighted by a flashing icon in the passenger side wing mirror, so that even if the vehicle is in a blind spot, its approach is made known.

2.6.4.1.4 Cooperative Adaptive Cruise Control. Adaptive cruise control is similar to conventional cruise control in that it maintains the vehicle's preset speed. Cooperative adaptive cruise control can automatically adjust speed in order to maintain a proper distance between vehicles in the same lane by communicating, using an ITS communications link, with the preceding vehicle. The ITS cooperative communications link to the vehicle in front obtains relevant information from that vehicle concerning its speed, acceleration, deceleration, set cruise speed, and incidence of manual override. If the lead vehicle slows down, or if another object is detected, the system sends a signal to the engine or braking system to decelerate. Then, when the road is clear, the system will reaccelerate the vehicle back to the set speed.

2.6.4.1.5 Cooperative Collision Warning. Collision warning systems assist the driver to read surrounding traffic by warning of the unreasonable approach of an adjacent vehicle. In-vehicle systems achieve this by the use of distance sensing radars and a video screen and audio alert when a vehicle is approaching too rapidly, indicating the direction of the intrusion, and possibly the time to impact and speed of oncoming vehicle. Cooperative collision warning systems collect accurate data from the approaching vehicle and provide warnings, not only to the driver, but to the approaching vehicle using an ITS communications link. The objective is to predict threatening collisions early enough to prevent them from occurring, and by warning the driver to take evasive action. When collision avoidance is not possible, the damages can be limited to a minimum by reducing the speed and stopping distance. Later systems may involve automatically stopping the vehicle. The warning is transmitted into the vehicle audio and/or visual display via an ITS communications interface.

2.6.4.1.6 Cooperative Glare Reduction. When an ITS equipped vehicle detects that an approaching vehicle has full headlights on or inappropriately positioned headlights, it advises the offending vehicle of the problem. It is possible to envisage subsequent generation systems automatically triggering headlight dipping using an ITS communications interface.

2.6.4.1.7 Cooperative Vehicle-Highway Automation System (Platooning)—Infrastructure Assisted. (See also "Platooning—Vehicle Assisted" below.) This system is a next generation to adaptive cruise control, in which vehicles, communicating with each other, form dense platoons at high speed.

With the support of data collected via the infrastructure, accurate intercar spacing in the platoon is achieved by a longitudinal control system through the use of radar and radio communication between cars and the facilities of adaptive cruise control (i.e., each car in the platoon uses radar to measure the distance to the preceding car). The radio communication system provides each car with broadcast or ITS communications between vehicles providing the velocity and acceleration of the preceding car and the lead car of the platoon. All of these signals are used by the longitudinal feedback control system to continuously determine the desired acceleration of each car. The throttle or the brake is then used to provide the desired acceleration. Knowledge of the dynamic behavior of the throttle and brake actuators ensures that they are expertly controlled so that the desired acceleration is achieved very accurately.

Since platooning enables vehicles to operate much closer together than is possible under manual driving conditions, each lane can carry at least twice as much traffic as it can today. This will greatly reduce highway congestion. At close spacing, aerodynamic drag is also significantly reduced and fuel consumption and exhaust emissions are improved.

2.6.4.1.8 Curve Speed Warning—Vehicle Based. This is the provision of warnings from preceding vehicles to drivers to adapt speed to curve (bend) conditions. This may be adaptive to current weather and road conditions. Vehicle based curve speed warning systems use signals from preceding vehicles to warn drivers, typically those in commercial trucks and other heavy vehicles, of potentially dangerous speeds in approach to curves on highways. The warning is transmitted into the vehicle audio and/or visual display via an ITS communications interface.

2.6.4.1.9 Highway Merge Assistant. This is the provision of warnings from preceding, adjacent, and following vehicles to drivers where highways or lanes merge. These in-vehicle electronic systems monitor the position of a vehicle within a roadway lane and warn a driver if it is unsafe to change lanes or merge into a line of traffic. The warning is transmitted into the vehicle audio and/or visual display via an ITS communications interface.

2.6.4.1.10 Lane Change Assistant. This is the provision of assistance from preceding, adjacent, and following vehicles to assist vehicles to change lanes. The assistance is transmitted into the vehicle audio and/or visual displays via an ITS communications interface.

2.6.4.1.11 Left Turn Assistant—Vehicle Assisted. This is the provision of assistance to drivers turning left from other vehicles in the vicinity. This countermeasure involves warning motorists making a left turn at a traffic signal of a potential conflict with vehicles approaching from the opposite direction (which causes a total of 192,000 crashes per year in the United States [Barr, 2001]). Data is collected directly using vehicle↔vehicle communications to assist basic sensing to identify potential conflicts by determining the speed and the acceleration or deceleration rate of each vehicle approaching the intersection from the opposite direction, including vehicles executing through and right-turn movements. Simple point measurements will not be sufficient, since vehicles can assume various trajectories and acceleration/deceleration/stopping movements, particularly when other vehicles are present. The vehicle-vehicle communications are achieved via an ITS communications interface.

2.6.4.1.12 Merge Assistant. This is the provision of assistance to drivers merging into traffic lanes using information transferred from adjacent vehicles. The information is transmitted via an ITS communications interface.

2.6.4.1.13 Platooning—Vehicle Assisted. (See also *Cooperative Vehicle-Highway Automation System (Platooning)-Infrastructure Assisted* above.) This system is a next generation to adaptive cruise control, in which vehicles, communicating with each other using an ITS communications air interface form dense platoons at high speed by using vehicle-vehicle communications using an ITS air interface.

Accurate intercar spacing in the platoon is achieved by a longitudinal control system through the use of radar and radio communication between cars, as well as adaptive cruise control (i.e., each car in the platoon uses radar to measure the distance to the preceding car). The radio communication system provides each car with broadcast or ITS communications between vehicles providing the velocity and acceleration of the preceding car and the lead car of the platoon. All of these signals are used by the longitudinal feedback control system to continuously determine the desired acceleration of each car. The throttle or the brake is then used to provide the desired acceleration. Knowledge of the dynamic behavior of the throttle and brake actuators ensures that they are expertly controlled so that the desired acceleration is achieved very accurately.

Since platooning enables vehicles to operate much closer together than is possible under manual driving conditions, each lane can carry at least twice as much traffic as it can today. This will greatly reduce highway congestion. At close spacing aerodynamic drag is also significantly reduced, and fuel consumption and exhaust emissions are improved.

2.6.4.1.14 Precrash Sensing—Vehicle↔Vehicle Assisted. (See also "Precrash Sensing—In-Vehicle" above.) Precrash sensing functionality is defined in functional steps that require a increasing situation analysis performance and a growing amount of application effort. Each functional step makes it necessary to define the appropriate range of view, the virtual barrier. It is subject to various constraints and the configurations possible with precrash sensing. Precrash sensing technology uses platform radar sensors designed for the functional integration of possible functions that rely on sensor information from the close surroundings of the vehicle.

The main focus of precrash sensing is to help passive safety devices in protecting the passenger in all crash situations. Thus, passive safety devices will no longer be limited to airbags, seatbelts, and active headrests, but will use vehicle↔vehicle communications via an ITS air interface, new restraint systems, and automatic brake and steering activation as well. The latter devices, traditionally being active safety devices where there is driver interaction, will become more and more partially automated to compensate for deficits in human reactions.

2.6.4.1.15 Post-Crash Warning. These are warnings provided to nearby vehicles that the vehicle has been debilitated/damaged by a recent crash event. They may include a request to pass on messages to emergency services. The information is transmitted via an ITS communications interface.

2.6.4.1.16 Right Turn Assistant—Vehicle Assisted. This is the provision of assistance to drivers turning right by receipt of data from adjacent vehicles. See *Left Turn Assistant* above. The assistance data is transmitted into the vehicle audio and/or visual display via an ITS communications interface.

2.6.4.1.17 Road Condition Warning—Vehicle Assisted. This is the provision of warnings from nearby vehicles to drivers where there are adverse road conditions (e.g., ice, flood, obstacle in road, pothole, spillage, surface deformation). The warning is transmitted into the vehicle audio and/or visual display via an ITS communications interface.

2.6.4.1.18 Road Feature Notification. This is information sent via an ITS vehicle↔vehicle communications network to following vehicles of abnormal road features being approached. This may include permanent features such as particularly severe bends where no infrastructure based warning has been received, road deformation and damage, or temporary features such as an obstacle on the road, ice, or flood. The location of the event is transmitted and the distance from the event and time to reach the event calculated. The received message sounds a first alert informing the driver of the upcoming event, distance to event, and time to event, audibly or via the driver's video screen, and provides a second audible/visual warning just before the location of the event is reached.

2.6.4.1.19 SOS Services–Vehicle Assisted. This is the provision of connections to vehicles from police and emergency service vehicles for the transmission of voice and/or data communications; or the forwarding of such communications from other vehicles within the locality. The warning is transmitted into the vehicle audio and/or visual display via an ITS communications interface.

2.6.4.1.20 Stop Sign Movement Assistance—Vehicle Assisted. (See also *Stop Sign Movement Assistance–Infrastructure Assisted* above.) This countermeasure involves warning motorists leaving a stop sign that their movement may conflict with another vehicle. The subject vehicle movement could be left turn, right turn, or through. (causing a total of 362,000 crashes per year in the United States [Barr, 2001].)

Data from adjacent vehicles is used to identify potential conflicts by determining the speed, acceleration, or deceleration rate of each vehicle approaching the intersection and the discharge from the stop line for vehicles at stop-controlled approaches. The data is transmitted into the vehicle audio and/or visual display via an ITS communications interface.

2.6.4.1.21 Vehicle Alert. This alerts drivers to the presence of vehicles moving too fast at blind intersections. For example, if the system determines that a car is approaching a driver too fast from the left, an audible alert will sound and a voice recording will call out: "Car approaching from left" or similar. At the same time, an image of an approaching vehicle will appear on the driver's navigation screen. The system may also alert a driver when it detects that he or she is approaching a stop sign or red traffic light too fast.

2.6.4.1.22 Visibility Enhancer—Vehicle Assisted. (See also *In-Vehicle Visibility Enhancer* above.) This is a system that uses information from nearby vehicles, collected via an ITS air interface, together with in-vehicle infrared or low light camera techniques to provide enhanced visibility of dark objects, persons, and animals in the roadway at night via a head-up display or enhanced display on the driver's video screen. This may be supported by audible warnings.

2.6.4.1.23 Wrong-Way Driver Warning—Vehicle Assisted. (See also *Wrong-Way Driver Warning-Infrastructure Assisted* above.) These are warnings from vehicles, transmitted via an ITS communications interface, to offenders, to the road operator, and other road users of vehicles traveling in the wrong direction on a controlled direction carriageway.

When the navigation system of a vehicle identifies that the vehicle is traveling on a controlled direction carriageway or one-way road, and vehicle sensors or cameras detect another vehicle driving opposite to the permitted direction, or the driver is proceeding in the wrong direction to that permitted, a warning is sent to the violating vehicle itself (or to the driver via audio/video display), to the traffic control center, and to nearby traffic. The warning is transmitted between vehicles via an ITS communications interface.

2.6.4.2 Vehicle↔Vehicle Commercial Services

Such services may include route optimization, sharing of information with and between drivers, and collation and forwarding of fleet management information.

2.6.5 Off-Trip Services

2.6.5.1 Pretrip Dynamic In-Vehicle Route Guidance and Navigation Programming/Setup

This involves the programming of a trip route prior to the start of the trip. The pretrip travel information user service allows travelers to access a complete range of real-time multimodal transportation information at home, work, and other major sites where trips originate. Information on road network conditions, incidents, weather, and transit services, are conveyed through these systems to provide travelers with the latest conditions and opportunities in order to plan their travel. Based on this information, the traveler can select the best departure time, route, and modes of travel, or perhaps decide not to make the trip at all.

2.6.5.2 Pretrip Integrated Multimodal Trip Guidance

This is the provision of route guidance advice from the infrastructure via land line or ITS air interface to account for road use, road condition, and congestion information. The pretrip travel information user service allows travelers to access a complete range of real-time multimodal transportation information at home, work, and other major sites where trips originate to enable multimodal routes to be planned.

2.6.5.3 Pretrip Pedestrian or Bicycle Route Guidance

This is the provision of route guidance advice from the infrastructure via land line, cellular, or wireless broadband to account for road use, road condition, and congestion information, adapted to the special needs of the mode of transport (bicycle or pedestrian in this case).

2.6.5.4 Trip Planning: Current Utilization Levels from Public Transport Information Systems

This is the provision of current utilization levels from public transport information systems, per trip, to the intending traveler, from the infrastructure via land line, cellular, or wireless broadband.

2.7 Services to Public Transport Users and Pedestrians

2.7.1 Dynamic Ridesharing

A ride matching and reservation user service provides real-time ride matching information and reservations to travelers in their homes, offices, or other locations, and assists transportation providers with vehicle assignments and scheduling. The user service provides one of the basic tools for altering the travel behavior of individuals who drive alone during congested periods.

For public transit, route deviation schemes, where vehicles would leave a fixed route for a short distance to pick up or discharge passengers, are another approach employed to improve the service. Vehicles providing this service can include small buses, taxicabs, or other small, shared-ride vehicles.

2.7.2 Emergency Call/Mayday Alert for Public Transport

Mayday requests from travelers, drivers, and transit vehicle operators are passed to the emergency system operator using an ITS or other communications link. The process involves acknowledgment of the emergency request from the system operator, which is passed on to the requestor. This process inputs processing parameters for biometric image matching analysis from the system operator, and outputs those parameters to another process. That process in turn provides image matching results to this process, which are then returned to the system operator.

2.7.3 Intrusion Detection

Intrusion detection (or motion detection) at transit facilities (e.g., transit yards and shops) remotely monitors sensor data collected in secure areas, including those frequented by travelers (transit stops, rest areas, park and ride lots, modal interchange facilities, on-board transit vehicles) and those typically away from travelers (tunnels, bridges, roadway infrastructure). Sensor data can also be collected from other emergency centers. The process returns collected sensor data (raw and processed) to those centers. The types of sensor data include threat sensors (such as chemical, biological, explosives, and radiological), object detection sensors (such as metal detectors), motion and intrusion sensors, and infrastructure integrity sensors. In addition to raw sensor inputs, this process also receives data preprocessed in the field, and provides additional processing if directed by processing parameters established by center personnel. The process can input threat information from field analysis functions and, together with its own processing, identify potential threats and verify those threats by correlating collected data. The process outputs identified threats to other processes for output to center personnel and to support further threat analysis.

2.7.4 Paratransit Fleet Dispatch

Paratransit is an alternative mode of flexible passenger transportation that does not follow fixed routes or schedules. Typically vans or mini-buses are used to provide paratransit service, but other important providers include share taxis and, in parts of Asia, "jitneys" and "Tuk-Tuks." Paratransit services may vary considerably on the degree of flexibility they provide their customers. At their simplest they may consist of a taxi or small bus that will run along a more or less defined route and then stop to pick up or discharge passengers on request. At the other end of the spectrum is fully demand-responsive transport: the most flexible paratransit systems offer on-demand call-up door-to-door service from any origin to any destination in a service area. Paratransit services are normally operated by public transit agencies, community groups, or not-for-profit corporations, and for-profit private companies or operators.

Such systems are optimized by sophisticated central planning and management. Typically users call in by phone and the central control locates the best unit to respond. This is more complicated than simple shared ride taxis, because the destinations will differ more significantly, and so efficiency is gained by simultaneous multidestination routing.

2.7.5 Public Transport Fares Management

This is the management of fare collection by fare (basis of charge) category and payment type, as well as the use of ITS/IT/ICT to efficiently manage traffic flows and collect fares. This includes management of fare rates according to demand levels across the day, or on particular days, or long-term price management to encourage/discourage usage patterns.

2.7.6 Public Transport Service Dispatch

This provides potential solutions to dispatchers and operators to optimize service dispatch and to facilitate quick responses to service delays, as well as to forecast service delays.

2.7.7 Public Transport Scheduling Services

This service maintains transportation schedules and assures transfer connections from vehicle to vehicle and between modes. This can be coupled with traffic control services to facilitate a quick response to service delays.

2.7.8 Public Transport Service Planning

Public transportation service planning automates the operations, planning, and subsequently the management functions of public transit systems. It provides real-time computer analysis of vehicles and facilities to improve transit operations and maintenance. This service helps to maintain transportation schedules and assure transfer connections from vehicle to vehicle and between modes Service schedulers will have timely data to adjust trips. Personnel management can be enhanced with the automatic recording and verification of operating and maintenance task performance.

2.7.9 Public Transport Surveillance

These are systems that monitor the environment in transit facilities, transit stations, parking lots, bus stops, and on-board transit vehicles, and which generate alarms (either automatically or manually) when necessary. The service also provides systems that monitor key infrastructure of transit (rail tracks, bridges, tunnels, bus guideways).

2.7.10 Public Transport Vehicle Fleet Tracking

This service monitors the location of transit vehicles, identifies deviations from the schedule, and offers potential solutions to dispatchers and operators. This service helps to maintain transportation schedules and assure transfer connections from vehicle to vehicle and between modes and can be coupled with traffic control services to facilitate quick responses to service delays.

2.7.11 Public Transport Vehicle Internal Systems Monitoring

This service involves, for example, measurement and reporting of tire pressures, the identity of towed trailers, suspension raising/lowering for improved access, engine efficiency and temperature, shock and vibration sensing, use of signal prioritization.

2.7.12 Safety Enhancements for Vulnerable Road Users

This service group covers the application of ITS functionality to the enhancement of safety levels for vulnerable road user groups (particularly the elderly or disabled, as well as road maintenance workers). User groups and service functions include:

- Motorcyclists;
- Pedal cyclists;
- Pedestrians;
- Smart pedestrian crossings (e.g., automatic warning of pedestrians for drivers, prolonging crossing times for elderly users, and changing pedestrian prioritization);
- Smart pedestrian crossings (e.g.; prolonging crossing times for elderly and disabled users);
- Vehicle presence detection (either to the vehicle or from the vehicle to the pedestrian);
- Automatic advice to drivers by disabled road users (e.g., presence of wheel chair);
- Intersection monitoring of specialized conveyances (e.g., wheelchairs, carts);
- Driver warnings for specialized conveyances;
- Clarification of right of way rules;
- Onboard echo of warning signs;
- Identification of the presence of oncoming vehicles by aural or tactile means;
- Warning of imminent signal phase change;
- Signal display advance warning;

- Oncoming vehicle advance warning (for junction without signals);
- In-vehicle signage and warning systems.

2.7.13 Silent Alarm

This is the provision of silent alarms via an ITS air interface or other means, including the provision of location information. This involves the silent activation of video cameras and the transmission of images, and includes systems on board public transit vehicles (bus, rail car) for video, audio (including covert microphones that can be triggered by the transit vehicle operator), and event recording (i.e., black box), as well as outputs to monitor activities, incidents, and potential threats. These systems and sensors are normally monitored by central dispatch.

2.7.14 Travel Services Information—Dedicated Location

This service provides travel services information specific to a particular location (e.g., train station, sports arena).

2.7.15 Travel Services Information—Personal Interactive

The traveler services information user service provides a business directory (or yellow pages) of information on travel-related services and facilities (e.g., the location, operating hours, and availability of food, lodging, parking, auto repair, hospitals, and police facilities). Traveler services information would be accessible from the home, office, or other public locations to help plan trips, and it would also be available en route. The service includes not only the traveler services information, but the capability to make reservations for many of the traveler services.

Personalized public transit user service supports flexibly routed transit vehicles. Small, publicly or privately operated vehicles provide on-demand routing to pick up passengers who have requested service and deliver them to their destinations.

2.8 Technology in Evolution

For many of the services briefly described above, it may seem to be too soon to be considering standardization—indeed, it may seem too soon even to identify which will need standards. Where work is known to be ongoing or completed, it has been identified in parenthesis. However, much of the content of these services are in conceptual design, research, or preliminary tests. It may be considered that, after the finalization of the communications platforms(s) to support ITS, these areas will form the bulk of ITS standardization in the years to come.

In order to enable such services to be developed and provided, standardized ITS communications interfaces and their behavior must first be defined and "future-proofed" as far as possible (because the long lead times for the development and implementation of some of these technologies will exceed the lifespan of many current telecommunications technologies), and methods to network between interfaces need to be defined and proved. This work forms the core of many current ITS standardization activities, and relevant standardization work is described in Chapter 4.

PART II
Standards to Achieve ITS Services

CHAPTER 3

An Introduction to Communication Technologies Standards for ITS

NOTE: Detailed consideration of specific wireless communications standards and regulations for ITS is provided in Chapter 4.

3.1 On-Board ITS Standards

3.1.1 Introduction to In-Vehicle ITS Systems

In Section 2.6.1 we described the concept of in-vehicle ITS services. These services are provided entirely within the vehicle, and require no bidirectional communication outside of the vehicle. As we have seen, they may be fully automatic, and can provide corrective action or in other cases a warning. As we observed in Section 2.6.1, the provision of an in-vehicle video screen and the human-machine interface to react with it also provides new capabilities for intelligent driver assistance.

Standards for these in-vehicle systems are usually developed within ISO/TC 22 (Road Vehicles), although, as we observed in Section 2.6.1, since they have no need for connection with the outside world, they may well be entirely proprietary and outside the realm of standardization.

However, it is both in the interest of vehicle manufacturers and road safety, that, where interaction with the driver is involved, whether active (as in connecting a seat belt) or semi-passive (as in providing a warning signal), a driver can step from one vehicle to another and interact with the system intuitively, and without having to read the operation manual.

We all change our vehicles from time to time, and many of us travel by plane or train and hire a vehicle at our destination. We get into a car, usually from a different manufacturer than that in our garage at home, pick up the keys, adjust the mirrors and seat position, and drive away. Some of us own more than one vehicle, and as these vehicles often have different main purposes, they are frequently from different manufacturers. We step from one to the other, with little adjustment usually required. For example, most vehicles in the world put the direction indicators on the left of the steering column, and the windshield wipers on the right, while some Asian manufacturers put direction indicators on the right and the wiper controls on the left. So when you are around an airport, watch out for someone operating their windshield wipers, it usually means that they are about to imminently turn left or right!

We mention this anomaly, because it actually highlights how little we have to think about these things, even when we change from a right-hand drive vehicle to a left-hand drive vehicle, or from a car to a van. This has not happened by accident. Motor vehicle manufacturers have worked assiduously over the years to make driving an interoperable, easy, and safe experience. With respect to in-vehicle ITS systems, they continue to do so, largely by developing standards. So while, as we observed in Section 2.6.1, the actual technology providing the ITS might vary, the human-machine interface (HMI), will be standardized or an industry practice.

In Section 2.6.1 we also used the example of tire pressure sensors requiring international standards, because tires will be replaced several times during the lifetime of the vehicle, and in a competitive world may be replaced by tires from a different manufacturer.

The general standards for driver interoperability, while being of very significant importance, are not the subject of this book, the subject of which is specifically ITS-related standards, rather than general vehicle operability and HMI standards. The reader who is interested in these generic standards is directed to the following standards.

For ISO/TC 22 Road Vehicles standards, see the following.

Downloadable FOC abstracts TC 22
http://isotc.iso.org/livelink/livelink?func=ll&objId=138072&objAction=browse&sort=name

For SAE Vehicles standards, see the following.

Downloadable FOC abstracts SAE
http://www.sae.org/technical/standards/ground_vehicle/

The individual standards for these systems, where they relate to ITS services, are to be found in Section 8.4.

3.2 Navigation Systems

3.2.1 SAT-NAV

A satellite navigation (SAT-NAV) system may be generically described as a "navigation system using location information from orbiting satellites." Although current generation navigation systems receive signals from satellites, the communication is one-way and so they are classified as *on-board systems*.

3.2.2 SAT-NAV+

While the majority of SAT-NAV systems are purely on-board or portable devices with a one-way communication link, there is a growing trend (at the time of this writing) to link the navigation data with congestion and route guidance data using 2G/3G mobile telephones. These current generation devices generally seek the presence of a telephone by Bluetooth short-range wireless communications links, and so long as the link is recognized and authorized, they will contact a subscribed information service (usually provided by the provider of the SAT-NAV equipment) for congestion information and will provide a hands-free telephone via the SAT-NAV speakers and an attached microphone.

This, then, moves the technology from being only an on-board device to a bidirectional communications device.

The future is clear to see. Either using the direct telephony links available, or in the future by linking into the vehicle's CALM (which stands for communications access for land mobiles; see Sections 4.3 and 4.6.2) or other communication link, or by directly accessing a mobile broadband service, various levels of driver guidance and advisory systems will evolve.

At the time of this writing, there is no direct standardization work being undertaken in this area, although it may be expected in the near future.

SAT-NAV and Standards

The standards for SAT-NAV operation are already provided in the operating specifications of GPS, GLONASS, and GALILEO. No SDO standardization is therefore required.

Specifications can be obtained from:

Downloadable FOC NAVCEN www.navcen.uscg.gov/gps/geninfo
Downloadable FOC GLONASS www.fas.org/spp/guide/russia/nav/glonass.htm
Downloadable FOC ESA www.esa.int
Downloadable FOC Europe-energy_transport http://ec.europa.eu/dgs/energy_transport/

Links from on-board SAT-NAV systems via 2G and 3G telephone systems are controlled by 3GPP and 3GPP2 standards for mobile telephones. These are general standards and are not ITS specific, and details can be obtained from:

Downloadable FOC 3GPP www.3gpp.org
Downloadable FOC 3GPP2 www.3gpp2.org

See also Section 4.2 for further information.

The links between wireless communications between mobile telephones and SAT-NAV are likely to be either Bluetooth or ZigBee. These are general standards and are not ITS specific, and details can be obtained from:

Downloadable FOC Bluetooth www.bluetooth.com

See also Section 3.11.

Downloadable FOC ZigBee www.zigbee.org

See also Section 3.11.

SDO standards become required in order to provide applications (ITS services) via such links. The reader is directed to the appropriate service domain in Part III of this book, and possibly Chapter 7 if your enquiry relates to an identification service.

3.3 Radar Systems

Ford initially introduced adaptive cruise control (ACC) on its Jaguar vehicles in Europe in the late 1990s. The system uses a high-performance, mechanically scanning, long-range radar sensor to help the driver detect objects up to 150m in the vehicle's forward path. If the lane ahead is clear, the system will maintain the cruising speed set by the driver. When slower traffic is detected, the system will maintain a driver-selected headway (adjustable between 1.0 and 2.2 seconds) using throttle control and limited braking. The system also can detect if another vehicle crosses into the path from an adjacent lane and respond appropriately. ACC became available from Mercedes Benz and Lexus (Toyota) in 2000, and from Infiniti (Nissan) in 2001. General Motors introduced ACC for the 2003 model year on their Cadillac XLR.

Radar systems are essentially on-board devices in that there is no bidirectional communication (in current generation devices).

An extension to adaptive cruise control technology is a service such as DELPHI's Forewarn Forward Alert system. When the driver has enabled the Forward Alert system (regardless of whether ACC is on or engaged), audible and visual alerts will be provided to the driver that braking is required due to slower-moving vehicles ahead.

Drivers can adjust system sensitivity to adapt alert levels to their preferred driving style and prevent nuisance alerts. It is important to note that the Forward Alert system will not warn of stopped vehicles or fixed objects ahead and should not be expected to prevent collisions in these situations.

Where standard cruise control provides added convenience for interstate highway travel, ACC gives similar levels of convenience on more congested roads and traffic conditions. The system may include driver information displays that indicate the set cruising speed and driver-selectable gap. With a Forward Alert type of system, drivers have the additional capability of receiving automatic alerts in nearly any type of road or traffic condition that they should brake for slower-moving traffic ahead, even when they are not operating with ACC on or engaged.

See Sections 5.4.1, 4.1.1.3, 4.1.2.1 (Part 15 -252/509/515), 8.4.3.1, and 12.2.1 (ARIB T48).

Some forward obstacle and rear obstacle alert systems use radar to detect objects (although ultrasound, optical, or infrared techniques are more common). See Section 8.4 for details of standards for these services, and also multiple references in the ITS services described in Chapter 2.

ACC and Forward Alert systems are generally proprietary, and there is no requirement for standards to enable interoperability, as interoperability as such is not required. However, as these systems become more common, standards are likely with respect to measuring the scope and effectiveness of the systems, and with respect to common driver interface operations. At the time of writing, no existing work is underway in any of the leading SDOs.

ITU Recommendations to national frequency regulators exist in "Low power short-range vehicular radar equipment at 60 GHz and 76 GHz" (May 2000), at http://www.itu.int/rec/R-REC-M.1452/en.

CEPT Recommendations to its member national frequency regulators exist at http://www.ero.dk/documentation/.

ETSI Communications standards for vehicle radar are developed by ETSI ERM. At the moment the principal developed ETSI standards for vehicle radar are:

- EN 302 288 24 GHz Automotive Radar (until June 30, 2013). (See Section 4.1.3.3.) http://webapp.etsi.org/exchangefolder/en_30228801 v010301o.pdf.
- EN 302 264 79 GHz Automotive Short Range Radar. http://europa.eu.int/eur-lex/lex/LexUriServ/LexUriServ.do?uri=OJ:L:2004:241:0066:0067:EN:PDF.

3.4 Optical Systems

3.4.1 On-Board Systems

On-board optical systems tend to present themselves as digital imaging devices (basically a digital camera that compares current image with previous image) or a video camera, usually installed to provide reversing views on the in-vehicle screen, collating data and providing appropriate warnings. Stand-alone systems are more likely to use ultrasound (see Section 3.5) or infrared (see Section 3.6). They require no interoperability and are generally proprietary devices, and therefore require no standards.

3.4.2 Enforcement Systems

Most enforcement systems capture a digital or film image of the offending vehicles. The authors are aware of no international standards with respect to the specification or performance of such devices; however, most national law enforcement agencies have their own minimum performance specifications.

3.4.3 Road Charging, Access Control, and Similar Systems

Access control and road charging systems may also use optical number plate recognition. The world's largest congestion pricing system (located in London) uses optical number plate recognition. The authors are aware of no international standards with respect to the specification or performance of such devices; however, most national law enforcement agencies have their own minimum performance specifications.

3.5 Ultrasound/Sonar Systems

Most forward and rear obstacle detection systems use ultrasound devices using the principal of SONAR (which is the acronym for SOund NAvigation and Ranging). The principle involves bouncing acoustic waves off of objects and determining their distances by measuring the time for the echoes to return. The technology (originally devised for submarines) is generic and the products are proprietary. There is no interoperability involved and no standards required, except to compare performance.

3.6 Infrared Systems

Infrared may be used as a bidirectional communication system. (See Sections 4.6.2.5, 4.6.2.6, and 4.6.3.) With respect to on-board systems, infrared, like ultrasound, can be used for object and movement detection. Infrared may also provide more capable obstacle detections systems, such as by collating data from a front-mounted radar and a twin-lens infrared stereo camera on the top edge of the windshield that picks up infrared radiation emitted from projected from the headlights and reflected off objects up to 25m away (LEXUS).

3.7 Wireless Systems Within a Vehicle

There is a recognition among vehicle manufacturers that:

- Communications units, such as 2G/3G phones and PDAs will be introduced to vehicles.
- Aftermarket products, such as SAT-NAV with communications capabilities (see Section 3.2), will be introduced to vehicles.
- Portable computers with wireless communications facilities will be used within vehicles.
- Vehicle users will want to bring their MP3 players and similar devices into vehicles and connect using plug or wireless technology (such as Bluetooth and ZigBee). (See Sections 3.11.2 and 4.2.7.)

Indeed, some automotive manufacturers see this as a way of simplifying or reducing the cost of automotive manufacture, by simply providing the wired and wireless access ports, instead of installing built-in radios, CD players, and phones.

These types of wireless networks are generically known as *personal area networks* (PANs). Further detail of the concepts and standards supporting PANs (such as Bluetooth and ZigBee) are provided in Sections 4.2.6 and 4.2.7.

It is also possible to use Wi-Fi to make a wireless local area network (WLAN), although this is a more complicated subject. For communication within a vehicle, the Wi-Fi protocols could be used either to provide a wireless access point to enable Wi-Fi-enabled devices within the vehicle to connect to the vehicle network, or as so-called "peer-to-peer" networks between the vehicle's WAN access point and that of a transient or permanent device within the vehicle (though we have not yet heard of this being proposed as a way of connecting permanent equipment within a vehicle). However, so long as a vehicle remains within a Wi-Fi hotspot—say, at a service station—Wi-Fi can be used in a limited way (there is no cell–cell handover) to support some ITS service provision (e.g., update of traffic information, bus/train/plane arrival and departure information, and so forth). Further details of the concepts and standards supporting WLAN are provided in Sections 3.10 and 4.2.3. At the time of this writing, ISO TC 204 was in the process of creating a new working group to consider standardization issues for nomadic devices in vehicles.

3.8 Infrastructure/Infrastructure Standards for ITS

3.8.1 Introduction

ITS is often envisaged as service provision that happens in the vehicle or for travelers on the move, and these aspects are an important facet of ITS service provision, but they are only part of ITS service provision.

One of the most important sectors of ITS has to do with traffic management and the exchange of information between control centers and the operation of street equipment—for example, to provide "green wave" traffic light control and traffic management systems such as SCOOT. The traveler information boards at bus stops and metro and rail stations, which are currently very popular, rely principally on hard wired systems.

Even when we consider services to vehicles, many of these require back office management and distribution to the roadside beacon/access point, which rely on infrastructure systems. Most of these infrastructure systems use generic infrastructure that is not purpose designed and is used by the ITS sector. However, at 64 to 66 GHz, possibly at 63 GHz, and at 5.9 GHz there are likely to be dedicated infrastructure systems for ITS.

As most countries are well equipped with fiber optic, wired, and wireless telecommunications systems, and considering the expected roll out of generic wireless broadband systems, generic telecommunications infrastructure architectures will be used and, where appropriate, interfaced to specific ITS systems—hence the potential significance of the CALM concept discussed in Section 4.3.

Within the context of this book, the impact of current infrastructure systems is very significant. In our review of systems that can be used to provide ITS service provision, although we do not attempt to list every telecommunications standard, we do have to include generic telecommunications infrastructure standards that can provide generic services (such as video and graphics), that may be used to support ITS service provision. This, therefore, forms a major chapter in this book (Chapter 6). Indeed, as this book is primarily designed to be a reference work for ITS specialists and students (rather than infrastructure telecommunications standards specialists), it is designed to provide an easy reference location to assist them to locate the relevant telecommunications standards that may be useful in the design and implementation of ITS service provision. We do not claim that this book provides references to *all* communications infrastructure standards—it does not; but it does provide pointers to, and in many case brief summaries of, generic/communications/infrastructure standards that are most likely to be used as part of or in support of ITS service provision.

3.8.2 Hard Wired Systems

Following the logic of Section 3.8.1, hard wired systems will operate according to the protocols defined in general telecommunications standards. Other than where these standards serve as possible support features for ITS service provision, we do not list these general wired standards. If knowledge of these standards is required, reference is made to the ITU Web site for further details (http://www.itu.int/rec/T-REC-G/e).

With respect to support features, we refer the reader to Chapter 6.

Because ITS services may often depend on the rapid transport of messages, a few words on the impact of the development of optical fibers in telecommunications networks is relevant. These developments have provided significant improvements in network performance, which has led to a fundamental change in the very nature of the use of telecommunications networks and has helped to enable the evolution of the communications backbone of the Internet (even if service delivery to the end user has remained largely dependent on copper cable). Communication systems have diversified from telephony applications to connecting a wide variety of multimedia information sources across the Internet.

Low-cost systems are now available even at transmission rates of more than 10 Gbps (gigabytes per second), a speed which until recently was regarded as possible only in high-cost national systems. Great advances have also occurred in standard twisted pair wire communications, with recent systems able to demonstrate fully broadband communication services over distances in excess of 1 km.

Additional substantial increases in transmission capability, as large as those recently delivered, can be envisaged within the next few years, allowing much greater levels of interactivity and functionality across the network. Innovations in optical fiber and wire based communication systems can be expected to allow the transmission of both digital and analog signals for both computer and mobile communication systems.

In order to have components that function compatibly in fiber optic communication systems, a number of standards have been developed. The International Telecommunication Union publishes several standards related to the characteristics and performance of fibers themselves, including: (1) ITU-T G.651 Characteristics of a 50/125 μm multimode graded index optical fiber cable, and (2) ITU-T G.652 Characteristics of a single-mode optical fiber cable. These are downloadable FOC via http://www.itu.int/rec/T-REC-G.651-199802-I/en.

More generally, ITU-T Recommendations can be downloaded via http://www.itu.int/rec/T-REC-G/e, and ITU-R (Radio) Recommendations can be downloaded via http://www.itu.int/ITU-R/index.asp?category=publications&link=rec-search&lang=en.

Other recommendations and standards that may be of interest in this area are:

IEEE 802.3-2005 Gigabit Ethernet.

Downloadable ITU Ethernet Recommendations
http://standards.ieee.org/getieee802/802.3.html

ISO/IEC 9314-9 FDDI – Fiber distributed data interface.

Downloadable ITU FDDI
http://www.iso.org/iso/en/CatalogueDetailPage.CatalogueDetail?CSNUMBER=22760

Fiber Channel. Standards here are messy and we recommend going to the Wikipedia site http://en.wikipedia.org/wiki/List_of_Fibre_Channel_standards for further details.

High-Performance Parallel Interface (HIPPI).

Downloadable HIPPI
http://www.hippi.org/

Synchronous Digital Hierarchy. ITU-T G783, G803, G813, G825 and others,

http://www.itu.int/rec/T-REC-G/e
Downloadable ITU SDH
http://www.itu.int/rec/T-REC-G/e

ETSI ETS 300 147.

Downloadable ETSI ETS 300 147
http://portal.etsi.org/docbox/EC_Files/EC_Files/ts_1021470102v010201p.pdf

EN 300 417 -6.

Downloadable ETSI ETS 300 147 -6
http://www.itu.int/ITU-T/2001-2004/com15/otn/SDH-rec.html

ITU-T G.707 and G.708 Synchronous Optical Networking.

Downloadable ITU-T G707/708
http://www.itu.int/rec/T-REC-G/e

Toslink- AES/EBU.

Downloadable IToslik – AES/EBU
http://webstore.iec.ch/webstore/webstore.nsf/artnum/031856

3.8.3 Wireless Systems

Whether vehicle↔vehicle, vehicle↔infrastructure, or infrastructure↔infrastructure, wireless communication plays a key element in ITS service provision. Note that communication to, from, with, and between vehicles is, by definition, wireless. Chapter 4 considers these issues in detail as they relate to the various communications types that may be encountered in ITS service provision.

BRAN standards are for equipment which provides broadband (25 Mbps or more) wireless access to wire-based networks in both private and public environments operating in either licensed or license-exempt spectrum. These systems address both business and residential applications. These are fixed wireless access systems which are intended as high performance, quick to set up, competitive alternatives for wire-based access systems.

The specifications address the physical (PHY) layer as well as the data link control (DLC) layer (with medium access and logical data link control sublayers as appropriate). Interworking specifications (convergence layers) that allow broadband radio systems to interface to existing wired networks, notably those based on ATM, TCP/IP protocol suites, and 3G mobile networks, are also specified.

Other bodies of interest in this area are the ATM Forum, the HiperLAN2 Global Forum, the IEEE Wireless LAN Committees P 802.11a and IEEE 802.16, the Internet Engineering Task Force, the MMAC-PC High Speed Wireless Access Systems Group, the International Telecommunication Union Radio sector (ITU-R), and a number of internal ETSI Technical Bodies.

ETSI BRAN currently produces specifications for three major standard areas:

- HiperLAN2, a mobile broadband short-range access network;
- HIPERACCESS, a fixed wireless broadband access network;
- HIPERMAN, a fixed wireless access network which operates below 11 GHz.

HiperLAN2

ETSI claims that HiperLAN/2 "is designed to give consumers in corporate, public and home environments wireless access to the Internet and future multimedia, as well as real time video services at speeds of up to 54 Mbit/s. The system will be quick and easy to install and provide interworking with several core networks including the Ethernet, IEEE 1394, and ATM."

The HiperLAN2 Web site, http://portal.etsi.org/bran/bran_tor.asp, provides more information about HiperLAN2.

HIPERACCESS

These are standards for broadband multimedia fixed wireless access. ETSI claims that the HIPERACCESS specifications "allow for a flexible and competitive alternative to wired access networks. It will be an interoperable standard, in order to promote a mass market and thereby low cost products."

HIPERACCESS targets high frequency bands. It will be particularly optimized for the 40.5 to 43.5-GHz band and is closely associated with IEEE-SA (Working Group 802.16) to harmonize the interoperability standards for broadband multimedia fixed wireless access networks.

The published HIPERACCESS specifications include:

- The HIPERACCESS system overview (ETSI TR 102 003), http://webapp.etsi.org/WorkProgram/Frame_WorkItemList.asp?qETSI_NUMBER=102+003;
- Te HIPERACCESS PHY specification (ETSI TS 101 999), http://webapp.etsi.org/WorkProgram/Frame_WorkItemList.asp?qETSI_NUMBER=101+999;
- The HIPERACCESS Data Link Control (DLC) Layer (ETSI TS 102 000), http://webapp.etsi.org/WorkProgram/Frame_WorkItemList.asp?qETSI_NUMBER=102+000;
- The Convergence Layers for the cell and packet based core networks (ETSI TS 102 115), http://webapp.etsi.org/WorkProgram/Frame_WorkItemList.asp?qETSI_NUMBER=102+115;
- ETSI TS 102 117, http://webapp.etsi.org/WorkProgram/Frame_WorkItemList.asp?qETSI_NUMBER=102+117).

Further information on HIPERACCESS can be found at http://portal.etsi.org/bran/bran_tor.asp.

HIPERMAN

HIPERMAN is an interoperable broadband fixed wireless access system operating at radio frequencies between 2 and 11 GHz. ETSI claims that the HIPERMAN standard "is designed for Fixed Wireless Access provisioning to SMEs and residences using the basic MAC (DLC and CLs) of the IEEE 802.16-2001 standard. It has been developed in very close cooperation with IEEE 802.16, such that the HIPERMAN standard and a subset of the IEEE 802.16a-2003 standard will interoperate seamlessly." HIPERMAN is capable of supporting ATM, though the main focus is on IP traffic. It offers various service categories, full quality of service, fast connection control management, strong security, fast adaptation of coding, modulation and transmit power to propagation conditions, and is capable of non-line-of-sight operation. HIPERMAN enables both PMP and Mesh network configurations. HIPERMAN also supports both FDD and TDD frequency allocations and H-FDD terminals. All this is achieved with a minimum number of options to simplify implementation and interoperability.

More information about HIPERMAN related work items can be found at:

http://webapp.etsi.org/WorkProgram/Frame_WorkItemList.asp?qPROJECT_CODE=HIPERMAN;685

The BRAN standards can be downloaded from:

Downloadable ETSI BRAN
http://portal.etsi.org/docbox/EC_Files/EC_Files/ts_1021470102v010201p.pdf
http://portal.etsi.org/docbox/EC_Files/EC_Files/ts_1021470202v010101p.pdf

While the standards listed in this section provide useful information regarding the "fixed" infrastructure, and are useful in that context of ITS service provision, it must be remembered that moving vehicles not only require wireless systems, but in most cases, because they are a moving location, require what we have come to call "cellular" wireless systems (i.e., access points with cell_cell handover). This requirement and its consequences feature significantly in Chapter 4.

3.8.4 Internet-Driven Systems

In the early days of development of the ITS sector, it was envisaged that a whole business infrastructure would have to be created to support the provision of commercial services for ITS. This, of course, had some lusting after potential new business opportunity (particularly with respect to location based services), while others argued pessimistically that the scale of such business could never justify the significant investment required in the early years, so would therefore never happen.

In the intervening years, however, the Internet has developed rapidly, and while many of the ITS-specific services described in this book require specific communications services, the commercial services can largely be provided via the Internet. This is, of course, achieved by the Internet technique of separating the provision of the service from the provision of the medium carrying the Internet connection.

Because different media will be available in different locations, or because vehicles will support only certain media, we have the same requirement for all but time-critical safety services-that is, to separate out the provision of the service from the provision of the communications media. This is achieved through a major initiative called CALM (see Section 4.3), which, for all but time-critical safety services, uses the principles of the Internet, and IPv6 in particular.

3.9 Sensors

If you read through the list of described ITS Services in Chapter 2, you will have seen the word "sensor" appear many times. Sensors, and the act of sensing, occur in many different ways; some of it is calculative based on known parameters or expected assumptions, and often it relies on specific sensing equipment (sensors). This may be, for example, a sensor to measure tire pressure, a biometric fingerprint sensor to authenticate a driver, a temperature sensor in a refrigerated truck, or a presence sensor at a junction.

Most commonly, sensors will be hard wired to some equipment or a vehicle data bus and therefore simply become "information"—data to be moved around an ITS system in order to provide some facet of the service.

There is also a role for stand-alone sensors that are not permanently attached to a piece of equipment, or, as in the case of tire pressure sensors for example, their location makes hard wiring impracticable. In these cases the link to the ITS system will most often be provided via an RFID link.

More information about the use of sensors in ITS, biometric, and RFID systems can be found in Sections 7.2, 7.4 (especially Sections 7.4.5 and 7.4.7), and 7.5, and in use in Section 8.4.4; and their roles in ITS service provision is best seen in Chapter 2.

3.10 Wi-Fi

Wi-Fi is the appellation given to compliant and approved systems of the members of the Wi-Fi Alliance, using the IEEE 802.11 series of standards for WLAN. Summaries of the Wi-Fi standards are given in Section 4.2.3.

The Wi-Fi Alliance has more than 300 members. The Wi-Fi Alliance develops universal specifications (based on IEEE 802.11 standards) and follow through with rigorous testing and Wi-Fi certification of wireless devices.

To date, more than 3,300 products have been certified. Wi-Fi Alliance also works to provide Wi-Fi users with the information they need to make decisions about Wi-Fi systems.

Unlike Bluetooth and Zigbee, Wi-Fi is based on standards (see Section 4.2.3). But standards alone do not guarantee interoperability and so the Wi-Fi Alliance (http://www.wi-fi.org/) provides certification programs (http://www.wi-fi.org/certification_programs.php), as well as common market representation, profile, and education.

3.11 Personal Area Networks

PANs differ from WLANs in that they are generally designed for 1:1 wireless connectivity over a short range (i.e., close to one's person).

The reach of a PAN is typically a few meters at most. PANs can be used for communication among the personal devices themselves (intrapersonal communication), or for connecting to a higher-level network—in the case of ITS service provision, this is normally an access point to the vehicle network, usually with a firewall between the PAN and the in-vehicle network. The most commonly found PANs are Bluetooth and ZigBee, although in the future Ultra Wide Band (UWB) and Near Field Communications (NFC) may perform similar functionality.

3.11.1 Bluetooth

Bluetooth is an industry interest group. The Bluetooth Special Interest Group (SIG) is a privately held, not-for-profit trade association. The Special Interest Group was founded in September 1998. The Bluetooth SIG itself does not make, manufacture, or sell Bluetooth-enabled products. The SIG has more than 8,000 member companies that are leaders in the telecommunications, computing, automotive, music, apparel, industrial automation, and network industries. SIG members drive development of Bluetooth wireless technology and implement and market the technology in their products. The SIG has a small group of dedicated staff in Hong Kong, Sweden, and the United States.

The Bluetooth SIG global headquarters are in Bellevue, Washington, and it has local offices in Hong Kong and Malmo, Sweden.

Members support a number of working groups and committees that focus on specific areas, such as engineering, qualification, and marketing.

The Bluetooth SIG includes *promoter* member companies Agere, Ericsson, Intel, Lenovo, Microsoft, Motorola, Nokia, and Toshiba, and thousands of *associate* and *adopter* member companies.

The Bluetooth Mission Statement. Strengthen the Bluetooth brand by empowering SIG members to collaborate and innovate, creating the preferred wireless technology to connect diverse devices.

Section 4.2.6 provides more information about Bluetooth standards and specifications.

3.11.2 ZigBee

The ZigBee Alliance is an association of companies working together to enable reliable, cost-effective, low-power, wirelessly networked, monitoring and control products based on what it describes as an "open global standard." However, no formal SDO standardization process has been undertaken.

The goal of the ZigBee Alliance is to provide the consumer with ultimate flexibility, mobility, and ease of use by building wireless intelligence and capabilities into everyday devices. ZigBee technology will be embedded in a wide range of products and applications across consumer, commercial, industrial, and government markets worldwide. For the first time, companies will have a standards-based wireless platform optimized for the unique needs of remote monitoring and control applications, including simplicity, reliability, low cost, and low power.

ZigBee technology is a wireless control technology designed to offer original equipment manufacturers (OEMs) and developers the ability to build reliable, cost-effective, low-power wireless control products used in residential, commercial, and industrial applications.

3.11.3 Next Generation Networks

There as been much talk of so-called "next generation networks" (NGN), largely but not always in relation to what follows 3G cellular telephony. The future ITS is intimately involved with the future of wireless communications networks, and therefore NGN is a subject area of great interest.

Next generation networking (NGN) is a broad term to describe some key architectural evolutions in telecommunication core and access networks that will be deployed over the next 5 to 10 years. It is an unfortunate name because "time" and "next" are moving boundaries. We may hopefully expect that "next" will become "current" and eventually will become "past" and be replaced. It seems that the original concept was to look at future concepts and then develop standards that would have appropriate names. Unfortunately in the rush to be the lead SDO, some of the standards (for example, see Section 6.2.2.73) incorporate the words "NGN" or "next generation network" into the title of their deliverables, which seems at best, rather naive and inappropriate.

The general concepts behind most NGN thinking is that one network can transport all information and services (voice, data, and all sorts of media such as video) by encapsulating these into packets (packet switching). NGNs are commonly built around the Internet Protocol (IPv6), and this is sometimes called IP based communications.

ITU-T defines an NGN as "a packet-based network able to provide services including Telecommunication Services and able to make use of multiple broadband, QoS-enabled transport technologies and in which service-related functions are independent from underlying transport-related technologies. It offers unrestricted access by users to different service providers. It supports generalized mobility which will allow consistent and ubiquitous provision of services to users."

There are three architectural concepts that separate NGN from its predecessors:

- A common protocol architecture for all types of service, implying, for example, the migration of voice from a switched architecture (PSTN) to VoIP, and also migration of legacy services such as X.25 and Frame Relay (either commercial migration of the customer to a new service like IP VPN, or the provision of an emulating capability in the NGN);
- In the wired access network, a converged setup where a digital subscriber line access multiplexer (DSLAM) integrates VoIP, enabling voice switching infrastructure from the exchange to be removed;
- In a cable access network, NGN convergence implies migration of constant bit rate voice to CableLabs PacketCable standards that provide VoIP and Session Initiation Protocol (SIP) services. Both services ride over DOCSIS as the cable data layer standard services are independent of transport details.

NGNs are envisaged as being based on Internet technologies including Internet Protocol (IP) and Multiprotocol Label Switching (MPLS). At the application level, Session Initiation Protocol seems to be taking over from ITU-T H.323. Most of the telcos are extensively researching and supporting IP Multimedia Subsystems, which gives SIP a major chance of being the most widely adopted protocol.

The IP Multimedia Subsystem (IMS) is a standardized NGN architecture for an Internet media-services capability defined by the ETSI and the Third Generation Partnership Project (3GPP). (See Sections 4.2.1 and 3.11.4.) NGNs therefore have much in common with the CALM concept described in Section 4.3.

The most active of the SDO activities on NGN seems to be ETSI's TISPAN group. See Section 3.11.4 for further information.

ITU-T has created a focus group to work on developing recommendations for NGN. It can be accessed via: http://www.itu.int/ITU-T/ngn/fgngn/index.html. Its standardization initiatives can be accessed via: http://www.itu.int/ITU-T/ngn/index.phtml.

How will the ITS sector deal with these as yet not much specified communications networks? The CALM concept (see Section 4.3), by separating the media provision from the ITS service provision, is designed to make accommodation of and transfer to these future networks.

Section 6.2.2.73 also has additional information on NGN.

3.11.4 TISPAN

TISPAN describes itself as "defining the next generation network." TISPAN is a major ETSI initiative, designed to build upon the work already done by 3GPP in creating the SIP-based IMS described in Section 3.11.4.1. TISPAN and 3GPP (see Section 4.2.1) are now working together to define a harmonized IMS-centric core for both wireless and wireline networks.

This harmonized all-IP network has the objective to provide a completely new telecom business model for both fixed and mobile network operators. Access independent IMS will be a key enabler for fixed/mobile convergence, reducing network installation and maintenance costs, and allowing new services to be rapidly developed and deployed to satisfy new market demands.

NGN Release 1 was launched by TISPAN in December 2005, providing the robust and open standards for industry to use as a reliable basis for the development and implementation of the first generation of NGN systems.

TISPAN is now working on Release 2, with a focus on enhanced mobility, new services, and content delivery with improved security and network management.

TISPAN is structured as a single Technical Committee, with eight Working Groups to deliver specifications back-up to the TISPAN Plenary meetings.

Building upon the work already done by 3GPP in creating the SIP-based IMS (see Section 3.11.4.1), TISPAN and 3GPP are now working together to define a harmonized all-IP network.

The list of TISPAN specifications can be found in Appendix D.

For more information, the reader is referred to http://www.etsi.org/tispan and portal.etsi.org/docbox/TISPAN/Open.

3.11.4.1 IMS

IMS, the IP Multimedia Core Network Subsystem, is a collection of different functions that are linked by standardized interfaces; these are linked to form one IMS administrative network.

The second phase of the IMS comprises all the core network elements for the provision of multimedia services. IMS has been designed to enable full IP-based communication, exploiting the benefits of IP for all types of traffic and providing for seamless operation between different systems and also with the Internet. Using SIP, defined by the IETF, IMS enables mobile operators to offer their subscribers multimedia services based on and built upon Internet applications, services, and protocols. This second phase of IMS ensures interworking with circuit-switched networks, non-IMS networks, and with the similar (but different) core network defined by 3GPP2 for CDMA systems.

A user can connect to an IMS network in various ways, all of which use the standard IPv6 (see Sections 6.2.14) and run SIP user agents.

Direct IMS terminals (such as mobile phones, PDAs, and computers) register directly on an IMS network, even when they are roaming in another network or country (i.e., the visited network).

The Home Subscriber Server (HSS), or User Profile Server Function (UPSF), is a master user database that supports the IMS network entities that actually handle calls. It contains the subscription-related information (user profiles), performs authentication and authorization of the user, and can provide information about the user's physical location. It is similar to the GSM Home Location Register (HLR) and Authentication Center (AUC). A Subscriber Location Function (SLF) is needed to map user addresses when multiple HSSs are used. Both the HSS and the SLF communicate through a protocol called DIAMETER.

IMS also requires IP Multimedia Private Identity (IMPI) and IP Multimedia Public Identity (IMPU). Both are uniform resource identifiers (URIs), which can be alphanumeric.

A variety of charging strategies can be supported.

For those requiring more detailed information, Motorola provides a good introduction to IMS at: http://mototracker.atomicserver.co.uk/files/Introduction%20to%20IMS.pdf.

As yet there are no direct links between ITS standards and IMS, although there is obvious synergy for this in the future.

CHAPTER 4

Wireless Communications Standards Used for ITS

4.1 Regulations

When dealing with ubiquitous wireless communications, varying national regulations must be considered. Radio spectrum is managed nationally and is generally considered to be a sovereign national issue. Control of *radio* (in its broadest sense of all radiated emissions) is regulated by national regulators: the Federal Communications Commission (FCC) in the United States, OFCOM in the United Kingdom, and ACMA in Australia. However, where regional trading blocks are established, there are additional complications. This results in considerable confusion in the market place. The following sections of this chapter will provide some explanation of current structures.

4.1.1 CEPT

The European Conference of Postal and Telecommunications Administrations (CEPT) was established on June 26, 1959, as a coordinating body for European state telecommunications and postal organizations. It was established by 19 countries, and has now expanded to 47 countries, which virtually covers the whole geographical area of Europe. It deals exclusively with regulatory matters in the field of posts and telecommunications. The acronym comes from the French version of its name *Conférence européenne des administrations des postes et des telecommunications.*

Until reforms were adopted in September 2001, CEPT had three committees:

- *European Radiocommunications Committee (ERC)*: The ERC has established the European Radiocommunications Office (ERO) in Copenhagen. ERC matters are dealt with by DTI (now DBERR), RA in the United Kingdom.
- *Committee on European Postal Regulation (CERP)*: Matters are dealt with by DTI (now DBERR) in the United Kingdom, CGBPS.
- *European Committee for Telecommunications Regulatory Affairs (ECTRA)*: ECTRA established the European Telecommunications Office (ETO) in Copenhagen. Their support functions have recently been transferred to the ERO and the ETO has been closed as of January 2001.

ERC and ECTRA have now been combined to form the Electronic Communications Committee (ECC).

Most of the radio related activities take place within four working groups of the ERC:

- Spectrum Engineering;
- Frequency Management;
- Radio Regulatory;
- Conference Preparatory Group.

The European Telecommunication Standards Institute (ETSI) was established by CEPT in 1985 when the CEPT telecommunication standardization activities were transferred from CEPT to ETSI, thus establishing a formal link between the "regulatory framework" and an SDO. How does this complex situation work in practice?

In Europe, CEPT may by consensus agree upon the parameters of a particular recommendation, but it is at the discretion of a particular nation as to whether to apply the recommendation by introducing regulation in its country. To achieve consensus or change within CEPT is a slow and tortuous process taking years; and even when achieved, the result is a "recommendation" that member countries are invited and encouraged, but not forced, to adopt.

For the European Commission, trying to establish a common market in Europe, this is most frustrating and it consequently creates organizations such as the ERO and the ECC to work towards common agreements that it can then mandate as requirements to the member states.

In general this just causes confusion with multiple committee structures, which can as well be used to obstruct or influence progress as to promote it.

The formal structure of these committees is as shown in Figures 4.1 and 4.2.

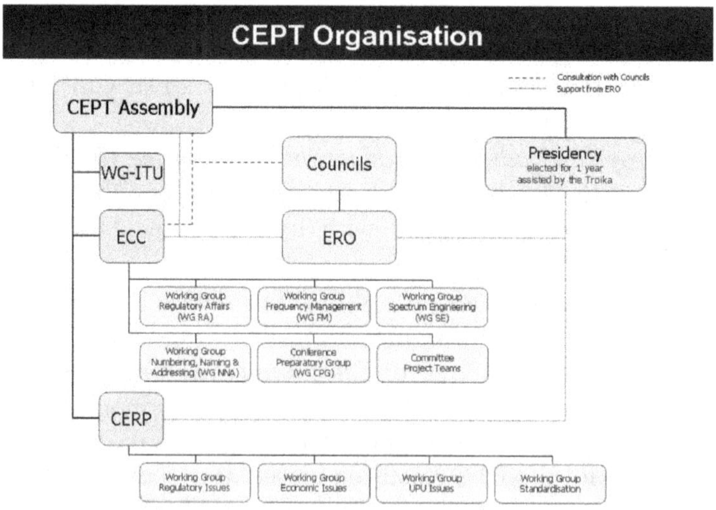

Figure 4.1 Organization of CEPT. (*Source:* ERO Web site.)

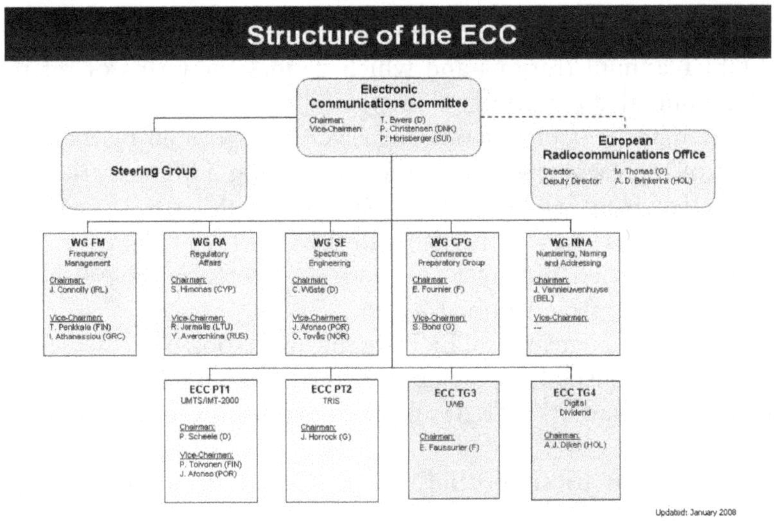

Figure 4.2 Organization chart of radio committees in Europe. (*Source:* ERO Web site.)

CEPT comprises 47 administrations, as follows: Albania, Andorra, Austria, Azerbaijan, Belarus, Belgium, Bulgaria, Bosnia and Herzegovina, Croatia, Cyprus, Czech Republic, Denmark, Estonia, Finland, France, Georgia, Germany, Greece, Hungary, Iceland, Ireland, Italy, Latvia, Liechtenstein, Lithuania, Luxembourg, Malta, Former Yugoslav Republic of Macedonia, Moldova, Monaco, Montenegro (application pending), Netherlands, Norway, Poland, Portugal, Romania, Russian Federation, San Marino, Serbia, Slovak Republic, Slovenia, Spain, Sweden, Switzerland, Turkey, Ukraine, United Kingdom, and Vatican City.

4.1.1.1 ECC

The objectives of the European Electronics Communication Committee are: the development of common electronic communications regulatory policies in a European context; the establishment within Europe of a harmonized plan for the efficient use of the radio spectrum, satellite orbits, and numbering resources, so as to satisfy requirements of users and industry; the promotion of European cooperation in the preparation of ITU fora; the encouragement of deregulation and liberalization; and the process of free circulation of radiocommunication equipment as a way of supporting the development of a more open and competitive market.

4.1.1.2 ERO

The European Radiocommunications Office was formally opened on May 6, 1991, and is located in Copenhagen, Denmark. ERO is the permanent office supporting the ECC of the CEPT. ECC is the committee that brings together the radio and telecommunications regulatory authorities of the 47 CEPT member countries.

ERO was established on the basis of a Memorandum of Understanding (MoU). In 1996 this MoU was replaced by the "Convention for the establishment of the

European Radiocommunications Office," which has to this date been signed by 30 CEPT administrations and which defines the terms of reference for ERO and the funding arrangement.

The management function for ERO is performed by the Office Councils, consisting of representatives from the contracting administrations.

In effect from January 2001, ERO's sister office, the European Telecommunications Office (ETO), was merged with ERO, which took over some of the functions previously carried out by ETO, and now performs these tasks on behalf of the ETO Administrative Council.

Another change in the CEPT organization took place in October 2001, when the ECC was established, replacing the European Committee for Telecommunications Regulatory Affairs (ECTRA) and the European Radiocommunications Committee (ERC).

In addition to supporting the ECC and its working groups, ERO has the following functions:

- To provide a center of expertise which shall act as a focal point to identify problem areas and new possibilities in the radio and telecommunications fields and advise the ECC accordingly;
- To draft long-term plans for future use of the radio frequency spectrum at a European level;
- To support and work together with national frequency management authorities;
- To conduct consultations on specific topics or parts of the frequency spectrum;
- To publish ECC Decisions and Recommendations and keep a record of the implementation;
- To identify and promote best practice in administration of national numbering schemes and number assignment procedures;
- To oversee the registrar service for the European Telephony Numbering Space.

ERO is the distribution point for all ECC documentation and also provides detailed information about the work of the ECC via the ERO Web site: http://www.ero.dk. The ERO Web site is an important element in the process where information is provided about the latest developments within the ECC with reports of recent meetings and approved texts of ECC decisions, recommendations, and reports. ERO also hosts the CEPT Web site: http://www.cept.org.

ERO has the overall objective of developing proposals for a European Table of Frequency Allocations and Utilizations for the frequency range 9 kHz to 275 GHz, scheduled to be implemented by 2008.

The European Common Allocation Table (ECA) with spectrum strategies for the current and future use of spectrum in Europe has been developed based on a major consultation activity called the Detailed Spectrum Investigation process (DSI). This consultation arrangement has been developed and organized by ERO with very active participation and involvement from European industry and users of the radio frequency spectrum.

The ECA is also to a large extent aligned with military use of frequency spectrum in Europe. Military frequency requirements included in the table have been supported by the NATO Joint Civil/Military Frequency Agreement (NJFA) and by ongoing agreements within the civil/military cooperation in Europe.

The European Common Allocation Table has been used by CEPT administrations as the basis for developing of National Frequency Allocation and Utilization tables. The ERO has developed an online information system (ERO Frequency Information System, EFIS) to provide easy access for industry and users to frequency utilization information across (ultimately) all 47 CEPT countries in Europe. This system also contains documentation and hyperlinks related to the use of frequency spectrum.

4.1.1.3 Principal Regulatory Regimes That Affect ITS Service Provision in Europe

There is a Commission Decision under the Radio Spectrum Decision (676/2002/EC) or a specific Directive (e.g., DECT, GSM) enforcing the standards. Other Commission Decisions are under development in the Radio Spectrum Committee (RSCOM).

Class 1 is divided into a number of subclasses, which are listed on the ERO Web site. The subclasses include the following (a nonexhaustive list). Provided that you comply, equipment may be marketed and used without restriction.

NOTE: The following tables are for guidance only. Certain bands are restricted in the applications that may be used.

The use of the frequency bands is not formally harmonized. You are required to inform the relevant member state of your intention to market the equipment, and inform the user of national usage restrictions. National administrations cooperate via CEPT, which may have adopted decisions or recommendations which can be used for guidance. However, they may not have been fully implemented in all member states. You should check with the relevant administration. A list of CEPT deliverables, as well as contact points in the CEPT administrations, are available from the ERO.

4.1.1.3.1 EN 300 422/EN 300 220 Aids for Hearing Impaired

ETSI	EN EN	300 422	169,4 MHz to 169,8125 MHz (frequency band previously designated for the ERMES paging system	Published Jan. 2007	ETSI
ETSI	EN EN	300 220		Published Jan. 2007	ETSI

This includes aids for the hearing impaired, social alarms, meter reading systems, tracking and asset tracing, and paging systems.

Downloadable FOC EN 300 422
http://webapp.etsi.org/exchangefolder/en_300422v010201p.pdf

2005/928/EC
http://europa.eu.int/eur-lex/lex/LexUriServ/LexUriServ.do?uri=CELEX:32005D0928:EN:NOT

Downloadable FOC EN 300 220
http://webapp.etsi.org/exchangefolder/en_30022001v020101p.pdf

4.1.1.3.2 EN 301 893 Wireless Access Systems Including RLAN

ETSI	EN	301 893	5 150 MHz to 5 350 MHz 5 470 MHz to 5 725 MHz Wireless Access Systems including Radio Local Area Networks	Published	ETSI

Downloadable FOC EN 301 893
http://webapp.etsi.org/exchangefolder/tr_102439v010101p.pdf
2005/513/EC
http://europa.eu.int/eur-lex/lex/LexUriServ/site/en/oj/2005/l_187/l_18720050719en 00220024.pdf

4.1.1.3.3 EN 302 288/EN 302 264 Automotive Radar

ETSI	EN	302 288	24 GHz Automotive Short Range Radar (until 30 June 2013)	Published	ETSI

Downloadable FOC EN 302 288
http://webapp.etsi.org/exchangefolder/en_30228801v010301o.pdf
2005/50/EC http://europa.eu.int/eur-lex/lex/LexUriServ/site/en/oj/2005/l_021/l_02120050125en 00150020.pdf

This involves road-vehicle based radar for collision mitigation and traffic safety applications.

ETSI	EN	302 264	79 GHz Automotive Short Range Radar	Under Development	ETSI

Downloadable FOC 302 264 not yet available
2004/454/EC http://europa.eu.int/eurlex/lex/LexUriServ/LexUriServ.do?uri=OJ:L:2004: 241:0066:0067:EN:PDF

This involves road-vehicle based radar for collision mitigation and traffic safety applications.

4.1.1.3.4 EN 300 330 Inductive SRDs

EN 300 330 is a standard that makes provisions for inductive short-range devices. (As this standard is in many parts with a common Web link, the link is shown first in this instance.)

Downloadable FOC EN 300 330
http://webapp.etsi.org/exchangefolder/en_300330v010202p.pdf

EN 300 330 ERM: Short Range Devices (SRD) Radio equipment in the frequency range 9 kHz–25 MHz and inductive loop-systems in the frequency range 9 kHz–30 MHz.

Part 1: Technical characteristics and test methods.

Part 2: Harmonized EN under article 3.2 of the R&TTE Directive.

The following band allocations/uses are determined within EN 300 330.

ETSI EN 300 330 20.05–59.75 kHz Published ETSI

Inductive applications (42 dBμA/m at 10m).

ETSI EN 300 330 59.750–60.250 kHz Published ETSI

Inductive applications (42 dBμA/m at 10m).

ETSI EN 300 330 60.250–67 kHz Published ETSI

Inductive applications (69 dBμA/m at 10m).

ETSI EN 300 330 67 – 70 kHz Published ETSI

Inductive applications (42 dBμA/m at 10m).

ETSI EN 300 330 70–119 kHz Published ETSI

Inductive applications (42 dBμA/m at 10m).

ETSI	EN	300 330	119–127 kHz	Published	ETSI

Inductive applications (66 dBμA/m at 10m).

ETSI	EN	300 330	127–135 kHz	Published	ETSI

Inductive applications (42 dBμA/m at 10m).

ETSI	EN	300 330	6765–6795 kHz	Published	ETSI

Inductive applications (42 dBμA/m at 10m).

ETSI	EN	300 330	7400 to 8800 kHz	Published	ETSI

Inductive applications (9 dBμA/m at 10m).

ETSI	EN	300 330	13.553–13.567 MHz	Published	ETSI

Inductive applications (42 dBμA/m at 10m).

4.1.1.3.5 EN 300 220 Nonspecific SRDs

EN 300 220 is a standard that makes provisions for nonspecific short-range devices. (As this standard is in many parts with a common Web link, the links are shown first in this instance.)

Downloadable FOC EN 300 220-1
http://webapp.etsi.org/exchangefolder/en_30022001v020101p.pdf
EN 300 220-2
http://webapp.etsi.org/exchangefolder/en_30022002v020102p.pdf
EN 300 220-3
http://webapp.etsi.org/exchangefolder/en_30022003v010101p.pdf

EN 300 220 ERM: Short Range Devices (SRD) Radio equipment to be used in the 25 MHz–1 GHz frequency range with power levels up to 500 mW.

Part 1: Technical characteristics and test methods.

Part 2: Supplementary parameters not intended for conformity purposes.

Part 3: Harmonized EN covering essential requirements under article 3.2 of the R&TTE Directive.

ETSI EN 300 220 26.995 MHz, 27.045 MHz 27.145 MHz 27.195 MHz Published ETSI

Nonspecific short-range devices (42 dBμA/m @ 10m, 1 mW erp).

ETSI EN EN 300 220 40.665 MHz, 40.675 MHz, 40.685 MHz, 40.695 MHz Published ETSI

Nonspecific short-range devices (10 mW).

ETSI EN EN 300 220 433.050 MHz–434.790 MHz Published ETSI

Nonspecific short-range devices (10 mW).

ETSI EN EN 300 220 868.0 MHz–868.6 MHz Published ETSI

Nonspecific short-range devices (25 mW).

ETSI EN EN 300 220 868.7 MHz–869.2 MHz Published ETSI

Nonspecific short-range devices (25 mW).

ETSI EN EN 300 220 869.2 MHz–869.25 MHz Published ETSI

Social alarms (10 mW).

ETSI EN EN 300 220 869.25 MHz–869.3 MHz Published ETSI

Alarms (10 mW).

ETSI EN EN 300 220 869.4 MHz–869.65 MHz Published ETSI

Nonspecific short-range devices (25 mW).

ETSI	EN EN	300 220	868.6 MHz–868.7 MHz	Published	ETSI

Alarms (10 mW).
Alarms (25 mW, 869.65 MHz–869.7 MHz).

ETSI	EN EN	300 220	869.7 MHz–870.0 MHz	Published	ETSI

Nonspecific short-range devices (5 mW).

4.1.1.3.6 EN 303 035 TETRA

ETSI	EN	303 035	380–385 MHz 390–395 MHz TETRA (power classes up to 45 dBm (30W))	Published	ETSI

Downloadable FOC EN 303 035-1
http://webapp.etsi.org/exchangefolder/en_30303501v010201p.pdf
Downloadable FOC EN 303 035-2
http://webapp.etsi.org/exchangefolder/en_30303502v010202p.pdf

EN 303 035 is a standard making provisions for Terrestrial Trunked Radio (TETRA).

4.1.1.3.7 EN 301 357 Wireless Audio

ETSI	EN	301 357	863–865 MHz Wireless Audio (10 mW)	Published	ETSI

Downloadable FOC EN 301 357
http://webapp.etsi.org/exchangefolder/en_301357v010101p.pdf

EN 301 357 is a standard making provisions for wireless audio.

4.1.1.3.8 EN 301 419 GSM

EN 301 419 is a standard in several parts making provisions for cellular telecommunications. (As this standard is in many parts with a common Web link, the links are shown first in this instance.)

Downloadable FOC EN 301 419-1
http://webapp.etsi.org/exchangefolder/en_30141901v040101p.pdf
Downloadable FOC EN 301 419-2
http://webapp.etsi.org/exchangefolder/en_30141902v050101p.pdf
Downloadable FOC EN 301 419-3
http://webapp.etsi.org/exchangefolder/en_30141903v050002p.pdf
Downloadable FOC EN 301 419-7
http://webapp.etsi.org/exchangefolder/en_30141907v050101p.pdf

ETSI	EN	301 418	Parts 1-3 880–890/925–935 MHz 890–915/ 935–960 Part 7 876–880/921–925 MHz	Published	ETSI

Parts 1–3 GSM (power classes up to 39 dBm (8W)); Part 7 GSM-R (Rail) (power classes up to 39 dBm (8W)).

ETSI	EN	301 418	1,710–1785/1805–1,880 MHz	Published	ETSI

GSM (power classes up to 36 dBm (4W)).

4.1.1.3.9 EN 301 444 / 301 426 / 301 427 Land Mobile Earth Satellite Stations

EN 301 444 / 301 426 / 301 427 are standards making provisions for Land Mobile Earth Satellite Stations. (As this standard is in many parts with a common Web link, the links are shown first in this instance.)

Downloadable FOC EN 301 444
http://webapp.etsi.org/exchangefolder/en_301444v010101p.pdf
Downloadable FOC EN 301 426
http://webapp.etsi.org/exchangefolder/en_301426v010201p.pdf
Downloadable FOC EN 301 427
http://webapp.etsi.org/exchangefolder/en_301427v010201p.pdf

ETSI	EN	301 444	1525.0–1544.0 MHz 1555.0–1559.0 MHz 1631.5–1634.5 MHz 1 656.5 MHz–1 660.5 MHz Land Mobile Earth Stations (LMES)	Published	ETSI
ETSI	EN	301 426	1525.0–1544.0 MHz 1555.0–1559.0 MHz 1 626.5 MHz–1 645.5 MHz 1 656.5 MHz–1 660.5 MHz Low data rate LMES	Published	ETSI
ETSI	EN	301 427	10.7–11.7 GHz 12.5–12.75 GHz 14.0–14.25 GHz Land Mobile Earth Stations	Published	ETSI

4.1.1.3.10 EN 301 441 / 301 442 Satellite-Personal

EN 303 441 / EN 301 442 are standards making provisions for the following.

ETSI	EN	301 441	1,610–1,613.5 MHz 1,613.8–1,626.5 MHz 2,483.5–2,500 MHz	Published	ETSI
ETSI	EN	301 442	1,980–2,010 MHz 2,170–2,200 MHz	Published	ETSI

Satellite-Personal Communications Earth stations.

Downloadable FOC EN 301 441
http://webapp.etsi.org/exchangefolder/en_301441v010101p.pdf

4.1.1.3.11 EN 301 406 DECT

ETSI	EN	301 406	1880 MHz–1900 MHz	Published	ETSI

DECT (250 mW).

EN 301 419 is a standard making provisions for Digitally Enhanced Cordless Telephony (DECT).

Downloadable FOC EN 301 406
http://webapp.etsi.org/exchangefolder/en_301406v010501p.pdf

4.1.1.3.12 EN 301 440 / EN 301 328 Nonspecific SRDs

ETSI	EN	301 440 301 328	2400–2483.5 MHz EN 300 440: Radio Equipment and Systems (RES); SRD; 1 GHz–25 GHz	Published	ETSI

EN 301 440 / EN 301 328 are standards making provisions for nonspecific short-range devices.

ETSI has developed specific standards for RFID operating at UHF frequencies and also has developed generic SRD standards for LF, HF, and microwave equipment which can be used for RFID.

Following the adoption of a Harmonized Standard EN 302 208 (September 2004) (Mandate M/284), CEPT has adopted a modified ECC Recommendation 70-03 to allow the use of 2W devices in the UHF band, using mitigation techniques recommended by ETSI in TR 102 313, 2004-07 (Listen Before Talk with Dynamic Frequency Agility) and TR 101 445, the ETSI System Reference Document for

UHF RFID. This document is now being revised to use channel plans instead of LBT/DFA for higher power devices.

See also EN 302 208, Section 4.1.1.3.13.

Downloadable FOC EN 301 440
http://webapp.etsi.org/exchangefolder/en_301440v010202p.pdf

Nonspecific short-range devices (10 mW).

Wideband Data Transmission Systems including RLANs (10 mW, 100 mW in the band 2,400 to 2,454 MHz).

Movement Detection (25 mW, 2,446 to 2,454 MHz).

ETSI	EN	301 440	5,725–5,875 MHz	Published	ETSI

Nonspecific short-range device (25 mW).

ETSI	EN	301 440	24.15–24.175 GHz	Published	ETSI

Movement Detection (100 mW eirp).

4.1.1.3.13 EN 302 208 UHF RFID up to 2W ERP

ETSI	EN	302 208	Electromagnetic compatibility and Radio spectrum Matters (ERM); Radio Frequency Identification Equipment operating in the band 865 MHz to 868 MHz with power levels up to 2 W	Published	ETSI

Downloadable FOC EN 302 208-1
http://pda.etsi.org/pda/home.asp?wki_id=zRbenU4K2Zz_'0,_CEs

Downloadable FOC EN 302 208-2
http://pda.etsi.org/pda/home.asp?wki_id=Oh1ms-DmSXdfhkihwX21S

Following the adoption of a Harmonized Standard EN 302 208 (September 2004) (Mandate M/284), CEPT has adopted a modified ECC Recommendation 70-03 to allow the use of 2W devices in the UHF band, using mitigation techniques recommended by ETSI in TR 102 313, 2004-07 (Listen Before Talk with Dynamic Frequency Agility), and TR 101 445, the ETSI System Reference Document for UHF RFID. This document is now being revised to use channel plans instead of

LBT/DFA for higher power devices and should be available from the links below when available.

4.1.1.3.14 EN 301 489 EMC Standard for Radio Equipment and Services

ETSI	EN	301 489	ElectroMagnetic Compatibility (EMC) standard for radio equipment and services	Published	ETSI

Downloadable FOC EMC for RES 301 489-1
http://pda.etsi.org/pda/home.asp?wki_id=rIWM21wT9kVXZWbVEg6df

Downloadable FOC EMC for RES 301 489-2
http://pda.etsi.org/pda/home.asp?wki_id=GX.T7b-vahABFIJJyyMh5

Downloadable FOC EMC for RES 301 489-3
http://pda.etsi.org/pda/home.asp?wki_id=ULGGfOjZNjmnrvmmW8lOP

Downloadable FOC EMC for RES 301 489-4
http://pda.etsi.org/pda/home.asp?wki_id=zbmNkzs_ohRSWaRSAR.Sv

Downloadable FOC EMC for RES 301 489-5
http://pda.etsi.org/pda/home.asp?wki_id=6P1AClWfaj',26'-ogLAA

Downloadable FOC EMC for RES 301 489-6 http://pda.etsi.org/pda/home.asp?wki_id=ceXIHGf1_gijnrilSzbEf

Downloadable FOC EMC for RES 301 489-7
http://pda.etsi.org/pda/home.asp?wki_id=pSr4p3JloiHJJPHL09@.@

Downloadable FOC EMC for RES 301 489-8
http://pda.etsi.org/pda/home.asp?wki_id=FiHCudR79gz@,2z,kSA0Q

Downloadable FOC EMC for RES 301 489-9
http://pda.etsi.org/pda/home.asp?wki_id=SVc,MQ6r_iYadZZgHhZok

Downloadable FOC EMC for RES 301 489-10
http://pda.etsi.org/pda/home.asp?wki_id=fJwruDqWok78CG7EvxyW-

Downloadable FOC EMC for RES 301 489-11
http://pda.etsi.org/pda/home.asp?wki_id=8KDn9rPUQ0MOMVQQ2m8bY

Downloadable FOC EMC for RES 301 489-12
http://pda.etsi.org/pda/home.asp?wki_id=eZjvEMYwr,128839m-Of,

Downloadable FOC EMC for RES 301 489-13
http://pda.etsi.org/pda/home.asp?wki_id=rN'hm8Cbe0deimdlKEnNI

Downloadable FOC EMC for RES 301 489-14
http://pda.etsi.org/pda/home.asp?wki_id=GdTprjL.,'IJPPKR,X,Ro

Downloadable FOC EMC for RES 301 489-15
http://pda.etsi.org/pda/home.asp?wki_id=URobJW-hr-vw@,v,cnN83

Downloadable FOC EMC for RES 301 489-16
http://pda.etsi.org/pda/home.asp?wki_id=zgEjO082C'ZaeiaaG0dCY

Downloadable FOC EMC for RES 301 489-17
http://pda.etsi.org/pda/home.asp?wki_id=6UYWwtsn—8ACFHHuF.@t

Downloadable FOC EMC for RES 301 489-18
http://pda.etsi.org/pda/home.asp?wki_id=ck,d_N@9P.rsw@stYYD,J

Downloadable FOC EMC for RES 301 489-19
http://pda.etsi.org/pda/home.asp?wki_id=pYIQTAetC,PQVRYQ5obmd

Downloadable FOC EMC for RES 301 489-20
http://pda.etsi.org/pda/home.asp?wki_id='LdC_'JX-0.'2,5,j'@Uy

Downloadable FOC EMC for RES 301 489-22
http://pda.etsi.org/pda/home.asp?wki_id=vNUrXBw4NwMNUSURP2CLO

Downloadable FOC EMC for RES 301 489-23
http://pda.etsi.org/pda/home.asp?wki_id=Lc@ycm-Vou1366AA3KTOu

Downloadable FOC EMC for RES 301 489-24
http://pda.etsi.org/pda/home.asp?wki_id=YQFl4YjAawegjkeehar69

Downloadable FOC EMC for RES 301 489-25
http://pda.etsi.org/pda/home.asp?wki_id='glt93sb_tIKINLMLt2Ae

Downloadable FOC EMC for RES 301 489-26
http://pda.etsi.org/pda/home.asp?wki_id=BU-fhwWGovvxv@y@z2Ryz

Downloadable FOC EMC for RES 301 489-27
http://pda.etsi.org/pda/home.asp?wki_id=tXqZEDJM_v89FGADAb0kj

Downloadable FOC EMC for RES 301 489-28
http://pda.etsi.org/pda/home.asp?wki_id=JnGhJnSoMsrszxzxvuHo9

Downloadable FOC EMC for RES 301 489-31
http://pda.etsi.org/pda/home.asp?wki_id=WbaTra6S9uPRPUYPS3gWU

Downloadable FOC EMC for RES 301 489-32
http://pda.etsi.org/pda/home.asp?wki_id=kOuGJNq7_w.,'.350J-Dp

4.1.1.4 CEPT/ERC Recommendation 70-03

Downloadable FOC CEPT 70.03
http://www.ero.dk/documentation/docs/doc98/official/pdf/REC7003E.PDF

ERC RECOMMENDATION 70-03 (Tromsø 1997 and subsequent amendments) RELATING TO THE USE OF SHORT RANGE DEVICES (SRD)

Recommendation adopted by the Frequency Management, Regulatory Affairs and Spectrum Engineering Working Groups, version of May 30, 2007.

This Recommendation sets out the general position on common spectrum allocations for short-range devices (SRDs) for countries within CEPT. It is also intended that it can be used as a reference document by the CEPT member countries when preparing their national regulations in order to keep in line with the provisions of the R&TTE Directive.

In using this recommendation, it should be remembered that it represents the most widely accepted position within the CEPT, but it should not be assumed that all allocations are available in all countries. An indication of where allocations are not available or where deviations from the CEPT position occur is to be found in Appendix 3 of the recommendation.

It should also be remembered that the pattern of radio use is not static. It is continuously evolving to reflect the many changes that are taking place in the radio environment; particularly in the field of technology. Spectrum allocations must reflect these changes and the position set out in this recommendation is therefore subject to continuous review.

Moreover, many administrations have designated additional frequencies or frequency bands for SRD applications on a national basis that do not conform to the CEPT position set out in this recommendation.

For these reasons, those wishing to develop or market SRDs based on this recommendation are advised to contact the relevant national administration to verify that the position set out herein still applies.

CEPT has adopted this recommendation to deal with SRDs, and ETSI has now developed harmonized standards for the majority of these devices. Other standards or technical specifications will be applicable within the framework of the R&TTE Directive for placing it on the market.

The term SRD is intended to cover the radio transmitters which provide either unidirectional or bidirectional communication and which have low capability of causing interference to other radio equipment. SRDs use either integral, dedicated, or external antennas, and all modes of modulation can be permitted subject to relevant standards.

This recommendation describes the spectrum management requirements for SRDs relating to allocated frequency bands, maximum power levels, channel spacing, and duty cycle.

For CEPT countries that have implemented the R&TTE Directive, Article 12 (CE-marking) and Article 7.2 (putting radio equipment into service) apply. Article 12 states that "any other marking may be affixed to the equipment provided that the visibility and legibility of the CE-marking is not hereby reduced." Article 7.2 states that "member states may restrict the putting into service of radio equipment only for reasons related to the effective and appropriate use of the radio spectrum, avoidance of harmful interference or matters relating to public health."

CEPT has considered the use of SRD devices on board aircraft and it has concluded that, from the CEPT regulatory perspective, such use is allowed under the same conditions provided in the relevant Annex of Recommendation 70-03. For aviation safety aspects, the CEPT is not the right body to address this matter which remains the responsibility of aircraft manufacturers or aircraft owners who should consult with the relevant national or regional aviation bodies before the installation and use of such devices on board aircraft.

For SRDs individual licenses are normally not required. Where licenses are required this is stated in the relevant Annex.

The following annexes define the regulatory parameters as well as additional information about harmonized standards, frequency issues, and important technical parameters. Other technical parameters are indicated in the relevant standard.

Appendix 2 covers the relevant ERC Decisions and ETSI standards.

For countries having implemented the R&TTE Directive, further details can be found on the relevant EC (http://europa.eu.int/comm/enterprise/rtte/index_en.htm) and ERO Web sites (http://www.ero.dk).

The European Conference of Postal and Telecommunications Administrations,

considering

a) that SRDs in general operate in shared bands and are not permitted to cause harmful interference to other radio services;

b) that in general SRDs cannot claim protection from other radio services;

c) that due to the increasing interest in the use of SRDs for a growing number of applications it is necessary to harmonise frequencies and regulations for these devices;

d) that there is a need to distinguish between different applications;

e) that additional applications and associated annexes will be added as necessary;

f) that for CEPT countries that have implemented the R&TTE Directive article 12 (CE marking) and article 7.2 on putting into service of radio equipment apply,

g) that equipment marketed before the adoption of this Recommendation marked with the abbreviation CEPT LPD Y according to the abrogated CEPT Recommendation T/R 01-04 should be allowed continuation of free circulation and use

h) that maintenance of Appendices 2 and 3 and also the related cross-references in the Annexes may be undertaken by the ERO based on information from Administrations,

i) that information about placing SRD equipment on the market and its use can be obtained by contacting individual administrations, especially with regard to equipment operating in frequencies or frequency bands that may be designated for SRDs by administrations in addition to those covered in this Recommendation;

j) that SRD equipment normally use either integral or dedicated antennas. In exceptional cases external antennas could be used which will be mentioned in the appropriate annex to this Recommendation;

k) that for those countries implementing the provisions of this Recommendation, national restrictions in respect of the annexes can be found in Appendix 3;

recommends

1) that CEPT administrations implement the parameters in accordance with the indications mentioned in the annexes;

2) that technical parameter limits should not be exceeded by any function of the equipment;

3) that CEPT administrations should allow visitors from other countries to carry and use their equipment temporarily without any further formalities unless there are national restrictions as shown in Appendix 3.

(The source of the above information is Recommendation 70.03, reproduced with the permission of ERO.)

4.1.2 FCC

The Federal Communications Commission (FCC) is an independent United States government agency, directly responsible to Congress. The FCC was established by the Communications Act of 1934 and is charged with regulating interstate and international communications by radio, television, wire, satellite, and cable. The FCC's jurisdiction covers the 50 states, the District of Columbia, and U.S. possessions.

However, Canada, Mexico, and several Central American countries generally adopt FCC regulations and sit on FCC committees.

The FCC is directed by five Commissioners appointed by the U.S. President and confirmed by the Senate for 5-year terms, except when filling an unexpired term. The President designates one of the Commissioners to serve as Chairperson. Only three Commissioners may be members of the same political party. None of them can have a financial interest in any Commission-related business.

As the chief executive officer of the Commission, the Chairman delegates management and administrative responsibility to the Managing Director. The Commissioners supervise all FCC activities, delegating responsibilities to staff units and bureaus.

Bureaus and Offices

The Commission staff is organized by function. There are seven operating bureaus and 10 staff offices. The bureaus' responsibilities include: processing applications for licenses and other filings; analyzing complaints; conducting investigations; developing and implementing regulatory programs; and taking part in hearings. Offices provide support services. Even though the bureaus and offices have their individual functions, they regularly join forces and share expertise in addressing Commission issues.

- *Consumer & Governmental Affairs Bureau:* Educates and informs consumers about telecommunications goods and services and engages their input to help guide the work of the Commission. CGB coordinates telecommunications policy efforts with industry and with other governmental agencies (federal, tribal, state, and local) in serving the public interest.
- *Enforcement Bureau:* Enforces the Communications Act, as well as the Commission's rules, orders, and authorizations.
- *International Bureau:* Represents the Commission in satellite and international matters.
- *Media Bureau:* Regulates AM and FM radio and television broadcast stations, as well as cable television and satellite services.
- *Wireless Telecommunications Bureau:* Oversees cellular and PCS phones, pagers, and two-way radios. This bureau also regulates the use of radio spectrum to fulfill the communications needs of businesses, aircraft and ship operators, and individuals.
- *Public Safety & Homeland Security Bureau:* Addresses public safety, homeland security, national security, emergency management and preparedness, disaster management, and other related issues.
- *Wireline Competition Bureau:* Responsible for rules and policies concerning telephone companies that provide interstate, and under certain circumstances intrastate, telecommunications services to the public through the use of wire-based transmission facilities (i.e., corded/cordless telephones).
- *Office of Administrative Law Judges:* Presides over hearings and issues initial decisions.
- *Office of Communications Business Opportunities:* Provides advice to the Commission on issues and policies concerning opportunities for ownership by small, minority, and women-owned communications businesses.
- *Office of Engineering and Technology:* Allocates spectrum for nongovernment use and provides expert advice on technical issues before the Commission.
- *Office of the General Counsel:* Serves as chief legal advisor to the Commission's various bureaus and offices.
- *Office of Inspector General:* Conducts and supervises audits and investigations relating to the operations of the Commission.
- *Office of Legislative Affairs:* The Commission's main point of contact with Congress.

- *Office of the Managing Director:* Functions as a chief operating official, serving under the direction and supervision of the Chairman.
- *Office of Media Relations:* Informs the news media of FCC decisions and serves as the Commission's main point of contact with the media.
- *Office of Strategic Planning & Policy Analysis:* Works with the Chairman, Commissioners, bureaus, and offices to develop strategic plans identifying policy objectives for the agency.
- *Office of Work Place Diversity:* Advises the Commission on all issues related to workforce diversity, affirmative recruitment, and equal employment opportunity.

4.1.2.1 FCC Regulations Part 15 – RFID

Downloadable FOC FCC Part 15
http://www.fcc.gov/oet/info/rules/part15/part15-2-16-06.pdf

The following Part 15 regulations contain all updates and changes adopted and released by the Commission as of August 14, 2006. However, changes to the rules do not become effective until at least 30 days after they are published in the *Federal Register*. It is possible that recent changes to these rules may not have been published in the *Federal Register* and may not yet be effective.

PART 15 – RADIO FREQUENCY DEVICES
Subpart A – General
Section 15.1 Scope of this Part.
Section 15.3 Definitions.
Section 15.5 General conditions of operation.
Section 15.9 Prohibition against eavesdropping.
Section 15.11 Cross reference.
Section 15.13 Incidental radiators.
Section 15.15 General technical requirements.
Section 15.17 Susceptibility to interference.
Section 15.19 Labelling requirements.
Section 15.21 Information to user.
Section 15.23 Home-built devices.
Section 15.25 Kits.
Section 15.27 Special accessories.
Section 15.29 Inspection by the Commission.
Section 15.31 Measurement standards.
Section 15.32 Test procedures for CPU boards and computer power supplies.
Section 15.33 Frequency range of radiated measurements.
Section 15.35 Measurement detector functions and bandwidths.
Section 15.37 Transition provisions for compliance with the rules.
Section 15.38 Incorporations by reference.
Subpart B – Unintentional Radiators
Section 15.101 Equipment authorization of unintentional radiators.

Section 15.102 CPU boards and power supplies used in personal computers.
Section 15.103 Exempted devices.
Section 15.105 Information to the user.
Section 15.107 Conducted limits.
Section 15.109 Radiated emission limits.
Section 15.111 Antenna power conducted limits for receivers.
Section 15.113 Power line carrier systems.
Section 15.115 TV interface devices, including cable system terminal devices.
Section 15.117 TV broadcast receivers.
Section 15.118 Cable ready consumer electronics equipment.
Section 15.119 Closed caption decoder requirements for analog television receivers.
Section 15.120 Program blocking technology requirements for television receivers.
Section 15.121 Scanning receivers and frequency converters used with scanning receivers.
Section 15.122 Closed caption decoder requirements for digital television receivers and converter boxes.
Section 15.123 Labeling of digital cable ready products.
Subpart C – Intentional Radiators
Section 15.201 Equipment authorization requirement.
Section 15.202 Certified operating frequency range.
Section 15.203 Antenna requirement.
Section 15.204 External radio frequency power amplifiers and antenna modifications.
Section 15.205 Restricted bands of operation.
Section 15.207 Conducted limits.
Section 15.209 Radiated emission limits, general requirements.
Section 15.211 Tunnel radio systems.
Section 15.213 Cable locating equipment.
Section 15.214 Cordless telephones.
Radiated Emission Limits, Additional Provisions
Section 15.215 Additional provisions to the general radiated emission limitations.
Section 15.217 Operation in the band 160–190 kHz.
Section 15.219 Operation in the band 510–1705 kHz.
Section 15.221 Operation in the band 525–1705 kHz.
Section 15.223 Operation in the band 1.705–10 MHz.
Section 15.225 Operation within the band 13.110–14.010 MHz.
Section 15.227 Operation within the band 26.96–27.28 MHz.
Section 15.229 Operation within the band 40.66–40.70 MHz.
Section 15.231 Periodic operation in the band 40.66–40.70 MHz and above 70 MHz.
Section 15.233 Operation within the bands 43.71–44.49 MHz, 46.60–46.98 MHz, 48.75–49.51 MHz and 49.66–50.0 MHz.
Section 15.235 Operation within the band 49.82–49.90 MHz.

Section 15.237 Operation in the bands 72.0–73.0 MHz, 74.6–74.8 MHz and 75.2–76.0 MHz.

Section 15.239 Operation in the band 88–108 MHz.

Section 15.240 Operation in the band 433.5–434.5 MHz.

Section 15.241 Operation in the band 174–216 MHz.

Section 15.242 Operation in the bands 174–216 MHz and 470–668 MHz.

Section 15.243 Operation in the band 890–940 MHz.

Section 15.245 Operation within the bands 902–928 MHz, 2435–2465 MHz, 5785–5815 MHz, 10500–10550 MHz, and 24075–24175 MHz.

Section 15.247 Operation within the bands 902–928 MHz, 2400–2483.5 MHz, and 5725–5850 MHz.

Section 15.249 Operation within the bands 902–928 MHz, 2400–2483.5 MHz, 5725–5875 MHz, and 24.0–24.25 GHz.

Section 15.250 Operation of wideband systems within the band 5925–7250 MHz.

Section 15.251 Operation within the bands 2.9–3.26 GHz, 3.267–3.332 GHz, 3.339–3.3458 GHz, and 3.358–3.6 GHz.

Section 15.252 Operation of wideband vehicular radar systems within the bands 16.2–17.7 GHz and 23.12–29.0 GHz.

Section 15.253 Operation within the bands 46.7–46.9 GHz and 76.0–77.0 GHz.

Section 15.255 Operation within the band 57–64 GHz.

Section 15.257 Operation within the band 92–95 GHz.

Subpart D – Unlicensed Personal Communications Service Devices

Section 15.301 Scope.

Section 15.303 Definitions.

Section 15.305 Equipment authorization requirement.

Section 15.307 Coordination with fixed microwave service.

Section 15.309 Cross reference.

Section 15.311 Labelling requirements.

Section 15.313 Measurement procedures.

Section 15.315 Conducted limits.

Section 15.317 Antenna requirement.

Section 15.319 General technical requirements.

Section 15.321 [Reserved]

Section 15.323 Specific requirements for devices operating in the 1920–1930 MHz sub-band.

Subpart E – Unlicensed National Information Infrastructure Devices

Section 15.401 Scope.

Section 15.403 Definitions.

Section 15.405 Cross reference.

Section 15.407 General technical requirements.

Subpart F – Ultra-Wideband Operation

Section 15.501 Scope.

Section 15.503 Definitions.

Section 15.505 Cross reference.

Section 15.507 Marketing of UWB equipment.

Section 15.509 Technical requirements for ground penetrating radars and wall imaging systems.

Section 15.110 Technical requirements for through-wall imaging systems.
Section 15.511 Technical requirements for surveillance systems.
Section 15.513 Technical requirements for medical imaging systems.
Section 15.515 Technical requirements for vehicular radar systems.
Section 15.517 Technical requirements for indoor UWB systems.
Section 15.519 Technical requirements for hand held UWB systems.
Section 15.521 Technical requirements applicable to all UWB devices.
Section 15.523 Measurement procedures.
Section 15.525 Coordination requirements.
Subpart G – Access Broadband Over Power Line (Access BPL)
Section 15.601 Scope.
Section 15.603 Definitions.
Section 15.605 Cross reference.
Section 15.607 Equipment Authorization of Access BPL equipment.
Section 15.609 Marketing of Access BPL equipment.
Section 15.611 General technical requirements.
Section 15.613 Measurement Procedures.
Section 15.615 General administrative requirements.

4.1.3 Other Countries

Throughout the world most countries simply have to operate within the regime of their national radio regulations. For example, Figure 4.3 shows the radio regulation regime in Japan.

In order to find information on the regulatory regimes for individual countries, the author suggests doing a Google search for "[countryname] Radio Regulations" or visit that country's government Web site and make a similar search.

However, although the radio regulations regime is more simple with respect to a single country, when it comes to achieving common bands in order to provide ITS services, countries outside Europe and North America face even more difficult challenges with coordination of frequency bands and services. For example: the frequencies proposed for ITS at circa 5 GHz are different in Japan than in the United States, and Europe is still in the process of deciding if it can adopt a regime similar to that in the United States. At the time of writing, Australia is currently considering what bands may be appropriate for the provision of ITS services.

Further, different countries, and even different road operators, considering the same scenarios, may not reach the same conclusions; indeed, it is unlikely that they will. For example, while most of Europe has developed and adopted a standard for plaza and free flow road tolling using 5-GHz technology, the largest country in Europe, Germany, has selected a system based on GPS/GSM.

Spending years either in EU Directorates or Standardization Committees, trying to get a single "best" global solution for continuous communications with vehicles, is unlikely to be productive, and by the time agreement is reached, if ever, technology will have moved on.

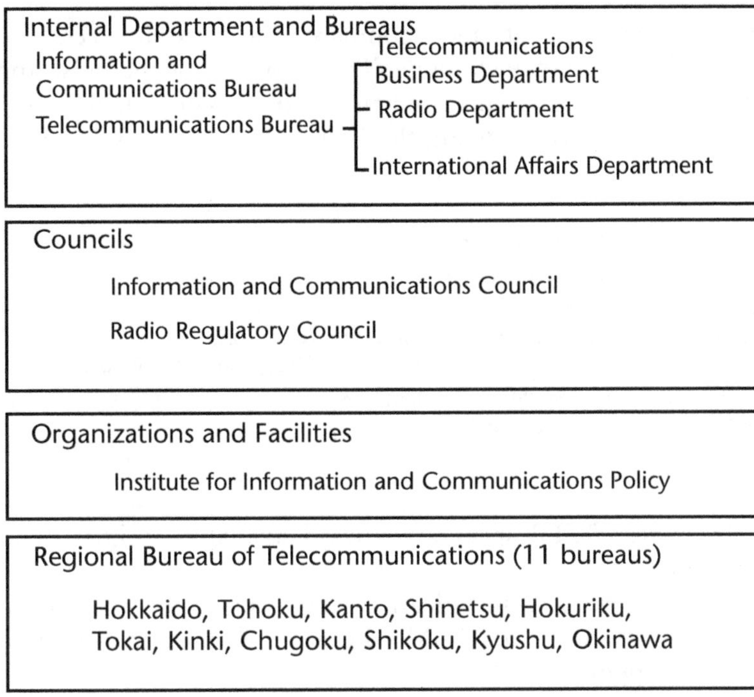

Figure 4.3 Radio regulation authorities in Japan. (*Source:* CSI from information supplied by the Telecommunications Bureau of the Ministry of Internal Affairs and Communication, Japan.)

4.2 Publicly Available Wireless Networks

In the context of this book, a *public network* is any telephony network that is available to the general public. The telephone in your home is a good example of such a network. Access to the network is available to anyone who has subscribed to the service—either by an on-going subscription, or in the case of a public phone box in the street, by giving money or credit card details or by using a pre-paid card at the time of use.

A public wireless network is a network, such as cellular telephony, satellite, or mobile wireless broadband, which is established for general communications purposes. In the context of ITS such networks may be available, where provided, so long as a contract exists with the service provider, for such wireless communications services to be utilized to provide or support ITS services (and indeed in some circumstances, such as crash notification, they may be required to provide some safety related services without a contract, much as we find today with telephone calls to emergency services).

4.2.1 3GPP

4.2.1.1 Overview

Although to the user, GSM and 3G cellular telephones seem like systems which are retrogradely interoperable in the case of 3G (to work over 2G), the standards that back them are extremely complex. There are in excess of 70 standards and deliverables for 2G/GPRS alone. This book is not the most appropriate place to cover these subjects deeply, and this chapter will provide only an overview. For those requiring further detail, the Web links listed may be useful.

Downloadable FOC 3GPP Home Page www.3gpp.org/

Downloadable FOC 3GPP Specifications
http://www.3gpp.org/specs/specs.htm

Downloadable FOC 3GPP Specifications Status database
http://www.3gpp.org/ftp/Information/Databases/Spec_Status/3GPP-Spec-Status.zip

Downloadable FOC 3GPP Specs Numbering
http://www.3gpp.org/specs/numbering.htm

The latest 3GPP specs are always available at http://www.3gpp.org/specs/specs.htm.

3GPP are writing specifications and reports for a Third Generation Mobile System based on evolved GSM core networks and the radio access technologies that they support.

All 3G and GSM specifications have a 3GPP specification number consisting of 4 or 5 digits. (e.g., 09.02 or 29.002). The first two digits define the series as listed in Table 4.1. They are followed by two additional further digits for the 01 to 13 series or three additional digits for the 21 to 55 series.

The term "3G" means a 3GPP system using a UTRAN radio access network; the term "GSM" means a 3GPP system using a GERAN radio access network. (Thus "GSM" includes GPRS and EDGE features.)

Table 4.1 is taken from the "3GPP Specifications–Numbering scheme" Web page.

The 3GPP Specifications are stored on the file server as zipped MS-Word files. More help on this subject is at http://www.3gpp.org/specs/specs.htm. Some of the more frequently used specifications can be downloaded from the following links:

Downloadable FOC

TR21.905 Vocabulary for 3GPP Specifications
http://www.3gpp.org/ftp/Specs/html-info/21905.htm

TS 22.101 Service Aspects; Service Principles
http://www.3gpp.org/ftp/Specs/html-info/22101.htm

TS 22.141 Presence Service; Stage 1
http://www.3gpp.org/ftp/Specs/html-info/22141.htm

TS22.228 Service requirements for the IP multimedia core network subsystem; Stage 1
http://www.3gpp.org/ftp/Specs/html-info/22228.htm

TS 22.250 IMS Group Management; Stage 1
http://www.3gpp.org/ftp/Specs/html-info/22250.htm

TS 22.340 IMS Messaging; Stage 1
http://www.3gpp.org/ftp/Specs/html-info/22340.htm

TR 22.800 IMS Subscription and access scenarios
http://www.3gpp.org/ftp/Specs/html-info/22800.htm

Table 4.1 3GPP Specifications

Subject of Specification Series	*3G/GSM R99 and Later*	*GSM Only (Rel-4 and Later)*	*GSM Only (Before Rel-4)*
General information (long defunct)			00 series
Requirements	21 series	41 series	01 series
Service aspects ("stage 1")	22 series	42 series	02 series
Technical realization ("stage 2")	23 series	43 series	03 series
Signaling protocols ("stage 3")–user equipment to network	24 series	44 series	04 series
Radio aspects	25 series	45 series	05 series
CODECs	26 series	46 series	06 series
Data	27 series		07 series
Signaling protocols ("stage 3")–(RSS-CN)	28 series	48 series	08 series
Signaling protocols ("stage 3")–intra-fixed-network	29 series	49 series	09 series
Program management	30 series	50 series	10 series
Subscriber Identity Module (SIM/USIM), IC Cards. Test specs.	31 series	51 series	11 series
OAM&P and Charging	32 series	52 series	12 series
Access requirements and test specifications		13 series	13 series
Security aspects	33 series		
UE and (U)SIM test specifications	34 series	–2	11 series
Security algorithms (3)	35 series	55 series	–4
Evolved UTRA aspects	36 series	—	—

Note (1) Note deleted for this presentation, see note (1) at http://www.3gpp.org/specs/numbering.htm
Note (2): The specifications of these aspects are spread throughout several series.
Note (3): Algorithms may be subject to export licensing conditions. See the relevant 3GPP page. See also the relevant ETSI pages.
Note (4): The original GSM algorithms are not published and are controlled by the GSM Association.

TS 23.002 Network Architecture
http://www.3gpp.org/ftp/Specs/html-info/23002.htm

TS 23.003 Numbering, Addressing and Identification
http://www.3gpp.org/ftp/Specs/html-info/23003.htm

TS 23.008 Organisation of Subscriber Data
http://www.3gpp.org/ftp/Specs/html-info/23008.htm

TS 23.107 Quality of Service (QoS) principles
http://www.3gpp.org/ftp/Specs/html-info/23107.htm

TS 23.125 Overall high level functionality and architecture impacts of flow based charging; Stage 2
http://www.3gpp.org/ftp/Specs/html-info/23125.htm

TS 23.141 Presence Service; Architecture and functional description; Stage 2
http://www.3gpp.org/ftp/Specs/html-info/23141.htm

TS 23.167 IMS emergency sessions
http://www.3gpp.org/ftp/Specs/html-info/23167.htm

TS 23.207 End-to-end QoS concept and architecture
http://www.3gpp.org/ftp/Specs/html-info/23207.htm

TS 23.218 IMS session handling; IM call model; Stage 2
http://www.3gpp.org/ftp/Specs/html-info/23218.htm

TS 23.221 Architectural Requirements
http://www.3gpp.org/ftp/Specs/html-info/23221.htm

TS 23.228 IMS stage 2
http://www.3gpp.org/ftp/Specs/html-info/23228.htm

TS23.234 WLAN interworking
http://www.3gpp.org/ftp/Specs/html-info/23234.htm

TS 23.271 Location Services (LCS); Functional description; Stage 2
http://www.3gpp.org/ftp/Specs/html-info/23271.htm

TS 23.278 Customized Applications for Mobile network Enhanced Logic (CAMEL)–IMS interworking; Stage 2
http://www.3gpp.org/ftp/Specs/html-info/23278.htm

TR 23.864 Commonality and interoperability between IMS core networks
http://www.3gpp.org/ftp/Specs/html-info/23864.htm

TR 23.867 IMS emergency sessions
http://www.3gpp.org/ftp/Specs/html-info/23867.htm

TR 23.917 Dynamic policy control enhancements for end-to-end QoS, Feasibility study
http://www.3gpp.org/ftp/Specs/html-info/23917.htm

TR 23.979 3GPP enablers for Push-to-Talk over Cellular (PoC) services; Stage 2
http://www.3gpp.org/ftp/Specs/html-info/23979.htm

TR 23.981 Interworking aspects and migration scenarios for IPv4-based IMS implementations (early IMS)
http://www.3gpp.org/ftp/Specs/html-info/23981.htm

TS 24.141 Presence Service using the IMS Core Network subsystem; Stage 3
http://www.3gpp.org/ftp/Specs/html-info/24141.htm

TS 24.147 Conferencing using the IMS Core Network subsystem
http://www.3gpp.org/ftp/Specs/html-info/24147.htm

TS 24.228 Signalling flows for the IMS call control based on SIP and SDP; Stage 3
http://www.3gpp.org/ftp/Specs/html-info/24228.htm

TS 24.229 IMS call control protocol based on SIP and SDP; Stage 3
http://www.3gpp.org/ftp/Specs/html-info/24229.htm

TS 24.247 Messaging using the IMS Core Network subsystem; Stage 3
http://www.3gpp.org/ftp/Specs/html-info/24247.htm

TS 26.235 Packet switched conversational multimedia applications; Default codecs
http://www.3gpp.org/ftp/Specs/html-info/26235.htm

TS 26.236 Packet switched conversational multimedia applications; Transport protocols
http://www.3gpp.org/ftp/Specs/html-info/26236.htm

TS 29.162 Interworking between the IMS and IP networks
http://www.3gpp.org/ftp/Specs/html-info/29162.htm

TS 29.163 Interworking between the IMS and Circuit Switched (CS) networks
http://www.3gpp.org/ftp/Specs/html-info/29163.htm

TC 29.198 Open Service Architecture (OSA)
http://www.3gpp.org/ftp/Specs/html-info/29198.htm

TS 29.207 Policy control over Go interface
http://www.3gpp.org/ftp/Specs/html-info/29207.htm

TS 29.208 End-to-end QoS signalling flows
http://www.3gpp.org/ftp/Specs/html-info/29208.htm

TC 29.209 Policy control over Gq interface
http://www.3gpp.org/ftp/Specs/html-info/29209.htm

TS 29.228 IMS Cx and Dx interfaces: signalling flows and message contents
http://www.3gpp.org/ftp/Specs/html-info/29228.htm

TS 29.229 IMS Cx and Dx interfaces based on the Diameter protocol; Protocol details
http://www.3gpp.org/ftp/Specs/html-info/29229.htm

TS 29.278 CAMEL Application Part (CAP) specification for IMS
http://www.3gpp.org/ftp/Specs/html-info/29278.htm

TS 29.328 IMS Sh interface : signalling flows and message content
http://www.3gpp.org/ftp/Specs/html-info/29328.htm

TS 29.329 IMS Sh interface based on the Diameter protocol; Protocol details
http://www.3gpp.org/ftp/Specs/html-info/29329.htm

TR 29.962 Signalling interworking between the 3GPP SIP profile and non-3GPP SIP usage

http://www.3gpp.org/ftp/Specs/html-info/29962.htm

TS 31.103 Characteristics of the IMS Identity Module (ISIM) application
http://www.3gpp.org/ftp/Specs/html-info/31103.htm

TS 32.240 Telecommunication management; Charging management; Charging architecture and Principles
http://www.3gpp.org/ftp/Specs/html-info/32240.htm

TS 32.260 Telecommunication management; Charging management; IMS charging
http://www.3gpp.org/ftp/Specs/html-info/32260.htm

TS 32.299 Telecommunication management; Charging management; Diameter charging applications
http://www.3gpp.org/ftp/Specs/html-info/32299.htm

TS 32.421 Telecommunication management; Subscriber and equipment trace: Trace concepts and requirements
http://www.3gpp.org/ftp/Specs/html-info/32421.htm

TS 33.102 3G security; Security architecture
http://www.3gpp.org/ftp/Specs/html-info/33102.htm

TS 33.108 3G security; Handover interface for Lawful Interception (LI)
http://www.3gpp.org/ftp/Specs/html-info/33108.htm

TS 33.141 Presence service; security
http://www.3gpp.org/ftp/Specs/html-info/33141.htm

TS 33.203 3G security; Access security for IP-based services
http://www.3gpp.org/ftp/Specs/html-info/33203.htm

TS 33.210 3G security; Network Domain Security (NDS); IP network layer security
http://www.3gpp.org/ftp/Specs/html-info/33210.htm

TR 33.978 Security aspects of early IP Multimedia Subsystem (IMS)
http://www.3gpp.org/ftp/Specs/html-info/33978.htm

2G and 3G telephony is important to ITS service provision in that, despite the charging business model not being too conducive to "always-on" systems, and the connection times therefore being too slow for time-critical ITS systems, the 2G/3G networks are ubiquitous to much of the Earth's surface, excluding a few countries and sparsely populated regions.

It is not therefore necessary to wait for ITS specific communications networks or mobile wireless broadband to be able to access some ITS services.

3GPP now manages both 2G/GRPRS/EDGE and 3G Standards, and manages the standards for migration and the operation of 3G over 2G networks where 3G is not available.

The term "3GPP specification" therefore covers all GSM (including GPRS and EDGE) and WCDMA specifications. The 3G specifications are known as UTRAN, UMTS (in Europe), and FOMA (in Japan).

4.2.1.2 GSM

Downloadable FOC GSM
http://www.3gpp.org/specs/specs.htm

GSM is the acronym for Global System for Mobile Communication, a globally accepted standard for digital cellular communication. GSM is the name of a standardization group established in 1982 to create a common European mobile telephone standard that would formulate specifications for a pan-European mobile cellular radio system operating at 900 MHz. GSM is ubiquitous in most of Europe, much of Asia, Central and South America, Australasia, and Pacific countries. Coverage is now widespread in Canada and in metropolitan areas of the United States.

GSM provides recommendations, not requirements. The GSM specifications define the functions and interface requirements in detail but do not address the hardware. The reason for this is to limit the designers as little as possible but still make it possible for the operators to buy equipment from different suppliers.

The architecture of GSM networks comprises three major elements:

- Switching system (SS);
- Base station system (BSS);
- Operation and support system (OSS).

Figure 4.4 shows the GSM network elements.

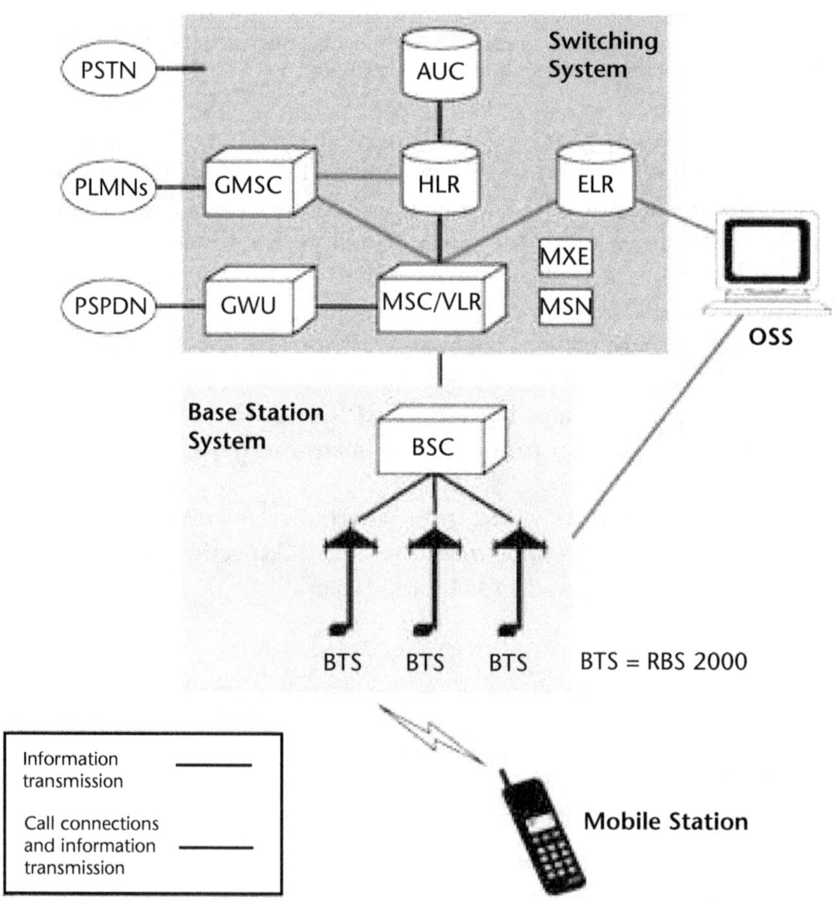

Figure 4.4 GSM network elements. (*Source:* CSI Library.)

4.2.1.3 SIM

Downloadable FOC GSM 11.11
http://pda.etsi.org/PDA/copy_file.asp?wki_id=igan562npJRVVaZ4AGuo&
Profile_id=PbOOv.6jd'@qcffcUgmVhchpe&page_id=CopyFile

Downloadable FOC GSM 11.14
http://pda.etsi.org/PDA/copy_file.asp?wki_id=aSs1-6.N.jgnjmi2tb7I&
Profile_id=PbOOv.6jd'@qcffcUgmVhchpe&page_id=CopyFile

A central and mandatory feature of the 2GGSM phone system is the Subscriber Identity Module (SIM). This is a miniaturized removable smart card held within the handset. SIM cards securely store a service-subscriber key (IMSI) and this is used to identify a subscriber. SIM cards can be moved by the user from one handset to another (unless this is blocked by the handset manufacturer/vendor), allowing users to access their same subscriber account and phone number by taking their SIM card out of one handset and putting it into a different handset.

In CDMA systems the equivalent is known as a Removable User Identity Module (RUIM) (see Section 4.2.1.4), and in 3G systems the equivalent is known as a Universal Integrated Circuit Card (UICC) (see Section 4.2.1.14).

SIM cards are 25 mm wide, with a height of 15 mm, and a thickness of 0.76 mm.

4.2.1.4 RUIM

Downloadable FOC RUIM
http://www.3gpp2.org/Public_html/specs/C.S0023-A_v3.0_041220.pdf

A Removable User Identity Module is the CDMA handset equivalent to the GSM SIM and 3G USIM (see Section 4.2.1.12). It can work in both CDMA and GSM phones and networks. It is physically compatible with GSM SIMs (see Section 4.2.1.3) and can fit into existing GSM phones and it is an extension of the GSM 11.11 standard (see Section 4.2.1.3).

4.2.1.5 IMS

Downloadable FOC

TS 22.250 IMS Group Management; Stage 1
http://www.3gpp.org/ftp/Specs/html-info/22250.htm

TS 22.340 IMS Messaging; Stage 1
http://www.3gpp.org/ftp/Specs/html-info/22340.htm

TR 22.800 IMS Subscription and access scenarios
http://www.3gpp.org/ftp/Specs/html-info/22800.htm

TS 23.167 IMS emergency sessions
http://www.3gpp.org/ftp/Specs/html-info/23167.htm

TS 23.218 IMS session handling; IM call model; Stage 2
http://www.3gpp.org/ftp/Specs/html-info/23218.htm

TS 23.228 IMS stage 2
http://www.3gpp.org/ftp/Specs/html-info/23228.htm

TR 23.864 Commonality and interoperability between IMS core networks
http://www.3gpp.org/ftp/Specs/html-info/23864.htm

TR 23.867 IMS emergency sessions
http://www.3gpp.org/ftp/Specs/html-info/23867.htm

TR 23.981 Interworking aspects and migration scenarios for IPv4-based IMS implementations (early IMS)
http://www.3gpp.org/ftp/Specs/html-info/23981.htm

TS 24.141 Presence Service using the IMS Core Network subsystem; Stage 3
http://www.3gpp.org/ftp/Specs/html-info/24141.htm

TS 24.147 Conferencing using the IMS Core Network subsystem
http://www.3gpp.org/ftp/Specs/html-info/24147.htm

TS 24.228 Signalling flows for the IMS call control based on SIP and SDP; Stage 3
http://www.3gpp.org/ftp/Specs/html-info/24228.htm

TS 24.247 Messaging using the IMS Core Network subsystem; Stage 3
http://www.3gpp.org/ftp/Specs/html-info/24247.htm

TS 29.162 Interworking between the IMS and IP networks
http://www.3gpp.org/ftp/Specs/html-info/29162.htm

TS 29.163 Interworking between the IMS and Circuit Switched (CS) networks
http://www.3gpp.org/ftp/Specs/html-info/29163.htm

TS 29.228 IMS Cx and Dx interfaces : signalling flows and message contents
http://www.3gpp.org/ftp/Specs/html-info/29228.htm

TS 29.229 IMS Cx and Dx interfaces based on the Diameter protocol; Protocol details
http://www.3gpp.org/ftp/Specs/html-info/29229.htm

TS 29.278 CAMEL Application Part (CAP) specification for IMS
http://www.3gpp.org/ftp/Specs/html-info/29278.htm

TS 29.328 IMS Sh interface : signalling flows and message content
http://www.3gpp.org/ftp/Specs/html-info/29328.htm

TS 29.329 IMS Sh interface based on the Diameter protocol; Protocol details
http://www.3gpp.org/ftp/Specs/html-info/29329.htm

TS 31.103 Characteristics of the IMS Identity Module (ISIM) application
http://www.3gpp.org/ftp/Specs/html-info/31103.htm

TS 32.260 Telecommunication management; Charging management; IMS charging
http://www.3gpp.org/ftp/Specs/html-info/32260.htm

TR 33.978 Security aspects of early IP Multimedia Subsystem (IMS)
http://www.3gpp.org/ftp/Specs/html-info/33978.htm

The IMS is designed to deliver IP multimedia over 3GPP systems. Its original formulation (3GPP R5) provided IP via GPRS but has been to support via EDGE and 3G protocols.

3GPP uses, as far as possible, IETF (standard Internet) protocols such as SIP, and, as the control layer is a common horizontal layer, services do not need their own control functions.

As you can see from the above list of deliverables, it has evolved through several iterations.

4.2.1.6 GPRS

Downloadable FOC GPRS
http://www.3gpp.org/specs/specs.htm

General Packet Radio Service (GPRS) is a packet-switched technology that enables data communications over GSM. This transmission technique does not set up a continuous channel from a portable terminal for the transmission and reception of data, but transmits and receives data in packets. It makes more efficient use of available radio spectrum. It provides a multislot packet data transmission protocol for GSM supporting speeds up to 170 Kbps. GPRS networks consist of two main elements: Service GPRS Support Node (SGSN) and Gateway GPRS Support Node (GGSN). EDGE (see Section 4.2.1.7) has generally replaced GPRS and offers higher data transmission speeds.

4.2.1.7 EDGE

Downloadable FOC EDGE
http://www.3gpp.org/specs/specs.htm

Enhanced Data rates for GSM Evolution (EDGE) is an evolution of GPRS with a more sophisticated modular scheme which increases data transmission rates to a range of 144 to 384 Kbps.

4.2.1.8 WCDMA

Wideband Code Division Multiple Access (WCDMA) is a 3G wireless access protocol that uses a wider channel band (5 MHz) to increase transmission bandwidth up to 384 Kbps in mobile environments and up to 2 Mbps in fixed settings.

4.2.1.9 UTRAN

Downloadable FOC UTRAN
http://www.3gpp.org/specs/specs.htm

UMTS Terrestrial Radio Access Network (UTRAN) is a generic European term for third-generation mobile cellular system. (See Section 4.2.1.10.)

4.2.1.10 UMTS

Downloadable FOC UMTS
http://www.3gpp.org/specs/specs.htm

Universal Mobile Telecommunications System (UMTS) is one of the third-generation cell phone technologies using W-CDMA as the underlying air interface, as standardized by the 3GPP. (See Section 4.2.1.)

4.2.1.11 FOMA

Downloadable FOC i-Mode
http://www.nttdocomo.co.jp/english/service/imode/index.htm

Freedom of Mobile Multimedia Access (FOMA) is the brand name for the 3G services being offered by Japanese mobile phone operator NTT DoCoMo. FOMA is compatible with standard UMTS, both in respect of additional protocols and the USIM card exchange, and hence provides several alternative options for global roaming: either with or without change of handset.

4.2.1.12 USIM

Downloadable FOC USIM Deliverables
http://www.3gpp.org/ftp/Specs/archive/31%5Fseries/31.102/

Universal Subscriber Identity Module is a UMTS function which runs on a UICC smart card within a 3G mobile handset. The USIM stores user subscriber information, authentication information, and provides storage space for text messages and maintains a phone book.

The USIM also holds a long-term secret code (known as "K") for the purposes of authentication, which is shared with the authentication center in the network. The USIM also provides a function to verify that a sequence number is within a range to avoid replay attacks, and generates the "CK" and "IK" session keys to maintain confidentiality using protection algorithms of a block cipher known as KASUMI.

4.2.1.13 ISIM

Downloadable FOC ISIM 3GPP TS.31.103 Characteristics of the IMS Identity Module (ISIM) application
http://www.3gpp.org/ftp/Specs/html-info/31103.htm

An IP Multimedia Services Identity Module (ISIM) is an application which runs on a (3G) UICC smart card and provides part of the IMS function. It provides parameters to identify and authenticate the user to the IMS. The ISIM application can coexist with SIM and USIM on the same UICC, making it possible to use the same smartcard in both GSM networks and earlier releases of UMTS.

ISIM holds a private user identity, one or more public user identities and a long-term code used as part of the calculation for authentication.

4.2.1.14 UICC

Downloadable FOC UICC
http://www.3gpp.org/specs/specs.htm

In the UMTS version of 3G, a UICC is an IC card which validates the integrity and security of the user and also holds additional user related data, typically up to several hundred kilobytes.

When used in a 2G GSM network, the UICC contains a SIM function application and in a UMTS network it provides the USIM function, allowing within one card access to both GSM and UMTS networks (which is most important until the UMTS 3G networks are ubiquitous) and performs the ISIM function. The IC card also holds a phone book and other applications. With the UMTS release 5, a new application, the IP Multimedia Services Identity Module is required for services in the IMS. The telephone book is a separate application and not part of either subscription information module.

4.2.1.15 KASUMI

Downloadable FOC KASUMI
http://portal.etsi.org/dvbandca/3GPPSPECIFICATIONS/3GTS35.202.pdf

KASUMI is a block cipher which is used in the confidentiality and integrity algorithms for 3GPP systems. A number of serious weaknesses in the cipher have been identified.

KASUMI is an optimized version of an existing lock cipher called MISTY1—indeed, KASUMI is the Japanese word for "misty." KASUMI maintains an efficient implementation in software.

4.2.1.16 CAMEL

Downloadable FOC
3GPP TS 02.78 Release 96 specification for the service aspects of CAMEL Phase 1
http://www.3gpp.org/ftp/Specs/archive/02_series/02.78/0278-560.zip

3GPP TS 22.078 Release 5 specification for CAMEL phase 4
http://www.3gpp.org/ftp/Specs/archive/22_series/22.078/22078-5f0.zip

3GPP TS 29.079 CAMEL Application Part specification for Release 5
http://www.3gpp.org/ftp/Specs/archive/29_series/29.078/29078-590.zip

TS 23.278 Customized Applications for Mobile network Enhanced Logic (CAMEL)–IMS interworking; Stage 2
http://www.3gpp.org/ftp/Specs/html-info/23278.htm

TS 29.278 CAMEL Application Part (CAP) specification for IMS
http://www.3gpp.org/ftp/Specs/html-info/29278.htm

Customized Applications for Mobile Networks Enhanced Logic (CAMEL) (ETSI TS 123 078) is a set of standards designed to work on either a GSM core network or UMTS network. They allow an operator to define services over and above standard GSM services/UMTS services. The CAMEL architecture is based on the Intelligent Network (IN) standards, and uses the CAP protocol.

CAMEL can be used to create many services and is particularly useful in a roaming situation.

4.2.1.17 IMSI

Downloadable FOC IMSI
http://www.3gpp.org/ftp/Specs/archive/23%5Fseries/23.003/

International Mobile Subscriber Identity (IMSI) is a unique phone identity that is stored in the SIM of all UMTS network users. It is sent by the phone to the network. To minimize the risk of the subscriber being identified and tracked by eavesdroppers on the radio interface, the IMSI is sent as rarely as possible and a randomly generated TMSI is sent instead.

An IMSI is usually 15 digits long, but can be shorter. The first 3 digits are the Mobile Country Code, and this is followed by the Mobile Network Code, which is either 2 digits (European standard) or 3 digits (North American standard). The remaining digits provide the mobile subscriber identification number (MSIN) within the network's customer base.

IMSI conforms to the ITU E.212 Numbering Recommendation

4.2.1.18 TMSI

Downloadable FOC TMSI
http://www.3gpp.org/ftp/Specs/archive/23%5Fseries/23.003/

The Temporary Mobile Subscriber Identity (TMSI) is the identity that is most commonly sent between the mobile and the network. A TMSI is generated for each geographical location to improve privacy. It is a randomly allocated number that is given to the mobile whenever it is switched on and registers with the network. The number is local to a location area, and so it has to be updated each time the mobile moves to a different cell.

The network can also change the TMSI of the mobile at any time. This is done frequently in order to avoid the subscriber from being identified and tracked by eavesdroppers. This makes it difficult to trace which mobile is which, except briefly, when the mobile is just switched on.

4.2.1.19 IMEI

The International Mobile Equipment Identity (IMEI) is a unique device identity and is phone specific to every GSM and UMTS mobile phone. It is usually found printed on the phone underneath the battery and can also be found by dialing the sequence *#06# into the phone.

The IMEI number is used by the GSM network to identify valid devices and can therefore be used to stop a stolen phone from accessing the network. This is done by using its IMEI number. This renders the phone useless, regardless of whether the SIM is changed.

4.2.1.20 MSISDN

The Mobile Subscriber ISDN Number (MSISDN) refers to the telephone number of a mobile subscriber. The MSISDN follows the numbering plan defined in the ITU-T Recommendation E.164. (ISDN = Integrated Services Digital Network.)

While IMSI/TMSI are used for user identification. The MSISDN is the telephone number of a user.

4.2.1.21 GSM/3G Evolution

Revised versions of many of these specifications are produced up to four times a year following the quarterly technical support group (TSG) plenary meetings. Following each TSG plenary meeting, a complete updated set of specifications is produced. This set includes not only the new specifications generated at that meeting, but also the latest versions of each specification that was not changed at that meeting.

The Status List (ZIPped files in either RTF or Word format) summarizes the current version number for every release of every 3GPP specification following each TSG plenary meeting. Also listed for each specification are:

- The 3GPP working group and rapporteur responsible for the specification;
- The project manager in the Mobile Competence Center (MCC) responsible for the specification;
- The meeting at which it was, or is expected to be, "frozen" (i.e., the point after which only corrections are allowed).

Full details of the specifications, their history and current status can be found in the 3GPP Specifications Status database.

Downloadable FOC 3GPP Specifications Status
http://www.3gpp.org/ftp/Information/Databases/Spec_Status/3GPP-Spec-Status.zip

4.2.2 3GPP2

The Third Generation Partnership Project 2 (3GPP2) is the standardization group for CDMA2000, the set of 3G standards based on earlier 2G CDMA technology.

The participating associations are ARIB/TTC (Japan), China Communications Standards Association, Telecommunications Industry Association (North America), and Telecommunications Technology Association (South Korea).

Downloadable FOC 3GPP2Specs
http://www.3gpp2.org/Public_html/specs/tsgcnumdesc.cfm

4.2.3 IEEE 802.11—WLAN, Wi-Fi, and Its Variants

IEEE	802.11	IEEE Std 802.11-1997 specifies a single Medium Access Control (MAC) sub layer and 3 Physical Layer Specifications.	Published	IEEE

ISO Available from
http://standards.ieee.org/getieee802/802.11.html

IEEE Std 802.11-1997 specifies a single medium access control (MAC) sublayer and three physical layer specifications for a single MAC that will support multiple PHY instantiations.

Introduction to IEEE 802.11

IEEE Std. 802.11-1997, was the first wireless LAN (WLAN) standard. It forms the backbone of a range of physical layer instantiations. There are a large number of variants both in protocols and frequency used. As the principal 802.11 standard provides the basis for all of the variants of 802.11, a greater level of detail is provided here to assist in understanding these subsequent options.

The origin of IEEE 802.11 can be traced back to IEEE 802 IEEE Standard for Local and Metropolitan Area Networks: Overview and Architecture, a standard whose origins go back to the 1970s to provide the common features for what were then, of course, wired networks. Each of the separate 802 standards deals with a different type of network. Standard 11 (IEEE 802.11) deals with wireless networks.

It is not that simple, however, because there are different approaches to the work, and the technology is evolving very rapidly. Hence, the first IEE 802.11 compliant products found in the market were the simpler 802.11b compliant products, whereas the more capable IEEE 802.11a products did not hit the market until some years later. Now there are more than 26 variants under development and so 802.^ standards will now include 802.^^ variants. In this short summary, we will consider only the major variants that we are likely to encounter in a vehicular environment.

The parent IEEE 802.11 standard defines the medium access control (MAC) and physical (PHY) layers for a LAN with wireless connectivity (see Figure 4.5). It addresses local area networking where the connected devices communicate over the air to other devices that are within close proximity to each other. This section provides an overview of the 802.11 architecture and the different topologies incorporated to accommodate the unique characteristics of the IEEE 802.11 wireless LAN standard.

The concepts for IEEE 802.11 originate with and are very similar to a wireless version of the IEEE 802.3 Ethernet standard (IEEE 802.3 IEEE Standard for Local and Metropolitan Area Networks: Overview and Architecture), but for a wireless environment. The IEEE 802.11 standard addresses:

- Functions required for an 802.11 compliant device to operate either in a peer-to-peer fashion or integrated with an existing wired LAN;
- Operation of the 802.11 device within possibly overlapping 802.11 wireless LANs and the mobility of this device between multiple wireless LANs;
- MAC level access control and data delivery services to allow upper layers of the 802.11 network;
- Several physical layer signaling techniques and interfaces;
- Privacy and security of user data being transferred over the wireless media.

Differences Between a Wireless LAN and a Wired LAN

The wireless environment is more challenging than for a wired LAN. The physical characteristics of a wireless LAN introduce range limitations and the consequential unreliability of the medium, dynamic topologies where stations move about, interference from outside sources, and lack of the ability for every device to hear every other device within the WLAN.

Limitations such as these create the need for the WLAN standard to create fundamental definitions for short-range LANs made up of components that are within close proximity of each other.

The mobility of wireless stations within a zone (more properly known as *nomadic devices*) is another important feature of a wireless LAN. However, in the base architecture the term "mobility" occurs. In the original IEEE 802 standards, the true concept of mobility (i.e., with homogeneous handover between cells, or

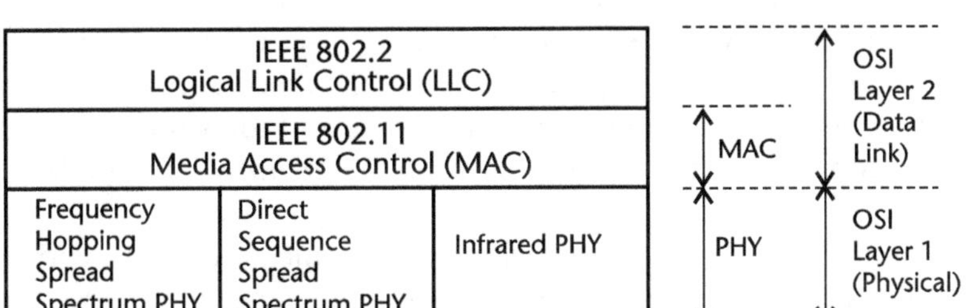

Figure 4.5 IEEE 802.11 standards mapped to the OSI reference model. (*Source:* CSI Library.)

heterogeneous handover between systems using different wireless technologies) was not considered. The furthest that 802.11 got into this concept was the ability to use a backbone wired network to transfer data from a nomadic network in one cell to a different access point in another cell. Until 802.11p, s, and x variants, the cell network handover techniques that we all take for granted to move from cell to cell were not contemplated.

Within 802.11 the concept of mobility was restricted to a nomadic device within the range of an access point being able to operate from different locations within that access zone. The concept assumes, indeed specifies, that even then, the device will remain stationary during any communication session. (However, we now know that the protocols are good enough that mobility within the access zone does in fact enable nomadic mobility during a session so long as the user remains within the zone of a single access point or peer-to-peer communication.)

IEEE 802.11 Wireless LAN Architecture

The 802.11 architecture is comprised of several components and services that interact to provide station mobility transparent to the higher layers of the network stack.

Wireless LAN Station

The station (STA) is the most basic component of the 802.11 wireless network. A station is any device that contains the functionality of the 802.11 protocol: the MAC and PHY. Typically the 802.11 functions are implemented in the hardware and software of a network interface card (NIC).

A station may be a desktop personal computer, a laptop personal computer, a handheld device such as a PDA, an access point, or similar device. Stations support the 802.11 station services of:

- Authentication;
- Deauthentication;
- Privacy;
- Data delivery.

Basic Service Set (BSS)

IEEE 802.11 defines the Basic Service Set (BSS) as the basic building block of an IEEE 802.11 wireless LAN. The BSS consists of a group of any number of stations.

802.11 Topologies

Independent Basic Service Set

The most basic wireless LAN topology is a set of stations, which have recognized each other and are connected via the wireless media in a peer-to-peer fashion. This form of network topology is referred to as an Independent Basic Service Set (IBSS) or, in common parlance, and as represented in Microsoft Windows, an ad hoc network.

In an IBSS, the mobile stations communicate directly with each other. Every mobile station may not be able to communicate with every other station due to the range limitations. There are no relay functions in an IBSS, therefore all stations need to be within range of each other and communicate directly. (See Figure 4.6.)

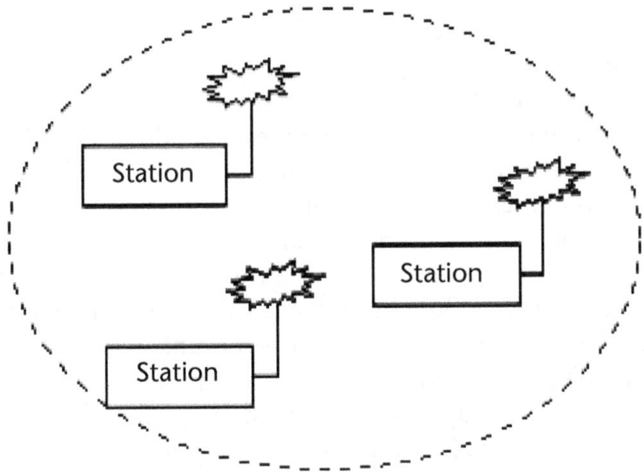

Figure 4.6 Independent Basic Service Set (IBSS). (*Source:* CSI Library.)

Infrastructure Basic Service Set

An Infrastructure Basic Service Set is a BSS with a component called an access point (AP). The access point provides a local relay function for the BSS. All stations in the BSS communicate with the access point and no longer communicate directly with each other. All frames are relayed between stations by the access point. This local relay function generally doubles the range of the IBSS.

The access point may also provide connection to a distribution system (a parent network or the Internet, for example).

802.11 Physical Layer

The 802.11 physical layer (PHY) manages the base layer "radio" aspects of the wireless transmission The PHY provides the enaction of physical interface protocols and protocol management to enable and make the exchange of frames with the upper MAC layer for transmission and receipt of data, and it provides a carrier sense indication back to the MAC to verify activity on the media. The PHY achieves this through modulation (and possibly spread spectrum) of the signal carrier to transmit binary data, organized in frames, across the medium.

The extensions to 802.11 – 802.11a – 802.11^^ provide techniques to achieve the physical transfer of data. (Key extensions that are likely to be encountered in a vehicular environment are summarized next.) Each PHY specification includes state diagrams to formally describe the protocols. One of the techniques, spread spectrum, which is used by several of the 802.11 variants, is worthy of a little more description.

Spread Spectrum. Spread spectrum is term used for a number of techniques which provide greater reliability and security of data transmission, and is therefore highly desirable in the vagaries of a wireless world, with its infinitely greater risk of interference and corruption of the signal as received, as well as eavesdropping by unauthorized parties, than a wired network. The origin of spread spectrum lies in military operations and support of covert operatives. Crudely, the techniques trade bandwidth for reliability. By chopping up the signal, and either transmitting it simultaneously at multiple frequencies to ensure the accuracy of the received data, or by rapidly changing frequencies in a predetermined sequence known only to the sender and receiver (which appears random to a third party) to prevent the message from interception, the message can achieve greater reliability and/or better resistance to interception by unauthorized listeners.

Spread spectrum is often touted as being a faster technique than traditional narrow band transmission, but the reality is that in order to achieve its reliability objectives, in either mode, more bandwidth has to be used. And the consequence of using more bandwidth is a higher achievable data rate. It is the bandwidth that affects the data rate, and not the spread spectrum technique. However, as more bandwidth is required to enable spread spectrum to work, the bands where it is permitted must be wide enough for the technique to work, and hence faster data rates are often achieved.

Another secondary benefit that was not part of the original conception, but is of increasing importance in a crowded wireless world, is that by providing the hopping management, which includes collision avoidance techniques, the result is more neighbor friendly and enables larger populations to coexist without interference. This is because, by spreading the data across more channels, although there is some overhead loss, the load on any one channel at any point in time is significantly lower. In a bidirectional wireless communication, even where high data volumes may be exchanged, much of the time available to transmit data is lost in any event, therefore by lowering the load on any one channel, and having efficient accessing/management techniques to send the data when the channel is quiet you make it easier to support more users of the channel, and the spread spectrum techniques manage this very well. In a BSS environment with multiple users to an access point, the techniques designed to enable spread spectrum to provide the reliability that they were designed for actually enable larger populations to use the access point.

Frequency Hopping Spread Spectrum. Frequency hopping utilizes a set of narrow channels and hops through all of them in a predetermined sequence. For example, the 2.4-GHz frequency band is divided into 70 channels of 1 MHz each. Every 20 to 400 msec, the system hops to a new channel following a predetermined cyclic pattern.

The 802.11 Frequency Hopping Spread Spectrum (FHSS) PHY variants such as IEEE802.11b typically uses the 2.4-GHz radio frequency band, operating with at 1 or 2 Mbps data rate.

Direct Sequence Spread Spectrum. The principle of direct sequence is to spread a signal on a larger frequency band by multiplexing it with a signature or code to minimize localized interference and background noise. To spread the signal, each bit is modulated by a code. In the receiver, the original signal is recovered by receiving the whole spread channel and demodulating with the same code used by the transmitter. The 802.11 Direct Sequence Spread Spectrum (DSSS) PHY also uses the 2.4-GHz radio frequency band.

802.11 Media Access Control

The 802.11 MAC provides a controlled access method to the shared wireless media called Carrier-Sense Multiple Access with Collision Avoidance (CSMA/CA). CSMA/CA is similar to the collision detection access method deployed by 802.3 Ethernet LANs. (See Sections 3.8.2 and 6.2.1.)

The 802.11 MAC layer provides functionality to allow reliable data delivery for the upper layers over the wireless PHY media. The data delivery itself is based on an asynchronous, best-effort, connectionless delivery of MAC layer data. There is no guarantee that the frames will be delivered successfully. The third function of the 802.11 MAC is to protect the data being delivered by providing security and privacy services. Security is provided by the authentication services and by WEP.

The MAC provides the following services:

- Authentication (station service);
- Deauthentication (station service);
- Privacy (station service);
- MSDU delivery (station service);
- Association (distribution system service);
- Disassociation (distribution system service);
- Distribution (distribution system service);
- Integration (distribution system service);
- Reassociation (distribution system service).

Stations can operate in two configurations:

- *Independent configuration:* The stations communicate directly to each other, so there is no infrastructure that needs to be installed—so-called ad hoc networks. It is easy to operate, but the disadvantage is that the coverage area is limited. Stations in such configuration are in a BSS. Without the ESS, the stations operate in an IBSS.
- *Infrastructure configuration:* The stations communicate to access points which are part of a distribution system. An access point serves the stations in a BSS. The set of BSSs are called ESS.

NOTE: 802.11 only specifies the air interface, which is the interface between stations and between stations and access points. With a distribution system, the coverage area can be extended to whatever the internals of the distribution system for instance with bridged wired LANs.

The standard provides the above-mentioned services with the following functionality: roaming within an ESS, multiple data rates in BSSs, and power management (i.e., stations can switch off their transceivers to conserve power).

The MAC protocol is CSMA/CA.

The standard includes a formal description of the MAC protocol using the SDL method standardized by the ITU-T (formerly CCITT).

CSMA/CA

The fundamental access method of 802.11 is Carrier Sense Multiple Access with Collision Avoidance. CSMA/CA works with a "listen before talk" scheme. This means that a station wishing to transmit must first sense the radio channel to determine if another station is transmitting. If the medium is not busy, the transmission may proceed.

The CSMA/CA protocol avoids collisions among stations sharing the medium by utilizing a random back-off time if the station's physical or logical sensing mechanism indicates a busy medium. The period of time immediately following a busy medium is the highest probability of collisions occurring, especially under high utilization.

The CSMA/CA scheme implements a minimum time gap between frames from a given user. Once a frame has been sent from a given transmitting station, that station must wait until the time gap is up to try to transmit again. Once the time has passed, the station selects a random amount of time (the back-off interval) to wait before listening again to verify a clear channel on which to transmit. If the channel is still busy, another back-off interval is selected that is less than the first. This process is repeated until the waiting time approaches zero and the station is allowed to transmit. This type of multiple access ensures judicious channel sharing while avoiding collisions.

Parts of 802.11

When purchasing 802.11 equipment, you will currently find a statement that reads something like "complies to IEEE 802.11b." The letter at the end refers to the responsibility of a task group to address a particular aspect which will form an amendment of the basic standard. The reader should note that it is the intention of IEEE that these will all be consolidated into a new revision of 802.11, simply entitled 802.11, with all of the amendments incorporated.

The purpose and status of all of the various task groups and amendments, as of mid-2007, are as listed in Table 4.2.

IEEE 802.11p is the IEEE 802 variant being designed for use in the vehicular environment (WAVE = wireless access for the vehicular environment). It is a further development of the 802.11a protocols, but with additions that provide for the very short latencies required. See also Section 4.6.6.1.

4.2.4 Mobile Wireless Broadband

Mobile wireless broadband (MWB) is a relatively recent technology that takes the principles of broadband and provides the service wirelessly. Many of the developments were initially created to replace broadband provision without the need for laying cables. Wireless broadband has similarities to IEEE 802.11, WLAN, but the range is measured in kilometers rather than meters. Mobile wireless broadband is wireless broadband with cell-to-cell handover, and it can be considered to be a sort of cross between WLAN and cellular telephony—indeed, some believe it will become 4G mobile telephony.

Apart from its range, it is significant that the business model philosophy evolves from broadband service provision, which is generally subscribed on a flat rate

Table 4.2 IEEE 802.11 Task Groups

Group	*Label*	*Description*	
IEEE 802.11 Working Group	WG	The Working Group is comprised of all of the Standing Committees, Task Groups, Study Groups, and Ad-Hoc Groups	
Task Group	TG	The committee(s) that are tasked by the WG as the author(s) of the standard or subsequent amendments via an approved PAR	
MAC Task Group	MAC	Scope of project	The scope of the project is to develop one common MAC for WLANs applications, in conjunction with the PHY Task Group work
		Status	Work has been completed and is now part of the original standard, published as IEEE Std. 802.11-1997
		Update status	Work has been completed on the ISO/IEC version of the original standard, published as IEEE Std. 802.11, 1999 (ISO/IEC 8802-11: 1999)
PHY Task Group	PHY	Scope of project	The scope of the project is to develop three PHYs for WLAN applications, using infrared (IR), 2.4-GHz FHSS, and 2.4-GHz DSSS, in conjunction with the one common MAC Task Group work
		Status	Work has been completed and is now part of the original standard, published as IEEE Std. 802.11-1997
		Update status	Work has been completed on the ISO/IEC version of the original standard, published as IEEE Std. 802.11, 1999 (ISO/IEC 8802-11: 1999)
Task Group a	TGa	Scope of project	The scope of the project is to develop a PHY to operate in the newly allocated UNII band
		Status	Work has been completed and is now part of the standard as an amendment, published as IEEE Std. 802.11a-1999
		Update status	Work has been completed on the ISO/IEC version of the original standard as an amendment, published as IEEE Std. 802.11a-1999 (ISO/IEC 8802-11: 1999 (E)/ Amd 1: 2000 (ISO/IEC)
Task Group b	TGb	Scope of project	The scope of the project is to develop a standard for a higher rate PHY in the 2.4-GHz band
		Status	Work has been completed and is now part of the standard as an amendment, published as IEEE Std. 802.11b-1999
Task Group b-Cor1	TGb-Cor1	Scope of project	The scope of this project is to correct deficiencies in the MIB definition of 802.11b
		Purpose of project	As the MIB is currently defined in 802.11b, it is not possible to compile an interoperable MIB. This project will correct the deficiencies in the MIB
		Status	Work has been completed and is now part of the standard as an amendment, published as IEEE Std. 802.11b-1999/Cor1-2001
Task Group c	TGc	Scope of project	To add a subclause under 2.5 Support of the Internal Sub-Layer Service by specific MAC Procedures to cover bridge operation with IEEE 802.11 MAC. This supplement to ISO/IEC 10038 (IEEE 802.1D) will be developed by the 802.11 Working Group in cooperation with the IEEE 802.1 Working Group
		Purpose of project	To provide the required 802.11 specific information to the ISO/IEC 10038 (IEEE 802.1D) standard
		Status	Work has been completed and is now part of the ISO/IEC 10038 (IEEE 802.1D) standard

Table 4.2 IEEE 802.11 Task Groups (continued)

Group	*Label*	*Description*	
Task Group d	TGd	Scope of project	This supplement will define the physical layer requirements: channelization, hopping patterns, new values for current MIB attributes, and other requirements to extend the operation of 802.11 WLANs to new regulatory domains (countries)
		Purpose of project	The current 802.11 standard defines operation in only a few regulatory domains (countries). This supplement will add the requirements and definitions necessary to allow 802.11 WLAN equipment to operate in markets not served by the current standard
		Status	Work has been completed and is now part of the standard as an amendment, published as IEEE Std. 802.11d-2001
Task Group e	TGe	Scope of project	Enhance the 802.11 MAC to improve and manage quality of service, provide classes of service, and enhanced security and authentication mechanisms. Consider efficiency enhancements in the areas of the Distributed Coordination Function (DCF) and Point Coordination Function (PCF)
		Purpose of project	To enhance the current 802.11 MAC to expand support for LAN applications with quality of service requirements. Provide improvements in security, and in the capabilities and efficiency of the protocol. These enhancements, in combination with recent improvements in PHY capabilities from 802.11a and 802.11b, will increase overall system performance, and expand the application space for 802.11. Example applications include transport of voice, audio, and video over 802.11 wireless networks, video conferencing, media stream distribution, enhanced security applications, and mobile and nomadic access applications
		Status	Active
Task Group f	TGf	Scope of project	To develop recommended practices for an Inter-Access Point Protocol (IAPP), which provides the necessary capabilities to achieve multivendor access point interoperability across a distribution system supporting IEEE P802.11 WLAN links. This IAPP will be developed for the following environment(s): 1) A distribution system consisting of IEEE 802 LAN components supporting an IETF IP environment 2) Others as deemed appropriate This Recommended Practices Document shall support the IEEE P802.11standard revision(s)
		Purpose of project	IEEE P802.11 specifies the MAC and PHY layers of a WLAN system and includes the basic architecture of such systems, including the concepts of access points and distribution systems. Implementation of these concepts where purposely not defined by P802.11 because there are many ways to create a Wireless LAN system. Additionally many of the

Table 4.2 IEEE 802.11 Task Groups (continued)

Group	*Label*	*Description*	
			possible implementation approaches involve concepts from higher network layers. While this leaves great flexibility in distributions system and access point functional design, the associated cost is that physical access point devices from different vendors are unlikely to interoperate across a distribution system due to the different approaches taken to distribution system design. As P802.11 based systems have grown in popularity, this limitation has become an impediment to WLAN market growth. At the same time it has become clear that there are a small number of distribution system environments that comprise the bulk of the commercial WLAN system installations This project proposes to specify the necessary information that needs to be exchanged between access points to support the P802.11 DS functions. The information exchanges required will be specified for one or more distribution systems; in a manner sufficient to enable the implementation of distribution systems containing access points from different vendors which adhere to the recommended practices
		Status	Work has been completed and is now part of the standard as a recommended practice, published as IEEE Std. 802.11F-2003
Task Group g	TGg	Scope of project	The scope of this project is to develop a higher speed(s) PHY extension to the 802.11b standard. The new standard shall be compatible with the IEEE 802.11 MAC. The maximum PHY data rate targeted by this project shall be at least 20 Mbps. The new extension shall implement all mandatory portions of the IEEE 802.11b PHY standard. The project will take advantage of the provisions for rate expansion that are in place on the current standard PHY. The 802.11 MAC defines a mechanism for operation of stations supporting different data rates in the same area. The current 802.11b standard already defines the basic rates of 1, 2, 5.5, and 11 Mbps. The proposed project targets further developing the provisions for enhanced data rate capability of 802.11b networks. The 802.11 MAC currently incorporates the interpretation of data rate information and the computation of expected packet duration even if the specific station does not support the rate at which the packet was sent.
		Purpose of project	To develop a new PHY extension to enhance the performance and the possible applications of the 802.11b compatible networks by increasing the data rate achievable by such devices. This technology will be beneficial for improved access to fixed network LAN and internetwork infrastructure (including access to other wireless LANs) via a network of access points, as well as creation of higher performance ad hoc networks
		Status	Work has been completed and is now part of the standard as an amendment, published as IEEE Std. 802.11g-2003

Table 4.2 IEEE 802.11 Task Groups (continued)

Group	*Label*	*Description*	
Task Group h	TGh	Scope of project	Enhance the 802.11 Medium Access Control (MAC) standard and 802.11a High Speed Physical Layer (PHY) in the 5-GHz band supplement to the standard; to add indoor and outdoor channel selection for 5-GHz license exempt bands in Europe; and to enhance channel energy measurement and reporting mechanisms to improve spectrum and transmit power management (per CEPT and subsequent EU committee or body ruling incorporating CEPT Recommendation ERC 99/23)
		Purpose of project	To enhance the current 802.11 MAC and 802.11a PHY with network management and control extensions for spectrum and transmit power management in 5-GHz license exempt bands, enabling regulatory acceptance of 802.11 5-GHz products. Provide improvements in channel energy measurement and reporting, channel coverage in many regulatory domains, and provide dynamic channel selection and transmit power control mechanisms
		Status	Work has been completed and is now part of the standard as an amendment, published as IEEE Std. 802.11h-2003
Task Group i	TGi	Scope of project	Enhance the 802.11 Medium Access Control (MAC) to enhance security and authentication mechanisms
		Purpose of project	To enhance the current 802.11 MAC to provide improvements in security
		Status	Work has been completed and is now part of the standard as an amendment, published as IEEE Std. 802.11i-2004
Task Group j	TGj	Scope of project	Enhance the 802.11 standard and amendments, to add channel selection for 4.9 and 5 GHz in Japan to additionally conform to the Japanese rules for radio operation
		Purpose of project	To obtain Japanese regulatory approval by enhancing the current 802.11 MAC and 802.11a PHY to additionally operate in newly available Japanese 4.9- and 5-GHz bands
		Status	Work has been completed and is now part of the standard as an amendment, published as IEEE Std. 802.11j-2004
Task Group k	TGk	Scope of project	To define radio resource measurement enhancements to provide interfaces to higher layers for radio and network measurements
		Purpose of project	The original standard has a basic set of radio resource measurements for internal use only. These measurements and others are required to provide services (such as roaming, coexistence, and others) to external entities. It is necessary to provide these measurements and other information in order to manage these services from an external source
		Status	Active

Table 4.2 IEEE 802.11 Task Groups (continued)

Group	*Label*	*Description*	
Task Group l	TGl	Scope of project	Task Group "Letter–L" not to be used by the IEEE 802.11 Working Group for inclusion into the published standard
		Purpose of project	N/A
		Status	N/A
Task Group m	TGm	Scope of project	Maintenance of the IEEE 802.11-1999 (reaff. 2003) standard
		Purpose of project	Maintenance of technical and editorial corrections to the 802.11-1999 (reaff. 2003) Wireless LAN Medium Access Control (MAC) and Physical Layer (PHY) specifications standard
		Status	Active
Task Group n	TGn	Scope of project	Investigating the possibility of improvements to the 802.11 standard to provide high throughput (> 100 Mbps)
		Purpose of project	The purpose of the project is to improve the 802.11 WLAN user experience by providing significantly higher throughput for current applications and to enable new applications and market segments.
		Status	Active
Task Group o	TGo	Scope of project	Task Group "Letter–O" not to be used by the IEEE 802.11 Working Group for inclusion into the published standard
		Purpose of project	N/A
		Status	N/A
Task Group p	TGp	Scope of project	To create an amendment of IEEE 802.11 to support communication between vehicles and the roadside and between vehicles while operating at speeds up to a minimum of 200 km/h for communication ranges up to 1,000m. The amendment will support communications in the 5-GHz bands; specifically 5.850- to 5.925-GHz band within North America with the aim to enhance the mobility and safety of all forms of surface transportation, including rail and marine. Amendments to the PHY and MAC will be limited to those required to support communications under these operating environments within the 5-GHz bands
		Purpose of project	To amend the existing IEEE 802.11 standard to make it suitable for interoperable communications to and between vehicles. The primary reasons for this amendment include the unique transport environments, and the very short latencies required (some applications must complete multiple data exchanges within 4 to 50 ms)
		Status	Active
Task Group q	TGq	Scope of project	Task Group "Letter–Q" not to be used by the IEEE 802.11 Working Group for inclusion into the published standard
		Purpose of project	N/A
		Status	N/A

Table 4.2 IEEE 802.11 Task Groups (continued)

Group	*Label*	*Description*	
Task Group r	TGr	Scope of project	Enhancements to the 802.11 Medium Access Control (MAC) layer to minimize or eliminate the amount of time data connectivity between the station and the distribution system is absent during a BSS transition, limited to the state necessary for the operation of the MAC
		Purpose of project	To improve BSS transitions within 802.11 ESSs and to support real-time constraints imposed by applications such as VoIP.
		Status	Active
Task Group s	TGs	Scope of project	To develop an IEEE 802.11 Extended Service Set (ESS) Mesh* with an IEEE 802.11 Wireless Distribution System (WDS) using the IEEE 802.11 MAC/PHY layers that supports both broadcast/multicast and unicast delivery over self-configuring multihop topologies
		Purpose of project	To provide a protocol for auto-configuring paths between APs over self-configuring multihop topologies in a WDS to support both broadcast/multicast and unicast traffic in an ESS Mesh using the four-address frame format or an extension
		Status	Active
Task Group t	TGt	Scope of project	To provide a set of performance metrics, measurement methodologies, and test conditions to enable measuring and predicting the performance of 802.11 WLAN devices and networks at the component and application level as a recommended practice
		Purpose of project	To enable testing, comparison, and deployment planning of 802.11 WLAN devices based on a common and accepted set of performance metrics, measurement methodologies and test conditions
		Status	Active
Task Group u	TGu	Scope of project	Amend the IEEE 802.11 MAC and PHY to support interworking with external networks
		Purpose of project	To provide amendments to the IEEE 802.11 PHY/MAC layers which enable interworking with other networks. This includes both enhanced protocol exchanges across the air interface and provision of primitives to support required interactions with higher layers for interworking
		Status	Active
Task Group v	TGv	Scope of project	Amendment to provide wireless network management enhancements to the 802.11 MAC, and PHY, to extend prior work in radio measurement to effect a complete and coherent upper layer interface for managing 802.11 devices in wireless networks
		Purpose of project	To provide amendments to the IEEE 802.11 PHY/MAC layers that enables management of attached stations in a centralized or in a distributed fashion (e.g., monitoring, configuring, and updating) through a layer 2 mechanism. While the 802.11k Task Group is defining messages to retrieve information from the station, the ability to configure the station is not in its scope. The proposed task group will also create an Access Port Management Information Base (AP MIB)
		Status	Active

Table 4.2 IEEE 802.11 Task Groups (continued)

Group	*Label*	*Description*	
Task Group x	TGy	Scope of project	Task Group "Letter–X" not to be used by the IEEE 802.11 Working Group for inclusion into the published standard
		Purpose of project	N/A
		Status	N/A
Task Group y	TGy	Scope of project	Contention based protocol standard for the IEEE 802.11 family of standards
		Purpose of project	To create a contention based protocol standard for the IEEE 802.11 family of standards
		Status	The PAR and Five Criteria for 802.11Y (Contention Based Protocols) were approved by ExCom in November 2005. However the item did not get on the NesCom agenda for their approval, so the group will continue to operate as the CBP SG through March 2006. The 802.11y standard is in its early proposal stages
Study Group	SG	Investigates the interest of placing something in the standard	
Study Group–ADS	ADS	Scope of project	Investigate the management frame security
		Purpose of project	To provide security for management frames and other control functions
		Status	Active
Standing Committee	SC	A group that reports directly to the WG Chair that investigates or provides assistance	
Standing Committee–Publicity	PSC	Looks at how IEEE 802.11 can better publicize the standard by collecting data related to its use and operation	
		Status	Ongoing
Standing Committee–Wireless Next Generations	WNG SC	Scope of project	Investigating the globalization and harmonization of WLANs jointly with ETSI-BRAN and MMAC, including revisions to the 802.11 standard
		Purpose of project	To provide one worldwide or common interface to WLAN standards acceptable to ETSI-BRAN, and MMAC, plus refinements to the existing 802.11 standard
		Status	Ongoing

monthly subscription, or a subscription with some usage level rates. For ITS service provision this potentially provides it with significant advantages over traditional 2G/3G cellular telephony, both in the cost to use and the operating speeds. However, it is not yet widely rolled out, and if successful, it will probably result in a change of the charging models for the traditionally call time–based 2G/3G services. A potential disadvantage is that the performance of MWB deteriorates in direct proportion to the number of users active at any one access point.

Although WIMAX and WIBRO (IEEE 802.16) have received most MWB publicity to date, the first, and at the time of writing, most successful of the MWB technologies is High Capacity, Spatial Division Multiple Access (HC-SDMA), although this requires new frequency allocation to allow it to operate. At the time of writing, it has achieved penetration in Australia, Africa, some parts of Northern Europe, some parts of Asia and the United States.

The mobile version of WIMAX was a relatively late addition to WIMAX, and has only recently become operational.

4.2.4.1 HC-SDMA Mobile Wireless Broadband

The use of HC-SDMA to provide ITS services is described in Section 4.5.7.1. The underpinning standards for HC-SDMA are:

ITU	ITU-R M.1801	Radio interface standards for broadband wireless access systems, including mobile and nomadic applications, in the mobile service operating below 6 GHz. ("RecDoc. 8/167 (Rev. 1) Annex 4 III")	Published	ITU
ANSI	0700004-2005	HC-SDMA protocol specification	Published	ANSI
ATIS	ATIS-0700004-2005	HC-SDMA protocol specification	Published	ATIS

ITU-R M.1801 Available from
http://www.itu.int/publications/bookshop/how-to-buy.html

ANSI.ATIS 0700004-2005 Available from
www.ansi.org

ATIS 0700004-2005 2 docs: Available from
[1.ATIS-0700004.2005] http://www.atis.org/docstore/product.aspx?id=22595
[2.ATIS-PP-0700004.2007] http://www.atis.org/docstore/product.aspx?id=22675

Whereas the standards and specifications for 2G cellular telephony are encapsulated in more than 70 standards, specifications and technical documents (of which there is no single comprehensive overview standard or summary), HC-SDMA is encapsulated in a single, very comprehensive document of several hundred pages. There are no technical differences nor layout differences (apart from SDO format pages) between the ATIS standard and the ANSI standard.

The HC-SDMA interface provides wide-area broadband wireless IP data connectivity for fixed, portable, and mobile computing devices and appliances. The protocol is designed to be implemented with smart antenna array techniques to substantially improve the radio frequency (RF) coverage, capacity, and performance for the system.

HC-SDMA operates on a similar premise as GSM or CDMA2000 for cellular phones, with hand-offs between HC-SDMA cells, providing the user with a seamless wideband wireless experience even when moving at the speed of a car or train.

The protocol is as follows:

- Specifies base station and client device RF characteristics, including output power levels, transmit frequencies and timing error, pulse shaping, in-band and out-of band spurious emissions, receiver sensitivity and selectivity;
- Defines associated frame structures for the various burst types including standard uplink and downlink traffic, paging and broadcast burst types;

- Specifies the modulation, forward error correction, interleaving, and scrambling for various burst types;
- Describes the various logical channels (broadcast, paging, random access, configuration and traffic channels) and their roles in establishing communication over the radio link;
- Specifies procedures for error recovery and retry.

The protocol also supports Layer 3 (L3) mechanisms for creating and controlling logical connections (sessions) between client device and base including registration, stream start, power control, handover, link adaptation, and stream closure, as well as L3 mechanisms for client device authentication and secure transmission on the data links.

4.2.4.2 IEEE 802.16 WIMAX and Its Variants

IEEE	802.16	Broadband Wireless Access	Published	IEEE

Various parts of IEEE 802.16 Available from
http://www.ieee.org/portal/innovate/products/standard/index.html

The history of IEEE 802.16 can be traced back to 1998, when the U.S. National Institute of Standards and Technology (NIST) started looking for a different wireless application and held the first initiative meeting. By the time of the first standardization of 802.16 (in December 2001), systems were already available. In fact, many basic concepts of 802.16 were borrowed from DOCSIS/HFC (a cable modem technology) and applied to the wireless setting, although the physical layer specification is different. (This is different from 802.11/Wi-Fi which is based on 802.3 Ethernet concepts. See Sections 3.8.2 and 6.2.1, also IEEE 802.11 Sections 3.10, 4.2.3, and 6.2.3.)

The IEEE 802.16 base standard is designed for line-of-sight applications utilizing the 10 to 66-GHz spectrum, but is not well suited to residential implementations where line of sight is often difficult.

IEEE 802.16 provides an amendment for nonline-of-sight utilizing 2 to 11 GHz. This solution is better suited to point-to-multipoint (PMP) and home applications. Orthogonal Frequency Division Multiplexing (OFDM) has been recently adopted. In mobile communication environments such as a moving vehicle, there is a Doppler shift effect which causes each subchannel to be shifted and orthogonality to be broken. To prevent this phenomenon, a cyclic prefix to each symbol is added. The cyclic prefix is used to implement the FFT when one needs to use circular convolution to emulate linear convolution. Instead of just sending blank during guard time, repeating the end of the symbol provides "guard time" (instead of the "guard band") to make sure all the subcarriers are orthogonal.

There is another benefit claimed by proponents of OFDM. Since each subcarrier of OFDM uses a narrow bandwidth, in the time domain each symbol is transmitted during a relatively long symbol time. This guarantees all the significant multipaths arrive within a symbol time minimizing intersymbol interference. This is good news, because it removes the need for an equalizer at a transmitter and a receiver side, reducing complexity and cost.

IEEE 802.16-2004 revises and replaces 802.16, 802.16a, and 802.16REVd. This revision represents the completion of the essential fixed wireless standard. As with most IEEE 802.16 standards, the principal IEEE 802.16 standard supports only fixed wireless and nomadic communications within the zone of one base station. It has no homogeneous (nor heterogeneous) hand-off provisions.

IEEE 802.16e is a MAC/PHY enhancement for supporting mobile communications at vehicular speeds. This supports a full hand-off. A user's session is maintained when he moves from one base station signal to another.

Network Architecture

Backplane management is achieved as a bridge, a router, or a host. When the service data unit (SDU) comes into the convergence sublayer, the classifier in the convergence sublayer can classify the type of the traffic (e.g., voice, Web surfing, ATM CBR) and map it into a particular connection ID (CID). In turn, each CID is referred to a certain specific profile. Each profile describes modulation scheme, forward error correction (FEC) scheme, and so on. Once the CID is well defined between both communication parties (such as a base station and a subscriber station), the nonchanging payload header information (such as ATM cell header or IP header) can be suppressed by payload header suppression (PHS). (See Figures 4.7 and 4.8.)

Scope of the IEEE 802.16 Standard

Having understood the principles of the architectural aspects, we now turn to the IEEE 802.16 standards. The scope of the 802.16 standards covers:

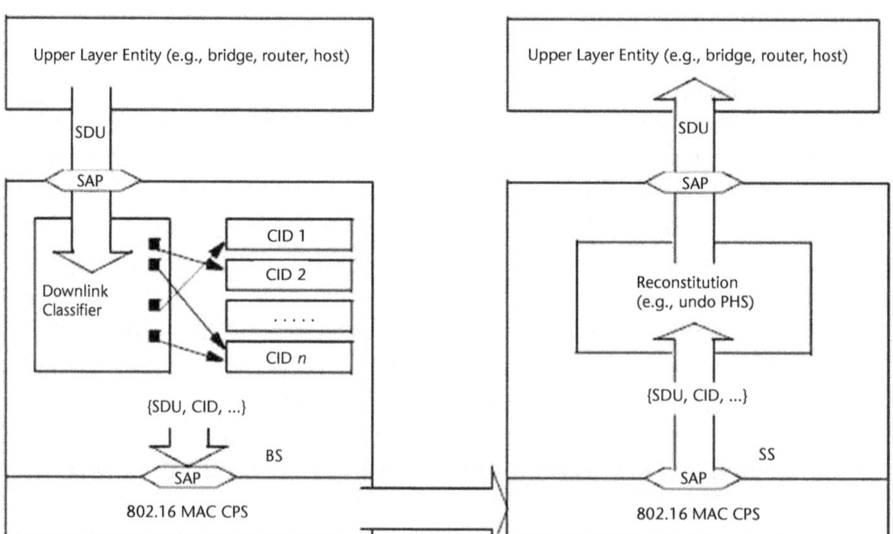

Figure 4.7 802.16 network architecture. (*Source:* http://www.eecs.berkeley.edu.)

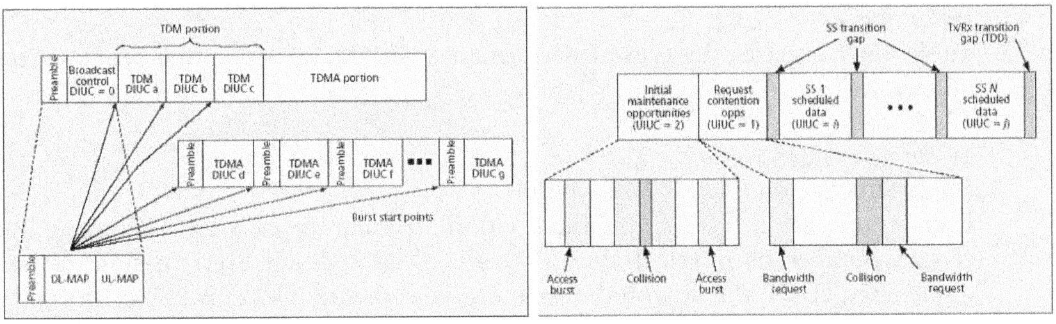

Figure 4.8 Downlink/uplink frame structures. (*Source:* www.eecs.berkeley.edu.)

- PHY;
- Service access point (SAP);
- Security sublayer;
- MAC common part sublayer (CPS);
- Service-specific convergence sublayer (CS).

Hand-Off Mechanism

The hand-off procedure of 802.16e is not radically different from that of other well-known wireless technologies. Depending on the signal-to-noise ratio, the subscriber station sends a hand-off request to the serving base station. The serving base station negotiates the bandwidth and QoS with possible target base stations on behalf of the subscriber station. Once it receives the confirmation from one of them, the serving base station notifies the subscriber station the hand-off response. The subscriber station then commences its session with the target base station.

Physical Layer

When using 2 to 11 GHz, utilizing OFDM gives several benefits including the use of adaptive antenna system (AAS), which is a directional signal processing technique that is able to highlight the signal from a certain direction when the system is sectorized. Table 4.3 summarizes the physical layer.

Table 4.3 IEEE 802.16 Physical Layer Summary

Designation	*Applicability*	*MAC*	*Duplexing*
Wireless MAN-SC	10–66 GHz Licensed	Basic	TDD, FDD, HFDD
	2–11 GHz Licensed	Basic, ARQ, STC, AAS	TDD, FDD
Wireless MAN-OFDM	2–11 GHz Licensed	Basic, ARQ, STC, AAS	TDD, FDD
	2–11 GHz License-exempt	Basic, ARQ, STC, DFS, MSH, AAS	TDD
Wireless MAN-OFDMA	2–11 GHz Licensed	Basic, ARQ, STC, AAS	TDD, FDD
	2–11 GHz License-exempt	Basic, ARQ, STC, DFS, MSH, AAS	TDD

MSH = mesh network; DFS = dynamic frequency selection.
Source: CSI Library.

Performance

Table 4.4 compares the typical performance of 802.16 with other wireless technology.

Media Access Control

Frame Structure and Access Mechanism

Every frame has a fixed frame time within an accuracy of a few milliseconds. A certain number of physical slots (PS) are given to each user. In time division duplexing (TDD), the downlink frame precedes the uplink frame. Since the end of downlink/start of uplink varies adaptively, this flexibility gives a good stream control in comparison to 802.11 protocols.

There are some basic notions. In the beginning of every frame, the downlink-MAP (DL-MAP) tells how the rest of the frame will be used; DL-MAP describes which part of the frame shall be used for whom with a certain downlink interval usage code (DIUC). DIUC is a scalar number indicating which specific burst profile will be used for a certain portion of the frame. A burst profile that is determined by DIUC is a preconfigured group of information specifying the FEC type and modulation type. The mapping between a DICU and a burst profile is done by the base station and is periodically broadcast to network in the downlink channel descriptor (DCD) message. The same applies to uplink, uplink-MAP (UL-MAP), uplink interval usage code (UIUC), and uplink channel descriptor (UCD) message.

Within a time division multiplexing (TDM) portion of a single frame, the modulation scheme gets more complex (equivalently less robust, equivalently again, more efficient) as TDM portions proceeds. By this clever strategy, the node that has better channel condition will successfully decode the received signal without losing synchronization, while the nodes with worse channel condition do not care about the synchronization under the fancier modulation after it successfully decodes the signal addressed to itself.

In the time division multiple access (TDMA) portion, every subframe carries its own preamble because some half-duplex nodes might be not able to listen to TDMA portion from the beginning because of possible prior transmissions.

In uplink, two initial parts of the uplink frame are subframes for contention based access. The contention based portion of the frame is used for the initial ranging of the newly joining stations or the best effort traffics/bandwidth requests of already associated stations. If a station wants to transmit only a single packet to the base station, it may well use this contention based portion of the frame instead of acquiring a dedicated bandwidth.

There is a complex power control scheme used in 802.16. When a subscriber station joins a network, it transmits its initial ranging request with the minimal

Table 4.4 Performance Comparison of Typical Wireless Technology

	Channel Bandwidth	*Maximum Data Rate*	*Maximum bps/Hz*
802.11a	20 MHz	54 Mbps	2.7
802.16a	10, 20, 3.5, 7, 14, 3, 6 MHz	70 Mbps	5
EDGE	200 kHz	384 Kbps	1.9
CDMA2000	1.25 MHz	2 Mbps	1.6

Source: CSI Library.

transmission power. If the subscriber station does not receive the ranging response from the base station, then it assumes the base station did not hear its request because the power was too weak. Then it repeats its ranging request with a slightly higher power. This process is repeated until the subscriber station gets a proper response from the base station. (See Figure 4.9.)

Network Entry Process

As with all IEEE 802 standards, a subscriber station has to enter and register itself to the network. Upon initialization, a subscriber station scans the possible channels of the downlink frequency band of operation until it finds a valid downlink signal. Once the PHY achieves connection, the subscriber station searches DL-MAP and obtains downlink parameters. After this stage, the subscriber station waits for UCD message to retrieve transmission parameters and perform initial ranging. Ranging is the process of acquiring the correct timing offset and power adjustments.

The ranging request (RNG-REQ) is performed on contention based slots. The ranging response (RNG-RSP) is sent to the subscriber station to adjust power level, frequency offset, and timing offset. Once successfully ranged, the subscriber station joins normal data traffic in the uplink. The subscriber station and the base station perform capability negotiation, authorization, key exchange, registration, IP version negotiation, DHCP address allocation, and operational parameter transfer. Finally they set up connections for preprovisioned service flows belonging to the subscriber station. (See Figure 4.10.)

Connection and Services

802.16 is a connection oriented protocol. Management messages are based on the preset connection ID (CID).

Every CID is associated with a service flow ID (SFID) that determines quality of service (QoS) parameters that the network tries to guarantee. Each CID is associated with a certain type of scheduling type and QoS parameters (see Table 4.5).

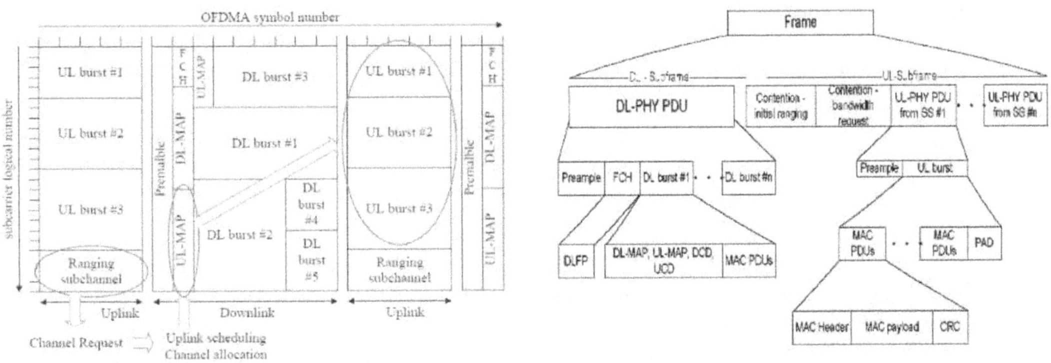

Figure 4.9 Frame structure of 802.16. Y axis is for subchannel; X axis is for symbol time in the left figure. (*Source:* www.eecs.berkeley.edu.)

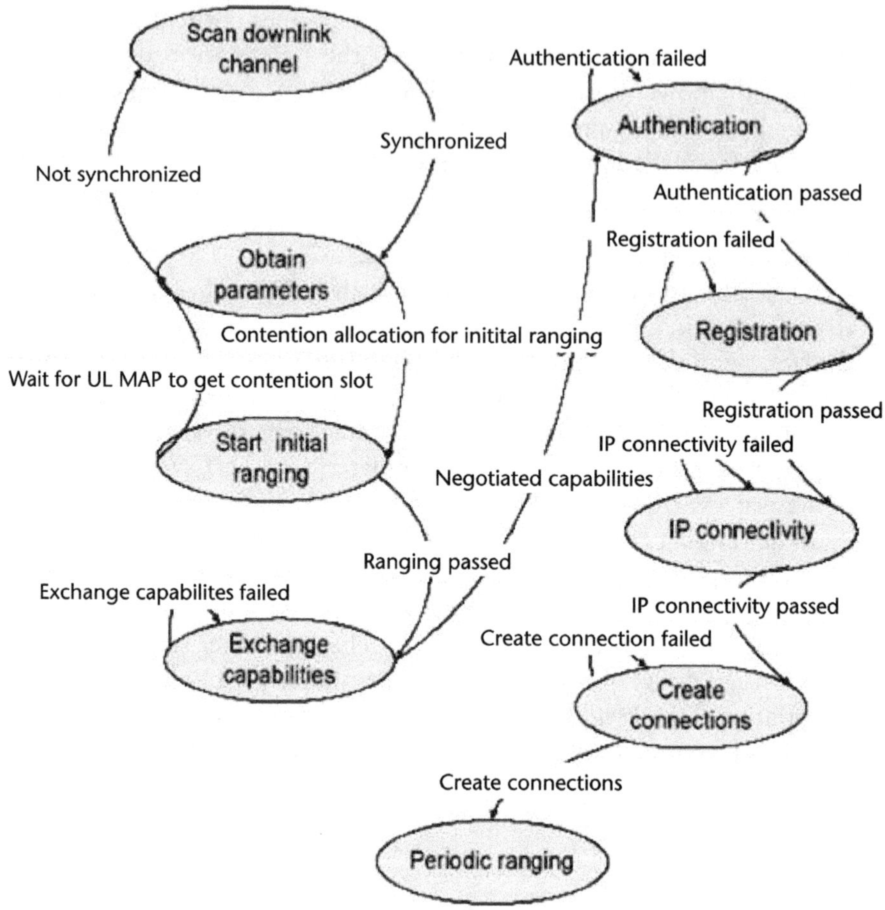

Figure 4.10 Network entry process of a newly joining station. (*Source:* CSI Library.)

Table 4.5 Scheduling Types and QoS

Scheduling Type	*QoS*
Unsolicited Grant Service (UGS)	Maximum sustained traffic rate, maximum latency, tolerated jitter
Real-Time Polling Service (rtPS)	Maximum sustained traffic rate, minimum reserved traffic rate, committed burst size, maximum latency
Nonreal-Time Polling Service (nrtPS)	Committed information rate, maximum information rate
Best Effort (BE)	Maximum information rate

Source: CSI Library.

Message Format

In 802.16, understanding the message format is straightforward. There are two types of header formats: (1) generic header, and (2) management header.

The significant header fields include Type, Length, CID. In a bandwidth request message, the BR (Bandwidth Request) field is also important. Currently 38 kinds of management messages are defined. They include UCD, DCD, DL-MAP, UL-MAP, ranging, registration, primary key management, dynamic service addition/change/deletion, TFT file exchange, ARQ, reports of channel status, and power control.

Advanced 802.16 Features

Some advanced features are significant. The first feature is MIMO channel capacity, which improves range and capacity.

The second feature is hybrid-ARQ. FEC has two major roles: error detection and error correction. Imaging the receiver station detects an error but is not able to correct it. In hybrid-ARQ mode, instead of discarding the error packet, the receiver sends a particular ARQ message to the transmitter and the transmitter sends back stronger FEC information. The receiver exploits both messages—the error packet and the stronger FEC information to recover the original packet.

Dynamic frequency selection (DFS) helps in minimizing interference, and space-time coding (STC) enhances performance in fading environments through spatial diversity.

4.2.4.3 IEEE 802.20 Mobile Broadband Wireless Access

IEEE	802.20	Mobile Broadband Wireless Access	Under Development	IEEE

IEEE 802.20 in development. Not currently available

On December 11, 2002, the IEEE Standards Board approved the establishment of IEEE 802.20, the Mobile Broadband Wireless Access (MBWA) Working Group. The mission of the work item is stated as follows: "The mission of IEEE 802.20 is to develop the specification for an efficient packet based air interface that is optimized for the transport of IP based services. The goal is to enable worldwide deployment of affordable, ubiquitous, always-on and interoperable multi-vendor mobile broadband wireless access networks that meet the needs of business and residential end user markets."

The scope of the work item 802.20 is as follows: "Specification of physical and medium access control layers of an air interface for interoperable mobile broadband wireless access systems, operating in licensed bands below 3.5 GHz, optimized for IP-data transport, with peak data rates per user in excess of 1 Mbps. It supports various vehicular mobility classes up to 250 km/h in a MAN environment and targets spectral efficiencies, sustained user data rates and numbers of active users that are all significantly higher than achieved by existing mobile systems."

The state of development as at the time of publication of this book is that a draft was expected for ballot by the end of 2007, with a view to a standard within 2008. A working group letter ballot was launched in July 2007 based on an initial draft decided in January 2006 combined with a new additional draft proposed in March 2007.

The overall structure proposed is:

Chapter 1: General structure including modes
Chapter 2–13: Wide band mode (FDD/TDD based on UMB in 3GPP2)
Chapter 14–25: 625k-MC mode (TDD based on ANSI HC- SDMA enhancement specification)

4.2.5 Satellite

Another form of communication that offers potential for communications with and between vehicles is the use of satellite communications. Advocates of satellites communications claim that satellite communications offer:

- Rural/desert region coverage;
- Usually global (but not necessarily continuous availability;
- Back-stop communications capability;
- Useful for polling vehicle when location not known;
- Very efficient broadcast mode;
- Satellite technology and performance improving, with cost reducing;
- Competitive service provision.

There is certainly interest in countries such as Australia, the very north of Europe, Canada, and parts of Africa, especially for emergency communications, such as eCall.

There is no standardization on base satellite technologies that are operated as proprietary systems.

The principal service providers (in no particular sequential order) are:

- S-DAB (Digital Audio Broadcast):
  - Worldspace;
  - ONDAS;
  - XM Radio;
  - SIRIUS.
- S-DMB (Multimedia Broadcast):
  - Eutelsat/SES Astra JV;
  - Mobile Broadcasting Corporation.
- Store and Forward:
  - Argos;
  - ORBCOM.
- MSS (Mobile Satellite Services) (GEO):
  - Iridium;
  - Space Segment Classification
  - Globalstar
  - MSS (Mobile Satellite Services) (non-GEO)
  - Inmarsat
  - Thuraya;
  - Aces;
  - Galileo;

- MSS/ATC (MSS with Ancillary Terrestrial Components);
- ICO GLOBAL;
- MSV (Mobile Satellite Ventures);
- Terrestar Networks;
- FSS (Fixed Satellite Services);
- SES Astra;
- Eutelsat;
- Telesat/Wildblue;
- IP STRAR;
- HYLAS.

However, service coverage spreads widely according to both the technology and the number of satellites supporting the technology. Some will claim to give continuous coverage to a geographic area; others only offer coverage when one of their satellites is in view. Geostationary satellites offer the most consistent coverage, but do not necessarily offer a full range of services.

With knowledge of the current location, service quality information can be predicted for satellites in high elliptical (e.g., Molniya) orbits, where the coverage pattern moves with the orbits. It is therefore possible to predict delay before connection. Low Earth orbit (LEO) satellites offer the potential of low-cost satellite communications, but penetration has not been as fast as expected.

The European Project SISTER is categorizing and grouping the applications according to common communications requirements:

- One-way communication link—from vehicle to service center;
- Bidirectional communication link;
- Broadcast applications;
- Broadcast with a return link;
- Voice and data link.

SISTER is a _10.5 million Integrated Project in the European Framework 6 collaborative R&D program and is contracted to run for 3 years, starting in November 2006.

Satellites can be used to page individual vehicles so that they can respond and create a call over networks without the need for massive HLR/VLR mobility management signaling.

Multihop use of a satellite system is possible from another vehicle—if payment can be collected.

The SISTER project is a major deliverable is to assist in the development of a standardized satellite interface for CALM. (See Section 4.5.8.) However, this standard will treat satellite as a form of public mobile wireless broadband access and will not provide standards for the satellite communications themselves. Other than Galileo, no standards therefore seem likely in this area.

4.2.6 Personal Area Networks: Bluetooth

4.2.6.1 Bluetooth Core Specifications

An introduction to personal area networks was given in Section 3.11.

There are no formal Bluetooth standards—the interest group was of the opinion that a quicker and more flexible route would be the publication of agreed specifications. In fact, the process has not proved much quicker and is less democratic than the SDO standardization processes. The Bluetooth specifications are available for (free) download (as PDFs) from:

Downloadable FOC Bluetooth technology Specifications
http://www.bluetooth.com/Bluetooth/Learn/Technology/Specifications/

The principal specifications are available from the link above and are described next.

Downloadable FOC Core Specification v2.0 + EDR
http://www.bluetooth.com/NR/rdonlyres/1F6469BA-6AE7-42B6-B5A1-65148B9DB238/840/Core_v210_EDR.zip

Downloadable FOC Volume 4: HCI Transports
http://www.bluetooth.com/NR/rdonlyres/24C24181-5B5D-4566-BB4F-D4C302910B9D/1766/Volume4_SPEC.pdf

Downloadable FOC Advanced Audio Distribution Profile
http://www.bluetooth.com/NR/rdonlyres/800D10CD-DE3D-4D51-ABC0-726C8DF26151/921/A2DPspecv10.pdf

Downloadable FOC Audio/Video Remote Control Profile
http://www.bluetooth.com/NR/rdonlyres/0CD6930C-302E-4BB8-ACA8-199066B47ADE/922/AVRCPSpecv10.pdf

Downloadable FOC Basic Imaging Profile (BIP)
http://www.bluetooth.com/NR/rdonlyres/182CCD92-3481-44F0-B901-9181BE573AFA/924/BIP_SPEC_V10.pdf

Downloadable FOC Basic Printing Profile 1.2 (BPP)
http://www.bluetooth.com/NR/rdonlyres/A73B2675-418C-4434-8B37-A3AD60035299/2940/BPP_SPEC_V12r00.pdf

Downloadable FOC Basic Printing Profile (BPP)
http://www.bluetooth.com/NR/rdonlyres/276DE16A-BDB9-48BF-8123-AF01E3730E5F/925/BPP_SPEC_V10.pdf

Downloadable FOC Common ISDN Access Profile (CIP)
http://www.bluetooth.com/NR/rdonlyres/A4A5D682-5AD8-4125-9ABB-946A7355F13F/926/CIP_SPEC_V10.pdf

Downloadable FOC Cordless Telephony Profile (CTP)
http://www.bluetooth.com/NR/rdonlyres/3D448C5B-E9C3-4485-A7F0-47F21A3C93B5/973/CTP_SPEC_V11.pdf

Downloadable FOC Device Identification Profile (DI)
http://www.bluetooth.com/NR/rdonlyres/9A211B3D-7A84-4B09-B131-806A59CE4F0E/2939/DI_SPEC_V12r00.pdf

Downloadable FOC Dial-Up Networking Profile (DUN)
http://www.bluetooth.com/NR/rdonlyres/447AB9BB-30F1-4F2B-9D75-562B56E16681/974/DUN_SPEC_V11.pdf

Downloadable FOC Fax Profile (FAX)
http://www.bluetooth.com/NR/rdonlyres/84D63C89-9B54-4F37-B440-838E5BACD63A/975/FAX_SPEC_V11.pdf

Downloadable FOC File Transfer Profile (FTP)
http://www.bluetooth.com/NR/rdonlyres/D769771D-460B-49C9-850A-068771251FD0/976/FTP_SPEC_V11.pdf

Downloadable FOC Generic Audio/Video Distribution Profile
http://www.bluetooth.com/NR/rdonlyres/1F14FA4E-9421-423B-93B4-62085876A349/873/GAVDPSpecv10.pdf

Downloadable FOC Generic Object Exchange Profile (GOEP)
http://www.bluetooth.com/NR/rdonlyres/170D5BFC-E6A5-4334-ABF7-5464E10577D1/977/GOEP_SPEC_V11.pdf

Downloadable FOC Hands-Free Profile 1.5 (HFP 1.5)
http://www.bluetooth.com/NR/rdonlyres/C0F90A55-BDE4-4FB3-A4FF-DAB0F137DBDF/1762/HFP15_SPEC_V10r00.pdf

Downloadable FOC Hands-Free Profile (HFP)
http://www.bluetooth.com/NR/rdonlyres/94EAD32B-3E70-4AEC-9CD5-1E3F887A15F4/917/HFPSpecVersion1_0.pdf

Downloadable FOC Hardcopy Cable Replacement Profile 1.2 (HCRP)
http://www.bluetooth.com/NR/rdonlyres/7B53802D-D1E3-43BA-A0BF-2C63BF69CE9C/2941/HCRP_SPEC_V12r00.pdf

Downloadable FOC Hardcopy Cable Replacement Profile
http://www.bluetooth.com/NR/rdonlyres/75127AC3-5BDD-48D4-9B76-2B7F8E05D5E2/979/HCRP_SPEC_V10.pdf

Downloadable FOC Headset Profile (HSP)
http://www.bluetooth.com/NR/rdonlyres/5C0DEE05-84CD-4D79-BD52-7ECA283430A0/981/HSP_SPEC_V11.pdf

Downloadable FOC Human Interface Device Profile (HID)
http://www.bluetooth.com/NR/rdonlyres/0BE438ED-DC1B-41D1-AAC0-1AAA956097A2/980/HID_SPEC_V10.pdf

Downloadable FOC Intercom Profile (ICP)
http://www.bluetooth.com/NR/rdonlyres/362A1E9C-637D-4CF0-B81A-B7CB13200AC2/982/ICP_SPEC_V11.pdf

Downloadable FOC Object Push Profile (OPP)
http://www.bluetooth.com/NR/rdonlyres/DB010C1F-6ADF-4F87-95EB-276AE321C59F/983/OPP_SPEC_V11.pdf

Downloadable FOC Personal Area Networking Profile (PAN)
http://www.bluetooth.com/NR/rdonlyres/279DC460-295E-42ED-8952-61B723620884/984/PAN_SPEC_V10.pdf

Downloadable FOC Serial Port Profile (SPP)
http://www.bluetooth.com/NR/rdonlyres/9C6DB2A4-A7D9-47A6-81B3-5F03981AE9C4/986/SPP_SPEC_V11.pdf

Downloadable FOC Service Discovery Application Profile
http://www.bluetooth.com/NR/rdonlyres/52B76D51-B8B9-44AA-879B-2E7D90060A23/985/SDAP_SPEC_V11.pdf

Downloadable FOC SIM Access Profile (SAP)
http://www.bluetooth.com/NR/rdonlyres/3FFE61A6-888E-49D1-BD3E-75C9164A796E/920/SAP_SPEC_V10.pdf

Downloadable FOC Synchronization Profile (SYNCH)
http://www.bluetooth.com/NR/rdonlyres/B677C19F-6603-4AF7-B826-05270AB7B9B5/987/SYNCH_SPEC_V11.pdf

Downloadable FOC Video Distribution Profile (VDP)
http://www.bluetooth.com/NR/rdonlyres/BF58FC4A-442C-419D-BBC2-479ABAC46E58/923/VDP_SPEC_V10.pdf

Downloadable FOC WAP Bearer (WAPB)
http://www.bluetooth.com/NR/rdonlyres/82A72AFB-21FC-4638-8BFA-7EAD4BFF7ACB/988/WAPB_SPEC_V11.pdf

Downloadable FOC Audio/Video Control Transport Protocol
http://www.bluetooth.com/NR/rdonlyres/40088B97-8E54-45DD-96EE-B6FED3F4C64E/914/AVCTPSpecv1_0.pdf

Downloadable FOC Audio/Video Distribution Transport
http://www.bluetooth.com/NR/rdonlyres/40162FD6-673D-4389-A979-4D73D7A9D01D/910/AVDTPSpecv10.pdf

Downloadable FOC Bluetooth Network Encapsulation Protocol
http://www.bluetooth.com/NR/rdonlyres/89B2BA02-D3FB-4717-97A1-A5B10D90F795/912/BNEPSpecification1.pdf

Downloadable FOC Object Exchange (OBEX)
http://www.bluetooth.com/NR/rdonlyres/D0600F99-8505-45A5-9635-518936123402/913/OBEX.pdf

Downloadable FOC Telephony Control Protocol (TCP)
http://www.bluetooth.com/NR/rdonlyres/CBB7B208-B257-485B-8648-127CFDB72709/915/TCSBinary.pdf

Downloadable FOC RFCOMM with TS 07.10
http://www.bluetooth.com/NR/rdonlyres/4C1E59CA-7E67-4126-8FE8-107C84A7B72C/916/rfcomm.pdf

Downloadable FOC HCI SD Transport
http://www.bluetooth.com/NR/rdonlyres/2836BFB8-42CA-47F1-BB65-3A9B6F689EE1/872/HCI_SD_Transport_SP_D10r3.pdf

Downloadable FOC HCI UART Transport
http://www.bluetooth.com/NR/rdonlyres/1D893C03-3218-4E3E-80C3-298CD0ADBFD7/853/HCI_UART.pdf

Downloadable FOC HCI USB Transport
http://www.bluetooth.com/NR/rdonlyres/81582E6C-216A-4657-A245-F01CA924946F/849/HCI_USB.pdf

Downloadable FOC Bluetooth Qualification and Testing
http://www.bluetooth.org/qualification/

The Bluetooth general site URL is:

General site : Bluetooth www.bluetooth.com

Although there are no formal international standards, some international and national standards organizations incorporate the use of Bluetooth specifications in their standards.

4.2.6.2 SAE Bluetooth Wireless Protocol for Automotive Applications

SAE	J2561_200112	Bluetooth Wireless Protocol for Automotive Applications	Published Dec. 2001	SAE

ISO Available from http://www.sae.org/technical/standards/J2561_200112

This SAE Information Report defines the functionality of typical Bluetooth applications used for remotely accessing in-vehicle automotive installations of electronic devices. Remote access may be achieved directly with on-board Bluetooth modules, or indirectly via a custom designed gateway that communicates with Bluetooth and non-Bluetooth modules alike. Access to the vehicle, in the form of two-way communications, may be made via a single master port, or via multiple ports on the vehicle. The Bluetooth technology may also be used in conjunction with other types of off-board wireless technology. This report recommends using a message strategy that is already defined in one or more of the documents listed. Those strategies may be used for some of the typical remote communications with a vehicle. It is recognized, however, that there may be specific applications requiring a unique message strategy or structure. This document depicts five different levels of security measures that may be required for the various types of communication. The EMC test requirements for the implementation of this protocol are recommended. This report makes no attempt to define all of the applications wherein the Bluetooth protocol may be utilized. A number of use cases are listed in Section B.1 of the appendix to serve as representative applications (e.g., credit card transactions, vehicle diagnostics, toll booth collection, flash reprogramming).

4.2.7 Personal Area Networks: ZigBee

The ZigBee specifications can be obtained from the following link:

Downloadable FOC Zigbee Specifications
http://www.zigbee.org/en/spec_download/download_request.asp

The general URL for ZigBee is:

General Site: Zigbee www.zigbee.org

ZigBee is another type of PAN that was introduced in Section 3.11. The principle offerings from the ZigBee Alliance are as follows.

SYNC. The ZigBee alliance offers a ZigBee Compliant Platform or the ZigBee Certified program. The ZigBee Certified testing program applies to modules and end products built on ZigBee Public Application profiles.

ZigBee Compliant Platform. The ZigBee Compliant Platform (ZCP) Program applies to modules or platforms which are intended to be used as building blocks for end products. Tested products will receive a confirmation letter granting the use of the term "ZigBee Compliant Platform."

NOTE: The scope of certification testing may differ if the system contains already certified components.

ZigBee Certified Products. The Certification Program applies to end products built upon a ZigBee Compliant Platform. These products can use ZigBee Public Application profiles or a Manufacturer Specific Application. After successful completion of this testing program, the ZigBee logo can be applied to the "ZigBee Certified" product as per the ZigBee Trademarks, Designations, and Logos Policy (previously referenced).

Although there are no formal international standards, some international and national standards organizations incorporate the use of ZigBee specifications in their standards.

Subsequent chapters in this book provide further details of such services and their standards in the context of ITS service provision.

4.3 CALM: Communications Access for Land Mobiles[1]

Downloadable FOC : http://www.thecalmforum.org

Car owners have noticed a significant improvement in the reliability and durability of vehicles in the past two decades. Indeed, car manufacturers now estimate that the average on-road life of cars is now in excess of 15 years, and expect that by 2010, the average on-road lifetime of a car will be up to 20 years. Heavy goods vehicles already frequently achieve a 20-year working lifespan.

1. The authors gratefully acknowledge the cooperation of The CALM Forum in the compilation of information for this and subsequent chapters in relation to the concept, architecture, and capabilities of CALM, and for their kind permission to use and reproduce any of their material in this book.

This has significant impact on any equipment, particularly cooperative equipment, fitted to vehicles. Systems fitted to vehicles sold in 2010–2020 must therefore still be able to operate in 2030–2040.

4.3.1 The Evolving Online World

Our personal world now expects online connectivity. We expect our computers to have ubiquitous access to the Internet, increasingly at broadband speeds. Current generations of cellular technology enable Internet access using our mobile telephones and PDAs. The introduction of wireless broadband will provide true broadband access in a mobile environment, together with its wide range of services. Today, we already expect to have ubiquitous online access. Such requirements include when we are in our vehicles.

Typical personal demands within a vehicle include the "are we there yet?" syndrome, keeping the family entertained/occupied during journeys by providing access to the Internet, downloading video, access to "yellow page" services for passengers, and online navigation, map update, and congestion avoidance navigation for the driver.

However, apart from the "demand pull" services called directly by the user, there are a large number of other services "pushed" by the requirements for safety and road management. Currently, governments are spending large amounts of money installing and managing variable messaging signs (VMS), but it is only practicable to have signs separated by one or many kilometers. Most of these messages are safety, congestion, or road management related (e.g., lane closures). It is preferable to bring these messages immediately in-vehicle, displayed on the drivers navigation screens, or given as audio messages.

Vehicles are increasingly controlled by on-board computer processors. These devices are complex and from time to time need to be upgraded, usually for safety reasons, although sometimes for efficiency reasons. At the moment, this requires a product recall, which is both expensive and bad publicity for vehicle manufacturers. Additionally, with ubiquitous online access, the opportunity arises that when a vehicle is malfunctioning, or not functioning efficiently, the vehicle can contact the service department, the problem analyzed remotely, often remedied remotely, or the driver can be alerted to take the vehicle to the service depot with an indication of the urgency of the situation.

The main roadway system requirements for continuous (quasi-continuous) communications are safety related. While simple RFID tags can provide vehicle traffic light priority, this is a useful, but limited, benefit and is only useful where there are traffic control lights, and work only for local vehicles equipped with appropriate tags. Most requirements of emergency vehicles are on open roads, and early systems will include notification of an approaching emergency vehicle, communications between emergency vehicles, and between emergency vehicles and their base stations, including, for example, medical monitoring systems in ambulances.

Further into the future, but the subject of very active current research and development activities, are those of collision avoidance and cooperative driving. For collision avoidance, vehicles need to communicate with adjacent vehicles and other vehicles in their proximity. Collaborative driving systems have similar requirements—for example, advising other vehicles when they are changing lanes or with adaptive cruise control systems where vehicles are adjusting their speed in dialog with the preceding vehicle.

Centrally controlled incident management systems envisage collecting data from vehicles involved in/passing the incident, and collating and disseminating information for approaching vehicles—for example, identifying the exact location of an object or stationary vehicle on the road pavement, and providing exact location and warning to approaching vehicles.

Other possibilities under discussion in some countries, but unlikely to be popular in all countries, include vehicle speed limitation (e.g., in the proximity of schools). Other proposed services include curve speed assistance, stop light assistant, railroad collision avoidance, vehicles as probes, road condition warning (ice, snow, surface irregularities), low bridge warning, public transport coordination (trains/trams/busses running on time or behind schedule), emergency vehicle video relay, border clearance, vehicle safety clearance, and driver verification, among others.

Other commercial services include access control, online payment systems (e.g., for parking or to acquire map updates), fleet management data, and rental car processing. More sophisticated route guidance systems linked to congestion data also require quasi-continuous access for congestion and road maintenance activities. Public transport vehicle fleet management is a particular example, especially in respect of on-board incidents or threat environments, enabling video coverage of busses and trams to be centrally monitored in real time. And, of course, there are many services made available by the provision of on-board access to the Internet.

All of these services need communications connectivity to the vehicle.

4.3.2 The Conundrum

While we may wish to see single choice types of decision, regulation, and standardization for the introduction and operation of ITS services, the situation presents a conundrum. In summary, we have to meet the following requirements:

- Provide systems that enable (quasi) continuous communications between vehicles and the infrastructure, and between vehicles, in an environment where the vehicle marketplace is global with a limited number of large manufacturers providing vehicle models on a global basis with the minimum of variation to meet local national requirements.
- Provide systems, with a useful standardization lifetime of 10 to 20 years, where vehicles may be operational for more than 20 years. Thus we are looking at systems functioning to 2040 and beyond.
- The lifecycle of telecommunications systems is getting shorter, and several future possibilities are now apparent, and it is clear that other new technologies, not yet envisaged, will be developed in the period of 2005–2040 and may be appropriate. Current communications systems, such as 2G and 3G cellular are unlikely to still be used in 2040.
- Governments and system providers will not make common decisions about the base media permitted in their countries.

- The characteristics of different media vary according to the properties of that media. Different media are better or less well suited to different applications.
- Vehicles frequently travel across national borders and operate in countries other than their home country. This is particularly true for commercial vehicles and for all vehicles within Europe.
- Manufacturers want to equip vehicles with a single global solution. They do not want the complexity, and risk, of too many uncoordinated communications systems providing different services.
- ITS Standards are being developed as the technology develops.
- It is the role of those that standardize to enable the market, not to determine it.

4.3.3 The CALM Concept

What then is CALM? The acronym, which started life as "Continuous Air-interface Long and Medium range" has evolved with the project to become "Communications Access for Land Mobiles" as its scope has evolved and developed.

The fundamental principles of the CALM concept, and the architecture and standards that embody it, are predicated on the principle of making "best" use of the resources available. The resources are the various communications media available, and "best" is defined by the objectives to be achieved and their relative cost. Flexibility, adaptability, and extensibility are the keys to its success.

The CALM concept is therefore developed to provide a layered solution that enables continuous or quasi-continuous communications between vehicles and the infrastructure, or between vehicles, using such (multiple) wireless telecommunications media that are available in any particular location, and have the ability to migrate to a different available media where required. Media selection is at the discretion of user determined parameters.

4.3.4 CALM "Application" Service Types

Application services that are likely to use CALM fall into two categories: safety and commercial, although some safety related services are of commercial interest as options on up-market vehicles. The lists given in Section 2.2 provide a selection of ITS services that have already been identified as probably suitable for support via CALM. The list is neither exhaustive nor complete and continues to expand as ITS evolves. However, it gives an indication of the types of service that will use CALM.

4.3.5 CALM Benefits

CALM combines several communication media in an open manner in accordance with international standards that provide:

- Openness, since the standards are available to everybody;
- Stability, since there is a formal body responsible;

- Visibility and credibility of the specifications;
- An open way to influence the next phases of standards;
- Extensibility.

Wherever possible, CALM is based on IPv6, which means that it is fully compatible with Internet services, while at the same time not being restricted by the addressing shortcomings of the current IPv4 protocols. For time-critical safety services, where processing or radio protocols are not rapid enough to support IPv6 protocols, there is a CALM-FAST mode of operation which enables very rapid transmission of short messages.

ERTICO assesses CALM's benefits to be[2]:

- CALM combines multiple communication media in an open way.
- IPv6 basis means that it is fully compatible with Internet services.
- Spanning multiple media, and open to more medias both broadcast and others—by the integration of a simple IPv6/Management convergence layer.
- CALM is based on well-tested standards and media that are adopted and optimized for the mobile environment. This means:
  - Low risk strategy;
  - Fast and relatively low cost implementation;
  - Limited "chicken-and-egg" problem—much of the infrastructure and service network is already available;
  - Allows new media to be easily introduced.
- CALM includes the spectrum protected, 5-GHz low-latency vehicle_vehicle/vehicle-roadside communications system needed for many vehicle-based public safety systems. This is a very high priority in North America, and also one of the findings of the recently concluded eSafety project.
- Supplements DSRC (European dedicated short range communication) with possibilities for higher speeds and longer distances, and manages spectrum so that units can, if required, support both CALM and DSRC in the same unit.
- Cooperation with ITS standardization Working Groups.
- Cooperation with IEEE 802.11 for mobile 5-GHz W-LAN.
- Cooperation with IETF for Internet functions.
- Cooperation with ITU-R and ETSI.
- Cooperation with ITS America.

4.3.6 The Role of CALM in the Provision of ITS Application Services

ITS will increasingly rely on data being passed across a wireless air interface in order to provide/maintain the service being provided. These wireless interfaces may be CALM type of service provision where the service provider is separated from the medium provision, or (particularly in early iterations) may be a dedicated medium air interface. Early electronic road tolling provides an example of such

2. This list is from: "ERTICO: CALM Fact Sheet Update," May 2004, Peter Van der Perre ERTICO SCRL.

dedicated air interfaces, and some early instantiations by automotive manufacturers will use existing media, for example, cellular communications or Wi-Fi. It is also important to remember that some intelligent transport systems, particularly early instantiations, may not require a bidirectional communication using any wireless air interface. Some of these systems may never require bidirectional communications with other vehicles or the infrastructure, others may evolve to do so.

4.3.7 CALM Architecture

While CALM can be described as an enabling concept and architecture, the concept itself does not define the ITS application services that can be provided via the CALM architecture. Its purpose is to provide the architecture and management entities to enable the networking of ITS services to their hosts, in a mobile environment using whatever wireless medium is available and usable by in vehicle equipment, and to separate the ITS service provider from any requirements regarding the specific wireless medium used to transfer the data.

Figure 4.11 provides a representation of the CALM Architecture. This architecture will evolve and develop over time but can be described as represented in Figure 4.12.

Figure 4.13 provides a view of the CALM kernel, including the non-IPv6 CALMFAST communications provisions. CALMFAST is used for short fast messages for time-critical safety applications, and unlike the IPv6 communications provisions, does not have heterogeneous handover capability.

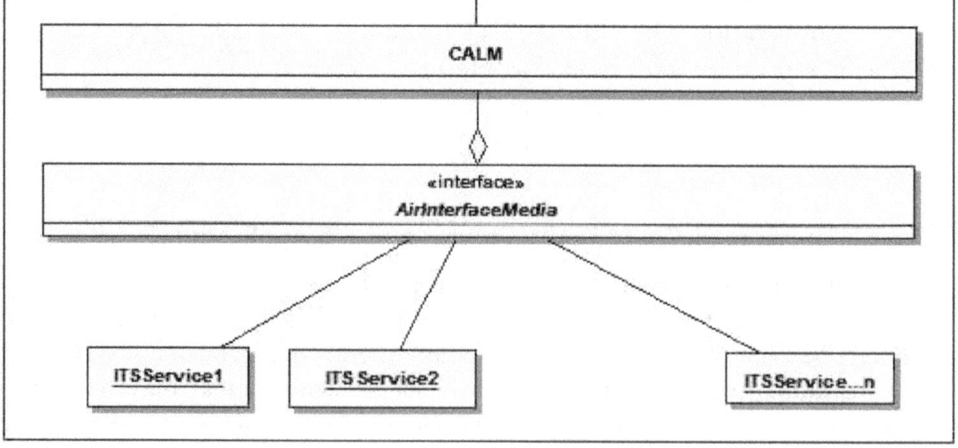

Figure 4.11 The CALM high level architecture. (*Source:* The CALM Forum.)

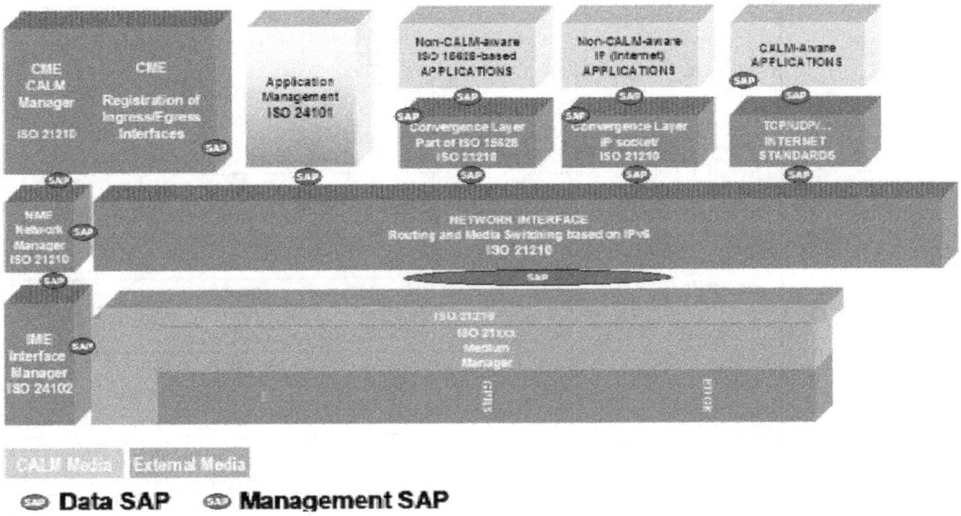

Figure 4.12 The CALM architecture showing media, SAPs, and management entities. (*Source:* The CALM Forum.)

The elements shown in Figure 4.13 inside the blocks "CALM User Services/ Applications," "Network/Transport Layer," and "Physical/Data Link Layer" are considered to be examples of optional elements.

The various functionalities contained in a CCK may be addressed for the purpose of station management by means of link-local IPv6 addresses and the simple network management protocol (SNMP).

This link-local addressing is affected within the station independent of the networking protocol used for wireless communications including situations where only CALMFAST is used in the wireless link.

4.3.8 CALM and Standards

While the scope of CALM is ambitious, it is probably the most robust and resilient to change, in that it is not fixed in a single media or service concept solution, does not require everyone in the world to use the same media or services, and has the capacity to migrate and evolve over time as technology moves on.

However, such interoperability does not happen by accident, and since 2000 a group within ISO, Working Group 16, has been evolving the concepts and developing the complex range of standards required to enable the CALM concept to function interoperably.

Figure 4.14 provides an architectural view of CALM with the key standards identified. The following chapters on wireless media standards for ITS include detail of the scope and provision of these standards.

4.4 Dedicated and Public Wireless Networks

Within this chapter, a dedicated wireless network is one designed and installed primarily for the provision of ITS services; it may be publicly available to all vehicles, but its prime purpose is the provision of ITS services.

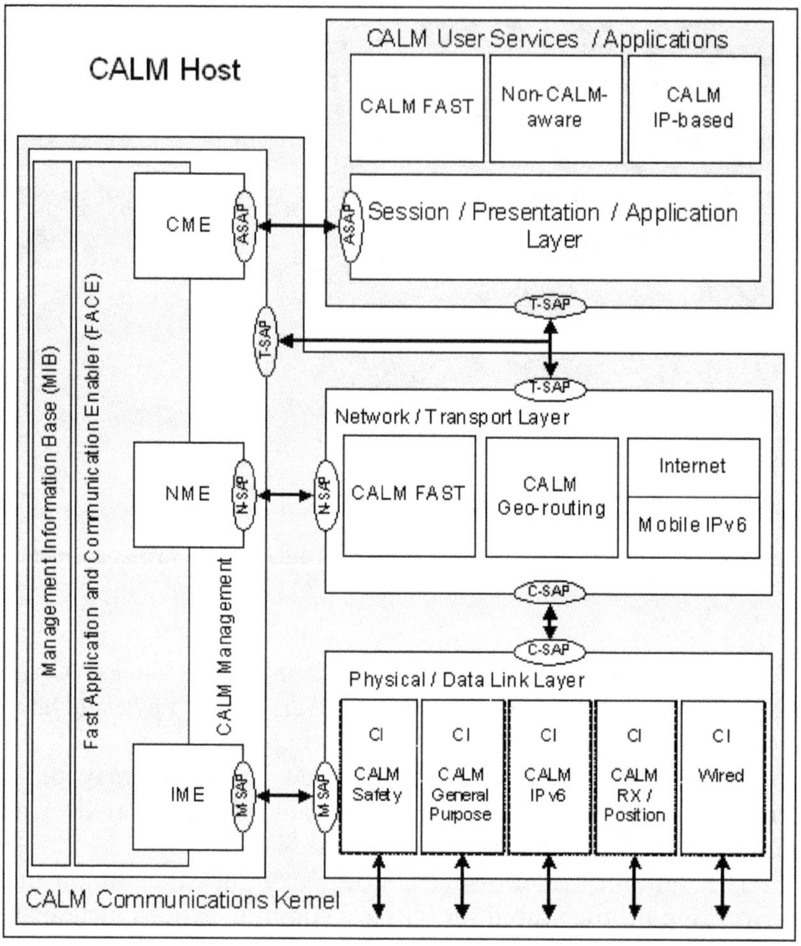

Figure 4.13 CALM communications kernel. (*Source:* The CALM Forum.)

A public wireless network, as described in Section 4.2, is a network, such as cellular telephony, satellite, or mobile wireless broadband, which is established for general communications purposes, but is available, where provided, so long as a contract exists with the service provider, to support ITS service provision.

The provision of ITS services to and between vehicles needs to be wireless. But where is it appropriate to use public wireless networks and when is it desirable or necessary to use networks dedicated to ITS?

It is clear that many services can technically be provided by both public and dedicated networks, and which network will provide the ITS service will depend to some extent on the comparative technical performance between the networks according to the properties their physical and operational behavior, and it will also depend on the business model provided.

Direct vehicle-vehicle services, even if provided across a mesh network along a highway, are, by the definition used here, dedicated wireless services for ITS.

Some services are time critical (e.g., collision avoidance and platooning), and these require bidirectional communication sessions that must respond in near real

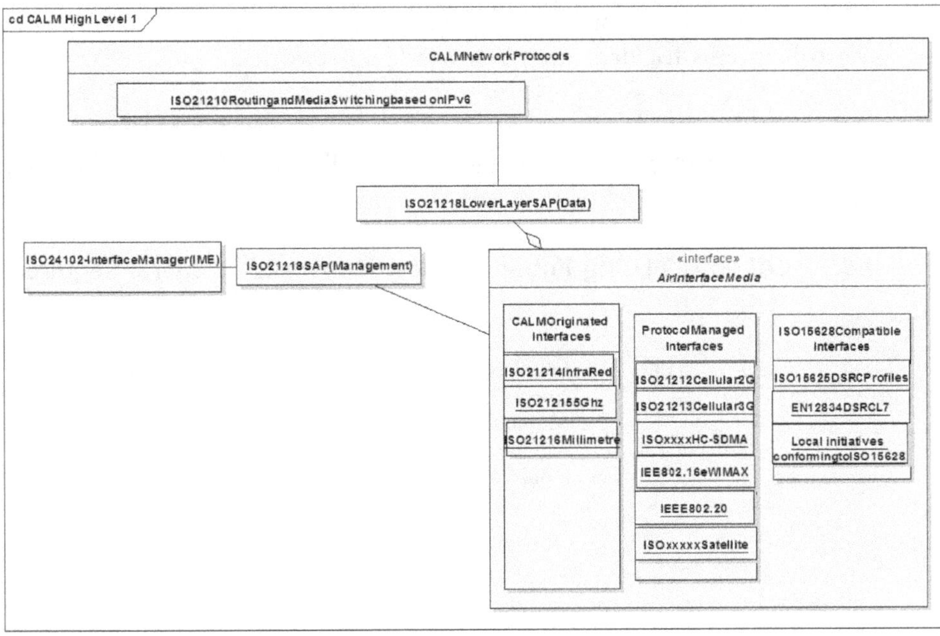

Figure 4.14 CALM architecture with key standards identified. (*Source:* CSI Library.)

time, with time management and response measured in milliseconds. The communications delays involved in communications via public networks make them unsuitable for reliable transactions in these timeframes. Even if the performance could be offered, it is unlikely, for liability reasons, that automotive manufacturers would permit their use.

Dedicated wireless networks for ITS are therefore being designed at 5 GHz, 63 GHz, and infrared. Where these networks will be installed remains an issue for the future, and it is not expected that all three will be ubiquitously available. One country may support one, another country or road operator another. This that may not be desirable, but it reflects the most likely reality, and the CALM concept will enable this to be workable wherever the appropriate communications equipment is in a vehicle.

A small or low cost car which will be overwhelmingly used within one country is likely only to support the dedicated network of that country and one or two public networks. A top-teir executive vehicle, which, in Europe at least, may cross borders with regularity, will support several or many media.

Software controlled radios combined with adaptive antennas will make this commercially feasible for automotive manufacturers, and the cost difference will drop significantly as take-up volume increases. (If this sounds unlikely, consider GPS receivers, which upon introduction cost many hundreds of dollars for the key components alone. Today a single chip solution exists and costs significantly less than $10 in volume.)

Within this book we group dedicated wireless networks for ITS separately from the use of public wireless for the provision of ITS services. However, in time, the separation may become academic, particularly if the provision of commercial services is seen as a way of funding the investment in installing the infrastructure and control systems for dedicated ITS wireless networks.

4.5 Standards Underpinning the Use of Public Wireless Networks for ITS

4.5.1 CALM: ITS Using Public Wireless Networks—General Requirements

ISO	CD	25111	Intelligent Transport Systems—Communications Access for Land Mobiles (CALM)—ITS using public wireless networks—General requirements	Under DIS Ballot	ISO TC204

Not Yet Available. Track Progress at:
http://www.iso.org/iso/iso_catalogue/catalogue_tc/catalogue_detail.htm?csnumber=42751

This work originally started in 2005 with respect to the advent of mobile wireless broadband systems, but in early 2007 it was realized that many of the features were common to all public wireless networks; in particular, the common differences between subscription based "continuously on" systems and "connection time managed" systems, which has more to do with business models rather than whether or not the providing system is technically broadband or not.

The implication on connecting into CALM is significant. A continuously-on system will operate much as a dedicated system, establishing the communications connection, and then opening the CALM session on top of it and accessing services immediately on demand as appropriate.

A connection time managed system, on the other hand, can only be registered as a possibility before the CALM session is opened, and each time a service is to be provided via such a medium, it subsequently has to establish the full connection with the medium.

ISO 25111 provides the architecture and common requirements for connection of both types of wireless communications systems. All common aspects including the general requirements for connection and disconnection are specified in ISO 25111. The specific CALM aspects need to use specific medium standards to require compliance to ISO 25111 and the other principal CALM management and networking standards, and provide the command and response sequences significant for that medium.

4.5.2 Analog Cellular Systems

Early on in the development of ITS service provision, some of the first services that were offered were emergency services using analog communications. However, as digital telephony became available, it became clear that the largely voice-based services, or very low data rate services possible using analog techniques, were so restricted that they hardly justified the appellation "ITS;" and for a number of other reasons, analog telecommunications were to be phased out. At this stage it was decided to abandon all analog options from ITS standardization.

4.5.3 Second Generation Cellular Systems

4.5.3.1 ISO 21212 Intelligent Transport Systems: Communications Access for Land Mobiles (CALM)—2G Cellular Systems

ISO	IS	21212	Intelligent Transport Systems—Communications Access for Land Mobiles (CALM)—2G Cellular systems	Approved. Awaiting Publication	TC204

ISO 21212 Available from
http://www.iso.org/iso/iso_catalogue/catalogue_tc/catalogue_detail.htm?csnumber=45378

This is one of a suite of standards developed as a CALM medium. Because of its endemic spread and ubiquitous availability throughout much of the world, it was seen as an early generation option for the provision of some ITS services. The low data rate and the time based charging model will count against it as alternatives evolve, but it remains an early exploiter medium for CALM and other ITS systems.

The development of ISO 21212 was commenced in 2000 and it became a fully approved International Standard published in 2008. It should be noted that the subsequent evolution of other CALM media, and particularly wireless mobile broadband and satellite, have caused the development working group to embark on a revision, to bring the format into line with other CALM media standards that use public wireless networks, under a general framework defined in ISO 25111 (CALM using public networks for land mobiles—General requirements); however, while the format will change, and indeed may have already changed by the date of publication of this book, the technical provisions will remain similar.

The ISO 21212 standard determines the air interface using second generation cellular networks. Second generation (e.g. using WAP and I-Mode type protocols) systems to be compliant to CALM, i.e., requirements that must be met before a 2G system can be incorporated into a CALM system. In particular, this International Standard specifies protocols and parameters that second generation systems shall include to support prolonged, long-range, high data rate wireless communication links in ITS environments where heterogeneous handovers or media independent handovers (MIH) are either necessary to maintain the link, or desirable as determined by media selection policies.

The standard provides protocols and parameters for long range, medium speed wireless communications in the ITS sector using second generation cellular communications.

Wherever practicable, this International Standard has been developed by reference to suitable extant standards, adopted by selection. Required regional variations are provided.

Specifically for this International Standard, extant 2G systems, as defined by various international and national standards, are adopted by reference.

Such links are required for quasi-continuous, prolonged, or short communications between vehicles and the roadside, between vehicles, and between mobile equipment and fixed infrastructure points, over medium and long ranges, specifically for this standard by adoption of second generation cellular protocols and standards.

Application-specific upper layers are not included in this standard, but they will be driven by application standards (which may not be technology specific).

The principal role and content of the standard is to provide the medium session initiation, registration, connect and disconnect requirements, and the requirements to establish a CALM session via the medium and to provide the requirements to utilize the CALM management, networking, and service access protocols to provide a viable media link in the CALM environment.

In order to claim conformance with this International Standard, cellular communication shall be established in full compliance with local telecommunications procedures and protocols for the appropriate 3GPP 2G standards, and shall comply with the requirements of ISO 21217 (CALM System architecture), ISO 21210 (CALM Networking protocols), ISO 21218 (CALM medium service access points), and CALM interface manager (ISO 24102).

Further information on the generic standards underpinning ISO 21212 can be found in Section 4.2.

4.5.4 ETSI DTS; Intelligent Transport Systems (ITS): CALM 2G/2.5G Cellular

ETSI	DTS	TG37-015	Intelligent Transport Systems: CALM 2G/ 2.5G Cellular	Under Development	ERM TG37

Downloadable FOC But not yet available

This work item of ETSI TC-ITS will be developed into a standard once ISO 21213 has been finalized and approved.

4.5.5 Third Generation Cellular Systems

4.5.5.1 ISO DIS 21213 Intelligent Transport Systems—Communications Access for Land Mobiles (CALM)—3G Cellular Systems

ISO	DIS	21213	Intelligent Transport Systems—Communications Access for Land Mobiles (CALM)—3G Cellular systems	Approved. Awaiting Publication	ISO TC204

ISO 21213 Available from
http://www.iso.org/iso/iso_catalogue/catalogue_ics/catalogue_detail_ics.htm?csnumber=45377&ICS1=35&ICS2=240&ICS3=60

This is the second of a suite of standards developed as a CALM medium. Because of the expectation of its endemic spread and ubiquitous availability throughout much of the world, it was seen as an early generation option for the provision of some ITS services. The medium data rate and the time based charging model will count against it as alternatives evolve, but it remains an early exploiter medium for CALM and other ITS systems.

The development of ISO 21213 was commenced in 2000 (at the same time as ISO 21212) and it became a fully approved International Standard published in 2008. It should be noted that the subsequent evolution of other CALM media, and particularly wireless mobile broadband and satellite, have caused the development working group to embark on a revision, to bring the format into line with other CALM media standards that use public wireless networks, under a general framework defined in ISO 25111 (CALM using public networks for land mobiles—General requirements); however, while the format will change, and indeed may have already changed by the date of publication of this book, the technical provisions will remain similar.

The ISO 21213 standard determines the air interface options applicable to CALM using third generation cellular networks. In particular, this International Standard specifies protocols and parameters that 3G systems shall include to support prolonged, long-range, high data rate wireless communication links in ITS environments where heterogeneous handovers or media independent handovers (MIH) are either necessary to maintain the link, or desirable as determined by media selection policies.

The standard provides protocols and parameters for long range, medium speed wireless communications in the ITS sector using third generation cellular communications.

Wherever practicable, this standard has been developed by reference to suitable extant standards, adopted by selection. Required regional variations are provided.

Specifically, for this standard, by adoption of "Third Generation" (3G), as defined by 3GPP and 3GPP2 and including IMT-2000, cellular protocols and standards.

Such links are required for quasi-continuous, prolonged, or short communications between vehicles and the roadside, between vehicles, and between mobile equipment and fixed infrastructure points, over medium and long ranges, specifically for this standard by adoption of third generation cellular protocols and standards.

Application-specific upper layers will not be included in this standard but will be driven by application standards (which may not be technology specific).

The principal role and content of the standard is to provide the medium session initiation, registration, connect and disconnect requirements, and the requirements to establish a CALM session via the medium and to provide the requirements to utilize the CALM management, networking, and service access protocols to provide a viable media link in the CALM environment.

In order to claim conformance with this International Standard, cellular communication shall be established in full compliance with local telecommunications procedures and protocols for the appropriate 3GPP/3GPP2 3G standards and shall comply with the requirements of ISO 21217 (CALM System architecture), ISO

21210 (CALM Networking protocols), ISO 21218 (CALM medium service access points), and CALM interface manager (ISO 24102).

Further information on the generic standards underpinning ISO 21213 can be found in Section 4.2.2.

4.5.6 ETSI DTS Intelligent Transport Systems (ITS); CALM 3G Cellular

ETSI	DTS	TG37-016	Electromagnetic compatibility and Radio spectrum Matters (ERM); Intelligent Transport Systems : CALM 3G Cellular	Under Development	ERM TG37

Downloadable FOC But not yet available

This work item of ETSI TC-ITS will be developed into a standard once ISO 21213 has been finalized and approved.

4.5.7 CALM Using Mobile Wireless Broadband Systems

It is now reasonably clear that wireless broadband will become ubiquitous within urban areas and in areas where wired broadband installation is difficult or not cost effective. Clearly, although wireless broadband cells are usually much larger than 3G cells, a traveling vehicle will soon move from one cell to another. One standard has already been developed to enable cell-cell handover protocols to provide mobile wireless broadband (HC-SDMA), and two others [802.16e (WIMAX/WiBro) and 802.20] will incorporate handover protocols once its fixed wireless broadband standards are finalized and proven.

4.5.7.1 ISO 25113 CALM Using HC-SDMA

ISO	DIS	25113	Intelligent Transport Systems—Calm using (HC-SDMA) CD	Under Development	ISO TC204

Not yet available. Track Progress at:
http://www.iso.org/iso/iso_catalogue/catalogue_tc/catalogue_detail.htm?csnumber=42753

WIMAX has stolen much of the initial publicity regarding the advent of mobile wireless broadband; it is an ANSI standard system, HC-SDMA, that has been the first to become available, fully functioning and operational. At the time of writing it is already available in most major cities in Australia, several countries in Africa, some parts of United States, and some northern European Countries (Norway, Denmark, and The Netherlands).

High capacity–spatial division multiple access (HC-SDMA) was adopted as an ANSI standard, and details of that generic WMB standard is given in Section 4.5.5. Standardization of the ITS use of HC-SDMA, within the CALM concept, is provided in ISO 25113, which, at the time of writing, was being submitted for CD Ballot.

ISO 25113 works within the general framework standard ISO 25111 described in Section 4.6.2.21.

This International Standard determines the air interface options applicable to CALM using mobile wireless broadband (MWB) techniques complying to ATIS-0700004-2005 HC-SDMA protocol specification and specifies the management interface requirements for long range, wireless communications in the ITS sector utilizing HC-SDMA techniques.

Wherever practicable, this International Standard has been developed by reference to suitable extant standards, adopted by selection. Required regional variations are provided.

Application-specific upper layers will not be included in this International Standard, but they will be driven by application standards (which may not be technology specific).

4.5.7.2 ISO 25112 CALM Using IEEE 802.16e/IEEE 802.16g

ISO	DIS	25112	Intelligent Transport Systems—Communications Access for Land Mobiles (CALM)—Mobile wireless broadband using IEEE 802.16e/ IEEE 802.16g	Under Development CD	ISO TC204

Not yet Available. Track Progress at:
http://www.iso.org/iso/iso_catalogue/catalogue_tc/catalogue_detail.htm?csnumber=42752

Capturing much of the attention regarding the advent of wireless mobile broadband, the launch of WIMAX has been somewhat protracted and slow. This is primarily because, at the time when it was originally devised, it was envisaged that the principal interest in wireless broadband was fixed, with the possibility of so-called nomadic roaming within the zone of a beacon. The significant advantage of the range of a mobile broadband cell, together with the higher data rates, was seen to be its advantage over 2G/3G, and its use was envisioned for both town center coverage and to provide broadband to more remote rural communities where the laying of wired infrastructure could not be cost justified. This resulted in the IEEE 802.16 Standard.

However, by the time that this standard was nearing completion, the importance of the MWB market was realized with the knowledge that others, including HC-SDMA, were moving fast to service this market and were selling mobility as one of their features. IEEE 802.16e was therefore somewhat hastily developed and published. In their haste there were some technical glitches that made the published version inoperable, and a revision process was launched and the glitches fixed during 2007.

At the time of this writing, there are only major deployments in Mauritius and some middle sized deployment in Korea, but it may be reasonably expected, probably by the time you read this book, to be quite widely deployed in Europe, North America, and significant parts of Asia and Africa.

Like ISO 25113, ISO 25112 uses the common layout and works within the general framework of standard ISO 25111, described in Section 4.6.2.21.

This International Standard selects the options appropriate for CALM using MWB techniques complying with IEEE 802.16e/802.16g, and specifies the management interface requirements.

CALM links are required for quasi-continuous, prolonged, and short duration communications between vehicles and the roadside, between vehicles, and between mobile equipment and fixed infrastructure points, over medium and long ranges.

Wherever practicable, this International Standard has been developed by reference to suitable extant standards, adopted by selection. Required regional variations are provided.

Application-specific upper layers will not be included in this International Standard, but they will be driven by application standards (which may not be technology specific).

4.5.7.3 ISO 29283 CALM Using IEEE 802.20

ISO	NP	29283	Intelligent Transport Systems—Communications Access for Land Mobiles (CALM)—Mobile wireless broadband using IEEE 802.20	WD Under Development	ISO TC204

Not Yet Available

Because of the weaknesses of the IEEE voting systems, IEEE 802.20, a more capable so-called "next generation" WMB system, has been the victim of commercially inspired machinations that at one time caused IEEE to suspend the work for 6 months while it investigated accusations of high-jacking the process, not following rules, and unfair play. As this seems to be a common occurrence in the IEEE process, the reader may assume that this was quite a serious situation. However, the development had moved back on track by the end of 2006 and the standard was expected to be published by the end of 2007.

In anticipation of the publication of IEEE 802.20, ISO/TC 204 has prepared a standard to use IEEE 802.20 WMB within the context of CALM. Like ISO 25113 (CALM using HC-SDMA) and ISO 25112 (CALM using IEEE 802.16e/g), ISO 29283 uses the common layout and works within the general framework standard ISO 25111, described at the start of Section 4.6.2.21.

This International Standard determines the air interface options applicable to CALM using MWB techniques complying with IEEE 802.20 – MBWA – 625K-MC

MODE protocol specification, and specifies the management interface requirements for long range, wireless communications in the ITS sector utilizing IEEE 802.20 – MBWA – 625K-MC MODE techniques.

CALM links are required for quasi-continuous, prolonged, and short duration communications between vehicles and the roadside, between vehicles, and between mobile equipment and fixed infrastructure points, over medium and long ranges.

Wherever practicable, this International Standard has been developed by reference to suitable extant standards, adopted by selection. Required regional variations are provided.

Application-specific upper layers will not be included in this International Standard, but they will be driven by application standards (which may not be technology specific).

See also Section 4.6.2.26.

4.5.8 ISO 29282 CALM Using Satellite

ISO	NP	29282	Intelligent Transport Systems—Communications Access for Land Mobiles (CALM)—CALM Using Satellite communications	WD Under Development	ISO TC204

Not Yet Available

While the focus of much of the work in the ITS domain has been centered on areas of dense traffic population, the issues or the use of ITS in remote areas has gained much less attention, but it is not forgotten. While advanced collision avoidance systems may be of limited use in the vast Australian outback, crashes, often not involving other vehicles, do occur, as do breakdowns and other incidents. Not being able to become aware of such situations (via eCall type emergency calls) is a major problem for the Australian government, and is a cause of death and injury totally disproportionate to the traffic volume and density. Indeed, some states require that a two-way radio be available in every vehicle in the Outback area under their control. One cannot envision the Outback, or the desert regions of the United States or the Russian Federation, being kitted out with variable message signs, but ITS provides the possibility of providing the information onto in-vehicle screens and getting assistance via eCall so long as a means of communication exists. In these remote areas, and even in the more remote areas of Northern Europe, there is often not even cellular network coverage.

Fortunately, the latest generations of satellite telephone networks have become both available and much more affordable—at this stage, not affordable enough for infotainment provision, though that may well come later.

Transmission delays mean that such services are not suitable for the so-called time-critical emergency services envisioned within ITS development, but this is not necessarily required in a desert region (however, investigation into ITS services that provide location of proximity vehicles in the event of an incident or to warn, for example, of the approach of a road-train are under investigation).

One of the advantages of CALM is the ability to use and switch networks, so a suitably equipped vehicle can use a 5-GHz or WMB network in the cities, but switch to satellite in the Outback or other remote region.

ISO/TC 204 is therefore developing a standard for CALM using satellite communication. Again, it is under the framework of ISO 25111 and in a similar format to the other "CALM Using. . ." standards.

At the time of writing this book, the work is at an intermediate working draft stage and currently provides the options appropriate for CALM using satellite communications and specifies the management interface requirements.

CALM links are required for quasi-continuous, prolonged, and short duration communications between vehicles and the roadside, between vehicles, and between mobile equipment and fixed infrastructure points, over medium and long ranges.

Wherever practicable, this International Standard has been developed by reference to suitable extant standards, adopted by selection. Required regional variations are provided.

At the time of this writing, it was undecided how many different satellite communications protocols will be supported within the standard.

4.5.9 ETSI WMB for ITS General Provisions

ETSI	DTR/ DTS	TG37-021	Intelligent Transport System nonfrequency-specific protocols and procedures capable of use up to very high speed operation (above 1 GB) across the range of 1 GHz–100 GHz, to enable ITS operations from and between: infrastructure–infrastructure, infrastructure–roadside, roadside–vehicles/mobiles, vehicle/ mobile–vehicle/mobile	Under Development	ERM TG37

Downloadable FOC But not yet available

4.6 ITS Specific Wireless Communications Networks

4.6.1 Dedicated Short Range Communications (European)

These are a set of European standards defining the framework of a dedicated short range communications (DSRC) link in the Road Transport and Traffic Telematics (RTTT) environment.

Originally envisaged as a set of CEN standards, they have been complemented by ETSI communications standards for the communication layer (to support the allocation of spectrum for this use), and by ETSI tests for conformance.

The original set of standards proclaimed that communication requirements of many RTTT applications can be fulfilled by DSRC. The DSRC standards enable compliant communication systems to serve multiple RTTT applications in parallel.

The small service areas and severe real-time constraints require a specific protocol architecture leading to the reduced protocol stack shown in Figure 4.15, consisting of the application layer, the data link layer, and the physical layer. Such an architecture is very common for real-time environments. This standard gives the architecture and services offered by the DSRC data link layer.

The following set of European standards for the DSRC link is issued by CEN:

- EN 12253 DSRC Physical Layer using Microwave at 5.8 GHz;
- EN 12795 DSRC Data Link Layer: MAC and LLC (this standard);
- EN 12834 DSRC Application Layer;
- EN 13372 DSRC Profiles for RTTT Applications.

The standards have been used widely and successfully around the world as the communications basis for road tolling.

4.6.1.1 ERM RTTT Part 1

ETSI	EN	200 674	Electromagnetic compatibility and Radio spectrum Matters (ERM); Road Transport and Traffic Telematics (RTTT); Part 1: Technical characteristics and test methods for High Data Rate (HDR) data transmission equipment operating in the 5,8 GHz Industrial, Scientific and Medical (ISM) band	Published	ERM TG37

Downloadable FOC ETSI 200 674
http://webapp.etsi.org/exchangefolder/es_20067401v020101p.pdf

This document specifies radio parameters, data link services and protocol data units, and application services and protocol data units which are necessary for the

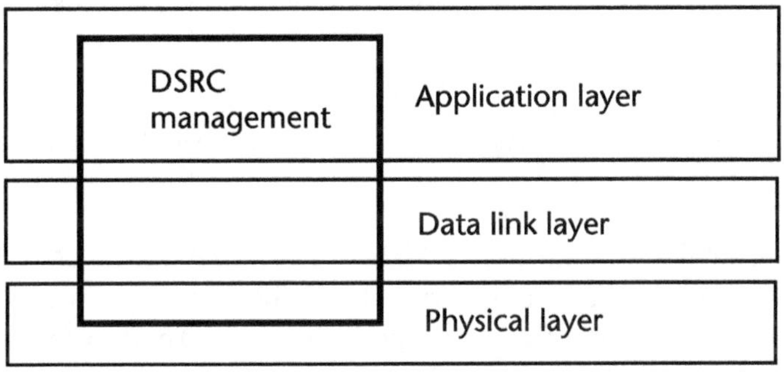

Figure 4.15 DSRC protocol stack. (*Source:* CSI Library.)

efficient use of the radio spectrum and for the purpose of DSRC based applications. This includes methods of measurements for verifying the limits stated in the document.

The present document applies to 5.8-GHz short range devices (SRDs) for use in RTTT:

- With a RF output connection and specified antenna or with an integral antenna;
- For data transmission only;
- Operating on radio frequencies in the 5.725- to 5.875-MHz Industrial, Scientific and Medical (ISM) band.

The applicability of the present document covers both the road side units (RSUs) and the on-board units (OBUs) with transceivers and transponders.

The present document does not necessarily include all the characteristics which may be required by a user, nor does it necessarily represent the optimum performance achievable.

The present document complies with CEPT/ERC/DEC/(92)02 and CEPT/ERC Recommendation 70-03. It is a specific standard covering various RTTT applications. The RTTT data rate specified in the present document is for RTTT systems using down and up link data rate of up to 1 Mbps.

For nonharmonized parameters, national regulatory conditions may apply regarding the type of modulation, channel/frequency separation, maximum transmitter output power/effective radiated power, equipment marking as a condition of the issue of an individual or general license or as a condition of use under license exemption.

Additional standards or specifications may be required for equipment such as that intended for connection to the Public Switched Telephone Network (PSTN) or other systems.

4.6.1.2 EN 12253 RTTT DSRC PHY Using 5.8 GHz

EN	12253	prENV 12253, Road Transport and Traffic Telematics (RTTT). -. Dedicated Short Range. Communication (DSRC). -. Physical Layer using Microwave at 5.8 GHz	Published	EN CEN TC278

EN 12253 Available from http://www.cen.eu/esearch/extendedsearch.aspx

The DSRC standards EN 12253, EN 12795, and EN 12834, which together form a three-layered architecture for DSRC, are designed to encompass a wide range of services for different purposes in order to make the basic DSRC architecture suited for many different applications and for a wide range of possible products and systems.

This European standard:

- Specifies a physical layer at 5.8 GHz for DSRC as applicable in the field of RTTT;
- Povides requirements for the communication medium to be used for exchange of information between RSUs and OBUs;
- Caters for a communication means to be used by several applications in the RTTT sector.

The contents of the standard include:

- Downlink parameters;
- Uplink parameters;
- DSRC link parameters:
  - General;
  - Downlink parameters;
  - Uplink parameters;
- Deviations.

4.6.1.3 EN 12795 RTTT DSRC Data Link Layer

EN	12795	Road Traffic and Transport Telematics (RTTT)–Dedicated Short-Range Communication (DSRC)–DSRC Data Link Layer: Medium Access and Logical Link Control	Published	EN CEN TC278

EN 12795 Available from http://www.cen.eu/esearch/extendedsearch.aspx

This European standard sets out to:

- Define the data link layer of DSRC;
- Be positioned with respect to other related standards by the layers defined in OSI Basic Reference Model (see ISO/IEC 7498-1, Section 6.2.2.8) as adapted for DSRC;
- Support broadcast and half-duplex transmission modes;
- Support a variety of fixed equipment configurations—it supports configurations where one fixed equipment unit communicates with one mobile equipment unit, as well as configurations where one fixed equipment unit can communicate with several mobile equipment units;
- Take into account that the mobile equipment communicates with the fixed equipment while passing through a limited communication zone;
- Define parameters to be used in negotiation procedures taking place between fixed equipment and mobile equipment.

The standard specifically does not define:

- To what extent different instances of fixed equipment, operating in the vicinity of each other, need to be synchronized with each other;
- Neither any specific configuration nor the layout of the communication zone.

By defining two distinct sublayers, namely, the medium access control sublayer and the logical link control sublayer, this standard defines:

- Medium access control procedures for the shared physical medium;
- Addressing rules and conventions;
- Data flow control procedures;
- Acknowledgment procedures;
- Error control procedures;
- Services provided to the application layer.

The MAC sublayer is specific to the DSRC. The LLC services offered are unacknowledged and acknowledged connectionless services based on ISO/IEC 8802-2.

Figure 4.16 illustrates the global data flow between the elements of the DSRC stack, (physical, data link, and application layers) and the application.

NOTE: For definitions of the terms used in Figure 4.16, see Section 6.2.2.8, ISO/IEC 7498-1.

4.6.1.4 EN 12834 RTTT DSRC Application Layer

EN	12834	Road Traffic And Transport Telematics (RTTT)–Dedicated Short-Range Communication (DSRC)–Application Layer	Published	IS/EN CEN TC278

CEN 12834 Available from
http://www.cen.eu/esearch/extendedsearch.aspx

This standard specifies the application layer core which provides communication tools for applications based on DSRC. These tools consist of kernels that can be used by application processes via service primitives. The application processes, including application data and application specific functions, are outside the scope of this standard.

The standard is named "Application Layer" although it does not cover all functionality of OSI Layer 7 and it includes functionality from lower layers.

This standard uses services provided by DSRC data link layer (EN 12795) and covers functionality of intermediate layers of the OSI Basic Reference Model (see Section 6.2.2.8, ISO/IEC 7498-1).

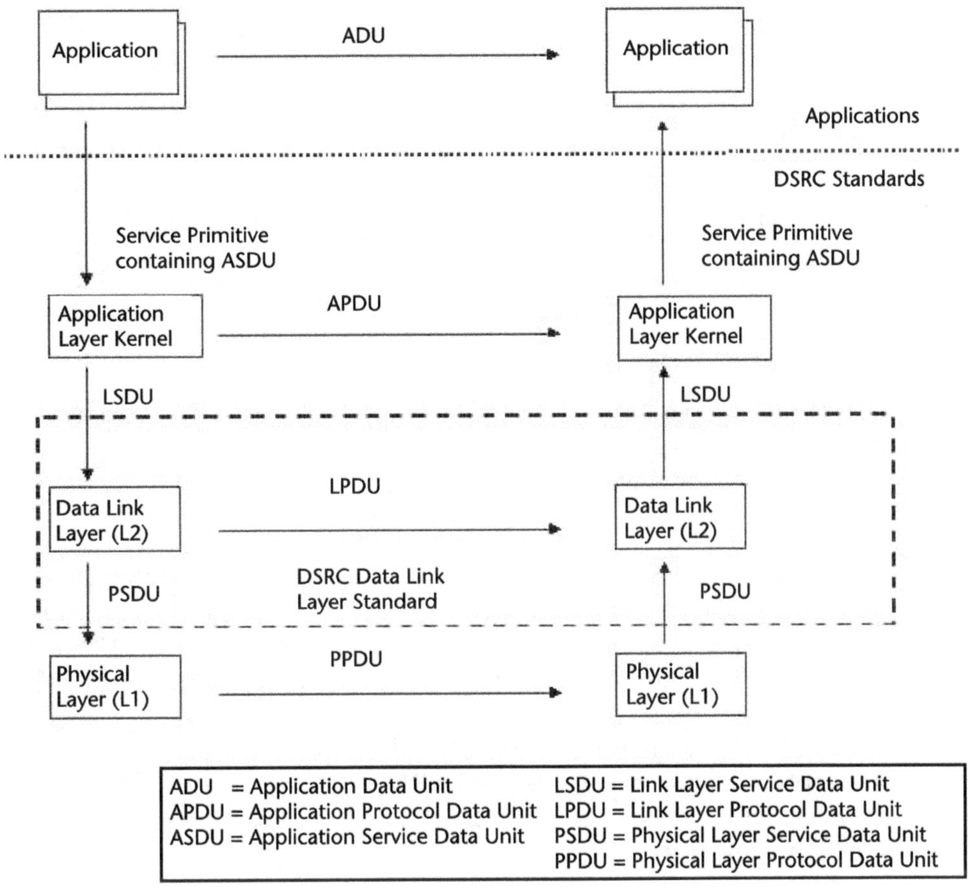

Figure 4.16 Architecture and data flow of the DSRC stack. (*Source:* CSI Library.)

Figure 4.18 in Section 4.6.1.3 illustrates the global data flow between the parts of the DSRC stack (physical, data link, and application layers) and the application.

The standard sets out to cover areas such as:

- Application Layer structure and framework;
- Services to enable data transfer and remote operations;
- Application multiplexing procedure;
- Fragmentation procedure;
- Concatenation and chaining procedures;
- Common encoding rules to translate data from abstract syntax ASN.1 (ISO/IEC 8824-1) into transfer syntax (ISO/IEC 8825-2), and vice versa;
- Communication initialization and release procedures;
- Broadcast service support;
- DSRC management support including communication profile handling.

It is outside the scope of the standard to define a security policy. Some transport mechanisms for security related data are provided.

NOTE: During the lifetime of the associated ENV, no implementation of the broadcast pool functionality became known. Broadcast pool functionality has to be considered untested and was kept in this standard for compatibility with the ENV only.

4.6.1.5 EN 13372 RTTT DSRC Profiles for RTTT Applications

CEN	EN	13372	Dedicated Short-Range Communication (DSRC)–DSRC profiles for RTTT applications	Published	CEN TC278

EN 13372 Available from
http://www.cen.eu/esearch/extendedsearch.aspx

The DSRC standards EN 12253, EN 12795, and EN 12834, which together form a three-layered architecture for DSRC, are designed to encompass a wide range of services for different purposes in order to make the basic DSRC architecture suited for many different applications and for a wide range of possible products and systems.

EN 13372 specifies DSRC profiles which provide coherent sets of communication tools for applications based on DSRC. These sets consist of subsets of functionality described in EN 12253, EN 12795, and EN 12834, out of which a minimum subset is mandatory.

DSRC systems can be built using one-way (downlink) or two-way (interactive) communication. The DSRC profiles described in this standard are intended for interactive DSRC systems based on two-way communication and DSRC systems using one-way communication (broadcast services).

While EN 12253 contains very little variation, EN 12795 and especially EN 12834 describe a wide range of communication services. It has not been considered feasible to assume that any one piece of equipment is designed to implement all of these services. It is often the case that the functionality of the on-board unit is more limited than the functionality of the road side unit.

EN 13372 is designed to increase compatibility of equipment belonging to DSRC-based systems. It defines only the communication aspects of the systems. In addition, also the application aspects need to be defined in order to ensure that on-board units belonging to one DSRC system can function meaningfully in another DSRC system for a similar application purpose (interoperability).

This standard makes provisions for:

- Physical layer parameter values;
- Data link layer subsets;
- Application layer subsets;
- Initialization procedures;
- Late response procedures;
- Termination procedures.

4.6.1.6 EN 300 674-1 ERM DSRC General Characteristics, Test Methods, and Essential Requirements for RSU and OBU

ETSI	EN	300 674-1	Electromagnetic compatibility and Radio spectrum Matters (ERM); Road Transport and Traffic Telematics (RTTT); Dedicated Short Range Communication (DSRC) transmission equipment (500 kbit/s / 250 kbit/s) operating in the 5,8 GHz Industrial, Scientific and Medical (ISM) band; General characteristics, test methods and essential requirements under the R& TTED for RSU and OBU; Part 1: General characteristics and test methods for RSU and OBU.	Published TS	ERM TG37

Downloadable FOC ETSI 300 674-1
http://webapp.etsi.org/exchangefolder/en_30067401v010201p.pdf

The present document applies to RTTT wireless systems: (1) with or without antenna connectors; (2) for digital data transmission; and (3) operating on radio frequencies in the 5,725 to 5,875-GHz ISM frequency band.

The applicability of the present document covers both the RSUs and the OBUs with transceivers and transponders.

The present document complies with ECC/DEC/(02)01 and CEPT/ERC/REC 70-03. It is a specific standard covering various RTTT applications.

Additional standards or specifications may be required for equipment such as that intended for connection to the PSTN or other systems.

4.6.1.7 EN 102 486-1-1 ERM RTTT DSRC Data Link and MAC PICS

ETSI	TS	102 486-1-1	Electromagnetic compatibility and Radio spectrum Matters (ERM); Road Transport and Traffic Telematics (RTTT); Test specifications for Dedicated Short Range Communication (DSRC) transmission equipment; Part 1: DSRC data link layer: medium access and logical link control; Sub-Part 1: Protocol Implementation Conformance Statement (PICS) pro forma specification	Published TS	ETSI ERM TG37

Downloadable FOC ETSI TS102 486-1-1
http://webapp.etsi.org/exchangefolder/ts_1024860101v010101p.pdf

The document specifies partial Protocol Implementation Conformance Statement (PICS) pro formas for the MAC and LLC layers of CEN DSRC as defined in EN

12795 in compliance with the relevant requirements, and in accordance with the relevant guidance given in ISO/IEC 9646-7 and ETS 300 406.

The pro forma is intended for use by suppliers of equipment which is claimed to conform to the CEN DSRC data link layer, as specified in EN 12795.

The document also contains profile implementation conformance statement (ICS) pro formas and profile requirements lists for the MAC and LLC layers for use by suppliers of equipment which is claimed to conform to EN 13372, DSRC Profiles for RTTT applications.

To evaluate conformance of a particular implementation, it is necessary to have a statement of which capabilities and options have been implemented for a telecommunication specification. Such a statement is called an implementation conformance statement. The present document provides pro forma ICS templates, to be filled in by equipment suppliers.

4.6.1.8 EN 102 486-1-2 ERM RTTT DSRC Data Link and MAC ATS & PIXIT

ETSI	TS	102 486-1-2	Electromagnetic compatibility and Radio spectrum Matters (ERM); Road Transport and Traffic Telematics (RTTT); Test specifications for Dedicated Short Range Communication (DSRC) transmission equipment; Part 1: DSRC data link layer: medium access and logical link control; Sub-Part 2: Test Suite Structure and Test Purposes (TSS&TP)	Published TS	ETSI ERM TG37

Downloadable FOC ETSI 102 486-1-2
http://webapp.etsi.org/exchangefolder/ts_1024860102v010101p.pdf

This technical specification contains the test suite structure (TSS) and test purposes (TP) to test the ERM/TG37 DSRC data link control (DLC) layer.

The objective of this test specification is to provide a basis for conformance tests for DSRC equipment giving a high probability of interoperability between different manufacturers' equipment.

The ISO standard for the methodology of conformance testing (ISO/IEC 9646-1 and ISO/IEC 9646-2) as well as the ETSI rules for conformance testing (ETS 300 406) are used as a basis for the test methodology.

4.6.1.9 EN 102 486-1-3 ERM RTTT DSRC MAC & LLC ATS and Partial PIXIT

ETSI	TS	102 486-1-3	Electromagnetic compatibility and Radio spectrum Matters (ERM); Road Transport and Traffic Telematics (RTTT); Test specifications for Dedicated Short Range Communication (DSRC)	Published TS	ETSI ERM TG37

transmission equipment; Part 1: DSRC data link layer: medium access and logical link control; Sub-Part 3: Abstract Test Suite (ATS) and partial PIXIT pro forma

Downloadable FOC ETSI 102 486-1-3
http://webapp.etsi.org/exchangefolder/ts_1024860103v010101p.pdf

This technical specification contains the abstract test suite (ATS) and partial PIXIT pro forma to test the ERM/TG37 DSRC data link control (DLC) layer.

The objective of this test specification is to provide a basis for conformance tests for DSRC equipment giving a high probability of interoperability between different manufacturers' equipment.

The ISO standard for the methodology of conformance testing (ISO/IEC 9646-1, ISO/IEC 9646-2, and ISO/IEC 9646-3) as well as the ETSI rules for conformance testing (ETS 300 406) are used as a basis for the test methodology.

4.6.1.10 EN 102 486-2-1 ERM RTTT DSRC Application Layer PICS

ETSI	TS	102 486-2-1	Electromagnetic compatibility and Radio spectrum Matters (ERM); Road Transport and Traffic Telematics (RTTT); Test specifications for Dedicated Short Range Communication (DSRC) transmission equipment; Part 2: DSRC application layer; Sub-Part 1: Protocol Implementation Conformance Statement (PICS) pro forma specification	Published TS	ETSI ERM TG37

Downloadable FOC ETSI 102 486-2-1
http://webapp.etsi.org/exchangefolder/ts_1024860201v010101p.pdf

This document specifies partial PICS pro forma for the application layer of CEN DSRC as defined in EN 12834, in compliance with the relevant requirements, and in accordance with the relevant guidance given in ISO/IEC 9646-7 and ETS 300 406.

The pro forma is intended for use by suppliers of equipment which is claimed to conform to the CEN DSRC application layer, as specified in EN 12834.

The document also contains profile ICS pro formas and profile requirements lists for the application layer for use by suppliers of equipment which is claimed to conform to EN 13372, DSRC Profiles for RTTT.

To evaluate conformance of a particular implementation, it is necessary to have a statement of which capabilities and options have been implemented for a telecommunication specification. Such a statement is called an implementation conformance statement. The present document provides pro forma ICS templates, to be filled in by equipment suppliers.

4.6.1.11 EN 102 486-2-2 ERM RTTT DSRC Application Layer TSS & TP

ETSI	TS	102 486-2-2	Electromagnetic compatibility and Radio spectrum Matters (ERM); Road Transport and Traffic Telematics (RTTT); Test specifications for Dedicated Short Range Communication (DSRC) transmission equipment; Part 2: DSRC application layer; Sub-Part 2: Test Suite Structure and Test Purposes (TSS&TP)	Published TS	ETSI ERM TG37

Downloadable FOC ETSI 102 486-2-2
http://webapp.etsi.org/exchangefolder/ts_1024860202v010101p.pdf

This technical specification (TS) contains the Test Suite Structure (TSS) and Test Purposes (TP) to test the ERM/TG37 Dedicated Short Range Communication (DSRC); Application layer.

The objective of this test specification is to provide a basis for conformance tests for DSRC equipment giving a high probability of inter-operability between different manufacturer's equipment.

The ISO standard for the methodology of conformance testing (ISO/IEC 9646-1) and ISO/IEC 9646-2) as well as the ETSI rules for conformance testing (ETS 300 406) are used as a basis for the test methodology.

4.6.1.12 EN 102 486-2-3 ERM RTTT DSRC Application Layer ATS & PIXIT Pro Forma

ETSI	TS	102 486-2-3	Electromagnetic compatibility and Radio spectrum Matters (ERM); Road Transport and Traffic Telematics (RTTT); Test specifications for Dedicated Short Range Communication (DSRC) transmission equipment; Part 2: DSRC application layer; Sub-Part 3: Abstract Test Suite (ATS) and partial PIXIT pro forma	Published TS	ETSI ERM TG37

Downloadable FOC ETSI 102 486-2-3
http://webapp.etsi.org/exchangefolder/ts_1024860203v010101p.pdf

This technical specification contains the abstract test suite and partial PIXIT pro forma to test the ERM/TG37 DSRC application layer.

The objective of this test specification is to provide a basis for conformance tests for DSRC equipment giving a high probability of interoperability between different manufacturers' equipment.

The ISO standard for the methodology of conformance testing (ISO/IEC 9646-1, ISO/IEC 9646-2, and ISO/IEC 9646-3) as well as the ETSI rules for conformance testing (ETS 300 406) are used as a basis for the test methodology.

4.6.1.13 EN Interoperability Application Profile for DSRC

CEN	EN	15509	Interoperability application profile for DSRC	Published	CEN TC278

EN 15509 Available from http://www.cen.eu/esearch/extendedsearch.aspx

This standard is designed to make it easier for operators to harmonize tolling systems. The standard is fully in line with application requirements being defined in the ongoing work on the European Electronic Toll Service (EETS).

EN 15509 "EFC interoperable application profile for DSRC" defines one unambiguous interoperable application profile (IAP) for DSRC. It provides a coherent set of parameter choices for the roadside and on-board equipment in EFC based on a set of underlying enabling base standards. This includes data elements and coding, security functions, and DSRC-functions. Having two security levels allows for the implementation of these requirements with or without protected access to the data in the OBU.

"This standard is based on the experiences from a large number of implementations throughout the world," says Jean-François Jouen (senior expert at ISIS) who coordinated the preparation of this standard. "It makes use of the results from European projects such as CARDME, PISTA and CESARE, the fruits of European EFC harmonization that have been used as the basis for several national implementations. It also provides for effective support for the work on the European Electronic Toll Service (EETS), called for by European EFC interoperability Directive."

Operators can implement the standard at their own pace and in coexistence with local or national EFC applications. Operators will be harmonized at a technical level using the standard as a tool when specifying requirements in procurements.

NOTE: Do not confuse this standard number with ISO 15509, which is not related to ITS.

This European standard provides for a coherent set of requirements of the EFC application that may serve as a common technical platform for EFC interoperability.

This European standard defines an interoperable application profile for DSRC-EFC transactions. The main objective is to support technical interoperability between EFC systems within the scope of the standard (as defined in Clause 1 below). A basic description of the EFC service and an EFC system can be found in CEN/ISO TS 17573 (see Section 8.8.1.4).

This European standard only defines a basic level of technical interoperability for EFC equipment (i.e., OBU and RSE) using DSRC. It does not provide a full solution for interoperability, and it does not define other parts of the EFC system, other services, other technologies, and nontechnical elements of interoperability.

The elaboration of this European standard is based on the experiences from a vast number of implementations and projects throughout Europe. The standard makes use of the results from European projects such as CARDME, PISTA, and CESARE, as they represent the fruit of European EFC harmonization and have been used as the basis for several national implementations. The development of a common EETS as a part of the European EFC Directive (2004/52/EC) also calls for the definition of an interoperable EFC service. This European standard provides for effective support for the work on the definition of EETS.

4.6.2 Standards for CALM

The CALM concept has been discussed in previous sections, especially Section 4.3. For completeness and reference, the whole suite of CALM standards (at the time of writing) is listed in this section.

Some of the standards have already been described; in these cases, this section provides only a reference to the section where the full description is to be found.

4.6.2.1 ISO 21210 CALM Networking

ISO	CD	21210	CALM—Medium and long range, high speed, air interface parameters and protocols for broadcast, point-point, vehicle–vehicle, and vehicle-point communications in the ITS sector–Networking Protocol	Under Development	ISO TC204

Not Yet Available. Track Progress at:
http://www.iso.org/iso/iso_catalogue/catalogue_tc/catalogue_detail.htm?csnumber=46549

The current scope of this International Standard specifies the networking protocols and parameters for CALM, initially using cellular communication, 5 GHz, millimeter, and infrared, but capable of supporting other media that are able to comply with the protocols.

Networking is a complex affair at any time; in a mobile vehicle environment with multiple media, it is particularly complex.

4.6.2.2 ETSI TS (CALM)/Network Service Access Point Definition

ETSI	DTS	TG37-010	Intelligent Transport Systems Continuous Air-interface for Long and Medium range (CALM) / Network Service Access Point definition	Under Development	ERM TG37

Downloadable FOC intended but not yet available (under development)
http://pda.etsi.org/PDA/queryform.asp?profile=rKGSWY0COczpbeebTflU@TyW0

4.6.2.3 ISO 21212 CALM Using 2G Cellular Networks

ISO	IS	21212	Intelligent Transport Systems—Communications Access for Land Mobiles(CALM)—2G Cellular systems	Approved Awaiting publication	ISO TC204

ISO 21212 Available from http://www.iso.org/iso/iso_catalogue/catalogue_tc/catalogue_detail.htm?csnumber=45378

See Section 4.5.3.

4.6.2.4 ISO 21213 CALM Using 3G Cellular Networks

ISO	CD	21213	Intelligent Transport Systems—Communications Access for Land Mobiles(CALM)—3G Cellular systems	Approved Awaiting Publication	ISO TC204

ISO 21213 Available from http://www.iso.org/iso/iso_catalogue/catalogue_tc/catalogue_detail.htm?csnumber=45377

See Section 4.5.5.

4.6.2.5 ISO 21214 CALM Using Infrared

ISO	DIS	21214	Intelligent Transport Systems—Communications Access for Land Mobiles (CALM)—Infra Red Systems	Published Standard	ISO TC204

IS0 21214 Available from http://www.iso.ch/iso/en/CatalogueDetailPage.CatalogueDetail?CSNUMBER=41104

See Section 4.6.3.

4.6.2.6 ETSI TS TG37-011

ETSI	DTS	TG37-011	Intelligent Transport Systems; Continuous Air Interface, Long and Medium Range (CALM), Test procedures for Infra-Red medium	Under Development	ERM TG37

Downloadable FOC Intended but final document not yet available

4.6.2.7 ISO 21215 CALM Using 5 GHz

ISO	WD	21215	Intelligent Transport Systems—Communications Access for Land Mobiles (CALM)—5 GHz Systems Not yet available	Under Development	ISO TC204

See Section 4.6.4.

4.6.2.8 ETSI TS TG37-007

ETSI	DEN	TG37-007	Electromagnetic compatibility and Radio spectrum Matters (ERM); Harmonized Standard covering Intelligent Transport Systems/Road Transport and Traffic Telematics communications systems operating in the 5,85 GHz to 5,925 GHz band.	Published	ERM TG37

Downloadable FOC

4.6.2.9 ISO 21216 CALM Using Millimeter Wave

ISO	NP	21216-1	Intelligent Transport Systems – Communications Access for Land Mobiles (CALM) – 63 GHz Systems Physical Layer	Under Development	CD ISO TC204

In ballot. Not Yet Available.

See Section 4.6.6.

4.6.2.10 ETSI TS TG37-018

ETSI	DTS	TG37-018	Electromagnetic compatibility and Radio spectrum Matters (ERM); Road Transport and Traffic Telematics (RTTT); CALM 63-64 GHz	Under Development	ERM TG37

Downloadable FOC Intended but final document not yet available

4.6.2.11 ETSI TS TG37-008

ETSI	DEN	TG37-008	Electromagnetic compatibility and Radio spectrum Matters (ERM); Candidate Harmonised Standard covering Intelligent Transport Systems/Road Transport and Traffic Telematics communications systems operating in the 63 GHz to 64 GHz band.	Published	ERM TG37

Downloadable FOC

4.6.2.12 ISO 3***** CALM Using WLAN

ISO	CD		Intelligent Transport Systems—Communications Access for Land Mobiles(CALM)—CALM using WLAN	Working Draft	ISO TC204

Number not yet allocated. Not Yet Available

This is the most recent addition to the family of media supported within the CALM context. Although in its BSS mode it is not suitable for vehicles on the move, it enables CALM functionality at service stations, and when stationary at hot spots. In peer-to-peer (IBS) mode it may have some limited uses in car-to-car communications. As with other public wireless media, it will operate within the context of ISO 25111.

This International Standard determines the air interface options applicable to CALM using WLANs within the concepts defined in IEEE 802.11 standards (excluding 802.11 WAVE) and specifies the management interface requirements for long range, wireless communications in the ITS sector utilizing IEEE 802.11 WLAN.

CALM links are required for quasi-continuous, prolonged, and short duration communications between vehicles and the roadside, between vehicles, and between mobile equipment and fixed infrastructure points, over medium and long ranges.

Wherever practicable, this International Standard has been developed by reference to suitable extant standards, adopted by selection. Required regional variations are provided.

Application-specific upper layers will not be included in this International Standard, but they will be driven by application standards (which may not be technology specific).

4.6.2.13 ISO 21217 CALM Architecture

ISO	CD	21217	Intelligent Transport Systems—Communications Access for Land Mobiles (CALM)—Architecture	Under Development	DIS ISO TC204

Not Yet Available. Track progress at:
http://www.iso.org/iso/iso_catalogue/catalogue_tc/catalogue_detail.htm?csnumber=50507

CALM architecture was introduced in Section 4.3.7. Subsequently it has been referred to in many standards. The standard defining and describing the CALM Architecture is ISO 21217.

This International Standard is part of a family of International Standards for CALM which specify a common architecture, network protocols, and communications interface definitions for wireless communications using different bearer technologies (e.g., cellular second generation, cellular third generation, infrared, 5-GHz microwave, 60-GHz millimeter-wave, and mobile wireless broadband communications). Other wireless communications interfaces may be added at a later date. These wireless communications interfaces are designed to provide parameters and protocols for broadcast, point-to-point, vehicle-to-vehicle, and vehicle-to-point communications in the ITS sector.

This International Standard determines the architectural framework for common usage by the CALM communications interface definition International Standards as described in ISO 21217, and network protocol International Standards determined by ISO21210.

The International Standard determines the architectural framework within which the CALM network protocols and CALM communications interface definition International Standards are required to operate.

Such links are required for safety related single hop, and for quasi-continuous, prolonged, or short communications between vehicles and the roadside, between vehicles, and between mobile equipment and fixed infrastructure points, over medium and long ranges.

The architecture provides for regional variations where regulations differ in different countries and regions.

The International Standard describes the architecture in an abstract way with several graphical views and examples. The graphical representation follows the ISO OSI principles and the UML principles, except where noted.

While CALM is primarily based on IPv6, it provides a route for limited "CALM-FAST" communications sessions, without handover, for time-critical safety applications.

In addition to the requirements specified within this International Standard, a number of notes and examples are provided to explain the CALM concept.

A major revision to this draft standard was undertaken in 2007, and at the time of this writing, it is undergoing DIS ballot. This shows the communications kernel described in Figure 4.17, and it incorporates the CALMFAST provisions.

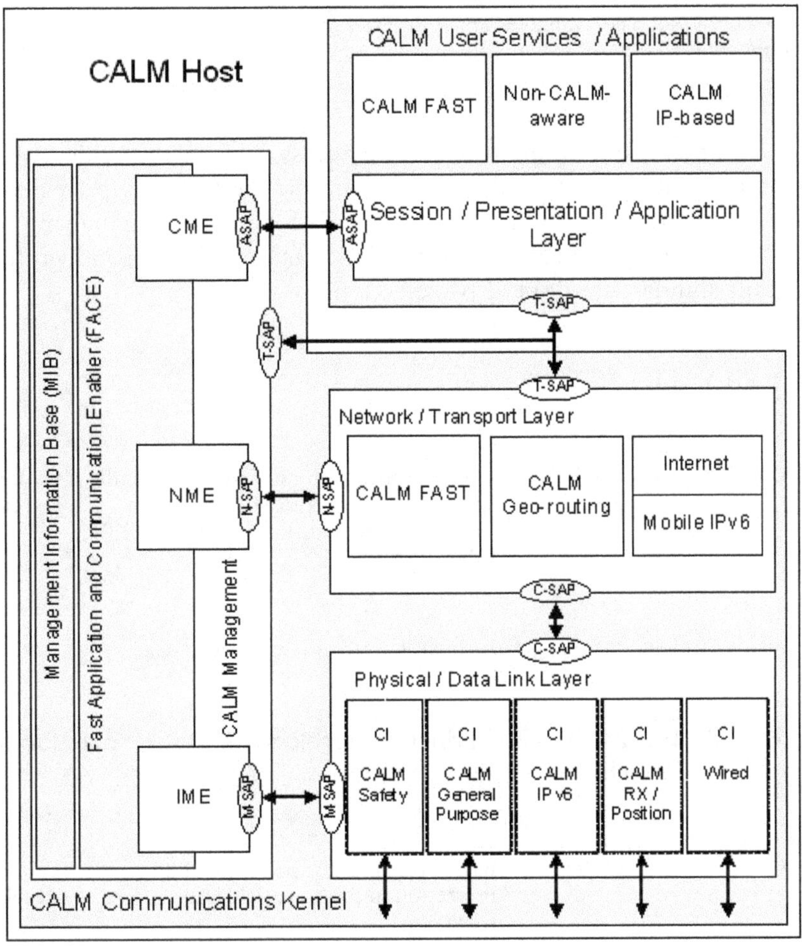

Figure 4.17 CALM communications kernel as presented in ISO 21217. (*Source:* Dr. H. J. Fischer, esf-gmbh.de.)

4.6.2.14 ETSI TS TG37-006

ETSI	DTS	TG37-006	Intelligent Transport Systems; Framework and architecture for (quasi-) continuous Communications between vehicles and the infrastructure.	Under Development	ERM TG37

Downloadable FOC Intended but final document not yet available

4.6.2.15 ISO 21218 CALM Networking – Lower Layer SAPs

ISO	IS	21218	CALM Networking–Lower Layer SAPs	Passed all ballots Awaiting publication	ISO TC204

ISO 21218 Not Yet Available. Track progress at:
http://www.iso.org/iso/iso_catalogue/catalogue_tc/catalogue_detail.htm?csnumber=35723

This standard determines the service access points as provided by the communication adaptation layer (CAL) for communication and as provided by the medium management adaptation entity (MMAE) for medium management.

Wherever practicable, this standard has been developed by reference to suitable extant standards, adopted by selection.

4.6.2.16 ETSI TS TG37-017

ETSI	DTS	TG37-017	Electromagnetic compatibility and Radio spectrum Matters (ERM); Road Transport and Traffic Telematics (RTTT); CALM SAPs	Under Development	ERM TG37

Downloadable FOC Intended but final document not yet available

4.6.2.17 ISO 24100 Basic Principles for Personal Data Protection in Probe Vehicle Information Services

ISO	DIS	24100	Basic principles for personal data protection in probe vehicle information services	Under Development	ISO TC204

ISO 24100 Available from
http://www.iso.ch/iso/en/CatalogueDetailPage.CatalogueDetail?CSNUMBER=42017&scopelist=PROGRAMME

At the time of this writing, this important deliverable in a difficult and complex area is only at a draft stage, and so it may evolve from its current scope and contents.

Probe vehicle systems are being investigated and deployed throughout the world. It is expected that the number of practical systems will grow steadily over the next few years. In general, probe data collection systems will incorporate extensive technical measures to minimize the use of personal data and protect any personal data that is used. Nevertheless, because technical measures cannot address every situation, we must address the possibility that situations may arise in which personal data becomes vulnerable to misuse. Since data collected by such systems can reveal sensitive personal information, it is critical to address consumer requirements for personal data protection through a formal policy for handling this data.

This protection is particularly important because it is difficult to completely eliminate any possibility of probe data being linked to a particular person or vehicle.

For example, consider a probe vehicle information service that does not include any personal data within the probe data, but uses personal data to authenticate the data source and ensure data integrity when collecting probe data. In this case, even if personal data is not contained in the collected probe data, probe data senders may still be identified. It is important to have both a system to protect personal data and a set of basic principles that are observed by the probe vehicle information service providers to reassure probe data senders about their personal data and facilitate the creation of information services using useful probe data.

The definition of personal data refers to the recommendation of the OECD council in 1980 concerning guidelines governing the protection of privacy and trans-border flows of personal data. The framework for describing the basic principles has adopted the eight principles of the recommendation of the OECD council in 1980 concerning guidelines governing the protection of privacy and trans-border flows of personal data. The basic principles in this standard are examined and developed on the basis of the results of the threat analysis

This standard states the basic rules to be observed by service providers who handle personal data in probe vehicle information services. This standard is aimed at protecting the personal data as well as the intrinsic rights and interests of probe data senders (i.e., owners and drivers of vehicles fitted with in-vehicle probe systems).

The standard specifies the following items related to probe vehicle systems that collect probe data from private vehicles and process the data statistically to generate useful information that is provided to various end users:

- A *reference architecture* for probe vehicle systems. This standard shows probe vehicle systems in which personal data are handled at the time of probe data collection. It should be specified in compliance with the reference architecture in ISO/CD 22837.
- The definition of *personal data* included in probe vehicle systems. This standard is defined in reference to the recommendation of the OECD Council in 1980 concerning guidelines governing the protection of privacy and trans-border flows of personal data (the OECD guidelines for personal data protection).
- The *basic principles* for personal data protection in probe vehicle systems. These principles specify the basic rules for handling personal data properly that should be observed when collecting probe data. They are stipulated in compliance with the eight principles described in the OECD guidelines for personal data protection.

  See Figure 4.18.

4.6.2.18 ISO 24101 CALM Application Management for ITS Communications

ISO	DIS	24101	CALM- Application Management for ITS Communications	Under Development DIS	ISO TC204

ISO Not Yet Available. Track Progress at:
http://www.iso.org/iso/iso_catalogue/catalogue_tc/catalogue_detail.htm?csnumber=50128

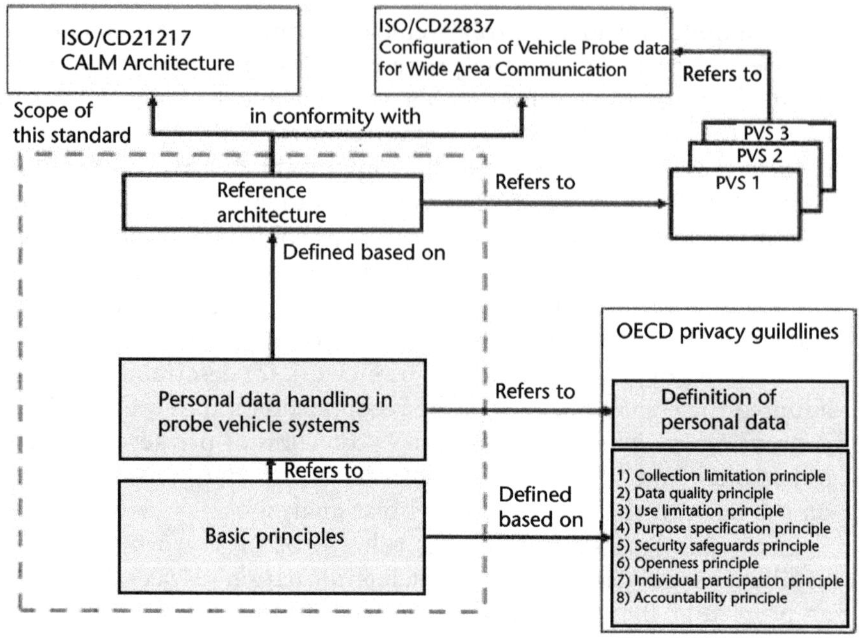

Figure 4.18 Scope of ISO 24100. (*Source:* ISO 24100.)

This document defines mechanisms, structures, and methods of application management, which means addition, update, and deletion of applications, for on-board and roadside equipment, providing multiapplication ITS by using CALM.

4.6.2.19 ISO 24102 CALM Interface Management

ISO	CD	24102	CALM Interface Management	Under Development CD	ISO TC204

Not Yet Available. Track Progress at:
http://www.iso.org/iso/iso_catalogue/catalogue_tc/catalogue_detail.htm?csnumber=42019

This standard determines the air interface manager as part of the CALM management. It serves CALM communication interfaces via the M-SAP.

Wherever practicable, this standard has been developed by reference to suitable extant standards, adopted by selection.

4.6.2.20 ISO 24103 CALM MAIL (Media Adapted Interface Layer)

ISO	CD	24103	CALM MAIL (Media Adapted Interface Layer)	Under Development DIS Ballot	ISO TC204

Not Yet Available. Track Progress at:
http://www.iso.org/iso/iso_catalogue/catalogue_tc/catalogue_detail.htm?csnumber=45806

This International Standard determines the logical structure of using DSRCs conformant to ISO 15628 (DSRC Layer 7) as CALM media. See Section 4.7.2.2.

DSRCs to which MAIL can be applied are the ones that are application layer compliant with ISO 15628, and they include the following:

- ARIB STD-T75 Dedicated Short-Range Communication (Japan);
- TTAS.KO-06.0025 Standard of DSRC Radio Communication between Road-side Equipment and On-board Equipment in 5.8 GHz band (Korea);
- CEN EN 12253 DSRC Physical Layer using Microwave at 5.8 GHz band (Europe);
- CEN EN 12795 DSRC Data Link Layer (Europe);
- CEN EN 12834 DSRC Application Layer (Europe).

This standard does not stipulate the detailed implementation but rather the logical structure of CALM MAIL.

To furnish related information, however, ARIB STD-T88 (DSRC Application Sub Layer), which is a standard with detailed definitions, is referred to and other standards may also be added as references as appropriate. As reference information, application of MAIL to non-IP protocols may be introduced.

See Figure 4.19.

4.6.2.21 ISO 25111 CALM Using Public Wireless Networks

ISO	CD	25111	Intelligent Transport Systems—Communications Access for Land Mobiles (CALM)—ITS using public wireless Networks—General requirements	Under DIS Ballot	ISO TC204

Not Yet Available. Track Progress at:
http://www.iso.org/iso/iso_catalogue/catalogue_tc/catalogue_detail.htm?csnumber=42751

See Section 4.5.1.

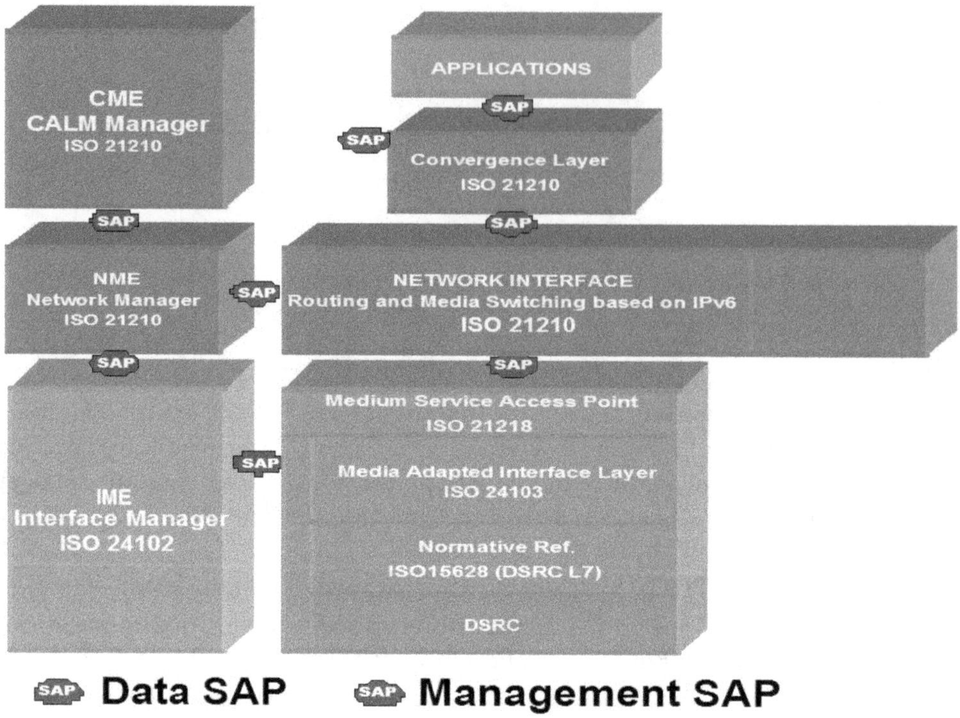

Figure 4.19 CALM-MAIL on CALM architecture. (*Source:* The CALM Forum.)

4.6.2.22 ISO 25112 CALM Using IEEE 802.16e/IEEE 802.16g (WIMAX)

ISO	NP	25112	Intelligent Transport Systems—Communications Access for Land Mobiles (CALM)—Mobile wireless broadband using IEEE 802.16e/ IEEE 802.16g	Under Development	ISO TC204

See Section 4.5.7.2.

4.6.2.23 ISO 25113 CALM Using HC-SDMA

ISO	NP	25113	Intelligent Transport Systems—Existing Systems (HC-SDMA)	Under Development	ISO TC204

See Section 4.5.7.1.

4.6.2.24 ISO 29281 CALM-FAST

ISO	NP	29281	Intelligent Transport System—CALM-CALM-Fast Subsystem	Under Development	ISO TC204

Not Yet Available

Under most circumstances, CALM is designed to use IPv6. However, in some time critical situations, and in certain other environments, the use of IPv6 is unsuited with current radio technology.

This standard determines the CALM fast subsystem. The fast subsystem is for:

- Specific applications (e.g., for road safety related applications) that need to have some knowledge of the MAC sublayer of a communication interface in order to directly control the behavior of the MAC;
- CALM communication interfaces that require a specific fast handling in the CALM system.

This standard specifies:

- The architecture of the CALM fast subsystem, being part of the overall CALM architecture;
- the CALM Management Entity (CME) functionality related to distributed CALM station (i.e., management communications inside a CALM station consisting of several router devices and host devices);
- The Network Management Entity (NME) functionality and protocol for CALM fast;
- The OSI network layer and transport layer protocol for CALM fast;
- CALM fast specific usage of the CALM communication interfaces.

Wherever practicable, this standard has been developed by reference to suitable extant standards, adopted by selection.

4.6.2.25 ISO 29282 CALM Using Satellite Communications

ISO	NP	29282	Intelligent Transport Systems—Communications Access for Land Mobiles (CALM)—CALM Using Satellite communications	Under Development	ISO TC204

Not Yet Available

See Section 4.5.8, and with respect to satellite navigations systems see Section 3.2.

4.6.2.26 ISO 29283 CALM Using IEEE 802.20

ISO	NP	29283	Intelligent Transport Systems—Communications Access for Land Mobiles (CALM)—Mobile wireless broadband using IEEE 802.20	Under Development	ISO TC204

Not Yet Available

See Section 4.5.7.3.

4.6.3 ISO 21214 Infrared Systems

ISO	IS	21214	Intelligent Transport Systems—Communications Access for Land Mobiles (CALM)—Infra Red Systems	Published Standard	ISO TC204

Available from
http://www.iso.org/iso/iso_catalogue/catalogue_tc/catalogue_detail.htm?csnumber=41104

Work on ISO 21214 was started in 2000, at the same time as work items for CALM using cellular communications and using 5 GHz. Thanks to the efforts and enthusiasm of the proponents of infrared technology for ITS, notably the renowned Dipl. Ing. Helmut Strasser, one of the three founders of the CALM concept, and with particular reference to CALM service provision, ISO 21214 was the first CALM standard to achieve the status of a full ISO International Standard. This was achieved towards the end of 2004. Unfortunately, Herr Strasser died a few months before its publication, and so this standard also stands as a testament to his pioneering work. It is indeed one of less than a handful of International Standards that is published with a dedication. The dedication reads:

> *This International Standard is dedicated to the Late Dipl. Ing. Helmut Strasser in grateful recognition of his leadership as the editor and project leader of project ISO 21214, and for his commitment and services to meeting the challenges of international standardization in the rapidly changing arena of ITS technology for more than a decade.*

Unlike the cellular standards, which operate over a public network and had only to define how to use that already standardized network, ISO 21214 had to define the complete set of protocols for the provision of the infrared media service in addition to the medium session initiation, registration, connect and disconnect requirements, and the requirements to establish a CALM session via the medium and to provide the requirements for utilizing the CALM management, networking, and service access protocols to provide a viable media link in the CALM environment.

The ISO 21214 standard determines the air interface using infrared systems at 820 to 1,010 nm.

The International Standard provides protocols and parameters for medium range, medium to high speed wireless communications in the ITS sector using infrared systems.

Conformance tests are currently being developed as follow-up work.

4.6.4 ISO 21215 5-GHz Systems

ISO	CD	21215	Intelligent Transport Systems—Communications Access for Land Mobiles (CALM)—5 GHz Systems	Under Development	ISO TC204

Not Yet Available

5-GHz communications technology, the band used for the successful European DSRC tolling systems, was, from the outset of CALM, seen as one of the key enabling technologies for CALM provision of ITS services. Work on the development of an International Standard, therefore, commenced at the same time as that for cellular communications and infrared.

At an early stage of the project, however, the United States decided to develop what it confusingly called a "DSRC" system for quasi-continuous communications with vehicles at 5.8 GHz. This was confusing because with a range approaching 1 km between beacons, it was neither short range, nor dedicated in the context previously used for DSRC (short range communications for toll collection) in Europe. Being a single medium project, it is different and more limited than CALM, but with many common features at 5 GHz. The United States decided to eschew the International Standards forum, ISO/TC 204 (for which it provides the chair and secretariat), and developed its 5-GHz standards for "DSRC" via IEEE, a body whose standards, whilst used globally with success for some areas (such as Wi-Fi), are essentially standards developed within the United States and balloted by people who turn up for meetings in the United States.

More than 2 years of discussion and negotiation at IEEE by the global CALM community were necessary before an agreement was reached that IEEE would develop the 5.9-GHz standards for ITS, with participation form the ISO community, specifically for 5.9 GHz, but with gateways to enable it to also function within the CALM context as its 5.9-GHz standard. It was therefore agreed that the ISO CALM standards would be developed once this determination was completed, and the expectation was that this would be done within 2 years.

The IEEE 802.11p/IEEE 1609 project, however, did not prove to be as simple as was first anticipated within IEEE. The initial designs were developed around existing 802.11a protocols and an existing "chip" that could be adapted. This proved a major limitation, and the project has gone through several iterations. On the positive side, in the meantime, the U.S. Department of Transportation has become more committed to the concept, has funded an extensive trial, and has declared that so long as the trial is successful, it will roll out the system nationally.

For the ISO community these delays have been frustrating and have delayed the development of the standards to the point that they have had to work to avoid automatic project cancellation.

The good news is that the evolution of the IEEE concepts has brought them much closer to those envisaged in the first place by the ISO community. The standard is now being written, like the other IEEE 802 standards as an "enabling" standard, and ISO 21215 will have to sit on top of this to select its interoperable method for CALM.

Meanwhile, now that the documents are more or less stable, the European contingent has formed European projects in part funded by the EC (CVIS, SAFE-SPOT, and COOPERS) and in part by private projects (The Car2Car Consortium) to test the technology in real environments. The projects CVIS, SAFESPOT, and COOPERS are also testing the media networking potential of CALM.

At the moment these are, therefore, only working draft status deliverables for CALM 5 GHz (ISO WD 21215).

This standard determines the CALM microwave medium operating in the 5-GHz frequency range, referred to as CALM M5. CALM M5 is based on the WLAN technology 802.11 standardized at IEEE for the automotive environment and on IEEE 1609 WAVE.

This standard specifies how the IEEE standards shall be used and amended in order to support the CALM architecture. Possible frequency allocations in regions other than North America are supported.

Wherever practicable, this standard has been developed by reference to suitable extant standards, adopted by selection.

NOTE: At the time of writing, this draft is awaiting the revision of IEEE 802.11p before it can be finalized. Additionally, as the latest drafts of IEEE 802.11p are more enabling and less prescriptive than previous drafts, further layers determining the CALM requirements on top of IEEE 802.11p and option selections within 802.11p will need to be added. The reader who requires further detail is therefore referred to http://www.iso.org and http://www.calm.hu for the latest drafts and the final word of contents.

4.6.5 ITU Dedicated Short Range Communications (DRSC) at 5.8 GHz

ITU	Rec	M.1453-2	Dedicated short range communications (DRSC) at 5.8 GHz	Published	ITU-R

ITU M.1453-2 available at http://www.itu.int/rec/R-REC-M.1453/en

This provides the ITU recommendation supporting the IEEE provisions.

4.6.6 ISO 21216 Millimeter Wave Systems

ISO	WD	21216-1	Intelligent Transport Systems—Communications Access for Land Mobiles (CALM)—63 GHz Systems Physical Layer	Under Development	ISO TC204

Not Yet Available

This work also began in 2000, although there was a recognition that, at that time, the technology was much less mature than 5 GHz, infrared, and cellular, particularly with respect to technology affordable for ITS.

Early work was began in Japan, but recent leadership has been much more oriented in Europe, especially the United Kingdom.

NP 21216 has now been divided into two parts: Part 1 dealing with the physical layer, and Part 2 dealing with the MAC and lower networking layers up to the SAP.

The current working draft for ISO 21216 Part 1 covers the following aspects:

- Adoption of other standards and internationally adopted practices;
- Physical (PHY) layer parameters;
- Regional and/or national limitations on radio parameters;
- Spectrum;
- Transmitted power;
- Access methods;
- Directivity;
- Data rate;
- Modulation;
- Marking, labeling, and packaging.

ISO 21216 Part 2 is under development at the time of this writing, but does not yet have a stable draft.

4.7 Regional and National Standards for ITS Communications

4.7.1 Long and Medium Range Systems—United States

4.7.1.1 IEEE 802.11 P WLAN -WAVE

IEEE	802.11p (WAVE)	Wireless Access for the Vehicular Environment (WAVE)	Under Ballot as work item	IEEE

Not Yet Available

At the time of this writing, this draft attained a 73% positive ballot but failed to reach the 75% of those voting required for acceptance. It is therefore currently being revised. The details of the contents may therefore change. However, the overall scope and contents are likely to remain similar. But if you require detailed content or knowledge of the latest version, you should consult http://www.ieee.org.

IEEE P802.11p, Draft Amendment to Standard for Wireless Access in Vehicular Environments (WAVE) – LAN/MAN Specific Requirements–Part 11: Wireless Medium Access Control (MAC) and physical layer (PHY) specifications) WAVE is the mode of operation used by IEEE 802.11 devices in the band allocated for ITS communications.

IMPORTANT: In order to understand IEEE 802.11p, it is important to understand that this is not a stand-alone standard, but an amendment to IEEE 802.11 (and all of its other amendments) and must be read together with these. Section 4.2.3 provides some details on 802.11.

The purpose of this specification is to provide wireless communications over short distances between information sources and transacting stations on the roadside with mobile radio units, between mobile units, and between portable units and mobile units. The communications generally occur over line-of-sight distances of less than 1,000m between roadside units and mostly high speed, but occasionally between stopped and slow moving vehicles or between high-speed vehicles. This specification also offers regulatory bodies a means of standardizing access to the ITS frequency band for the purpose of interoperable communications to and between vehicles at line-of-sight distances on the roadway.

This specification makes provision for the following:

- Describes the functions and services required by WAVE-conformant stations to operate in a high-speed mobile environment;
- Refers to IEEE 802.11 MAC procedures;
- Defines the WAVE signaling technique and interface functions that are controlled by the IEEE 802.11 MAC;
- Permits the operation of a WAVE-conformant station within a DSRC zone that may coexist with multiple overlapping DSRC communication zones;
- Describes the requirements and procedures to provide privacy of user information being transferred over the wireless medium and authentication of the WAVE-conformant stations.

This amendment to IEEE 802.11 is based on extensive testing and analysis of wireless communications in a mobile environment. The results of these efforts were documented in ASTM E 2213-03, "Standard Specification for Telecommunications and Information Exchange Between Roadside and Vehicle Systems—5.9 GHz Band Wireless Access in Vehicular Environments (WAVE) / Dedicated Short Range Communications (DSRC) Medium Access Control (MAC) and Physical Layer (PHY) Specifications."

In this amendment a new clause, Clause 20, was created that specifically defines the requirements of the extension of the PHY for a WAVE implementation.

4.7.1.2 IEEE 1609-1 DSRC Resource Manager

The IEEE 1609 series of standards are designed to sit on top of the IEEE 1609.11p standard to provide functionality essentially for the United States' VII program. However, they may also provide some of the functionality required for the CALM standards at 5.9 GHz.

IEEE 1609-1/2/3/4 Available from www.ieee.org

IEEE	1609-1	Standard for Dedicated Short Range Communications (DSRC) Resource Manager	Published IEEE Trial Standard

The status of this deliverable is: "IEEE Trial-Use Standard for Wireless Access in Vehicular Environments (WAVE) – Resource Manager." The standard asserts that are two types of wireless access in vehicular environments (WAVE) devices.

The first type, referred to as the roadside unit (RSU), is stationary while in operation and usually permanently mounted along the roadside. The second type, known as the on-board unit (OBU), may operate while mobile and is usually mounted onboard a vehicle. Typically, the stationary devices host an application that provides a service, and the mobile devices host a peer application that uses this service. There may also be applications on devices remote from the RSU, whose purpose is to provide services to the OBU. This standard describes a WAVE application that resides on the RSU but is designed to multiplex requests from remote applications, providing them access to the OBU.

This standard specifies the WAVE application known as the *resource manager* (RM), which resides on, for example, the RSU, and its peer known as the *resource command processor*, which resides on the OBU. Remote from the RSU, the other applications, known as *resource manager applications* (RMAs), communicate with the resource command processor through the resource manager. This standard describes how the RM multiplexes requests from multiple RMAs, each of which is communicating with multiple OBUs hosting a resource command processor. The purpose of the communication is to provide the RMAs access to "resources," such as memory, user interfaces, and interfaces to other on-board equipment controlled by the resource command processor in a consistent, interoperable, and timely manner to meet the requirements of RMAs.

The RM uses the concept of all of the communication being initiated from an entity known as a *provider*, which issues requests to an entity known as a *user*, which responds only to requests that it receives. Within this standard, the RM is the provider of a service (as a representative of the RMAs), and the resource command processor is the user of a service (representing the resources to be managed). Either the RSU or OBU can operate as the provider; in other words, either device type can host the RM. The device, either RSU or OBU, that is hosting the RM will be referred to as the *provider device.*

The scope of the standard is to specify the services and interfaces of the WAVE RM, including protective mechanisms for security and privacy, applicable and available to all users of DSRC and WAVE mode operations in the 5.9-GHz band authorized by the FCC for ITS.

NOTE: This version of the standard does not specify explicitly the details of the security interface. Security provisions are in IEEE 1609.2.

NOTE: The interpretation of the role of the RSU in this standard is more restricted than that envisaged in ISO CALM standards, and also has not been designed with a vehicle-vehicle configuration in its design concept.

The purpose of this standard is to enable complete interoperability of applications using WAVE in a manner that simplifies the on-board vehicle systems, reducing cost and improving performance. Effective use of the memory pages by applications can also minimize configuration management issues over the life of a system.

The standard is intended to enable a wide range of applications to be supported by an OBU of the lowest possible cost. The low cost is enabled by removing the need for the OBU to interpret application messages.

There is no OBU software representing applications using RM; thus the processing, memory, and configuration management requirements are removed from the OBU. Instead of putting such processing requirements on the OBU, they are placed on the RSU or an application processor remote from the RSU.

The only processing requirement is that of interpreting the specific commands and message headers defined herein, which is application independent. The OBU merely serves as a mobile mailbox to carry application messages and data from one RSU to another or as a common interface point to transfer data to other on-board systems.

By allowing memory to be assigned to an application at any time during the life of the OBU, future applications can be developed and deployed without on-board hardware or software modification.

For applications using RM as a mobile mailbox, with no on-board use of the data stored in memory, there are significant security advantages. By having the OBU treat each application's messages as a bit-stream to be saved and later retrieved from memory, such data can be encrypted in a manner that is not known to the OBU. There is no need for the OBU to support the encryption schemes used by these applications, and such security schemes can be under the total and absolute control of each of these applications.

4.7.1.3 IEEE 1609-3 Standard for Dedicated Short Range Communications (DSRC) Application Layer

IEEE	1609-2	Standard for Dedicated Short Range Communications (DSRC) Application Layer	Published IEEE Trial Standard

IEEE 1609-1/2/3/4 Available from www.ieee.org

The status of this deliverable is: "IEEE Trial-Use Standard for Wireless Access in Vehicular Environments – Security Services for Applications and Management Messages."

WAVE is a radio communications system intended to provide interoperable wireless networking services for transportation. These services include those recognized for DSRC by the U.S. National Intelligent Transportation Systems Architecture (NITSA) and many others not specifically identified in the architecture. The system enables vehicle-to-vehicle (V2V) and vehicle-to-roadside or vehicle-infrastructure (V2I) communications, generally over line-of-sight distances of less than 1,000m, where the vehicles may be moving at speeds up to 140 km/h.

The physical layer and MAC use elements of the IEEE 802.11 PHY and MAC and were under development at the time this standard was issued. Channelization and the upper layers of the network stack are defined in IEEE P1609.4 and IEEE P1609.3, respectively. IEEE P1609.1 defines an application, the resource manager, which uses the network stack for communications. IEEE Std. 1609.2, specifies security services for the WAVE networking stack and for applications that are intended to run over that stack. Services include encryption using another party's public key, and nonanonymous authentication.

The scope of the standard is to define secure message formats, and the processing of those secure messages, within the DSRC/WAVE system. The standard covers methods for securing WAVE management messages and application messages, with the exception of vehicle-originating safety messages. It also describes administrative functions necessary to support the core security functions.

The safety-critical nature of many DSRC/WAVE applications makes it vital that services be specified that can be used to protect messages from attacks such as eavesdropping, spoofing, alteration, and replay.

Additionally, the fact that the wireless technology will be deployed in personal vehicles, whose owners have a right to privacy, means that in as much as possible the security services must be designed to respect that right and not leak personal, identifying, or linkable information to unauthorized parties. This standard describes security services for WAVE management messages and application messages, with the exception of vehicle-originating safety messages, to meet these requirements. It is anticipated that vehicle originating safety messages will be added in an amendment to this standard.

4.7.1.4 IEEE 1609-3 Standard for IP Interface for Dedicated Short Range Communications

IEEE	1609-3	Standard for IP Interface for Dedicated Short Range Communications (DSRC)	Published Trial Standard	IEEE

IEEE 1609-1/2/3/4 Available from www.ieee.org

The status of this deliverable is: "IEEE Trial-Use Standard for Wireless Access in Vehicular Environments (WAVE)—Standard for IP Interface for Dedicated Short Range Communications (DSRC)."

A WAVE system is a radio communications system intended to provide seamless, interoperable services to transportation. These services include those recognized by the U.S. NITSA and many others contemplated by the automotive and transportation infrastructure industries. These services include vehicle-to-roadside as well as vehicle-to-vehicle communications. WAVE networking services provide data delivery services between WAVE devices and management services to all layers. This is but one component in the overall WAVE architecture, which includes IEEE Std 1609-2006, IEEE Std 1609.1-2006, IEEE Std 1609.4-2006, IEEE Std 802.11, and IEEE P802.11p.

The scope of the standard is to define services, operating at the network and transport layers, in support of wireless connectivity among vehicle-based devices, and between fixed roadside devices and vehicle-based devices using the 5.9-GHz DSRC/WAVE mode.

WAVE networking services represent layers 3 and 4 of the OSI communications stack. The purpose of this standard is to provide addressing and routing services within a WAVE system, enabling multiple stacks of upper layers above WAVE networking services and multiple lower layers beneath WAVE networking services. Upper layer support includes in-vehicle applications offering safety and convenience to their users.

WAVE networking services carry user data through the system. Clause 6 specifies the management plane functions that support system operations. Clause 7 defines the primitives used to communicate between WAVE networking services and other system entities. Annex A and Annex B contain a description, and formal definition, of the management information employed by WAVE networking services. Annex C provides an informative bibliography and definitions. Annex D provides a protocol implementation conformance statement (PICS) pro forma.

The system described in this standard supports high-rate, low latency communications between WAVE devices.

Generic IPv6 traffic is supported (not IPv4), as well as a specialized short message service. A control channel provides a common channel for signaling. IP application data is restricted to service channels; short message application data may be sent on either type of channel. Applications benefiting from the WAVE communications can reside on the WAVE devices, or reside on generic devices located on other networks connected to these devices.

The protocol supports wireless communications between any and all WAVE devices. These devices may be mobile, portable, or stationary. The mobile devices include vehicles operating at the high speeds occurring on open highways. A common characteristic of WAVE systems is the need for extremely low communications latency (measured in milliseconds) from initially encountering a device that provides services to completing a set of data transfers.

The intent of the standards is to ensure interoperability and robust safety/public safety communications among these WAVE devices.

4.7.1.5 IEEE 1609-4 Standard for Dedicated Short Range Communications (DSRC) Medium Access Control (MAC) Layer

IEEE	1609-4	Standard for Dedicated Short Range Communications (DSRC) Medium Access Control (MAC) Layer	Published IEEE Trial Standard

IEEE 1609-1/2/3/4 Available from www.ieee.org

The status of this deliverable is: "IEEE Trial-Use Standard for Wireless Access in Vehicular Environments (WAVE) – Multi-channel Operation."

The standard describes multichannel wireless radio operations, WAVE mode, MAC, and physical layers, including the operation of control channel and service channel interval timers, parameters for priority access, channel switching and routing, management services, and primitives designed for multichannel operations.

Its purpose is to enable effective mechanisms that control the operation of upper layers across multiple channels, without requiring knowledge of PHY parameters, and to describe the multichannel operation channel routing and switching for different scenarios.

4.7.1.6 SAE J2735_200612 Dedicated Short Range Communications (DSRC) Message Set Dictionary

SAE	J2735_ 200612	Dedicated Short Range Communications (DSRC) Message Set Dictionary	Published Dec. 2006	SAE

SAE J2735_200612 Available from
http://www.sae.org/technical/standards/J2735_200612

This SAE Recommended Practice is intended as a guide toward standard practice and is subject to change to keep pace with experience and technical advances. This SAE Recommended Practice specifies standard message sets, data frames, and data elements for use by applications intended to utilize the 5.9-GHz DSRC for wireless access in vehicular environments (DSRC/WAVE, referenced in this document simply as DSRC) communication systems. The scope is limited to specifying initial representative message structure and providing sufficient background information to allow readers to properly interpret the DSRC standards and message definitions from the point of view of an application developer.

The contents of the standard define:

- 58 Data Elements. These data elements are new or uniquely defined for J2735.
- 29 External Data Entries. These data elements are defined in other standards and included in J2735.
- 13 Data Concepts. These data elements are still in development in this edition of the standard. They are presented as thought starters with the anticipation that the subcommittees will refine them and present recommendations back to the technical committee.

All data elements and messages are defined as ASN.1 along with the equivalent XML representation.

4.7.2 Long and Medium Range Systems—Japan

4.7.2.1 ARIB T55/T75 Long and Medium Range Systems (Japanese DSRC)

The Japanese have developed their own set of DSRC standards, published as ARIB T55. This was followed in 2003 by ARIB T75. These include standards for L1, L2, and L7. The ARIB L7 is based on and is compatible with the CEN L7 standard.

T55 and T75 form the main platform for current generation Japanese ITS (which is either complementary to or instead of other international ITS standards). ARIB standards are developed or revised at the Standard Assembly of ARIB. ARIB standards approved by the Standard Assembly are made publicly available in hard copy, on CDs, or through Web posting, generally within 1 month of the date of approval. Many are translated into English. The general reference site is: http://www.arib.or.jp/english.

ARIB	T55	Electronic Toll Collection System (and revision from v1 to v2.0)	Approved	ARIB

ARIB T55 Available from
http://www.arib.or.jp/english/html/overview/img/arib_std-t55_e.pdf

In Japan, the Association of Radio Industries and Businesses (ARIB) approved in November 1997 a standard entitled, "Dedicated Short Range Communication (DSRC) for Transport Information and Control Systems (TICS)," an English version of which was published in August 1998 under the reference "ARIB STD-T55 Version 1.0."

The standard was developed based on the participation of and discussion with the various radio equipment manufacturers, operators, and users. The scope of application of the standard covers the minimum requirements for communications between on-board equipment and roadside unit. The purpose is to ensure the quality and compatibility of radio facilities and equipment.

The standard adopts the three-layer structure of the OSI basic reference model, and the standardized objects are Layer 1, Layer 2, and Layer 7.

The services provided by the system have the following attributes:

- Information transfer capability: unrestricted digital information;
- Information transfer rate: 1,024 Kbps;
- Communication configuration: point-to-point, point-to-multipoint.

The basic services provided through the information transfer channel are:

- Exchange of the information about the toll collection, and the reading/writing of the data performed through the radio facility (lane based antenna) installed at the toll gate;
- Transmission of the guiding information about the lanes, in order to guide the vehicles equipped with OBU to the dedicated lane and to achieve smooth operation of traffic lanes, notification of the route information.

ARIB	T75	Dedicated Short-Range Communication System (and from Ver.1.2 to Ver.1.3)	Approved	ARIB

ARIB T75 Available from
http://www.arib.or.jp/english/html/overview/img/arib_std-t75_e.pdf

This standard specifies the radio communication interface between a land mobile station and a base station for a DSRC system.

The system is designed to operate within the framework and regulation of Article 49-26 (including related notifications) of Japanese Radio Facility Regulations when the system is used in Japan.

Mobile stations and the radio station for testing communication to the mobile stations are required to be in accord with Article 4-3 of the Japanese Radio Act and Article 6-4-7 of the Enforcement Regulations.

The system consists of a roadside unit installed on the roadside (termed "base station") and on-board equipment (termed "land mobile station" or simply "mobile station"), as well as equipment for testing the mobile station (termed "station for testing" or simply "test equipment").

In terms of mutual connectivity and compatibility, the standard defines the minimum level of specifications required for basic connections and services as the mandatory requirement, and the specifications required for what free choice is permitted, such as protocols, as optional standards to provide for future expansion.

Further, in order to provide options and future expansion capabilities as much as possible, care has been taken not to place restrictions on nonstandardized specifications.

The standard adopts the three-layer structure of the OSI basic reference model and the standardized objects are Layer 1, Layer 2, and Layer 7.

As for functions in Layer 3, Layer 4, Layer 5, and Layer 6 defined in the OSI basic reference model, they are specified in the Layer 7 if they are needed for the system, taking into account the fact that the transaction occurs in a short period each time a mobile station passes through a small communication zone of the base station.

The application sublayer should accommodate as an option when processing near to the application level is required, and portion where Layer 2 cannot make in order to be compatible with ARIB STD-55 (Electronic Toll Collection System), described earlier.

The standard provides a framework for a system that consists of a roadside unit installed at the roadside (termed "base station" hereunder) and on-board equipment installed in the vehicle (termed "mobile station").

The base station performs land mobile radio communication with mobile station(s). The base station is composed of radio equipment with antenna(s), a transmitter and receiver, a control unit, and a display unit.

Depending on the radio communication range, the base station is classified as follows:

- Class 1: radio communication range is below 10m;
- Class 2: radio communication range exceeds 10m, but within 30m.

The mobile station performs land mobile radio communication with the base station. The mobile station consists of radio equipment with antennae(s), a transmitter and receiver, and optional equipment such as an IC card, a control unit, and a display unit.

The standard should be read in conjunction with ARIB T88, the Japanese DSRC Application Layer Standard (see Section 4.6.6.4).

Other ARIB telecommunications standards are not detailed here since, at the time of writing, they are not often referred to in an ITS context. However, they and their references are summarized in Section 12.2.

4.7.2.2 ISO 15628 Application Interface Standards to Enable Non-CALM Media to Interface with CALM

ISO	15628	Road transport and traffic telematics—Dedicated short range communication (DSRC)—DSRC application layer	Published Standard	ISO TC204

ISO Available from
http://www.iso.org/iso/iso_catalogue/catalogue_tc/catalogue_detail.htm?csnumber=44432

The communication requirements of many ITS applications can be fulfilled by DSRC. The DSRC standards enable compliant communication systems to serve multiple ITS applications in parallel.

The small service areas and severe real-time constraints require a specific protocol architecture leading to the reduced protocol stack shown in Figure 4.20, built up by the application layer, the data link layer, and the physical layer. Such architecture is very common for real-time environments.

This International Standard gives the architecture and services offered by the DSRC application layer.

This International Standard contains, besides the normative main body, three normative annexes: "Data Structures," "Naming and Registration," and "Declara-

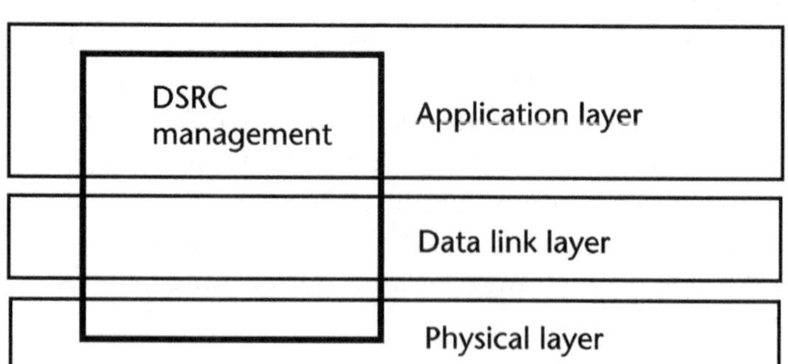

Figure 4.20 European DSRC protocol stack. (*Source:* CSI Library.)

tion of Application Layer Features Supported," respectively; plus two informative annexes: "Examples of Coding" and "Lower Layer Services."

The International Standard specifies the application layer core, which provides communication tools for applications based on DSRC. These tools consist of Kernels that can be used by application processes via service primitives. The application processes, including application data and application-specific functions, are outside the scope of this International Standard.

The standard is named "Application Layer" although it does not cover all functionality of OSI Layer 7 and it includes functionality from lower layers.

The standard uses services provided by DSRC data link layer, and covers functionality of intermediate layers of the OSI basic reference model (see Section 6.2.2.8, ISO/IEC 7498-1).

4.7.2.3 ARIB T88 Japanese DSRC Application Sublayer

ARIB STD-T88	DSRC Application Sub-Layer	2004.5.25	Ver.1.0

ISO Available from
http://www.arib.or.jp/english/html/overview/img/arib_std-t88_e.pdf

ARIB Standard T88 is essentially the Japanese instantiation of ISO 15628 DSRC application sublayer. However, whereas the ISO standard has only been proven in the context of EFC, Japan sees ISO 15628/ARIB T88 as its portal to CALM and has implemented a much wider range of services, or its equivalent of IEEE 1609 series. At the GSC meeting in Sophia Antipolis, and subsequently on the GSC Web site, they explained their use of ARIB T88 as follows (the following summary is extracted from the Japanese presentation to GSC and posted for public dissemination and information on the GSC Web site, which provides the source of this description):

ARIB T88/ISO 15628 is used to provide the following ITS services:

- ETC in Japan;
- Development of DSRC applications;
- Application sublayer (ASL);
- Study of a basic application interface to extend application in vehicles;
- Standardization related to DSRC.

CHAPTER 5

Technical Support Standards for ITS

This chapter provides a summary of general tools that have been developed by the ITS standards community, which are not specific to the provision of any particular ITS service (for these see Part III, Chapter 8) but may be of general use in the sector to help understand and further its evolution, as well as to develop ITS service specific standards. Chapter 6 will then consider generic standards developed outside of the sector that may provide similar support; and then Chapter 7 will consider standards for identification.

5.1 General Architecture Development Standards and Specifications for ITS

5.1.1 ISO 14813-1 ITS Service Domains, Service Groups, and Services

ISO	IS	14813-1	Intelligent transport systems—Reference model architecture(s) for the ITS sector-Part 1—ITS service domains, service groups and services	Revision Published 2007	ISO TC204

ISO 14813-1 Available from
http://www.iso.org/iso/iso_catalogue/catalogue_tc/catalogue_detail.htm?csnumber=43664

Originally referred to in the prior version of TR 14813-1 as "Fundamental Services," ITS service domains and groups reflect the evolution of technology-oriented transportation practices and applications. This has become of increasing importance and interest as the scope of ITS expands beyond its original applications in road traffic management, traveler information, and electronic payment systems. ITS is now also expected to address:

- Transport network operations and maintenance activities;
- Freight mobility and intermodal connectivity;
- Multimodal travel including both pretrip and on-trip information and journey planning;
- Variable road pricing strategies for freight and personal travel;

- Emergency and natural disaster related response activities and coordination;
- National security needs related to transportation infrastructure.

Additionally, ITS activities as described earlier also will interface with more generalized activities and environments outside the transportation sector. For example, road pricing and revenue systems activities may interface with electronic commerce (e-commerce) activities, and may thus utilize standards and principles associated with the banking industry along with generally accepted accounting principles. National security and coordination issues also require addressing specific national standards related to civil defense, emergency communications, and other procedures. These interfaces, while largely outside the scope of TC204, are nevertheless critical external influences on the functionality of the various services supported by ITS service domains and groups.

The standards that have been developed within TC204 may all be mapped to one or more of the ITS service groups defined in this document. At the same time, the applications for these standards are becoming broader. Additionally, the development of a standard international data dictionary and registry for ITS requires the ability to address both current and emerging applications.

To this end, the ITS service domains and groups presented in this document serve as a framework for developing ITS-related concepts of operation, which in turn lead to the definition of the appropriate requirements and standards necessary to deploy specific ITS applications. As the definition of transportation activities that utilize ITS tools has broadened, the original *fundamental services* developed by TC204 are now revised and expanded into *ITS service domains and groups*.

Figure 5.1 illustrates the hierarchy of functional definitions, and the basis upon which user-oriented architecture views (known as *use cases*) are derived (refer to the other TR 14813 parts, which further define the ITS applications of Unified Modeling Language, including use cases, to document the architecture).

In order to develop a cohesive reference architecture, and in order to establish the relationship and interdependencies of the various ITS services, it is beneficial to first determine the underlying ITS services. Thus, the purpose of this international standard is to identify the ITS service groups and the domains within which the service groups reside within the current perception of the ITS sector.

ITS service domains and groups, while they build upon existing U.S., E.U., Japanese, and other international and national taxonomies (or classification systems), can also *provide a common descriptive basis for comparing* these taxonomies, as well as others being developed throughout the world.

ISO 14813-1 (2007) provides a definition of the primary services and application areas that can be provided to ITS users. Those with a common purpose can be collected together in ITS service domains, and within these there can be a number of ITS service groups for particular parts of the domain. This international standard identifies 11 service domains, within which numerous groups are then defined. Within this framework, there are varying levels of detail related to definition of different services. These details differ from nation to nation, depending on whether the specific national architecture building blocks are based directly upon services or on groups of functions. Thus, the intent here is to address *groups* of services and the respective *domains* within which they fit. As these domains and service

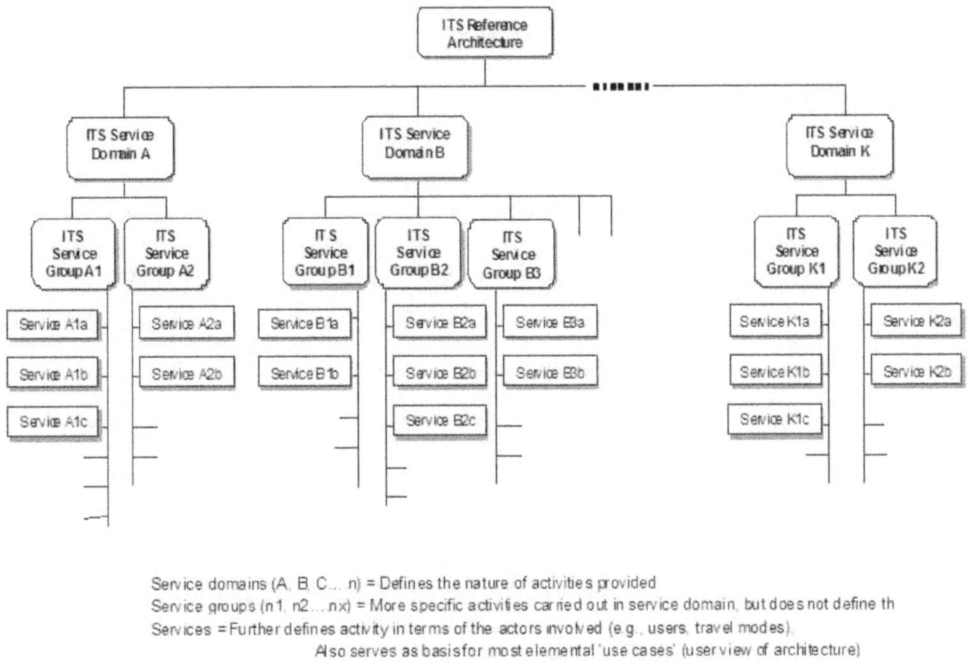

Figure 5.1 ITS services: hierarchy of definitions for ITS reference architecture. (*Source:* CSI Library.)

groups evolve over time, it is intended that this international standard shall be revised to include them.

This international standard is applicable to the working groups of ISO/TC 204 and other TCs which are developing international standards for the ITS sector and associated sectors whose boundaries cross into the ITS sector (such as some aspects of urban light railways, intermodal freight, and fleet). This international standard is designed to provide information and explanation to those developing ITS international standards and to those developing specifications, implementations, and deployments for ITS.

The standard is in itself, by its nature, advisory and informative. It is designed to assist the integration of services into a cohesive reference architecture, as well as to assist interoperability and common data definition. Specifically, services defined within the service groups will be the basis for definition of use cases and the resultant reference architecture functionality, along with definition of applicable data within data dictionaries, as well as applicable communications and data exchange standards.

The standard provides descriptions of the concepts by domain, service group, and service.

Since the publication of the standard, and indeed, partly because of the research work undertaken for this book, ISO TC204 WG1 has declared that while it fully supports this standard as effective work, the finalized standard may be too oriented toward the institutions behind ITS, and that the broader scope (i.e., as provided in this book) justifies the commencement of the next round of revision.

As the ITS sector evolves, resources permitted, such reviews and revisions to this deliverable are seen as an inevitably regular process.

As 14813-1 is fundamental to any interoperable provision of ITS services, and therefore to many of the other standards referenced in this book, some additional description, provided courtesy of CSI Ltd.'s summary of the objective and method of ISO 14813-1, is given here[1]:

> It is central to the understanding of the nature of ITS systems to understand that they vary in their implementation either because of the organization within a specific geographic region, or may vary according to the perception of any actor.
>
> Therefore, and regardless of any specific implementation services and groups of services may be provided in functionally different (although often interrelated) application areas. These application areas are the highest level of abstraction in an ITS architecture, and are called service domains.
>
> ITS services do not represent the technology or the functionality of an ITS system included in any ITS applications.
>
> [Examples of ITS service domains are given in the summary of contents above.]
>
> A service domain comprises one or more types of ITS service. Each type of ITS service may comprise several instances of related services. These collations of related ITS service instances are called "ITS service groups." Therefore, an ITS service group consists of one or more similar or complementary services provided to ITS users.
>
> There are several characteristics of "ITS service groups" and services contained within ISO 14813-1:
>
> a) Each 'ITS service group' is oriented to a specific activity related to management or information of the road transport network that is divided into specific services that should address particular users or modes.
>
> b) The name of each service group reflects the type of activities supported (e.g., "pre-trip information").
>
> c) Each service within the service group references both the service group activity and the nature of the users or modes supported by the service (e.g., "pre-trip information – public transport").
>
> d) Each level of the hierarchy should be at an equivalent level of granularity.
>
> The concept of "an ITS service" consists of a product or activity provided to a specific ITS user. ITS services may therefore be considered as the elemental building blocks of any ITS architecture/system.
>
> It must also be noted that different countries do and will continue to partition their reference architectures in different ways – some through more granular service or needs definition, others at a higher level of abstraction, others emphasizing the aspects that require interoperability and exchange of data. Therefore elaboration of specific "ITS services" is inappropriate in any "architecture" Standard (although instantiations may well require application standards if they are to operate internationally or even nationally).
>
> Elaboration of specific ITS services needs to be undertaken in a consistent manner throughout any specific architecture. There are several methodologies that assist the development of consistent elaboration. Elementary services can be elaborated by using particularization about service groups. Different viewpoints may result in different elaborations.
>
> The provision of an ITS service requires both a service provider and a user. In general one may say that the ITS user is a "partner," one who receives an ITS

1. This text was taken from the CSI input document at the creation of the 14813-1 revision process.

service through interaction with the ITS system. They have elsewhere been described as "Those who want the benefit of ITS systems, those who use ITS systems, those who manufacture or operate ITS systems, and those who regulate, control Transport using ITS systems."... By definition, all human interaction with ITS systems involves external actors interfacing at the boundary of the system. Humans interacting with an ITS system are sometimes called "external ITS users."

In order to study, consider and effect interoperability, it is first necessary to categorise ITS activities This serves to delineate different sectors of the ITS industry (and are roughly parallel with current TC204 working group activities).

The lists and descriptions in ISO14813-1 describe eleven "ITS service domains." [See Section 5.1.1.]

Of course, the categorization of the services into 11 domains is designed to help with architecture design and creation of architecture models, but it does not also imply that all ITS Architectures should always be required to follow this construction. The construction that they use should be that which is best suited to their ultimate use and should be independent of the services that they support.

Instantiation of Services are often interdependent on, or providers to, other services within a service group or are key enablers for the provision of services in other service groups. In architecture elaborations based on these services it is important that the proposed classification schema identify WHO is responsible for the provision of the service.

ITS Service Domains are comprised of 'ITS service groups,' and these 'ITS Service Groups' represent a further delineation of the various sectors of ITS activity The delineation specifically addresses different types of activities carried out within the domain. The service groups do not necessarily address specific users, modes or audiences for these activities; that level of detail is provided by the specific services (which will be defined as separate services).Those separate services required for an interoperable instantiation may be the subject of separate service specific standards, but will not be included in an architecture at anything other than an abstracted functional level. This abstracted functional level permits the transport operator (state, municipality or authority) to select which specific services within each service group are relevant to their needs. This also permits a more modular definition of specific architecture needs that may be applicable to a particular project or system. And it permits experts in the provision of specific services to devise appropriate services, rather than try to fit to a concept designed by an architect who is not experienced in the application of a particular subject.

5.1.2 ISO 14813-2 TICS System Architecture—Core TICS Reference Architecture

ISO	TR	14813-2	Transport Information and Control Systems—Transport information and control systems—Reference model architecture(s) for the TICS sector—Part 2: Core TICS reference architecture	Published TR ISO	ISO TC204

ISO 14813-2 Available from
http://www.iso.org/iso/iso_catalogue/catalogue_tc/catalogue_detail.htm?csnumber=25615

NOTE: Transport Information and Control Systems (TICS) was the previous name for ISO/TC 204 (Intelligent Transport Systems).

ISO TR 14813-2 identifies that there may well be more than one single ITS/TICS architecture approach to be considered and documented, and that existing architecture approaches will have previously produced documentation developed according to disparate standards and conventions.

Full documentation of all possible architectural approaches is obviously not feasible given the high level of resources required to carry this out. Indeed, full documentation and description of all possible approaches is undesirable as an item for standardization.

A defined and consistent approach, however, is required to facilitate the specification of architecture requirements in order to enable a clear view of the work of each participant group to be developed and presented. This technical report is one of a set of documents intended to respond to stated objectives regarding the production of a ITS/TICS reference architecture.

In order to document an architecture, graphical and textual components of a model are required. This technical report has adopted a methodology based on the unified modeling language (UML) for documenting the ITS/TICS reference architecture. A tutorial on the UML is provided in ISO 14813-4. UML (ISO 19501) is a visual modeling language for building object-oriented and component-based systems. A commercially available computer aided software engineering (CASE) tool has been used to document the architecture. While the tool is a commercial product, UML is open and nonproprietary.

NOTE: The following paragraphs appear in ISO 14813-2, ISO 14813-3, and ISO 14813-4.

The architecture of an information and control system merges hardware and software considerations into a coordinated and integrated system view. The system architecture is a high level abstraction (or model) of the system. A system architecture should embrace both today's applications and the applications that are expected in the future. Architecture begins with the definition of the conceptual services (e.g., Part 1—ITS/TICS Fundamental Services). There are several identifiable stages of system architecture development:

- Reference architecture;
- Logical architecture;
- Physical architecture.

A reference architecture is the first of all architectures. It is a concise generic framework which guides the development of more concrete system architectures. It is large enough that distinct concepts are not merged out of necessity and small enough that it does not become unwieldy.

A most significant example of a reference architecture in information systems is the Reference Model of Open Systems Interconnection (often called the seven layer model) developed by ISO in the 1970s. This model has underpinned the development of all modern computer networks, allowing services such as global networking (of which the prime example is the Internet) to become a reality.

A reference architecture is generic and nonprescriptive and captures the concepts of the system. A logical architecture elaborates the conceptual behavior, and in so doing it provides more detail about the modularity. A physical architecture is reached when the actual distribution of the system modules is defined, thus leading to important implications for communications.

There is no firm demarcation between a reference architecture and a logical architecture. Thus, the essence of behavior and modularity is present in a reference architecture. The ITS/TICS Reference Architecture developed by WG1 shows important interrelationships that arise in the provision of the services of the sector. However the ITS/TICS Reference Architecture is more abstract than, for example, the logical architecture of the U.S. National Architecture.

It is envisioned that the ITS/TICS Reference Architecture will be used by the TC204 Working Groups to develop their own logical and physical architectures in a cohesive manner.

Some ITS/TICS fundamental services are already well developed by the industry, while others are less mature. Therefore the ITS/TICS Reference Architecture does not have a uniform granularity across all services. This characteristic is a direct result of the fore mentioned requirement that architecture embrace the applications that are intended in the future. This suggests one of the ways in which the architecture will undergo change in the future.

Architectures may present static characteristics only, or both static and dynamic characteristics. Dynamic characteristics may be seen as belonging solely to the design/implementation stages of system development. However, by including dynamic characteristics at the reference architecture stage, one gains important insights into the static architecture. Thus, two orthogonal views of architecture are presented:

- Static relationship view (class diagram);
- Dynamic interactive view (sequence diagram).

NOTE: End of text is repeated in ISO 14813-2, ISO 14813-3, and ISO 14813-4.

This part of ISO TR 14813 develops a core reference architecture. The static scope is determined by deriving the system boundary and the use cases from an analysis of the ITS/TICS fundamental services (ISO TR 14813 Part 1).

The core reference architecture is a reference for the development of national architectures.

Part 3 of ISO TR 14813 elaborates the core reference architecture by refinement of two orthogonal views. The elaboration calls upon domain expertise that would be provided by other TC204 Working Groups in the development of ISO standards or by national groups developing national architectures and standards.

The core reference architecture is described in clauses 5 to 8 of the Technical Report. Clause 5 introduces the architecture at a highly abstract level. Clause 6 defines all the actors. Clause 7 derives all the use cases from the ITS/TICS fundamental services and develops eight use case diagrams. Clause 8 defines an abstract collection of classes and develops a set of sequence diagrams, one per use case diagram.

Readers should refer to Part 4 of ISO TR 14813 (Tutorial) for an introduction to the modeling views used in this part and the methodology applied.

ISO 14813-2, -3, -4 provide one methodology using ISO 19501-1, Unified Modeling Language. However, the ISO ITS architecture standards do not require the use of a specific methodology, and other techniques, such as process decomposition, may also be used.

5.1.3 ISO 14813-3 TICS System Architecture—Example Elaboration

ISO	TR	14813-3	Transport Information and Control Systems- System Architecture- Example Elaboration	Published TR	ISO TC204

ISO 14813-3 Available from
http://www.iso.org/iso/en/CatalogueDetailPage.CatalogueDetail?CSNUMBER=29534

NOTE: Transport Information and Control Systems (TICS) was the previous name for ISO/TC 204 (Intelligent Transport Systems).

It is important to read the introductory text in Section 5.1.2, which is repeated in ISO 14813-2, ISO 14813-3, and ISO 14813-4.

ISO TR 14813-2 develops a core ITS/TICS reference architecture. The static scope is determined by deriving the system boundary and the use cases from an analysis of the ITS/TICS fundamental services (ISO 14813-1).The core reference architecture is a reference for the development of national architectures.

This technical report elaborates the core by refinement of the two orthogonal views. The elaboration calls upon domain expertise which would be provided by other TC204 Working Groups in the development of ISO standards, or by national groups developing national architectures and standards.

The core reference architecture is elaborated in this technical report which also introduces the elaboration method employed. The technical report further elaborates the classes, the sequence diagrams, the elaborated packages, and some of the main dependencies between the packages.

Readers of this technical report should refer to ISO 14813-4 (Tutorial) for an introduction to the modeling views used in this technical report and the overall methodology.

ISO 14813-2, -3, -4 provide one methodology using ISO 19501-1, Unified Modeling Language. However, the ISO ITS architecture standards do not require the use of a specific methodology and other techniques, such as process decomposition, may also be used.

ISO 14813-5 provides guidance as to the requirements for architecture description in ITS standards; ISO 24529, Using UML in Developing ITS Standards (see Section 5.2.4) and ISO 17452, Using UML for ITS Data Registries and Data Dictionaries (see Section 5.2.3) provide further assistance and guidance.

5.1.4 ISO 14813-4 TICS System Architecture—Reference Model Tutorial

ISO	TR	14813-4	Transport Information and Control Systems—System Architecture—Reference Model Tutorial	Published TR	ISO TC204

ISO 14813-4 Available from
http://www.iso.org/iso/iso_catalogue/catalogue_tc/catalogue_detail.htm?csnumber=25613

NOTE: Transport Information and Control Systems (TICS) was the previous name for ISO/TC 204 (Intelligent Transport Systems).

It is important to read the introductory text in Section 5.1.2, which is repeated in ISO 14813-2, ISO 14813-3, and ISO 14813-4.

5.1.5 ISO 14813-5 TICS System Architecture—Requirements for Architecture Description in TICS Standards

ISO	TR	14813-5	Transport Information and Control Systems—System Architecture—Requirements for Architecture Description in TICS Standards	Published TR Revised in DIS Ballot	ISO TC204

ISO 14813-5 Available from
http://www.iso.org/iso/iso_catalogue/catalogue_tc/catalogue_detail.htm?csnumber=29535

NOTE: Transport Information and Control Systems (TICS) was the previous name for ISO/TC 204 (Intelligent Transport Systems).

ISO 14813-5 provides guidance as to the requirements for architecture description in ITS standards.

"Architecture" can be defined as "design; the way components fit together" (from: "Interoperability Clearinghouse Glossary of Terms," at http://www.ichnet.-org/glossary.htm). Architecture is implicit in any construction, be it of a physical entity (such as a building), an operational entity (such as a company or organization), a system entity (such as a software system), or a business entity (such as a commercial business operation).

While it may be stated that every entity has an architecture, that architecture may be an explicit construction as a result of a deliberate design process, or it may be the implicit result of an unplanned series of events, and sometimes it may be the combination of both.

In physical construction, it is generally recognized that a deliberate design process will produce a better and more efficient building than one where a group of individuals have collected whatever materials happened to be nearby.

ITS is a group of services (see Section 5.1.1) which can be implemented to improve both the driving experience and the safety and security of drivers, passengers, and pedestrians. ITS systems can also assist in the labor, energy, environment, and cost efficiency of transportation systems. It is a feature of most ITS systems that their architecture involves the collection, use, and exchange of information/data within and between software systems which affect or control the behavior of physical equipment in order to provide a service to the actors involved in, or interacting with, the transport sector.

In order to maximize the efficiency of coexisting ITS systems, and in order to obtain compatibility and/or interoperability and to eliminate contention, the systems need to coexist and operate within a known and supportive architectural framework.

The ITS sector is still emerging and developing and is still close to the start of its evolution and application. The technology is developing and changing rapidly, and ITS services have to generally make provision not only for its interaction with other services, but with migration from one technology generation to later iterations.

This international standard is designed to ensure that, in order to obtain maximum interoperability, efficiency, and migration capability, architecture is an explicit process in the development of and specifications defined within ITS international standards.

The word "architecture" has been used in an informal manner to mean a variety of different concepts, and in formal architecture design there are differing methodologies and opinions as to their suitability for use in ITS system and standards design. This has limited the effective communication in the ITS sector by causing uncertainty as to the meaning of the word "architecture" when it is used. A second function of this international standard is to provide consistent terminology to be used in describing architectural aspects of ITS standards, and to provide a consistent form for ITS architecture description in standards in the ITS sector.

This international standard does not give preference to any one methodology for architecture development and description; it requires only that the consideration of architecture is an explicit process that takes into account the interrelationships and interoperability of ITS systems and that an architecture description is provided within ITS standards.

The standard requires that the architecture aspects of ITS standards are described explicitly in each and every ITS standard and that all standards are related to the (one or more) ITS service domains, service groups and service(s) (ISO 14813-1), that they are designed to enable or support.

An ITS architecture is a framework for ITS deployments. It is a high level description of the major elements and the interconnections among them. It provides the framework around which the interfaces, specifications, and detailed ITS systems designs can be defined. An ITS architecture is not a product design, nor a detailed specification for physical deployment, and it is not specific to any one location.

The title "Systems Architecture" is perhaps the closest general terminology, but that term is sometimes too specific to include the conceptual aspects included in the terminology "ITS Architecture" and also often implies a location specific solution.

The purpose of an ITS architecture is to maximize efficiency, interoperability, and multimodality of multiple interacting ITS systems in a complex and developing sector.

This international standard defines:

- Terminology to be used when documenting or referencing aspects of architecture description in ITS standards;
- The requirement that aspects of system architecture are to be documented and described in ITS standards.

In compiling this standard, the authors have assumed that contemporary systems engineering practices are used. Such practices are not defined within this standard.

5.1.6 ITS Data Dictionaries

At the time of this writing, there are no standardized data dictionaries for ITS. However, see Section 8.7 where (draft) CEN 15722 (was 24977) determines a minimum set of data for eCall services, and Section 8.3.2.3 where ISO 24978 establishes the framework for an emergency message data registry.

Section 5.1.7 provides details of the principal standard that determines the framework for ITS data registries and ITS glossaries. Several instantiations have already appeared, most notably in the United Kingdom (TIH), Australia, and the Czech Republic.

See also Section 7.3 for details of vehicle identification schemes.

5.1.7 ISO 14817 ITS Data Registries

ISO	14817	Transport information and control systems – Requirements for an ITS/TICS central Data Registry and ITS/TICS Data Dictionaries	Published Standard	ISO TC204

ISO 14817 Available from
http://www.iso.org/iso/iso_catalogue/catalogue_tc/catalogue_detail.htm?csnumber=36030

This international standard defines the framework, formats, and procedures used to define information and information exchanges within the ITS/TICS sector. The standard is designed to be used by the ITS/TICS community at large, but should be of special interest to application developers, equipment providers, and data registry managers.

The international standard specifies a set of meta attributes for ITS/TICS data dictionaries, as well as associated conventions and schemes, which enables the description, standardization, and management of all exchanged ITS/TICS data.

Through consistent use of these common structures and associated conventions and schemes, interchange of data and information among the various ITS/TICS functional subsystems via their specific application systems can be maximized. This international standard also supports reuse of data elements and other data concepts across various ITS/TICS functional subsystems and their specific application systems.

The data registry process defined within ISO 14817 is consistent with implementation(s) of the ISO ITS/TICS system architecture defined in the ISO 14813 standardization deliverables, particularly ISO14813 Parts 2 and 3. This does not preclude the application of the data registry using alternative international, regional, or national system architecture methodologies or techniques; indeed, a common data registry will ease migration and interoperability between such approaches.

The ITS/TICS data concepts that populate the ITS/TICS data registry may originate from a CASE tool implementation of the ISO 14813 TICS reference architecture, from international standards for ITS, from national implementations for ITS, or from the submission by relevant users. Data dictionary entries are not limited to those generated by object oriented methodologies.

The annexes to the international standard describe the specific details for implementing the requirements. Annex A details ITS/TICS functional operating procedures for registration and harmonization of data concepts. Annex B prescribes the detailed definitions and descriptions of the ITS/TICS data registry and data dictionary meta attributes. Annex C prescribes the meta attribute requirements for data concepts contained in the ITS/TICS data registry and ITS/TICS data dictionaries. Annex D specifies the naming and name abbreviation conventions and the process for converting ITS/TICS descriptive names to ASN.1 names. Annex E contains the rules for data representation in an information model, along with examples. Annex F describes the ASN.1 information object specification for an ITS/TICS data concept with examples.

The international standard specifies the framework, formats, and procedures used to define information exchanges within the ITS/TICS sector. It defines the content of the ITS/TICS central data registry and data dictionaries, as well as the registration process to enter data concepts into the data registry. Throughout the text, "data registry" should be taken to mean the ITS/TICS central data registry.

This international standard specifies:

- The framework used to identify and define all information exchanges;
- The framework used to extend standardized information exchanges to support local customizations and combinations;
- The information modeling method for defining ITS/TICS data concepts, when used;
- Meta attributes used to describe, standardize and manage each of the data concepts defined within this framework;
- The requirements used to record these definitions;
- The formal procedures used to register these definitions within the data registry.

The data registry described in the standard supports, and is designed to include, data concepts using alternative international, regional, or national system architecture methodologies or techniques. A common data registry will ease migration and interoperability between such approaches.

5.1.8 ISO TR 25102 TICS System Architecture—ITS Use Case pro forma Template

ISO	TR	25102	ITS Use Case pro forma Template	Approved Published	TR	ISO TC204

ISO 25102 Available from
http://www.iso.org/iso/iso_catalogue/catalogue_tc/catalogue_detail.htm?csnumber=42743

The objective of this technical report is to propose a standard pro forma template for use cases for ITS and to provide guidance on how the template should be used.

While this technical report provides a pro forma template, the elements may be augmented or omitted as applicable. The technical report provides guidance to develop use cases and is a guide rather than a prescription to be followed without variation.

A "use case model" is simply a term to describe, and in many cases define, a user's view of interactions with (and within) the system. Use cases show how entities interact and are usually presented as structured text or diagrammatically.

Use cases are a means to define requirements for a system in terms of the primary users (known as actors) that interact with the system and the scenarios or activities that are performed by the system in response to stimuli from the actors or from other system entities. Each use case has a starting state and conditions, a series of activity steps that together comprise a scenario, and a finishing state and conditions. There may be more than one scenario in a use case. The use case should also include exceptional cases with alternate outcomes.

In many situations, including in some international standards, there has been more attention paid to the definition of "actors" and "use cases" and the relationships between them, than to the detail of each use case, especially the explanatory text that goes with the use case.

The identification of use cases is most frequently associated with use case model diagrams using UML (ISO 19501). In this document, for consistency, we use this methodology throughout. However use cases can be elaborated and developed for any system methodology and are as appropriate for process-oriented methodology as object-oriented methodology, and, indeed, there is no requirement to use any technical architecture methodology at all. A use case can often be elaborated simply with pen and paper.

The benefits of applying use cases to the development of ITS include:

- A common, standardized approach available for the first segment of software system development, namely, requirements elicitation and definition;
- Requirements are related to each other informally, thus providing some assurance of compatibility and consistency.

This technical report discusses the application of use cases for requirements and related aspects of a software intensive system such as ITS.

The scope of this technical report is to provide a pro forma template for the consistent consideration and development of use cases within ITS international standards and associated deliverables.

NOTE: This technical report provides a pro forma template; the elements may be augmented or omitted as applicable. The technical report provides guidance to develop use cases and is a guide rather than a prescription to be followed without variation.

5.1.9 ISO 24098 Procedures for Developing ITS Deployment Plans Utilizing ITS System Architecture

ISO	TR	24098	Procedures for developing ITS deployment plans utilizing ITS system architecture	Approved Published	ISO TC204

ISO 24098 Available from
http://www.iso.org/iso/iso_catalogue/catalogue_tc/catalogue_detail.htm?csnumber=42015

It is well perceived in the ITS community that understanding the system architecture is significant for the realization of ITS, which extends wide areas on service domains, system sizes, and relating stakeholders and users. Those who deploy ITS systems within a specific region need to consider the coverage area as well as existing inventories.

This technical report facilitates the introduction of ITS at regional levels through the development of the ITS deployment plans by utilizing regional ITS architectures.

The scope of this technical report is to describe the procedure for developing ITS deployment plans utilizing ITS system architectures. The document consists of the basic policy of ITS deployment and the procedure for developing ITS deployment plans. Framework, procedure, and requirements for the developing regional ITS deployment plans utilizing regional ITS architecture are reported.

5.1.10 ISO 25100 Intelligent Transport Systems—System Architecture—User Guide for Harmonization of Data Concepts

ISO	TR	25100	User guide for harmonization of data concepts	Approved Published	ISO TC204

ISO 25100 Available from
http://www.iso.org/iso/iso_catalogue/catalogue_tc/catalogue_detail.htm?csnumber=42741

The objective of this technical report is to provide user guidance for the harmonization of data concepts where there are similarities in definitions, including semantics.

Harmonization has been discussed by several groups and there have already emerged some preliminary guidance and principles for the effective harmonization of data concepts for ITS.

It should be clearly recognized that harmonization is not essential for interoperability, which can usually be achieved given sufficient investment of knowledge and resources. Nevertheless this generally leads to duplication and other unnecessary, futile, and even useless work being undertaken. This also assumes that there are unlimited resources available to achieve the desired interoperability, whereas, in practice, time, budget, and a shortage of skilled resources often cause compromise. Additionally interoperability in one aspect is sometime achieved by the lack or loss of interoperability in another. Harmonization is intended to reduce the nugatory work, increase efficiency, and thereby reduce the incidence of errors and faults.

This technical report describes a proposed process for harmonization of data concepts to arrive at preferred definitions for use in formal standards, specifications, technical reports, and information architecture (data) models. The proposal is based on consideration of harmonization processes used by three international groups involved in transport and logistics information and control systems.

Harmonization provides a means to improve efficiency and effectiveness of ITS by helping to remove duplication, inefficiency, ambiguity, and confusion, and thereby improve clarity, comprehension, safety, and efficiency.

The scope of this technical report is harmonization of data concepts that are being managed by data registry and data dictionaries such as those described in ISO 14817:2002

The technical report describes a proposed process for harmonization of such data concepts to arrive at preferred definitions for use in formal standards, specifications, technical reports, and information architecture (data) models. The proposal is based on consideration of harmonization processes used by three international groups involved in the ITS sector and in the wider sector of transport and logistics information and control systems.

5.1.11 ISO 25106 Procedures and Formats for ITS Glossaries

ISO	CD	25106	Procedures and format for ITS Glossaries	Ballot in process	ISO TC204

ISO 25106 Not Yet Available. Track Progress at:
http://www.iso.org/iso/iso_catalogue/catalogue_tc/catalogue_detail.htm?csnumber=42748

A technical report providing an ITS glossary of terms was prepared during the 1990s, and was approved; however, during the publication stage it became clear that, firstly, updating would be a major problem in the fast evolving ITS sector, and secondly, there were potential intellectual property rights (IPR) problems with the use of some terms. It had not occurred to the developers that terms would be the subject of IPR, and so this situation had not been checked during development. Reluctantly, this TR (14812) was abandoned.

In its place, ISO 25106 is being developed. This standard provides a regime for the operation of a data registry set up to operate in accordance with ISO 14817, but it is specifically tailored as a framework for supporting either an online or off-line ITS glossary.

The consistent use of terminology is a major component for interoperability. However, the ITS sector is rapidly evolving and expanding. Rather than publishing a paper based "Glossary of ITS Terms," which will be out of date before it is published, and in these circumstances extremely difficult to maintain, this international standard sets the procedures and format for providing and maintaining one or more instances of a Web-based ITS glossary.

Submission to the glossary can be made by any contributor, but inclusion in the Web glossary of terms shall be determined by the ITS glossary registrar in accordance with the provisions of ITS standards for data registry management and this international standard.

A further advantage of the methodology determined in this international standard is that the submitter takes the responsibility for ensuring that terms submitted are free of intellectual property rights or that the submitter has dispensation to make the terms freely available and useable.

The scope of the draft is to provide a standardized set of protocols, parameters, and a method of management of an updateable "Glossary of ITS Terms." The procedures defined in this international standard determine how to create and an manage an ITS glossary of terms in accordance with the provisions of ISO 14817:2002, Intelligent Transport Systems, Requirements for an ITS/TICS Central Data Registry and ITS/TICS Data Dictionaries.

NOTE: Terms and conditions of access to any such glossary are outside of the scope of this international standard.

5.1.12 ISO 20452 Requirements and a Logical Data Model for PSF and API Used in ITS Database Technologies and Logical Data Organization for a PSF

ISO	TS	20452	Intelligent Transport systems—Requirements and a Logical Data Model for PSF and API used in ITS Database Technologies and Logical Data Organization for a PSF	Published	ISO TC204

ISO 20452 Available from
http://www.iso.org/iso/iso_catalogue/catalogue_tc/catalogue_detail.htm?csnumber=39447

This specification describes the functional requirements and logical data model for physical storage format (PSF) and application programming interface (API), as well as the logical data organization for PSF, which were completed under ISO 14826. It does not specify a physical data organization.

ISO/TS 20452 and ISO 17267 replace the following:

ISO	TS	14826	Physical Storage	Published	ISO TC204

Also refer to Sections 8.2.1.1 and 8.2.4.1.

See Section 8.2.4.1 for information regarding:

- ISO14826/17267-2:2000, Physical Storage Format and Application Program Interface for Navigation Applications - Part 2: Physical Storage Format.
- ISO14826/17267-3:2000, Physical Storage Format and Application Program Interface for Navigation Applications - Part 3: Application Program Interface.

For information concerning eXtended Geographic Data Files (XGDF), see Section 5.5.3.

5.1.13 ISO 15662 Wide Area Communications—Protocol Management Information

ISO	IS	15662	Intelligent Transport Systems – Wide Area Communications – Protocol Management Information	Published Standard	ISO TC204

ISO 15662 Available from
http://www.iso.org/iso/iso_catalogue/catalogue_tc/catalogue_detail.htm?csnumber=44322

Most of the application services in the ITS sector use a variety of wide area communication systems in order to connect user terminals and *service centers*. In addition, the application services that are currently being provided connect specific user terminals to specific service centers using specific wide area communications systems. In other words, the various conditions that must be established to provide services are fixed. However, when the future modes of service use are considered, it is assumed that a user will utilize the same terminal to access Service Center A in some cases and Service Center B in other cases. It can also be assumed that in some cases the user may be on foot and in others he or she may be traveling in a vehicle. It can also be assumed that some users may access the service center from smart phones, while others may do so from navigation systems, while still others may do so using interactive TVs.

When a variety of user terminals utilize a variety of wide area communications systems to connect to a variety of service centers in this manner, the type and content of the conditions that must be established will differ for each individual service usage. In order to provide appropriate service based on these conditions, it is crucial to establish a mechanism by which the type of conditions and the content established for them are transmitted to an appropriate entity and can be interpreted.

This international standard summarizes information as a checklist to consider internal processing in communication systems, terminals, and so forth (protocol management information), suitable for providing ITS application services utilizing wide area communication systems.

The document provides information as a checklist to consider handling messages that are defined by the application working groups of ISO TC204, installing systems and selecting suitable wide area communication systems for providing ITS application services.

The uses of this information are for the framework of message headers, pay load items for initializing communication link, checklists for system design, and so on. Thus, these information items are not necessarily contained in message instances and/or headers that are actually transmitted.

For example, this information is used to organize the characteristics of messages such as those requesting handover based on the relationship between the size of the information service area and the communication range of each communication system (see Figure 5.2).

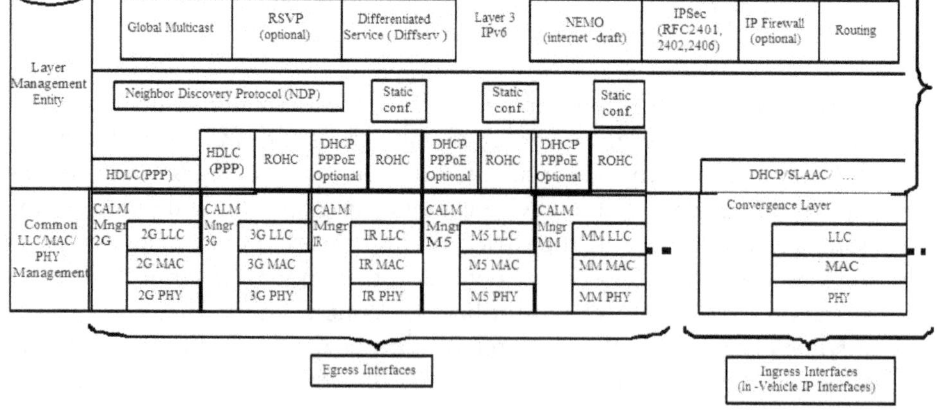

Figure 5.2 Context of ISO 15662. (*Source:* ISO 15662.)

5.1.14 ISO 21707 Data Quality

ISO	NP	21707	Quality of input data for ITS systems	Under Development	ISO TC204

ISO 21707 Not yet available. Track Progress at:
http://www.iso.org/iso/iso_catalogue/catalogue_tc/catalogue_detail.htm?csnumber=34668

This standard, which is under development, will specify the format for defining the quality of quantitative data being exchanged between systems (including users) in a transport information application.

In the past, many users have wished to allocate obtained data to their own system. This is not possible unless the quality of the data is specified. The standard is intended to define the minimum quality requirements for 10 items, including accuracy, reliability, and timeliness, regarding transportation information.

5.2 Technology Specific Standards and Guides

5.2.1 ISO 14813-6 TICS System Architecture—Use of ASN.1 in ITS Standards, Systems, and Services

ISO	TR/ DIS	14813-6	Transport Information and Control Systems- System Architecture- Use of ASN.1 in ITS Standards & Systems	Published TR Revised. In DIS Ballot	ISO TC204

ISO 14813-6 Available from
http://www.iso.org/iso/iso_catalogue/catalogue_tc/catalogue_detail.htm?csnumber=30622

The generic standards for ASN.1 are described in Section 6.2.2.20. This section describes specific standards requirements that determine the use of ASN.1 in ITS standards. ITS service provision is based largely on the transfer, use, and exchange of data, often between different systems. For interoperability within a system many standards define the data used in the system. However, data interoperability is only limited to use within that system, or specific systems that adopt the same definitions. As there are many standards, there tend to be many definitions, mostly developed without awareness of others, and therefore they are inconsistent and not interoperable. Another of the problems with data definition is that it tends to be biased to the technology initially envisaged for its transfer, or to the concepts behind a particular system.

Once the data is to be comprehended outside of the system of origin, complex look-up is required, and even where possible this is often inadequate. SC6 of ISO/

IEC JTC1 having studied this problem has developed Abstract Syntax Notation One and encapsulated it in the ISO 8824 series of standards. The advantage of ASN.1 it is that it is a completely neutral, free-standing notation that is capable of being encoded and transferred by almost any means, while retaining its comprehension and interoperability. It is now widely used. The ISO 8825 series of standards provides standard means to encode and transfer ASN.1 data, although use of ISO 8824 does not require transfer by any of the means defined in ISO 8825. (However, with the current popularity of XML/XSD it is worth noting that one of the parts of 8825 provides a means of ASN.1/XML data encoding.)

ISO 8824 is particularly useful for consistent data definition in data repositories and data dictionaries, and is specified by several data registry standards, including ISO 14817, which is the data registry standard for ISO/TC 204.

In most cases the use of ASN.1 is embedded within an application standard. However, there are some standards where the use of ASN.1 is a principal function of the standard. These are listed later in this chapter.

This international standard is one of a series of documents to provide a form and structure to the reference architectures for ITS. Specifically, this international standard enables conformance with a resolution of ISO/TC 204, a determination for the consistent use and elaboration of abstract syntax notation one (ASN.1) (ISO/IEC 8824-1 through ISO/IEC 8824-4) within ITS international standards.

The previous version of this document was developed following the decision of the ISO TC204 to adopt ASN.1 as its normal syntax notation for data definitions within ITS international standards and to provide instructions and rules to facilitate interoperability and mobility of data. This revised document clarifies its predecessor and updates it by setting the use of ASN.1 in context with the use of other notations within ITS international standards.

To be explicit, the TC204 decision does not *require* that all ITS international standards and systems use ASN.1 as their only means of encoding and transfer. In many cases other methods will be used because of industry practices or efficiency in certain situations. The TC204 decision requires only that: (1) where data is defined within an ITS international standard or data registry, it is elaborated in a consistent form within all ITS international standards as an ASN.1 module to promote interoperability and reuse; and (2) where ASN.1 is the chosen method for encoding, that it is consistently defined according to ISO/IEC 8825.

ASN.1 and its encoding rules provide a means of achieving interoperability of otherwise incompatible data concepts. In order to achieve this, levels of identification are required to precede certain data elements, to enable the comprehension of data messages.

Within the ASN.1 data definition there may be a requirement to use other notations or encoding rules in the transfer of information within a system specified within an ITS international standard.

ISO 14817 defines the format of data dictionaries and data registries and is consistent with this document in requiring the definition of data according to ASN.1.

This document is designed to provide an enabling structure for use in the ITS sector. It provides a formal means to enact the ISO/TC 204 decision by resolution to use ASN.1 for data definitions within ITS international standards. This provides a common message form to enable interoperability and reuse. It provides consistency of use so that where other aspects of ASN.1 (defined within ISO 8824 & ISO 8825), such as transfer rules, are selected to be used, they are used in a common and consistent way in order to maximize interoperability and reuse.

It is important to be clear that this international standard does not require the use of ASN.1 for anything other than providing a common and flexible form of data definition, and this document makes specific provision for the support of use of other extant standardized syntax notations (such as EDIFACT, XML) while maintaining interoperability and reuse by defining these practices within an ASN.1 data definition.

Specific implementation requirements, other than those determined in the syntax notations identified earlier, are beyond the scope of this book.

This document also provides a means where particular ITS sector requirements, or existing international standards, that require particular message forms and procedures that are expressed in other notations (e.g., EDIFACT, XML) may be referenced and reused by other ITS applications. Thus, it presents an unambiguous system for identifying all the different data types and describing them in ITS international standards in a common way.

5.2.2 Web Services in ITS Standards, Systems, and Services

"Web services" is a generic name given to Web applications that interact with other Web applications for the purpose of exchanging data. Initially used for the exchange of data on large private enterprise networks, Web services are evolving to include transactions over the public Internet.

A Web service is a service that is provided automatically between computers to provide a generic service to a third party that is linked to one of the computers provided by one or more service providers. Because it is generic and not custom, the same routine can be used by many service providers and clients.

Web services require a standardized way to integrate Web-based applications, and this is achieved using a combination of extended mark up language (XML), Simple Object Access Protocol (SOAP), Web Services Description Language (WSDL) and Universal Description, Discovery and Integration (UDDI), which is a Web-based distributed directory that enables businesses to list themselves on the Internet and discover each other over the Internet. XML is used to tag the data, SOAP is used to transfer the data, WSDL is used for describing the services available, and UDDI is used for listing what services are available. Web services allow organizations to communicate data without intimate knowledge of each other's IT/ICT systems behind the firewall.

Unlike traditional client/server models, Web services do not provide the user with a graphical user interface (GUI). Web services work by sharing business logic, data, and processes through a programmatic interface across the network. This enables the applications to interface (not the users to interface). Developers can then add the Web service to a GUI (such as a Web page or an executable program) to offer specific functionality to users.

Web services allow different applications from different sources to communicate with each other without time-consuming custom coding, and because all communication is in XML, Web services are not tied to any one operating system or programming language. For example, Java can talk with Perl, and Windows applications can talk with UNIX applications.

Web services do not require the use of browsers or HTML.

In addition to the use of standards, Web services are characterized by the presence of the following actors:

- A service requester (client);
- A service provider;
- A registry containing the Web service.

Web services also enact some common processes between the actors to provide and validate the service:

- Service providers make the services available (this is called *publishing*).
- The client requesting the service has to first "discover" that the service is available.
- There needs to be a connection made between the service provider and the client (this is called *binding*) using standard protocols.

In an ITS environment, where services are being provided remotely, and their information is being exchanged by parties that do not know each other, Web services are expected to play an increasingly important role.

Simple ITS examples may be: identify location, transmit speed and direction, send message board information, and give this vehicle priority. These generic services and many other examples are components of many services.

See also ISO 25437, Section 6.2.2.71, 6.2.2.72, Sections 6.2.17 (SOAP), 6.2.18 (WSDL), 6.2.19 (UDDI), and 6.2.20 (XML).

5.2.2.1 ISO 24097 Using Web Services (Machine-Machine Delivery) for ITS Service Delivery

ISO/	FDIS 24097	Using Web Services (machine-machine delivery) for ITS service delivery	Under Development	ISO TC204

ISO 24097 Not Yet Available. Track Progress at:
http://www.iso.org/iso/iso_catalogue/catalogue_tc/catalogue_detail.htm?csnumber=42014

NOTE: At the time of writing this deliverable is being revised to WSDL 2.0.

ITS services have been evolving from single functional and limited area services to services in which many systems cooperate to provide effectiveness and efficient services across a wide area. In today's world, ITS services are required to communicate not just with other parts of the same ITS service provision, but between different ITS services and even with non-ITS services or a user's system directly.

(Some examples of these systems are communication between traffic management systems, route guidance systems, homeland security systems, environment protection systems, and private freight management systems).

These systems (even those limited to ITS services) are usually deployed in a heterogeneous circumstance, and they use different hardware, different operating systems (OS), middleware, or development languages. This, therefore, creates a challenge in order to realize system coordination across the organizations in a way that is flexible, quick, and at reasonable cost. Web services are a recent methodology that overcomes these difficulties. Using Web services technology for ITS services can significantly simplify and reduce the cost of Internet-based service provision, which may well affect the level and speed of take-up of use of ITS services.

The World Wide Web Consortium (W3C) defines Web services as follows:

> A web service is a software system designed to support interoperable machine-to-machine interaction over a network. It has an interface described in a machine-processable format (specifically WSDL {web services Description Language}). Other systems interact with the web service in a manner prescribed by its description using SOAP-messages, typically conveyed using HTTP with an XML serialization in conjunction with other web-related standards.

Web services require quite a lot of functionalities, and as a result, architecture is indispensable. Web services standardization organizations construct standards by service-oriented architecture (SOA). SOA is evolutional form of distributed computing and object orientation.

Applying SOA-based standards to the ITS services, the following effects can be expected.

From a business viewpoint:

- Increased service value;
- Internationalization;
- Expansion of the business automation.

From a systems developing viewpoint:

- Easy and quick development of ITS service coordination and service area expansion;
- Web services enables system developer to focus not on "how" but rather "what." "How" is covered by standard base tools. This enables quick and easy system software development;
- Web services standards have a composable structure and so promote reusability of software;
- Easy connection to a legacy system.

In the ITS sector, message standardization of many applications has already been completed, is well advanced, or is determined regionally. Message standardization is intended to improve system coordination, interoperability, and reuse, and so the conditions for Web services are considered already mature. In addition, the use of Web services will increase the flexibility of ITS services to interoperate and communicate beyond the ITS sector and in areas where the delineation between ITS services and general commercial services converge.

This technical specification has therefore been developed. The rationale for this deliverable to be a technical specification (as opposed to a full international standard) is because some of the upper parts of the SOA stack are still being developed. Further iterations of this specification, or of subsequent specifications, may be expected in the future as these developments evolve. However, in order to assist with the introduction and use of Web services in a consistent manner, guidance has been requested without waiting until this lengthy process has been completed. This technical specification therefore provides specification guidelines for core Web services (WSDL, SOAP, and UDDI).

The scope of this technical specification is to provide specification guidelines for core Web services (WSDL, SOAP, and UDDI) developed and used within the ITS sector, in order to support ITS services provided in an Internet environment.

Figure 5.3, which is provided within the scope of the standard, shows a high level overview architecture for the provision of Web services.

A second figure (Figure 5.4) provided within the scope of the standard shows the evolution of and how current WSDL and SOAP versions relate, and also identifies the correct combination of each version. In the figure, only (WSDL 1.1, SOAP 1.1), (WSDL 2.0, SOAP 1.2), and (WSDL 2.0, SOAP 1.1) provide an acceptable combination. SOAP 1.2 is not backwards-compatible to SOAP 1.1, and also WSDL 2.0 is not backwards-compatible to WSDL 1.1. Therefore, WSDL 1.1 is the candidate recommendation for this technical specification. In this technical specification, WSDL is therefore taken to imply WSDL 1.1 and SOAP is taken to imply SOAP 1.1.

Even if WSDL 2.0 is significantly superior to WSDL 1.1, and WSDL1.1 users will migrate to WSDL 2.0, the demise of WSDL 1.1 will take considerable time (see

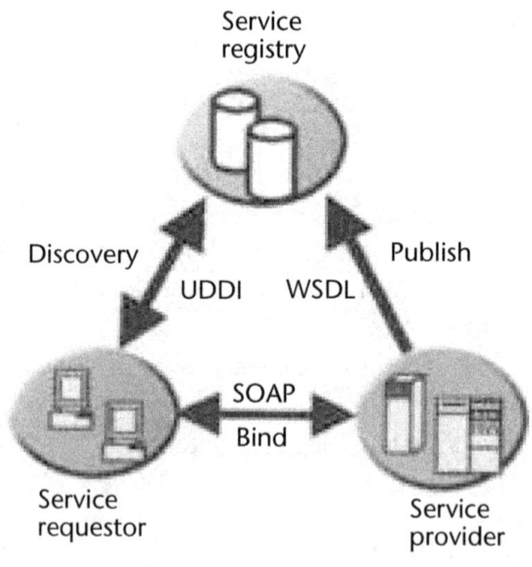

Figure 5.3 High level overview architecture for the provision of Web services. (*Source:* ISO 24097 Scope.)

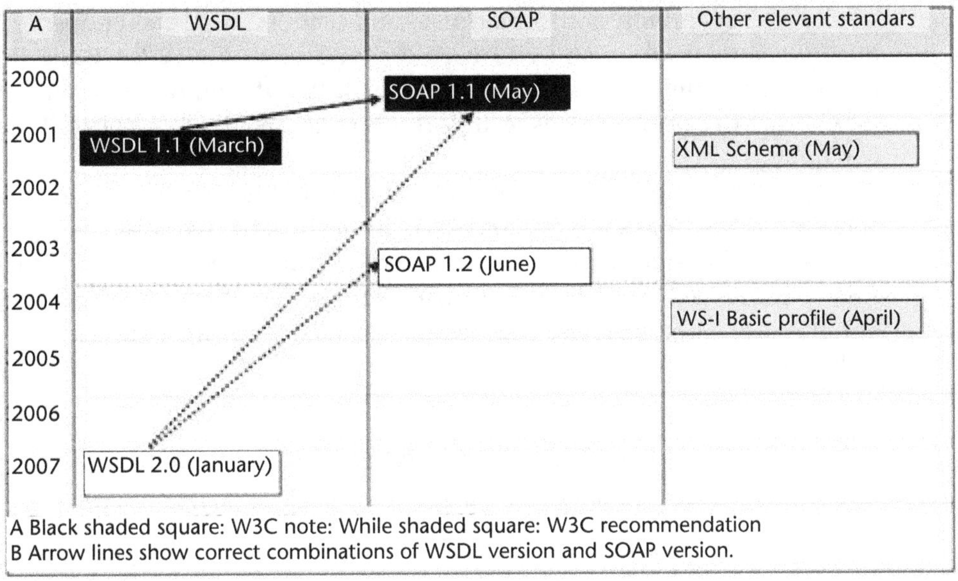

Figure 5.4 WSDL and SOAP correct version combination. (*Source:* ISO 24097 Scope.)

Figure 5.5), especially because Web services are usually effected between different organizations.

This technical specification is limited to two service combinations:

- ITS sector service provider, ITS sector service consumer;
- ITS sector service provider, non-ITS sector service consumer.

In the second case, it may be expected that the non-ITS sector service consumer will expect ITS services to be offered in a manner that conforms to the WS-I Basic Profile 1.1 (referred to as WS-I Basic Profile, called BP in this document).

It is to be noted that core Web service standards (SOAP, WSDL, and UDDI) still contain some ambiguities and conflicts among core standards. One reason for this is because SOAP and WSDL were developed, as W3C notes, before the recommendation to use XML schema (see Figure 5.5), and this has subsequently become a base component of Web services standards. In order to promote Web

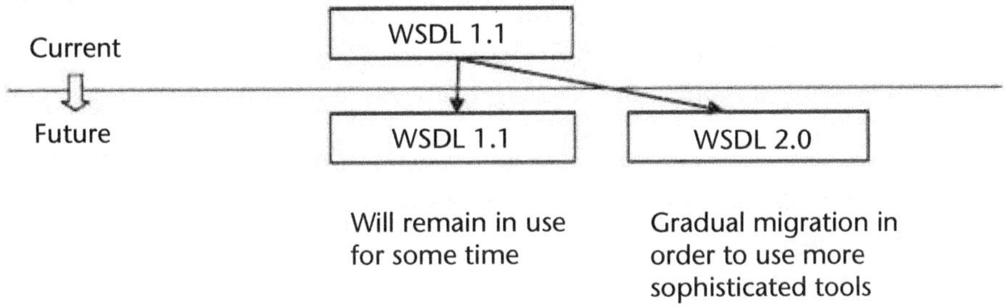

Figure 5.5 WSDL upgrade scenario. (*Source*: ISO 24097 Scope.)

service interoperability across platforms and vendors, WS-I was cooperatively developed by both vendors and end user organizations. The WS-I Basic Profile (BP) provides an outline of requirements to which WSDL and the Web service protocol (SOAP) are to comply. In this technical specification, WS-I Basic Profile compliance is required.

Figure 4 of the scope of the document, reproduced as Figure 5.6, depicts the concept of WS-I Basic Profile and its relationships to this technical specification.

NOTE: Considering the wide range of user applicability of ITS Web services, ASN.1-based Web services are not within the scope of this technical specification, but this does not prohibit the use of ASN.1-based Web services between ITS stakeholders.

5.2.2.2 NIST Guide to Secure Web Services Special Publication 800-95

NIST	800-95	Guide to Secure Web Services Special Publication 800-95	Published 2006	NIST

NIST 800-95 Available from
www.csrc.nist.gov/publications/drafts/Draft-SP800-95.pdf

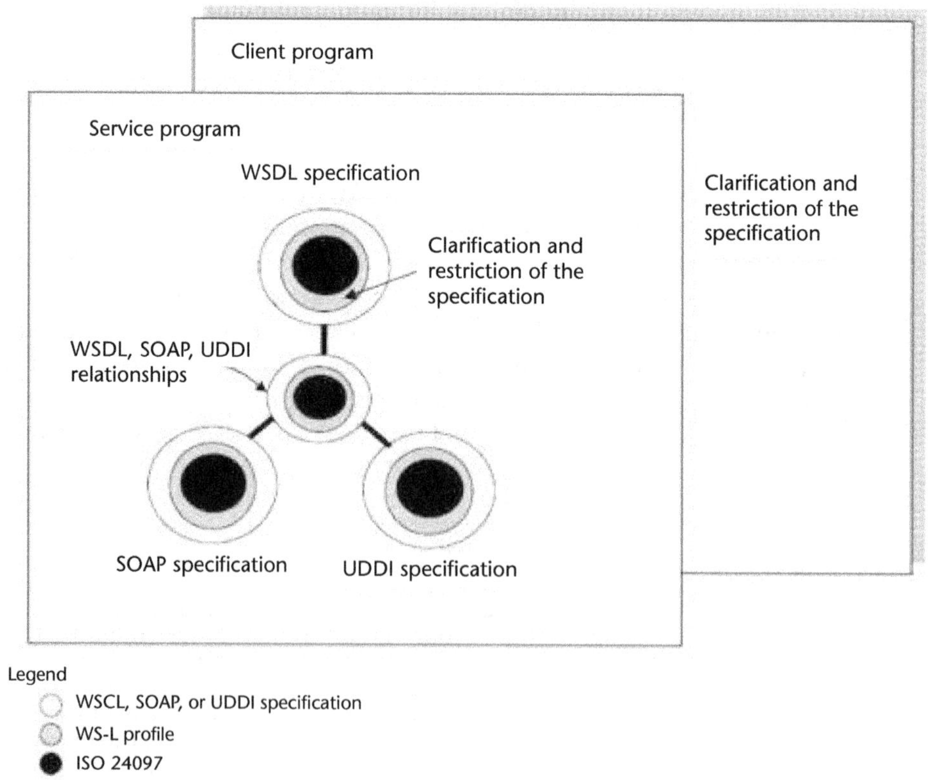

Figure 5.6 ISO TS 24097 Conceptual Framework. (*Source:* ISO 24097 Scope.)

This document is not a standard, but rather a NIST equivalent of a technical report. Because of its relevance in the field of provision of Web services, we have included reference to it in this book.[2]

Some background information about NIST and this deliverable is therefore useful at this stage.

The Information Technology Laboratory (ITL) at the National Institute of Standards and Technology (NIST) promotes the U.S. economy and public welfare by providing technical leadership for the nation's measurement and standards infrastructure. ITL develops tests, test methods, reference data, proof of concept implementations, and technical analysis to advance the development and productive use of information technology. ITL's responsibilities include the development of technical, physical, administrative, and management standards and guidelines for the cost-effective security and privacy of sensitive unclassified information in federal computer systems. This Special Publication 800-series reports on ITL's research, guidance, and outreach efforts in computer security and its collaborative activities with industry, government, and academic organizations.

The advance of Web services technologies promises to have far-reaching effects on the Internet and enterprise networks. Web services based on XML, SOAP, and related open standards, and deployed in SOA allow data and applications to interact without human intervention through dynamic and ad hoc connections. Web services technology can be implemented in a wide variety of architectures, can coexist with other technologies and software design approaches, and can be adopted in an evolutionary manner without requiring major transformations to legacy applications and databases.

The security challenges presented by the Web services approach are formidable and unavoidable. Many of the features that make Web services attractive, including greater accessibility of data, dynamic application-to-application connections, and relative autonomy (lack of human intervention) are at odds with traditional security models and controls. Difficult issues and unsolved problems exist, such as the following:

- Confidentiality and integrity of data that is transmitted via Web services protocols in service-to-service transactions, including data that traverses intermediary (pass-through) services;
- Functional integrity of the Web services that requires both establishment in advance of the trustworthiness of service in orchestrations or choreographies, and the establishment of trust between services on a transaction-by-transaction basis;
- Availability in the face of denial-of-service attacks that exploit vulnerabilities unique to Web service technologies, especially targeting core services, such as discovery service, on which other services rely.

Perimeter-based network security technologies (e.g., firewalls, intrusion detection) are inadequate to protect SOAs for the following reasons:

2. Singhal, A., and T. Winograd, "Guide to Secure Web Services Special Publication 800-95," Computer Security Division, Information Technology Laboratory, National Institute of Standards and Technology, Gaithersburg, MD, 2006.

- SOAs are dynamic and can seldom be fully constrained to the physical boundaries of a single network.
- SOAP is transmitted over HTTP, which is allowed to flow without restriction through most firewalls. Moreover, TLS, which is used to authenticate and encrypt Web-based messages, is unsuitable for protecting SOAP messages because it is designed to operate between two endpoints.
- TLS cannot accommodate Web services' inherent ability to forward messages to multiple other Web services simultaneously.

The SOA processing model requires the ability to secure SOAP messages and XML documents as they are forwarded along potentially long and complex chains of consumer, provider, and intermediary services. The nature of Web services processing makes those services subject to unique attacks, as well as variations on familiar attacks targeting Web servers.

Ensuring the security of Web services involves implementation of new security frameworks based on use of authentication, authorization, confidentiality, and integrity mechanisms. This document describes how to implement those security mechanisms in Web services. It also discusses how to make Web services and portal applications robust against the attacks to which they are subject.

5.2.2.3 SAE Converting ATIS Message Standards from ASN.1 to XML

SAE	J2630_ 200312	Converting Atis Message Standards From Asn.1 to Xml	Published Dec. 2003	SAE

SAE J2630_200312 Available from
http://www.sae.org/technical/standards/J2630_200312

This SAE standard presents a set of rules for transforming an ASN.1 message set definition into an XML schema. The result is intended to be a stand-alone XML schema that is fully consistent with an existing ASN.1 information model. This is a different goal from other related work by other standards bodies developing a set of XML encoding rules for ASN.1 or ASN.1 encodings for XML schema. These rules were initially developed in order to produce an XML schema for the SAE ATIS standard. While other standards may also choose to use these rules, the rules may not be applicable for all environments. The goal for these transformation rules is twofold.

The first goal is to provide a uniform set of such rules that all interested parties can use. The second goal is to use such rules to define an adopted schema for traveler information that reflects the preferred translation of ASN.1 message sets to XML for use by ITS system implementers. The first goal is met by this document.

The second goal is met by employing this document to produce XML information as part of the periodic reballoting of the SAE ATIS standard. This is a parallel standards effort with this document. These rules were developed as part of the process to draft an XML version of the ATIS data element and message set standards.

The original effort focused upon the needs found in the currently adopted October 2000 SAE J2354 standard for ATIS message sets. SAE J2354 makes extensive use of elements from the ITETMDD work and from ITE-TCIP work. In addition, the draft Event Report Message (ERM) portion of the Message Sets for External Traffic Management Center Communication (MS/ETMCC) standard was also examined and translated. Message and data elements from the IEEE Incident Management standard (IEEE 1512-2000) were also examined. By this effort, every major message set of ITS was considered to some degree to ensure that the resulting translations could be successfully employed by others toward a common result. It is intended that the resulting ATIS schema will be voted on as an SAE ATIS standard.

Other standards-developing organizations may also choose to use these rules; however, it is recognized that translations may not be necessary in some environments and that these rules may not be applicable to all environments within ITS. It is left to each standards-development organization to make such determinations for their specific environments.

Section 4 of the document is the complete set of conversion rules. An example of use for each rule is given, showing an original ASN.1 definition, the resulting XML schema definition, and a sample XML document element where applicable. Section 5 of this document discusses the background of some nonobvious conversion rules presented in Section 4. These include methods of handling the XML representation of enumerations, octet strings, and bit strings.

Section 5 also describes how XML namespaces can be used to point to types defined in the Traffic Management Data Dictionary (TMDD), the Transit Communications Interface Profiles (TCIP) standard, the Location Reference Message Set (LRMS), or the International Traveler Information Systems (ITIS) phrase lists. It also discusses additional translation refinements proposed for the conversion of the ATIS message standard. Some omissions in the SAE standards required new types to be defined in order to have a valid XML schema. In addition, some new types must be defined to implement the representation of octet strings and bit strings. These changes are documented in Section 6.

In the currently adopted revision of the standard the use of in-line definitions rather than the more proper use of formal type definitions is also a challenge for conversion. Section 6 also documents a number of common types that were defined globally at the start of the schema rather than being redefined identically, in-line multiple times in the schema.

Section 7 presents a sizable sample ASN.1 definition and the corresponding XML schema. The sample illustrates a large number of the conversion rules.

5.2.2.4 24824-3 Information Technology—Generic Applications of ASN.1: Fast Infoset Security

ISO/IEC	FCD	24824-3	Information technology—Generic applications of ASN.1: Fast infoset security	Under Development FCD	JTC1/SC6

See Section 6.2.2.71.

5.2.3 ISO 17452 Using UML (Unified Language) for Defining and Documenting ITS Interfaces

ISO	TR	17452	Using UML (Unified Language) for defining and documenting ITS Interfaces	In Ballot	ISO TC204

ISO 17452 Available from
http://www.iso.org/iso/iso_catalogue/catalogue_tc/catalogue_detail.htm?csnumber=38373

ISO 14817 specifies the formats and procedures used to define information exchanges within the ITS/TICS sector. Such information arises through the development of the architecture for a particular application standard and the subsequent specification of the detailed data concept instances that arise in association with the architecture's interfaces. This document illustrates the steps involved in such development.

In the development of standards, it is often the case that working groups have a well-formed perception of the conceptual context within which their standard is to be applied. This is the case because many standards are the result of a refinement and consensus of requirements based on recent practice. The formal process for the identification of the requirements is streamlined to capitalize on this available body of knowledge.

For completeness it begins with the capture of requirements. These requirements need be only those which directly affect the standard. The context of a real-world system that incorporates the standard would include a much wider range of requirements; however, we are focusing on that aspect of standards which produces data elements and other concept instances which will be registered in a data dictionary/registry. The methodology is derived from processes used in the development of software intensive systems.

This technical report gives guidelines for using UML for defining and documenting ITS/TICS interfaces. It presents these guidelines in the context of a case study for the creation of an ITS/TICS data dictionary and submissions to the ITS/TICS data registry.

In UML, an interface is a collection of operations that are used to specify a service of a class or component. ITS/TICS data registry, as defined in ISO 14817, builds on this definition by mapping an operation to a message, and then it extends the definition of an interface to be a dialog (i.e., a collection of messages with in an implied protocol). This technical report conforms to these steps.

5.2.4 ISO TR 24529 Using UML in ITS Standards

ISO	TR	24529	Using UML in ITS Standards	Approved Published	ISO TC204

ISO 24529 Available from:
http://www.iso.org/iso/iso_catalogue/catalogue_tc/catalogue_detail.htm?csnumber=39591

The objective of this technical report is to provide guidance on the use of UML in the development of standards for ITS.

The advantages of applying UML to the development of ITS include the following:

- UML provides an internationally standardized form of system model that should be readily interpreted anywhere worldwide.
- UML enables cohesive description from multiple user views.
- There is available extensive training and tool support for UML.
- UML is capable of manipulation by a metadata registry for ITS.
- UML tools enable conversion directly to computer coding.
- UML is very widely used in the architecture, design, and development of software-intensive systems.

The disadvantages of using UML include the following:

- UML is not understood by many stakeholders who are not also software developers.
- UML uses a larger amount of unfamiliar language and jargon which, while it may be necessary for precision, is daunting and off-putting to the nonspecialist and lay reader.
- UML is not yet developed enough to support the full scope of systems engineering.
- UML is still under active development and therefore the compatibility of UML models may be an issue.

There are therefore some risks in using UML, but nevertheless the benefits are widely judged as exceeding the disadvantages. This document is intended to provide guidance to stakeholders who are considering the use of UML for ITS.

The scope of this technical report is the use of UML within international standards technical specifications and technical reports and related documents.

This technical report discusses the application of UML to the development of standards within the context of ITS.

5.2.5 ISO 24531 Using XML in ITS Standards, Data Registries, and Data Dictionaries

ISO	DIS	24531	Intelligent Transport Systems—System Architecture, taxonomy and terminology—Using XML in ITS standards, data registries and data dictionaries	Under Development	ISO TC204

ISO 24531 Available from
http://www.iso.org/iso/iso_catalogue/catalogue_tc/catalogue_detail.htm?csnumber=37503

As the exchange of information via the Internet and other wired and wire-free networks develops and expands, the use of XML and its variants continue to grow and develop. XML will be an important tool in the development and operation of ITS services.

However, within XML and its variants there are options. In order to obtain maximum benefit, interoperability and reuse of data, within the ITS sector, it is important to implement XML and its variants in a consistent manner.

This international standard provides definition of how to use XML and its variants in a consistent and interoperable manner within the ITS sector.

The international standard has been developed to assist ITS standards developers and users of ITS standards who wish to use XML, by providing a consistent definition of the rules and rule references for the use of XML within ITS systems. The scope of the international standard is to define consistent rules and rule references to provide a framework to be used when implementing XML based applications in ITS, and particularly, in specifying XML in ITS standards, ITS data registries, and ITS data dictionaries. This international standard also provides guidance and examples with respect to the use of XML in ITS and the elaboration of XML within the ASN.1 data definitions required by ISO 14813-6 (Section 5.2.1).

This document defines:

- Rules concerning the creation of XML schemas for ensuring interoperability in various types of ITS applications that use XML. –Normative– (clause 7);
- Rules for using XML for the purpose of reusing XML schemas. –Normative– (clause 7);
- Rules concerning registration and management of XML components in data dictionaries and data registries. –Normative– (clause 8);
- Examples of the use of XML in ITS applications. –Informative– (Annex A);
- Representation of international resource identifiers (IRIs) and/or ID-related constructs of this standard. –Informative– (Annex B);
- Schema header template. –Informative– (Annex C);
- Example of registering XML constructs. –Informative– (Annex D);
- Example of automatic generation of an XML schema from UML. –Informative– (Annex E);
- Applying ASN.1 encoding for XML document. –Informative– (Annex F);
- ASN.1 transformation to XML schema example. –Informative– (Annex G).

5.2.6 ISO 24532 Common Object Request Broker Architecture (CORBA)

ISO	TR	24532	Intelligent transport systems—Systems Architecture, taxonomy and terminology—Using CORBA (Common Object Request Broker Architecture) in ITS Standards, data registries and data dictionaries	Published TR	ISO TC204

ISO 24532 Available from
http://www.iso.org/iso/iso_catalogue/catalogue_tc/catalogue_detail.htm?csnumber=37504

CORBA is one of many software technologies involved in distributed systems and system integration, There are a significant number of existing CORBA deployments in ITS, and discussions on best practice and standardization have naturally emerged. Discussion can often lead to comparisons between different technologies, confusion, and even apparent competition between different software technologies.

The objective of this technical report is to identify the role of CORBA in ITS and to provide guidelines for the use of CORBA in ITS.

This technical report clarifies the purpose of CORBA and its role in ITS; it provides some broad guidance on usage, and it prepares the way for further ISO deliverables on the use of CORBA in ITS.

5.2.7 ISO TR 26999 Rules and Guidance for the Use of Process (Functional) Oriented Methodology in ITS Standards, Data Registries, and Data Dictionaries

ISO	NP	26999	Rules and guidance for the use of Process (Functional) Oriented Methodology in ITS Standards, Data Registries and Data Dictionaries	TR Under Development	ISO TC204

ISO 26999 Not yet available. Track Progress at :
http://www.iso.org/iso/iso_catalogue/catalogue_tc/catalogue_detail.htm?csnumber=46068

At the time of writing, there are no deliverables or working drafts. The first working draft is expected at the end of 2007.

The justification for this standard is that the process (functional) oriented methodology is being used by many in the ITS/TICS community, particularly for high-level ITS architectures at the national and project level.

Having studied and discussed the subject within the WG and within the ITS community, it is apparent that the application of this methodology would benefit from the availability of rules and guidance. Without a technical report, technical specification, or IS there is a significant risk that there will be diversity of method of implementations that will restrict interoperability.

The specific aim of this TR/TS/IS will be to define the rules and rule references that should be used when implementing the process (functional) oriented methodology in ITS. Particularly, for specifying functionality and data elements in ITS standards, data registries, and data dictionaries and to provide guidance and examples in respect of the use of the process (functional) oriented methodology in ITS, and the elaboration of ITS/TICS interfaces.

The main interests that will benefit from this work are the developers of ITS standards and those producing system architectures that are above the design level. Users and students will also benefit from the clarifications, explanations, and examples provided in the TR/TS/IS.

- *Feasibility:* Process (functional) oriented methodology is already developed, documented, and in widespread use. There is no aspect of novelty or risk.

- *Timeliness:* This work arises because of the widespread and growing use of process (functional) oriented methodology for high-level ITS architectures across Europe and many other parts of the world.
- *Benefits:* These include improved ease of use of the process (functional) oriented methodology in a consistent manner, improved understanding of these issues, and interrelationships by developers of ITS standards, leading to faster development of ITS standards.

5.2.8 IEEE 1455 IEEE Standard for Message Sets for Vehicle/Roadside Communications

IEEE	1455	IEEE Standard for Message Sets for Vehicle/Roadside Communications	Published 1999	IEEE

IEEE 1455 Available from www.ieee.org

IEEE 1455-1999 is applicable to dedicated short-range communications (DSRC). Within the overall context of DSRC operations, this standard specifies the message set, data dictionary, and communications protocols above the OSI data link layer for the DSRC wireless interface. These communications protocols define the low-level commands used to control transponder resources and thereby enable message transfer. This standard also specifies the resources that may be present on a vehicle's transponder and the means by which the roadside equipment can control those on-board equipment resources.

This standard, in conjunction with related standards governing the data link (ASTM PS105-99) and physical (ASTM E2158-01) layers of the DSRC protocol stack, provides the basis for interoperable, noninterfering DSRC implementations using equipment from multiple vendors. These implementations include commercial vehicle, toll collection, and border crossing applications.

The standard is intended for equipment manufacturers, system integrators, toll and turnpike agency engineers and procurement specialists, research consultants, and other interested individuals and groups.

IEEE 1455-1999 may be used in several ways by various DSRC equipment stakeholders, including manufacturers that develop compliant transponders. In addition, it can be used by roadside equipment manufacturers to develop equipment that communicates with such transponders. Transponder and beacon manufacturers should note the internal resources that must be provided within, and commands that are recognized by, compliant transponders. Manufacturers of roadside equipment utilize protocols found in this standard to complete the communications stack between the data link layer and application-level processing. Finally, ITS application developers (such as engineers for toll or turnpike agencies) who design and implement systems that use DSRC can utilize the messages that transfer information between related applications.

The standard specifies:

- Overall DSRC architecture;
- Scope and organization of the resources provided on compliant transponders;
- Commands that are used to control transponders and memory areas used to store information;
- Roadside resource manager that arbitrates requests for transponder usage;
- ITS application-specific messages that may be stored in the transponder memory areas;
- Services that are used to connect the upper layer processing with the DSRC data link layer.

This standard also indicates the manner in which new systems achieve backward compatibility or noninterference with existing systems.

5.2.9 IEEE P1556 Standard for Security and Privacy of Vehicle/Roadside Communication Including Smart Card Communications

IEEE	P1556	Standard for Security and Privacy of Vehicle/Roadside Communication Including Smart Card Communications	IEEE

IEEE P1556 Available from www.ieee.org

This standard defines the logical and physical interfaces, as well as the performance attributes for the interface between the rail subsystem and the highway subsystem at a highway rail intersection.

The standard describes a resource manager that arbitrates requests for transponder usage.

5.2.10 EN 12896 Transmodel

CEN	EN	12896	Reference Data Model For Public Transport	CEN	TC278

Text Downloadable FOC Transmodel Documentation www.transmodel.org

Transmodel is the European Reference Data Model for Public Transport; it provides an abstract model of common public transport concepts and structures that can be used to build many different kinds of public transport information systems, including for timetabling, fares, operational management, and real-time data. It is described here for ease of comparison with NTCIP (7.7 above)

ENV 12896 was prepared by the work area Transmodel of the EuroBus project (1992–1994) and by the DRIVE II task force Harpist (1995). The EuroBus/Transmodel and Harpist kernel team was considered as a subgroup of Working Group 3 of CEN TC278. The results of these projects were based upon earlier results reached within the Drive I Cassiope project and the ÖPNV data model for public transport, a German national standard. They have been approved by Topic Group 6 (TG6, entitled "Public Transport Data Models") of the EC Drive II program and by two validation seminars open to participants in various Drive II projects, to European public transport operators and to other public transport experts. The standard reflected the contents of deliverable C1 of the HARPIST task force, published in May 1995, with modifications resulting from the discussion process in CEN TC278/WG3 between May and October 1995. It was first published as a pre-standard (ENV) and later ratified as a European standard (EN).

The standard is composed of two parts:

- Normative part (main document and the normative appendix);
- Informative appendix.

The main document presents the history (1. Foreword) and the rationale (2. Introduction) of the proposed standard:

- The executive summary of the reference data model (3. Scope);
- the definitions of the terms as they are used in this document (5. Definitions);
- The technical requirements in form of detailed textual descriptions and diagrams.

The normative appendix contains:

- The definitions of the concepts (entity definitions);
- The main properties of the concepts (main attributes, identifiers, super-types).

The informative appendix is composed of the following parts:

- Consistency and integrity conditions (Appendix B1);
- Introduction to data modeling and the methodology used for the reference data model (Appendix B2);
- Functional model (Appendix B3);
- Modifications to ENV 12896 (Appendix C);
- Synthetic view of the reference data model for public transport (Appendix D).

5.2.11 National Transportation Communications for ITS Protocol (NTCIP)

The NTCIP is a (U.S.) joint standardization project of AASHTO, ITE, and NEMA, with funding from the FHWA. It is therefore a U.S. program. However, the organizers have aspirations to wider, perhaps global implementation. This may cause problems and will probably not succeed, in that the European "Transmodel" provides a different and probably more sophisticated option. However, it is relevant to provide a short summary of its provisions. At the time of this writing, NTCIP has support in the United States and Japan.

From 1996 to 1999, NTCIP was referenced as the NTCIP Guide. However, to provide an organized numbering scheme for the NTCIP documents, this document is now referenced as NTCIP 9001.

Downloadable FOC The NTCIP guide
http://www.ntcip.org/library/documents/pdf/9001v0302b.pdf

Further documents links table can be found at http://www.ntcip.org/library/documents/.

For more detail go to http://www.ntcip.org/.

5.3 Transport Systems Security

Security issues have three facets in ITS. First, there is the physical security of the traveler and vehicles. These issues tend to be dealt with on an application service by application service basis. The reader is referred to Chapter 8 to find such standards. Although it should be stated that, perhaps with the exception of security aspects dealt with in ISO CD 24533 (see Section 7.3.4.1), the development of such standards is largely still in its infancy, as in most cases where communications are concerned, the sector is still largely focused on the basic framework standards that will enable service provision, rather than the services themselves.

The second security aspect concerns security of and secure access to data. At the time of writing this book, this is an area of significant interest and activity in the ITS sector, but coverage is patchy and needs further work. There are provisions in the electronic fee collection sector. The electronic fee collection standards are to be found in Section 8.8. With respect to probe vehicle data, see Section 8.2.8; and with respect to the logistical aspects of national security, ISO CD 24533 (see Section 7.3.4.1) also address some of these issues. There is an interesting dilemma of being able to electronically identify a vehicle for the provision of ITS services, yet at the same time preventing a terrorist identifying when a target vehicle reaches a particular spot where he may detonate an explosive device. However, we may expect many standards to be developed to protect services and control security of data once the application services become better defined.

The third security aspect is the provision of data protection security for users of ITS systems. Here there is interest from two opposing positions: privacy with respect to the individual, and protection of the majority of users through what is called "lawful intercept," the interpretation of which depends on where you are coming from. Lawful intercept as an anti-terrorist measure has widespread support; however, there are others who believe that lawful intercept should include all access by legitimate authorities where laws are, or are believed to be, being broken. As with all human rights versus society rights issues, we may imagine that there will be many interesting and lively debates over the coming years. So far, the only deliverable to emerge is ISO CD 24100, "Basic Principles for Personal Data Protection in Probe Vehicle Information Services" (see Section 8.2.8.2).

Fortunately, issues of generic data security have been addressed over the years and continue to be addressed, and these standards will be of significant use to developers of ITS systems. Because of the potential significance of Web services in ITS service provision, readers are also referred to a deliverable from NIST - Guide to Secure Web Services Special Publication 800-95, which is summarized in Section 5.2.2.2.

5.3.1 ISO 15408 Evaluation Criteria for IT Security

ISO/IEC	IS	15408	Evaluation criteria for IT Security	Under Development	ISO TC204
ISO/IEC	IS	15408-1	Information technology—Security techniques—Evaluation criteria for IT security—Part 1: Introduction and general model	Published	ISO/IEC JTC1

ISO 15408-1 Available from
http://www.iso.org/iso/iso_catalogue/catalogue_tc/catalogue_detail.htm?csnumber=40612

NOTE: The electronic version of this international standard can be downloaded from the ISO/IEC Information Technology Task Force (ITTF) Web site.

http://isotc.iso.org/livelink/livelink/fetch/2000/2489/Ittf_Home/PubliclyAvailableStandards.htm

ISO/IEC 15408-1:2005 defines two forms for expressing IT security functional and assurance requirements. The protection profile (PP) construct allows the creation of generalized reusable sets of these security requirements. The PP can be used by prospective consumers for specification and identification of products with IT security features which will meet their needs. The security target (ST) expresses the security requirements and specifies the security functions for a particular product or system to be evaluated, which is called the target of evaluation (TOE). The ST is used by evaluators as the basis for evaluations conducted in accordance with ISO/IEC 15408.

ISO/IEC	IS	15408-2	Information technology — Security techniques — Evaluation criteria for IT security — Part 2: Security functional requirements	Published	ISO/IEC JTC1

ISO 15408-2 Available from
http://www.iso.org/iso/iso_catalogue/catalogue_tc/catalogue_detail.htm?csnumber=40613

NOTE: The electronic version of this international standard can be downloaded from the ISO/IEC Information Technology Task Force (ITTF) Web site.

http://isotc.iso.org/livelink/livelink/fetch/2000/2489/Ittf_Home/PubliclyAvailableStandards.htm

ISO/IEC 15408-2:2005 defines the required structure and content of security functional components for the purpose of security evaluation. It includes a catalog of functional components that will meet the common security functionality requirements of many IT/ICT products and systems.

ISO/IEC	IS	15408-3	Information technology — Security techniques — Evaluation criteria for IT security — Part 2: Security functional requirements	Published	ISO/IEC JTC1

ISO 15408-3 Available from
http://www.iso.org/iso/iso_catalogue/catalogue_tc/catalogue_detail.htm?csnumber=40614

NOTE: The electronic version of this international standard can be downloaded from the ISO/IEC Information Technology Task Force (ITTF) Web site.

http://isotc.iso.org/livelink/livelink/fetch/2000/2489/Ittf_Home/PubliclyAvailableStandards.htm

ISO/IEC 15408-3:2005 defines the assurance requirements of ISO/IEC 15408. It includes the evaluation assurance levels (EALs) that define a scale for measuring assurance, the individual assurance components from which the assurance levels are composed, and the criteria for evaluation of protection profiles and security targets.

5.3.2 ISO 15446 Guide for Production of Protection Profiles and Security Targets

ISO/IEC	TR	15446	Guide for production of protection profiles and security targets	Published	ISO TC204

ISO 15446 Available from
http://www.iso.org/iso/iso_catalogue/catalogue_tc/catalogue_detail.htm?csnumber=39690

NOTE: The electronic version of this International Standard can be downloaded from the ISO/IEC Information Technology Task Force (ITTF) Web site.

http://isotc.iso.org/livelink/livelink/fetch/2000/2489/Ittf_Home/PubliclyAvailableStandards.htm

ISO/IEC TR 15446:2004 (see Section 5.3.2) provides guidance relating to the construction of protection profiles (PPs) and security targets (STs) that are intended to be compliant with ISO/IEC 15408 (the "Common Criteria") (see Section 5.3.1).

ISO/IEC TR 15446:2004 gives suggestions on how to develop each section of a PP or ST. It is supported by an annex that contains generic examples of each type of PP and ST component, and by other annexes that contain detailed worked examples.

ISO/IEC TR 15446:2004 (see Section 5.3.2) is primarily aimed at those who are involved in the development of PPs and STs. However, it is also likely to be useful to evaluators of PPs and STs and to those who are responsible for monitoring PP and ST evaluation. It may also be of interest to consumers and users of PPs and STs who wish to understand what guidance the PP/ST author used, and which parts of the PP or ST are of principal interest.

5.3.3 IEEE P1556 Standard for Security and Privacy of Vehicle/Roadside Communication Including Smart Card Communications

See Section 5.2.9.

5.3.4 ISO 9160 Data Encipherment—Physical Layer Interoperability Requirements

See Section 6.2.2.24.

5.3.5 ISO 9591 Information Technology—Open Systems Interconnection, Part 2--The Directory: Models; Part 10—The Directory: Use of Systems Management for Administration of the Directory

See Section 6.2.2.30.

5.3.6 ISO 10736 Telecommunications and Information Exchange Between Systems—Transport Layer Security Protocol

See Section 6.2.2.41.

5.3.7 ISO 11577 Open Systems Interconnection—Network Layer Security Protocol

See Section 6.2.2.48.

5.3.8 ISO 13594 Information Technology—Lower Layers Security

See Section 6.2.2.53.

5.3.9 ISO 26927 Corporate Telecommunication Networks—Mobility for Enterprise Communications

See Section 6.2.2.76.

5.3.10 ISO 19773-12 Information Technology—Metadata Modules (MM) Data Structure for Entity-Person-Group (EPG) Security Credentials Data

See Section 6.2.9.12.

5.3.11 Specification and Standardization of the Internet Protocol Version 6 (IPv6)—Encapsulating Security Payload Header

See Section 6.2.14.2.

5.3.12 UDDI

See Section 6.2.19.

5.3.13 7816-8 Personal Identification (Including IC Cards/Smart Cards)—Identification Cards—Integrated Circuit Cards—Part 8: Security Related Interindustry Commands

See Section 7.1.5.

5.3.14 ISO 24534-2 Road Transport and Traffic Telematics—Automatic Vehicle and Equipment Identification—Electronic Registration Identification (ERI) for Vehicles—Part 2: Operational Requirements

See Section 7.3.3.2.

5.3.15 ISO 24534-4 Road Transport and Traffic Telematics—Automatic Vehicle and Equipment Identification—Electronic Registration Identification (ERI) for Vehicles—Part 4: Secure Communications Using Asymmetric Techniques

See Section 7.3.3.4.

5.3.16 ISO 24534-5 Road Transport and Traffic Telematics—Automatic Vehicle and Equipment Identification—Electronic Registration Identification (ERI) for Vehicles—Part 5: Secure Communications Using Symmetric Techniques

See Section 7.3.3.5.

5.3.17 ISO FDIS 24535 Intelligent Transport Systems—Automatic Vehicle Identification—Basic Electronic Registration (Basic ERI)

See Section 7.3.3.6.

5.3.18 ISO 24533 Intelligent Transport Systems—Data Dictionary and Message Set for Tracking of Freight and its Intermodal Transfer

See Section 7.3.4.1.

5.3.19 ISO 18000-1 Radio Frequency Identification for Item Management—Part 1: Reference Architecture and Definition of Parameters to Be Standardized

See Section 7.4.7.

5.3.20 ISO 18000-2 Radio Frequency Identification for Item Management—Part 2: Parameters for Air Interface Communications Below 135 kHz

See Section 7.4.8.

5.3.21 ISO 18000-6 Radio Frequency Identification for Item Management—Part 6: Parameters for Air Interface Communications at 860 MHz to 960 MHz

See Section 7.4.11.

5.3.22 ISO 18000-7 Radio Frequency Identification for Item Management—Part 7: Parameters for Active Air Interface Communications at 433 MHz

See Section 7.4.12.

5.3.23 ISO 15961 Radio Frequency Identification (RFID) for Item Management—Data Protocol: Application Interface

See Section 7.4.7.

5.3.24 J15674 Road Vehicles—Extended Data Link Security

See Section 8.4.15 ISO 15674.

5.3.25 J1760_200112 Data Security Services

See Section 8.4.5.21/12.1.2.

5.3.26 J2186_200506 E/E Data Link Security

See Section 8.4.5.32/12.1.2.

5.3.27 Transport Related Emergency Notification and Personal Security

See Section 8.7.1.

5.3.28 ISO 17574 Electronic Fee Collection (EFC)—Guidelines for EFC Security Protection Profiles

See Section 8.8.1.5.

5.3.29 Public Travel Security

See Section 8.9.1.

5.3.30 National Security Service Groups

See Section 8.12.

5.3.31 ARIB RCR STD-30 Security Radio Equipment for Low Power Radio Station

See Section 12.2.1.

5.3.32 ISO 24100 Basic Principles for Personal Data Protection in Probe Vehicle Information Services

See Section 4.6.2.18.

5.3.33 24824-3 Information Technology—Generic Applications of ASN.1: Fast Infoset Security

See Section 6.2.2.20.3

5.4 Transport Systems Safety

If you read the introductory chapters of this book, and particularly if you study the list of ITS evaluated services in Chapter 2, you will see the word "safety" occur on many occasions. You may therefore find it surprising that there are only a few instances of ITS standards that expressly use the word "safety" in their title, or even in their texts.

In Section 5.1.1, where we describe how ISO 14813-1 divides the ITS sector into service domains and groups, the standard identifies some of these as specific to the provision of safety. The standards for these groupings are described in Chapter 8, where you will find headings such as "Safety Readiness" (Section 8.4.3.4), "Automated Roadside Safety Inspection" (Section 8.5.5), "Commercial Vehicle On-Board Safety Monitoring" (Section 8.5.6), "Transport Related Emergency Notification and Personal Safety" (Section 8.7.1), and "Road Transport Related Personal Safety" (Section 8.9), but you will find only a few standards yet developed. If you study Chapter 7 (Automatic Identification), you will see references to safety, but no specific standards for safety. Why is this?

Safety is implicit to the provision of many ITS services. The improvement of safety, the saving of life, and the reduction in the number and severity of injuries are probably the single most important factors in ITS service provision. However, because the *objective* is safety, safety does not occur in the standard title, scope, or specifications. For example, a lane departure warning system, or a forward or rear obstacle warning system, have the sole objective of improving safety by warning the driver when he is veering from the lane, or on a collision path with an object. However, the standards specify procedures and protocols: types of warnings, situations in which the warning is activated, and so on. (so that driver response is intuitive regardless of which car he happens to be driving that day), and the word "safety" will probably not occur in the standard at all.

So to establish what standards are available to assist safety, I regret that you will have to trawl through the whole list and assess the objectives of each standard. I did consider identifying safety related standards in the numerical lists at the end of the book. But almost all direct ITS standards could in some way claim to be safety related, albeit in a very obtuse way, so I abandoned this idea.

The few standards that specifically mention safety may be found as follows.

5.4.1 EN 302 288/EN 302 264 Automotive Radar

This standard describes road-vehicle based radar for collision mitigation and traffic safety applications. See Section 4.1.1.3.3.

5.4.2 TR 24714-1Cross-Jurisdictional and Societal Aspects of Implementation of Biometric Technologies—Part 1: Guide to the Accessibility, Privacy, and Health and Safety Issues in the Deployment of Biometric Systems for Commercial Application

See Section 7.2.7.

5.4.3 J2189_200112 Guidelines for Evaluating Child Restraint System Interactions with Deploying Airbags

See Section 8.4.3.5.

5.4.4 ISO/CD TS 22240 Road Vehicles—Vehicles Safety Information Model (VSIM)

See Section 8.4.4.18

5.4.5 ISO 24978 Emergency and Safety Message Data Registry

See Section 8.3.2.3.

5.4.6 CEN TS/15722 (Was 24977) e-Call Minimum Set of Data

See Section 8.3.2.1.

5.4.7 CEN WI 00278220 eCall Operating Requirements

See Section 8.3.2.2.

5.4.8 SAE J2313_199909 On-Board Land Vehicle Mayday Reporting Interface

See Section 8.7.1.6.

5.5 Geographic and Location Based Services for ITS

Location based services (LBS) are services that exploit knowledge about where a user is located in order to provide services specific to that location. LBS therefore need to identify geographic location. Section 6.1 summarizes the generic standards which are used in relation to identifying location and aspects related to it. This section deals with ITS-specific aspects, which will be used in coordination with the generic standards.

GNSS (GPS, Galileo, GLONASS) is the most accurate means of identifying location (where the sight of enough of their satellites is known). However, an access point that a vehicle is transacting with (such as a roadside beacon) may provide adequate location data in order to provide or offer a service, or it can be provided by trilateration, for example, with a cellular phone based on the signal-strength of the closest cell-phone access points. See the standards in Section 6.1.

Many ITS services are, or will be, location based. Time-critical and many other safety services, including collision avoidance of whatever nature, all depend on accurate positioning of the location of the vehicle(s) involved.

Other such services may advertise local restaurants or recreation facilities and provide other nonessential services, and this may be used to provide the business case to provide access points that may be used in addition to providing safety services. However, it is the role that LBS plays in the provision of safety services where knowledge of exact location is a crucial requirement.

See Section 6.1 for general geographic and location identification standards and for related standards in the area known as geodetics. The sections that follow summarize ITS-specific aspects of the subject.

5.5.1 ISO 14825 Intelligent Transport Systems—Geographic Data Files (GDF)—Overall Data Specification

CEN/ISO	ENV/ ISO	14825	Intelligent transport systems—Geographic Data Files (GDF)—Overall data specification	Published	ISO TC204/ CEN TC278

ISO 14825 Available from
http://www.iso.ch/iso/en/CatalogueDetailPage.CatalogueDetail?CSNUMBER=30763&
scopelist=PROGRAMME

This is the standard for exchange of data from geographical databases serving as the basis for map data used for navigation.

As the file is not used directly for navigation, emphasis is placed on ease of editing—that is, genre-by-genre compilation of data. With respect to XGDF, however, consideration is also given to methods for providing information.

Thanks to the preceding work for standardization by CEN, work proceeded relatively smoothly compared with other items, and GDF was announced as ISO 14825 in February 2004.

The scope and potential use of GDF is huge. As discussions went on, new ideas were proposed resulting in a new PWI being approved at the TC204 conference in November 2000. The purpose was to revise the current GDF and accommodate the latest developments in information exchange, such as the Internet. The work entered the NP phase in August 2004 after compiling requests for improvement (see ISO 22953 in Section 5.5.3).

5.5.2 ISO 17572 Intelligent Transport Systems (ITS)—Location Referencing for Geographic Databases

ISO	CD	17572	Intelligent Transport Systems (ITS) - Location Referencing for Geographic Databases	Published	ISO TC204/ WG3

ISO	CD	17572-1	Location Referencing for Geographic Databases - Part 1: General Requirements and Conceptual Model	Published 2007	ISO TC204/ WG3

ISO 17572-1 Available from
http://www.iso.org/iso/en/CatalogueDetailPage.CatalogueDetail?CSNUMBER=45960&
scopelist=PROGRAMME

ISO	CD	17572-2	Location Referencing for Geographic Databases - Part 2: Pre-coded Location References (Pre-coded Profile)	Published 2007	ISO TC204/ WG3

ISO 17572-2 Available from
http://www.iso.org/iso/en/CatalogueDetailPage.CatalogueDetail?CSNUMBER=45961&
scopelist=ALL

ISO	CD	17572-3	Location Referencing for Geographic Databases - Part 3: Dynamic Location References (Dynamic Profile)	Published 2007	ISO TC204/ WG3

ISO 17572-3 Available from
http://www.iso.org/iso/en/CatalogueDetailPage.CatalogueDetail?CSNUMBER=45962&scopelist=PROGRAMME

This draft has had a difficult history but is now back on track. It is designed to provide methods for location referencing when information is exchanged between different applications and geographical databases and to enable the establishment of location on a different map database when traffic information is exchanged between systems.

Initially, it was decided that a method based on coordinate systems and road descriptors would be adopted as options, pending the results of demonstration experiments in Europe and the United States. However, activities in this field were stalled for some time because the results were not readily available. During the stalemate, the need for standardization of general-purpose location referencing increased as the information community moved rapidly toward standardization. Two profiles were subsequently proposed:

- *Profile 1. Precoded Location Referencing:* a method premised on regulation in a system, including a method using precoded location tables;
- *Profile 2. Geographic Object Referencing:* a method using coordinates (required) and descriptors (optional).

Profile 1 has been completed, and CD comment has been finished. Full-scale discussions have just started on Profile 2 after receiving proposals from Europe.

Coordination efforts for the details are being carried out in response to Japan's proposal for including a method that uses geometrical shapes with an emphasis on coordinate arrays.

5.5.3 ISO 22953 Intelligent Transport Systems (ITS) eXtended Geographic Data Files (XGDF)

ISO	NP	22953	Intelligent Transport Systems (ITS) eXtended Geographic Data Files (XGDF)	Under Development	ISO TC204

ISO 22953 is not yet available

See also Sections 5.5.3 and 8.2.3.8, ISO 14825. ISO 22953 is a further development from this work.

This deliverable is a description of the GDF standard. This is a specification of how to store geographic information for intelligent transport systems. The structures in GDF, although primarily developed for storing data, are already a sophisticated ontology for transport networks.

GDF is the leading business-to-business (B2B) map delivery format for high-end map-dependent ITS applications and services that require detailed, rich, and accurate geographic data. Its primary use is for car navigation systems, but it is also very usable for location-based services and many other transport and traffic applications like fleet management, dispatch management, traffic analysis, and traffic management. The European Union promotes the use of the GDF format. Maps in GDF are typically provided by map vendors such as Tele Atlas and NAVTEQ. Bosch, Philips, and Volvo use GDF for their car navigation systems.

GDF version 4.0 was published as ISO standard ISO 22953 in the early 2004. Through its extensive dictionary of standardized semantic definitions, GDF 4.0 forms an excellent foundation for interoperability between map-based ITS applications and services.

The purpose of ISO 22953 is to revise the current GDF and accommodate the latest developments in information exchange, such as the Internet. The work entered the NP phase in August 2004 after compiling requests for improvement.

Discussions on XGDF are underway with themes on required performance and models. After appropriate consultation with TC 211, which handles geographical information on a generic basis, TC 211 decided to adopt UML for the concept model. The latest model has been circulated in TC 211.

5.5.4 SAE J1698 Location Referencing Message Specification (LRMS)

SAE	J1698_ 200502	Location Referencing Message Specification (LRMS)	Published Oct. 2004	SAE

SAE J1698_200502 Available from
http://www.sae.org/technical/standards/J1698_200502

This recommended practice aims to establish a common format for displaying and presenting crash-related data recorded and stored within certain electronic components currently installed in many light-duty vehicles. This recommended practice pertains only to the post-download format of such data and is not intended to standardize the format of the data stored within any on-board storage unit, or to standardize the method of data recording, storing, or extraction. Historically, crash data recording technology in light-duty vehicles has developed and evolved based on differing technical needs of manufacturers and their customers without industry standards or government regulation. As a result, wide variations currently exist among vehicle manufacturers regarding the scope and extent of recorded data. For this reason, this recommended practice is not intended to standardize or mandate the recording of any specific data element or to specify a minimum data set. Rather, it is intended to be a compilation of data elements and parameters that various manufacturers are currently recording, as well as those elements reasonably predicted to be recorded in the foreseeable future, and to establish a common format for display and presentation of that data so recorded.

This version of the recommended practice is limited in application to vehicular data recorded in single frontal impact events. Provisions for multiple-impact events may be included in the next version. Side-impact and rollover events may be addressed at a later time.

SAE	J1698/1_ 200503	Vehicle Event Data Interface—Output Data Definition	Published Oct. 2004	SAE

SAE J1698/1_200503 Available from
http://www.sae.org/technical/standards/J1698_200502

This document is part of the J1698 document family, and provides the definitions for event-related data items.

5.5.5 SAE J2374 Location Referencing Message Specification

SAE	J2374_ 199907	Location Referencing Message Specification	Published July 1999	SAE

SAE J2374_199907 Available from
http://www.sae.org/technical/standards/J2374_199907

The LRMS is intended to provide a practical approach to standardization for location referencing within a mixed data set environment (i.e., where more than one kind of spatial data set exists, and where spatial references between these data sets must be made). Although some ITS applications in local areas may be satisfied by having one common data set—for which location references may be implemented in any number of ways—many ITS applications will have broad interoperability requirements within the nation or a region.

For example, a vehicle driven from California to Florida should be able to receive and understand spatial references for traffic information or routing instructions throughout the trip. Similarly, information sent from a vehicle to a central site should be understood in any city regardless of the kinds of data sets in use, whether they are public or private, or how locations are referenced internally to particular data sets.

The LRMS can be applied to ITS systems involving mobile vehicles on roads, rails, and waterways. It can also be applied to location references to and from central sites to nonmobile sites such as kiosks, other central sites, or pedestrians.

The broadest scope of the LRMS is therefore intermodal spatial data set interoperability at the national level and across all of ITS. Given the great variety of ITS systems, it is expected that individual LRMS profiles will generate location referencing standards for subsets of ITS applications, such as ISP→vehicle→ISP, or center→center.

5.6 Human-Machine Interface Standards

5.6.1 J1757 Standard Metrology for Vehicle Displays

SAE	J1757/1_ 200704	Standard Metrology for Vehicular Displays	Published April 2007	SAE

SAE J1757/1_200704 Available from
http://www.sae.org/technical/standards/J1757/1_200704

The scope of this SAE standard is to provide methods to determine display optical performance in all typical automotive ambient light illumination, with a focus on high ambient contrast ratio, which is critical for display legibility in a sunshine environment. It covers indoor measurements and simulated outdoor lighting. It is not the scope of this document to set threshold values for automotive compliance; however, some recommended values are presented for reference.

CHAPTER 6

General Standards for Information Technology That May Be Used to Support ITS Service Provision

This chapter provides summaries and references to standards which have been developed generically for information technology, and not specifically for ITS service provision, but which may be useful to enable, or as a building block for, the provision of ITS services. Many, but not all, of these standards are available for download free of charge.

6.1 Geographic and Location Based Standards

Location based services (LBS) are services that exploit knowledge about where a user is located in order to provide services specific to that location. Location based services therefore need to identify geographic location. This section summarizes the generic standards which are used in relation to identifying location and aspects related to it.

As we observed in Section 5.5, many ITS services are, or will be, location based, critically, many safety services, and collision avoidance of whatever nature, all depend on accurate positioning of the location of the vehicle(s) involved.

For geographic and location based standards developed specifically for ITS, the reader is directed to Section 5.5 This section details generic geographic and location based standards (that may be well suited to supporting ITS service provision).

Most of these standards are developed by ISO/TC 211. The current work program of ISO/TC 211 can be downloaded from http://www.isotc211.org/pow.htm.

6.1.1 eXtended Geographic Data File (XGDF)

ISO	NP	22953	Intelligent Transport Systems (ITS) eXtended Geographic Data Files (XGDF)	Under Development	ISO TC204

Although this section deals with generic geographic standards, because of its generic application when developed, this deliverable is cross referenced here. For details, see Section 5.5.3.

6.1.2 ISO 6709:2006 Standard Representation of Latitude, Longitude, and Altitude for Geographic Point Locations

ISO	6709	Standard representation of latitude, longitude and altitude for geographic point locations	Published 1983	ISO TC211

ISO 6709 Available from
http://www.iso.org/iso/en/CatalogueDetailPage.CatalogueDetail?CSNUMBER=13152

ISO 6709 dates back to 1983 and is the basis of most geographic/geodetic standards. It describes a variable-length format for the representation of latitude, longitude, and altitude for use in data interchange. It allows the use of normal sexagesimal notations involving degrees, minutes, and seconds as well as various combinations of sexagesimal and decimal notations. It uses numeric characters 0 to 9, graphic characters plus (+), minus (–), full stop (.), and comma (,).

Unusually for an ISO standard, permission is granted to copy, distribute, and/or modify this document under the terms of the GNU Free Documentation License, Version 1.2 or any later version published by the Free Software Foundation; with no Invariant Sections, no Front-Cover Texts, and no Back-Cover Texts. A copy of the license "GNU Free Documentation License" can be found at http://www.gnu.org/copyleft/fdl.html.

6.1.3 ISO 19100 Series of Standard—Geographic Information

ISO	19100 Series Geographic Information	ISO TC211

ISO 19100 is frequently referred to generically for a series of standards (starting at 19101) for defining, describing, and managing geographic information (i.e., information concerning objects or phenomena that are directly or indirectly associated with a location relative to the Earth). This series of standard specifies methods, tools, and services for management of information, including the definition, acquisition, analysis, access, presentation, and transfer of such data in digital/electronic form between different users, systems, and locations. This series of standards make it possible to define profiles in order to facilitate the development of geographic information systems and application systems that will be used for specific purposes; profiling consists of putting together "packages/subsets" of the total set of standards to fit individual application areas or users.

6.1.4 ISO 19101 Geographical Information—Reference Model

ISO	19101	Geographical Information -Reference model	Published 2002	ISO TC204

ISO 19101 Available from
http://www.iso.org/iso/en/CatalogueDetailPage.CatalogueDetail?CSNUMBER=26002

This International Standard defines the framework for standardization in the field of geographic information and sets forth the basic principles by which this standardization takes place.

This framework identifies the scope of the standardization activity being undertaken and the context in which it takes place. The framework provides the method by which what is to be standardized can be determined and describes how the contents of the standards are related.

Although structured in the context of information technology and information technology standards, this International Standard is independent of any application development method or technology implementation approach.

ISO	TS	19101-2	Geographical Information—Reference model—Part 2: Imagery	Published 2002	ISO TC211

ISO 19101-2 Available from
http://www.iso.org/iso/en/CatalogueDetailPage.CatalogueDetail?CSNUMBER=39983&scopelist=PROGRAMME

This technical specification defines a reference model for standardization in the field of geographic imagery. This reference model identifies the scope of the standardization activity being undertaken and the context in which it takes place. It will include gridded data with an emphasis on imagery. Although structured in the context of information technology and information technology standards, this technical specification will be independent of any application development method or technology implementation approach.

This technical specification provides a reference model for the open distributed processing of geographic imagery. The motivating themes addressed in this reference model are as follows:

- In terms of volume, imagery is the dominant form of geographic information.
- Stored geographic imagery volume will grow to the order of an exabyte
- National imagery archives are multiple petabytes in size, ingesting a terabyte per day.
- Individual application data centers are archiving hundreds of terabytes of imagery.

- Tens of thousands of datasets have been catalogued but are not yet on-line.
- Most geographic imagery will never be directly accessed by humans.
- Human attention is a scarce resource and is insufficient to view petabytes of data.
- Semantic processing will be required: automatic detection of features and mining based on geographic concepts.
- Geographic imagery is a key input to support policy makers in decision support.

The ultimate challenge is to enable the geographic imagery collected from different sources to become an integrated digital representation of the Earth, widely accessible for humanity's critical decisions.

NOTE: Project 19102 was cancelled.

6.1.5 ISO 19103:2004 Geographic Information—Conceptual Schema Language

ISO	TS	19103	Geographical Information –Conceptual Schema Language	Published 2005	ISO TC211

ISO 19103 Available from
http://www.iso.org/iso/en/CatalogueDetailPage.CatalogueDetail?CSNUMBER=37800&scopelist=PROGRAMME

The technical specification ISO/TS 19103:2005 provides rules and guidelines for the use of a conceptual schema language within the ISO geographic information standards. The chosen conceptual schema language is the Unified Modeling Language (UML)—ISO 19501 (see Section 6.2.12.1).

ISO/TS 19103:2005 provides a profile of UML for use with geographic information. In addition, it provides guidelines on how UML should be used to create standardized geographic information and service models.

6.1.6 ISO 19104 Geographical Information—Terminology Introduction

ISO	TS	19104	Geographical Information—Terminology Introduction		ISO TC211

This work item has not yet resulted into a published deliverable. ISO/TC 211 has authorized the ISO 19104 Terminology editing committee to review the ISO 19104 document and to make recommendations as follows:

- The ISO 19104 Terminology editing committee is authorized to meet periodically until the formation of a maintenance agency for purposes of conducting the activities listed.
- The ISO 19104 Terminology editing committee will make recommendations for amendments under ISO Directives 1, clauses 2.10.2 and 2.10.3.
- The ISO 19104 Terminology editing committee shall receive recommendations from other ISO/TC 211 ECs on terminology, report progress of harmonization efforts, and recommend terminology maintenance procedures.
- The efforts will be guided by the scope and purpose statements for ISO 19104 Terminology as approved by ISO/TC 211. The ISO 19104 Terminology editing committee may recommend revision of its terms of reference and scope after 6 months.
- The ISO 19104 Terminology editing committee will establish a Web database of the terminology of geographic information prior to the elevation of ISO 19104 to DIS. This to be done in cooperation with the secretariat.
- The ISO 19104 Terminology editing committee may propose terminology maintenance procedures for ISO/TC 211.

6.1.7 ISO 19105 Geographical Information—Conformance and Testing

ISO	TS	19105	Geographic information—Conformance and testing	Published 2000	ISO TC211

ISO 19105 Available from
http://www.iso.org/iso/en/CatalogueDetailPage.CatalogueDetail?CSNUMBER=26010

This International Standard specifies the framework, concepts, and methodology for testing and criteria to be achieved to claim conformance to the family of ISO geographic information standards. It provides a framework for specifying abstract test suites (ATS) and for defining the procedures to be followed during conformance testing. Conformance may be claimed for data or software products or services or by specifications including any profile or functional standard.

Standardization of test methods and criteria for conformance to geographic information standards will allow verification of conformance to those standards. Verifiable conformance is important to geographic information users, in order to achieve data transfer and sharing.

This International Standard is applicable to all the phases of conformance and testing. These phases are characterized by the following major activities:

- The definition of ATS for conformance to the ISO geographic information standards;
- The definition of test methods for conformance to the ISO geographic information standards;

- The conformance assessment process carried out by a testing laboratory for a client, culminating in the production of a conformance test report.

This International Standard specifies the requirements for, and gives guidance on, the procedures to be followed in conformance testing for the ISO geographic information standards. It includes only such information as is necessary to meet the following objectives:

- To achieve confidence in the tests as a measure of conformance;
- To achieve comparability between the results of corresponding tests applied in different places at different times;
- To facilitate communication between the parties responsible for the activities described in the first two bullets.

This standard provides a framework for certification (i.e., an administrative procedure which may follow conformance testing) in an informative annex (B).

The International Standard does not include provisions for:

- The description of requirements for procurement and contracts;
- Testing by means of test methods which are specific to particular applications or systems;
- Acceptance testing, performance testing, and robustness testing.

The framework established by this International Standard includes the concept of executable test suites (ETS). These, by their very nature, cannot be standardized; consequently, standardization of executable test suites is outside the scope of the International Standard.

6.1.8 ISO 19106:2003 Geographic Information—Profiles

ISO	TS	19106	Geographic information—Profiles	Published 2003	ISO TC211

ISO 19106 Available from
http://www.iso.org/iso/en/CatalogueDetailPage.CatalogueDetail?CSNUMBER=26011

ISO 19106:2004 is intended to define the concept of a profile of the ISO geographic information standards developed by ISO/TC 211 and to provide guidance for the creation of such profiles. Only those components of specifications that meet the definition of a profile contained herein can be established and managed through the mechanisms described in this International Standard. These profiles can be standardized internationally using the ISO standardization process. This document also provides guidance for establishing, managing, and standardizing at the national level (or in some other forum).

6.1.9 ISO 19107 Geographic information—Spatial Schema

ISO	IS	19107	Geographic information—Spatial Scheme	Published 2003	ISO TC211

ISO 19107 Available from
http://www.iso.org/iso/en/CatalogueDetailPage.CatalogueDetail?CSNUMBER=26012

ISO 19107:2003 specifies conceptual schemas for describing the spatial characteristics of geographic features, as well as a set of spatial operations consistent with these schemas. It treats vector geometry and topology up to three dimensions. It defines standard spatial operations for use in access, query, management, processing, and data exchange of geographic information for spatial (geometric and topological) objects of up to three topological dimensions embedded in coordinate spaces of up to three axes.

6.1.10 ISO 19108:2004 Geographic Information—Temporal Schema

ISO	IS	19108	Geographic information—Temporal Schema	Published 2002 Corrigenda 2006	ISO TC211

ISO 19108 Available from
http://www.iso.org/iso/en/CatalogueDetailPage.CatalogueDetail?CSNUMBER=26013

ISO 19108:2002 defines concepts for describing temporal characteristics of geographic information. It depends upon existing information technology standards for the interchange of temporal information. It provides a basis for defining temporal feature attributes, feature operations, and feature associations, and for defining the temporal aspects of metadata about geographic information. Since this International Standard is concerned with the temporal characteristics of geographic information as they are abstracted from the real world, it emphasizes valid time rather than transaction time.

6.1.11 ISO 19109:2004 Geographic Information—Rules for Application Schema

ISO	IS	19109	Geographic information—Rules for Application Schema	Published 2004	ISO TC211

ISO 19109 Available from
http://www.iso.org/iso/en/CatalogueDetailPage.CatalogueDetail?CSNUMBER=39891&
scopelist=PROGRAMME

ISO 19109:2005(E) defines rules for creating and documenting application schemas, including principles for the definition of features. Its scope includes the following:

- Conceptual modeling of features and their properties from a universe of discourse;
- Definition of application schemas;
- Use of the conceptual schema language for application schemas;
- Transition from the concepts in the conceptual model to the data types in the application schema;
- Integration of standardized schemas from other ISO geographic information standards with the application schema.

The following are outside the scope of the standard:

- Choice of one particular conceptual schema language for application schemas;
- Definition of any particular application schema;
- Representation of feature types and their properties in a feature catalog;
- Representation of metadata;
- Rules for mapping one application schema to another;
- Implementation of the application schema in a computer environment;
- Computer system and application software design;
- Programming.

6.1.12 ISO 19110:2004 Geographic Information—Methodology for Feature Cataloguing

ISO	IS	19110	Geographic information—Methodology for feature cataloguing	Published 2004	ISO TC211

ISO 19110 Available from
http://www.iso.org/iso/en/CatalogueDetailPage.CatalogueDetail?CSNUMBER=39965

ISO 19110:2005 defines the methodology for cataloging feature types and specifies how the classification of feature types is organized into a feature catalog and presented to the users of a set of geographic data. ISO 19110:2005 is applicable to creating catalogs of feature types in previously uncataloged domains and to revising existing feature catalogs to comply with standard practice. ISO 19110:2005 applies to the cataloging of feature types that are represented in digital form. Its principles can be extended to the cataloging of other forms of geographic data.

The standard is applicable to the definition of geographic features at the type level. ISO 19110:2005 is not applicable to the representation of individual instances of each type and excludes spatial, temporal, and portrayal schemas as specified in ISO 19107, ISO 19108, and the future ISO 19117, respectively. It also excludes collection criteria for feature instances.

ISO 19910:2005 may be used as a basis for defining the universe of discourse being modeled in a particular application, or to standardize general aspects of real-world features being modeled in more than one application.

6.1.13 ISO 19111 Geographic Information—Spatial Referencing by Coordinates

ISO	IS	19111	Geographic information—Spatial referencing by coordinates	Published 2003: Rev 2007	ISO TC211

ISO 19111 Available from
http://www.iso.org/iso/en/CatalogueDetailPage.CatalogueDetail?CSNUMBER=26016

ISO 19111:2007 defines the conceptual schema for the description of spatial referencing by coordinates, and is optionally extended to spatial-temporal referencing. It describes the minimum data required to define one-, two- and three-dimensional spatial coordinate reference systems with an extension to merged spatial-temporal reference systems. It allows additional descriptive information to be provided. It also describes the information required to change coordinates from one coordinate reference system to another.

In the standard, a coordinate reference system does not change with time. For coordinate reference systems defined on moving platforms such as cars, ships, aircraft, and spacecraft, the transformation to an Earth-fixed coordinate reference system can include a time element.

ISO 19111:2007 is applicable to producers and users of geographic information. Although it is applicable to digital geographic data, its principles can be extended to many other forms of geographic data such as maps, charts, and text documents.

The schema described can be applied to the combination of horizontal position with a third nonspatial parameter which varies monotonically with height or depth. This extension to nonspatial data is beyond the scope of ISO 19111:2007 but can be implemented through profiles.

Work item "19111-2 Geographic information—Spatial referencing by coordinates—Part 2: Extension for parametric value" is under development.

6.1.14 ISO 19112 Geographic Information—Spatial Referencing by Geographic Identifiers

ISO	IS	19112	Geographic information—Spatial referencing by geographic identifiers	Published 2003: Rev 2007	ISO TC211

ISO 19112 Available from
http://www.iso.org/iso/en/CatalogueDetailPage.CatalogueDetail?CSNUMBER=26017

ISO 19912:2003 defines the conceptual schema for spatial references based on geographic identifiers. It establishes a general model for spatial referencing using geographic identifiers, defines the components of a spatial reference system, and defines the essential components of a gazetteer. Spatial referencing by coordinates is not addressed in this document; however, a mechanism for recording complementary coordinate references is included.

The standard assists users in understanding the spatial references used in datasets. It enables gazetteers to be constructed in a consistent manner and supports the development of other standards in the field of geographic information. It is applicable to digital geographic data, and its principles may be extended to other forms of geographic data such as maps, charts and textual documents.

6.1.15 ISO 19113 Geographic Information—Quality Principles

ISO	IS	19113	Geographic information—Quality principles	Published 2002 In Revision	ISO TC211

ISO 19113 Available from
http://www.iso.ch/iso/en/CatalogueDetailPage.CatalogueDetail?CSNUMBER=26018&ICS1=35&ICS2=240&ICS3=70

ISO 19113:2002 establishes the principles for describing the quality of geographic data and specifies components for reporting quality information. It also provides an approach to organizing information about data quality.

The standard is applicable to data producers providing quality information to describe and assess how well a dataset meets its mapping of the universe of discourse as specified in the product specification, formal or implied, and to data users attempting to determine whether or not specific geographic data is of sufficient quality for their particular application. This International Standard should be considered by organizations involved in data acquisition and purchase, in such a way that it makes it possible to fulfill the intentions of the product specification.

It can additionally be used for defining application schemas and describing quality requirements.

In addition to being applicable to digital geographic data, the principles of ISO 19113:2002 can be extended to identify, collect, and report the quality information for a geographic dataset; its principles can be extended and used to identify, collect, and report quality information for a dataset series or smaller groupings of data that are a subset of a dataset.

Although ISO 19113:2002 is applicable to digital geographic data, its principles can be extended to many other forms of geographic data such as maps, charts, and textual documents.

ISO 19113:2002 does not attempt to define a minimum acceptable level of quality for geographic data.

Work has been concluded on a revision and at the time of writing this was in an internal ballot process.

6.1.16 ISO 19114 Geographic Information—Quality Evaluation Procedures

ISO	IS	19114	Geographic information—Quality evaluation procedures	Published 2003	ISO TC211

ISO 19114 Available from
http://www.iso.ch/iso/en/CatalogueDetailPage.CatalogueDetail?CSNUMBER=26019&scopelist=ALL

ISO 19114:2003 provides a framework of procedures for determining and evaluating quality that is applicable to digital geographic datasets, consistent with the data quality principles defined in ISO 19113. It also establishes a framework for evaluating and reporting data quality results, either as part of data quality metadata only or also as a quality evaluation report.

The standard is applicable to data producers when providing quality information on how well a dataset conforms to the product specification, and to data users attempting to determine whether or not the dataset contains data of sufficient quality to be fit for use in their particular applications.

Although ISO 19114:2003 is applicable to all types of digital geographic data, its principles can be extended to many other forms of geographic data such as maps, charts, and textual documents.

6.1.17 ISO 19115 Geographic Information—Metadata

ISO	IS	19115	Geographic information—Metadata	Published 2003	ISO TC211

ISO 19115 Available from
http://www.iso.ch/iso/en/CatalogueDetailPage.CatalogueDetail?CSNUMBER=26020

ISO 19115:2003 defines the schema required for describing geographic information and services. It provides information about the identification, the extent, the quality, the spatial and temporal schema, spatial reference, and distribution of digital geographic data.

This standard is applicable to:

- The cataloging of datasets, clearinghouse activities, and the full description of datasets;
- Geographic datasets, dataset series, and individual geographic features and feature properties.

ISO 19115:2003 defines:

- Mandatory and conditional metadata sections, metadata entities, and metadata elements;
- The minimum set of metadata required to serve the full range of metadata applications (data discovery, determining data fitness for use, data access, data transfer, and use of digital data);
- Optional metadata elements in order to allow for a more extensive standard description of geographic data, if required;
- A method for extending metadata to fit specialized needs.

Though ISO 19115:2003 is applicable to digital data, its principles can be extended to many other forms of geographic data such as maps, charts, and textual documents as well as nongeographic data.

NOTE: Certain mandatory metadata elements may not apply to these other forms of data.

6.1.18 ISO 19115 Geographic Information—Metadata Extensions for Imagery and Gridded Data

ISO	WD	19115 pt 2	Geographic information—Metadata Part 2 Extensions for imagery and gridded data	Under development	ISO TC211

ISO 19115 part 2 is Not Yet Available

ISO 19115-2:2006 "Geographic Information—Metadata—Part 2: Extensions for imagery and gridded data" is a complementary standard to ISO 19115 "Geographic information—Metadata" that will define metadata elements to support imagery and gridded data and will extend the UML model for metadata to include the following:

- It will support the collection and processing of natural and synthetic imagery produced by remote sensing and other imaging processes.

- It will support the collection and processing of geospatial metadata for imagery, gridded, and coverage data.
- It will define a data model for information describing geographic imagery and gridded data, establishing the names, definitions, and permissible values for new data elements including new classes relevant to imagery and gridded data.

The standard is of particular relevance to the following sectors:

- Developers of GIS products;
- Developers of GIS application systems;
- Producers/suppliers of geographic data;
- Users of geographic data and GIS;
- Developers of standards.

Further information on this standard and its implementation is available from the ISO/TC 211 secretariat via http://www.isotc211.org.

6.1.19 ISO 19116:2004 Geographic Information—Positioning Services

ISO	IS	19116	Geographic information—Positioning services	Published 2004	ISO TC211

ISO 19116 Available from
http://www.iso.org/iso/en/CatalogueDetailPage.CatalogueDetail?CSNUMBER=37805

ISO 19116:2004 specifies the data structure and content of an interface that permits communication between position-providing device(s) and position-using device(s) so that the position-using device(s) can obtain and unambiguously interpret position information and determine whether the results meet the requirements of the use. A standardized interface of geographic information with position allows the integration of positional information from a variety of positioning technologies into a variety of geographic information applications, such as surveying, navigation, and intelligent transportation systems. ISO 19116:2004 is designed to benefit a wide range of applications for which positional information is important.

6.1.20 ISO 19117:2004 Geographic Information—Portrayal

ISO	IS	19117	Geographic information—Portrayal	Published 2004	ISO TC211

ISO 19117 Available from
http://www.iso.org/iso/en/CatalogueDetailPage.CatalogueDetail?CSNUMBER=40395

ISO 19117:2005 defines a schema describing the portrayal of geographic information in a form understandable by humans. It includes the methodology for describing symbols and mapping of the schema to an application schema. It does not include standardization of cartographic symbols nor their geometric and functional description.

6.1.21 ISO 19118:2004 Geographic Information—Encoding

ISO	IS	19118	Geographic information—Encoding	Published 2004	ISO TC211

ISO 19118 Available from
http://www.iso.org/iso/en/CatalogueDetailPage.CatalogueDetail?CSNUMBER=37796&scopelist=PROGRAMME

ISO 19118:2005 specifies the requirements for defining encoding rules to be used for interchange of geographic data within the ISO 19100 series of International Standards.

ISO 19118:2005 specifies:

- Requirements for creating encoding rules based on UML schemas;
- Requirements for creating encoding services;
- An informative XML-based encoding rule for neutral interchange of geographic data.

ISO 19118:2005 does not specify any digital media, it does not define any transfer services or transfer protocols, nor does it specify how to encode inline large images.

NOTE: OGC and ISO/TC 211 have developed geography markup language (GML) together, which is an XML encoding scheme in compliance with ISO 19118 for the transport and storage of geographic information modeled according to the conceptual modeling framework used in the ISO 19100 series and including both the spatial and nonspatial properties of geographic features. The GML specification defines the XML schema syntax, mechanisms, and conventions that provide an open, vendor-neutral framework for the definition of geospatial application schemas and objects

All specifications use UML according to ISO19103 to allow for XML encoding according to ISO 19118.

Other than the network itself, the standards do not specify any specific features related to the road application domain. Instead, the specifications are actually frameworks and it is up to users of the specifications to define the specific features for a certain application domain.

6.1.22 ISO 19119:2005 Geographic Information—Services

ISO	IS	19119	Geographic information—Services	Published 2005	ISO TC211

ISO 19119 Available from
http://www.iso.org/iso/en/CatalogueDetailPage.CatalogueDetail?CSNUMBER=39890

ISO 19119:2005 identifies and defines the architecture patterns for service interfaces used for geographic information, defines its relationship to the Open Systems Environment model, and presents a geographic services taxonomy and a list of example geographic services placed in the services taxonomy. It also prescribes how to create a platform-neutral service specification, how to derive conformant platform-specific service specifications, and it provides guidelines for the selection and specification of geographic services from both platform-neutral and platform-specific perspectives.

6.1.23 ISO/TR 19120 Geographic Information—Functional Standards

ISO	TR	19120	Geographic information—Functional Standards	Published 2001	ISO TC211

ISO 19120 Available from
http://www.iso.org/iso/en/CatalogueDetailPage.CatalogueDetail?CSNUMBER=28928

Within the context of this technical report, a functional standard has been identified as an existing geographic information standard, in active use within the international community. National standards have not been considered within this report.

This technical report seeks to identify the components of those recognized functional standards and to identify elements that can be harmonized between these standards and with the ISO/TC 211 base standards. This technical report provides a starting point for a feedback cycle between the functional standards communities and the ISO 19100 series component project teams.

6.1.24 ISO/TR 19121 Geographic Information—Imagery and Gridded Data

ISO	TR	19121	Geographic information—Functional Standards	Published 2001	ISO TC211

ISO 19121 Available from
http://www.iso.org/iso/en/CatalogueDetailPage.CatalogueDetail?CSNUMBER=29775

This technical report reviews the manner in which raster and gridded data is currently being handled in the geomatics community in order to propose how this type of data should be supported by geographic information standards.

This technical report identifies those aspects of imagery and gridded data that have been standardized or are being standardized in other ISO committees and external standards organizations, and that influence or support the establishment of raster and gridded data standards for geographic information. It also describes the components of those identified ISO and external imagery and gridded data standards that can be harmonized with the ISO 19100 series of geographic information/geomatics standards.

A plan is presented for ISO/TC 211 to address imagery and gridded data in an integrated manner within the ISO 19100 series of geographic information standards.

6.1.25 ISO/TR 19122 Geographic Information—Qualifications and Certification of Personnel

ISO	TR	19122	Geographic information—Qualifications and Certification of personnel	Published 2004	ISO TC211

ISO 19122 Available from
http://www.iso.org/iso/en/CatalogueDetailPage.CatalogueDetail?CSNUMBER=31088

ISO/TR 19122:2004 is applicable to the following aspects of the field of geographic information/geomatics:

- To develop a Type 3 report, which describes a system for the qualification and certification, by a central independent body, of personnel in the field of geographic information/geomatics;
- To define the boundaries between geographic information/geomatics and other related disciplines and professions;
- To specify technologies and tasks pertaining to geographic information/geomatics;
- To establish skill sets and competency levels for technologists, professional staff, and management in the field;
- To research the relationship between this initiative and other similar certification processes performed by existing professional associations;
- To develop a plan for the accreditation of candidate institutions and programs, for the certification of individuals in the workforce, and for collaboration with other professional bodies.

6.1.26 ISO 19123 Geographic Information—Schema for Coverage Geometry and Functions

ISO	IS	19123	Geographic information—Schema for coverage geometry and functions	Published 2005	ISO TC211

ISO 12193 Available from
http://www.iso.org/iso/en/CatalogueDetailPage.CatalogueDetail?CSNUMBER=40121&scopelist=PROGRAMME

ISO 19123:2005 defines a conceptual schema for the spatial characteristics of coverages. Coverages support mapping from a spatial, temporal, or spatio-temporal domain to feature attribute values where feature attribute types are common to all geographic positions within the domain. A coverage domain consists of a collection of direct positions in a coordinate space that may be defined in terms of up to three spatial dimensions as well as a temporal dimension. Examples of coverages include rasters, triangulated irregular networks, point coverages, and polygon coverages. Coverages are the prevailing data structures in a number of application areas, such as remote sensing, meteorology, and mapping of bathymetry, elevation, soil, and vegetation.

The International Standard defines the relationship between the domain of a coverage and an associated attribute range. The characteristics of the spatial domain are defined, whereas the characteristics of the attribute range are not part of ISO 19123:2005.

6.1.27 ISO 19124 Geographic Information—Imagery and Gridded Data Components

ISO	Summary	19124	Geographic information—Imagery and gridded data components	ISO TC211

This work is a follow-up to ISO/TR 19120. It started in 1999 and was completed in 2000. The work produced a Review Summary containing the following:

- Framework of imaging and gridded (I&G) data:
- Term definitions and conceptual hierarchy;
- UML model of I&G data;
- Five components:
  - Data model;
  - Metadata;
  - Encoding;
  - Services;
  - Spatial registration;
  - Plus impact on ISO 2111 standards/projects;
- Proposed two new work items:
- Stage 1 of I&G project;
- IS on sensor and data modes.

Stage 0 work on imagery, gridded and coverage data has resulted in this Review Summary and a new work item proposal (NWIP) to progress to Stage 1, as well a second NWIP proposal for sensor data. The NWIP to progress to Stage 1 was for a technical specification, because it is easier to achieve this goal since it has one fewer step than the preparation of an International Standard, and because it is expected to use this development as a vehicle to prepare material for future technical amendments to other standards.

This document is not available for sale, nor is it on the ISO/TC 211 Web site. For more details, contact ISO via http://www.isotc211.org.

6.1.28 ISO 19125-1 Geographic Information—Simple Feature Access—Part 1: Common Architecture

ISO	IS	19125-1	Geographic information—Simple feature access—Part 1: Common architecture	Published 2004	ISO TC211

ISO 19125-1 Available from
http://www.iso.org/iso/en/CatalogueDetailPage.CatalogueDetail?CSNUMBER=40114

ISO 19125-1:2004 establishes a common architecture for geographic information and defines terms to use within the architecture. It also standardizes names and geometric definitions for types for geometry.

ISO 19125-1:2004 does not place any requirements on how to define the geometry types in the internal schema nor does it place any requirements on when, how, or who defines the geometry types. ISO 19125-1:2004 does not attempt to standardize and does not depend upon any part of the mechanism by which types are added and maintained.

6.1.29 ISO 19125-2 Geographic Information—Simple Feature Access—Part 2: SQL Option

ISO	IS	19125-2	Geographic information—Simple feature access—Part 2: SQL option	Published 2004	ISO TC211

ISO 19125-2 Available from
http://www.iso.org/iso/en/CatalogueDetailPage.CatalogueDetail?CSNUMBER=40115

This part of ISO 19125:2004 specifies an SQL schema that supports storage, retrieval, query, and update of simple geospatial feature collections via the SQL Call Level Interface (SQL/CLI), and it establishes an architecture for the implementation of feature tables.

ISO 19125-2:2004 defines terms to use within the architecture for geographic information and it defines a simple feature profile of ISO 19107. In addition, this part of ISO 19125:2004 describes a set of SQL geometry types together with SQL functions on those types. The geometry types and functions described represent a profile of ISO 13249-3.

This part of ISO 19125:2004 standardizes the names and geometric definitions of the SQL types for geometry and the names, signatures, and geometric definitions of the SQL functions for geometry.

This part of ISO 19125:2004 does not attempt to standardize and does not depend upon any part of the mechanism by which Types are added and maintained in the SQL environment, including the following:

- The syntax and functionality provided for defining types;
- The syntax and functionality provided for defining SQL functions;
- The physical storage of type instances in the database;
- Specific terminology used to refer to user defined types, for example, UDT.

6.1.30 ISO 19126 Geographic Information—Profile—FACC Data Dictionary

ISO	DELETED	19126	Geographic information—Profile—FACC Data Dictionary	DELETED Being reactivated	ISO TC211

ISO 19126 set out to specify schemas for geographic feature concept dictionaries and for managing feature concept dictionaries as registers. These registers are in accordance with ISO 19135 "Geographic information—Procedures for registration of items of geographic information." Due to lack of progress, and ISO's current policy regarding projects that do not deliver to time, the project was deleted. It is being reactivated. However, it is worthwhile spending a little time considering the aspects that this work item set out to achieve.

As described in ISO 19101 "Geographic information—Reference Model," geographic features are real-world phenomena associated with a location relative to the surface of the Earth, about which data is collected, maintained, and disseminated. This International Standard defines schemas for geographic feature concept dictionaries.

This International Standard also defines schemas for registers conformant to ISO 19135 "Geographic Information—Procedures for registration of items of geographic information." ISO 19135 specifies that a technical standard is required to define the item classes in any conformant register. This International Standard serves as the technical standard that defines the item classes required for feature concept dictionary registers.

6.1.31 ISO 19127 Geographic Information—Geodetic Codes and Parameters

ISO	IS	19127	Geographic information—Profile—FACC Data Dictionary	Published 2005	ISO TC211

ISO 19127 Available from
http://www.iso.org/iso/en/CatalogueDetailPage.CatalogueDetail?CSNUMBER=41784

ISO/TS 19127:2005 defines rules for the population and maintenance of registers of geodetic codes and parameters and identifies the data elements, in compliance with ISO 19135 and ISO 19111, required within these registers. Recommendations for the use of the registers, the legal aspects, the applicability to historic data, the completeness of the registers, and a mechanism for maintenance are specified by the registers themselves.

6.1.32 ISO 19128 Geographic Information—Web Map Server Interface

ISO	IS	19128	Geographic information—Profile—Web Map server interface	Published 2005	ISO TC211

ISO 19128 Available from
http://www.iso.org/iso/en/CatalogueDetailPage.CatalogueDetail?CSNUMBER=32546

ISO 19128:2005 specifies the behavior of a service that produces spatially referenced maps dynamically from geographic information. It specifies operations to retrieve a description of the maps offered by a server, to retrieve a map, and to query a server about features displayed on a map. ISO 19128:2005 is applicable to pictorial renderings of maps in a graphical format; it is not applicable to retrieval of actual feature data or coverage data values.

6.1.33 ISO 19129 Geographic Information—Imagery, Gridded, and Coverage Data Framework

ISO	DELETED	19129	Geographic information—Profile—Imagery, gridded and coverage data framework	DELETED	ISO TC211

This project team was characterized by the lack of experts participating in the work. Since the project team took too long a time in ISO procedures, it has been deleted. Nevertheless, some parts of ISO 19129 have been put into ISO 19101-2 and the rest of the standard will be reproposed as a NWIP soon.

The new document lacks thorough work and parts of its scope is still not covered, particularly what concerns the way of using ISO 19109 for imagery. The work item may be reactivated at some time in the future if additional resources can be found.

6.1.34 ISO 19130 Geographic Information—Sensor and Data Models for Imagery and Gridded Data

ISO	TS	19130	Geographic information—Profile—Sensor and data models for imagery and gridded data	In Development CD	ISO TC211

ISO 19130 is Not Yet Available

In March 2001 ISO/TC 211 approved a project to develop International Standard ISO 19130 "Sensor and Data Models for Imagery and Gridded Data." Since August 2002, the project has made significant progress. The standard has been advanced from the working draft (WD) stage to the committee draft (CD) stage.

The scope of the standard includes: (1) specifying a sensor model which describes the physical and geometrical properties of each kind of photogrammetric, remote sensing, and other sensors that produce imagery data; and (2) defining a conceptual data model that specifies, for each kind of sensor, the minimum content requirement and the relationship among the components of the content for the raw data that was measured by the sensor and provided in an instrument-based coordinate system, to make it possible to geo-locate and analyze the data. Meanwhile, a project team, consisting of 37 experts nominated by 16 ISO/TC 211 national members and liaison organizations, was formed to undertake the task of developing the standard.

6.1.35 ISO 19131 Geographic Information—Data Product Specifications

ISO	IS	19131	Geographic information—Data product specifications	Published 2007	ISO TC211

ISO 18131 Available from
http://www.iso.org/iso/en/CatalogueDetailPage.CatalogueDetail?CSNUMBER=36760&scopelist=PROGRAMME

ISO 19131:2007 specifies requirements for the specification of geographic data products, based upon the concepts of other ISO 19100 International Standards. It also provides help in the creation of data product specifications, so that they are easily understood and fit for their intended purpose.

6.1.36 ISO 19132 Geographic Information—Location Based Services Possible Standards

ISO	DIS	19132	Geographic information—Location based services possible standards	Under Development DIS	ISO TC211

ISO 18131 Available from
http://www.iso.org/iso/iso_catalogue/catalogue_tc/catalogue_detail.htm?csnumber=40601

This deliverable seeks to identify the manner by which ISO/TC 211 should handle the development of standards for location based services and aims to identify areas in which standards for LBS could be required. The increasing popularity and availability of LBS without a standards framework could lead to a digression of services and service quality, which may ultimately be critical for a mobile user.

ISO 19132:2007 defines a reference model and a conceptual framework for LBS and describes the basic principles by which LBS applications may interoperate. This framework references or contains an ontology, a taxonomy, a set of design patterns, and a core set of LBS service abstract specifications in UML. ISO 19132:2007 further specifies the framework's relationship to other frameworks, applications, and services for geographic information and to client applications.

6.1.37 ISO19133 Geographic Information—Location Based Services Tracking and Navigation

ISO	IS	19133	Geographic information—Location based services tracking and navigation	Published 2005	ISO TC211

ISO 19133 Available from
http://www.iso.org/iso/en/CatalogueDetailPage.CatalogueDetail?CSNUMBER=32551&scopelist=PROGRAMME

ISO 19133:2005 describes the data types, and operations associated with those types, for the implementation of tracking and navigation services. It is designed to specify Web services that can be made available to wireless devices through Web-resident proxy applications, but is not restricted to that environment.

6.1.38 ISO 19134 Geographic Information—Multimodal Location Based Services for Routing and Navigation

ISO	IS	19134	Geographic information—Multimodal location based services for routing and navigation	Published 2006	ISO TC211

ISO 19134 Available from
http://www.iso.org/iso/en/CatalogueDetailPage.CatalogueDetail?CSNUMBER=32552

ISO 19134:2006 specifies the data types and their associated operations for the implementation of multimodal LBS for routing and navigation. It is designed to specify Web services that may be made available to wireless devices through Web-resident proxy applications, but is not limited to that environment.

6.1.39 ISO 19135 Geographic Information—Procedures for Registration of Geographical Information Items

ISO	IS	19135	Geographic information—Procedures for registration of geographical information items	Published 2005	ISO TC211

ISO 19135 Available from
http://www.iso.org/iso/en/CatalogueDetailPage.CatalogueDetail?CSNUMBER=32553&scopelist=PROGRAMME

ISO 19135:2005 specifies procedures to be followed in establishing, maintaining, and publishing registers of unique, unambiguous, and permanent identifiers, as well as meanings that are assigned to items of geographic information. In order to accomplish this purpose, ISO 19135:2005 specifies elements of information that are necessary to provide identification and meaning to the registered items and to manage the registration of these items (see Figure 6.1).

NOTE: The general definition of registration has been narrowed so that registration is the assignment of linguistically independent identifiers, rather than names, to items of geographic information.

6.1.40 ISO 19136 Geographic Information—Geography Markup Language (GML)

ISO	CD	19136	Geographic information—Geography Markup Language	CD	ISO TC211

Downloadable FOC ISOtc211 Outreach 19136
www.isotc211.org/Outreach/Overview/Factsheet_19136.pdf

ISO 19136 Available from
http://www.iso.org/iso/en/CatalogueDetailPage.CatalogueDetail?CSNUMBER=32554&ICS1=35&ICS2=240&ICS3=70

The membership of the Open GIS Consortium (OGC) has approved the release of the *OpenGIS Geography Markup Language (GML) Implementation Specification* Version 3.1.0 as a publicly available OpenGIS Recommendation Paper. The GML specification is now being edited jointly in the OGC GML Revision Working Group and in ISO/TC 211/WG 4 (Geographic Information/Geomatics). The ISO version is now Committee Draft level (ISO/CD 19136), while the OGC version is character-

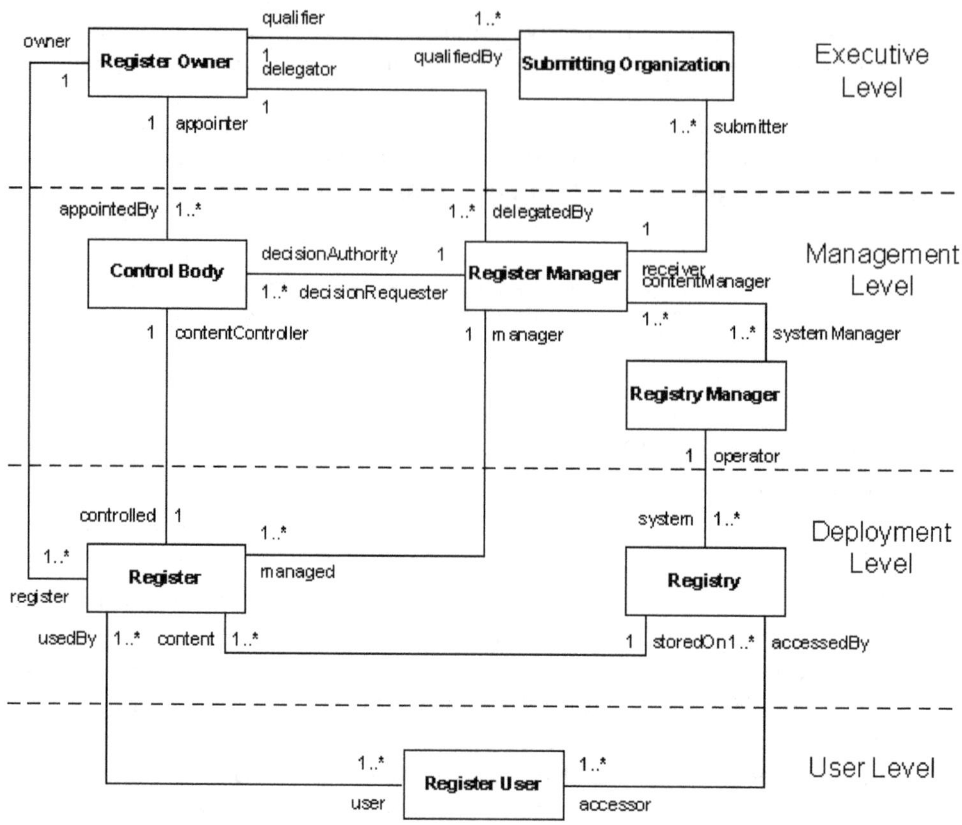

Figure 6.1 Organizational relationships between different registry "actors." (*Source:* ISO 19135 Scope.)

ized as a Recommendation Paper in order to maintain alignment with the ISO editing process.

The Geography Markup Language is the most widely supported open specification for representation of geographic (spatial and location) information. It defines XML encoding for the transport and storage of geographic information, including both the geometry and properties of geographic features. In keeping with OGC's IPR policies for OpenGIS standards, GML is freely available for use on royalty-free terms. The 601-page prose document is supported by 33 separate XML schema files.

The specification is based upon a large number of other W3C, IETF, ISO, and OpenGIS standards; it normatively references the XML Linking Language (XLink) Version 1.0 and The Schematron Assertion Language 1.5.

GML Version 3.1.0 adds new geometries, is more compliant with the ISO/TC 211 family of specifications, and contains some items for increased efficiency and simplicity. It "maintains backward compatibility for GML version 3.0.0 and 2.1.2 instance documents by preserving, but deprecating, some schema components that have been replaced by different constructs in the current version."

The OGC is "an international, member-driven, non-profit industry consortium of 258 companies, government agencies and universities participating in a consensus process to develop publicly available interface specifications and geoprocessing interoperability computing standards. OGC supports interoperable solutions that 'geo-enable' the Web, wireless and location-based services, and mainstream IT. The specifications empower technology developers to make complex spatial information and services accessible and useful with all kinds of applications."

6.1.41 ISO 19137 Geographic Information—Generally Used Profiles of the Spatial Schema and of Similar Important Other Schemas

ISO	IS	19137	Geographic information—Generally used profiles of the spatial schema and of similar important other schemas	Published 2007	ISO TC211

ISO 19137 Available from
http://www.iso.org/iso/en/CatalogueDetailPage.CatalogueDetail?CSNUMBER=32555&ICS1=35&ICS2=240&ICS3=70

ISO 19137:2007 defines a core profile of the spatial schema specified in ISO 19107 that specifies, in accordance with ISO 19106, a minimal set of geometric elements necessary for the efficient creation of application schemata.

It supports many of the spatial data formats and description languages already developed and in broad use within several nations or liaison organizations.

6.1.42 ISO 19138 Geographic Information—Data Quality Measures

ISO	IS	19138	Geographic information—Data quality measures	Published 2006	ISO TC211

ISO 19138 Available from
http://www.iso.ch/iso/en/CatalogueDetailPage.CatalogueDetail?CSNUMBER=32556&scopelist=PROGRAMME

ISO/TS 19138:2006 defines a set of data quality measures. These can be used when reporting data quality for the data quality subelements identified in ISO 19113. Multiple measures are defined for each data quality subelement, and the choice of which to use will depend on the type of data and its intended purpose.

The data quality measures are structured so that they can be maintained in a register established in conformance with ISO 19135.

ISO/TS 19138:2006 does not attempt to describe every possible data quality measure, only a set of commonly used ones.

6.1.43 ISO 19139 Geographic Information—Metadata—Implementation Specification

ISO	IS	19139	Geographic information—Metadata—Implementation specification	Published 2007	ISO TC211

ISO 19139 Available from
http://www.iso.org/iso/en/CatalogueDetailPage.CatalogueDetail?CSNUMBER=32557&scopelist=PROGRAMME

ISO/TS 19139:2007 defines Geographic MetaData XML (GMD) encoding, an XML schema implementation derived from ISO 19115.

6.1.44 ISO 19140 Geographic Information—Technical Amendment to the ISO 191** Geographic Information Series of Standards for Harmonization and Enhancements

ISO	WD	19140	Geographic information—Technical amendment to the ISO 191** Geographic information series of standards for harmonization and enhancements	Under Development WD	ISO TC211

ISO 19140 is Not Yet Available

This is a technical amendment to the ISO 191** geographic information series of standards for harmonization and enhancements. This project will develop technical amendments to the ISO 191** geographic information series of standards to achieve harmonization between them. This will include issues of consistency, cross-references, terminology, data model, and presentation. Other amendments necessary to achieve the objectives of these standards will also be included, subject to ensuring consistency with the other standards.

6.1.45 ISO 19141 Geographic Information—Schema for Moving Features

ISO	WD	19141	Geographic information—Schema for moving features	Under Development CD	ISO TC211

ISO 19141 is Not Yet Available. Track Progress at
http://www.iso.org/iso/iso_catalogue/catalogue_tc/catalogue_detail.htm?csnumber=41445

6.1.46 ISO 19142 Geographic Information—Web Feature Service

ISO	WD	19142	Geographic information—Web feature service	Under Development CD	ISO TC211

ISO 19142 is Not Yet Available. Track Progress at
http://www.iso.org/iso/iso_catalogue/catalogue_tc/catalogue_detail.htm?csnumber=42136

6.1.47 ISO 19143 Geographic Information—Filter Encoding

ISO	WD	19143	Geographic information—Filter encoding	Under Development CD	ISO TC211

ISO 19143 is Not Yet Available. Track Progress at
http://www.iso.org/iso/iso_catalogue/catalogue_tc/catalogue_detail.htm?csnumber=42137

6.1.48 ISO 19144-1 Geographic Information—Classification Systems—Part 1: Classification System Structure

ISO	WD	19144-1	Geographic information—Classification Systems—Part 1: Classification system structure	Under Development CD	ISO TC211

ISO 19144-1 is Not Yet Available. Track Progress at
http://www.iso.org/iso/iso_catalogue/catalogue_tc/catalogue_detail.htm?csnumber=32562

6.1.49 ISO 19144-2 Geographic Information—Classification Systems—Part 2: Land Cover Classification System LCCS

ISO	WD	19144-2	Geographic information—Classification Systems—Part 2: Land Cover Classification System LCCS	Under Development CD	ISO TC211

ISO 19144-2 is Not Yet Available. Track Progress at
http://www.iso.org/iso/iso_catalogue/catalogue_tc/catalogue_detail.htm?csnumber=44342

6.1.50 ISO 19145 Geographic Information—Registry of Representations of Geographic Point Locations

ISO	WD	19145	Geographic Information—Registry of representations of geographic point locations	Under Development WD	ISO TC211

ISO 19145 is Not Yet Available.

6.1.51 ISO 19146 Geographic Information—Cross-Domain Vocabularies

ISO	WD	19146	Geographic information—Cross-domain vocabularies	Under Development CD	ISO TC211

ISO 19146 is Not Yet Available. Track Progress at
http://www.iso.org/iso/iso_catalogue/catalogue_tc/catalogue_detail.htm?csnumber=44873

6.1.52 ISO 19147 Geographic Information—Location Based Services—Linear Referencing System

ISO	WD	19147	Geographic information—Location based services—Linear referencing system	Under Development WD	ISO TC211

ISO 19147 is Not Yet Available.

6.1.53 ISO 19148 Geographic Information—Location Based Services—Linear Referencing System

ISO	WD	19148	Geographic information—Location based services—Linear referencing system	Under Development WD	ISO TC211

ISO 19148 is Not Yet Available. Track Progress at
http://www.iso.org/iso/iso_catalogue/catalogue_tc/catalogue_detail.htm?csnumber=32566

6.1.54 ISO 19149 Geographic Information—Rights Expression Language for Geographic Information—GeoREL

ISO	WD	19149	Geographic information—Rights expression language for geographic information—GeoREL	Under Development WD	ISO TC211

ISO 19149 is Not Yet Available. Track Progress at
http://www.iso.org/iso/iso_catalogue/catalogue_tc/catalogue_detail.htm?csnumber=32567

6.1.55 ISO 19150 Geographic Information—Ontology

ISO	WD	19150	Geographic information—Ontology	Under Development WD	ISO TC211

ISO 19150 is Not Yet Available.

6.1.56 ISO 19151 Geographic Information—Dynamic Position Identification Scheme for Ubiquitous Space (u-position)

ISO	WD	19151	Dynamic position identification scheme for Ubiquitous space (u-position)	Under Development WD	ISO TC211

ISO 19151 is Not Yet Available. Track Progress at
http://www.iso.org/iso/iso_catalogue/catalogue_tc/catalogue_detail.htm?csnumber=32569

6.2 Data Specification, Management, and Transmission

6.2.1 General Data Specification, Management, and Transmission Standards

This reference work acknowledges that ITS services may often need to use generic data specification, management, and transmission techniques, but the subject of general data specification, management, and transmission is an area that would justify a reference work of its own, and is far too complex to be relevantly explained in this book. This section, therefore, simply provides a reference table (Table 6.1), which provides pointers to the leading relevant standards.

Open Systems Interconnect (OSI) was a series of projects started in 1982 between ISO and ITU-T to standardize networking. The core standard is generally regarded as ISO 7498, the seven-layer communications architecture model, but as can be seen from Table 6.1, there are now a number of standards in the series.

The remainder of this section deals with general communications standards that may be used to support one or more aspects of ITS service provision and may specifically be referred to in ITS standards.

6.2.2 ISO/IEC JTC1 SC6 Information Technology—Telecommunications and Information Exchange Between Systems

SC6 is an important provider of generic standards with respect to telecommunications and information exchange between systems—standards which may be utilized to support ITS service provision in both wired and wireless environments. SC6 deals with standardization in the field of telecommunications and deals with the

Table 6.1 OSI Standards for Open VMS

Layer	*Standard*	*Description*	*Observation*	*Committee*
	ISO 7498	Basic Reference Model	This is the basic 7 layer communications model that underpins most communications architecture description	See Section 6.2.2.8
Application	ISO 8571	File Transfer, Access and Management (FTAM)		
	ISO 8649	Service Definition—Association Control		
	ISO 8650	Association Control Service Element (ACSE)		
	ISO 9041	Virtual Terminal Protocol		
	ISO 9072	Remote Operations Service Element		
Presentation	ISO 8822	Presentation Service		
	ISO 8823	Presentation (Kernel)		
Session	ISO 8326	Connection-Oriented Network Service (CONS)		
	ISO 8327	Connection-Oriented Network Service Protocol		
	ISO 9595	Common Management Information Service Element (CMISE)		
Transport	ISO 8072	Connection-Oriented Transport Service (COTS) Definition		
	ISO 8073	Connection-Oriented Transport Service (COTS) Protocol		
	RFC 1006		Defines how to implement ISO 8073 Transport Class 0 on top of TCP	
	RFC 1006 Extension (Internet Draft)		Defines how to implement ISO 8073 Transport Class 2 nonuse of Explicit Flow Control on top of TCP	
Network	ISO 7498 Addendum 1	Connectionless-Mode Transmission		
	ISO 8208	X.25 Packet Level Protocol (PLP)		
	ISO 8348	Service definition: Connection-Oriented Network Service (CONS)		
	ISO 8348 Addendum 2	OSI addressing formats		
	ISO 8473,8348 Addendum 1	Connectionless-Mode Network Service (CLNS)		
	ISO 8878	X.25/Connection-Oriented Network Service (CONS)		
	ISO 8881	X.25 Packet Level Protocol in local area networks		
	ISO 9542	to intermediate-system routing exchange protocol		

Table 6.1 OSI Standards for Open VMS (continued)

Layer	*Standard*	*Description*	*Observation*	*Committee*
Data Link	ISO 3309, 4335, 7809, 8471, 8885	Point-to-point data links (HDLC)		
	ISO 7776	Point-to-point X.25 data links (only with an X.25 license; LAPB, one possible user)		
	ISO 8802-1 (IEEE 802.1)	Ethernet support (CSMA-CD) (*NOTE: DECnet-Plus supports only the addressing specifications in this standard*)	Published	JTC1/SC6
	ISO 8802-2 (IEEE 802.2)	Frame formats for 8802-3 LANs (CSMA-CD logical link control (LLC1)) X.25 logical link control (LLC2) (one possible user)	Published	JTC1/SC6
	ISO 8802-3 (CSMA/CD)	LAN support (CSMA-CD)	Published/ Under Revision	JTC1/SC6
	ISO 9314	Fiber Distributed Data Interface (FDDI)		
Physical	ISO 8802-3 (IEEE 802.3)	CSMA-CD devices	Published/Under Revision	JTC1/SC6
	ISO 9314	Fiber Distributed Data Interface (FDDI)		
	EIA RS-232-C	HDLC point-to-point devices		
	EIA RS-422	HDLC point-to-point devices		
	EIA RS-423	HDLC point-to-point devices		
	CCITT V.35	HDLC point-to-point devices		

Source: CSI Library.

exchange of information between open systems including system functions, procedures, parameters and equipment, as well as the conditions for their use. This standardization includes both the lower layers that support the physical, data link, network, and transport services, including private integrated services networking, as well as the upper layers that support the application protocols and services such as directory and ASN.1. A vital aspect of this work is done in effective cooperation with ITU-T and other worldwide and regional standardization bodies (e.g., IEEE and IETF).

Some of the standards that may be relevant to ITS service provision are detailed in this section.

The work program of ISO/IEC JTC1 SC6 can be downloaded from:

http://www.iso.org/iso/iso_catalogue/catalogue_tc/catalogue_tc_browse.htm?commid=45072&development=true

6.2.2.1 ISO 1155 Use of Longitudinal Parity to Detect Errors in Information Messages

ISO/IEC	IS	1155	Information processing—Use of longitudinal parity to detect errors in information messages	Published 1978	JTC1/SC6

ISO 1155 Available from
http://www.iso.org/iso/iso_catalogue/catalogue_tc/catalogue_detail.htm?csnumber=5723

6.2.2.2 ISO 1177 Character Structure for Start/Stop and Synchronous Character Oriented Transmission

ISO/IEC	IS	1177	Information processing—Character structure for start/stop and synchronous character oriented transmission	Published 1985	JTC1/SC6

ISO 1177 Available from
http://www.iso.org/iso/iso_catalogue/catalogue_tc/catalogue_detail.htm?csnumber=5762

6.2.2.3 ISO 1745 Basic Mode Control Procedures for Data Communication Systems

ISO/IEC	IS	1745	ISO: Information processing—Basic mode control procedures for data communication systems	Published 1975	JTC1/SC6

ISO 1745 Available from
http://www.iso.org/iso/iso_catalogue/catalogue_tc/catalogue_detail.htm?csnumber=6365

6.2.2.4 ISO 2628 Basic Mode Control Procedures—Complements

ISO/IEC	IS	2628	Basic mode control procedures—Complements	Published 1973	JTC1/SC6

ISO 2628 Available from
http://www.iso.org/iso/iso_catalogue/catalogue_tc/catalogue_detail.htm?csnumber=7605

6.2.2.5 ISO 2629 Basic Mode Control Procedures—Conversational Information Message Transfer

ISO/IEC	IS	2629	Basic mode control procedures—Conversational information message transfer	Published 1973	JTC1/SC6

ISO 2629 Available from
http://www.iso.org/iso/iso_catalogue/catalogue_tc/catalogue_detail.htm?csnumber=7606

6.2.2.6 ISO 7478 Data Communication—Multilink

ISO/IEC	IS	7478	Information processing systems—Data communication—Multilink	Published 1987 Cor 1 1989	JTC1/SC6

ISO 7478 Available from
http://www.iso.org/iso/iso_catalogue/catalogue_tc/catalogue_detail.htm?csnumber=14216
ISO 7478 Cor 1Available from
http://www.iso.org/iso/iso_catalogue/catalogue_tc/catalogue_detail.htm?csnumber=14218

This standard defines multilink procedures where multiple parallel data links at the data link layer are used to provide a variable bandwidth data link between network layer entities. It does not specify the way in which single data link protocol functions indicate to multilink procedures that the transmission of a multilink frame has been successful.

ISO/IEC	IS	7478 Cor 1	Information processing systems—Data communication—Multilink	ISO:1987/ Cor 1:1989	JTC1/SC6

Page 10, clause A.2 MLP reset initiated by a single station: replace the existing diagram by that given in this Corrigendum.

6.2.2.7 ISO 7480 Telecommunications and Information Exchange Between Systems—Start-Stop Transmission Signal Quality at DTE/DCE Interfaces

ISO/IEC	IS	7480	Information technology—Telecommunications and information exchange between systems—Start-stop transmission signal quality at DTE/DCE interfaces	Published 1991	JTC1/SC6

ISO 7480 Available from
http://www.iso.org/iso/iso_catalogue/catalogue_tc/catalogue_detail.htm?csnumber=14221

This standard specifies signal quality requirements for serial data transmission. The interface of concern conforms to CCITT Recommendations V.24 (telephone networks) and X.24 (data networks). Annex A is for information only.

6.2.2.8 ISO 7498 Information Technology—Open Systems Interconnection

ISO/IEC	IS	7498	Information technology—Open Systems Interconnection		JTC1/SC6

This standard is in several parts.

ISO/IEC	IS	7498-1	Information technology—Open Systems Interconnection—Basic Reference Model: The Basic Model	Published 1994 2nd Edition	JTC1/SC6

Downloadable FOC ISO/IEC 7498-1:1994
http://standards.iso.org/ittf/PubliclyAvailableStandards/s020269_ISO_IEC_7498-1_1994(E).zip

This is the basic communications systems architecture model that underpins most modern communication architecture descriptions.

The current 1994 version cancels and replaces the first edition (1984). The model provides a common basis for the coordination of standards development for the purpose of systems interconnection, while allowing existing standards to be placed into perspective within the overall reference model. The model identifies areas for developing or improving standards. It does not intend to serve as an implementation specification.

ISO/IEC	IS	7498-3	Information technology—Open Systems Interconnection—Basic Reference Model: Part 3 : Naming and addressing	Published 1994 2nd Edition	JTC1/SC6

Downloadable FOC ISO/IEC 7498-3:1997
http://standards.iso.org/ittf/PubliclyAvailableStandards/s025022_ISO_IEC_7498-3_1997(E).zip

ISO/IEC	IS	7498-4	Information technology—Open Systems Interconnection—Basic Reference Model: Part 4 : Management framework	Published 1994 1st Edition	JTC1/SC6

Downloadable FOC ISO/IEC 7498-4:1989
http://standards.iso.org/ittf/PubliclyAvailableStandards/s014258_ISO_IEC_7498-4_1989(E).zip

6.2.2.9 ISO 7776 Telecommunications and Information Exchange Between Systems—High-Level Data Link Control Procedures—Description of the X.25 LAPB-Compatible DTE Data Link Procedures

ISO/IEC	IS	7776	ISO/IEC: Information technology—Telecommunications and information exchange between systems—High-level data link control procedures—Description of the X.25 LAPB-compatible DTE data link procedures	Published 1995 Amd1 1996	JTC1/SC6

ISO 7776 Available from
http://www.iso.org/iso/iso_catalogue/catalogue_tc/catalogue_detail.htm?csnumber=22172
ISO 7776 Amd 1 Available from
http://www.iso.org/iso/iso_catalogue/catalogue_tc/catalogue_detail.htm?csnumber=23304

This cancels and replaces the first edition (1986). It defines an application of certain HDLC standards, and also defines the structure, elements, and procedures for the operation of a DTE using the X.25 LAPB protocol. The procedures are applicable to data interchange between a DTE and a DCE, or between two DTEs. The procedures are defined for use on duplex links, using synchronous transmission or start/stop transmission.

Amd 1 adds the multiselective reject option (option 3.3) and modulo 32 768 (option 10.2) from ISO/IEC 7809:1993.

6.2.2.10 ISO 8072 Open Systems Interconnection (OSI)—Transport Service Definition

ISO/IEC	IS	8072	Information technology—Open systems interconnection—Transport service definition	Published 1996	JTC1/SC6

ISO 8072 Available from
http://www.iso.org/iso/iso_catalogue/catalogue_tc/catalogue_detail.htm?csnumber=27497

6.2.2.11 ISO 8073 Open Systems Interconnection (OSI)—Protocol for Providing the Connection-Mode Transport Service

ISO/IEC	IS	8073	Information technology—Open Systems Interconnection—Protocol for providing the connection-mode transport service	Published 1997 4th Edition	JTC1/SC6

Downloadable FOC ISO/IEC 8073:1997
http://standards.iso.org/ittf/PubliclyAvailableStandards/s024077_ISO_IEC_8073_1997(E).zip

ISO/IEC	IS	8073 Amd 1	Relaxation of class conformance requirements and expedited data service feature negotiation	1997/Amd 1:1998	JTC1/SC6

Downloadable FOC ISO/IEC 8073:1997/Amd.1:1998 (E)
http://standards.iso.org/ittf/PubliclyAvailableStandards/c028666_ISO_IEC_8073_1997_Amd_1_1998(E).zip

6.2.2.12 ISO 8348 Open Systems Interconnection—Network Service Definition

ISO/IEC	IS	8348	Information technology—Open Systems Interconnection—Network service definition	Published 2002 3rd Edition	JTC1/SC6

Downloadable FOC ISO/IEC 8348:2002 (E)
http://standards.iso.org/ittf/PubliclyAvailableStandards/c035872_ISO_IEC_8348_2002(E).zip

ISO/IEC 8348:2002 defines the set of capabilities, in terms of abstract service definition, provided by the network layer to the transport layer. For designers of transport layer protocols, it provides a definition of the network service to allow design and implementation independent of details of the Network Layer protocol. For designers of network layer protocols, it defines the set of capabilities to be made available through the action of the protocol.

6.2.2.13 ISO 8473 Protocol for Providing the Connectionless-Mode Network Service

ISO/IEC	IS	8473-1	Information technology—Protocol for providing the connectionless-mode network service: Protocol specification	Published 1998 2nd Edition	JTC1/SC6

Downloadable FOC ISO/IEC 8473-1:1998
http://standards.iso.org/ittf/PubliclyAvailableStandards/c030931_ISO_IEC_8473-1_1998(E).zip

ISO/IEC	IS	8473-2	Information technology—Protocol for providing the connectionless-mode network service—Part 2: Provision of the underlying service by an ISO/IEC 8802 Sub-network	Published 1996	JTC1/SC6

Downloadable FOC ISO/IEC 8473-2:1996
http://standards.iso.org/ittf/PubliclyAvailableStandards/s023306_ISO_IEC_8473-2_1996(E).zip

This differs from the other related International Standards by the layers defined in ISO/IEC 7498-1. In particular, it defines the way in which a local area network conforms.

ISO/IEC	IS	8473-3	Information technology—Protocol for providing the connectionless-mode network service—Part 3: Provision of the underlying service by an X.25 subnetwork	Published 1995	JTC1/SC6

Downloadable FOC ISO/IEC 8473-3:1995
http://standards.iso.org/ittf/PubliclyAvailableStandards/s023305_ISO_IEC_8473-3_1995(E).zip

ISO/IEC	IS	8473-4	Information technology—Protocol for providing the connectionless-mode network service: Provision of the underlying service by a Sub-network that provides the OSI data link service	Published 1995	JTC1/SC6

Downloadable FOC ISO/IEC 8473-4:1995
http://standards.iso.org/ittf/PubliclyAvailableStandards/s023335_ISO_IEC_8473-4_1995(E).zip

This specifies the way in which the underlying service assumed by the protocol defined by ITU-T Rec. X.233 ISO/IEC 8473-1 is provided by a subnetwork which provides the OSI data link service defined by CCITT Rec. X.212 ISO/IEC 8886, through the operation of a subnetwork dependent convergence function (SNDCF) as described in ISO/IEC 8648. It also provides the PICS pro forma for this protocol, in compliance with the relevant requirements and in accordance with the relevant guidance.

ISO/IEC	IS	8473-5	Information technology—Protocol for providing the connectionless-mode network service: Provision of the underlying service by ISDN circuit-switched B-channels	Published 1997	JTC1/SC6

Downloadable FOC ISO/IEC 8473-5:1997
http://standards.iso.org/ittf/PubliclyAvailableStandards/s020905_ISO_IEC_8473-5_1997(E).zip

6.2.2.14 ISO 8602 Protocol for Providing the OSI Connectionless-Mode Transport Service

ISO/IEC	IS	8602	Information technology—Protocol for providing the OSI connectionless-mode transport service	Published 2nd edition	JTC1/SC6

Downloadable FOC ISO/IEC 8602:1995
http://standards.iso.org/ittf/PubliclyAvailableStandards/s015907_ISO_IEC_8602_1995(E).zip

ISO/IEC	IS	8602 Amd 1	Information technology—Protocol for providing the OSI connectionless-mode transport service Amendment 1 : Addition of connectionless-mode multicast capability	Published	JTC1/SC6

Downloadable FOC ISO/IEC 8602:1995/Amd.1:1996
http://standards.iso.org/ittf/PubliclyAvailableStandards/s024883_ISO_IEC_8602_1995_Amd_1_1996(E).zip

6.2.2.15 ISO 8480 DTE/DCE Interface Back-Up Control Operation Using ITU-T Recommendation V.24 Interchange Circuits

ISO/IEC	IS	8480	Information technology—Telecommunications and information exchange between systems—DTE/DCE interface back-up control operation using ITU-T Recommendation V.24 interchange circuits	Published 1995	JTC1/SC6

ISO 8480 Available from
http://www.iso.org/iso/iso_catalogue/catalogue_tc/catalogue_detail.htm?csnumber=24074

This specifies the procedures for back-up operation applicable to data transmission on a leased line when implemented through a single interface, controlling both the leased and the back-up facilities, using ITU-T Recommendation V.24 interchange circuits in conjunction with either the 25-pole connector (ISO 2110) or the 26-pole connector (ISO/IEC 11569). This cancels and replaces the first edition, which has been technically revised.

6.2.2.16 ISO 8481 Telecommunications and Information Exchange Between Systems—DTE to DTE Direct Connections

ISO/IEC	IS	8481	Information technology—Telecommunications and information exchange between systems—DTE to DTE direct connections	Published 1996	JTC1/SC6

ISO 8481 Available from
http://www.iso.org/iso/iso_catalogue/catalogue_tc/catalogue_detail.htm?csnumber=26137

This describes an arrangement for interconnection of data terminal equipment (DTE), without intermediate data circuit-terminating Equipment (DCE), in terms of electrical, mechanical, and functional characteristics. It applies to DTEs with interface circuits standardized in CCITT Rec. X.24 for transmission over public data networks (PDN). This replaces the first edition.

6.2.2.17 ISO 8602 Protocol for Providing the OSI Connectionless-Mode Transport Service

ISO/IEC	IS	8602	Information technology—Protocol for providing the OSI connectionless-mode transport service	Published 1995	JTC1/SC6

ISO 8602 Available from
http://www.iso.org/iso/iso_catalogue/catalogue_tc/catalogue_detail.htm?csnumber=15907

NOTE: The electronic version of this International Standard can be downloaded from the ISO/IEC Information Technology Task Force (ITTF) Web site.

http://isotc.iso.org/livelink/livelink/fetch/2000/2489/Ittf_Home/PubliclyAvailableStandards.htm

This cancels and replaces the first edition (1987). It specifies: procedures for the connectionless-mode transmission of data and protocol control information; procedures for the correct interpretation of transport protocol control information; the functional requirements for implementations claiming conformance to this standard; and the encoding of the transport protocol data units used for the transmission of data and control information.

ISO/IEC	IS	8602 Amd 1	Information technology—Protocol for providing the OSI connectionless-mode transport service Amd 1:Addition of connectionless-mode multicast capability	Amd 1 1996	JTC1/SC6

ISO 8481 Available from
http://www.iso.org/iso/iso_catalogue/catalogue_tc/catalogue_detail.htm?csnumber=24883

NOTE: The electronic version of this International Standard can be downloaded from the ISO/IEC Information Technology Task Force (ITTF) Web site.

http://isotc.iso.org/livelink/livelink/fetch/2000/2489/Ittf_Home/PubliclyAvailableStandards.htm

This defines additional assumptions concerning the services optionally provided by the network layer and adds no new functions of its own. The identical text is published as ITU-T Rec. X.234/Amd. 1.

6.2.2.18 ISO 8648 Open Systems Interconnection—Internal Organization of the Network Layer

ISO/IEC	IS	8648	ISO: Information processing systems—Open Systems Interconnection—Internal organization of the Network Layer	Published 1988	JTC1/SC6

ISO 8648 Available from
http://www.iso.org/iso/iso_catalogue/catalogue_tc/catalogue_detail.htm?csnumber=16011
ISO 8648:1988/Cor 1:1991 Available from
http://www.iso.org/iso/iso_catalogue/catalogue_tc/catalogue_detail.htm?csnumber=16012

This provides an architectural model of the OSI network layer as a frame for OSI network layer standardization. Is intended for use in the design and application of network layer protocols. It is to be used: to provide a common set of concepts and terminology for network layer standards; to analyze network layer functionality and classify network layer protocols; to specify the use of "real network" in supporting or providing the OSI network service, above all, in cases where multiple "real network" are to be interconnected and used.

Cor 1: includes changes to Subclauses: 3.3, 5.2.2, 5.4, and 5.5 (of the standard): "relay system" will be replaced by "network-relay system."

6.2.2.19 ISO 8802 Telecommunications and Information Exchange Between Systems—Local and Metropolitan Area Networks

ISO/IEC/ IEEE	IS	8802	ISO/IEC 8802-11:2005 Information technology—Telecommunications and information exchange between systems—Local and metropolitan area networks		ISO JTC1/ SC6

This standard is ostensibly in 11 parts, but some parts do not exist and some have been withdrawn.

See also Section 6.2.1 and IEEE 802.11 Sections 3.8.3, 4.2.3, and 4.6.7.1.1. This should be viewed in coordination with (confusingly) ISO 11802 (see Section 6.2.2.49).

ISO/IEC/ IEEE	IS	8802-1	Information technology—Telecommunications and information exchange between systems—Local and metropolitan area networks—Specific requirements—Part 1: Overview of Local Area Network Standards	Published 2001	JTC1 SC6

ISO 8802-1 Available from
http://www.iso.org/iso/iso_catalogue/catalogue_tc/catalogue_detail.htm?csnumber=34385

ISO/IEC/ IEEE	IS	8802-2	Information technology—Telecommunications and information exchange between systems—Local and metropolitan area networks—Technical reports and guidelines—Part 2: Logical link control	Published 2005	JTC1 SC6

ISO 8802-2 Available from
http://www.iso.org/iso/iso_catalogue/catalogue_tc/catalogue_detail.htm?csnumber=30174
ISO 8802-2 Cor 1 Available from
http://www.iso.org/iso/iso_catalogue/catalogue_tc/catalogue_detail.htm?csnumber=34553

ISO/IEC/ IEEE	IS	8802-3	Information technology—Telecommunications and information exchange between systems—Local and metropolitan area networks—Specific requirements—Part 3: Carrier sense multiple access with collision detection (CSMA/CD) access method and physical layer specifications	Published Draft Amds 1,2,3,4 Under development	JTC1 SC6

ISO 8802-3 Available from
http://www.iso.org/iso/iso_catalogue/catalogue_tc/catalogue_detail.htm?csnumber=31002
ISO/IEC 8802-3/DAmd 1
http://www.iso.org/iso/iso_catalogue/catalogue_tc/catalogue_detail.htm?csnumber=38885
ISO/IEC 8802-3/DAmd 2
http://www.iso.org/iso/iso_catalogue/catalogue_tc/catalogue_detail.htm?csnumber=39543
ISO/IEC 8802-3/DAmd 3
http://www.iso.org/iso/iso_catalogue/catalogue_tc/catalogue_detail.htm?csnumber=40655
ISO/IEC 8802-3/DAmd 4
http://www.iso.org/iso/iso_catalogue/catalogue_tc/catalogue_detail.htm?csnumber=40656

Draft Amd 1: Media Access Control (MAC) Parameters, Physical Layers, and Management Parameters for 10 Gb/s Operation.

Draft Amd 2: Data Terminal Equipment (DTE) power via Media Dependent Interface (MDI).

Draft Amd 3: Maintenance.

Draft Amd 4: Physical layer and management parameters for 10Gb/s operation, type 10GBASE-CX4.

ISO/IEC/ IEEE	IS	8802-4	Information technology—Telecommunications and information exchange between systems—Local and metropolitan area networks—Specific requirements—Information processing systems—Local area networks—Part 4: Token-passing bus access method and physical layer specifications	withdrawn	JTC1 SC6

ISO 8802-4 is withdrawn

ISO/IEC/ IEEE	IS	8802-5	ISO/IEC 8802-11:2005 Information technology(Telecommunications and information exchange between systems—Local and metropolitan area networks—Specific requirements—Part 5: Token ring access method and physical layer specifications	Published	JTC1 SC6

ISO 8802-5 Available from http://www.iso.org/iso/iso_catalogue/catalogue_tc/catalogue_detail.htm?csnumber=29923

ISO/IEC/ IEEE	IS	8802-6	Information technology—Telecommunications and information exchange between systems—Local and metropolitan area networks—Specific requirements—Part 6: Distributed Queue Dual Bus (DQDB) access method and physical layer specifications	Withdrawn	JTC1 SC6

ISO 8802-6 is withdrawn

ISO/IEC/ IEEE	IS	8802-7	Information technology—Telecommunications and information exchange between systems—Local and metropolitan area networks—Specific requirements—Part 7: Slotted ring access method and physical layer specification	Withdrawn	JTC1 SC6

ISO 8802-7 is withdrawn

8802-8 does not exist as a published standard, withdrawn standard, standard under development, or deleted work item.

ISO/IEC/ IEEE	IS	8802-9	Information technology—Telecommunications and information exchange between systems—Local and metropolitan area networks—Specific requirements—Part 9: Integrated Services (IS) LAN Interface at the Medium Access Control (MAC) and Physical (PHY) Layers	Withdrawn	JTC1 SC6

ISO 8802-9 is withdrawn

8802-10 does not exist as a published standard, withdrawn standard, standard under development, or deleted work item.

ISO/IEC/ IEEE	IS	8802-11	Information technology—Telecommunications and information exchange between systems—Local and metropolitan area networks—Specific requirements—Part 11: Wireless LAN Medium Access Control (MAC) and Physical Layer (PHY) specifications	Published 2005 Amd 4,5,6 2006	JTC1 SC6

ISO 8802-11 Available from
http://www.iso.org/iso/iso_catalogue/catalogue_tc/catalogue_detail.htm?csnumber=39777
ISO 8802-11 Amd 4 Available from
http://www.iso.org/iso/iso_catalogue/catalogue_tc/catalogue_detail.htm?csnumber=41476
ISO 8802-11 Amd 5 Available from
http://www.iso.org/iso/iso_catalogue/catalogue_tc/catalogue_detail.htm?csnumber=41477
ISO 8802-11 Amd 6 Available from
http://www.iso.org/iso/iso_catalogue/catalogue_tc/catalogue_detail.htm?csnumber=41243

The medium access control (MAC) and physical characteristics for wireless local area networks (WLANs) are specified in ISO/IEC 8802-11:2005, which is part of a series of standards for local and metropolitan area networks. The MAC unit in ISO/IEC 8802-11:2005 is designed to support physical layer units, as they may be adopted dependent on the availability of spectrum. ISO/IEC 8802-11:2005 contains five physical layer units: four radio units, operating in the 2,400 to 2,500-MHz band and in the bands comprising 5.15 to 5.25 GHz, 5.25 to 5.35 GHz, and 5.725 to 5.825 GHz, and one baseband infrared (IR) unit. One radio unit employs the frequency-hopping spread spectrum (FHSS) technique, two employ the direct sequence spread spectrum (DSSS) technique, and another employs the orthogonal frequency division multiplexing (OFDM) technique.

Amds 1–3 are incorporated into the current version.

Amd 4: Further Higher Data Rate Extension in the 2.4 GHz Band.

Amd 5: Spectrum and Transmit Power Management Extensions in the 5 GHz band in Europe.

Amd 6: Medium Access Control (MAC) Security Enhancements.

6.2.2.20 ISO 8824/5 Abstract Syntax Notation.1

ASN.1 has particular significance in the provision and support of ITS services because ISO/TC 204 has mandated that, for interoperability, the definition of data in ITS standards is always defined in ASN.1 in accordance with the specifications in ISO 8824 (but not necessarily defined only in ASN.1). Several ITS standards also use the encoding specified in ISO 8825, but this is not mandatory and other encoding tools such as XML are widely used.

6.2.2.20.1 Information Technology—Abstract Syntax Notation.1 (ASN.1)

ISO/TC 204 mandates the definition of data to be specified in ASN.1 (in addition to any other formats) in order to establish interoperability. As listed in Section

6.1.5.3, the standards for ASN.1 have been defined by ISO/IECJTC1/SC6. Because of the significance of the use of ASN.1 in ITS, it is shown in a separate subsection for easy reference. The standards for ASN.1 are specified in ISO/IEC 8824 (4 Parts) and ISO/IEC 8825 (4 Parts). ISO/IEC 8824 specifies basic notation, while ISO/IEC 8825 specifies several alternative encoding rules (including encoding for XML/XSD).

ISO/IEC/ ITU	IS/Rec	8824-1 X.680/ 8824-1:2002	Information technology—Abstract Syntax Notation One (ASN.1): Specification of basic notation	Published 2002 Cor 1 2006	JTC1/ SC6 ITU

ISO 8824-1 Available from
http://www.iso.org/iso/iso_catalogue/catalogue_tc/catalogue_detail.htm?csnumber=35684
ISO 8824-1 Cor 1 Available from
http://www.iso.org/iso/iso_catalogue/catalogue_tc/catalogue_detail.htm?csnumber=43764
Downloadable FOC as ITU X.680
http://www.itu.int/ITU-T/studygroups/com17/languages/X.680-0207.pdf

ISO/IEC 8824-1:2002 provides a notation called Abstract Syntax Notation One (ASN.1) for defining the syntax of information data. It defines a number of simple data types and specifies a notation for referencing these types and for specifying values of these types.

The ASN.1 notations can be applied whenever it is necessary to define the abstract syntax of information without constraining in any way how the information is encoded for transmission.

ISO/IEC/ ITU	IS/Rec	8824-2 X.681:2002 (2002)	Information technology—Abstract Syntax Notation One (ASN.1): Information object specification	Published 2002 Amd1 2004	JTC1/ SC6 ITU

ISO 8824-2 Available from
http://www.iso.org/iso/iso_catalogue/catalogue_tc/catalogue_detail.htm?csnumber=35685
ISO 8824-2 Amd 1 Available from
http://www.iso.org/iso/iso_catalogue/catalogue_tc/catalogue_detail.htm?csnumber=40232
Downloadable FOC as ITU X.681
http://www.itu.int/ITU-T/studygroups/com17/languages/X.681-0207.pdf

ISO/IEC 8824-2:2002 provides the ASN.1 notation which allows information object classes as well as individual information objects and sets thereof to be defined and given reference names. An information object class defines the form of a conceptual table (an information object set) with one column for each field in the information object class, and with each complete row defining an information object.

Amd 1: Support for EXTENDED-XER.

ISO/IEC/ ITU	IS/Rec	8824-3 X.682 2002	Information technology—Abstract Syntax Notation One (ASN.1): Constraint specification	Published 2002	JTC1/ SC6 ITU

ISO 8824-3 Available from
http://www.iso.org/iso/iso_catalogue/catalogue_tc/catalogue_detail.htm?csnumber=35686

Downloadable FOC as ITU X.682
http://www.itu.int/ITU-T/studygroups/com17/languages/X.682-0207.pdf

ISO/IEC 8824-3:2002 provides the ASN.1 notation for the general case of constraint and exception specification by which the data values of a structured data type can be limited. The notation also provides for signaling if and when a constraint is violated.

ISO/IEC/ ITU	IS/Rec	8824-4 X.683:2002	Information technology—Abstract Syntax Notation One (ASN.1): Parameterization of ASN.1 specifications	Published 2002	JTC1/ SC6 ITU

ISO 8824-4 Available from
http://www.iso.org/iso/iso_catalogue/catalogue_tc/catalogue_detail.htm?csnumber=35687

Downloadable FOC as ITU X.683
http://www.itu.int/ITU-T/studygroups/com17/languages/X.683-0207.pdf

ISO/IEC 8824-4:2002 defines the provisions for parameterized reference names and parameterized assignments for data types which are useful for the designer when writing specifications where some aspects are left undefined at certain stages of the development to be filled in at a later stage to produce a complete definition of an abstract syntax.

6.2.2.20.2 Information Technology—ASN.1 Encoding Rules

ISO/IEC/ ITU	IS/Rec	8825-1 X.690 (2002)	Information technology—ASN.1 encoding rules: Specification of Basic Encoding Rules (BER), Canonical Encoding Rules (CER) and Distinguished Encoding Rules (DER)	Published 2002 Amd 1 2004	JTC1/ SC6 ITU

ISO 8825-1 Available from
http://www.iso.org/iso/iso_catalogue/catalogue_tc/catalogue_detail.htm?csnumber=35688

ISO 8825-1 Amd 1 Available from
http://www.iso.org/iso/iso_catalogue/catalogue_tc/catalogue_detail.htm?csnumber=40233

Downloadable FOC as ITU X.690
http://www.itu.int/ITU-T/studygroups/com17/languages/X.690-0207.pdf

ISO/IEC 8825-1:2002 defines a set of Basic Encoding Rules (BER) that may be applied to values of types defined using the ASN.1 notation. Application of these encoding rules produces a transfer syntax for such values. It is implicit in the specification of these encoding rules that they are also used for decoding. ISO/IEC 8825-1:2002 also defines a set of Distinguished Encoding Rules (DER) and a set of Canonical Encoding Rules (CER) both of which provide constraints on the BER. The key difference between them is that DER uses the definite length form of encoding while CER uses the indefinite length form. DER is more suitable for the small encoded values, while CER is more suitable for the large ones. It is implicit in the specification of these encoding rules that they are also used for decoding.

Amd 1: Support for EXTENDED-XER.

ISO/IEC/ ITU	IS/Rec	8825-2 X.691 (2002)	Information technology—ASN.1 encoding rules: Specification of Packed Encoding Rules (PER)	Published 2002 Amd 1 2004 Cor 1 2006 Amd 2 2007 Cor 2 2006	JTC1/ SC6 ITU

ISO 8825-2 Available from
http://www.iso.org/iso/iso_catalogue/catalogue_tc/catalogue_detail.htm?csnumber=35689
ISO 8825-2 Amd 1 Available from
http://www.iso.org/iso/iso_catalogue/catalogue_tc/catalogue_detail.htm?csnumber=40234
ISO 8825-2 Cor 1 Available from
http://www.iso.org/iso/iso_catalogue/catalogue_tc/catalogue_detail.htm?csnumber=43765
ISO 8825-2 Amd 2 Available from
http://www.iso.org/iso/iso_catalogue/catalogue_tc/catalogue_detail.htm?csnumber=43803
ISO 8825-2Cor 2 Available from
http://www.iso.org/iso/iso_catalogue/catalogue_tc/catalogue_detail.htm?csnumber=44410
Downloadable FOC as ITU X.691
http://www.itu.int/ITU-T/studygroups/com17/languages/X.691-0207.pdf

ISO/IEC 8825-2:2002 describes a set of encoding rules that can be applied to values of all ASN.1 types to achieve a much more compact representation than that achieved by the BER and its derivatives (described in ITU-T Rec.X.690 | ISO/IEC 8826-1).

Amd 1: Support for EXTENDED-XER.

Amd 2: Time type support.

ISO/IEC/ ITU	IS/Rec	8825-3 X.692 (200)	ISO/IEC: Information technology—ASN.1 encoding rules: Specification of Encoding Control Notation (ECN)	Published 2002 Amd 1 2005 Cor 1 2006 Amd 2 2006	JTC1/SC6

ISO 8825-3 Available from
http://www.iso.org/iso/iso_catalogue/catalogue_tc/catalogue_detail.htm?csnumber=34390

ISO 8825-3 Amd 1 Available from
http://www.iso.org/iso/iso_catalogue/catalogue_tc/catalogue_detail.htm?csnumber=40054

ISO 8825-3 Cor 1 Available from
http://www.iso.org/iso/iso_catalogue/catalogue_tc/catalogue_detail.htm?csnumber=43766

ISO 8825-3 Amd 2 Available from
http://www.iso.org/iso/iso_catalogue/catalogue_tc/catalogue_detail.htm?csnumber=43800

Downloadable FOC as ITU X.692
http://www.itu.int/ITU-T/studygroups/com17/languages/X.692-0203.pdf

Downloadable FOC ISO/IEC 8825-3:2003 Annex E
http://standards.iso.org/ittf/PubliclyAvailableStandards/c034390_ISO_8825-3_2003(E)_Annex_E.zip

ISO/IEC 8825-3:2002 defines the Encoding Control Notation (ECN) used to specify encodings (of ASN.1 types) that differ from those provided by standardized encoding rules such as the BER and the Packed Encoding Rules (PER).

Amd 1: Extensibility support.

ISO/IEC/ ITU	IS/Rec	8825-4 X.693 (2002)	ISO/IEC: Information technology—ASN.1 encoding rules: XML Encoding Rules (XER)	Published 2002 Amd 1 2004 Amd 2 2006	JTC1/SC6

ISO 8825-4 Available from
http://www.iso.org/iso/iso_catalogue/catalogue_tc/catalogue_detail.htm?csnumber=34389

ISO 8825-4 Amd 1 Available from
http://www.iso.org/iso/iso_catalogue/catalogue_tc/catalogue_detail.htm?csnumber=37232

ISO 8825-4 Amd 2 Available from
http://www.iso.org/iso/iso_catalogue/catalogue_tc/catalogue_detail.htm?csnumber=43804

Downloadable FOC as ITU X.693
http://www.itu.int/ITU-T/studygroups/com17/languages/X.693-0112.pdf

ISO/IEC 8825-4:2002 specifies rules for encoding values of ASN.1 types using XML.

Amd 1: EXTENDED-XER

Amd 2: Time type support.

ISO/IEC/ ITU	IS/Rec	8825-5 X.694 (2002)	Information technology—ASN.1 encoding rules: Mapping W3C XML schema definitions into ASN.1	Published 2004 Cor 1 2006	JTC1/SC6

ISO 8825-5 Available from
http://www.iso.org/iso/iso_catalogue/catalogue_tc/catalogue_detail.htm?csnumber=37979

ISO 8825-5 Cor 1 Available from
http://www.iso.org/iso/iso_catalogue/catalogue_tc/catalogue_detail.htm?csnumber=44409

Downloadable FOC as ITU X.693
http://www.itu.int/ITU-T/studygroups/com17/languages/X694.pdf

ISO/IEC 8825-5:2004 defines rules for mapping an XSD schema (a schema conforming to the W3C XML schema specification) to an ASN.1 schema in order to use ASN.1 encoding rules such as BER, DER, PER, or the XML Encoding Rules (XER) for the transfer of information defined by the XSD schema.

The use of ISO/IEC 8825-5:2004 with the ASN.1 Extended XML Encoding Rules (EXTENDED-XER) provides the same XML representation of values as that defined by the original XSD schema.

6.2.2.20.3 Generic Applications of ASN.1: Fast Infoset

At the time of this writing, this standard comprised four parts.

ISO/IEC	FDIS	24824	Information technology—Generic applications of ASN.1: Fast Infoset.		JTC1/SC6

ISO/IEC	IS	24824-1	Information technology—Generic applications of ASN.1: Fast infoset	Published 2007	JTC1/SC6

ISO/IEC 24824-1:200X
http://www.iso.org/iso/iso_catalogue/catalogue_tc/catalogue_detail.htm?csnumber=41327
(to be downloadablew FOC at some time in near future
http://standards.iso.org/ittf/PubliclyAvailableStandards/

ISO/IEC 24824-1:2007 specifies a representation of an instance of the W3C XML Information Set using binary encodings. These binary encodings are specified using the ASN.1 notation and the ASN.1 Encoding Control Notation (ECN).

The technology specified in ISO/IEC 24824-1:2007 is called Fast Infoset. This technology provides an alternative to W3C XML syntax as a means of representing instances of the W3C XML Information Set. This representation generally provides smaller encoding sizes and faster processing than a W3C XML representation.

ISO/IEC 24824-1:2007 specifies the use of several techniques that minimize the size of the encodings and that maximize the speed of creating and processing Fast Infoset documents.

These techniques include the use of dynamic tables (for both character strings and qualified names), initial vocabularies, and external vocabularies.

ISO/IEC 24824-1:2007 also specifies a Multipurpose Internet Mail Extensions (MIME) media type that identifies a Fast Infoset document.

ISO/IEC	IS	24824-2	Information technology—Generic applications of ASN.1: Fast Web Services	Published 2006	JTC1/SC6

Downloadable FOC ISO/IEC 24824-2:2006
http://standards.iso.org/ittf/PubliclyAvailableStandards/c041328_ISO_IEC_24824-2_2006(E).zip

ISO/IEC 24824-2:2006 specifies the messages required for Fast Web Services. It provides the specification of ASN.1 SOAP messages which carry the same semantics as W3C SOAP messages. The exchange of ASN.1 SOAP messages provides Fast Web Services.

ISO/IEC 24824-2:2006 references other ASN.1 International Standards and the Web services specification W3C SOAP 1.2 to fully define application exchanges that use Web protocols and W3C SOAP functionality. This is an extension to the provision of Web services using W3C SOAP messages, without any change to the functionality of W3C SOAP and service description languages. The main change is to the use of compact and easily processed binary encodings of XML data, rather than character encodings.

The use of ISO/IEC 24824-2:2006 (with the ASN.1 SOAP messages encoded using the ASN.1 PER) allows applications to provide Web services using messages that require less network bandwidth and less processing power (and hence provide a higher transaction processing rate) than the use of the character encoding of XML data.

ISO/IEC 24824-2:2006 also specifies a MIME media type that identifies an ASN.1 SOAP message encoded in PER. It also specifies a MIME media type that identifies a W3C SOAP message infoset encoded as a Fast Infoset document (see ISO/IEC 24824-1, Section 6.2.2.71). Both of these encodings are used by ISO/IEC 24824-2:2006.

ISO/IEC	FCD	24824-3	Information technology—Generic applications of ASN.1: Fast infoset security	Under Development FCD	JTC1/SC6

ISO 24824-3 Available from
http://www.iso.org/iso/iso_catalogue/catalogue_tc/catalogue_detail.htm?csnumber=44076

This standard is currently under development and led by the ASN.1 Group. It will become an ITU-T Recommendation, International Standard, entitled "Information Technology—Generic Applications of ASN.1—Fast Infoset and Fast Web Services Security" and will specify the application of standard security methods to the Fast Web Services specification.

In order for applications to apply standard security methods to the Fast Infoset specification, it is necessary to specify transformations to be applied to Fast Infoset documents (or parts of them) to support the application of those security methods.

6.2.2.21 ISO 8878 Telecommunications and Information Exchange Between Systems—Use of X.25 to Provide the OSI Connection-Mode Network Service

ISO/IEC	IS	8878	Information technology—Telecommunications and information exchange between systems—Use of X.25 to provide the OSI Connection-mode Network Service	Published 1992 2nd Edition	JTC1/SC6

Downloadable FOC ISO/IEC 8878:1992
http://standards.iso.org/ittf/PubliclyAvailableStandards/s016385_ISO_IEC_8878_1992(E).zip

Three technical corrigendum to this standard have been published.

Downloadable FOC ISO/IEC 8878:1992/Cor.1:1993
http://standards.iso.org/ittf/PubliclyAvailableStandards/s022300_ISO_IEC_8878_1992_Cor_1_1993(E).zip

Downloadable FOC ISO/IEC 8878:1992/Cor.2:1993
http://standards.iso.org/ittf/PubliclyAvailableStandards/s022685_ISO_IEC_8878_1992_Cor_2_1993(E).zip

Downloadable FOC ISO/IEC 8878:1992/Cor.3:1996
http://standards.iso.org/ittf/PubliclyAvailableStandards/s026240_ISO_IEC_8878_1992_Cor_3_1996(E).zip

6.2.2.22 ISO 8881 Data Communications—Use of the X.25 Packet Level Protocol in Local Area Networks

ISO/IEC	IS	8881	Information processing systems—Data communications—Use of the X.25 packet level protocol in local area networks	Published 1989	JTC1/SC6

ISO 8881 Available from
http://www.iso.org/iso/iso_catalogue/catalogue_tc/catalogue_detail.htm?csnumber=16394

This deals with the use of the X.25 Packed Level Protocol (PLP) as defined in ISO 8280 operating over ISO 8802 Local Area Networks (LANs). The operation of the X.25/PLP using the Logical Link Control (LLC) Type 2 procedures is described in Section 2. The operation of the X.25/PLP using the LLC Type 1 procedures is specified in Section 3. It is to be used in compliance with the standards: ISO 7498:1984; ISO 8208:1987; ISO 8208/Add. 1: ISO 8343/Add. 2:1988; ISO 8802-2; ISO 8878:1987; ISO/IEC/TR 10029:1989.

6.2.2.23 ISO 8886 Open Systems Interconnection—Data Link Service Definition

ISO/IEC	IS	8886	Information technology—Open Systems Interconnection—Data link service definition	Published 1996	JTC1/SC6

ISO 8886 Available from
http://www.iso.org/iso/iso_catalogue/catalogue_tc/catalogue_detail.htm?csnumber=26140

The purpose of this standard is to specify the characteristics of a conceptual data link service and thus supplement the OSI reference model in guiding the development of the data link.

6.2.2.24 ISO 9160 Data Encipherment—Physical Layer Interoperability Requirements

ISO/IEC	IS	9160	Information processing—Data encipherment—Physical layer interoperability requirements	Published 1988	JTC1/SC6

ISO 9160 Available from
http://www.iso.org/iso/iso_catalogue/catalogue_tc/catalogue_detail.htm?csnumber=16757

This defines interoperability and security related requirements for using encipherment at the physical layer of the ISO OSI reference model in telecommunication systems conveying automatic data processing (ADP) information. It facilitates interoperation of data encipherment equipment. It is applicable mainly to various encipherment algorithms. Annex B provides additional requirements for using DEA and applies to synchronous and asynchronous operation in full and half duplex modes. It specifies delayed option and immediate option as incompatible modes of synchronous operation.

6.2.2.25 ISO 9542 End System to Intermediate System Routing Exchange Protocol for Use in Conjunction with the Protocol for Providing the Connectionless-Mode Network Service (ISO 8473)

ISO/IEC	IS	9542	Information processing systems—Telecommunications and information exchange between systems—End system to Intermediate system routing exchange protocol for use in conjunction with the Protocol for providing the connectionless-mode network service (ISO 8473)	Published 1988	JTC1/SC6

Downloadable FOC ISO/IEC 9542:1988
http://standards.iso.org/ittf/PubliclyAvailableStandards/s017285_ISO_IEC_9542_1988(E).zip

ISO/IEC	IS	9542 Amnd 1	Information processing systems—Telecommunications and information exchange between systems—End system to Intermediate system routing exchange protocol for use in conjunction with the Protocol for providing the connectionless-mode network service (ISO 8473)—Amendment 1: Addition of group composition information	Amnd 1999	JTC1/SC6

Downloadable FOC ISO 9542:1988/Amd 1:1999
http://standards.iso.org/ittf/PubliclyAvailableStandards/c030377_ISO_9542_1988_Amd_1_1999(E).zip

This specifies a protocol used by network layer entities in end systems (ES) and intermediate systems (IS) to maintain routing information. It defines: transmission procedures for configuration and routing information between ES and IS; the encoding of protocol data units; procedures for protocol control information interpretation; and functional requirements for implementations conforming with this standard. It references: ISO 7498; 7498 Add. 1 and 4; 8208; 8348 Add. 1 and 2; ISO 8473; 8648; 8802; CCITT X.25.

6.2.2.26 ISO 9543 Information Exchange Between Systems—Synchronous Transmission Signal Quality at DTE/DCE Interfaces

ISO/IEC	IS	9543	Information processing systems—Information exchange between systems—Synchronous transmission signal quality at DTE/DCE interfaces	Published 1989	JTC1/SC6

ISO 9543 Available from
http://www.iso.org/iso/iso_catalogue/catalogue_tc/catalogue_detail.htm?csnumber=17288

This prescribes quality requirements for serial data transmission at the interface between synchronous transmission data terminal equipment and data circuit-terminating equipment in conformance with the CCITT Recommendations V.24 (telephone networks) and X.24 (data networks) as specified in DCEs of the corresponding CCITT Recommendations. It takes into account the need to have performance categories of signal quality depending on the type of interchange circuits.

6.2.2.27 ISO 9574 Provision of the OSI Connection-Mode Network Service by Packet Mode Terminal Equipment to an Integrated Services Digital Network (ISDN)

ISO/IEC	IS	9574	Information technology—Provision of the OSI connection-mode network service by packet mode terminal equipment to an integrated services digital network (ISDN)	Published 1992	JTC1/SC6

ISO 9574 Available from
http://www.iso.org/iso/iso_catalogue/catalogue_tc/catalogue_detail.htm?csnumber=17337

This specifies the method of providing service in accordance with the procedures described in Recommendation X.31. This is done by specifying the mapping of the CONS primitives and parameters to and from the elements of the protocols used by two types of packet mode terminal equipment: an X.25 DTE (TE2) and a packet mode ISDN terminal (TE1).

6.2.2.28 ISO 9575 Telecommunications and Information Exchange Between Systems—OSI Routing Framework

ISO/IEC	TR	9575	Information technology—Telecommunications and information exchange between systems—OSI Routing Framework	TR 1995	JTC1/SC6

ISO 9575 Available from
http://www.iso.org/iso/iso_catalogue/catalogue_tc/catalogue_detail.htm?csnumber=25981

This provides a framework in which OSI protocols for routing may be developed and to expedite the progression of routing protocols through the standardization process. It reflects the current state of OSI routing and does not preclude future extensions and developments. It replaces the first edition, which has been technically revised.

6.2.2.29 ISO 9577 Protocol Identification in the Network Layer

ISO/IEC	TR	9577	Information technology—Protocol identification in the network layer	TR 1999	JTC1/SC6

Downloadable FOC ISO/IEC TR 9577:1999
http://standards.iso.org/ittf/PubliclyAvailableStandards/c031053_ISO_IEC_TR_9577_1999(E).zip

6.2.2.30 ISO 9591 Information Technology—Open Systems Interconnection

Parts 1 through 10 of this standard have been published as of this writing.

ISO/IEC	IS	9594		Published/ Under Revision	JTC1/SC6
ISO/IEC	IS	9594-1	Information technology—Open Systems Interconnection—The Directory: Overview of concepts, models and services	Published 2005	JTC1/SC6

ISO 9594-1 Available from
http://www.iso.org/iso/iso_catalogue/catalogue_tc/catalogue_detail.htm?csnumber=41884

ISO/IEC 9594-1:2005 includes specifications for how information about objects (e.g., persons) is organized, created, maintained, and retrieved. It provides provisions for protecting stored information through authentication and access control specifications.

ISO/IEC 9594-1:2005 introduces the concepts of the directory and the directory information base (DIB), and overviews the services and capabilities which they provide. It is intended to give an introduction to the other parts of ISO/IEC 9594. It is not an implementation specification.

ISO/IEC	IS	9594-2	Information technology—Open Systems Interconnection—The Directory: Models	Published 2005	JTC1/SC6

ISO 9594-2 Available from
http://www.iso.org/iso/iso_catalogue/catalogue_tc/catalogue_detail.htm?csnumber=41885

ISO/IEC 9594-2:2005 includes specifications for how information about objects (e.g., persons) is organized, created, maintained, and retrieved. It also gives provisions for protecting stored information through authentication and access control specifications.

ISO/IEC 9594-2:2005 provides a number of different models for the directory as a framework for the other ISO/IEC 9594 parts. It specifies the basis for directory protocol specifications.

ISO/IEC	IS	9594-3	Information technology—Open Systems Interconnection—The Directory: Abstract service definition	Published 2005	JTC1/SC6

ISO 9594-3 Available from
http://www.iso.org/iso/iso_catalogue/catalogue_tc/catalogue_detail.htm?csnumber=41886

ISO/IEC 9594-3:2005 provides specifications for how information about objects (e.g., persons) is organized, created, maintained, and retrieved. It also gives provision for protecting stored information through authentication and access control specifications.

ISO/IEC 9594-3:2005 defines in an abstract way the externally visible service provided by the directory, including bind and unbind operations, read operations, search operations, modify operations, and errors. It also defines interworking with LDAP.

ISO/IEC	IS	9594-4	Information technology—Open Systems Interconnection—The Directory: Procedures for distributed operation	Published 2005	JTC1/SC6

ISO 9594-4 Available from
http://www.iso.org/iso/iso_catalogue/catalogue_tc/catalogue_detail.htm?csnumber=43794

ISO/IEC 9594-4:2005 provides specifications for how information about objects (e.g., persons) is organized, created, maintained, and retrieved. It also gives provisions for protecting stored information through authentication and access control specifications.

ISO/IEC 9594-4:2005 specifies the procedures by which the distributed components of the directory interwork in order to provide a consistent service to its users. This includes defining procedures for interworking with LDAP.

ISO/IEC	IS	9594-5	Information technology—Open Systems Interconnection—The Directory: Protocol specifications	Published 2005	JTC1/SC6

ISO 9594-5 Available from
http://www.iso.org/iso/iso_catalogue/catalogue_tc/catalogue_detail.htm?csnumber=43795

ISO/IEC 9594-5:2005 provides specifications for how information about objects (e.g., persons) is organized, created, maintained, and retrieved. It also gives provisions for protecting stored information through authentication and access control specifications.

ISO/IEC 9594-5:2005 specifies the Directory Access Protocol, the Directory System Protocol, the Directory Information Shadowing Protocol, and the Directory Operational Binding Management Protocol, fulfilling the abstract services specified in ISO/IEC 9594-2, ISO/IEC 9594-3, ISO/IEC 9594-4 and ISO/IEC 9594-9. It defines the use of both an OSI and an TCP/IP underlying protocol stack.

ISO/IEC	IS	9594-6	Information technology—Open Systems Interconnection—The Directory: Selected attribute types	Published 2005	JTC1/SC6

ISO 9594-6 Available from
http://www.iso.org/iso/iso_catalogue/catalogue_tc/catalogue_detail.htm?csnumber=43796

ISO/IEC 9594-6:2005 provides specifications for how information about objects (e.g., persons) is organized, created, maintained, and retrieved. It also gives provisions for protecting stored information through authentication and access control specifications.

ISO/IEC 9594-6:2005 defines a number of attribute types, matching rules and context types which may be found useful across a range of applications of the directory. One particular use for many of the attributes defined is in the formation of names, particularly for the classes of object defined in ISO/IEC 9594-7.

ISO/IEC	IS	9594-7	Information technology—Open Systems Interconnection—The Directory: Selected object classes	Published 2005	JTC1/SC6

ISO 9594-7 Available from
http://www.iso.org/iso/iso_catalogue/catalogue_tc/catalogue_detail.htm?csnumber=43797

ISO/IEC 9594-7:2005 provides specifications for how information about objects (e.g., persons) is organized, created, maintained, and retrieved. It also gives provision for protecting stored information through authentication and access control specifications.

ISO/IEC 9594-7:2005 defines a number of selected object classes and name forms which may be useful across a range of applications of the directory. An object class definition specifies the attribute types which are relevant to the objects of that class. A name form definition specifies the attributes to be used in forming names for the objects of a given class.

ISO/IEC	IS	9594-8	Information technology—Open Systems Interconnection—The Directory: Public-key and attribute certificate frameworks	Published 2005	JTC1/SC6

ISO 9594-8 Available from
http://www.iso.org/iso/iso_catalogue/catalogue_tc/catalogue_detail.htm?csnumber=43793

ISO/IEC 9594-8:2005 provides specifications for how information about objects (e.g., persons is organized, created, maintained, and retrieved. Multiple entities are likely deployed to provide the directory service. Communication amongst these entities is authenticated and/or encrypted.

ISO/IEC 9594-8:2005 specifies three frameworks and a number of data objects that can be used to authenticate and secure the communication between two entities (e.g., between two directory service entities or between a Web browser and Web server). The data objects can also be used to prove the source and integrity of data structures such as digitally signed documents.

ISO/IEC	IS	9594-9	Information technology—Open Systems Interconnection—The Directory: Replication	Published 2005	JTC1/SC6

ISO 9594-10 Available from
http://www.iso.org/iso/iso_catalogue/catalogue_tc/catalogue_detail.htm?csnumber=43798

ISO/IEC 9594-9:2005 provides specifications for how information about objects (e.g., persons) is organized, created, maintained, and retrieved. It also gives provisions for protecting stored information through authentication and access control specifications.

ISO/IEC 9594-9:2005 specifies a shadow service which DSAs may use to replicate directory information. The service allows directory information to be replicated among DSAs to improve service to directory users, and provides for the automatic updating of this information.

ISO/IEC	IS	9594-10	Information technology—Open Systems Interconnection—The Directory: Use of systems management for administration of the Directory	Published 2005	JTC1/SC6

ISO 9594-11 Available from
http://www.iso.org/iso/iso_catalogue/catalogue_tc/catalogue_detail.htm?csnumber=43799

ISO/IEC 9594-10:2005 provides specifications for how information about objects (e.g., persons) is organized, created, maintained, and retrieved. It also gives provisions for protecting stored information through authentication and access control specifications.

ISO/IEC 9594-10:2005 describes the requirements for directory management and analyzes these requirements to identify those that may be realized by OSI Systems Management services (and protocols), those that are realized by directory services (and protocols), and those that are realized by local means.

This purpose of directory management is to assure that needed, accurate directory information is available to users as scheduled with the expected response time, integrity, security, and level of consistency. Furthermore, systems management may be accomplished with the minimum burden on processing time and memory on platforms and the communications system.

6.2.2.31 ISO 9834 Open Systems Interconnection—Procedures for the Operation of OSI Registration Authorities

ISO/IEC	IS	9834	ISO/IEC Information technology—Open Systems Interconnection—Procedures for the operation of OSI Registration Authorities		JTC1/SC6

At the time of this writing, there are eight parts to this standard.

ISO/IEC	IS	9834-1	Information technology—Open Systems Interconnection—Procedures for the operation of OSI Registration Authorities: General procedures and top arcs of the ASN.1 Object Identifier tree	Published 2005	JTC1/SC6

ISO 9834-1 Available from
http://www.iso.org/iso/iso_catalogue/catalogue_tc/catalogue_detail.htm?csnumber=40286

ISO/IEC 9834-1:2005:

- Specifies a registration-hierarchical-name-tree (RH-name-tree), which is a generic tree structure for allocations made by registration authorities, and the ASN.1 object identifier tree, which is a specific instance of the RH-name-tree;
- Registers the three top-level arcs of the ASN.1 object identifier tree;
- Specifies procedures which are generally applicable to registration in the context of an RH-name-tree;
- Provides guidelines for the establishment and operation of International registration authorities;
- Provides guidelines for additional International Standards which reference the procedures in ISO/IEC 9834-1:2004.

ISO/IEC 9834-1:2005 does not exclude or disallow the use of any syntactic forms of names or any naming domains for registration purposes provided that the domains ensure nonambiguity within their scope. It is intended to cover those cases in which the registration-hierarchical-name is appropriate. Information about registration for specific objects is contained in separate International Standards.

ISO/IEC 9834-1:2005 applies to registration by International Standards, by international registration authorities, and by any other registration authority.

ISO/IEC	ISO/ IEC	9834-2	Information technology—Open Systems Interconnection—Procedures for the operation of OSI Registration Authorities—Part 2: Registration procedures for OSI document types		JTC1/SC6

ISO 9834-2 Available from
http://www.iso.org/iso/iso_catalogue/catalogue_tc/catalogue_detail.htm?csnumber=17706

This specifies the contents of register entries, recording information about OSI document types and assigning an unambiguous name of ASN.1 type object identifier to OSI document type definitions. It specifies the procedures for the operation of an international registration authority for OSI document types.

ISO/IEC	IS	9834-3	ISO/IEC: Information technology—Open Systems Interconnection—Procedures for the operation of OSI Registration Authorities : Registration of Object Identifier arcs beneath the top-level arc jointly administered by ISO and ITU-T	Published 2005	JTC1/SC6

ISO 9834-3 Available from
http://www.iso.org/iso/iso_catalogue/catalogue_tc/catalogue_detail.htm?csnumber=37230

ISO/IEC 9834-3:2004 specifies the procedures for operating the international registration authority for assignment of values to ASN.1 Object identifier arcs beneath the top-level arc are jointly administered by ISO and ITU-T.

ISO/IEC	IS	9834-4	Information technology—Open Systems Interconnection—Procedures for the operation of OSI Registration Authorities—Part 4: Register of VTE Profiles	Published 1991	JTC1/SC6

ISO 9834-4 Available from
http://www.iso.org/iso/iso_catalogue/catalogue_tc/catalogue_detail.htm?csnumber=17709

This standard specifies the contents of register entries, recording information about VTE-profiles and assigning unambiguous names of ASN.1 type object identifier to VTE-profile definitions.

ISO/IEC	IS	9834-5	Information technology—Open Systems Interconnection—Procedures for the operation of OSI Registration Authorities—Part 5: Register of VT Control Object Definitions	Published 1991	JTC1/SC6

ISO 9834-5 Available from
http://www.iso.org/iso/iso_catalogue/catalogue_tc/catalogue_detail.htm?csnumber=17711

This standard specifies the contents of register entries, recording information about VT control object definitions and assigning unambiguous names of ASN.1 type object identifier to VT CO definitions.

ISO/IEC	IS	9834-6	ISO/IEC: Information technology—Open Systems Interconnection—Procedures for the operation of OSI Registration Authorities: Registration of application processes and application entities	Published 2005	JTC1/SC6

ISO 9834-6 Available from
http://www.iso.org/iso/iso_catalogue/catalogue_tc/catalogue_detail.htm?csnumber=40287

ISO/IEC 9834-6:2005 specifies the procedures applicable to the registration of application processes and application entities. No requirement for an international registration authority has been identified; therefore, these procedures apply to registration at any point in the ASN.1 object identifier tree.

ISO/IEC 9834-6:2005 does not cover the registration of application-process types or application-entity types. No requirement for such registration has been identified.

ISO/IEC	IS	9834-7	Information technology—Open Systems Interconnection—Procedures for the operation of OSI Registration Authorities: Joint ISO and ITU-T Registration of International Organizations	Published 2005	JTC1/SC6

ISO 9834-7 Available from
http://www.iso.org/iso/iso_catalogue/catalogue_tc/catalogue_detail.htm?csnumber=37231

ISO/IEC 9834-7:2005 specifies procedures for registration authorities responsible for the assignment to organizations of names that are globally unambiguous in the context of:

- O/R addresses, as defined in ISO/IEC 10021-2;
- Directory names, as defined in ISO/IEC 9594-2; and
- ASN.1 object identifiers, as defined in ISO/IEC 8824-1.

ISO/IEC	IS	9834-8	Information technology—Open Systems Interconnection—Procedures for the operation of OSI Registration Authorities: Generation and registration of Universally Unique Identifiers (UUIDs) and their use as ASN.1 Object Identifier components	Published 2005	JTC1/SC6

ISO 9834-8 Available from
http://www.iso.org/iso/iso_catalogue/catalogue_tc/catalogue_detail.htm?csnumber=36775

ISO/IEC 9834-8:2005 specifies the format and generation rules that enable users to produce 128-bit identifiers that are either guaranteed or have a high probability of being globally unique.

The UUIDs generated in conformance with ISO/IEC 9834-8:2005 are suitable either for transient use, with generation of a new UUID every 100 nanoseconds, or as persistent identifiers.

ISO/IEC 9834-8:2005 is derived from earlier nonstandard specifications of UUIDs and their generation, and it is technically identical to those earlier specifications. It specifies the procedures for the operation of a Web-based registration authority for UUIDs.

ISO/IEC 9834-8:2005 also specifies and allows the use of UUIDs (registered or not registered) as OID components under the arc {joint-iso-itu-t uuid(25)}. This enables users to generate OIDs without any registration procedures. It also specifies and allows the use of UUIDs (registered or not registered) to form a URN.

6.2.2.32 ISO 10000 Information Technology—Framework and Taxonomy of International Standardized Profiles

ISO/IEC	IS	10000	Information technology—Framework and taxonomy of International Standardized Profiles		JTC1/SC6

This standard is in three parts.

ISO/IEC	IS	10000-1	Information technology—Framework and taxonomy of International Standardized Profiles—Part 1: General principles and documentation framework	Published 4th Edition	JTC1/SC6

Downloadable FOC ISO/IEC TR 10000-1:1998
http://standards.iso.org/ittf/PubliclyAvailableStandards/c030726_ISO_IEC_TR_10000-1_1998(E).zip

ISO/IEC	IS	10000-2	Information technology—Framework and taxonomy of International Standardized Profiles Part 2: Principles and Taxonomy for OSI Profiles	Published 5th Edition	JTC1/SC6

Downloadable FOC ISO/IEC TR 10000-2:1998
http://standards.iso.org/ittf/PubliclyAvailableStandards/c030728_ISO_IEC_TR_10000-2_1998(E).zip

ISO/IEC	IS	10000-3	Information technology—Framework and taxonomy of International Standardized Profiles— Part 3: Principles and Taxonomy for Open System Environment Profiles	Published 2nd Edition	JTC1/SC6

Downloadable FOC ISO/IEC TR 10000-3:1998
http://standards.iso.org/ittf/PubliclyAvailableStandards/c030727_ISO_IEC_TR_10000-3_1998(E).zip

6.2.2.33 ISO 10021 Information Technology—Message Handling Systems (MHS)

ISO/IEC	IS	10021	Information technology—Message Handling Systems (MHS)		JTC1/SC6

This series of standards is comprised of 10 parts at the time of this writing.

This Recommendation | International Standard is one of a series on message handling. ITU-T Rec. X.402 | ISO/IEC 10021-2 constitutes the introduction to the series and identifies the other documents in it.

The architectural basis and foundation for message handling are defined in still other Recommendations | International Standards. ITU-T Rec. X.402 | ISO/IEC 10021-2 identifies those documents as well.

The purpose of an MHS is to enable users to exchange messages on a store-and-forward basis. A message submitted on behalf of one user, the originator, is conveyed by the Message Transfer System (MTS) and subsequently delivered to the agents of one or more additional users, the recipients. Access Units (AU) link the MTS to communication systems of other kinds (e.g., postal systems). A user is assisted in the preparation, storage, and display of messages by a User Agent (UA). Optionally, it is assisted in the storage of messages by a Message Store (MS). The MTS comprises a number of Message Transfer Agents (MTA) which collectively perform the store-and-forward message transfer function.

ISO/IEC	IS	10021-1	Information technology—Message Handling Systems (MHS)—Part 1: System and service overview	Published 2003	JTC1/SC6

ISO 10021-1 Available from
http://www.iso.org/iso/iso_catalogue/catalogue_tc/catalogue_detail.htm?csnumber=31632

ISO/IEC 10021-1:2003 defines the overall system and service of an MHS and serves as a general overview of MHS.

Other aspects of Message Handling Systems and Services are defined in other parts of ISO/IEC 10021. The structure of ISO/IEC 10021 (all parts) defining the Message Handling System and Services is given in Table 1 of the standard.

The technical aspects of MHS are defined in other parts of ISO/IEC 10021. The overall system architecture of MHS is defined in ISO/IEC 10021-2:1996.

ISO/IEC	IS/ITU	10021-2 ITU-T Rec. X.402	Information technology—Message Handling Systems (MHS): Overall architecture	Published 2003	JTC1/SC6

ISO10021-2 Available from
http://www.iso.org/iso/iso_catalogue/catalogue_tc/catalogue_detail.htm?csnumber=33058

ISO/IEC 10021-2:2003 | ITU-T Rec. X.402 (1999) defines the overall architecture of the Message Handling System (MHS) and serves as a technical introduction to it. ISO/IEC 10021-2:2003 | ITU-T Rec. X.402 (1999) first presents abstract models of Message Handling, then specifies how one can configure the MHS to satisfy any of a variety of functional, physical, and organizational requirements. It describes the naming and addressing of users and distribution lists and the routing of information objects to them. It describes the uses the MHS may make of the directory and how the MHS is realized by means of OSI. The conventions used in the definition of the abstract services provided by MHS components are defined. Annexes provide important supplemental information.

ISO/IEC	IS		Information technology—Text Communication—Message-Oriented Text Interchange Systems (MOTIS)—Part 3: Abstract Service Definition Conventions	Withdrawn	JTC1/SC6

ISO 10021-3				Withdrawn	JTC1/SC6

ISO/IEC	IS	10021-4	Information technology—Message Handling Systems (MHS): Message transfer system—Abstract service definition and procedures	Published 2003	JTC1/SC6

ISO 10021-4 Available from
http://www.iso.org/iso/iso_catalogue/catalogue_tc/catalogue_detail.htm?csnumber=33059

ISO/IEC 10021-4:2003 | ITU-T Rec. X.411 defines the Message Transfer System (MTS) Abstract Service and the MTS Model, and provides an overview of the MTS Abstract Service. It defines both the semantics of the parameters of the MTS Abstract Service and the abstract-syntax of the MTS Abstract Service. ISO/IEC 10021-4:2003 | ITU-T Rec. X.411 defines the Message Transfer Agent (MTA) Abstract Service and refines the model of the MTS to show that the MTS comprises a number of MTAs that interwork with one another to provide the MTS Abstract Service. An overview of the MTA Abstract Service is provided. ISO/IEC 10021-4:2003 | ITU-T Rec. X.411 defines both the semantics of the parameters of the MTA Abstract Service and the abstract-syntax of the MTA Abstract Service. It specifies the procedures performed by MTAs to ensure the correct distributed operation of the MTS.

ISO/IEC	IS	10021-5	Information technology—Message Handling Systems (MHS): Message store: Abstract service definition	Published 1999	JTC1/SC6

ISO 10021-5 Available from
http://www.iso.org/iso/iso_catalogue/catalogue_tc/catalogue_detail.htm?csnumber=33060

This part of ISO/IEC 10021 defines the message store abstract service. This abstract service is provided by the message store access protocol (specified in ISO/IEC 10021-6) in conjunction with the MTS abstract service (defined in ISO/IEC 10021-4), together with the remote operations service element (ROSE) services (defined in ISO/IEC 9072-1). The abstract syntax notation for the application-layer protocols used in this part of ISO/IEC 10021 is defined in ISO 8824.

Other parts of ISO/IEC 10021 define other aspects of the MHS. ISO/IEC 10021-1 defines the user-oriented services provided by the MHS. ISO/IEC 10021-2 provides an architectural overview of the MHS. ISO/IEC 10021-3 provides a description of the abstract service definition conventions used in MHS. ISO/IEC 10021-7 defines the abstract service for interpersonal messaging and defines the format of interpersonal messages.

ISO/IEC	IS	10021-6	Information technology—Message Handling Systems (MHS): Protocol specifications	Published 2003	JTC1/SC6

ISO 10021-6 Available from
http://www.iso.org/iso/iso_catalogue/catalogue_tc/catalogue_detail.htm?csnumber=39564

ISO/IEC 10021-6:2003 specifies the MTS Access Protocol (P3) used between a remote user agent and the MTS to provide access to the MTS abstract service defined in ITU-T Rec. X.411 | ISO/IEC 10021-4. It also specifies the MS Access Protocol (P7) used between a remote user agent and a message store (MS) to provide access to the MS abstract service defined in ITU-T Rec. X.413 | ISO/IEC 10021-5. This document also specifies the MTS Transfer Protocol (P1) used between MTAs to provide the distributed operation of the MTS as defined in ITU-T Rec. X.411 | ISO/IEC 10021-4 and identifies the other Recommendations/International Standards which define other aspects of Message Handling Systems.

ISO/IEC	IS	10021-7	Information technology—Message Handling Systems (MHS): Interpersonal messaging system	Published 2003	JTC1/SC6

ISO 10021-7 Available from
http://www.iso.org/iso/iso_catalogue/catalogue_tc/catalogue_detail.htm?csnumber=31635

ISO/IEC 10021-7:2003 defines interpersonal messaging, a form of message handling tailored for ordinary interpersonal business or private correspondence. It is one of a series on message handling. ISO/IEC 10021-2 constitutes the introduction to the series and identifies the other documents in it. ISO/IEC 10021-7:2003 defines the kinds of information objects exchanged in interpersonal messaging, defines the associated abstract service, and specifies how it is provided. Annexes provide important supplemental information.

ISO/IEC	IS	10021-8	Information technology—Message Handling Systems (MHS)—Part 8: Electronic Data Interchange Messaging Service	Published 1999	JTC1/SC6

ISO 10021-8 Available from
http://www.iso.org/iso/iso_catalogue/catalogue_tc/catalogue_detail.htm?csnumber=34072

ISO/IEC	IS	10021-9	Information technology—Message Handling Systems (MHS): Electronic Data Interchange Messaging System	Published 1999	JTC1/SC6

10021-9 Available from
http://www.iso.org/iso/iso_catalogue/catalogue_tc/catalogue_detail.htm?csnumber=33079

ISO 10021-9 defines the message handling application called EDI messaging (EDIMG), a form of message handling tailored for exchange of EDI information, a new message content type and associated procedures known as PEDI. It is designed to meet the requirements of users of ISO 9735 (EDIFACT), and other commonly used EDI systems.

ISO/IEC	IS	10021-10	Information technology—Message Handling Systems (MHS): MHS routing	Published 1999	JTC1/SC6

ISO 10021-10 Available from
http://www.iso.org/iso/iso_catalogue/catalogue_tc/catalogue_detail.htm?csnumber=33063

ISO 10021-10 specifies the means by which messages are routed through the MHS, and it supplements the procedures defined in 14.3 of ITU-T Rec. X.411 | ISO/IEC 10021-4.

ISO/IEC	TR	10021-11	Information technology—Message Handling Systems (MHS): MHS Routing—Guide for messaging systems managers	Published 1999	JTC1/SC6

ISO 10021-11 Available from
http://www.iso.org/iso/iso_catalogue/catalogue_tc/catalogue_detail.htm?csnumber=17968

As its title implies, this technical report provides guidance for messaging systems managers.

6.2.2.34 ISO 10022 Open Systems Interconnection—Physical Service Definition

ISO/IEC	IS	10022	Information technology—Open Systems Interconnection—Physical Service Definition	Published 1996	JTC1/SC6

ISO 10022 Available from
http://www.iso.org/iso/iso_catalogue/catalogue_tc/catalogue_detail.htm?csnumber=24898

The purpose of this standard is to specify the characteristics of a conceptual physical service and thus supplement the OSI basic reference model in guiding the development of physical layer protocols.

6.2.2.35 ISO 10028 Telecommunications and Information Exchange Between Systems—Definition of the Relaying Functions of a Network Layer Intermediate System

ISO/IEC	IS	10028	Information technology—Telecommunications and information exchange between systems—Definition of the relaying functions of a Network layer intermediate system	Published 1993	JTC1/SC6

ISO 10028 Available from
http://www.iso.org/iso/iso_catalogue/catalogue_tc/catalogue_detail.htm?csnumber=17987

This is intended for use in guiding the design and application of real interworking units and real subnetworks (e.g., local area networks and private packet switched networks) which are intended to support the OSI network service. As the principal means for expressing the definition, the concept of the network internal layer service is used. The definition includes the invocation of network routing functions as a necessary element of the network relaying functions. The definition of network relaying functions applies both to a subnetwork supporting all elements of the network service and to a network relay system interconnecting two subnetworks.

6.2.2.36 ISO 10030 End System Routing Information Exchange Protocol for Use in Conjunction with ISO/IEC 8878

ISO/IEC	IS	10030	Information technology—Telecommunications and information exchange between systems—End System Routing Information Exchange Protocol for use in conjunction with ISO/IEC 8878	Published 1995	JTC1/SC6

ISO 10030 Available from
http://www.iso.org/iso/iso_catalogue/catalogue_tc/catalogue_detail.htm?csnumber=23965

This cancels and replaces the first edition (1990). It defines a protocol for the exchange of routing information between an end system and a subnetwork address resolution entity, and between an intermediate system and a subnetwork address resolution entity.

It is applicable to: end systems which operate according to the main body of ISO/IEC 8878 to provide and support the OSI connection-mode network service using ISO/IEC 8208; subnetwork address resolution entities which operate ISO/IEC 8208; and intermediate systems which operate ISO/IEC 8208.

6.2.2.37 ISO 10177 Provision of the Connection-Mode Network Internal Layer Service by Intermediate Systems Using ISO/IEC 8208, the X.25 Packet Layer Protocol

ISO/IEC	IS	10177	Information technology—Telecommunications and information exchange between systems—Provision of the connection-mode Network internal layer service by intermediate systems using ISO/IEC 8208, the X.25 Packet Layer Protocol	Published 1993	JTC1/SC6

ISO 10177 Available from
http://www.iso.org/iso/iso_catalogue/catalogue_tc/catalogue_detail.htm?csnumber=18194

This specifies the method by which a network-layer interworking unit (IWU) uses the X.25 packet layer protocol to support the OSI connection-mode network service. The specification is expressed in terms of a mapping between the network internal layer service defined in ISO/IEC 10028 and the virtual call and permanent virtual circuit services of the X.25 packet layer protocol. It provides the Protocol Implementation Conformance Statement (PICS) pro forma in compliance with the relevant requirements, and in accordance with the relevant guidance, given in ISO/IEC 9646-2.

6.2.2.38 ISO 10589 Intermediate System to Intermediate System Intra-Domain Routing Information Exchange Protocol for Use in Conjunction with the Protocol for Providing the Connectionless-Mode Network Service (ISO 8473)

ISO/IEC	IS	10589	Information technology—Telecommunications and information exchange between systems—Intermediate System to Intermediate System intra-domain routing information exchange protocol for use in conjunction with the protocol for providing the connectionless-mode network service (ISO 8473)	Published 2002	JTC1/SC6

Downloadable FOC ISO/IEC 10589:2002
http://standards.iso.org/ittf/PubliclyAvailableStandards/c030932_ISO_IEC_10589_2002(E).zip

6.2.2.39 ISO 10611 International Standardized Profiles AMH1n—Message Handling Systems—Common Messaging

ISO/IEC	IS/ISP	10611	Information technology—International Standardized Profiles AMH1n—Message Handling Systems—Common Messaging		JTC1/SC6

There are six parts to this series of standards at the time of publication of this book.

ISO/IEC	IS/ISP	10611-1	ISO/IEC ISP: Information technology—International Standardized Profiles AMH1n—Message Handling Systems—Common Messaging—Part 1: MHS Service Support	Published 2003	JTC1/SC6

ISO 10611-1 Available from
http://www.iso.org/iso/iso_catalogue/catalogue_tc/catalogue_detail.htm?csnumber=30717

ISO/IEC ISP 10611-1:2003(E) contains the overall specifications of the support of MHS Elements of Service and associated MHS functionality, which are generally not appropriate for consideration only from the perspective of a single MHS protocol. These specifications form part of the Common Messaging application functions, as defined in ISO/IEC ISP 10611 (all parts), which form a common basis for content type-dependent International Standardized Profiles for MHS that will be developed. Such specifications are in many cases applicable to more than one

MHS protocol or are otherwise concerned with component functionality which, although it can be verified via protocol, is not just related to protocol support. They are therefore designed to be referenced in the MHS Common Messaging application profiles ISO/IEC ISP 10611-3 (AMH11), ISO/IEC ISP 10611-4 (AMH12 and AMH14), ISO/IEC ISP 10611-5 (AMH13), and ISO/IEC ISP 10611-6 (AMH15), which specify the support of specific MHS protocols and associated functionality.

The specifications in ISO/IEC ISP 10611-1:2003(E) cover the provision and use of features associated with the Message Transfer Service (MTS), together with those features associated with intercommunication with Physical Delivery (PD) Services. Features which are associated with the MS and UA which are content type independent are also covered. Features which are specific to a particular content type (including the provision of services by a UA to an MHS user) are covered in separate content type dependent ISPs.

The specifications in ISO/IEC ISP 10611-1:2003(E) are divided into basic requirements, which are required to be supported by all MHS implementations, and a number of optional functional groups, which cover significant discrete areas of related functionality which are not required to be supported by all implementations.

ISO/IEC	IS/ISP 10611-2	Information technology—International Standardized Profiles (ISP) AMH1n—Message Handling Systems—Common Messaging—Part 2: Specification of ROSE, RTSE, ACSE, Presentation and Session Protocols for use by MHS	Published 1997	JTC1/SC6

ISO 10611-2 Available from
http://www.iso.org/iso/iso_catalogue/catalogue_tc/catalogue_detail.htm?csnumber=28445

This part of ISO/IEC ISP 10611 specifies how the Remote Operations Service Element, the Reliable Transfer Service Element, the Association Control Service Element, the Presentation Layer, and the Session Layer standards shall be used to provide the required OSI upper layer functions for MHS (see also Figure 1 of the standard). These specifications are therefore the common basis for the common messaging application functions, as defined in the other parts of ISO/IEC ISP 10611, and for content type dependent International Standardized Profiles for MHS that will be developed.

The OSI upper layer services and protocols to support the MHS functions covered by the AMH1 set of profiles are specified in the set of standards identified in Table 1 of the standard.

ISO/IEC	IS/ISP 10611-3	Information technology—International Standardized Profiles (ISP) AMH1n—Message Handling Systems—Common Messaging—Part 3: AMH11—Message Transfer (P1)	Published 2003	JTC1/SC6

ISO 10611-3 Available from
http://www.iso.org/iso/iso_catalogue/catalogue_tc/catalogue_detail.htm?csnumber=30718

ISO/IEC ISP 10611-3:2003(E) (AMH11) covers message transfer between MTAs using the P1 Message Transfer Protocol. These specifications form part of the common messaging application functions, as defined in ISO/IEC ISP 10611 (all parts), which form a common basis for content type dependent International Standardized Profiles for MHS that will be developed.

An MTA which conforms to profiles AMH11n as specified in ISO/IEC ISP 10611-3:2003(E) shall support a "normal mode" OSI protocol infrastructure (AMH111) as required by both ISO/IEC 10021-6 and the ITU-T X.400 Recommendations, and may additionally support an X.410 mode OSI protocol infrastructure (AMH112) as required, for ADMDs, by the ITU-T X.400 Recommendations.

ISO/IEC	IS/ISP	10611-4	Information technology—International Standardized Profiles AMH1n—Message Handling Systems—Common Messaging—Part 4: AMH12 and AMH14—MTS Access (P3) and MTS 94 Access (P3)	Published 2003	JTC1/SC6

ISO 10611-4 Available from
http://www.iso.org/iso/iso_catalogue/catalogue_tc/catalogue_detail.htm?csnumber=30719

ISO/IEC ISP 10611-4:2003(E) covers access to an MTS using the P3 MTS Access Protocol. These specifications form part of the common messaging application functions, as defined in ISO/IEC ISP 10611 (all parts), which form a common basis for content type dependent International Standardized Profiles for MHS that will be developed.

An MTA or an MTS-user which conforms to profile AMH12 as specified in ISO/IEC ISP 10611-4:2003(E) shall support an "mts-access" application context and for the MTA also an "mts-forced-access" application context. The MTA or the MTS-user may additionally conform to profile AMH14 as specified in ISO/IEC ISP 10611-4:2003(E) and shall support the "mts-access-94" and "mts-forced-access-94" application contexts.

ISO/IEC	IS/ISP	10611-5	Information technology—International Standardized Profiles AMH1n—Message Handling Systems—Common Messaging—Part 5: AMH13—MS Access (P7)	Published 2003	JTC1/SC6

ISO 10611-5 Available from
http://www.iso.org/iso/iso_catalogue/catalogue_tc/catalogue_detail.htm?csnumber=30720

ISO/IEC ISP 10611-5:2003(E) covers access to a message store using the P7 MS Access Protocol. These specifications form part of the common messaging application functions, as defined in ISO/IEC ISP 10611 (all parts), which form a common basis for content type dependent International Standardized Profiles for MHS that will be developed.

ISO/IEC	IS/ISP	10611-6	Information technology—International Standardized Profiles AMH1n—Message Handling Systems—Common Messaging—Part 6: AMH15—MS 94 Access (P7)	Published 2003	JTC1/SC6

ISO 10611-6 Available from
http://www.iso.org/iso/iso_catalogue/catalogue_tc/catalogue_detail.htm?csnumber=30724

ISO/IEC ISP 10611-6:2003(E) covers access to a message store using the P7 MS Access Protocol as defined in ISO/IEC 10021-5:1994 and ISO/IEC 10021-6:1996. These specifications form part of the common messaging application functions, as defined in ISO/IEC ISP 10611 (all parts), which form a common basis for content type dependent International Standardized Profiles for MHS that will be developed.

6.2.2.40 ISO 10733 Elements of Management Information Related to the OSI Network Layer

ISO/IEC	IS	10733	Information technology—Elements of management information related to the OSI Network Layer	Published 1998	JTC1/SC6

ISO 10733 Available from
http://www.iso.org/iso/iso_catalogue/catalogue_tc/catalogue_detail.htm?csnumber=30999

ISO 10733 provides the specification of management information within an Open System related to those operations of the OSI Network Layer. Specifics on how network layer management is accomplished are beyond the scope of this International Standard. Network layer management information is defined by specifying:

- The managed object class definition of Network Layer Managed Objects following guidelines put forth by the Structure of Management Information (see Recommendations X.720-X.724 and ISO/IEC 10165);
- The relationship of the managed objects and attributes to both the operation of the layer and to other objects and attributes of the layer;

- The action type operations on the attributes of Network Layer Managed Objects that are available to OSI Systems Management.

Annexes D, E, F, and G, which are integral parts of this Recommendation/International Standard, provide ICS pro forma associated with network layer management information.

6.2.2.41 ISO 10736 Telecommunications and Information Exchange Between Systems—Transport Layer Security Protocol

ISO/IEC	IS	10736	ISO/IEC: Information technology—Telecommunications and information exchange between systems—Transport layer security protocol	Published 1995	JTC1/SC6

ISO 10736 Available from
http://www.iso.org/iso/iso_catalogue/catalogue_tc/catalogue_detail.htm?csnumber=18828

This defines the transport layer security protocol. It does not specify the management functions and protocols needed to support this security protocol. It defines a protocol which may be used for security association establishment. It specifies one algorithm for authentication and key distribution which is based on public key crypto systems.

6.2.2.42 ISO 10742 Elements of Management Information Related to OSI Data Link Layer Standards

ISO/IEC	IS	10742	Information technology—Telecommunications and information exchange between systems—Elements of management information related to OSI Data Link Layer standards	Published 1994	JTC1/SC6

ISO 10742 Available from
http://www.iso.org/iso/iso_catalogue/catalogue_tc/catalogue_detail.htm?csnumber=20897

This provides the specification of management information within an Open System related to those operations of the OSI data link layer specified by the specifications in this document.

6.2.2.43 ISO 10747 Protocol for Exchange of Inter-Domain Routing Information Among Intermediate Systems to Support Forwarding of ISO 8473 PDUs

ISO/IEC	IS	10747	Information technology—Telecommunications and information exchange between systems—Protocol for exchange of inter-domain routing information among intermediate systems to support forwarding of ISO 8473 PDUs	Published 1994	JTC1/SC6

ISO 10747 Available from
http://www.iso.org/iso/iso_catalogue/catalogue_tc/catalogue_detail.htm?csnumber=21417

This specifies a protocol to be used by boundary intermediate systems to acquire and maintain information for the purpose of routing NPDUs between different routing domains. It lays down the procedures for the exchange of inter-domain reachability and path information between BISs, the procedures for maintaining inter-domain routing information bases within a BIS, the encoding of protocol data units used to distribute inter-domain routing information between BISs, and the functional requirements for implementations that claim conformance to this standard. The protocol described operates at the level of individual routing domains. It does not cover the establishment of administrative domains.

6.2.2.44 ISO/IEC 11179 Metadata Registries

ISO/IEC	IS	11179	Information technology — Metadata Registries (MDR)		JTC1/ SC32

The ISO/IEC 11179 series of standards is particularly significant for ITS standardization. These are the metadata registry standards (ISO 11179) on which the ITS Metadata Registry Standard ISO 14817 and Emergency Message Data Registry Standard ISO 24978 are based.

With ITS data moving from system to system, there needs to be repositories where the data and metadata can be stored, retrieved, and understood. This requires a registration process and a data repository/data dictionary. Data registry standards have been developed to enable this efficiently, consistently, and interoperably.

It should be noted that ISO 14817, which was approved in 2002, and ISO 24978, in which it was felt that consistency with ISO 14817 was paramount, were based on the version of ISO 11179 concurrent in 2000–2002, and not the current versions of ISI/IEC 11179.

The ISO/IEC 11179 standard is a multipart standard that includes the following six parts.

ISO/IEC	IS	11179-1	Information technology—Metadata Registries (MDR)—Part 1: Framework	2004 (2nd edition)	JTC1/ SC32

Downloadable FOC ISO/IEC 11179-1:2004
http://standards.iso.org/ittf/PubliclyAvailableStandards/c035343_ISO_IEC_11179-1_2004(E).zip

Part 1: Framework, introduces and discusses fundamental ideas of data elements, value domains, data element concepts, conceptual domains, and classification schemes essential to the understanding of this set of standards, and it provides the context for associating the individual parts of ISO/IEC 11179.

This edition replaces ISO/IEC 11179-1:1999(E) (1st edition) "Information technology—Specification and standardization of data elements—Part 1: Framework for the standardization of data elements."

ISO/IEC	IS	11179-2	Information Technology—Metadata Registries (MDR)—Part 2: Classification	2005 (2nd edition)	JTC1/ SC32

Downloadable FOC ISO/IEC 11179-2:2005
http://standards.iso.org/ittf/PubliclyAvailableStandards/c035345_ISO_IEC_11179-2_2005(E).zip

Part 2: Classification, provides a conceptual model for managing classification schemes. There are many structures used to organize classification schemes and there are many subject matter areas that classification schemes describe. So, this part also provides a two-faceted classification for classification schemes themselves.

This edition replaces ISO/IEC 11179-2:2000 (1st edition) "Information technology—Specification and standardization of data elements—Part 2: Classification for data elements."

ISO/IEC	IS	11179-3	Information Technology—Metadata Registries (MDR)—Part 3: Registry metamodel and basic attributes.	2003 (2nd edition)	JTC1/ SC32

Downloadable FOC ISO/IEC 11179-3:2003
http://standards.iso.org/ittf/PubliclyAvailableStandards/c031367_ISO_IEC_11179-3_2003(E).zip

Part 3: Registry Metamodel and Basic Attributes, specifies a conceptual model for a metadata registry. It is limited to a set of basic attributes for data elements, data element concepts, value domains, conceptual domains, classification schemes, and other related classes, called administered items. The basic attributes specified for data elements in ISO/IEC 11179-3:1994 are provided in this revision.

This edition replaces ISO/IEC 11179-3:1994(E) (1st edition) "Information technology—Specification and standardization of data elements—Part 3: Basic attributes of data elements."

ISO/IEC	IS	11179-4	Information Technology—Metadata Registries (MDR)—Part 4: Formulation of data definitions	2004 (2nd edition)	JTC1/ SC32

Downloadable FOC ISO/IEC 11179-4:2004
http://standards.iso.org/ittf/PubliclyAvailableStandards/c035346_ISO_IEC_11179-4_2004(E).zip

Part 4: Formulation of Data Definitions, provides guidance on how to develop unambiguous data definitions. A number of specific rules and guidelines are presented in ISO/IEC 11179-4 that specify exactly how a data definition should be formed. A precise, well-formed definition is one of the most critical requirements for shared understanding of an administered item; well-formed definitions are imperative for the exchange of information. Only if every user has a common and exact understanding of the data item can it be exchanged without trouble.

This edition replaces ISO/IEC 11179-4:1995(E) (1st edition) "Information technology—Specification and standardization of data elements—Part 4: Rules and guidelines for the formulation of data definitions."

ISO/IEC	IS	11179-5	Information Technology—Metadata Registries (MDR)—Part 5: Naming and identification principles	2005 (2nd edition)	JTC1/ SC32

Downloadable FOC ISO/IEC 11179-5:2005
http://standards.iso.org/ittf/PubliclyAvailableStandards/c035347_ISO_IEC_11179-5_2005(E).zip

Part 5: Naming and Identification Principles, provides guidance for the identification of administered items. Identification is a broad term for designating, or identifying, a particular data item. Identification can be accomplished in various ways, depending upon the use of the identifier. Identification includes the assignment of numerical identifiers that have no inherent meanings to humans; icons (graphic symbols to which meaning has been assigned); and names with embedded meaning, usually for human understanding, that are associated with the data item's definition and value domain.

This edition replaces ISO/IEC 11179-5:1995 (1st edition) "Information technology—Specification and standardization of data elements—Part 5: Naming and identification principles for data elements."

ISO/IEC	IS	11179-6	Information Technology—Metadata Registries (MDR)—Part 6: Registration	2005 (2nd edition)	JTC1/ SC32

Downloadable FOC ISO/IEC 11179-6:2005
http://standards.iso.org/ittf/PubliclyAvailableStandards/c035348_ISO_IEC_11179-6_2005(E).zip

Part 6: Registration, provides instruction on how a registration applicant may register a data item with a central registration authority and the allocation of unique identifiers for each data item. Maintenance of administered items already registered is also specified in this document.

This edition replaces ISO/IEC 11179-6:1997 (1st edition) "Information technology—Specification and standardization of data elements—Part 6: Registration of data elements."

6.2.2.45 ISO 11570 Open Systems Interconnection—Transport Protocol Identification Mechanism

ISO/IEC	IS	11570	Information technology—Telecommunications and information exchange between systems—Open Systems Interconnection—Transport protocol identification mechanism	Published 1992	JTC1/SC6

ISO 11570 Available from
http://www.iso.org/iso/iso_catalogue/catalogue_tc/catalogue_detail.htm?csnumber=20325

The procedures specified do not prevent communication between transport entities conforming to ISO/IEC 8073 only and those conforming to ISO/IEC 8073 as well as to this standard. The use of a protocol identification procedure allows transport entities to be implemented which can support both the OSI transport protocols and non-OSI transport protocols above the OSI network layer.

6.2.2.46 Telecommunications and Information Exchange Between Systems—Private Integrated Services Networks (PISN)

This is an important series of standards, of largely unrelated numbering, spanning a number of years in development. Their role in ITS is difficult to define at this stage and may only be in the handling of certain back office processing; however, some of these standards may turn out to be relevant in certain types of direct ITS communications. They are listed next, but are not detailed. All are directly downloadable without cost.

Private Integrated Services Networks (PISN) are private networks providing services to a specific set of users (this is different from a public network which provides services to the general public).

The series of standards provides specifications, based on a reference configuration (RC) for private integrated services network exchanges (PINX) for their interconnection to form PISNs. By combining multiple PINXs to a PISN the RC becomes applicable to a PISN.

The configuration is not intended to require any specific implementation of a PINX, but only to provide guidance for the specification of PINX capabilities. This RC is sufficient to support ISDN-like applications. It can be extended to also support non-ISDN-like applications. There are many aspects to provide reference specification, and hence there are many standards developed over a period of more than a decade.

There are a large number of (free to download) standards in this series, under the general context:

ISO/IEC	IS	xxxxx	Information technology—Telecommunications and information exchange between systems	JTC1/SC6

Downloadable FOC ISO/IEC 11571:1998—Private Integrated Services Networks—Addressing
http://standards.iso.org/ittf/PubliclyAvailableStandards/c029520_ISO_IEC_11571_1998(E).zip

Downloadable FOC ISO/IEC 11572:2000—Circuit mode bearer services—Inter-exchange signalling procedures and protocol
http://standards.iso.org/ittf/PubliclyAvailableStandards/c031613_ISO_IEC_11572_2000(E).zip

Downloadable FOC ISO/IEC 11574:2000—Circuit-mode 64 kbit/s bearer services—Service description, functional capabilities and information flows
http://standards.iso.org/ittf/PubliclyAvailableStandards/c031612_ISO_IEC_11574_2000(E).zip

Downloadable FOC ISO/IEC 11579-1:1994—Part 1: Reference configuration for PISN Exchanges (PINX)
http://standards.iso.org/ittf/PubliclyAvailableStandards/s020893_ISO_IEC_11579-1_1994(E).zip

Downloadable FOC ISO/IEC 11579-1:1994/Cor 1:1996—Part 1: Reference configuration for PISN Exchanges (PINX) Technical Corrigendum 1
http://standards.iso.org/ittf/PubliclyAvailableStandards/s027237_ISO_IEC_11579-1_1994_Cor_1_1996(E).zip

Downloadable FOC ISO/IEC 11579-3:1999—Part 3: Reference configuration for PINX extension lines
http://standards.iso.org/ittf/PubliclyAvailableStandards/c031128_ISO_IEC_11579-3_1999(E).zip

Downloadable FOC ISO/IEC 11582:2002—Generic functional protocol for the support of supplementary services—Inter-exchange signalling procedures and protocol
http://standards.iso.org/ittf/PubliclyAvailableStandards/c036049_ISO_IEC_11582_2002(E).zip

Downloadable FOC ISO/IEC 13863:1998—Specification, functional model and information flows—Path replacement additional network feature
http://standards.iso.org/ittf/PubliclyAvailableStandards/c29666_ISO_IEC_13863_1998(E).zip

Downloadable FOC ISO/IEC 13864:1995—Specification, functional model and information flows—Path replacement additional network feature
http://standards.iso.org/ittf/PubliclyAvailableStandards/s023103_ISO_IEC_13864_1995(E).zip

Downloadable FOC ISO/IEC 13865:2003—Specification, functional model and information flows—Call Transfer supplementary service
http://standards.iso.org/ittf/PubliclyAvailableStandards/c036843_ISO_IEC_13865_2003(E).zip

Downloadable FOC ISO/IEC 13866:1995—Specification, functional model and information flows—Call Transfer supplementary service
http://standards.iso.org/ittf/PubliclyAvailableStandards/s023105_ISO_IEC_13866_1995(E).zip

Downloadable FOC ISO/IEC 13868:2003—Inter-exchange signalling protocol—Name identification supplementary services
http://standards.iso.org/ittf/PubliclyAvailableStandards/c036839_ISO_IEC_13868_2003(E).zip

Downloadable FOC ISO/IEC 13869:2003—Inter-exchange signalling protocol—Call Transfer supplementary service
http://standards.iso.org/ittf/PubliclyAvailableStandards/c036844_ISO_IEC_13869_2003(E).zip

Downloadable FOC ISO/IEC 13870:2003—Inter-exchange signalling protocol—Call Completion supplementary services
http://standards.iso.org/ittf/PubliclyAvailableStandards/c036845_ISO_IEC_13870_2003(E).zip

Downloadable FOC ISO/IEC 13872:2003—Specification, functional model and information flows—Call Diversion supplementary services
http://standards.iso.org/ittf/PubliclyAvailableStandards/c036840_ISO_IEC_13872_2003(E).zip

Downloadable FOC ISO/IEC 13873:2003—Inter-exchange signalling protocol—Call Diversion supplementary service
http://standards.iso.org/ittf/PubliclyAvailableStandards/c036841_ISO_IEC_13873_2003(E).zip

Downloadable FOC ISO/IEC 13874:2003—Inter-exchange signalling protocol—Path Replacement additional network feature
http://standards.iso.org/ittf/PubliclyAvailableStandards/c036842_ISO_IEC_13874_2003(E).zip

Downloadable FOC ISO/IEC 14136:1995—Specification, functional model and information flows—Identification supplementary services
http://standards.iso.org/ittf/PubliclyAvailableStandards/s023656_ISO_IEC_14136_1995(E).zip

Downloadable FOC ISO/IEC 14474:1998—Functional requirements for static circuit-mode inter-PINX connections
http://standards.iso.org/ittf/PubliclyAvailableStandards/c023547_ISO_IEC_14474_1998(E).zip

Downloadable FOC ISO/IEC TR 14475:2001—Architecture and scenarios for Private Integrated Services
http://standards.iso.org/ittf/PubliclyAvailableStandards/c034118_ISO_IEC_TR_14475_2001(E).zip

Downloadable FOC ISO/IEC 14841:1996—Specification, functional model and information flows—Call offer supplementary service
http://standards.iso.org/ittf/PubliclyAvailableStandards/s025738_ISO_IEC_14841_1996(E).zip

Downloadable FOC ISO/IEC 14842:1996—Specification, functional model and information flows—Do not disturb and do not disturb override supplementary services
http://standards.iso.org/ittf/PubliclyAvailableStandards/s025739_ISO_IEC_14842_1996(E).zip

Downloadable FOC ISO/IEC 14843:2003—Specification, functional model and information flows—Do not disturb and do not disturb override supplementary services
http://standards.iso.org/ittf/PubliclyAvailableStandards/c036846_ISO_IEC_14843_2003(E).zip

Downloadable FOC ISO/IEC 14844:2003—Inter-exchange signalling protocol—Do Not Disturb and Do Not Disturb Override supplementary services
http://standards.iso.org/ittf/PubliclyAvailableStandards/c036847_ISO_IEC_14844_2003(E).zip

Downloadable FOC ISO/IEC 14845:1996—Specification, functional model and information flows—Call intrusion supplementary service
http://standards.iso.org/ittf/PubliclyAvailableStandards/s025743_ISO_IEC_14845_1996(E).zip

Downloadable FOC ISO/IEC 14846:2003—Inter-exchange signalling protocol—Call intrusion supplementary service
http://standards.iso.org/ittf/PubliclyAvailableStandards/c036848_ISO_IEC_14846_2003(E).zip

Downloadable FOC ISO/IEC 15049:1997—Specification, functional model and information flows—Advice of charge supplementary services
http://standards.iso.org/ittf/PubliclyAvailableStandards/s026329_ISO_IEC_15049_1997(E).zip

Downloadable FOC ISO/IEC 15050:2003—Inter-exchange signalling protocol—Advice Of Charge supplementary services
http://standards.iso.org/ittf/PubliclyAvailableStandards/c036849_ISO_IEC_15050_2003(E).zip

Downloadable FOC ISO/IEC 15051:2003—Specification, functional model and information flows—Recall supplementary service
http://standards.iso.org/ittf/PubliclyAvailableStandards/c036850_ISO_IEC_15051_2003(E).zip

Downloadable FOC ISO/IEC 15052:2003—Inter-exchange signalling protocol—Recall supplementary service
http://standards.iso.org/ittf/PubliclyAvailableStandards/c036851_ISO_IEC_15052_2003(E).zip

Downloadable FOC ISO/IEC 15053:2003—Specification, functional model and information flows—Call interception additional network feature
http://standards.iso.org/ittf/PubliclyAvailableStandards/c036852_ISO_IEC_15053_2003(E).zip

Downloadable FOC ISO/IEC 15054:2003—Inter-exchange signalling protocol—Call interception additional network feature
http://standards.iso.org/ittf/PubliclyAvailableStandards/c036853_ISO_IEC_15054_2003(E).zip

Downloadable FOC ISO/IEC 15055:1997—Specification, functional model and information flows—Transit counter additional network feature
http://standards.iso.org/ittf/PubliclyAvailableStandards/s026335_ISO_IEC_15055_1997(E).zip

Downloadable FOC ISO/IEC 15056:1997—Inter-exchange signalling protocol—Transit counter additional network feature
http://standards.iso.org/ittf/PubliclyAvailableStandards/s026336_ISO_IEC_15056_1997(E).zip

Downloadable FOC ISO/IEC 15428:1999—Specification, functional model and information flows—Wireless Terminal Location Registration supplementary service and Wireless Terminal Information Exchange additional network feature
http://standards.iso.org/ittf/PubliclyAvailableStandards/c027671_ISO_IEC_15428_1999(E).zip

Downloadable FOC ISO/IEC 15429:2003—Inter-exchange signalling protocol—Wireless Terminal Location Registration supplementary service and Wireless Terminal Information exchange additional network feature
http://standards.iso.org/ittf/PubliclyAvailableStandards/c036861_ISO_IEC_15429_2003(E).zip

Downloadable FOC ISO/IEC 15430:1999—Specification, functional model and information flows—Wireless terminal call handling additional network features
http://standards.iso.org/ittf/PubliclyAvailableStandards/c027673_ISO_IEC_15430_1999(E).zip

Downloadable FOC ISO/IEC 15431:2003—Inter-exchange signalling protocol—Wireless terminal call handling additional network features
http://standards.iso.org/ittf/PubliclyAvailableStandards/c036862_ISO_IEC_15431_2003(E).zip

Downloadable FOC ISO/IEC 15432:1999—Specification, functional model and information flows—Wireless Terminal Authentication supplementary services (WTAT and WTAN)
http://standards.iso.org/ittf/PubliclyAvailableStandards/c027675_ISO_IEC_15432_1999(E).zip

Downloadable FOC ISO/IEC 15433:2003—Inter-exchange signalling protocol—Wireless Terminal Authentication supplementary services
http://standards.iso.org/ittf/PubliclyAvailableStandards/c036863_ISO_IEC_15433_2003(E).zip

Downloadable FOC ISO/IEC 15505:2003—Specification, functional model and information flows—Message Waiting Indication supplementary service
http://standards.iso.org/ittf/PubliclyAvailableStandards/c036960_ISO_IEC_15505_2003(E).zip

Downloadable FOC ISO/IEC 15506:2003—Inter-exchange signalling protocol—Message Waiting Indication supplementary service
http://standards.iso.org/ittf/PubliclyAvailableStandards/c036854_ISO_IEC_15506_2003(E).zip

Downloadable FOC ISO/IEC 17309:2000—Mapping functions for the employment of a circuit mode basic service and the supplementary service user-to-user signalling as a pair of on-demand inter-PINX connections
http://standards.iso.org/ittf/PubliclyAvailableStandards/c032966_ISO_IEC_17309_2000(E).zip

Downloadable FOC ISO/IEC 17310:2000—Mapping functions for the employment of a circuit mode basic service and the supplementary service user-to-user signalling as a pair of on-demand inter-PINX connections
http://standards.iso.org/ittf/PubliclyAvailableStandards/c032970_ISO_IEC_17310_2000(E).zip

Downloadable FOC ISO/IEC 17311:2000—Mapping functions for the employment of 64 kbit/s circuit mode connections with 8 kbit/s sub-multiplexing
http://standards.iso.org/ittf/PubliclyAvailableStandards/c032971_ISO_IEC_17311_2000(E).zip

Downloadable FOC ISO/IEC 17875:2000—Specification, functional model and information flows—Private User Mobility (PUM)—Registration supplementary service
http://standards.iso.org/ittf/PubliclyAvailableStandards/c031600_ISO_IEC_17875_2000(E).zip

Downloadable FOC ISO/IEC 17876:2003—Inter-exchange signalling protocol—Private User Mobility (PUM)—Registration supplementary service
http://standards.iso.org/ittf/PubliclyAvailableStandards/c036858_ISO_IEC_17876_2003(E).zip

Downloadable FOC ISO/IEC 17877:2000—Specification, functional model and information flows—Private User Mobility (PUM)—Call handling additional network features
http://standards.iso.org/ittf/PubliclyAvailableStandards/c031602_ISO_IEC_17877_2000(E).zip

Downloadable FOC ISO/IEC 17878:2003—Inter-exchange signalling protocol—Private User Mobility (PUM)—Call handling additional network features
http://standards.iso.org/ittf/PubliclyAvailableStandards/c036859_ISO_IEC_17878_2003(E).zip

Downloadable FOC ISO/IEC 19459:2001—Specification, functional model and information flows—Single Step Call Transfer Supplementary Service
http://standards.iso.org/ittf/PubliclyAvailableStandards/c033847_ISO_IEC_19459_2001(E).zip

Downloadable FOC ISO/IEC 19460:2003—Inter-exchange signalling protocol—Single Step Call Transfer supplementary service
http://standards.iso.org/ittf/PubliclyAvailableStandards/c036860_ISO_IEC_19460_2003(E).zip

Downloadable FOC ISO/IEC 20113:2004—Specification, functional model and information flows—Make call request supplementary service
http://standards.iso.org/ittf/PubliclyAvailableStandards/c039223_ISO_IEC_20113_2004(E).zip

Downloadable FOC ISO/IEC 20114:2004—Inter-exchange signalling protocol—Make call request supplementary service
http://standards.iso.org/ittf/PubliclyAvailableStandards/c039224_ISO_IEC_20114_2004(E).zip

Downloadable FOC ISO/IEC 20115:2004—Use of QSIG for Message Centre Access (MCA) profile standard
http://standards.iso.org/ittf/PubliclyAvailableStandards/c039225_ISO_IEC_20115_2004(E).zip

Downloadable FOC ISO/IEC 20116:2004—Specification, functional model and information flows—Message centre monitoring and mailbox identification supplementary services
http://standards.iso.org/ittf/PubliclyAvailableStandards/c039226_ISO_IEC_20116_2004(E).zip

Downloadable FOC ISO/IEC 20117:2004—Inter-exchange signalling protocol—Message centre monitoring and mailbox identification supplementary services
http://standards.iso.org/ittf/PubliclyAvailableStandards/c039227_ISO_IEC_20117_2004(E).zip

Downloadable FOC ISO/IEC 20161:2001—Use of QSIG at the C reference point between a PINX and an Interconnecting Network
http://standards.iso.org/ittf/PubliclyAvailableStandards/c035305_ISO_IEC_20161_2001(E).zip

Downloadable FOC ISO/IEC 21407:2001—Specification, functional model and information flows—Simple dialog supplementary service
http://standards.iso.org/ittf/PubliclyAvailableStandards/c034499_ISO_IEC_21407_2001(E).zip

Downloadable FOC ISO/IEC 21408:2003—Inter-exchange signalling protocol—Simple dialog supplementary service
http://standards.iso.org/ittf/PubliclyAvailableStandards/c036864_ISO_IEC_21408_2003(E).zip

Downloadable FOC ISO/IEC 21888:2001—Specification, functional model and information flows—Call Identification and Call Linkage Additional Network Feature
http://standards.iso.org/ittf/PubliclyAvailableStandards/c034764_ISO_IEC_21888_2001(E).zip

Downloadable FOC ISO/IEC 21889:2001—Inter-exchange signaling protocol—Call Identification and Call Linkage Additional Network Feature
http://standards.iso.org/ittf/PubliclyAvailableStandards/c034765_ISO_IEC_21889_2001(E).zip

Downloadable FOC ISO/IEC TR 21890:2001—Telecommunications and information exchange between systems—Interoperation of PISNs with IP networks
http://standards.iso.org/ittf/PubliclyAvailableStandards/c034767_ISO_IEC_TR_21890_2001(E).zip

Downloadable FOC ISO/IEC 21989:2002—Specification, functional model and information flows—Short message service
http://standards.iso.org/ittf/PubliclyAvailableStandards/c036050_ISO_IEC_21989_2002(E).zip

Downloadable FOC ISO/IEC 21990:2002—Inter-exchange signalling protocol—Short message service
http://standards.iso.org/ittf/PubliclyAvailableStandards/c036051_ISO_IEC_21990_2002(E).zip

Downloadable FOC ISO/IEC 21992:2003—Mapping functions for the tunnelling of QSIG through IP networks
http://standards.iso.org/ittf/PubliclyAvailableStandards/c037519_ISO_IEC_21992_2003(E).zip

Downloadable FOC ISO/IEC 23290:2004—Mapping functions for the tunneling of QSIG through H.323 networks
http://standards.iso.org/ittf/PubliclyAvailableStandards/c040165_ISO_IEC_23290_2004(E).zip

6.2.2.47 ISO 11575 Protocol Mappings for the OSI Data Link Service

ISO/IEC	IS	11575	Information technology—Telecommunications and information exchange between systems—Protocol mappings for the OSI Data Link service	Published 1995	JTC1/SC6

ISO 11 575 Available from
http://www.iso.org/iso/iso_catalogue/catalogue_tc/catalogue_detail.htm?csnumber=20916

This specifies general principles for the mappings between the OSI data link service, both connection-mode (CO-DLS) and connectionless-mode (CL-DLS), and standard data link protocols. It specifies the detailed mappings M1, M2, M5, and M6; it also specifies the main features of the mappings M3 and M4. It does not specify individual implementations or products, nor does it constrain the implementation of data link entities and interfaces within an information processing system.

6.2.2.48 ISO 11577 Open Systems Interconnection—Network Layer Security Protocol

ISO/IEC	IS	11577	Information technology—Open Systems Interconnection—Network layer security protocol	Published 1995	JTC1/SC6

ISO 11577 Available from
http://www.iso.org/iso/iso_catalogue/catalogue_tc/catalogue_detail.htm?csnumber=22084

This specifies a protocol to be used by end systems and intermediate systems in order to provide security services in the network layer, which is defined by CCITT

Rec. X.213, ISO/IEC 8348 and ISO 8648. The protocol defined herein is called the Network Layer Security Protocol (NLSP).

6.2.2.49 ISO TRs 11802 Telecommunications and Information Exchange Between Systems—Local and Metropolitan Area Networks—Technical Reports and Guidelines

ISO/IEC	TR	TR 11802	Information technology—Telecommunications and information exchange between systems—Local and metropolitan area networks—Technical reports and guidelines		JTC1/SC6

This subject area has been divided into five work items and should be viewed in conjunction with ISO (confusingly) 8802 (see Section 6.2.2.19).

ISO/IEC	TR	11802-1	Information technology—Telecommunications and information exchange between systems—Local and metropolitan area networks—Technical reports and guidelines—Part 1: The structure and coding of Logical Link Control addresses in Local Area Networks	Published 2005	JTC1/SC6

ISO 11802-1 Available from
http://www.iso.org/iso/iso_catalogue/catalogue_tc/catalogue_detail.htm?csnumber=40637

ISO/IEC TR 11802-1:2005 provides:

- A description of the ISO/IEC 8802-2 LLC addressing conventions;
- The consideration for the manner in which new LLC address uses are assigned a value.

It is outside the scope of this TR to provide architectural judgments regarding the entities which are identified by particular address value(s).

ISO/IEC	TR	11802-2	Information technology—Telecommunications and information exchange between systems—Local and metropolitan area networks—Technical reports and guidelines—Part 2: Standard Group MAC Addresses	Published 2005	JTC1/SC6

ISO 11802-2 Available from
http://www.iso.org/iso/iso_catalogue/catalogue_tc/catalogue_detail.htm?csnumber=40636

ISO/IEC TR 11802-2:2005 provides:

- A description of the binary and hexadecimal representation of ISO/IEC 8802 LAN MAC addresses;
- A description of the subdivision of the Universally Administered Standard Group MAC Address Block into: ISO/IEC 10038 MAC Bridge Filtered MAC Group Addresses and Standard MAC Group Addresses.

11802-3 does not appear on the ISO Web site as a current standard, development work, or deleted work item.

ISO/IEC	TR	11802-4	Information technology—Telecommunications and information exchange between systems—Local and metropolitan area networks—Technical reports and guidelines—Part 4: Token ring access method and physical layer specifications—Fiber optic station attachment	Withdrawn	JTC1/SC6

ISO 11802-4 is withdrawn

ISO/IEC	TR	11802-5	Information technology—Telecommunications and information exchange between systems—Local and metropolitan area networks—Technical reports and guidelines—Part 5: Media Access Control (MAC) Bridging of Ethernet V2.0 in Local Area Networks	Published 1997	JTC1/SC6

ISO 11802-5 Available from
http://www.iso.org/iso/iso_catalogue/catalogue_tc/catalogue_detail.htm?csnumber=29274

6.2.2.50 ISO 13236 Information Technology—Quality of Service: Framework

ISO/IEC	IS	13236	Information technology—Quality of service: Framework	Published 1998	JTC1/SC6

ISO 13236 Available from
http://www.iso.org/iso/iso_catalogue/catalogue_tc/catalogue_detail.htm?csnumber=27993

ISO 13236 provides to the upper-layer applications enhanced transport services over the current OSI transport service; major enhancements include multicast services and enhanced QoS.

6.2.2.51 ISO 13239 Telecommunications and Information Exchange Between Systems—High-Level Data Link Control (HDLC) Procedures

ISO/IEC	IS	13239	Information technology—Telecommunications and information exchange between systems—High-level data link control (HDLC) procedures	Published 2002	JTC1/SC6

ISO 13239 Available from
http://www.iso.org/iso/iso_catalogue/catalogue_tc/catalogue_detail.htm?csnumber=37010

This International Standard specifies the frame structures, the elements of procedures, the classes of procedures, the content and format of the general purpose exchange identification (XID) frame, and a means for resolution/negotiation of a data link layer address in switched environments for data communication systems using bit-oriented high-level data link control (HDLC) procedures.

NOTE: The use of the phrase "bit-oriented," referring to the HDLC control procedures, pertains to the allocation of a nonintegral number of bits to various subfields used for HDLC control purposes. However, the frame as an entirety may be constructed from octet-oriented units (e.g., start-stop mode) for transmission purposes.

6.2.2.52 Telecommunications and Information Exchange Between Systems—Broadband Private Integrated Services Network

These standards are extensions to the PISN standards summarized in Section 6.2.2.46 above, in order to add broadband capabilities.

ISO/IEC	IS	13246	Information technology—Telecommunications and information exchange between systems—Broadband Private Integrated Services Network—Inter-exchange signalling protocol—Signalling ATM adaptation layer	Published 1997	JTC1/SC6

Downloadable FOC ISO/IEC 13246:1997
http://standards.iso.org/ittf/PubliclyAvailableStandards/s021481_ISO_IEC_13246_1997(E).zip

ISO/IEC	IS	13247	Information technology—Telecommunications and information exchange between systems—Broadband Private Integrated Services Network—Inter-exchange signaling protocol—Basic call/connection control	Published 1997	JTC1/SC6

Downloadable FOC ISO/IEC 13247:1997
http://standards.iso.org/ittf/PubliclyAvailableStandards/s021482_ISO_IEC_13247_1997(E).zip

ISO/IEC	IS	19058	Information technology—Telecommunications and information exchange between systems—Broadband Private Integrated Services Network—Inter-exchange signalling protocol—Generic functional protocol	Published 2001	JTC1/SC6

Downloadable FOC ISO/IEC 19058:2001
http://standards.iso.org/ittf/PubliclyAvailableStandards/c033736_ISO_IEC_19058_2001(E).zip

6.2.2.53 ISO 13594 Information Technology—Lower Layers Security

ISO/IEC	TR	13594	Information technology—Lower layers security	TR:1995	JTC1/SC6

ISO 13594 Available from
http://www.iso.org/iso/iso_catalogue/catalogue_tc/catalogue_detail.htm?csnumber=23330

This describes the cross layer aspects of the revision of security services in the lower layers of the OSI reference model (transport, network, data link, physical). It describes the architectural concepts common to these layers, the basis for interactions relating to security between layers, and the placement of security protocols in the lower layers.

6.2.2.54 ISO 13642 Elements of Management Information Related to the OSI Physical Layer

ISO/IEC	IS	13642:	Information technology—Elements of management information related to the OSI Physical Layer	Published 1999	JTC1/SC6

ISO 13642 Available from
http://www.iso.org/iso/iso_catalogue/catalogue_tc/catalogue_detail.htm?csnumber=33022

6.2.2.55 ISO 14476 Enhanced Communications Transport Protocol: Specification of Simplex Multicast Transport

So far there is only one published part to this standard.

ISO/IEC	IS	14476-1	Information technology—Enhanced communications transport protocol: Specification of simplex multicast transport	Published 2002	JTC1/SC6

NOTE: "Transport" is used here in the context of transporting data via a medium, not "transport" in the context of ITS.

ISO 14476-1 Available from
http://www.iso.org/iso/iso_catalogue/catalogue_tc/catalogue_detail.htm?csnumber=35679

ISO 14476-1 specifies the Enhanced Communications Transport Protocol (ECTP), which is a transport protocol designed to support Internet multicast applications over multicast-capable IP networks.

This Recommendation/International Standard specifies the ECTP for the simplex multicast transport connection that consists of one sender and many receivers. This International Standard specifies the protocol procedures for the following protocol operations:

- Connection creation with tree creation;
- Multicast data transmission;
- Tree-based reliability control with error detection, retransmission request, and retransmission;
- Late join and leave;
- Tree membership maintenance;
- Connection termination.

NOTE: ISO/IEC 14476 consists of the following parts, under the general title "Information technology—Enhanced Communications Transport Protocol:"

- Part 1: Specification of simplex multicast transport;
- Part 2: Specification of QoS management for simple multicast transport;
- Part 3: Specification of duplex multicast transport;
- Part 4: Specification of QoS management for duplex multicast transport;
- Part 5: Specification of n-plex multicast transport;
- Part 6: Specification of QoS management for n-plex multicast transport.

Other parts are available at:

http://www.iso.org/iso/search.htm?qt=iso+14476&sort=rel&type=simple&published=true

6.2.2.56 ISO 14699 Open Systems Interconnection—Transport Fast Byte Protocol

ISO/IEC	IS	14699	Information technology—Open Systems Interconnection—Transport Fast Byte Protocol	Published 1997	JTC1/SC6

ISO 14699 Available from
http://www.iso.org/iso/iso_catalogue/catalogue_tc/catalogue_detail.htm?csnumber=25266

6.2.2.57 ISO 14700 Open Systems Interconnection—Network Fast Byte Protocol

ISO/IEC	IS	14700	Information technology—Open Systems Interconnection—Network Fast Byte Protocol	Published 1997	JTC1/SC6

ISO 14700 Available from
http://www.iso.org/iso/iso_catalogue/catalogue_tc/catalogue_detail.htm?csnumber=25267

6.2.2.58 ISO 14765 Framework for Protocol Identification and Encapsulation

ISO/IEC	IS	14765	Information technology—Framework for protocol identification and encapsulation	Published 1997	JTC1/SC6

ISO 146765 Available from
http://www.iso.org/iso/iso_catalogue/catalogue_tc/catalogue_detail.htm?csnumber=25501

6.2.2.59 ISO 14766 Use of OSI Applications over the Internet Transmission Control Protocol (TCP)

ISO/IEC	IS	14766	Information technology—Telecommunications and information exchange between systems—Use of OSI applications over the Internet Transmission Control Protocol (TCP)	Published 1997	JTC1/SC6

ISO 14766 Available from
http://www.iso.org/iso/iso_catalogue/catalogue_tc/catalogue_detail.htm?csnumber=25502

6.2.2.60 ISO 15802 Local and Metropolitan Area Networks—Common Specifications

ISO/IEC	IS	15802	Information technology—Telecommunications and information exchange between systems—Local and metropolitan area networks—Common specifications		JTC1/SC6

At the time of publication of this book, this standard has four parts.

ISO/IEC	IS	15802-1	Information technology—Telecommunications and information exchange between systems—Local and metropolitan area networks—Common specifications—Part 1: Medium Access Control (MAC) service definition	Published 1995	JTC1/SC6

ISO 15802-1 Available from
http://www.iso.org/iso/iso_catalogue/catalogue_tc/catalogue_detail.htm?csnumber=21507

ISO/IEC 15802-1 forms one of a set of International Standards produced to facilitate the interconnection of information processing systems. It is related to other International Standards in the set as defined by ISO/IEC 7498-1. The reference model described by ISO/IEC 7498-1 subdivides the area of standardization for OSI into a series of layers and allows for each layer to be further divided into sublayers.

ISO/IEC	IS	15802-2	Information technology—Telecommunications and information exchange between systems—Local and metropolitan area networks—Common specifications—Part 2: LAN/MAN management	Published 1995	JTC1/SC6

ISO 15802-2 Available from
http://www.iso.org/iso/iso_catalogue/catalogue_tc/catalogue_detail.htm?csnumber=23332

This defines an OSI management-compatible architecture, as well as service and protocol elements for use in a LAN/MAN environment for the purpose of performing remote management of LAN-based or MAN-based devices.

ISO/IEC	IS IEEE ANSI	15802-3 802.1D	Information technology—Telecommunications and information exchange between systems—Local and metropolitan area networks—Common specifications—Part 3: Media Access Control (MAC) Bridges	Published 1998	JTC1/SC6

ISO 15802-3 Available from
http://www.iso.org/iso/iso_catalogue/catalogue_tc/catalogue_detail.htm?csnumber=25021

The concept of MAC bridging, introduced in the 1993 edition of this standard, has been expanded to define additional capabilities in bridged LANs aimed at providing for expedited traffic capabilities, to support the transmission of time-critical information in a LAN environment, and providing filtering services that support the dynamic use of group MAC addresses in a LAN environment.

This is also published as ANSI/IEEE Std 802.1D, 1998 Edition.

ISO/IEC	IS	15802-4	Information technology—Telecommunications and information exchange between systems—Local and metropolitan area networks—Common specifications—Part 4: System load protocol	Published 1994	JTC1/SC6

ISO 15802-4 Available from
http://www.iso.org/iso/iso_catalogue/catalogue_tc/catalogue_detail.htm?csnumber=22115

This defines a protocol to load the memory of data processing equipment installed on IEEE 802 networks. It discusses the LS in only as much detail as is necessary to define the loading protocol. Provides the PICS pro forma for the System Load Protocol in compliance with the relevant requirements, and in accordance with the relevant guidance, given in ISO/IEC 9646-2.

6.2.2.61 ISO 16512 Relayed Multicast Control Protocol (RMCP)

At the time of publication of this book, there is only one published part to this standard.

ISO/IEC	IS	16512-1	Information technology—Relayed Multicast Control Protocol (RMCP)—Part 1: Framework	Published 2005	JTC1/SC6

ISO 165 12-1 Available from
http://www.iso.org/iso/iso_catalogue/catalogue_tc/catalogue_detail.htm?csnumber=37964

ISO/IEC 16512-1:2005 addresses the basic concepts needed to specify RMCP for relayed multicast. It defines the related terminology and proposes a framework for the future development of RMCP. The framework covers network topology, including network entities and relationships between them, service scenarios, basic operations, and message encoding rules. RMCP is a protocol which is used to realize a relayed multicast data transport scheme. Differently than the conventional IP multicast, RMCP can configure a relayed multicast path where multicast traffic flows by using intermediate end hosts. RMCP can be applied to the current unicast-based Internet where IP multicast has not been deployed completely without any modifications.

6.2.2.62 ISO 16513 Information Technology—Group Management Protocol

ISO/IEC	IS	16513	Information technology—Group management protocol	Published 2005	JTC1/SC6

ISO 16513 Available from
http://www.iso.org/iso/iso_catalogue/catalogue_tc/catalogue_detail.htm?csnumber=37965

ISO/IEC 16513:2005 provides a specification of a Group Management Protocol (GMP), which is an application-layer control protocol for creating a group session and for managing the group's participating members.

The GMP consists of session management (SM), membership management (MM), and the function of exchanging information between SM and MM. SM is responsible for session creation and deletion. MM manages the member lists based on session information retrieved from SM.

6.2.2.63 ISO 18016 Message Handling Systems (MHS): Interworking with Internet E-Mail

ISO/IEC	TR	18016	Information technology—Message Handling Systems (MHS): Interworking with Internet e-mail	Published 2003	JTC1/SC6

ISO Available from
http://www.iso.org/iso/iso_catalogue/catalogue_tc/catalogue_detail.htm?csnumber=30801

ISO/IEC TR 18016:2003 identifies documents that specify how implementations of the MHS defined in the ITU-T X.400 series Recommendations | ISO/IEC 10021 may interwork with implementations of Internet e-mail. No requirements for conformance to this technical report are imposed.

6.2.2.64 ISO 18051 Telecommunications and Information Exchange Between Systems—Services for Computer Supported Telecommunications Applications (CSTA) Phase III

ISO/IEC	IS	18051	Information technology—Telecommunications and information exchange between systems—Services for Computer Supported Telecommunications Applications (CSTA) Phase III	Published 2005 3rd Edition	JTC1/SC6

Downloadable FOC ISO/IEC 18051:2005
http://standards.iso.org/ittf/PubliclyAvailableStandards/c041245_ISO_IEC_18051_2005(E).zip

6.2.2.65 ISO 18052 Telecommunications and Information Exchange Between Systems—Protocol for Computer Supported Telecommunications Applications (CSTA) Phase III

ISO/IEC	IS	18052	Information technology—Telecommunications and information exchange between systems—Protocol for Computer Supported Telecommunications Applications (CSTA) Phase III	Published 2000	JTC1/SC6

Downloadable FOC ISO/IEC 18052:2000
http://standards.iso.org/ittf/PubliclyAvailableStandards/c031681_ISO_IEC_18052_2000(E.zip

6.2.2.66 ISO 18053 Telecommunications and Information Exchange Between Systems—Glossary of Definitions and Terminology for Computer Supported Telecommunications Applications (CSTA) Phase III

ISO/IEC	IS	18053	Information technology—Telecommunications and information exchange between systems—Glossary of definitions and terminology for Computer Supported Telecommunications Applications (CSTA) Phase III	Published 2000	JTC1/SC6

Downloadable FOC ISO/IEC TR 18053:2000
http://standards.iso.org/ittf/PubliclyAvailableStandards/c031680_ISO_IEC_TR_18053_2000(E).zip

6.2.2.67 ISO 18056 Telecommunications and Information Exchange Between Systems—XML Protocol for Computer Supported Telecommunications Applications (CSTA) Phase III

ISO/IEC	IS	18056	Information technology—Telecommunications and information exchange between systems—XML Protocol for Computer Supported Telecommunications Applications (CSTA) Phase III	Published 2000	JTC1/SC6

Downloadable FOC ISO/IEC 18056:2005
http://standards.iso.org/ittf/PubliclyAvailableStandards/c041246_ISO_IEC_18056_2005(E).zip

6.2.2.68 Telecommunications and Information Exchange Between Systems—Near Field Communication (NFC)

This is a series of standards of unlinked reference numbers dealing with near field communications within telephony and information exchange between systems. More standards may be expected in the future.

ISO/IEC	IS	18092	Information technology—Telecommunications and information exchange between systems—Near Field Communication—Interface and Protocol (NFCIP-1)	Published 2004	JTC1/SC6

Downloadable FOC ISO/IEC 18092:2004
http://standards.iso.org/ittf/PubliclyAvailableStandards/c038578_ISO_IEC_18092_2004(E).zip

ISO/IEC TR 18016:2003 identifies documents that specify how implementations of the MHS defined in the ITU-T X.400 series Recommendations | ISO/IEC 10021 may interwork with implementations of Internet e-mail. No requirements for conformance to this Technical Report are imposed.

ISO/IEC	IS	21481	Information technology—Telecommunications and information exchange between systems—Near Field Communication Interface and Protocol -2 (NFCIP-2)	Published 2005	JTC1/SC6

Downloadable FOC ISO/IEC 21481:2005
http://standards.iso.org/ittf/PubliclyAvailableStandards/c040261_ISO_IEC_21481_2005(E).zip

ISO/IEC 21481:2005 specifies the communication mode selection mechanism, designed not to disturb any ongoing communication at 13.56 MHz, for devices implementing ISO/IEC 18092, and the reader functionality for integrated circuit cards compliant with ISO/IEC 14443 or ISO/IEC 15693 (and ISO 18000-3).

ISO/IEC 21481:2005 requires implementations to enter the selected communication mode, as specified in the respective International Standard. The communication mode specifications, however, are outside the scope of this NFCIP-2 International Standard.

ISO/IEC	IS	22536	Information technology—Telecommunications and information exchange between systems—Near Field Communication Interface and Protocol (NFCIP-1)—RF interface test methods	Published 2005	JTC1/SC6

Downloadable FOC ISO/IEC 22536:2005
http://standards.iso.org/ittf/PubliclyAvailableStandards/c041001_ISO_IEC_22536_2005(E).zip

ISO/IEC 22536:2005 is part of a suite of standards that specify tests for ISO/IEC 18092. It defines test methods for the RF interface. ISO/IEC 22536:2005 specifies RF test methods for NFC devices with antennas fitting within the rectangular area of 85 mm by 54 mm.

This test standard, the first of two parts, specifies compliance tests for the RF interface of ISO/IEC 18092 devices. The companion test standard specifies protocol tests for ISO/IEC 18092.

ECMA purposefully aligned this International Standard with ISO/IEC 10373-6 to allow testing laboratories to reuse equipment and expertise.

ISO/IEC	IS	23917	Information technology—Telecommunications and information exchange between systems—NFCIP-1—Protocol Test Methods	Published 2005	JTC1/SC6

Downloadable FOC ISO/IEC 23917:2005
http://standards.iso.org/ittf/PubliclyAvailableStandards/c041928_ISO_IEC_23917_2005(E).zip

ISO/IEC 23917:2005 specifies protocol test methods for ISO/IEC 18092 in addition to those specified in ISO/IEC 22536.

ISO/IEC	DIS	28361	Information technology—Near Field Communication Wired Interface (NFC-WI)	Under Development	JTC1/SC6

ISO 28361 Available from
http://www.iso.org/iso/iso_catalogue/catalogue_tc/catalogue_detail.htm?csnumber=44659

6.2.2.69 Telecommunications and Information Exchange Between Systems—Application Session Services

ISO/IEC	IS	22534	Information technology—Telecommunications and information exchange between systems—Application session services	Published 2005	JTC1/SC6

Downloadable FOC ISO/IEC 22534:2005
http://standards.iso.org/ittf/PubliclyAvailableStandards/c040999_ISO_IEC_22534_2005(E).zip

The services defined in ISO/IEC 22534:2005 are used to establish and maintain a relationship between an application and a server for the purpose of exchanging application messages. For the purpose of ISO/IEC 22534:2005, this relationship is called an application session.

Application protocols, such as ECMA-323, require that an application session be established before application messages are exchanged. ECMA-269 specifies several mechanisms for establishing an application context. One possible mechanism is ACSE (ISO/IEC 8649), but since ASCE uses ASN.1 encoding for its services, it is not desirable for use with XML-based protocols such as ECMA-323.

ISO/IEC 22534:2005 provides an XML-based alternative for establishing application sessions.

6.2.2.70 ISO 24771 MAC/PHY Standard for Ad Hoc Wireless Network to Guarantee QoS in an Industrial Work Environment

ISO/IEC	IS	24771	MAC/PHY standard for ad hoc wireless network to guarantee QoS in an industrial work environment	Published 2005	JTC1/SC6

ISO CD24771 Available from
http://www.iso.org/iso/iso_catalogue/catalogue_tc/catalogue_detail.htm?csnumber=43472

6.2.2.71 ISO 24824 Generic Applications of ASN.1: Fast Infoset

ISO/IEC	FDIS	24824	Information technology—Generic applications of ASN.1: Fast Infoset.		JTC1/SC6

See also Sections 5.2.2.4, 5.3.3.3, and 6.2.2.20.3.

Downloadable FOC as ITU RecX.891
http://www.itu.int/publications/bookshop/how-to-buy.html#free

ITU-T Rec. X.891 | ISO/IEC 24824-1 (Fast Infoset) specifies a representation of an instance of the W3C XML information set using binary encodings. These binary encodings are specified using the ASN.1 notation and the ASN.1 ECN.

The Fast Infoset technology provides an alternative to W3C XML syntax as a means of representing instances of the W3C XML information set. This representation generally provides smaller encoding sizes and faster processing than a W3C XML representation.

Fast Infoset specifies the use of several techniques that minimize the size of the encodings (called Fast Infoset documents) and that maximize the speed of creating and processing Fast Infoset documents. These techniques include the use of dynamic tables (for both character strings and qualified names), initial vocabularies, and external vocabularies.

ISO/IEC	IS	24824-1	Information technology—Generic applications of ASN.1: Fast infoset	Published 2007	JTC1/SC6

ISO 24824-1 Available from
http://www.iso.org/iso/iso_catalogue/catalogue_tc/catalogue_detail.htm?csnumber=41327

ISO/IEC 24824-1:2007 specifies a representation of an instance of the W3C XML information set using binary encodings. These binary encodings are specified using the ASN.1 notation and the ASN.1 ECN.

The technology specified in ISO/IEC 24824-1:2007 is called Fast Infoset. This technology provides an alternative to W3C XML syntax as a means of representing instances of the W3C XML information set. This representation generally provides smaller encoding sizes and faster processing than a W3C XML representation.

ISO/IEC 24824-1:2007 specifies the use of several techniques that minimize the size of the encodings and that maximize the speed of creating and processing Fast Infoset documents. These techniques include the use of dynamic tables (for both character strings and qualified names), initial vocabularies, and external vocabularies.

ISO/IEC	IS	24824-2	Information technology—Generic applications of ASN.1: Fast Web Services	Published 2006	JTC1/SC6

ISO 24824-2 Available from
http://www.iso.org/iso/iso_catalogue/catalogue_tc/catalogue_detail.htm?csnumber=41328

ISO/IEC 24824-2:2006 specifies the messages required for fast Web services. It provides the specification of ASN.1 SOAP messages which carry the same semantics as W3C SOAP messages. The exchange of ASN.1 SOAP messages provides fast Web services.

ISO/IEC 24824-2:2006 references other ASN.1 International Standards and the Web services specification W3C SOAP 1.2 to fully define application exchanges that use Web protocols and W3C SOAP functionality. This is an extension to the provision of Web services using W3C SOAP messages, without any change to the functionality of W3C SOAP and service description languages.

The main change is to the use of compact and easily processed binary encodings of XML data, rather than character encodings.

The use of ISO/IEC 24824-2:2006 (with the ASN.1 SOAP messages encoded using the ASN.1 PER) allows applications to provide Web services using messages that require less network bandwidth and less processing power (and hence provide a higher transaction processing rate) than use of the character encoding of XML data.

ISO/IEC 24824-2:2006 also specifies a MIME media type that identifies an ASN.1 SOAP message encoded in PER. It also specifies a MIME media type that identifies a W3C SOAP message infoset encoded as a Fast Infoset document (see ISO/IEC 24824-1). Both of these encodings are used by ISO/IEC 24824-2:2006.

ISO/IEC		24824-3	Information technology—Generic applications of ASN.1: Fast infoset security	Under Development FCD	JTC1/SC6

ISO 24824-3 FCD Track Status at
http://www.iso.org/iso/iso_catalogue/catalogue_tc/catalogue_detail.htm?csnumber=44076

The ASN.1 Group has begun to work on a new ITU-T Recommendation |International Standard, entitled "Information Technology—Generic Applications of ASN.1—'Fast Infoset' and Fast Web Services Security," that will specify the application of standard security methods to the Fast Infoset specification.

In order for applications to apply standard security methods to the Fast Infoset specification, it is necessary to specify transformations to be applied to Fast Infoset documents (or parts of them) to support the application of those security methods.

6.2.2.72 ISO 25437 WS-Session—Web Services for Application Session Services

ISO/IEC	IS	25437	Information technology—Telecommunications and information exchange between systems—WS-Session—Web Services for Application Session Services	Published 2006	JTC1/SC6

ISO 25437 Available from
http://www.iso.org/iso/iso_catalogue/catalogue_tc/catalogue_detail.htm?csnumber=42921

ISO/IEC 25437:2006 is a general standard which specifies generic Web services (in WSDL) and a SOAP binding for the application session services defined in ISO/IEC 22534 (ECMA-354). The application session services allow applications to create and maintain a relationship with servers, termed application session. The Web services specified herein allow service subscribers [applications in ISO/IEC 22534 (ECMA-354)] and service providers [servers in ISO/IEC 22534 (ECMA-354)] to create and maintain such application sessions. See also Section 6.2.2.71.

6.2.2.73 ISO 26905 Enterprise Communication in Next Generation Corporate Networks (NGCN) Involving Public Next Generation Networks (NGN)

ISO/IEC	TR	26905	ISO/IEC Information technology—Telecommunications and information exchange between systems—Enterprise Communication in Next Generation Corporate Networks (NGCN) involving Public Next Generation Networks (NGN)	TR 2006	JTC1/SC6

Downloadable FOC ISO/IEC TR 26905:2006
http://standards.iso.org/ittf/PubliclyAvailableStandards/c043893_ISO_IEC_TR_26905_2006(E).zip

ISO/IEC TR 26905:2006 identifies key use cases for communication with or between IP-based next generation corporate networks (NGCN) involving public next generation networks (NGN), analyzes these use cases in terms of available or planned standardized technology, and identifies requirements that will have to be met.

ISO/IEC TR 26905:2006 investigates configurations involving NGCNs and NGNs and their interoperating requirements. Non-IP-based interoperation (i.e., using circuit-switched technology) between NGCNs and NGNs is outside the scope of ISO/IEC TR 26905:2006.

ISO/IEC TR 26905:2006 does not discriminate between wireless and wired access technology. The terminal equipment (TE) interface within an NGCN is outside the scope of ISO/IEC TR 26905:2006.

All mobility aspects are outside the scope of ISO/IEC TR 26905:2006. They are covered by a companion technical report, ISO/IEC TR 26927.

Application considerations such as IP Centrex and Computer Supported Telecommunications Applications (CSTA) are outside the scope of ISO/IEC TR 26905:2006.

6.2.2.74 ISO 26907 High Rate Ultra Wideband PHY and MAC Standard

ISO/IEC	IS	26907	Information technology—Telecommunications and information exchange between systems—High Rate Ultra Wideband PHY and MAC Standard	Published 2007	JTC1/SC6

Downloadable FOC ISO/IEC 26907:2007
http://standards.iso.org/ittf/PubliclyAvailableStandards/c043900_ISO_IEC_26907_2007(E).zip

ISO/IEC 26907:2007 specifies a distributed MAC sublayer and a physical layer for wireless networks.

6.2.2.75 ISO 26908 MAC-PHY Interface for High Rate Ultra Wideband (ISO/ IEC 26907)

ISO/IEC	PRF	26908	Information technology—Telecommunications and information exchange between systems—MAC-PHY Interface for ISO/IEC 26907	Published 2007	JTC1/SC6

Downloadable FOC ISO/IEC 26908:2007
http://standards.iso.org/ittf/PubliclyAvailableStandards/c043901_ISO_IEC_26908_2007(E).zip

ISO/IEC 26908:2007 specifies the interface between implementations of the PHY and MAC as specified in ISO/IEC 26907.

6.2.2.76 ISO 26927 Corporate Telecommunication Networks—Mobility for Enterprise Communications

ISO/IEC	TR	26927	Information technology—Telecommunications and information exchange between systems—Corporate Telecommunication Networks—Mobility for Enterprise Communications	Published TR Under development to IS	JTC1/SC6

Downloadable FOC ISO/IEC TR 26927:2006
http://standards.iso.org/ittf/PubliclyAvailableStandards/c043914_ISO_IEC_TR_26927_2006(E).zip

ISO/IEC TR 26927:2006 identifies key mobility issues for access to IP-based enterprise information and communications services. It defines a taxonomy of mobility terms and then explores use cases and connectivity scenarios involving mobile enterprise users. From these, it derives requirements for mobility management, architecture, and security.

Mobility for enterprise communication is about making desktop communication and information resources available at different locations and while on the move. ISO/IEC TR 26927:2006 encompasses both wired and wireless connectivity using enterprise and public all-IP networks for voice, data, and converged services.

ISO/IEC TR 26927:2006 is intended as an aid to analyzing gaps in standardization that prevent or hinder mobility in enterprise communications and information access. More general interworking issues of enterprise communications are covered by a companion technical report, ISO/IEC TR 26905.

6.2.2.77 Further SC6 Standards

Further SC6 standards can be found via the following link:

http://www.iso.org/iso/en/CatalogueListPage.CatalogueList?ICS1=33&ICS2=&ICS3=&scopelist=

6.2.3 ISO 802-11 Information Technology—Telecommunications and Information Exchange Between Systems—Local and Metropolitan Area Networks—Specific Requirements—Part 11: Wireless LAN Medium Access Control (MAC) and Physical Layer (PHY) Specifications

ISO/IEC	IS	802-11	Information technology—Telecommunications and information exchange between systems—Local and metropolitan area networks—Specific requirements—Part 11: Wireless LAN Medium Access Control (MAC) and Physical Layer (PHY) specifications	Published/ Under Revision	JTC1/SC6
ISO/IEC	IS	8602 Amd 1	Information technology—Protocol for providing the OSI connectionless-mode transport service Amd 1: Addition of connectionless-mode multicast capability	Amd 1 1996	JTC1/SC6

Downloadable FOC ISO/IEC 8602:1995
http://standards.iso.org/ittf/PubliclyAvailableStandards/s015907_ISO_IEC_8602_1995(E).zip
Downloadable FOC ISO/IEC 8602:1995/Amd.1:1996
http://standards.iso.org/ittf/PubliclyAvailableStandards/s024883_ISO_IEC_8602_1995_Amd_1_1996(E).zip

The MAC and physical characteristics for WLANs are specified in ISO/IEC 8802-11:2005, which is part of a series of standards for local and metropolitan area networks. The MAC unit in ISO/IEC 8802-11:2005 is designed to support PHY units as they may be adopted dependent on the availability of spectrum. ISO/IEC 8802-11:2005 contains five physical layer units: four radio units, operating in the 2,400 to 2,500-MHz band and in the bands comprising 5.15 to 5.25 GHz, 5.25 to 5.35 GHz, and 5.725 to 5.825 GHz, and one baseband infrared (IR) unit. One radio unit employs the frequency-hopping spread spectrum (FHSS) technique, two employ the direct sequence spread spectrum (DSSS) technique, and another employs the orthogonal frequency division multiplexing (OFDM) technique.

Amd 1 1996 defines additional assumptions concerning the services optionally provided by the network layer and adds no new functions of its own. The identical text is published as ITU-T Rec. X.234/Amd 1. See Section 4.2.3, IEEE 802.11.

6.2.4 ISO/IEC JTC 1/SC7 Information Technology—Software Engineering

The declared scope of SC7 is the "Standardization of processes, supporting tools and supporting technologies for the engineering of software products and systems." The vision of SC7 is to provide a unified set of software and system engineering standards widely accepted by the intended class of users.

The list of published standards and the work program of SC7 may be a useful place to start any search for software engineering standards, and ITS standards developers and systems designers and implementers are recommended to search for existing software engineering standards before reinventing the wheel in the ITS context. This section lists some of the deliverables of SC7 that may be relevant in some aspects of ITS service provision. Most are downloadable without charge using the links that provided here.

6.2.4.1 ISO 10476 Information Technology—Open Distributed Processing

ISO/IEC	IS	10476	Information technology—Open Distributed Processing	Published/ Under Revision	JTC1/SC7

This series of standards is in four parts.

ISO/IEC	IS	10476-1	Information technology—Open Distributed Processing—Reference model: Overview	Published 1998	JTC1/SC7

Downloadable FOC ISO/IEC 10746-1:1998
http://standards.iso.org/ittf/PubliclyAvailableStandards/c020696_ISO_IEC_10746-1_1998(E).zip

ISO/IEC	IS	10476-2	Information technology—Open Distributed Processing—Reference model: Foundations	Published 1996	JTC1/SC7

Downloadable FOC ISO/IEC 10746-2:1996
http://standards.iso.org/ittf/PubliclyAvailableStandards/s018836_ISO_IEC_10746-2_1996(E).zip

ISO/IEC	IS	10476-3	Information technology—Open Distributed Processing—Reference Model: Architecture	Published 1996	JTC1/SC7

Downloadable FOC ISO/IEC 10746-3:1996
http://standards.iso.org/ittf/PubliclyAvailableStandards/s020697_ISO_IEC_10746-3_1996(E).zip

ISO/IEC	IS	10476-4	Information technology—Open Distributed Processing—Reference Model: Architectural semantics	Published 1998	JTC1/SC7

Downloadable FOC ISO/IEC 10746-4:1998
http://standards.iso.org/ittf/PubliclyAvailableStandards/c020698_ISO_IEC_10746-4_1998(E).zip

Further SC7 standards can be found via the following link:

http://www.iso.org/iso/en/CatalogueListPage.CatalogueList?ICS1=33&ICS2=&ICS3=&scopelist=

6.2.5 ISO/IEC JTC 1/SC22: Information Technology—Programming Languages, Their Environments, and Systems Software Interfaces

The scope of SC6 is the "Standardization of programming languages, their environments and systems software interfaces such as:

- Specification techniques
- Common facilities and interfaces"

As such, it is, of course, the place to look for standards that support programming languages. This in itself is a little too generic a summary for this book on ITS. However, some of the "common interfaces and facilities" standards are worth mentioning.

6.2.5.1 ISO 10967 Language Independent Arithmetic

ISO/IEC	IS	10967	Information technology—Language independent arithmetic semantics	Published	JTC1/SC22

At the time of this writing, this standard was in three parts.

ISO/IEC	IS	10967-1	Information technology—Language independent arithmetic—Part 1: Integer and floating point arithmetic	Published 1994	JTC1/SC22

Downloadable FOC ISO/IEC 10967-1:1994
http://standards.iso.org/ittf/PubliclyAvailableStandards/c018939_ISO_IEC_10967-1_1994(E).zip

This defines integer and floating point data types (bounded, unbounded, and modulo integer types, as well as normalized and denormalized floating point types) on computer systems and their properties to ensure that the processing of arithmetic data can be undertaken in a reliable and predictable manner. The requirements given shall be in addition to those that may be specified in other standards (e.g., those for programming languages).

ISO/IEC	IS	10967-2	Information technology—Language independent arithmetic—Part 2: Elementary numerical functions	Published 2001	JTC1/ SC22

Downloadable FOC ISO/IEC 10967-2:2001
http://standards.iso.org/ittf/PubliclyAvailableStandards/c024427_ISO_IEC_10967-2_2001(E).zip

ISO/IEC	IS	10967-3	Information technology—Language independent arithmetic—Part 3: Complex integer and floating point arithmetic and complex elementary numerical functions	Published 2006	JTC1/ SC22

Downloadable FOC ISO/IEC 10967-3:2006
http://standards.iso.org/ittf/PubliclyAvailableStandards/c037994_ISO_IEC_10967-3_2006(E).zip

ISO/IEC 10967-3:2005 is based on ISO/IEC 10967-2 and ISO/IEC 10967-1, as well as IEC 60559.

ISO/IEC 10967-3:2005 specifies the properties of complex and imaginary integer datatypes and floating point datatypes, basic operations on values of these datatypes, as well as some numerical functions for which operand or result values are of imaginary or complex integer datatypes or imaginary or complex floating point datatypes constructed from integer and floating point datatypes satisfying the requirements of ISO/IEC 10967-1. These operations and functions are available in a variety of programming languages in common use for mathematical and numerical applications.

6.2.5.2 ISO 10176 Guidelines for the Preparation of Programming Language Standards

ISO/IEC	IS	10176	Guidelines for the preparation of programming language standards	Published 2003	JTC1/ SC22

Downloadable FOC ISO/IEC TR 10176-2003
http://standards.iso.org/ittf/PubliclyAvailableStandards/c037765_ISO_IEC_TR_10176_2003(E).zip

ISO/IEC TR 10176:2003(E) provides guidelines for the preparation of programming language standards. Standards for programming languages are developed by many committees from many countries, with many different editors supporting the effort. ISO thus considered it necessary to develop guidelines.

6.2.5.3 ISO 10182 Programming Languages, Their Environments, and System Software Interfaces—Guidelines for Language Bindings

ISO/IEC	IS	10182	Information technology—Programming languages, their environments and system software interfaces—Guidelines for language bindings	Published 1993	JTC1/ SC22

Downloadable FOC ISO/IEC TR 10182:1993
http://standards.iso.org/ittf/PubliclyAvailableStandards/s018201_ISO_IEC_TR_10182_1993(E).zip

Section 2 contains the results of a survey of current methods used for language binding development. Characteristics of each method are given, followed by reasons for the selection of the method. Application of the methods has suggested some guidelines that are presented in Section 3. Sections 2 and 3 contain documentation of the current state of language binding efforts; Section 4 addresses future directions for language bindings.

6.2.5.4 ISO 11017 Framework for Internationalization

ISO/IEC	IS	11017	Information technology—Framework for internationalization	Published 1998	JTC1/ SC22

Downloadable FOC ISO/IEC TR 11017:1998
http://standards.iso.org/ittf/PubliclyAvailableStandards/s019001_ISO_IEC_TR_11017_1998(E).zip

6.2.5.5 ISO 11404 Programming Languages, Their Environments and System Software Interfaces—Language-Independent Datatypes

ISO/IEC	IS	11404	Information technology—Programming languages, their environments and system software interfaces—Language-independent datatypes	Published 1996	JTC1/ SC22

Downloadable FOC ISO/IEC TR 11404:1996
http://standards.iso.org/ittf/PubliclyAvailableStandards/s019346_ISO_IEC_TR_11404_1996(E).zip

This specifies the nomenclature and shared semantics for a collection of datatypes commonly occurring in programming languages and software interfaces, referred to as the language-independent (LI) data types.

6.2.5.6 ISO 13886 Language-Independent Procedure Calling (LIPC)

ISO/IEC	IS	13886	Information technology—Language-Independent Procedure Calling (LIPC)	Published 1996	JTC1/ SC22

Downloadable FOC ISO/IEC 13886:1996
http://standards.iso.org/ittf/PubliclyAvailableStandards/s023183_ISO_IEC_13886_1996(E).zip

This specifies a model for procedure calls and a reference syntax for mapping to and from the model. The syntax is referred to as the interface definition notation. The model includes procedure invocation, parameter passing, completion status, and environmental issues.

6.2.5.7 Further SC22 Standards

Further SC6 standards can be found via the following link:

http://www.iso.org/iso/en/CatalogueListPage.CatalogueList?ICS1=33&ICS2=&ICS3=&scopelist=

6.2.6 Audio, Video, and Graphics Standards

There are many standards for audio, video, and graphics. This is another area where a whole book could be devoted to the description, detail, and evolution of this subject. It is periphery to ITS, but ITS standards and systems developers may need to define the use of audio, video, or graphics in ITS service provision. We therefore list the more significant standards, mostly developed by ISO/IEC JTC1 SC6 (Telecommunications and Information Exchange Between Systems) and SC24 (Computer Graphics, Image Processing, and Environmental Data Representation). Further SC6/SC24 standards can be found via the following link:

http://www.iso.org/iso/en/CatalogueListPage.CatalogueList?ICS1=33&ICS2=&ICS3=&scopelist=

As this is another, though important, periphery area, which could once again provide the material for a book in its own right, we do not provide detailed summaries, but rather list the most relevant standards and short summaries for the most relevant. Most of this group of standards can be downloaded without charge using the links provided, in the event that they are relevant to your work. All of the standards are generic, but they may be used as part of ITS service provision.

6.2.6.1 ISO 8602 Computer Graphics—Protocol for Providing the OSI Connectionless-Mode Transport Service

ISO/IEC	IS	8602	Information technology—Protocol for providing the OSI connectionless-mode transport service	Published 2nd Edition 1995:	JTC1/SC6
ISO/IEC	IS	8602	Amendment 1: Addition of connectionless-mode multicast capability	Amd 1 1996	JTC1/SC6

Downloadable FOC ISO/IEC 8602:1995
http://standards.iso.org/ittf/PubliclyAvailableStandards/s015907_ISO_IEC_8602_1995(E).zip
Downloadable FOC ISO/IEC 8602:1995/Amd.1:1996
http://standards.iso.org/ittf/PubliclyAvailableStandards/s024883_ISO_IEC_8602_1995_Amd_1_1996(E).zip

This cancels and replaces the first edition (1987). It specifies: procedures for the connectionless-mode transmission of data and protocol control information; procedures for the correct interpretation of transport protocol control information; the functional requirements for implementations claiming conformance to this standard; and the encoding of the transport protocol data units used for the transmission of data and control information.

6.2.6.2 ISO 8632 Computer Graphics—Metafile for the Storage and Transfer of Picture Description Information

ISO/IEC	IS	8632	Information technology—Computer graphics—Metafile for the storage and transfer of picture description information		JTC1/SC6

ISO 8632 is currently in three parts: Parts 1, 3, and 4. ISO 8632-2:1987 Information processing systems—Computer graphics—Metafile for the storage and transfer of picture description information—Part 2: Character encoding was withdrawn in November 1992.

ISO/IEC	IS	8632-1	Information technology—Computer graphics—Metafile for the storage and transfer of picture description information—Part 1: Functional specification	Published 1999 2nd Edition	JTC1/SC6
ISO/IEC	IS	8632-1	Technical Corrigendum 1	Published 2006	JTC1/C6

Downloadable FOC ISO/IEC 8632-1:1999
http://standards.iso.org/ittf/PubliclyAvailableStandards/c032378_ISO_IEC_8632-1_1999(E).zip

Downloadable FOC ISO/IEC 8632-1:1999/Cor.1:2006
http://standards.iso.org/ittf/PubliclyAvailableStandards/c044114_ISO_IEC_8632-1_1999_Cor_1_2006(E).zip

ISO/IEC	IS	8632-3	Information technology—Computer graphics—Metafile for the storage and transfer of picture description information—Part 3: Binary encoding	Published 1999 2nd Edition	JTC1/SC6

Downloadable FOC ISO/IEC 8632-3:1999
http://standards.iso.org/ittf/PubliclyAvailableStandards/c032380_ISO_IEC_8632-3_1999(E).zip

ISO/IEC	IS	8632-4	Information technology—Computer graphics—Metafile for the storage and transfer of picture description information—Part 4: Clear text encoding	Published 1999 2nd Edition	JTC1/SC6

Downloadable FOC ISO/IEC 8632-4:1999
http://standards.iso.org/ittf/PubliclyAvailableStandards/c032381_ISO_IEC_8632-4_1999(E).zip

Further standards can be found through the following link:

http://www.iso.org/iso/en/CatalogueListPage.CatalogueList?ICS1=33&ICS2=&ICS3=&scopelist

6.2.7 ISO/IEC JTC 1/SC 25 Information Technology—Interconnection of Information Technology Equipment

This important SC has the scope to develop standards relating to: "Standardization of microprocessor systems; and of interfaces, protocols architectures and associated interconnecting media for information technology equipment and networks, generally for commercial and residential environments, to support for embedded and distributed computing environments, storage systems, and other input/output components."

Again, this is a periphery area but one that may be of relevance to some aspects of ITS service provision. Its work program can be found at:

http://isotc.iso.org/livelink/livelink/fetch/2000/2122/327993/755080/1054033/2787945/JTC001-N-8327.pdf?nodeid=5801448&vernum=0

ISO/IEC	FDIS	24475	Information technology—Storage management	Under Development	JTC1/SC25

ISO 24475 Not yet available

6.2.8 ISO/IEC JTC 1/SC 29 Information Technology—Coding of Audio, Picture, Multimedia, and Hypermedia Information

The scope of SC29 is the

> Standardization of coded representation of audio, picture, multimedia and hypermedia information and sets of compression and control functions for use with such information such as:
>
> - Audio information
> - Bi-level and Limited Bits-per-pixel Still Pictures
> - Digital Continuous-tone Still Pictures
> - Computer Graphic Images
> - Moving Pictures and Associated Audio
> - Multimedia and Hypermedia Information for Real-time Final Form Interchange
> - Audio Visual Interactive Script ware.

Many of its standards will be used in ITS service provision where audio and video material is transferred. Again, this is an important peripheral area where we provide the titles of available standards, but do not provide summaries. If these standards are of relevance to your work, most of the standards are downloadable, without charge, from the Web hyperlinks provided next. The work program of SC29 can be downloaded from the following link:

http://www.iso.org/iso/iso_catalogue/catalogue_tc/catalogue_tc_browse.htm?commid=45316&development=true

6.2.8.1 Coding of Moving Pictures and Associated Audio for Digital Storage Media at Up to About 1.5 Mbit/s

ISO/IEC	IS	11172:1995	Information technology—Coding of moving pictures and associated audio for digital storage media at up to about 1.5 Mbit/s	Published IS & Corrigendum	JTC1/ SC29

There are many parts to this standard, and some parts have been withdrawn.

ISO/IEC	IS	11172-4	Information technology—Coding of moving pictures and associated audio for digital storage media at up to about 1.5 Mbit/s—Part 4: Compliance	Published 1995	JTC1/ SC29

Downloadable FOC ISO/IEC 11172-4:1995 Compliance testing bitstreams
http://standards.iso.org/ittf/PubliclyAvailableStandards/s022691_ISO_IEC_11172-4_1995(E)_Compliance_Testing.zip

This specifies how tests can be designed to verify whether bit streams and decoders meet requirements specified in Parts 1, 2, and 3 of ISO/IEC 11172. It summarizes the requirements, cross references them to characteristics, and defines how compliance with them can be tested. It gives guidelines how to construct tests and determine their outcome. It defines some actual tests only for audio.

ISO/IEC	IS	11172-5	Information technology—Coding of moving pictures and associated audio for digital storage media at up to about 1.5 Mbit/s—Part 5: Software simulation	Published 1998	JTC1/ SC29

Downloadable FOC ISO/IEC TR 11172-5:1998 Reference Software
http://standards.iso.org/ittf/PubliclyAvailableStandards/c025029_ISO_IEC_TR_11172-5_1998(E)_Software_Simulation.zip

6.2.8.2 ISO 13818 Generic Coding of Moving Pictures and Associated Audio Information

ISO/IEC	IS	13818	Information technology—Generic coding of moving pictures and associated audio information	Published IS & Corrigenda	JTC1/ SC29

Parts 4 and 5 are freely downloadable. The status of other parts is unknown.

ISO/IEC	IS	13818-4	Information technology—Generic coding of moving pictures and associated audio information Part 4: Conformance testing	Published IS 2004 & Corrigenda	JTC1/ SC29

Downloadable FOC ISO/IEC 13818-4:2004 Conformance testing bitstreams
http://standards.iso.org/ittf/PubliclyAvailableStandards/ISO_IEC_13818-4_2004_Conformance_Testing/

ISO/IEC	IS	13818-4 Amd 1	Information technology—Generic coding of moving pictures and associated audio information Part 4: Conformance testing Amendment 1: MPEG-2 IPMP conformance testing	Amd Published 2005	JTC1/ SC29

Downloadable FOC ISO/IEC 13818-4:2004/Amd.1:2005 Conformance testing bitstreams
http://standards.iso.org/ittf/PubliclyAvailableStandards/c040093_ISO_IEC_13818-4_2004_Amd_1_2005_Conformance_Testing.zip

ISO/IEC	IS	13818-4 Amd 2	Information technology—Generic coding of moving pictures and associated audio information Part 4: Conformance testing Amendment 2: Additional audio conformance test sequences	Amd Published 2005	JTC1/ SC29

Downloadable FOC ISO/IEC 13818-4:2004/Amd.2:2005 Conformance testing bitstreams
http://standards.iso.org/ittf/PubliclyAvailableStandards/ISO_IEC_13818-4_2004_Amd_2_2005_Conformance_Testing/

ISO/IEC	IS	13818-5	Information technology—Generic coding of moving pictures and associated audio information Part 5: Software simulation	Published 2005	JTC1/ SC29

ISO/IEC TR 13818-5:2005 Downloadable FOC Reference Software
http://standards.iso.org/ittf/PubliclyAvailableStandards/c039486_ISO_IEC_13818-5_2005_Reference_Software.zip

6.2.8.3 ISO 14496 Coding of Audio-Visual Objects

ISO/IEC	IS	14496	Information technology—Coding of audio-visual objects	Published IS & Corrigenda	JTC1/ SC29

NOTE: there are many parts, amendments, and corrigendum to this standard, and some parts are withdrawn.

ISO/IEC	IS	14496-4	Information technology—Coding of audio-visual objects—Part 4: Conformance testing	Published 2004	JTC1/ SC29

Downloadable FOC ISO/IEC 14496-4:2004 Conformance testing bitstreams
http://standards.iso.org/ittf/PubliclyAvailableStandards/ISO_IEC_14496-4_2004_Conformance_Testing/

ISO/IEC	IS	14496-4/ Cor 1	Information technology—Coding of audio-visual objects—Part 4: Conformance testing—Technical corrigendum 1	Cor 1 2005	JTC1/ SC29

Downloadable FOC ISO/EC 14496-4:2004/Cor.1:2005 Conformance testing bitstreams
http://standards.iso.org/ittf/PubliclyAvailableStandards/c041934_ISO_IEC_14496-4_2004_Cor_1_2005_Bitstreams.zip

ISO/IEC	IS	14496-4/ Cor.3	Information technology—Coding of audio-visual objects—Part 4: Conformance testing—Technical Corrigendum 3	Cor 3 2006	JTC1/ SC29

Downloadable FOC ISO/EC 14496-4:2004/Cor.3:2006 Conformance testing bitstreams
http://standards.iso.org/ittf/PubliclyAvailableStandards/c043569_ISO_IEC_14496-4_2004_Cor_3_2006_Bitstreams.zip

ISO/IEC	IS	14496-4 Amd1/ Cor 1	Information technology—Coding of audio-visual objects—Part 4: Conformance testing—Amd.1:2005/Cor.1	Published 2005	JTC1/ SC29

Downloadable FOC ISO/EC 14496-4:2004/Amd.1:2005/Cor.1:2005 Conformance testing bitstreams
http://standards.iso.org/ittf/PubliclyAvailableStandards/c041935_ISO_IEC_14496-4_2004_Amd_1_2005_Cor_1_2005_Bitstreams.zip

ISO/IEC	IS	14496-4 Amd 2	Information technology—Coding of audio-visual objects—Part 4: Conformance testing—Amendment 2: MPEG-4 conformance extensions for XMT and media nodes	Amd 2 2005	JTC1/ SC29

Downloadable FOC ISO/IEC 14496-4:2004/Amd.2:2005 Conformance testing bitstreams
http://standards.iso.org/ittf/PubliclyAvailableStandards/ISO_IEC%2014496-4_2004_Amd_2_2005_Bitstreams/

ISO/IEC	IS	14496-4 Amd 3	Information technology—Coding of audio-visual objects—Part 4: Conformance testing—Amendment 3: Visual new levels and tools	Amd 3 2005	JTC1/ SC29

Downloadable FOC ISO/IEC 14496-4:2004/Amd.3:2005 Conformance testing bitstreams
http://standards.iso.org/ittf/PubliclyAvailableStandards/c038547_ISO_IEC%2014496-4_2004_Amd%203_2005_Bitstreams.zip

ISO/IEC	IS	14496-4 Amd 4	Information technology—Coding of audio-visual objects—Part 4: Conformance testing—Amendment 4: IPMPX conformance extensions	Amd 2005	JTC1/ SC29

Downloadable FOC ISO/IEC 14496-4:2004/Amd.4:2005 Conformance testing bitstreams
http://standards.iso.org/ittf/PubliclyAvailableStandards/c039864_ISO_IEC%2014496-4_2004_Amd%204_Bitstreams.zip

ISO/IEC	IS	14496-4 Amd 5	Information technology—Coding of audio-visual objects—Part 4: Conformance testing—Amendment 5: Conformance extensions for error-resilient simple scalable profile	Amd 2005	JTC1/ SC29

Downloadable FOC ISO/IEC 14496-4:2004/Amd.5:2005 Conformance testing bitstreams http://standards.iso.org/ittf/PubliclyAvailableStandards/ISO_IEC%2014496-4_2004_Amd_5_2005_Bitstreams/

ISO/IEC	IS	14496-4 Amd 6	Information technology—Coding of audio-visual objects—Part 4: Conformance testing—Amendment 6: Advanced Video Coding conformance	Amd 2005	JTC1/ SC29

Downloadable FOC ISO/IEC 14496-4:2004/Amd.6:2005 Conformance testing bitstreams http://standards.iso.org/ittf/PubliclyAvailableStandards/ISO_IEC%2014496-4_2004_Amd_6_2005_Bitstreams/

ISO/IEC	IS	14496-4 Amd 7	Information technology—Coding of audio-visual objects—Part 4: Conformance testing—Amendment 7: AFX conformance extensions	Amd 2005	JTC1/ SC29

Downloadable FOC ISO/IEC 14496-4:2004/Amd.7:2005 Conformance testing bitstreams http://standards.iso.org/ittf/PubliclyAvailableStandards/ISO_IEC%2014496-4_2004_Amd_7_2005_Bitstreams/

ISO/IEC	IS	14496-4 Amd 8	Information technology—Coding of audio-visual objects—Part 4: Conformance testing—Amendment 8: High Efficiency Advanced Audio Coding, audio BIFS, and structured audio conformance	Amd 2005	JTC1/ SC29

Downloadable FOC ISO/IEC 14496-4:2004/Amd.8:2005 Conformance testing bitstreams http://standards.iso.org/ittf/PubliclyAvailableStandards/ISO_IEC%2014496-4_2004_Amd_8_2005_Bitstreams/

ISO/IEC	IS	14496-4 Amd 9	Information technology—Coding of audio-visual objects—Part 4: Conformance testing—Amendment 9: AVC fidelity range extensions conformance	Amd 2006	JTC1/ SC29

Downloadable FOC ISO/IEC 14496-4:2004/Amd.9:2006 Conformance testing bitstreams http://standards.iso.org/ittf/PubliclyAvailableStandards/ISO_IEC%2014496-4_2004_Amd_9_2006_Bitstreams/

ISO/IEC	IS	14496-4 Amd 10	Information technology—Coding of audio-visual objects—Part 4: Conformance testing—Amendment 10: Conformance extensions for simple profile levels 4a and 5	Amd 2005	JTC1/ SC29

Downloadable FOC ISO/IEC 14496-4:2004/Amd.10:2005 Conformance testing bitstreams
http://standards.iso.org/ittf/PubliclyAvailableStandards/c041491_ISO_IEC%20_14496-4_2004_Amd_10_2005_Conformance_Testing.zip

ISO/IEC	IS	14496-4 Amd 11	Information technology—Coding of audio-visual objects—Part 4: Conformance testing—Amendment 11: Parametric stereo conformance	Amd 2006	JTC1/ SC29

Downloadable FOC ISO/IEC 14496-4:2004/Amd.11:2006 Conformance testing bitstreams
http://standards.iso.org/ittf/PubliclyAvailableStandards/c042023_ISO_IEC_14496-4_2004_Amd_11_2006_Conformance_Testing.zip

ISO/IEC	IS	14496-4 Amd 12	Information technology—Coding of audio-visual objects—Part 4: Conformance testing—Amendment 12: Morphing & Textures conformance	Amd 2007	JTC1/ SC29

Downloadable FOC ISO/IEC 14496-4:2004/Amd.12:2007 Conformance testing bitstreams
http://standards.iso.org/ittf/PubliclyAvailableStandards/c043586_ISO_IEC_14496-4_2004_Amd_12_2007_Bitstreams.zip

ISO/IEC	IS	14496-4 Amd 13	Information technology—Coding of audio-visual objects—Part 4: Conformance testing—Amendment 13: Parametric coding for high quality audio conformance	Amd 2007	JTC1/ SC29

Downloadable FOC ISO/IEC 14496-4:2004/Amd.13:2007 Conformance testing bitstreams
http://standards.iso.org/ittf/PubliclyAvailableStandards/c043177_ISO_IEC%20_14496-4_2004_Amd_13_2007_Conformance_Testing.zip

ISO/IEC	IS	14496-4 Amd 17	Information technology—Coding of audio-visual objects—Part 4: Conformance testing—Amendment 17: Advanced text and 2D graphics conformance	Amd 2007	JTC1/ SC29

Downloadable FOC ISO/IEC 14496-4:2004/Amd.17:2007 Conformance testing bitstreams
http://standards.iso.org/ittf/PubliclyAvailableStandards/c044864e_ISO_IEC_14496-4_2004_Amd_17_2007_Conformance_Testing.zip

ISO/IEC IS 14496-5 Information technology—Coding of audio-visual objects—Part 5: Reference software Published 2001 JTC1/SC29

Downloadable FOC ISO/IEC 14496-5:2001 Reference Software
http://standards.iso.org/ittf/PubliclyAvailableStandards/ISO_IEC_14496-5_2001_Software_Reference/

ISO/IEC IS 14496-5 Amd 1 Information technology—Coding of audio-visual objects—Part 5: Reference software—Amendment 1: Reference software for MPEG-4 Amd 2002 JTC1/SC29

Downloadable FOC ISO/IEC 14496-5:2001/Amd.1:2002 Reference Software
http://standards.iso.org/ittf/PubliclyAvailableStandards/c036089_ISO_IEC_14496-5_2001_Amd_1_2002_Reference_Software.zip

ISO/IEC IS 14496-5 Amd 2 Information technology—Coding of audio-visual objects—Part 5: Reference software—Amendment 2: MPEG-4 reference software extensions for XMT and media nodes Amd 2003 JTC1/SC29

Downloadable FOC ISO/IEC 14496-5:2001/Amd.2:2003 Reference Software
http://standards.iso.org/ittf/PubliclyAvailableStandards/c036355_ISO_IEC_14496-5_2001_Amd_2_2003_Reference_Software.zip

ISO/IEC IS 14496-5 Amd 3 Information technology—Coding of audio-visual objects—Part 5: Reference software—Amendment 3: Visual new level and tools Amd 2003 JTC1/SC29

Downloadable FOC ISO/IEC 14496-5:2001/Amd.3:2003 Reference Software
http://standards.iso.org/ittf/PubliclyAvailableStandards/c038058_ISO_IEC_14496-5_2001_Amd_3_2003(E)_Reference_Software.zip

ISO/IEC IS 14496-5 Amd 4 Information technology—Coding of audio-visual objects—Part 5: Reference software—Amendment 4: IPMPX reference software extensions Amd 2004 JTC1/SC29

Downloadable FOC ISO/IEC 14496-5:2001/Amd.4:2004 Reference Software
http://standards.iso.org/ittf/PubliclyAvailableStandards/c039417_ISO_IEC_14496-5_2001_Amd_4_2004(E)_Reference_Software.zip

ISO/IEC	IS	14496-4 Amd 5	Information technology—Coding of audio-visual objects—Part 5: Reference software—Amendment 5: Reference software extensions for error resilient simple scalable profile	Amd 2004	JTC1/ SC29

Downloadable FOC ISO/IEC 14496-5:2001/Amd.5:2004 Reference Software http://standards.iso.org/ittf/PubliclyAvailableStandards/c039392_ISO_IEC_14496-5_2001_Amd%205_2004(E)_Reference_Software.zip

ISO/IEC	IS	14496-4 Amd 6	Information technology—Coding of audio-visual objects—Part 5: Reference software—Amendment 6: Advanced Video Coding (AVC) and High Efficiency Advanced Audio Coding (HE AAC) reference software	Amd 2006	JTC1/ SC29

Downloadable FOC ISO/IEC 14496-5:2001/Amd.6:2005 Reference Software http://standards.iso.org/ittf/PubliclyAvailableStandards/ISO_IEC_14496-5_2001_Amd_6_Reference_Software/

ISO/IEC	IS	14496-4 Amd 6/ Cor 1	Information technology—Coding of audio-visual objects—Part 5: Technical Corrigendum 1	Amd 6 Cor 1 2005	JTC1/ SC29

Downloadable FOC ISO/IEC 14496-5:2001/Amd.6:2005/Cor.1:2006 Reference Software http://standards.iso.org/ittf/PubliclyAvailableStandards/c044371_ISO_IEC_14496-5_2

ISO/IEC	IS	14496-5 Amd 7	Information technology—Coding of audio-visual objects ing of audio-visual objects—Part 5: Reference software—Amendment 7: AFX reference software extensions	Amd 2005	JTC1/ SC29

Downloadable FOC ISO/IEC 14496-5:2001/Amd.7:2005 Reference Software Http://standards.iso.org/ittf/PubliclyAvailableStandards/c040361_ISO_IEC%2014496-5_2001_Amd%207_2005_Reference_Software.zip

ISO/IEC	IS	14496-5 Amd 8	Information technology—Coding of audio-visual objects—Part 5: Reference software—Amendment 8: AVC fidelity range extensions reference software	A,d 2006	JTC1/ SC29

Downloadable FOC ISO/IEC 14496-5:2001/Amd.8:2006 Reference Software http://standards.iso.org/ittf/PubliclyAvailableStandards/c041955_ISO_IEC_14496-5_2001_Amd_8_Reference_Software.zip

ISO/IEC	IS	14496-5 amd 9	Information technology—Coding of audio-visual objects—Part 5: Reference software—Amendment 9: Morphing & Textures reference software	Amd 2007	JTC1/ SC29

Downloadable FOC ISO/IEC 14496-5:2001/Amd.9:2007 Reference Software http://standards.iso.org/ittf/PubliclyAvailableStandards/c043931_ISO_IEC_14496-5_2001_Amd_9_2007_Reference_Software.zip

ISO/IEC	IS	14496-5 Amd 10	Information technology—Coding of audio-visual objects—Part 5: Reference software—Amendment 10: SSC, DST, ALS and SLS reference software	Amd 2007	JTC1/ SC29

Downloadable FOC ISO/IEC 14496-5:2001/Amd.10:2007 Reference Software http://standards.iso.org/ittf/PubliclyAvailableStandards/c043465_ISO_IEC_14496-5_2001_Amd_10_2007_Reference_Software.zip

ISO/IEC	IS	14496-7	Information technology—Coding of audio-visual objects—Part 7: Optimized reference software for coding of audio-visual objects	Published 2004 2nd Edition	JTC1/ SC29

Downloadable FOC ISO/IEC TR 14496-7:2004 Reference Software http://standards.iso.org/ittf/PubliclyAvailableStandards/c040728_ISO_IEC_TR_14496-7_2004_Reference_Software.zip

ISO/IEC	IS	14496-12	Information technology—Coding of audio-visual objects—Part 12: ISO base media file format	Published 2005	JTC1/ SC29

Downloadable FOC ISO/IEC 14496-12:2005 http://standards.iso.org/ittf/PubliclyAvailableStandards/c041828_ISO_IEC_14496-12_2005(E).zip

ISO/IEC	IS	14496-12 Cor 1	Information technology—Coding of audio-visual objects—Part 12: ISO base media file format—Technical Corrigendum 1	Cor 2005	JTC1/ SC29

Downloadable FOC ISO/IEC 14496-12:2005/Cor.1:2005 http://standards.iso.org/ittf/PubliclyAvailableStandards/c042292_ISO_IEC_14496-12_2005_Cor_1_2005(E).zip

ISO/IEC	IS	14496-12 Cor 2	Information technology—Coding of audio-visual objects—Part 12: ISO base media file format—Technical Corrigendum 2	Cor 2006	JTC1/ SC29

Downloadable FOC ISO/IEC 14496-12:2005/Cor.2:2006
http://standards.iso.org/ittf/PubliclyAvailableStandards/c043226_ISO_IEC_14496-12_2005_Cor_2_2006(E).zip

ISO/IEC	IS	14496-12 Amd 1	Information technology—Coding of audio-visual objects—Part 12: ISO base media file format—Amendment 1: Support for timed metadata, non-square pixels and improved sample groups	Amd 2007	JTC1/ SC29

Downloadable FOC ISO/IEC 14496-12:2005/Amd.1:2007
http://standards.iso.org/ittf/PubliclyAvailableStandards/c043689_ISO_IEC_14496-12_2005_Amd_1_2007(E).zip

ISO/IEC	IS	14496-20	Information technology—Coding of audio-visual objects—Part 20: Lightweight Application Scene Representation (LASeR) and Simple Aggregation Format (SAF)	Published 2006	JTC1/ SC29

Downloadable FOC ISO/IEC 14496-20:2006
http://standards.iso.org/ittf/PubliclyAvailableStandards/c041650_ISO_IEC_14496-20_2006(E).zip

Downloadable FOC Electronic Attachments
http://standards.iso.org/ittf/PubliclyAvailableStandards/c041650_ISO_IEC_14496-20_2006(E)_Electronic_Attachements.zip

6.2.8.4 ISO 15444 JPEG 2000 Image Coding System

ISO/IEC	IS	15444	Information technology—JPEG 2000 image coding system	Published IS & Corrigenda	JTC1/ SC29

NOTE: There are many parts to this standard; some parts have been withdrawn, and others have not yet been published.

ISO/IEC	IS	15444-4	Information technology—JPEG 2000 image coding system	Published 2004	JTC1/ SC29

Downloadable FOC ISO/IEC 15444-4:2004
http://standards.iso.org/ittf/PubliclyAvailableStandards/c039079_ISO_IEC_15444-4_2004(E).zip

Downloadable FOC Conformance testing bitstreams
http://standards.iso.org/ittf/PubliclyAvailableStandards/ISO_IEC_15444-4_2004_Conformance_Testing/

ISO/IEC	IS	15444-5	Information technology—JPEG 2000 image coding system: Reference software	Published 2003	JTC1/ SC29

Downloadable FOC ISO/IEC 15444-5:2003
http://standards.iso.org/ittf/PubliclyAvailableStandards/c033877_ISO_IEC_15444-5_2003(E).zip
Downloadable FOC Reference software
http://standards.iso.org/ittf/PubliclyAvailableStandards/c033877_ISO_IEC_15444-5_2003(E)_Software.zip

ISO/IEC	IS	15444-5	Information technology—JPEG 2000 image coding system: Reference software Part 5—Amendment 1: Reference software for the JP2 file format	Published 2003	JTC1/ SC29

Downloadable FOC ISO/IEC 15444-5:2003/Amd.1:2003
http://standards.iso.org/ittf/PubliclyAvailableStandards/c036080_ISO_IEC_15444-5_2003_Amd_1_2003(E).zip
Downloadable FOC Reference software
http://standards.iso.org/ittf/PubliclyAvailableStandards/c036080_ISO_IEC_15444-5_2003_Amd_1_2003(E)_Software.zip

ISO/IEC	IS	15444-12	Information technology—JPEG 2000 image coding system: Reference software—Part 12: ISO base media file format	Published 2005 2nd Edition	JTC1/ SC29

Downloadable FOC ISO/IEC 15444-12:2005
http://standards.iso.org/ittf/PubliclyAvailableStandards/c041827_ISO_IEC_15444-12_2005(E).zip

ISO/IEC	IS	15444-12 Cor 1	Information technology—JPEG 2000 image coding system: Reference software—Part 12: ISO base media file format Technical Coorigendum 1	Cor 1 2005	JTC1/ SC29

Downloadable FOC ISO/IEC 15444-12:2005/Cor.1:2005
http://standards.iso.org/ittf/PubliclyAvailableStandards/c042293_ISO_IEC_15444-12_2005_Cor_1_2005(E).zip

ISO/IEC	IS	15444-12 Cor 2	Information technology—JPEG 2000 image coding system: Reference software—Part 12: ISO base media file format Technical Corrigendum 2	Cor 1 2005	JTC1/ SC29

Downloadable FOC ISO/IEC 15444-12:2005/Cor.2:2006
http://standards.iso.org/ittf/PubliclyAvailableStandards/c043227_ISO_IEC_15444-12_2005_Cor_2_2006(E).zip

ISO/IEC	IS	15444-12 Amd 1	Information technology—JPEG 2000 image coding system: Reference software—Part 12: ISO base media file format Amendment 1: Support for timed metadata, non-square pixels and improved sample groups	Amd 1 2007	JTC1/ SC29

Downloadable FOC ISO/IEC 15444-12:2005/Amd.1:2007
http://standards.iso.org/ittf/PubliclyAvailableStandards/c043690_ISO_IEC_15444-12_2005_Amd_1_2007(E).zip

6.2.8.5 ISO 15938 Multimedia Content Description

ISO/IEC	IS	15938	Information technology—Multimedia content Description	Published IS & Corrigenda	JTC1/ SC29

NOTE: There are many parts to this standard; the following parts are available for download.

ISO/IEC	IS	15938 Part 6	Information technology—Multimedia content Description interface—Part 6: Reference software	Published 2003	JTC1/ SC29

Downloadable FOC ISO/IEC 15938-6:2003 Reference Software
http://standards.iso.org/ittf/PubliclyAvailableStandards/c035364_ISO_IEC_15938-6(E)_Reference_Software.zip

ISO/IEC	IS	15938 Part 6 Amd 1	Information technology—Multimedia content Description interface—Part 6: Reference software—Amendment 1: Reference software extensions	Amd 1 2003	JTC1/ SC29

Downloadable FOC ISO/IEC 15938-6:2003/Amd.1:2006 Reference Software
http://standards.iso.org/ittf/PubliclyAvailableStandards/c040493_ISO_IEC_15938-6_2003_Amd_1_2006_Reference_Software.zip

ISO/IEC	IS	15938 Part 7	Information technology—Multimedia content Description interface—Part 7: Conformance testing	2003	JTC1/ SC29

Downloadable FOC ISO/IEC 15938-7:2003 Conformance testing bitstreams
http://standards.iso.org/ittf/PubliclyAvailableStandards/ISO_IEC_15938-7_2003_Conformance_Testing/

ISO/IEC	IS	15938 Part 7 Amd 1	Information technology—Multimedia content Description interface—Part 7—Amendment 1: Conformance extensions	Amd 1: 2005	JTC1/ SC29

Downloadable FOC ISO/IEC 15938-7:2003/Amd.1:2005 Conformance testing bitstreams
http://standards.iso.org/ittf/PubliclyAvailableStandards/ISO_IEC_15938-7_2003_Amd_1_2005_Conformance_Testing/

ISO/IEC	IS	15938 Part 7 Amd 2	Information technology—Multimedia content Description interface—Part 7—Amendment 2: Fast access extensions conformance	Amd 2: 2007	JTC1/ SC29

Downloadable FOC ISO/IEC 15938-7:2003/Amd.2:2007 Conformance testing bitstreams
http://standards.iso.org /ittf/PubliclyAvailableStandards/c043589_ISO_IEC_15938-7_2003_Amd_2_2007_Conformance_Testing.zip

ISO/IEC	IS	15938 Part 11	Information technology—Multimedia content Description interface—Part 11: MPEG-7 profile schemas	2005	JTC1/ SC29

Downloadable FOC ISO/IEC TR 15938-11:2005
http://standards.iso.org/ittf/PubliclyAvailableStandards/c042114_ISO_IEC_TR_15938-11_2005(E).zip

ISO/IEC	IS	15938 Part 10 & 11	Information technology—Multimedia content Description—Part 10 & 11	Published IS	JTC1/ SC29

Downloadable FOC ISO/IEC 15938: MPEG-7 Schema Files
http://standards.iso.org/ittf/PubliclyAvailableStandards/MPEG-7_schema_files/

6.2.8.6 ISO 21000 Multimedia Framework (MPEG-21)

ISO/IEC	IS	21000	Information technology—Multimedia framework (MPEG-21)	Published IS & Corrigenda	JTC1/ SC29

NOTE: There are many parts to this standard.

ISO/IEC	IS	21000-1	Information technology—Multimedia framework (MPEG-21)—Part 1: Vision, Technologies and Strategy	Published 2004 2nd Edition	JTC1/ SC29

Downloadable FOC ISO/IEC TR 21000-1:2004
http://standards.iso.org/ittf/PubliclyAvailableStandards/c040611_ISO_IEC_TR_21000-1_2004(E).zip

ISO/IEC	IS	21000 Parts 2-17	Information technology—Multimedia framework (MPEG-21)—Part 2 to 17 *Contain the electronic form of the XML Schema files for all standards within ISO/IEC 21000.*	Published	JTC1/ SC29

Downloadable FOC ISO/IEC 21000: MPEG-21Schema Files
http://standards.iso.org/ittf/PubliclyAvailableStandards/MPEG-21_schema_files/

ISO/IEC	IS	21000 Part 2	Information technology—Multimedia framework (MPEG-21)—Part 2: Digital Item Declaration	Published 2005 2nd Edition	JTC1/ SC29

Downloadable FOC ISO/IEC 21000-2:2005
http://standards.iso.org/ittf/PubliclyAvailableStandards/c041112_ISO_IEC_21000-2_2005(E).zip

ISO/IEC	IS	21000 Part 8	Information technology—Multimedia framework (MPEG-21)—Part 2: Digital Item Declaration—Part 8: Reference Software	Published 2006	JTC1/ SC29

Downloadable FOC ISO/IEC 21000-8:2006 Reference Software
http://standards.iso.org/ittf/PubliclyAvailableStandards/ISO_IEC_21000-8_2006_Reference_Software/

6.2.8.7 ISO 23000 Multimedia Application Format (MPEG-A)

ISO/IEC	TR	23000	Information technology—Multimedia application format (MPEG-A)	TR under Development to IS	JTC1/ SC29

ISO 23000 Available from
http://www.iso.org/iso/iso_catalogue/catalogue_tc/catalogue_detail.htm?csnumber=42010

MPEG-A facilitates the swift development of innovative, standards-based multimedia applications and services. To this end, the MPEG-A introduces a set of standardized multimedia application formats (MAFs), together with their related software implementation. The normative software demonstrates how MAFs are used and offers vendors a head start for developing multimedia products based on MAFs. MPEG's ultimate objective for MAFs is to stimulate the increased use of MPEG technology through additional interoperability of different media types at the application level. Although in the past MPEG has specified profiles, subsets of technologies within specific parts of the standards, such as MPEG-2 Video, it has never recommended the use of specific combinations of profiles across different parts of a standard, such as MPEG-2 Audio and MPEG-2 Video, or across different standards, such as MPEG-4 and MPEG-7. It is this combination of technologies across MPEG standards which is the topic of MPEG-A.

ISO/IEC TR 23000-1:2007 contains a description of the original motivation for this new family of standards and the concept underlying the construction of all other parts (MAFs) of ISO/IEC 23000. Furthermore, it provides a short overview of all MAFs that have been standardized to date.

6.2.8.8 ISO 23001 MPEG Systems Technologies

ISO/IEC	FPD	23001	Information technology—MPEG systems technologies	Under Development	JTC1/ SC29

ISO 23001 Available from
http://www.iso.org/iso/iso_catalogue/catalogue_tc/catalogue_detail.htm?csnumber=35417

ISO/IEC 23001-1:2006 provides a standardized set of generic technologies for encoding XML documents. It addresses a broad spectrum of applications and requirements by providing generic methods for transmitting and compressing XML documents.

ISO/IEC 23001-1:2006 provides a specification which gives rules for the preparation of XML documents for efficient transport and storage, and enables the development of ISO/IEC 23001-1 terminals to receive, decode, and assemble possibly partitioned and compressed XML documents.

6.2.8.9 ISO 23002 MPEG Video Technologies

ISO/IEC	FCD	23002	Information technology—MPEG video technologies	Under Development	JTC1/ SC29

ISO 23002 Available from
http://www.iso.org/iso/iso_catalogue/catalogue_tc/catalogue_detail.htm?csnumber=42030

A number of image and video coding related standards include a requirement for decoders to implement an integer-output 8×8 inverse discrete cosine transform (IDCT) for the generation of inverse-transformed sample differences with a nominal range from –2B to (2B) –1 for some integer number of bits B, where B is greater than or equal to 8. ISO/IEC 23002-1:2006 specifies conformance requirements for establishing sufficient accuracy in such an integer-output IDCT implementation. It is intended to be suitable for reference to establish partial or complete requirements for IDCT accuracy for conformance to other standards that require IDCT use.

The accuracy requirements specified in the main body of ISO/IEC 23002-1:2006 are essentially the same as those previously specified in IEEE 1180-1990 (which has since been withdrawn), in Annex A of ITU-T Rec. H.261, and in Annex A of ITU-T Rec. H.263. These requirements have been specified herein to resolve normative references to IEEE 1180-1990 in MPEG standards after its withdrawal and to provide improved clarity for the specification of IDCT accuracy requirements.

6.2.8.10 ISO 23003 MPEG Audio Technologies

ISO/IEC	FDIS	23003	Information technology—MPEG audio technologies	Under Development	JTC1/ SC29

ISO 23003 Available from
http://www.iso.org/iso/iso_catalogue/catalogue_tc/catalogue_detail.htm?csnumber=44159

ISO/IEC 23003-1:2007 (MPEG Surround) is efficient technology for multichannel audio compression. Rather than performing a discrete coding of the individual audio input channels, MPEG Surround captures the spatial image of a multichannel audio signal into a compact set of parameters that are used to synthesize a high quality multichannel representation from a transmitted downmix signal. MPEG Surround extends traditional techniques for coding of two or more channels in a way that provides several significant advantages in terms of compression efficiency and user benefits. First, it allows the transmission of multichannel audio at bit rates, which so far have been used for the transmission of monophonic audio. Second, by its underlying structure, the multichannel audio signal is transmitted in a backward-compatible way—that is, the technology can be used to upgrade existing distribution infrastructures for stereo or mono audio content (radio channels, Internet streaming, music downloads, and so forth) towards the delivery of multichannel audio while retaining full compatibility with existing receivers.

6.2.8.11 ISO 23004 MPEG Multimedia Middleware

ISO/IEC	FCD	23004	Information technology—MPEG Multimedia middleware	Under Development	JTC1/ SC29

ISO 23004 Available from
http://www.iso.org/iso/iso_catalogue/catalogue_tc/catalogue_detail.htm?csnumber=44353

ISO/IEC 23004-4:2007 specifies the interfaces of the support application programming interface and the realization technology used for resource management in MPEG Multimedia Middleware (M3W). Resource management is an optional framework for M3W platforms. ISO/IEC 23004-4:2007 specifies:

- Entities and interfaces for resource budget creation, assignment and removal;
- Entity and interfaces for assessing the feasibility and selecting resource configurations (resource configuration = set of assigned budgets);
- Interfaces implemented by quality-aware entities (quality-aware entities can provide multiple quality levels and know the resource needed to provide each quality level);
- Entity and interfaces for coordination of the "budget–quality-level" negotiation (includes interfaces for registration and setting priorities).

6.2.9 ISO/IEC JTC 1/SC 32 Information Technology Data Management and Interchange

SC32 is another periphery but important SC with respect to supporting ITS service provision. The scope of SC32 is as follows:

> Standards for data management within and among local and distributed information systems environments. SC 32 provides enabling technologies to promote harmonization data management facilities across sector-specific areas. Specifically, SC 32 standards include:
>
> 1. reference models and frameworks for the coordination of existing and emerging standards;
> 2. definition of data domains, data types and data structures, and their associated semantics;
> 3. languages, services and protocols for persistent storage, concurrent access, concurrent update and interchange of data;
> 4. methods, languages, services and protocols to structure, organize and register metadata and other information resources associated with sharing and interoperability, including electronic commerce.

The more relevant standards are listed next, because of the relevance of this subject to ITS service provision. A short summary for each is also provided for the more closely relevant standards. The work program for SC32 can be downloaded from:

http://www.iso.org/iso/iso_catalogue/catalogue_tc/catalogue_tc_browse.htm?commid=45342&development=true

6.2.9.1 Metadata Registries (MDR)

Perhaps the most significant family of standards for ITS that are based on ISO/IEC JTC1/SC32 standards are the metadata registry standards on which the ITS metadata registry Standard ISO 14817 and Emergency Message Data Registry Standard are based. For details of these standards, see Sections 5.1.7 and 8.7.1.1.

ISO 2382 will comprise 26 parts. The following paragraphs summarize the parts that may be more significant to ITS.

6.2.9.2 ISO 2382 Information Technology—Vocabulary Part 4: Organization of Data

ISO/IEC	IS	2382-4	Information technology—Vocabulary Part 4: Organization of data	1999	JTC1/ SC32

Downloadable FOC ISO/IEC 2382-4:1999
http://standards.iso.org/ittf/PubliclyAvailableStandards/c027922_ISO_IEC_2382-4_1999(Bil).zip

Information technology gives rise to numerous international exchanges of both an intellectual and a material nature. These exchanges often become difficult, either because of the great variety of terms used in various fields or languages to express the same concept, or because of the absence or imprecision of the definitions of useful concepts.

To avoid misunderstandings and to facilitate such exchanges, it is essential to clarify the concepts, to select terms to be used in various languages or in various countries to express the same concept, and to establish definitions providing satisfactory equivalents for the various terms in different languages.

ISO 2382 was initially based mainly on the usage to be found in the Vocabulary of Information Processing which was established and published by the International Federation for Information Processing and the International Computation Centre, and in the American National Dictionary for Information Processing Systems and its earlier editions published by the American National Standards Institute (formerly known as the American Standards Association). Published and Draft International Standards relating to information technology of other international organizations (such as the International Telecommunication Union and the International Electrotechnical Commission) as well as published and draft national standards have also been considered.

The purpose of ISO/IEC 2382 is to provide definitions that are rigorous, uncomplicated, and which can be understood by all concerned. The scope of each concept defined has been chosen to provide a definition that is suitable for general application. In those circumstances, where a restricted application is concerned, the definition may need to be more specific.

However, while it is possible to maintain the self-consistency of individual parts, the reader is warned that the dynamics of language and the problems associated with the standardization and maintenance of vocabularies may introduce duplications and inconsistencies among parts.

This part of ISO/IEC 2382 is intended to facilitate international communication in information technology. It presents, in two languages, terms and definitions of selected concepts relevant to the field of information technology and identifies relationships among the entries.

In order to facilitate their translation into other languages, the definitions are drafted so as to avoid, as far as possible, any peculiarity attached to a language.

6.2.9.3 ISO 2382 Information Technology—Vocabulary Part 5: Representation of Data

ISO/IEC	IS	2382-5	Information technology—Vocabulary Part 5: Representation of data	1999	JTC1/ SC32

Downloadable FOC ISO/IEC 2382-5:1999
http://standards.iso.org/ittf/PubliclyAvailableStandards/c030851_ISO_IEC_2382-5_1999(Bil).zip

Information technology gives rise to numerous international exchanges of both an intellectual and a material nature. These exchanges often become difficult, either because of the great variety of terms used in various fields or languages to express the same concept, or because of the absence or imprecision of the definitions of useful concepts.

To avoid misunderstandings and to facilitate such exchanges, it is essential to clarify the concepts, to select terms to be used in various languages or in various countries to express the same concept, and to establish definitions providing satisfactory equivalents for the various terms in different languages.

ISO 2382 was initially based mainly on the usage to be found in the Vocabulary of Information Processing which was established and published by the International Federation for Information Processing and the International Computation Centre, and in the American National Dictionary for Information Processing Systems and its earlier editions published by the American National Standards Institute (formerly known as the American Standards Association). Published and Draft International Standards relating to information technology of other international organizations (such as the International Telecommunication Union and the International Electrotechnical Commission) as well as published and draft national standards have also been considered.

The purpose of ISO/IEC 2382 is to provide definitions that are rigorous, uncomplicated and which can be understood by all concerned. The scope of each concept defined has been chosen to provide a definition that is suitable for general application. In those circumstances, where a restricted application is concerned, the definition may need to be more specific.

However, while it is possible to maintain the self-consistency of individual parts, the reader is warned that the dynamics of language and the problems associated with the standardization and maintenance of vocabularies may introduce duplications and inconsistencies among parts.

6.2.9.4 Information Technology—Vocabulary Part 6: Preparation and Handling of Data

ISO/IEC	IS	2382-6	Information technology—Vocabulary Part 6: Preparation and handling of data	1987 2nd Edition	JTC1/ SC32

Downloadable FOC ISO 2382-6:1987
http://standards.iso.org/ittf/PubliclyAvailableStandards/s007238_ISO_2382-6_1987(Bil).zip

This part of IS0 2382 deals in particular with input and output of data, transfer and conversion methods, and also with search techniques.

6.2.9.5 ISO 10032 Information Technology—Reference Model of Data Management

ISO/IEC	IS	10032	Information technology—Reference Model of Data Management	2003	JTC1/ SC32

Downloadable FOC ISO/IEC TR 10032:2003
http://standards.iso.org/ittf/PubliclyAvailableStandards/c038607_ISO_IEC_TR_10032_2003(E).zip

6.2.9.6 ISO 14957 Notation of Format for Data Elements

ISO/IEC	IS	14957	Information Technology—Notation of Format for Data Elements	1996	JTC1/ SC32

ISO 14957 Available from
http://www.iso.org/iso/iso_catalogue/catalogue_tc/catalogue_detail.htm?csnumber=2365

The following is an excerpt from the Introduction to 1996 edition of standard 14957:

> Data interchange is experiencing rapid expansion, in the commercial, technical, and public sectors. It gives rise to inter working between different communities which often have developed independently information processing applications and telecommunication networks which meet specific needs. Hence an overall situation which suffers from a lack of homogeneity.
>
> In order to remedy this situation, an urgent standardization effort focused in particular on the representation of data elements is necessary.
>
> The representation of a data element supposes in the first place that the format, i.e., the type of characters used in the representation and in the length of the latter

is specified. So that these specifications have the same significance for everyone involved, it is necessary to express them in accordance with standardized conventions.

Such rules are likely to eliminate any and all risk of ambiguity, lack of understanding and error; they also facilitate the comparison of data element dictionaries, the design and creation of information systems as well as electronic data interchange (EDI).

These notations have been partially and variously expressed in different International Standards according to the specific contexts in which they have been defined, e.g., EDIFACT (ISO 9735), Banking Standards (as ISO 7982-1), Character sets (ISO 8859), Information processing (ISO 6093).

Therefore, the objective of this International Standard is to provide a unique source of reference on this issue for all Standards utilizing these types of notations independently of their environments.

6.2.9.7 ISO 14662 Information Technology—Open-EDI Reference Model

ISO/IEC	IS	14662	Information technology—Open-EDI reference model	Published 2004	JTC1/ SC32

Downloadable FOC ISO/IEC 14662:2004 (E)
http://standards.iso.org/ittf/PubliclyAvailableStandards/c037354_ISO_IEC_14662_2004(E).zip

ISO/IEC 14662:2004 specifies the framework for coordinating the integration of existing International Standards and the development of future International Standards for the interworking of Open-EDI Parties via Open-EDI and provides a reference for those International Standards. As such, it serves to guide the work necessary to accomplish Open-EDI by providing the context to be used by developers of International Standards to ensure the coherence and integration of related standardized modeling and descriptive techniques, services, service interfaces, and protocols.

This International Standard describes, through two perspectives of business transactions, significant aspects relevant to the interoperability of information technology systems used by Open-EDI Parties engaging in Open-EDI.

6.2.9.8 ISO 15944 Information Technology—Business Agreement Semantic Descriptive Techniques—Part 1: Operational Aspects of Open-EDI for Implementation

ISO/IEC	IS	15944-1	Information technology—Business agreement semantic descriptive techniques—Part 1: Operational aspects of Open-EDI for implementation	Published 2002	JTC1/ SC32

Downloadable FOC ISO/IEC 15944-1:2002
http://standards.iso.org/ittf/PubliclyAvailableStandards/c033322_ISO_IEC_15944-1_2002(E).zip

6.2.9.9 ISO 19502 Meta Object Facility (MOF) 1.4 Specification (OMG PAS)

ISO/IEC	IS	19502	Meta Object Facility (MOF) 1.4 specification (OMG PAS)	2005	JTC1/ SC32

Downloadable FOC The OMG version may be accessed from the OMG site
http://jtc1sc32.org/doc/recent/32N1328T-ISO-IEC-PAS-IS19502-MOF-05-06-01.pdf

6.2.9.10 ISO 19503 XML Metadata Interchange (XMI) 2.0 Specification (OMG PAS)

ISO/IEC	IS	19503	XML Metadata Interchange (XMI) 2.0 specification (OMG PAS)	2005	JTC1/ SC32

Downloadable FOC The OMG version may be accessed from the OMG site
http://www.omg.org/docs/formal/02-04-03.pdf

6.2.9.11 ISO 19763 Framework for Metamodel Interoperability

ISO/IEC	IS	19763	Information Technology—Framework for Metamodel Interoperability	Published 2007	JTC1/ SC32
ISO/IEC	IS	19763-1	Information Technology—Framework for Metamodel Interoperability Part 1: Reference Model	Published 2007	JTC1/ SC32

ISO 19763-1 Available from
http://www.iso.org/iso/iso_catalogue/catalogue_tc/catalogue_detail.htm?csnumber=38635

This part of ISO/IEC 19763 describes the concepts and an overall architecture of the metamodel framework standard to be applied in the development and the registration of the following individual metamodel frameworks.

ISO/IEC	IS	19763-2	Information Technology—Framework for Metamodel Interoperability Part 2: Core Model	Under Development	JTC1/ SC32

ISO 19763-2 is not yet available

This part of ISO/IEC 19763 specifies the core model which is required to describe metamodel items, and which may be used in situations where a complete metadata/metamodel registry is appropriate.

ISO/IEC	IS	19763-3	Information Technology—Framework for Metamodel Interoperability Part 3: Metamodel for Ontology Registration	Published 2007	JTC1/ SC32

ISO 19763-3 Available from
http://www.iso.org/iso/iso_catalogue/catalogue_tc/catalogue_detail.htm?csnumber=38637

This part of ISO/IEC 19763 specifies the metamodel that provides a facility to register administrative information about ontologies.

ISO/IEC	IS	19763-4	Information Technology—Framework for Metamodel Interoperability Part 4: Metamodel for Model Mapping	Published 2007	JTC1/ SC32

ISO 19763-4 Available from
http://www.iso.org/iso/iso_catalogue/catalogue_tc/catalogue_detail.htm?csnumber=38638

This part of ISO/IEC 19763 provides a normative metamodel for describing differences regarding formats and types of objects to be exchanged or shared. This metamodel framework also provides a capability for describing transformation rules between different objects in term of a metamodel instance.

6.2.9.12 ISO 19773 Information Technology—Metadata Modules (MM)

ISO/IEC 19773 is a multipart standard that initially includes the following parts. It is intended to consolidate some of these parts and add others.

ISO/IEC	IS	19773	ISO/IEC: Information Technology—Metadata modules (MM)	working draft	JTC1/ SC32
ISO/IEC	IS	19773-01	Framework	working draft	JTC1/ SC32
ISO/IEC	IS	19773-02	Data structure for reference-or-literal (reflit)	working draft	JTC1/ SC32

ISO/IEC	IS	19773-03	Data structure for internationalized/ localized multivalue/multidata	working draft	JTC1/ SC32
ISO/IEC	IS	19773-04	Data structure for internationalized/ localized multistring/multitext	working draft	JTC1/ SC32
ISO/IEC	IS	19773-05	Data structure for designation-kind-value (DKV) tuple	working draft	JTC1/ SC32
ISO/IEC	IS	19773-06	Data structure for unstructured array of designation-kind-value (DKV) tuples	working draft	JTC1/ SC32
ISO/IEC	IS	19773-07	Data structure and syntax for selections of designation-kind-value (DKV) tuples	working draft	JTC1/ SC32
ISO/IEC	IS	19773-08	Data structure for UPU postal data	working draft	JTC1/ SC32
ISO/IEC	IS	19773-09	Data structure for ITU-T E.164 phone number data	working draft	JTC1/ SC32
ISO/IEC	IS	19773-10	Data structure for who-what-where-when-why-how (W5H) event data	working draft	JTC1/ SC32
ISO/IEC	IS	19773-11	Data structure for entity-person-group (EPG) contact data	working draft	JTC1/ SC32
ISO/IEC	IS	19773-12	Data structure for entity-person-group (EPG) security credentials data	working draft	JTC1/ SC32
ISO/IEC	IS	19773-13	Data structure for entity-person-group (EPG) relationships and grouping data	working draft	JTC1/ SC32

ISO 19773 WDs Available from
http://www.iso.org/iso/search.htm?qt=19773&sort=rel&type=simple&published=true

6.2.9.13 ISO 20943 Achieving Metadata Registry Content Consistency

ISO/IEC	TR	20943	Information Technology—Achieving metadata registry content consistency	JTC1/ SC32

The 20943 standard is a multipart technical report that includes the following parts.

NOTE: At the time of this writing, two parts of this work item had been published as technical reports.

ISO/IEC	TR	20943-0	Information Technology—Achieving metadata registry content consistency—Part 0: Overview	Working Draft	JTC1/SC32

ISO 20943-0 Not Yet Available

Regarding metadata registries, this contains an overview and collection of frequently answers questions (FAQs) and answers.

ISO/IEC	TR	20943-1	Information Technology—Achieving metadata registry content consistency—Part 1: Data Elements	TR 2003	JTC1/SC32

Downloadable FOC ISO/IEC TR 20943-1:2003
http://standards.iso.org/ittf/PubliclyAvailableStandards/c034343_ISO_IEC_TR_20943-1_2003(E).zip

This provides guidance and recommendations on the registration of data elements within a metadata registry.

ISO/IEC	IS	20943-2	Information Technology—Achieving metadata registry content consistency—Part 2: XML Structured Data,	Working Draft	JTC1/SC32

ISO 20943-2 Not Yet Available

This provides guidance and recommendations on the registration of XML features within a metadata registry.

ISO/IEC	IS	20943-3	Information Technology—Achieving metadata registry content consistency—Part 3: Value Domains	TR 2004	JTC1/SC32

Downloadable FOC ISO/IEC TR 20943-3:2004
http://standards.iso.org/ittf/PubliclyAvailableStandards/c035339_ISO_IEC_TR_20943-3_2004(E).zip

This provides guidance and recommendations on the registration of value domains within a metadata registry.

6.2.9.14 ISO 20944 Metadata Registry Interoperability and Bindings (MDRIB)

ISO/IEC	IS	20944	Information Technology—Metadata Registry Interoperability and Bindings (MDRIB)	All parts are WD or CD at the date of publication of this book	JTC1/ SC32

Scheduled To be Available FOC once finalized. ISO/IEC 20944 series
http://www.iso.org/iso/search.htm?qt=20944&sort=rel&type=simple&published=true

NOTE: At the time of publishing this book, the link provided by ISO does not work. We suggest you contact SC32 directly for download.

The ISO/IEC 20944 standard is complex and is comprised of many parts. The main subject areas are:

- General Provisions (Parts 01–19);
- Coding Bindings (Parts 20–39);
- API Bindings (Parts 40–59);
- Protocol Bindings (Parts 60–79);
- Profiles (Parts 80–99).

The ISO/IEC 20944 series was developed for providing interoperability among metadata registries (11179-3), such as reading/writing attributes from/to a metadata registry. However, the ISO/IEC 20944 series may be used generically, such as for applications that are unrelated to 11179-3 metadata registries, or for applications that extend 11179-3 metadata registry attributes (attributes outside of the 11179-3 specification).

6.2.9.15 ISO 24706 Metadata for Technical Standards

ISO/IEC	IS	24706	Metadata for Technical Standards	IS Expected Dec 07	JTC1/ SC32

ISO 24706 Draft Available from
http://www.iso.org/iso/iso_catalogue/catalogue_tc/catalogue_detail.htm?csnumber=46208

6.2.9.16 ISO 24707 Metadata for Technical Standards—Common Logic

ISO/IEC	IS	24707	Metadata for Technical Standards–Common Logic	IS expected Dec 07	JTC1/ SC32

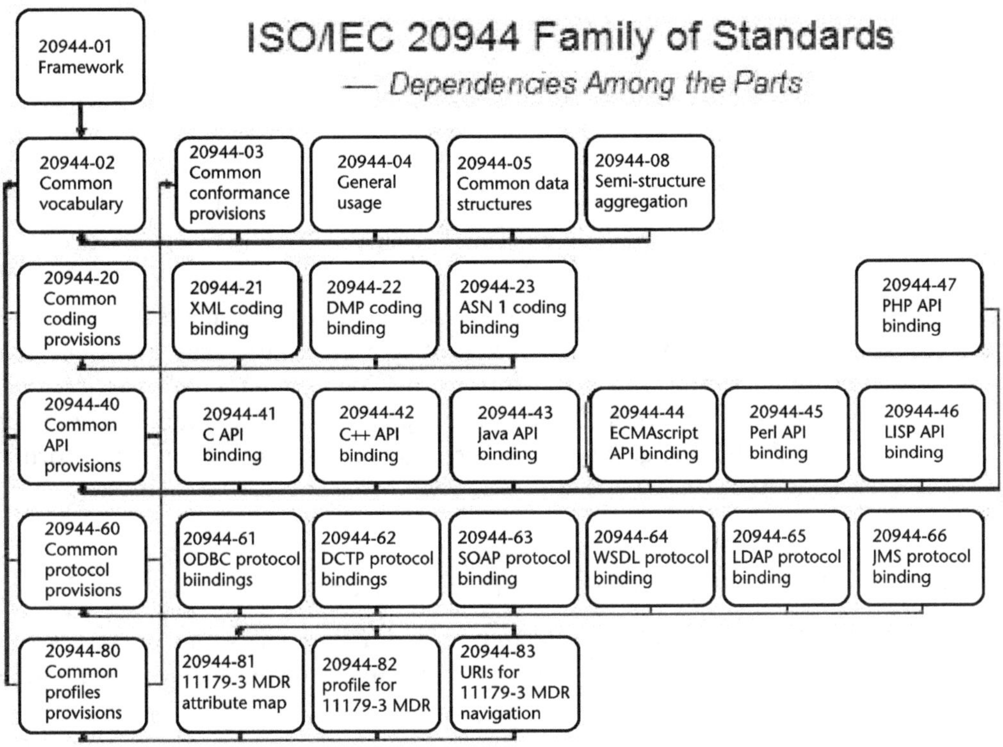

Figure 6.2 ISO/IEC 20944 Metadata Registry Interoperability and Bindings Family of Standards. (*Source:* ISO SC32 Web site.)

ISO 24707 Available from
http://www.iso.org/iso/iso_catalogue/catalogue_tc/catalogue_detail.htm?csnumber=39175

This standard specifies a family of languages designed for use in the representation and interchange of knowledge among disparate computer systems.

The following features are essential to the design of this standard:

- The language has declarative semantics. It is possible to understand the meaning of expressions in the language without appeal to an interpreter for manipulating those expressions.
- The language is logically comprehensive. At its most general, it provides for the expression of arbitrary logical sentences.

The following are within the scope of this standard:

- Interchange of knowledge among heterogeneous computer systems;
- Representation of knowledge in ontologies and knowledge bases;
- Specification of expressions that are the input or output of inference engines.

The following are outside the scope of this standard:

- Specification of proof theory or inference rules;
- Specification of translators between the notations of heterogeneous computer systems;
- Free logics;
- Conditional logics;
- Methods of providing relationships between symbols in the logical "universe" and individuals in the "real world."

This document describes common logic's syntax and semantics.

The standard defines an abstract syntax and an associated model-theoretic semantics for a specific extension of first-order logic. The intent is that the content of any system using first-order logic can be represented in the standard. The purpose is to facilitate interchange of first-order logic-based knowledge and information between systems.

6.2.10 JTC 1/SC34 JTC 1/SC34 Markup Languages

The scope of SC34 is to produce standards for languages and resources for the description and processing of compound and hypermedia documents, including:

- Standard generalized markup language and support facilities;
- Document processing architecture and formatting for documents represented in SGML and XML;
- Final-form document architecture and standard page description language;
- Font architecture, interchange format, and services;
- Glyph architecture and registration procedures;
- Hypermedia document structuring language and application resources;
- Content architectures and content notations for document architectures defined or supported by "Document Description and Processing Languages" standards.

Some of the standards of SC34 that are likely to be of use in ITS service provision are summarized next. The Work Program of SC34 can be downloaded from:

http://www.iso.org/iso/iso_catalogue/catalogue_tc/catalogue_tc_browse.htm?commid=45374&development=true.

6.2.10.1 ISO 8879 Standard Generalized Markup Language (SGML)

ISO/IEC	IS	8879	Information technology—Text and office systems—Standard Generalized Markup Language (SGML)	Published	JTC1/ SC34

ISO 8879 Available from
http://www.iso.org/iso/iso_catalogue/catalogue_tc/catalogue_detail.htm?csnumber=16387

Both SGML and XML are "meta" languages because they are used for defining markup languages. A markup language defined using SGML or XML has a specific vocabulary (labels for elements and attributes) and a declared syntax (grammar defining the hierarchy and other features).

6.2.11 ISO/IEC JTC1/SC35 Generic IT User Interfaces

ISO/IEC JTC1 SC35 has the following scope: "Standardization in the field of User-system interfaces between users (including people with special needs) and systems encompassing input and output devices in information technology environments, with a priority of meeting the JTC1 requirements for cultural and linguistic adaptability."

Included in the standard are the following related areas:

- Interfaces between users and devices such as keyboard, mice, pointers, pens, as well as visual displays, and forms of audio and tactile input/output, with the emphasis on functionality;
- Rules for system control by voice, vision, movement, gestures;
- Presentations of technical mechanisms, icons, graphical symbols;
- Dialog control and navigation in interactions between humans and systems assistance and tutoring.

The full work program for SC35 can be downloaded from:

http://www.iso.org/iso/iso_catalogue/catalogue_tc/catalogue_tc_browse.htm?commid=45382&development=true

The following standards, developed by SC35, may be useful in the context of ITS service provision.

6.2.11.1 ISO 11580 Model for Describing User Interface Objects, Actions, and Attributes

ISO/IEC	CD TR	11580	Information technology—Model for describing user interface objects, actions and attributes	Under Development	JTC1/ SC35

ISO 11580 Available from
http://www.iso.org/iso/iso_catalogue/catalogue_tc/catalogue_detail.htm?csnumber=42779

ISO/IEC TR 11580:2007 defines a format for describing user interface objects, actions, and attributes. It provides a basis for standardizing the names and properties of user interface objects, actions, and attributes across multiple applications and platforms.

ISO/IEC TR 11580:2007 contains guidance both on the standardization of user interface objects, actions, and attributes and on the implementation of these objects, actions, and attributes in any or all modalities. It is primarily intended for developers of standards, style guides, and architectures involving user interface objects, actions, and attributes.

ISO/IEC TR 11580:2007 also provides software developers with a range of functionalities to be considered in the design of objects, actions, and attributes within user interfaces.

6.2.11.2 ISO 19765 Survey of Existing Icons and Symbols for Elderly and Disabled Persons

ISO/IEC	DTR	19765	Information technology—Survey of existing icons and symbols for elderly and disabled persons	Under Development	JTC1/ SC35

ISO 19765 Available from
http://www.iso.org/iso/iso_catalogue/catalogue_tc/catalogue_detail.htm?csnumber=33903

Different users of information technology products possess different sets of abilities. Some abilities may not ever be present in a user, as they may have been born without them. Some abilities are acquired or developed, or deteriorate over time due to education, maturity, injury, illness, or age. Just as it is possible that a user possesses a combination of abilities, it is also possible that they may lack a combination of abilities.

ISO/IEC TR 19765:2007 presents icons and symbols currently used to provide access to facilities and tools to support the needs of elderly and disabled users of information technology (IT) products, and could form the basis of a future International Standard which would provide a recommended collection of icons and symbols.

These icons and symbols have been collected from a variety of sources including other standards, contemporary software products, Web sites, and hardware devices. These sources are cross-referenced and listed in a bibliography.

The icons and symbols presented in ISO/IEC TR 19765:2007 are categorized by modality and method of use.

6.2.11.3 ISO 19766 Guidelines for the Design of Icons and Symbols Accessible to All Users, Including the Elderly and Persons with Disabilities

ISO/IEC	CTR	19766	Information Technology—Guidelines for the design of icons and symbols accessible to all users, including the elderly and persons with disabilities	TR 2007	JTC1/ SC35

ISO 19766 Available from
http://www.iso.org/iso/iso_catalogue/catalogue_tc/catalogue_detail.htm?csnumber=42128

ISO/IEC TR 19766:2007 provides recommendations relating to the design of icons to support accessibility by the elderly and people with disabilities. These recommendations assist accessible implementation of all icons for users. While these recommendations were developed to meet the needs of the elderly and people with disabilities, they can also provide greater accessibility to a wider range of users in a variety of different contexts.

ISO/IEC TR 19766:2007 introduces a set of attributes and operations that can be implemented as features of graphic icons to make the functionality of these icons accessible to the widest possible range of users. Textual attributes are emphasized because they can be rendered in various alternate modalities. ISO/IEC 11581-1 provides guidance on the graphic aspects of icons. Specific renderings of these attributes (or of icons in general) are not dealt with as part of ISO/IEC TR 19766:2007.

6.2.11.4 ISO 24738 Icon Symbols and Functions for Multimedia Link Attributes

ISO/IEC	IS	24738	Information technology—Icon symbols and functions for multimedia link attributes	Published	JTC1/ SC35

ISO 24738 Available from
http://www.iso.org/iso/iso_catalogue/catalogue_tc/catalogue_detail.htm?csnumber=38848

ISO/IEC 24738:2006 defines a consistent set of icon symbols and related attributes that are presented on a computer screen and with which users interact to decide whether or not to take the associated link. These symbols represent attributes of the link and/or the destination of the link.

6.2.11.5 ISO 24752 User interfaces—Universal Remote Console

ISO/IEC	FCD	24752	Information technology—User interfaces—Universal remote console		JTC1/ SC35

This is a multipart series of standards under development comprising five parts at the date of publication of this book.

ISO/IEC	FCD	24752-1	Information technology—User interfaces—Universal remote console—Part 1: Framework	Under Development	JTC1/ SC35

ISO 24572-1 Not Yet Available

ISO/IEC FCD 24752-2	Information technology—User interfaces—Universal remote console—Part 2: User Interface Socket Description	Under Development	JTC1/SC35

ISO 24572-2 Not Yet Available

ISO/IEC FCD 24752-3	Information technology—User interfaces—Universal remote console—Part 3: Presentation Template	Under Development	JTC1/SC35

ISO 24572-3 Not Yet Available

ISO/IEC FCD 24752-4	Information technology—User interfaces—Universal remote console—Part 4: Target Description	Under Development	JTC1/SC35

ISO 24572-4 Not Yet Available

ISO/IEC FCD 24752-5	Information technology—User interfaces—Universal remote console—Part 5: Resource Description	Under Development	JTC1/SC35

ISO 24572-5 Not Yet Available

6.2.11.6 Screen Icons and Symbols for Personal Mobile Communication Device

ISO/IEC FCD 24755	Information Technology—Screen icons and symbols for personal mobile communication device	Under Development FDIS	JTC1/SC35

ISO 24755 Available from
http://www.iso.org/iso/iso_catalogue/catalogue_tc/catalogue_detail.htm?csnumber=41525

ISO/IEC 24755:2007 defines a consistent set of screen icons and symbols, together with their related functions, that are presented by personal mobile communications devices (e.g., mobile phones and personal digital assistants). These devices have touch screens accessible by stylus pen, finger, or button, with personalized application.

ISO/IEC 24755:2007 provides a consistent set of icon graphics for controlling these devices and for using personal information management related applications. These icons and symbols represent typical functions and statuses by their association with conventional controls and functions on real-world objects.

ISO/IEC 24755:2007 applies to all icon graphics displayed with a resolution of 32 x 32 pixels or higher. The graphic presentation can be either dynamic or fixed.

6.2.11.7 Algorithmic Framework for Determining Accessibility for Individual Users of Interactive Systems

ISO/IEC	FCD	24756	Information technology—Algorithmic framework for determining accessibility for individual users of interactive systems	Under Development FDIS	JTC1/ SC35

ISO 24756 Available from
http://www.iso.org/iso/iso_catalogue/catalogue_tc/catalogue_detail.htm?csnumber=41526

6.2.11.8 Taxonomy of Cultural and Linguistic Adaptability User Requirements

ISO/IEC	CD	24785	Information Technology—Taxonomy of cultural and linguistic adaptability user requirements	Under Development	JTC1/ SC35

ISO CD24785 Available from
http://www.iso.org/iso/iso_catalogue/catalogue_tc/catalogue_detail.htm?csnumber=41555

6.2.11.9 User Interfaces—Accessible User Interface for Accessibility Setting on Information Devices—Part 1: General and Methods to Start

At the time of developing this book, there is only one work item that has made ballot progress.

ISO/IEC	CD	24786-1	Information Technology—User Interfaces —Accessible User Interface for Accessibility Setting on Information Devices—Part 1: General and methods to start	Under Development	JTC1/ SC35

ISO 24786-1 Not Yet Available

6.2.12 Unified Modeling Language (UML)

6.2.12.1 ISO 19501 Information Technology—Open Distributed Processing—Unified Modeling Language (UML)

ISO/IEC	IS	19501	Information technology—Open Distributed Processing—Unified Modeling Language (UML) Version 1.4.2	Published	ISO/IEC JTC1

ISO 19501 Available from
http://www.iso.org/iso/iso_catalogue/catalogue_tc/catalogue_detail.htm?csnumber=32620

The UML is a graphical language for visualizing, specifying, constructing, and documenting the artifacts of a software-intensive system. The UML offers a standard way to write a system's blueprints, including conceptual things such as business processes and system functions, as well as concrete things such as programming language statements, database schemas, and reusable software components.

6.2.13 Terminology

ISO/TC 37 has also produced some useful standards with respect to terminology. These are summarized in this section. The full work program of TC 37 can be downloaded from:

http://www.iso.org/iso/standards_development/technical_committees/list_of_iso_technical_committees/iso_technical_committee.htm?commid=48104.

6.2.13.1 ISO 860 Harmonization of Concepts and Terms

ISO	IS	860	Terminology work—Harmonization of concepts and terms	Published 1996	ISO TC37

ISO 860 Available from
http://www.iso.org/iso/iso_catalogue/catalogue_tc/catalogue_detail.htm?csnumber=5240

This standard deals with the principles which are the basis upon which concept systems can be harmonized and with the development of harmonized terminologies, in order to improve the efficiency in interlinguistic communication.

It has been prepared by ISO/TC 37/SC 1 "Principles of terminology," whose scope is "Standardization of basic principles and methods for developing scientific and technical terminologies." The objective of ISO/TC 37/SC 1 is to prepare documents covering principles and methods of terminology work, describing basic concepts in the field of terminology as well as practical aspects of terminography, specifically concerning:

- Terminological principles and methods;
- Vocabulary of terminology;
- Terminology documentation;
- Coding and codes in the field of terminology.

6.2.13.2 ISO 704 Principles and Methods of Terminology

ISO	IS	704	Principles and methods of terminology	Published 1987 Under revision	ISO TC37

ISO 704 Available from
http://www.iso.org/iso/iso_catalogue/catalogue_tc/catalogue_detail.htm?csnumber=31696

ISO 704:1987 is currently under revision. This standard outlines the principles of conceptual analysis, definition writing, and term formation, as well as the methods used in terminology work. An informative annex provides examples of term formation methods.

6.2.13.3 Harmonization of Terminology

ISO	IS	860	Terminology work—Harmonization of concepts and terms	Published revised 2007	ISO TC37

ISO 860 Available From
http://www.iso.org/iso/iso_catalogue/catalogue_tc/catalogue_detail.htm?csnumber=40130&commid=48104

ISO 860:2007 specifies a methodological approach to the harmonization of concepts, concept systems, definitions, and terms. It applies to the development of harmonized terminologies, at either the national or international level, in either a monolingual or a multilingual context.

6.2.13.4 ISO 1087 Terminology—Vocabulary

ISO	IS	1087	Principles and methods of terminology	Published 1990	ISO TC37

ISO 1087:1990 is revised to become ISO 1087-1 and 1087-2. This standard provides a set of terminological entries covering the basic concepts in the field of theory and application of terminology work.

ISO	IS	1087-1	Terminology—Vocabulary—Part 1: Theory and application	Published 1990	ISO TC37

ISO 1087-1 Available from
http://www.iso.org/iso/iso_catalogue/catalogue_tc/catalogue_detail.htm?csnumber=20057

ISO	IS	1087-2	Terminology—Vocabulary—Part 2: Computer applications	Published 1990	ISO TC37

ISO 1087-2 Available from
http://www.iso.org/iso/iso_catalogue/catalogue_tc/catalogue_detail.htm?csnumber=32819

6.2.13.5 Principles, Methods, and Vocabulary

ISO	IS	1087	Vocabulary—Theory and Application	Published Under revision	ISO TC37

ISO 1087 Available from
http://www.iso.org/iso/iso_catalogue/catalogue_tc/catalogue_detail.htm?csnumber=20057

ISO 1087-1: 2000, which specifies fundamental terms and concepts for applied terminology work, is being revised by ISO/TC 37/SC 1/WG 3 "Principles, methods and vocabulary." Its scope is the standardization of principles, methods, and vocabulary of terminology work, terminology management, and language resources.

6.2.13.6 Terminology in Sociolinguistic Application

ISO	CD	22134	Terminology in sociolinguistic application	Under development	ISO TC37

ISO 22134 Available from
http://www.iso.org/iso/iso_catalogue/catalogue_tc/catalogue_detail.htm?csnumber=36143

ISO/TR 22134:2007 proposes guidelines for socio-terminology principles, methods, and vocabularies.

6.2.14 Internet Protocol (IP)

Internet standards are not in fact standards in the traditional sense, but are specifications developed by the Internet Engineering Task Force (IETF) (see Section 11.4). Principal IETF specifications used in ITS are summarized in this chapter.

The Internet has become pervasive in most developed societies. More than this, it has become an efficient means by which to move data without creating application-specific architecture instantiations. Whereas, just a decade ago, ITS prophets foresaw whole new sectors of ITS-specific business activities, the commonality to non-ITS specific services (such as hotel, restaurant, and entertainment booking), which are now generically provided via the Internet, mean that these services are available in the vehicle so long as Internet is made available within the vehicle.

Commercially, the Internet has already become the most efficient way of moving data in many cases, particularly in distributed and mobile networks. The standards supporting Internet provision and behavior are therefore of crucial importance to the provision of ITS services.

See also Section 5.2.2 regarding the use of Web services in an ITS environment.

6.2.14.1 Specification and Standardization of the Internet Protocol Version 4 (IPv4)

IETF	Specification and standardization of the Internet Protocol version 4 (IPv4)	IETF

Downloadable FOC IPv4 (original DARPA Spec)
http://www.ietf.org/rfc/rfc791.txt

Internet Protocol version 4 is the fourth iteration of the Internet Protocol (IP) and it is the first version of the protocol to be widely deployed. IPv4 is the dominant network layer protocol on the Internet and apart from IPv6 it is the only standard internetwork-layer protocol used on the Internet.

It is described in IETF RFC 791 (September 1981) which made obsolete RFC 760 (January 1980). The United States Department of Defense also standardized it as MIL-STD-1777.

IPv4 is a data-oriented protocol to be used on a packet switched internetwork (e.g., Ethernet). It is a best effort protocol in that it does not guarantee delivery. It does not make any guarantees on the correctness of the data; it may result in duplicated packets and/or packets that are out of order. These aspects are addressed by an upper layer protocol (e.g., TCP, and partly by UDP).

The entire purpose of IP is to provide unique global computer addressing to ensure that two computers communicating over the Internet can uniquely identify one another.

IPv4 has a 32-bit address (compare with IPv6 in Section 6.2.14.2).

6.2.14.2 Specification and Standardization of the Internet Protocol Version 6 (IPv6)

IETF	RFC	2460	Specification and standardization of the Internet Protocol version 6 (IPv6)		IETF
IETF	RFC	2373	IP Version 6 Addressing Architecture		IETF
IETF	RFC		Default Address Selection for IPv6	draft	IETF
IETF	RFC	2463	Internet Control Message Protocol (ICMPv6) for IPv6		IETF
IETF	RFC	1981	Path MTU Discovery For IPv6		IETF
IETF	RFC	2461	Neighbor Discovery for IPv6		IETF

Downloadable FOC IETF 2460
http://www.ietf.org/rfc/rfc2460.txt

Downloadable FOC IETF 2373
http://www.ietf.org/rfc/rfc2373.txt

Downloadable FOC IETF 2463
http://www.ietf.org/rfc/rfc2463.txt

Downloadable FOC IETF 1981
http://www.ietf.org/rfc/rfc1981.txt

Downloadable FOC IETF 2461
http://www.ietf.org/rfc/rfc2461.txt

The number of addresses is limited in IPv4. IETF has addressed the issues of wireless networks connecting 6 billion humans and then 6 trillion computers in IPv6. Whereas the address for IPv4 has 32 bits, the address for IPv6 has 128 bits, providing an exponentially larger population capacity (see Figure 6.3).

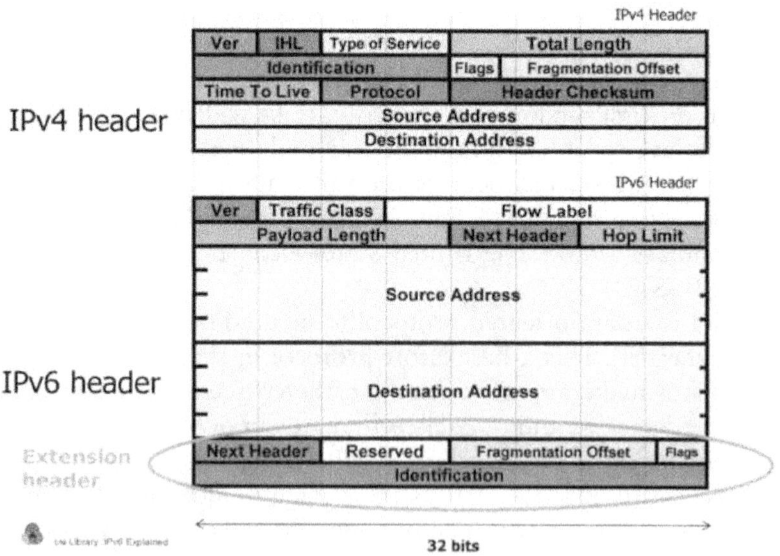

Figure 6.3 IPv6 compared to IPv4. (*Source:* CSI Library.)

Version 6 of the Internet Protocol [Version 5 was allocated to Internet Stream Protocol (RFC1190)] is being standardized by IETF IPv6 Working Group and Next Generation Transition Working Group. Work started in 1991 and the base protocol was standardized in 1998 as RFC2460.

Whereas in IPv4 there is "address exhaustion," IPv6 introduces a Network Address Translator (NAT).

The design philosophy for IPv6 features the following:

- Expanded addressing capabilities; {blistHeader format simplification;
- Improved support for extensions and options;
- Flow labeling capability;
- Authentication and privacy capabilities.

IPv6 provides an evolutionary (not revolutionary) step from IPv4. IPv6 includes a transition mechanism which is designed to allow users to adopt and deploy IPv6 in a highly diffuse fashion and to provide direct interoperability between IPv4 and IPv6 hosts. The transition to a new version of the Internet Protocol must be incremental, with few or no critical interdependencies, if it is to succeed. The IPv6 transition allows the users to upgrade their hosts to IPv6, and the network operators to deploy IPv6 in routers, with very little coordination between the two.

Simple Internet Transition Mechanisms

The key transition objective is to allow IPv6 and IPv4 hosts to interoperate. A second objective is to allow IPv6 hosts and routers to be deployed in the Internet in a highly diffuse and incremental fashion, with few interdependencies. A third objective is that the transition should be as easy as possible for end-users, system administrators, and network operators to understand and carry out.

The Simple Internet Transition (SIT) is a set of protocol mechanisms implemented in hosts and routers, along with some operational guidelines for addressing and deployment, designed to make transitioning the Internet to IPv6 work with as little disruption as possible.

SIT provides a number of features, including:

- Incremental upgrade and deployment. Individual IPv4 hosts and routers may be upgraded to IPv6 one at a time without requiring any other hosts or routers to be upgraded at the same time. New IPv6 hosts and routers can be installed one by one.
- Minimal upgrade dependencies. The only prerequisite to upgrading hosts to IPv6 is that the DNS server must first be upgraded to handle IPv6 address records. There are no prerequisites to upgrading routers.
- Easy Addressing. When existing installed IPv4 hosts or routers are upgraded to IPv6, they may continue to use their existing address. They do not need to be assigned new addresses. Administrators do not need to draft new addressing plans.
- Low start-up costs. Little or no preparation work is needed in order to upgrade existing IPv4 systems to IPv6, or to deploy new IPv6 systems.

The mechanisms employed by SIT include:

- An IPv6 addressing structure that embeds IPv4 addresses within IPv6 addresses, and encodes other information used by the transition mechanisms;
- A model of deployment where all hosts and routers upgraded to IPv6 in the early transition phase are "dual" capable (i.e., implement complete IPv4 and IPv6 protocol stacks);
- The technique of encapsulating IPv6 packets within IPv4 headers to carry them over segments of the end-to-end path where the routers have not yet been upgraded to IPv6.
- The header translation technique to allow the eventual introduction of routing topologies that route only IPv6 traffic, and the deployment of hosts that support only IPv6. Use of this technique is optional, and would be used in the later phase of transition if it is used at all.

SIT ensures that IPv6 hosts can interoperate with IPv4 hosts anywhere in the Internet up until the time when IPv4 addresses run out, and it allows IPv6 and IPv4 hosts within a limited scope to interoperate indefinitely after that. This feature protects the huge investment users have made in IPv4. SIT ensures that IPv6 does not render IPv4 obsolete. Hosts that need only a limited connectivity range (e.g., printers) never need to be upgraded to IPv6.

The incremental upgrade features of SIT allow the host and router vendors to integrate IPv6 into their product lines at their own pace, and allow the end-users and network operators to deploy IPng on their own schedules.

For more information on the IPng transition mechanisms, see the RFC 1933, Transition Mechanisms for IPv6 Hosts and Routers.

Downloadable RFC 1933, Transition Mechanisms for IPv6 Hosts and Routers
http://www.ietf.org/rfc/rfc1933.txt

Numerous products (mature) and initial commercial deployment now exist using IPv6.

IPv6 addresses feature different address types, as follows:

- Anycast: one-to-nearest communication (similar specification to Unicast Address);
- Unicast: one-to-one communication (Subnet Prefix is equivalent to the "subnet field" in an IPv4 address; Interface ID is equivalent to the "host field" in an IPv4 address);
- Multicast: one-to-many communication.

Within these address types there are the following address features:

- Interface-local;
- Link-local;

- Subnet-local (Subnet ID is an identifier of a link);
- Admin-local;
- Site-local;
- Organization-local;
- Global (global routing prefix is hierarchical).

A good source for further information is http://playground.sun.com/ipv6.

6.2.14.3 Network Mobility (NEMO)

IETF	RFC	3963	Network Mobility (NEMO) Basic Support	IETF
IETF	RFC	Draft 4885	Network Mobility Support Terminology	IETF

Downloadable FOC IETF RFC 3963
http://www.ietf.org/rfc/rfc3963.txt
Downloadable FOC IETF RFC DRAFT 4885
http://www.ietf.org/rfc/rfc4885.txt

Mobile Nodes and Multiple Interfaces in IPv6 (NEMO) offers the following features:

- Host mobility support for end systems that change point of attachment;
- Network mobility support;
- Entire networks that change point of attachment;
- Ad hoc networks: routing protocols for infrastructureless networks;
- Multihoming;
- Stepwise approach;
- NEMO Basic Support: session maintenance (now);
- NEMO Extended Support: performances issues (may be later);
- Standards track RFC 3963 (Jan. 2005);
- NEMO addresses;
- Migration transparency: permanent connectivity and session continuity;
- Performance transparency/Seamless mobility;
- Minimum signaling overhead, packet loss and delay;
- Network mobility support transparency;
- Operational transparency: implementation at IP layer;
- Arbitrary configurations: nested, multihomed;
- Local and global mobility: roaming across various admin domains;
- Optimal routing between arbitrary CNs and MNNs;
- Scalability: large number of mobile networks, CNs;
- Backward compatibility: MIPv6, AAA, IPSec, @ allocation, multicast;
- Secure signaling: authentication, authorization, confidentiality;

- Location privacy;
- IPv4 and NAT traversal.

IETF is working directly in liaison with ISO/TC 204 to address ITS-specific issues.

6.2.15 UTC—Coordinated Universal Time

BIPM	UTC Co-ordinated Universal Time

Downloadable FOC BIPM UTC
http://www.bipm.org/en/scientific/tai/time_server.html

Time is a critical data concept in the provision of many ITS services, particularly emergency and collision avoidance systems. The most widely accepted identification/measurement of time is so-called UTC time. Because of its importance in a wide number of ITS services, UTC is discussed in this section.

UTC is a standard for time, but it is not defined in the usual terms of an SDO generated standard. The full definition of UTC is contained in CCIR Recommendation 460-4.

The ITU sought a single agreed abbreviation for this time measurement, and the term coordinated universal time (UTC) as a universally accepted compromise was agreed upon.

The acronym UTC also fits with legacy system abbreviations of variants of *universal time* such as UT1 and UT1R.

UTC is a high-precision atomic time standard initially developed by the U.S. Department of Defense, and now maintained by the Bureau International de Poids et Measures (BIPM) (which translates to the International Bureau of Weights and Measures) who publishes monthly tables of differences between canonical International Atomic Time (TAI)/UTC and TAI/UTC as estimated in real time by participating laboratories.

6.2.16 WGS84—World Geodetic System

UN	WGS84	World Geodetic System release 1984	Last revised 2005

Download FOC WGS84
http://earth-info.nga.mil/GandG/wgs84/

Another important aspect for ITS systems is the two- and three-dimension location of objects in relation to the surface of the Earth and the relative positioning of one object or feature to another object. In ITS service provision this has an obvious use in navigation systems, and is also important for a wide range of current and projected systems.

Geodesy is applied mathematics concerned with the determination of the size and shape of the Earth (geoid). Direct measurements (triangulation, leveling, and gravimetric observations) determine the exact location of points on the Earth's surface. Geodetics are methods to achieve the measurement and representation of the Earth, its gravitational field, and its geodynamic phenomena (polar motion, Earth tides, and crustal motion) in three-dimensional time-varying space.

The World Geodetic System defines a reference frame for the Earth for use in geodesy and navigation.

6.2.17 Simple Object Access Protocol (SOAP)

W3C	Simple Object Access Protocol (SOAP)	Version 1.2 Published

Downloadable FOC SOAP
http://www.w3.org/TR/soap/

SOAP is a lightweight protocol intended for exchanging structured information in a decentralized, distributed environment. It is used particularly in providing Web services. The SOAP standard was developed by the World Wide Web Consortium (W3C). It is one of three major interlinked Web services protocols (see also WSDL and UDDI). For an explanation of the role and relationship of these standards, see also Section 5.2.2.

This standard is in multiple parts.

W3C	Simple Object Access Protocol (SOAP) SOAP Version 1.2 Part0: Primer	Version 1.2 Published

Downloadable FOC SOAP Primer
http://www.w3.org/TR/soap12-part0/

SOAP Version 1.2 Part 0: Primer (Second Edition) is a non-normative document intended to provide an easily understandable tutorial on the features of SOAP Version 1.2. In particular, it describes the features through various usage scenarios and is intended to complement the normative text contained in Parts 1 and 2 of the SOAP 1.2 specifications. This second edition includes additional material on the SOAP Message Transmission Optimization Mechanism (MTOM), the XML-binary Optimized Packaging (XOP), and the Resource Representation SOAP Header Block (RRSHB) specifications.

W3C		Simple Object Access Protocol (SOAP) SOAP Version 1.2 Part1: Messaging Framework	Version 1.2 Published

Downloadable FOC SOAP Messaging Framework
http://www.w3.org/TR/soap12

SOAP Part 1: Messaging Framework defines, using XML technologies, an extensible messaging framework containing a message construct that can be exchanged over a variety of underlying protocols.

W3C		Simple Object Access Protocol (SOAP) SOAP Version 1.2 Part2: Adjuncts	Version 1.2 Published

Downloadable FOC SOAP Adjuncts
http://www.w3.org/TR/soap12-part2/

SOAP Version 1.2 Part 2: Adjuncts defines a set of adjuncts that may be used with SOAP Version 1.2 Part 1: Messaging Framework. This specification depends on SOAP Version 1.2 Part 1: Messaging Framework [SOAP Part 1].

W3C		Simple Object Access Protocol (SOAP) SOAP Version 1.2 Specification Assertions and Test Collection	Version 1.2 Published

Downloadable FOC SOAP Assertions and Test Collection
http://www.w3.org/TR/2003/REC-soap12-testcollection-20030624/

This document draws on assertions found in the SOAP Version 1.2 specifications [SOAP Part 1], [SOAP Part 2], and provides a set of tests in order to show whether the assertions are implemented in a SOAP processor.

A SOAP 1.2 implementation that passes all of the tests specified in this document may claim to conform to the SOAP 1.2 Test Suite, 2003 06 24. It is incorrect to claim to be compliant with the SOAP Version 1.2 specifications merely by passing successfully all the tests provided in this test suite. It is also incorrect to claim that an implementation is noncompliant with the SOAP Version 1.2 specifications based on its failure to pass one or more of the tests in this test suite.

6.2.18 Web Services Description Languages (WSDL)

W3C	WGS84	Web Services Description Language (WSDL)	Published 2001

Downloadable FOC WSDL
http://www.w3.org/TR/wsdl

WSDL is an XML format for describing network services as a set of endpoints operating on messages containing either document-oriented or procedure-oriented information. The operations and messages are described abstractly, and are then bound to a concrete network protocol and message format to define an endpoint. Related concrete endpoints are combined into abstract endpoints (services). WSDL is extensible to allow description of endpoints and their messages regardless of what message formats or network protocols are used to communicate; however, the only bindings described in this document describe how to use WSDL in conjunction with SOAP 1.1, HTTP GET/POST, and MIME. For an explanation of the role and relationship of these standards, see also Section 5.2.2.

6.2.19 Universal Description, Discovery, and Integration (UDDI)

OASIS	Universal Description, Discovery, and Integration (UDDI)

Downloadable FOC UDDI
http://www.oasis-open.org/committees/uddi-spec/doc/tcspecs.htm#uddiv3

Organization for the Advancement of Structured Information Standards (OASIS) is a not-for-profit consortium that drives the development, convergence, and adoption of open standards for the global information society. The consortium produces more Web services standards than any other organization, along with standards for security, e-business, and standardization efforts in the public sector and for application-specific markets. Founded in 1993, OASIS has more than 5,000 participants representing more than 600 organizations and individual members in 100 countries.

UDDI Version 3 builds on the vision of UDDI: that is, a "meta service" for locating Web services by enabling robust queries against rich metadata. Expanding on the foundation of versions 1 and 2, version 3 offers the industry a specification for building flexible, interoperable XML Web services registries which are useful in private and public deployments. For an explanation of the role and relationship of these standards, see also Section 5.2.2.

The UDDI Version 3 specification consists of the following documents.

UDDI Version 3.0 Features List. OASIS recommends first reading the UDDI Version 3.0 Features List.

Downloadable FOC UDDI Version 3.0 Features List
http://uddi.org/pubs/uddi_v3_features.htm

UDDI Version 3.0.2. This is an OASIS Standard, which represents the second errata to the UDDI Version 3 specification.

Downloadable FOC OASIS Standard
http://www.oasis-open.org/specs/index.php#uddiv3

UDDI Version 3.0.2, UDDI Spec. Technical Committee Draft, Dated 20041019.

Downloadable FOC HTML / PDF
http://uddi.org/pubs/uddi-v3.0.2-20041019.pdf

OASIS	Universal Description, Discovery and Integration (UDDI)—UDDI Version 3.0.2 XML Schema.

UDDI uses the XML schema language to formally describe its data structures. UDDI Version 3.0.2 XML schema is provided by the following files.

Downloadable FOC UDDI API Schema uddi_v3.xsd
http://uddi.org/schema/uddi_v3.xsd

Downloadable FOC UDDI Custody Schema uddi_v3custody.xsd
http://uddi.org/schema/uddi_v3custody.xsd

Downloadable FOC UDDI Subscription Schema uddi_v3subscription.xsd
http://uddi.org/schema/uddi_v3subscription.xsd

Downloadable FOC UDDI Subscription Listener Schema uddi_v3subscriptionListener.xsd
http://uddi.org/schema/uddi_v3subscriptionListener.xsd

Downloadable FOC UDDI Replication Schema uddi_v3replication.xsd
http://uddi.org/schema/uddi_v3replication.xsd

Downloadable FOC UDDI Value Set Validation Schema uddi_v3valueset.xsd
http://uddi.org/schema/uddi_v3valueset.xsd

Downloadable FOC UDDI Value Set Caching uddi_v3valuesetcaching.xsd http://uddi.org/schema/uddi_v3valuesetcaching.xsd

Downloadable FOC UDDI Policy uddi_v3policy.xsd
http://uddi.org/schema/uddi_v3policy.xsd

Downloadable FOC UDDI Policy Instance Parameters uddi_v3policy_instanceParms.xsd
http://uddi.org/schema/uddi_v3policy_instanceParms.xsd

UDDI Version 3.0.2 WSDL Service Interface Descriptions.

OASIS	Universal Description, Discovery and Integration (UDDI) —. complete set of UDDI Version 3.0.2 WSDL definitions

The complete set of UDDI Version 3.0.2 WSDL definitions is provided by the following files.

Downloadable FOC UDDI API Binding uddi_api_v3_binding.wsdl
http://uddi.org/wsdl/uddi_api_v3_binding.wsdl

Downloadable FOC UDDI API Port Type uddi_api_v3_portType.wsdl
http://uddi.org/wsdl/uddi_api_v3_portType.wsdl

Downloadable FOC UDDI Custody Binding uddi_custody_v3_binding.wsdl
http://uddi.org/wsdl/uddi_custody_v3_binding.wsdl

Downloadable FOC UDDI Custody Port Type uddi_custody_v3_portType.wsdl
http://uddi.org/wsdl/uddi_custody_v3_portType.wsdl

Downloadable FOC UDDI Replication Binding uddi_repl_v3_binding.wsdl
http://uddi.org/wsdl/uddi_repl_v3_binding.wsdl

Downloadable FOC UDDI Replication Port Type uddi_repl_v3_portType.wsdl
http://uddi.org/wsdl/uddi_repl_v3_portType.wsdl

Downloadable FOC UDDI Subscription Binding uddi_sub_v3_binding.wsdl
http://uddi.org/wsdl/uddi_sub_v3_binding.wsdl

Downloadable FOC UDDI Subscription Port Type uddi_sub_v3_portType.wsdl
http://uddi.org/wsdl/uddi_sub_v3_portType.wsdl

Downloadable FOC UDDI Subscription Listener Binding uddi_subr_v3_binding.wsdl
http://uddi.org/wsdl/uddi_subr_v3_binding.wsdl

Downloadable FOC UDDI Subscription Listener Port Type uddi_subr_v3_portType.wsdl
http://uddi.org/wsdl/uddi_subr_v3_portType.wsdl

Downloadable FOC UDDI Value Set Validation Binding uddi_vs_v3_binding.wsdl
http://uddi.org/wsdl/uddi_vs_v3_binding.wsdl

Downloadable FOC UDDI Value Set Validation Port Type uddi_vs_v3_portType.wsdl
http://uddi.org/wsdl/uddi_vs_v3_portType.wsdl

Downloadable FOC UDDI Value Set Caching Binding uddi_vscache_v3_binding.wsdl
http://uddi.org/wsdl/uddi_vscache_v3_binding.wsdl

Downloadable FOC UDDI Value Set Caching Port Type uddi_vscache_v3_portType.wsdl
http://uddi.org/wsdl/uddi_vscache_v3_portType.wsdl

6.2.20 Extensible Markup Language (XML)

W3C	Extensible Markup Language (XML) 1.0 (Fourth Edition)	Sep. 2006

Downloadable FOC xml 1.0 Fourth Edition
http://www.w3.org/TR/xml

Extensible Markup Language (XML) is a simple, very flexible text format derived from SGML (ISO 8879). Originally designed to meet the challenges of large-scale electronic publishing, XML is also playing an increasingly important role in the exchange of a wide variety of data on the Web and elsewhere.

XML is a subset of SGML that is completely described in XML 1.0 Fourth Edition. Its goal is to enable generic SGML to be served, received, and processed on the Web in the way that is now possible with HTML. XML has been designed for ease of implementation and for interoperability with both SGML and HTML.

CHAPTER 7

Identification Technology Standards

Identification, whether it is identification of a vehicle, a driver, or some other type of identification, plays a crucial role in ITS service provision; and as systems that require higher levels of security are designed and implemented, and as the integration, intermodality, and multimodality of ITS systems increase, standardized means of identification will play an increasingly important role. Some of the identification standards have been designed specifically for vehicle identification, some specifically for ITS service provision, while others are general identification standards that may have application in ITS (e.g., fingerprint recognition to identify an authorized driver). ITS will also combine with so-called smart card technology (e.g., a road charging system where the credits are held on and decremented from a smart card), and with radio frequency identification (RFID) technology (e.g., in road tolling systems). Where appropriate, in addition to summarizing the available standards we have described some of the features of these technologies which will either make them highly suited or unsuited as part of the provision of ITS services.

Vehicles are at the heart of most of the ITS sector, so we also provide greater detail of vehicle identification standards, which is openly available information from many sources, and so not protected by standards copyrights.

7.1 Personal Identification (Including IC Cards/Smart Cards)

7.1.1 Background to Smart Cards

The term "smart card" is synonymous with "chip card" and "IC card" (integrated circuit card). Smart cards are increasingly used to support ITS service provision in areas such as electronic ticketing for public transport, access control, and electronic payment applications.

Smart cards have generally adopted the form of other plastic cards because they substituted existing applications such as credit cards which were already in use and familiar to users. Later uses, with no predisposition as to size but with a requirement to minimize size, such as SIM cards for mobile phones, mount the chip on a smaller form factor to reduce size and weight. The trend has moved from contact cards to noncontact cards, which have both ease of use and maintenance advantages in that there are no mechanical parts to go wrong in card readers and less physical damage to cards themselves where there is not contact.

Since these cards have become noncontact, and the reading range has been increased, it has become increasingly difficult to see the difference between smart cards and RFID devices. (See Section 7.4.) Indeed, the protocols for of ISO/IEC 18000-3 Mode 1, RFID for item management, are identical to the protocols 15693 Long range contactless smart cards, upon which they depend.

Because of the evolution from non-IC contact cards, ISO 7813 is also relevant. However, contactless smart card technology is based on ISO 10373 (Close Coupled), ISO 14443 (Proximity), and ISO 15693 (Vicinity) standards. Cards that comply with these standards are intelligent, read/write devices capable of storing different kinds of data and operating at different ranges. Standards-based contactless smart cards can authenticate an identity, determine the appropriate level of access, and admit the cardholder to a facility, all from data stored on the card. These cards can include additional authentication factors (such as biometric templates or PINs) and other card technologies, including a contact smart card chip, to satisfy the requirements of legacy applications or applications for which a different technology is more appropriate.

Standards-based contactless smart cards offer organizations the flexibility to select appropriate technologies driven by business requirements, rather than implementation constraints. This allows organizations to implement and enforce a wide range of security policies by deploying a system best suited to the application.

The reader for a smart card is generally a simple interface to a computer. Smart card readers are frequently built into larger service providing units such as metro transit systems and ATM machines. Figure 7.1 shows an ITS example of an electronic ticket access control system with a contactless smart card reading capability and examples of a telephone smartcard or a SIM card.

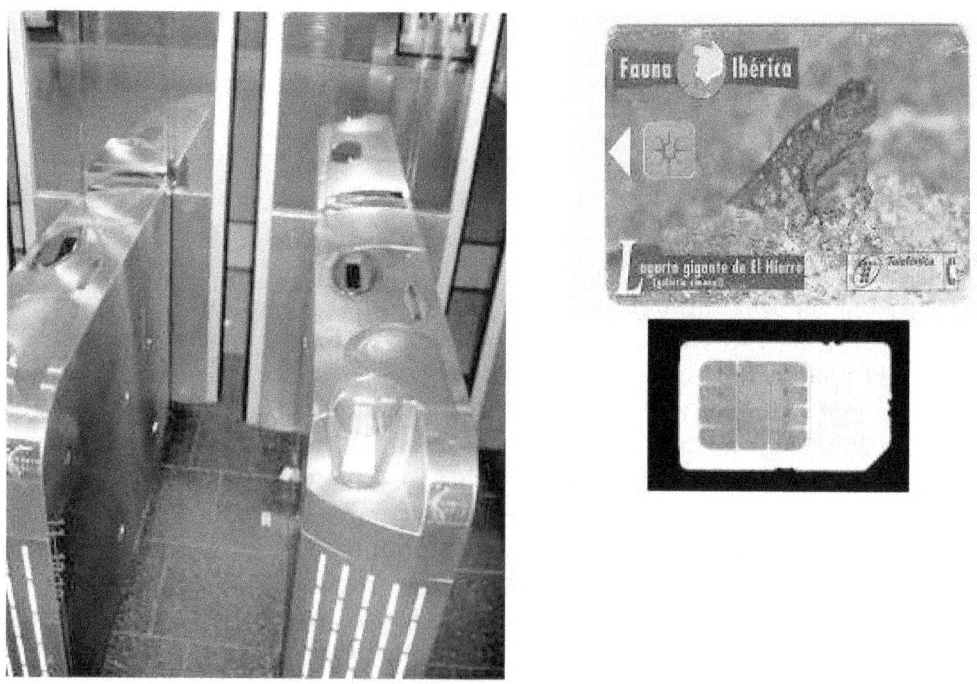

Figure 7.1 Metro electronic ticket reader with smart card reading; a typical example of a smart card; and a smart card for a GSM telephone. (*Source:* CSI Library.)

7.1.2 History of Smart Card/Personal Identification Standards

Standardization for this subject is developed by ISO/IEC JTC1/SC17. In the area of identification cards (the original role of SC17), and its expanded role of personal Identification) there are many developed standards, and for a more detailed investigation of this area, readers should visit http://www.iso.org site and search for JTC1/SC17. This section considers those personal identification standards that are most likely to be used in an ITS context.

The full work program of SC17 can be downloaded from:

http://www.iso.org/iso/iso_catalogue/catalogue_tc/catalogue_tc_browse.htm?commid=45144&development=true

The most widespread use of magnetic stripe cards and so-called smart cards is for banking cards, both credit cards and debit cards. The majority of loyalty cards also adhere to banking card standards.

Such systems adhere to the ISO/IEC 7810, 7811, 7812, and 7813 series of standards. These standards deal with all aspects of bank cards. ISO/IEC 7812 and 7816 define low and high coercivity magnetic stripes, respectively. Almost all banking cards used in the United States, United Kingdom, and Europe conform to the ISO 7810-7813 standards for magnetic cards.

While these aspects may be used in some ITS solutions where a smart card is used to provide an access or payment means, they are more relevant to equipment design and not fundamental to ITS. This section therefore provides only a list of the titles of the parts of ISO/IEC 7812 and 7816 for reference purposes, and provides detail in areas which may have direct relevance to ITS service provision.

7.1.3 Radio Regulations Environment for Smart Cards

It is important to be aware that in addition to the standards, contactless cards are affected by national and regional radio regulations.

Most modern noncontact smart cards operate at either low frequency (ISO 14443) or VHF—13.56 MHz (ISO 15693), and both use inductive coupling (i.e., the magnetic component of the emission).

The readers are therefore radio emitters and the cards, effectively, are passive transponders (see RFID below). As radio emitters they fall under radio regulations for short-range devices.

In Europe, for example, the following standards apply: (1) CEPT/ERC 70-03 Recommendation 70-03, relating the use of short range devices (SRD); and (2) EN 300 330 ERM: Short Range Devices (SRD) Radio equipment in the frequency range 9 kHz—25 MHz and inductive loop-systems in the frequency range 9 kHz—30 MHz; Part 1: Technical characteristics and test methods, and Part 2: Harmonized EN under article 3.2 of the R&TTE Directive.

In the United States they operate within FCC part 15 (see Section 4.1.2).

7.1.4 ISO 7812 Identification Cards—Identification of Issuers

ISO/IEC	IS	7812	Identification cards—Identification of issuers	Published Revised Version 2006	ISO JTC1 SC17

This standard is in three parts and is an architecture component of all personal ID cards.

ISO/IEC	IS	7812-1	Identification cards—Identification of issuers—Part 1	Published Revised Version 2006	ISO JTC1 SC17

ISO 7812-1 Available from
http://www.iso.org/iso/iso_catalogue/catalogue_tc/catalogue_detail.htm?csnumber=39698

ISO/IEC 7812-1:2006 specifies a numbering system for the identification of issuers of cards that requires an issuer identification number to operate in international, interindustry and/or intraindustry interchange.

ISO/IEC	IS	7812-2	Identification cards—Identification of issuers—Part 2: Application and registration procedures (Revision of ISO/IEC 7812-2:2000)	IS Revised Version 2007	ISO JTC1 SC17

ISO 7812-2 Available from
http://www.iso.org/iso/iso_catalogue/catalogue_tc/catalogue_detail.htm?csnumber=39699

ISO/IEC 7812-2:2007 is one of a series of International Standards describing the parameters for identification cards and the use of such cards for international and/or interindustry interchange. It describes the application and registration procedures for numbers issued in accordance with ISO/IEC 7812-1. ISO/IEC 7812-1 specifies the numbering system for the identification of issuers of identification cards used in international and/or interindustry interchange.

7.1.5 ISO 7816 Identification Cards—Integrated Circuit Cards

ISO/IEC	IS	7816	Identification cards—Integrated circuit cards		ISO JTC1 SC17

ISO 7816 is a multipart standard, much of which has to do with the physical form and layout of plastic identification cards.

ISO 7816 part 1 describes the physical characteristics of integrated circuit cards. It includes the accommodation of exposure limits for a number of electromagnetic phenomena such as x-rays, ultraviolet (UV) light, electromagnetic fields, static electrical fields, as well as the ambient temperature of the card. It also describes the characteristics of a card when it is bent or flexed and how connections between the surface connectors and the I/O pins of the embedded silicon die must be maintained to withstand mechanical stress

ISO 7816 part 2 deals with dimensions and location of contacts.

ISO 7816 Part 3 deals with smart card electronic signals and transmission protocols.

ISO 7816 Part 4 covers interindustry command for interchange including data structures, file organization, file referencing, file structures, data referencing methods, security architecture on the card, message structures, coding conventions, secure messaging, and basic interindustry commands.

The entire set of ISO 7816 standards is detailed below because even if only some parts are directly relevant to smart card use in ITS, there will be cross-references between the parts.

Parts of ISO/IEC 7816 are as follows.

ISO/IEC	IS	7816-1	Identification cards—Integrated circuit cards—Part 1: Physical characteristics	Published 1998	ISO JTC1 SC17

ISO 7816-1 Available from
http://www.iso.org/iso/iso_catalogue/catalogue_tc/catalogue_detail.htm?csnumber=29257

ISO/IEC	IS	7816-2	Identification cards—Integrated circuit cards—Part 2: Dimensions and location of the contacts	Published 1998	ISO JTC1 SC17

ISO 7816-2 Available from
http://www.iso.org/iso/iso_catalogue/catalogue_tc/catalogue_detail.htm?csnumber=26536

ISO/IEC	IS	7816-3	Identification cards—Integrated circuit cards—Part 3: Electronic signals and transmission protocols	Published 2006	ISI JTC1 SC17

ISO 7816-3 Available from
http://www.iso.org/iso/iso_catalogue/catalogue_tc/catalogue_detail.htm?csnumber=38770

ISO/IEC 7816-3:2006 specifies the power and signal structures, and information exchange between an integrated circuit card and an interface device such as a terminal. It also covers signal rates, voltage levels, current values, parity convention, operating procedure, transmission mechanisms and communication with the card.

It does not cover information and instruction content, such as identification of issuers and users, services and limits, security features, journaling and instruction definitions.

ISO/IEC	IS	7816-4	Identification cards—Integrated circuit cards—Part 4: Industry commands for interchange		ISO JTC1 SC17

ISO 7816-4 Available from
http://www.iso.org/iso/iso_catalogue/catalogue_tc/catalogue_detail.htm?csnumber=36134

ISO/IEC 7816-4:2005 specifies:

- Contents of command-response pairs exchanged at the interface;
- Means of retrieval of data elements and data objects in the card;
- Structures and contents of historical bytes to describe operating characteristics of the card;
- Structures for applications and data in the card, as seen at the interface when processing commands;
- Access methods to files and data in the card;
- A security architecture defining access rights to files and data in the card;
- Means and mechanisms for identifying and addressing applications in the card;
- Methods for secure messaging;
- Access methods to the algorithms processed by the card. It does not describe these algorithms.

It does not cover the internal implementation within the card or the outside world.

ISO/IEC 7816-4:2005 is independent from the physical interface technology. It applies to cards accessed by one or more of the following methods: contacts, close coupling, and radio frequency.

ISO/IEC	IS	7816-5	Identification cards—Integrated circuit cards—Part 5: Number system and registration procedure for application identifiers	Published 2004	ISO JTC1 SC17

ISO 7816-5 Available from
http://www.iso.org/iso/iso_catalogue/catalogue_tc/catalogue_detail.htm?csnumber=34259

ISO/IEC 7816-4 defines how to use an application identifier to ascertain the presence of and/or perform the retrieval of an application in a card.

ISO/IEC 7816-5:2004 shows how to grant the uniqueness of application identifiers through the international registration of a part of this identifier, and defines:

- The registration procedure;
- The authorities in charge thereof;
- The availability of the register which links the registered parts of the identifiers and the relevant application providers.

ISO/IEC	IS	7816-6	Identification cards—Integrated circuit cards—Part 6: Interindustry data elements	Published 2004	ISO JTC1 SC17

ISO 7816-6 Available from
http://www.iso.org/iso/iso_catalogue/catalogue_tc/catalogue_detail.htm?csnumber=38780

ISO/IEC 7816-6:2004 specifies the data elements (DEs) used for interindustry interchange based on integrated circuit cards (ICCs), both with contacts and without contacts. It gives the identifier, name, description, format, coding, and layout of each DE and defines the means of retrieval of DEs from the card.

ISO/IEC	IS	7816-7	Identification cards—Integrated circuit cards—Part 7: Interindustry commands for Structured Card Query Language (SCQL)	Published 1998	ISO JTC1 SC17

ISO 7816-7 Available from
http://www.iso.org/iso/iso_catalogue/catalogue_tc/catalogue_detail.htm?csnumber=28869

ISO/IEC	IS	7816-8	Identification cards—Integrated circuit cards—Part 8: Security related interindustry commands	Published 2004	ISO JTC1 SC17

ISO 7816-8 Available from
http://www.iso.org/iso/iso_catalogue/catalogue_tc/catalogue_detail.htm?csnumber=37989

ISO/IEC 7816-8:2004 specifies interindustry commands for integrated circuit cards (either with contacts or without contacts) that may be used for cryptographic operations. These commands are complementary to and based on the commands listed in ISO/IEC 7816-4.

Annexes are provided that give examples of operations related to digital signatures, certificates, and the import and export of asymmetric keys.

The choice and conditions of use of cryptographic mechanisms may affect card exportability. The evaluation of the suitability of algorithms and protocols is outside the scope of ISO/IEC 7816-8.

ISO/IEC	IS	7816-9	Identification cards—Integrated circuit cards—Part 9: Additional interindustry commands and security attributes (not published yet)	Published 2004	ISO JTC1 SC17

ISO 7816-9 Available from
http://www.iso.org/iso/iso_catalogue/catalogue_tc/catalogue_detail.htm?csnumber=37990

ISO/IEC 7816-9:2004 specifies interindustry commands for integrated circuit cards (both with contacts and without contacts) for card and file management (e.g., file creation and deletion). These commands cover the entire life cycle of the card, and therefore some commands may be used before the card has been issued to the cardholder or after the card has expired.

An annex is provided that shows how to control the loading of data (secure download) into the card, by means of verifying the access rights of the loading entity and protection of the transmitted data with secure messaging. The loaded data may contain, for example, code, keys, and applets.

ISO/IEC	IS	7816-10	Identification cards—Integrated circuit cards—Part 10: Electronic signals and answer to reset for synchronous cards	Published 1999	ISO JTC1 SC17

ISO 7816-10 Available from
http://www.iso.org/iso/iso_catalogue/catalogue_tc/catalogue_detail.htm?csnumber=30558

ISO/IEC	IS	7816-11	Identification cards—Integrated circuit cards—Part 11: Framework for dynamic handling of multiple application in integrated circuit cards	Published 2004	ISO JTC1 SC17

ISO 7816-11 Available from
http://www.iso.org/iso/iso_catalogue/catalogue_tc/catalogue_detail.htm?csnumber=31419

ISO/IEC 7816-11:2004 specifies the usage of interindustry commands and data objects related to personal verification through biometric methods in integrated circuit cards. The interindustry commands used are defined in ISO/IEC 7816-4. The data objects are partially defined in this International Standard, and partially imported from ISO/IEC 19785-1.

ISO/IEC 7816-11 also presents examples for enrollment and verification and addresses security issues.

ISO/IEC	IS	7816-12	Identification cards—Integrated circuit cards—Part 12—Identification cards—Integrated circuit cards—Part 12: Cards with contacts—USB electrical interface and operating procedures		ISO JTC1 SC17

ISO 7816-12 Available from
http://www.iso.org/iso/iso_catalogue/catalogue_tc/catalogue_detail.htm?csnumber=40604

ISO/IEC 7816-12:2005 specifies the operating conditions of an integrated circuit card that provides a USB interface. An integrated circuit card with a USB interface is named USB-ICC.

ISO/IEC 7816-12:2005 specifies:

- The electrical conditions when a USB-ICC is operated by an interface device—for those contact fields that are not used, when the USB interface is applied;
- The USB standard descriptors and the USB-ICC class specific descriptor;
- The data transfer between host and USB-ICC using bulk transfers or control transfers;
- The control transfers which allow two different protocols named version A and version B;
- The (optional) interrupt transfers to indicate asynchronous events;
- Status and error conditions.

ISO/IEC 7816-12:2005 provides two protocols for control transfers. This is to support the protocol T=0 (version A) or to use the transfer on APDU level (version B). ISO/IEC 7816-12:2005 provides the state diagrams for the USB-ICC for each of the transfers (bulk transfers, control transfers version A and version B). Examples of possible sequences which the USB-ICC must be able to handle are given in an informative annex.

ISO/IEC	FDIS	7816-13	Identification cards—Integrated circuit cards—Part 13: Commands for application management in a multi-application environment	Under Ballot	ISO JTC1 SC17

ISO 7816-13 Not Yet Available

There is no part 14 with any recorded status.

ISO/IEC	IS	7816-15	Identification cards—Integrated circuit cards—Part 15: Cryptographic information application	Approved Standard under Revision	ISO JTC1 SC17
ISO/IEC	CD	7816-15: 2004	Identification cards—Integrated circuit cards—Part 15: Cryptographic information application—Amendment 1: Informative Annex for ISO/IEC 7816-15 including examples of integrated use with other ISO/IEC 7816 Standards	Under Development	ISO JTC1 SC17

ISO 7816-15: Available from
http://www.iso.org/iso/iso_catalogue/catalogue_tc/catalogue_detail.htm?csnumber=35168
ISO 7816-15 Cor 1: Available from
http://www.iso.org/iso/iso_catalogue/catalogue_tc/catalogue_detail.htm?csnumber=40743
ISO 7816 Amd 1 : Not Yet Available

ISO/IEC 7816-15:2004 specifies a card application. This application contains information on cryptographic functionality. Further, ISO/IEC 7816-15:2004 defines a common syntax (in ASN.1) and format for the cryptographic information and mechanisms to share this information whenever appropriate.

ISO/IEC 7816-15:2004 supports the following capabilities:

- Storage of multiple instances of cryptographic information in a card;
- Use of the cryptographic information;
- Retrieval of the cryptographic information;
- Cross-referencing of the cryptographic information with DOs defined in ISO/IEC 7816 when appropriate;
- Different authentication mechanisms; and
- Multiple cryptographic algorithms.

7.1.6 ISO 10373 Identification Cards—Contact Cards

ISO/IEC	IS	ISO/IEC 10373	Identification cards—Test methods		ISO JTC1 SC17

Early generation IC cards are still in production for many applications and are based on ISO/IEC 10373 contact IC cards. These cards have physical electrical contacts that make a physical contact to achieve access to the card.

This series of standards comprises several parts.

ISO/IEC	IS	10373-1	Identification cards—Test methods—Part 1: General characteristic tests	Published 2006	ISO JTC1 SC17

ISO 10373-1 Available from
http://www.iso.org/iso/iso_catalogue/catalogue_tc/catalogue_detail.htm?csnumber=40682

ISO/IEC 10373-1:2006 specifies the nontechnology-specific test methods required to establish conformance of identification cards to the base (requirements) standards, for which the fundamental properties are defined in ISO/IEC 7810.

ISO/IEC	IS	10373-2	Identification cards- Test methods—Part 2: Identification cards—Test methods—Part 2: Cards with magnetic stripes	Published	ISO JTC1 SC17

ISO 10373-2 Available from
http://www.iso.org/iso/iso_catalogue/catalogue_tc/catalogue_detail.htm?csnumber=39497

This standard will only be of use if connecting within a vehicle using a magnetic IC card to identify the user.

ISO/IEC 10373-2:2006 defines test methods for the magnetic recording characteristics of identification cards according to the definitions given in base standards ISO/IEC 7811-2, ISO/IEC 7811-6, and ISO/IEC 7811-7.

ISO/IEC	IS	10373-3	Identification cards—Test methods—Part 3: Integrated circuit(s) cards	Published	ISO JTC1 SC17

ISO 10373-3 Available from
http://www.iso.org/iso/iso_catalogue/catalogue_tc/catalogue_detail.htm?csnumber=31083

ISO/IEC	IS	10373-4	Identification cards—Test methods—Part 4: Contactless integrated circuit cards	No formal status	ISO JTC1 SC17

There is no record in ISO of a part 4 with any status.

ISO/IEC	IS	ISO/IEC 10373-5	Identification cards—Test methods—Part 5:—Part 5: Optical memory cards	Published 2006	ISO JTC1 SC17

ISO 10373-5 Available from
http://www.iso.org/iso/iso_catalogue/catalogue_tc/catalogue_detail.htm?csnumber=41013

ISO 10373 defines test methods for characteristics of identification cards as defined in ISO/IEC 7810. Each test method is cross-referenced to one or more base standards, which may be ISO/IEC 7810 or one or more of the supplementary standards that define the information storage technologies employed in identification cards applications.

Criteria for acceptability do not form part of ISO 10373 but will be found in the International Standards mentioned above.

Test methods defined in ISO 10373 are intended to be performed separately. A given card is not required to pass through all the tests sequentially.

ISO 10373-1 deals with test methods which are common to one or more card technologies and other parts deal with other technology-specific tests.

ISO 10373-5:2006 deals with test methods which are specific to optical memory card technology.

ISO/IEC	IS	10373-6	Identification cards—Test methods—Part 6: Proximity cards	Published 2001	ISO JTC1 SC17

ISO 10373-6 Available from
http://www.iso.org/iso/iso_catalogue/catalogue_tc/catalogue_detail.htm?csnumber=31438

ISO/IEC	IS	ISO/IEC 10373-7	Identification cards—Test methods—Part 7: Vicinity Cards	Published 2001	ISO JTC1 SC17

ISO10373-7 Available from
http://www.iso.org/iso/iso_catalogue/catalogue_tc/catalogue_detail.htm?csnumber=31439

7.1.7 ISO 10536 Identification Cards—Contactless Integrated Circuit(s) Cards—Close-Coupled Cards

ISO/IEC	IS	10536	Identification cards—Contactless integrated circuit(s) cards—Close-coupled cards	Published 2001	ISO JTC1 SC17

This standard is in three parts.

ISO/IEC	IS	10536-1	Identification cards—Contactless integrated circuit(s) cards—Close-coupled cards—Part 1: Physical characteristics	Published 2000	ISO JTC1 SC17

ISO 10536-1 Available from
http://www.iso.org/iso/iso_catalogue/catalogue_tc/catalogue_detail.htm?csnumber=26737

ISO/IEC	IS	10536-2	Identification cards—Contactless integrated circuit(s) cards—Close-coupled cards—Part 2: Dimensions and location of coupling areas	Published 2005	ISO JTC1 SC17

ISO 10536-2 Available from
http://www.iso.org/iso/iso_catalogue/catalogue_tc/catalogue_detail.htm?csnumber=23131

This specifies the dimensions, location, nature, and assignment of each of the coupling areas to be provided for interfacing slot or surface card coupling devices (CCDs) with contactless integrated circuit(s) cards (CICCs) of the ID-1 card type.

ISO/IEC	IS	10536-3	Identification cards—Contactless integrated circuit(s) cards—Close-coupled cards—Part 3: Electronic signals and reset procedures	Published 2001	ISO JTC1 SC17

ISO 10536-3 Available from
http://www.iso.org/iso/iso_catalogue/catalogue_tc/catalogue_detail.htm?csnumber=24184

This specifies the nature and characteristics of the fields to be provided for power and bidirectional communications between card coupling devices and contactless integrated circuit(s) cards of the ID-1 card type in slot or surface operation. It is to be used in conjunction with ISO/IEC 10536-1 and ISO/IEC 10536-2.

ISO/IEC	IS	10536-4	Identification cards—Contactless integrated circuit(s) cards—Close-coupled cards—Part-4: Answer to reset and transmission protocols	Under development or withdrawn	ISO JTC1 SC17

ISO 10536-4 Not Available

This work item was created but does not exist on the ISO database. It may be in protracted development or may have been withdrawn. There are no further details available.

7.1.8 ISO 14443 Identification Cards—Contactless Integrated Circuit(s) Cards—Proximity [Contactless] Cards

ISO/IEC	IS	ISO/IEC 14443	Identification cards—Contactless integrated circuit(s) cards—Proximity cards	Published	ISO JTC1 SC17

A second generation of cards (ISO/IEC 14443) relies on optical connection, or on the presence of a magnetic coupling. These cards have the advantage that they do not need physical electrical contact; they just have to be in close proximity. This series of standards is the most widely used of the smart card standards, and in ITS is very widely used for electronic public transport ticketing and in some electronic fee collection systems.

ISO/IEC	IS	ISO/IEC 14443-1	Identification cards—Contactless integrated circuit(s) cards—Proximity cards—Part 1: Physical characteristics	Published 2001	ISO JTC1 SC17

ISO 14443-1 Available from
http://www.iso.org/iso/iso_catalogue/catalogue_tc/catalogue_detail.htm?csnumber=28728

ISO/IEC	IS	ISO/IEC 14443-2	Identification cards—Contactless integrated circuit(s) cards—Proximity cards—Part 2: Radio frequency interface power and signal interface	Published 2001	ISO JTC1 SC17

ISO 14443-2 Available from
http://www.iso.org/iso/iso_catalogue/catalogue_tc/catalogue_detail.htm?csnumber=28729

ISO/IEC	IS	ISO/IEC 14443-3	Identification cards—Contactless integrated circuit(s) cards—Proximity cards—Part 3: Initialization and anti-collision	Published 2001	ISO JTC1 SC17

ISO 14443-3 Available from
http://www.iso.org/iso/iso_catalogue/catalogue_tc/catalogue_detail.htm?csnumber=28730

ISO/IEC	IS	ISO/IEC 14443-4	Identification cards—Contactless integrated circuit(s) cards—Proximity cards—Part 4: Transmission protocols	Published 2001	ISO JTC1 SC17

ISO 14443-4 Available from
http://www.iso.org/iso/iso_catalogue/catalogue_tc/catalogue_detail.htm?csnumber=31425

7.1.9 ISO 15457 Identification Cards—Thin Flexible Cards

ISO/IEC	IS	15457-1	Identification cards—Thin flexible cards—Part 1: Physical characteristics	Published 2001	ISO JTC1 SC17

ISO 15457-1 Available from
http://www.iso.org/iso/iso_catalogue/catalogue_tc/catalogue_detail.htm?csnumber=27700

Only one part of this standard has ever been published (2001).

7.1.10 ISO 15693 Identification Cards—Contactless Integrated Circuit(s) Cards—Vicinity Cards

ISO/IEC	IS	15693	Identification cards—Contactless integrated circuit(s) cards—Vicinity cards	Approved	ISO JTC1/ SC17

ISO/IEC 15693 forms part of a series of International Standards that specify a contactless smart card. The card can be carried by members of the public in a purse or wallet and when presented at or near a terminal device can give access to places, goods, or services. In addition, the card can be attached to objects like bags and valuable items, which can then be tracked while in the vicinity of a reading device.

The ISO/IEC 15693 series of International Standards takes the contactless option further and extends the effective range of IC cards to several centimeters, and in some cases to 0.75m. Extension beyond this range is not possible because of power emission regulations around the world combined with the decay curve of the magnetic field.

This series of standards is the most capable of the IC card standards with respect to read range. In an RFID environment it is also encapsulated in ISO/IEC 18000-3 MODE 1. See Section 7.4.9.

ISO/IEC 15693 is one of a series of International Standards describing the parameters for identification cards as defined in ISO/IEC 7810 and the use of such cards for international interchange.

Contactless card standards cover a variety of types as embodied in ISO/IEC 10536 (Close-coupled cards), ISO/IEC 14443 (Proximity cards), and ISO/IEC 15693 (Vicinity cards). These are intended for operation when very near, nearby, and at a longer distance from associated coupling devices, respectively.

ISO/IEC 15693 is intended to allow operation of vicinity cards in the presence of other contactless cards conforming to ISO/IEC 10536 and ISO/IEC 14443 standards.

ISO/IEC	IS	15693-1	Identification cards—Contactless integrated circuit(s) cards—Vicinity cards—Part 1: Physical characteristics	Published 2000	ISO JTC1 SC17

ISO 15693-1 Available from
http://www.iso.org/iso/iso_catalogue/catalogue_tc/catalogue_detail.htm?csnumber=30995

This part of ISO/IEC 15693 specifies the physical characteristics of vicinity cards (VICC). It applies to identification cards of the card type ID-1 operating in vicinity of a coupling device.

ISO/IEC	IS	15693-2	Identification cards—Contactless integrated circuit cards—Vicinity cards—Part 2: Air interface and initialization	Published Revised 2006	ISO JTC1 SC17

ISO 15693-2 Available from
http://www.iso.org/iso/iso_catalogue/catalogue_tc/catalogue_detail.htm?csnumber=39695

ISO/IEC 15693-2:2006 defines the power and communications interface between the vicinity card and the reading device. Other parts of ISO/IEC 15693 define the physical dimensions of the card and the commands interpreted by the card and reader.

Power is coupled to the vicinity card by an ac field produced in the reader, also known as a coupler; the powering field has a frequency of 13.56 MHz and is one of the industrial, scientific, and medical (ISM) frequencies available for worldwide use. When sufficient power is received by the card, it is able to respond to commands sent from the coupler. The coupler sends commands to the card by modulating the powering field and by using a modulation system known as pulse position modulation, whereby the position of a single pulse relative to a known reference point codes the value of a nibble or byte of data. This allows the card to draw the maximum energy from the field almost continuously. Vicinity cards, which have no power source, can be energized at ranges of up to 1m from a coupler that can only transmit power within the limits permitted by international radio frequency (RF) regulations.

A vicinity card only responds when it receives a valid command that selects a single card from a possible collection of cards within range of the coupler. This process of collision detection and selection, also known as anticollision, is made possible by detecting the unique identification number encoded into every card. Anticollision and the commands used are defined in ISO/IEC 15693-3. The card responds to the coupler by drawing more or less power from the field and generates one or two subcarriers of around 450 kHz. These are switched on and off to provide Manchester-encoded data that are then detected by the coupler.

Thus, both power and bidirectional communications form the air interface between the vicinity card and the coupler. It is the flexibility of the interface to select one or two subcarriers when communicating from card to coupler, while also using slow or fast data rates from the coupler to the card, that allows systems to be tuned to suit different operational requirements ranging from use with high RF noise at short range to low RF noise at long range.

This part of ISO/IEC 15693 describes the electrical characteristics of the contactless interface between a vicinity card and a vicinity coupling device. The interface includes power and bidirectional communications.

This International Standard does not preclude the incorporation of other standard technologies on the card.

This part of ISO/IEC 15693 specifies the nature and characteristics of the fields to be provided for power and bidirectional communications between vicinity coupling devices and vicinity cards.

ISO/IEC 15693-2 does not specify the means of generating coupling fields nor the means of compliance with electromagnetic radiation and human exposure regulations, which can vary according to country regulations and/or standards.

ISO/IEC	IS	15693-3	Identification cards—Contactless integrated circuit(s) cards—Vicinity cards—Part 3: Anti-collision and transmission protocol	Published	ISO JTC1 SC17

ISO 15693-3 Available from
http://www.iso.org/iso/iso_catalogue/catalogue_tc/catalogue_detail.htm?csnumber=31428

This part of ISO/IEC 15693 describes the anti-collision and transmission protocols.

This International Standard does not preclude the incorporation of other standard technologies on the card.

This part of ISO/IEC 15693 describes:

- Protocol and commands;
- Other parameters required to initialize communications between a VICC and a VCD;
- Methods to detect and communicate with one card among several cards (anti-collision);
- Optional means to ease and speed up the selection of one among several cards based on application criteria.

7.1.11 ISO 18013 Personal Identification—ISO-Compliant Driving License

ISO/IEC	FCD	18013	Information technology—Personal identification—ISO-compliant driving license	Published 2005	ISO JTC1 SC17

This work item is being developed as a three-part standard.

ISO/IEC	FCD	18013-1	Information technology—Personal identification—ISO-compliant driving license—Part 1: Physical characteristics and basic data set	Published 2005	ISO JTC1 SC17

ISO 18013-1 Available from
http://www.iso.org/iso/iso_catalogue/catalogue_tc/catalogue_detail.htm?csnumber=41920

ISO/IEC 18013-1:2005 establishes the design format and data content of an ISO-compliant driving license (IDL) with regard to the human-readable (visual) features and the placement of ISO machine-readable technologies on the card. It creates a common basis for international use and mutual recognition of the IDL without restricting individual domestic or regional driver licensing authorities from incorporating their specific needs on the IDL.

The intent of the ID-1 sized IDL is to allow one document to serve the purpose of both what is currently known amongst driver licensing authorities as a domestic driving permit and an international driving permit (IDP). Thus the IDL replaces the need for two separate documents. Alternatively, those countries that choose to maintain their individual domestic design can issue a second card (with or without ISO machine-readable technologies), a domestic driving license (DDL), while the IDL serves to replace the current IDP paper document only. ISO/IEC 18013-1:2005 also specifies an explanatory booklet with sleeve insert pocket that may optionally accompany an IDL to facilitate its worldwide interpretation when used instead of an IDP.

ISO/IEC	FCD	18013-2	Information technology—Personal identification—ISO-compliant driving license—Part 2: Machine readable technologies	Under Development FCD	ISO JTC1 SC17

ISO 18013-2 FCD Track Status at
http://www.iso.org/iso/iso_catalogue/catalogue_tc/catalogue_detail.htm?csnumber=42199

ISO 18013-2 describes the technologies that may be used for the standard, including the logical data structure and data mapping for each technology.

This International Standard establishes guidelines for the design format and data content of an ISO compliant driving license (IDL) in regard to both human-readable features and ISO machine readable technologies. It creates a common basis for international use and mutual recognition of the IDL without impeding individual national/community/regional motor vehicle authorities in taking care of their specific needs.

ISO/IEC	FCD	18013-3	Information technology—Personal identification—ISO-compliant driving license—Part 3: Access control, authentication and integrity validation.	Under Development FCD	ISO JTC1 SC17

ISO 18013-3 FCD Track Status at
http://www.iso.org/iso/iso_catalogue/catalogue_tc/catalogue_detail.htm?csnumber=50351

While this work item is officially recorded as "Access control, authentication and integrity validation," internally, the SC17 work program describes the work item as "Biometrics, Image Processing and Cryptography." No documents are publicly available.

7.1.12 ISO 24727 Identification Cards—Integrated Circuit Card Programming Interfaces

ISO/IEC	FDIS	24727	Identification cards—Integrated circuit card programming interfaces	Under Development	ISO JTC1 SC17

ISO/IEC 24727 provides a set of programming interfaces for interactions between integrated circuit cards and applications to include multisector use of generic services for identification, authentication, and signature. ISO/IEC 24727 is specifically relevant to identity management applications desiring interoperability among diverse application domains. This standard defines interfaces such that independent implementations are interoperable. Card application and associated services are discoverable without the need for proprietary information. The organization and the operation of the ICC conform to ISO/IEC 7816-4.

Although not yet finalized, this standard has been publicly adopted by the European community for the European Citizens Card and by Australia for their citizen social services card. It also may be used for the U.S. NIST Personal Identity Verification Program (NPIVP). The mission of the NPIVP is to validate personal identity verification (PIV) components required for interfaces for personal identity verification.

ISO/IEC	FDIS	24727-1	Identification cards—Integrated circuit card programming interfaces Part 1: Architecture	Published 2007	ISO JTC1 SC17

ISO 24727-1 Available from
http://www.iso.org/iso/iso_catalogue/catalogue_tc/catalogue_detail.htm?csnumber=38837

ISO/IEC 24727-1 specifies the framework and supporting mechanisms and interfaces. It provides essential background information for the subsequent parts.

ISO/IEC 24727-1:2007 specifies:

- System architecture and principles of operation;
- A capabilities discovery mechanism;
- Security rationale.

It does not cover the internal implementation within the card or the outside world.

ISO/IEC 24727-1:2007 is independent of the physical interface technology. It applies to cards accessed by one or more of the following methods: contacts, close coupling, and radio frequency.

ISO/IEC	FCD	24727-2	Identification cards—Integrated circuit card programming interfaces—Part 2: Generic card interface	Under Development FCD	ISO JTC1 SC17

ISO 24727-2 Available from
http://www.iso.org/iso/iso_catalogue/catalogue_tc/catalogue_detail.htm?csnumber=43308

ISO/IEC 24727-2 provides a generic card interface. It details the functionality and related information structures available to the implementation of the application interface defined in ISO/IEC 24727-3.

ISO/IEC	CD	24727-3	Identification cards—Integrated circuit card programming interfaces—Part 3: Application interface	Under Development CD	ISO JTC / SC17

ISO 24727-3 CD Track Status at
http://www.iso.org/iso/iso_catalogue/catalogue_tc/catalogue_detail.htm?csnumber=43809

ISO/IEC 24727-3 details service access mechanisms for use by any application to include authentication protocols that are in use by identity systems (i.e., PIN, biometric, symmetric key). It provides a common application programming interface (API) and interoperable authentication protocols.

ISO/IEC	NP	24727-4	Identification Cards—Programming Interfaces for Integrated Circuit Cards—Part 4: API administration	Under Development CD	ISO JTC1 SC17

ISO 24727-4 Available from
http://www.iso.org/iso/iso_catalogue/catalogue_tc/catalogue_detail.htm?csnumber=43358

ISO/IEC 24727-4 details the security model and interface for secure messaging within the framework. It provides API administration between Part 2 and Part 3.

The scope of 24727-4 is to provide "Interoperable, functional and operational interconnectivity between host-based client applications and card-based services, including consistent API interconnectivity, end-to-end security policy specification, realization and administration, and administration of interoperable discovery and operational mechanisms."

ISO/IEC	NP	24727-5	Identification Cards—Programming Interfaces for Integrated Circuit Cards—Part 5: Testing	Under Development NP/WD	ISO JTC1 SC17

ISO 24727-5 Not Yet Available

ISO/IEC 24727-5 will contain testing requirements for ensuring compliance.

7.1.13 ISO 24749 Identification Cards—Secure and Interoperable IC Card Transaction Device

ISO/IEC	NP	24749	Identification cards—Secure and interoperable IC card transaction device	Under Development NP/WD	ISO JTC1 SC17

ISO 24749 Not Yet Available

Little information is available on this work item other than its title, and that to date, only one part, 24749-1, Architecture, is under development. It will be important as security becomes an ever more important issue in wireless communications

7.1.14 ISO 24787 On-Card Fingerprint Matching

ISO/IEC	NP	NP 24787	On-Card finger print matching	Under Development	ISO JTC/ SC17

ISO CD Track Status at
http://www.iso.org/iso/iso_catalogue/catalogue_tc/catalogue_detail.htm?csnumber=41557

This document was undergoing its first (CD) ballot in late 2007. This standard will also be important as security becomes an ever more important issue in wireless communications.

7.2 Biometric Identification

The relevance of biometric information to ITS service provision is primarily concerned with driver identification, user identification, and access control. However, as the ITS sector evolves, other application services may well require biometric identification or monitoring of biometrics.

Traditionally, two major types of automatic personal identification approaches have been widely used:

- Token-based;
- Knowledge-based.

Token-based approaches use *something that you have* to make a personal identification, such as a passport, driver's license, ID card, credit card, or keys.

Knowledge-based approaches use *something that you know* to make a personal identification, such as a password and personal identification number (PIN).

Since these traditional approaches are not based on any inherent attributes of an individual to make a personal identification, they suffer from a number of disadvantages. Tokens may be lost, stolen, forgotten, or misplaced; a PIN may be forgotten or guessed by impostors. All of these approaches are also unable to differentiate between an authorized person and an impostor who fraudulently acquires the token or knowledge of the authorized person. Therefore, knowledge-based and token-based approaches are unable to satisfy the security requirements of our electronically interconnected information society.

By contrast, "biometrics" refers to identifying an individual based on his physiological or behavioral characteristics (biometric identifiers), and relies on "who you are" or "what you do" to make a positive personal identification. It is inherently more reliable and more capable than knowledge-based and token-based techniques in differentiating between an authorized person and a fraudulent impostor, because many of the physiological or behavioral characteristics are distinctive to each person.

Regardless of which biometrics technology we refer, a biometric system is essentially a pattern recognition system which makes a personal identification by establishing the authenticity of a specific physiological or behavioral characteristic possessed by the user.

Generically a biometric system is depicted in two phases:

- Enrollment module;
- Identification module.

The enrollment module is responsible for enrolling individuals into the biometric system. During the enrollment phase, the biometric characteristic of an individual is first scanned by a biometric sensor to acquire a digital representation of the characteristic.

In order to facilitate matching and to reduce the storage requirements, the digital representation is further processed by a feature extractor to generate a compact but expressive representation, called a template. Depending on the application, the template may be stored in the central database of the biometric system or be recorded on a magnetic card or smart card issued to the individual. The identification module is responsible for identifying individuals at the point-of-access. During the operation phase, the biometric reader captures the characteristic of the individual to be identified and converts it to a digital format, which is further processed by the feature extractor to produce the same representation as the template. The resulting representation is fed to the feature matcher, which compares it against the template(s) to establish the identity of the individual.

An ideal biometric should meet the following requirements:

- *Universality:* each person should have the characteristic;
- *Uniqueness:* no two persons should be the same in terms.

While typically more secure, biometric techniques can add significant complexity to an installation, as follows:

- The most obvious consideration is up-front cost for the hardware and initial implementation.
- The database of biometric information and its security is another important consideration. If compromised, fake biometric data could be added to allow inappropriate accesses.
- The level of security needed is important. The greater the security, the less tolerance there will be for false positives and the more complex the biometric mapping will likely be.
- Where the biometric template for an individual user is stored will also influence system development. If stored in a central database, there may be processing delays as entire databases get searched. If stored on local machines there are security considerations as well as problems when those machines go down or individuals move to a new location. Some implementations store the template on a smart card; but then the loss of that card must be considered in the implementation.

Another consideration in implementing a biometric system is the "privacy" consideration. The common belief is that biometric data about an individual consists of things like their fingerprint or a picture of their retina. In actual fact, biometric systems typically only store a mathematical representation of the item. Users often must be convinced of this before they will accept and use a biometric identification system.

A biometric system may be either a *verification* (authentication) system or an *identification* system.

A verification system authenticates a person's identity by comparing the captured biometric characteristic with her own biometric template(s) prestored in the system. In a verification (authentication) system, the target individual to be identified submits a claim to an identity to the system usually via a magnetic stripe card, login name, or smart card, and the system either rejects or accepts the submitted claim of identity.

An identification system recognizes an individual by searching the entire template database for a match. In an identification system, the system establishes a subject's identity (or fails if the subject is not enrolled in the system database) without the subject having to claim an identity.

The overall performance of a biometric system is assessed in terms of its

- Accuracy;
- Speed;
- Storage.

Other factors such as cost and ease-of-use also affect the practicability of biometric identification systems.

7.2.1 Fingerprint Recognition

A fingerprint is the pattern of ridges and furrows on the surface of a fingertip. Its formation is determined during the fetal period. Fingerprints of identical twins are different and there is no correlation between the prints on the different fingers of an individual. Fingerprints are one of the most understood and studied biometrics. Humans have used fingerprints for personal identification for centuries and the validity of fingerprint identification has been well established. With the development of solid-state sensors, the marginal cost of incorporating a fingerprint-based biometric system is now affordable in many applications, such as a replacement driver validation in place of an ignition key, or for other access control systems. Fingerprints have sufficient information to allow large scale identification. One problem with fingerprint technology is its acceptability by a typical user, because fingerprints have traditionally been associated with criminal investigations and police work. People may feel uncomfortable in using fingerprints in civilian applications.

Another problem with fingerprint technology is that automatic fingerprint identification generally requires a large amount of computational resources. Fingerprints of a small fraction of the population may be unusable for automatic identification because of genetic, aging, environmental, or occupational reasons.

Typically, 90 or more reference points are determined when the fingerprint is first registered. These points are linked via a mathematical formula and the results are stored; not the fingerprint itself. When reading a fingerprint, the system runs the same calculations to verify the user's identification. Some sort of heat sensor is also usually used to make certain the hand is alive.

7.2.2 Eye Recognition

The same sort of calculations used for fingerprints are used for eye scanners. The most common eye scanner scans the iris of the eye and is implemented via a camera about 18-inches from the eye. However, this may limit the applicability of this technology with respect to many ITS applications.

Iris scanning does not require the person to interact close coupled with a device; a video image of the eye can be taken from 1 foot away. This has obvious benefits in applications like the one to positively identify for access control. The user's iris pattern is reflected back to the camera, which captures the unique pattern and stores it using less than 35 bytes of information. However, the success rate is directly relational to the closeness of the reader to the eye.

Retinal scanners, on the other hand, usually require the user to place their eye into some sort of device and then ask the user to look at a particular spot so the retina can be clearly imaged.

With current generation equipment, software running in a handheld computer can search for a suitable image at the rate of about 20 frames/second and captures the first image that meets the criteria for focus, freedom from artifacts, proper orientation, and sufficient information for analysis. The retinal vessels can be analyzed by simple algorithms, thus speeding search times and reducing costs. As processors become faster, the response time improves.

7.2.2.1 Retinal Pattern

Retinal vascular patterns have been known for some time as among the most distinctive, robust, and easily measured biometrics in human beings. It has been proven that the same applies to other mammals and many species of animals.

Every retinal vascular pattern is highly distinct, making the likelihood of misidentification essentially zero, even in very large populations. These patterns are easily captured via a digital video camera and handheld computer system designed especially for use in livestock. Additionally the vessels are quite easy to visualize and digitize, making small, readily manipulated files. All members of each species share a common structure that can be used both to align the images in the cameras and also to normalize each image for comparison purposes.

The pattern formed by veins beneath the retinal surface in an eye is stable and unique. Digital images of retinal patterns can be acquired by projecting a low-intensity beam of visual or infrared light into the eye and an image of the retina thus illuminated is captured using optics similar to a retinascope. In order that a fixed portion of the retinal vasculature is used for identification, the subject is required to closely gaze into an eye-piece and focus on a predetermined spot in the visual field. Retinal pattern based identification is very accurate. The degree of user cooperation/involvement required in imaging a retina may not be acceptable to the subjects undergoing identification. Retinal scan is currently perceived to be the most secure biometric technique. A large number of retinal scan-based biometric systems have been installed in several highly secure environments (e.g., prisons). A disadvantage of this biometric is that retinal scanners are expensive.

7.2.2.2 Iris

Iris is the annular region of eye bounded by pupil and sclera (white of the eye) on either side. Visual texture of iris stabilizes very early in life (first 2 years) and its complex structure carries very distinctive information useful for identification of individuals. Initial available results on accuracy and speed of iris-based identification are extremely promising and point to the feasibility of a large-scale identification using iris information. Each iris is unique and even irises of identical twins are different. Iris is more readily imaged than retina. Although the early iris-based identification systems needed considerable user participation and were expensive, efforts are underway to build more user-friendly and cost-effective versions. It remains to be seen how this relatively recently discovered biometric matures and gains public acceptance. It is extremely difficult to surgically tamper iris texture information and it is easy to detect artificial irises (e.g., designer contact lenses).

7.2.3 Facial Recognition

Like iris scanning, facial feature identification systems can capture images from a distance (several meters) by using video equipment. As in other more complex systems, the challenge is achieving high levels of performance as the size of the database increases. The potential of these systems is generating much interest. Increased development efforts are needed in the areas of multimedia video technology and the complex software that facial identification requires.

Facial images are probably the most common biometric characteristic used by humans to make a personal identification. Face recognition is one of the most active areas of research, with applications ranging from static, controlled mug shot verification to dynamic, uncontrolled face identification in a cluttered background.

Face recognition is a nonintrusive technique and people generally do not have any problem in accepting face as a biometric characteristic.

During the past quarter century, a substantial amount of research effort has been devoted to face recognition. Approaches to face recognition are typically based on either on the location and shape of facial attributes such as eyes, eyebrows, nose, lips, chin shape, and so on, and their spatial relationships or are based on an overall (global) analysis of the face image and the decomposition of face into a number of canonical faces.

A number of commercial face recognition systems are available. While the performance of these systems is reasonable, it is questionable whether the face itself, without any contextual information, is sufficiently effective to make a personal identification with a high level of confidence. Further, current face recognition systems impose a number of restrictions on how the facial images are acquired (e.g., simple background, uniform and fixed illumination). In order for the face recognition to be widely adopted, they should automatically detect whether there exists a face in the acquired image, locate the face if there is one, and recognize the face from a general viewpoint.

7.2.4 Facial Thermogram

The underlying vascular system in the human face produces a unique facial signature when heat passes through the facial tissue and is emitted from the skin. Such facial signatures can be captured using an infrared camera, resulting in an image called a face thermogram. It is believed that a face thermogram is unique to each individual and they are not vulnerable to disguises. Even plastic surgery, which does not reroute the flow of blood through the veins, cannot change the formation of the face thermogram. An infrared camera can capture the face thermogram in low/changing ambient light or in the absence of any light at all, which greatly reduces the restrictions on how face thermograms are acquired.

Face thermogram is a nonintrusive biometric technique which can verify an identity without contact, without full camera view, and without the cooperation of subjects. It is claimed that face thermogram-based recognition is superior to face recognition using CCD cameras.

Although it may be true that face thermograms are unique to each individual, it has not been proven that face thermograms are sufficiently discriminative. Face thermograms depend heavily on a number of factors such as the emotion of the subjects and body temperature, and as with face recognition, face thermogram recognition is view-dependent.

7.2.5 Hand Geometry

A variety of measurements of the human hand including its shape and lengths and widths of the fingers can be used as biometric characteristics. Hand geometry-based biometric systems have already been installed at thousands of locations around the world.

The technique is very simple, relatively easy to use, and is inexpensive. Operational environmental factors (e.g., dry weather) or individual anamolies (e.g., dry skin) generally have virtually no negative effects on the identification accuracy. It also does not appear to be a problem for people to accept this technology.

However, a significant disadvantage of this technique is its low discriminative capability (it is very difficult for a hand geometry-based biometric system to achieve a very high identification accuracy especially for a large population). The physical size of a hand geometry-based system is also large, which may restrict it from certain applications such as local access control systems.

The existing systems also have problems sensing the hand geometry measurements from individuals with limited dexterity (e.g., due to arthritis). Hand geometry information may not remain invariant permanent biometric characteristic over the lifespan of an individual, especially during childhood. Different types of rings may pose additional challenges in extracting the correct hand geometry information.

7.2.6 Hand Vein

Hand veins provide a very robust and stable pattern that can be used as a biometric characteristic to make a personal identification. Digitized images of hand vein patterns can be easily captured with an infrared camera. Hand vein patterns are unique to each individual. It is very difficult to change the formation of the hand vein pattern of an individual by surgery. Thus, hand vein–based technique is very efficient in circumventing fraudulent attempts. A hand vein–based biometric system has the potential to achieve reasonable identification accuracy and people are normally willing to accept it. However, there is no hand vein–based biometric system available that is able to demonstrate its superior capability in conducting automatic personal identification. Like hand geometry, it might be very difficult for a hand vein–based biometric system to achieve a very high identification accuracy. The physical size of a hand vein–based system is large. Hand vein patterns could be obliterated due to medical conditions (e.g., obesity, aging).

7.2.7 Signature

Each person has a unique style of handwriting. No two signatures of a person are exactly identical; the variations from a typical signature also depend upon the physical and emotional state of a person. Despite the variations in an individual's signatures, a few successful systems for signature-based authentication systems have been designed. The authentication accuracy of signature-based biometric systems is reasonable but does not appear to be sufficiently high to lead to large scale identification.

There are two approaches to signature verification: static and dynamic.

Static signature verification uses only the geometric (shape) features of a signature. Dynamic signature verification uses both the geometric (shape) features and the dynamic features such as acceleration, velocity, and trajectory profiles of the signature. An inherent advantage of a signature-based biometric system is that the signature has been established as an acceptable form of personal identification method and can be "transparently" incorporated into the existing business processes requiring signatures (e.g., credit card transactions). Another advantage of signature is that it is impossible for an impostor to obtain the dynamics information from a written signature.

7.2.8 Voice Recognition

One of the easiest methods to implement, voice recognition is often not particularly useful. People's voices change over time with the onset of colds or other diseases and background noise can hamper the identification task.

The characteristics of human speech are determined by the shape/size of the appendages (e.g., vocal tracts, mouth, nasal cavities, lips) synthesizing the sound. Speech of a person is distinctive but may not contain sufficient information to offer speech-based identification.

Speech-based verification could be either a text-dependent verification or a text-independent verification. A text-dependent verification authenticates the identity of an individual based on utterance of a fixed predetermined phrase. A text-independent verification verifies the identity of a speaker independent of the phrase, which is more difficult than a text-dependent verification but offers more protection against fraud. Existing commercial speaker verification systems can achieve reasonable identification accuracy.

Generally, people are willing to accept a speech based biometric system. However, speech-based features are sensitive to a number of factors such as background noise as well as the emotional and physical state of the speaker. In addition, some people seem to be extraordinarily skilled in mimicking other's voice which may be a reason why speech-based authentication is perceived to be of low-security.

7.2.9 DNA Techniques

Deoxyribonucleic acid (DNA) is found in every cell of every creature, and it contains the information for carrying out the activities of the cell. Since every person's DNA structure is completely unique, DNA analysis is a very accurate way of proving identification. Due to the extensive testing and advanced technology required, it is not the most cost efficient biometric science, but when a positive identification is needed it is the most reliable. For practical commercial automatic identification in an ITS context, it has to be discounted, at least for the time being. DNA is a laboratory technique, and within today's technology not something that can be employed in the day-to-day commercial field.

7.2.10 Principal Biometrics Standards

Biometric equipment is largely stand-alone and proprietary. However, many of the basic building blocks of biometrics are in the process of being standardized because of the complexity and because there is a risk that de facto standards will emerge based simply on the content of large databases or widely deployed biometric devices.

Globally, the ISO Joint Technical Committee has formed a biometric standards subcommittee. Known as Subcommittee 37 (SC37), it first met in December 2002. About 20 countries have joined SC37 through their national standards bodies.

The full work program for SC37 can be downloaded from:

http://www.iso.org/iso/iso_catalogue/catalogue_tc/catalogue_tc_browse.htm?commid=313770&development=true

7.2.10.1 ISO 24787 On-Card Fingerprint Matching

ISO/IEC	NP	24787	On-Card fingerprint matching	Under Development	ISO JTC1 SC17

See Section 7.1.14.

7.2.10.2 ISO 19784 Information Technology—Biometric Application Programming Interface

ISO/IEC	FPD	19784	Information technology—Biometric application programming interface	Under Development	ISO JTC1 SC37

This series of standards is in three parts.

ISO/IEC	FPD	19784-1	Information technology—Biometric application programming interface—Part 1: BioAPI specification—Amendment 1: BioGUI	Published 2006	ISO JTC1 SC37

ISO 19784-1 Available from
http://www.iso.org/iso/iso_catalogue/catalogue_tc/catalogue_detail.htm?csnumber=33922

ISO/IEC 19784-1:2006 provides a defined interface that allows a software application to communicate with (utilize the services of) one or more biometric technologies. It includes a high-level generic biometric authentication model suited to a broad range of biometrically enabled applications and to most forms of biometric technology.

An architectural model is described which enables components of a biometric system to be provided by different vendors and to interwork through fully defined Application Programming Interfaces (APIs), corresponding Service Provider Interfaces (SPIs), and associated data structures. ISO/IEC 19784-1:2006 covers the basic biometric functions of enrollment, verification, and identification, and includes a database interface to allow an application to manage the storage of biometric records.

Conformance requirements are identified and informative annexes, including sample code, are provided. ISO/IEC 19784-1:2006 specifies a biometric data structure which is compatible with ISO/IEC 19785 and 19794.

A proposal for an amendment to provide improvement of the Application-Controlled GUI of the BioAPI specification and harmonize the GUI specification with the latest revision of the BioAPI and Common Biometric Exchange Formats Framework (CBEFF) is under discussion, but no deliverables have yet appeared.

ISO/IEC	FDIS	19784-2	Information technology—Biometric application programming interface—Part 2: Biometric archive function provider interface	Published 2007	ISO JTC1 SC37

ISO 19784-2 Available from
http://www.iso.org/iso/iso_catalogue/catalogue_tc/catalogue_detail.htm?csnumber=43303

ISO/IEC 19784-2:2007 defines the interface between a biometric service provider (BSP) and a biometric archive function provider (BAFP) for BioAPI. A BAFP encapsulates all functionality for the storage, search, and management of biometric reference data regardless of the kind of physical storage media. Using a BAFP, a BSP does not have to provide special handling of different storage media like database servers, smartcards, and database Web services. Whatever media is used, the BSP in all cases handles the same interface for a BAFP.

The interface description contains management functions to attach and detach different BAFPs, to query biometric data records, and to store biometric data records.

ISO/IEC	NP	19784-3	Information technology—Biometric application program interface—Part 3: BioAPILite	Under Development	ISO JTC1 SC37

ISO 19784-3 Not Yet Available. Track Progress at:
http://www.iso.org/iso/iso_catalogue/catalogue_tc/catalogue_detail.htm?csnumber=43614

No other information is known at this stage.

7.2.10.3 ISI 19785 Information Technology—Common Biometric Exchange Formats Framework

ISO/IEC	FCD	19785-1	Information technology—Common Biometric Exchange Formats Framework—Part 1: Data element specification	Published 2006	ISO JTC1 SC37

ISO 19785-1 Available from
http://www.iso.org/iso/iso_catalogue/catalogue_tc/catalogue_detail.htm?csnumber=41047

ISO/IEC 19785-1:2006 defines a basic structure for standardized biometric information records (BIRs) within the Common Biometric Exchange Formats Framework (CBEFF). This structure consists of three parts: the standard biometric header

(SBH), the biometric data block (BDB), and the security block (SB). CBEFF also defines several data elements and their standardized abstract values that can be used in SBHs and SBs (CBEFF treats the BDB as opaque data). CBEFF also establishes mechanisms by which organizations, called "patrons" by CBEFF, can specify and publish BIR format specifications, which are in turn called "patron formats." CBEFF enables patrons to develop BIR specifications that are fully standardized and interoperable yet are specifically adapted to the requirements of a particular application environment.

CBEFF defines rules for BIRs that contain only one BDB (simple BIR) and that contain at least one BDB (complex BIR). CBEFF defines mandatory data elements that identify the format of a BDB and its security attributes (encryption and integrity). All the other CBEFF-defined data elements and abstract values are optional. CBEFF enables patrons to define additional data elements and abstract values as required by the application environment.

ISO/IEC	FCD	19785-2	Information technology—Common Biometric Exchange Formats Framework—Part 2: Procedures for the operation of the Biometric Registration Authority	Under Development	ISP JTC1 SC37

ISO 19785-2 Available from
http://www.iso.org/iso/iso_catalogue/catalogue_tc/catalogue_detail.htm?csnumber=41048

ISO/IEC 19785-2:2006 specifies the requirements for the operation of the Biometric Registration Authority within the CBEFF.

The Registration Authority is responsible for assigning and publishing, via its Web site, unique biometric organization identifier values to organizations that own or are otherwise responsible for standardized or proprietary format specifications for biometric data blocks, biometric information record security blocks, and/or CBEFF patron formats, and to organizations that intend to assign biometric product identifier values to their products.

ISO/IEC	FCD	19785-3	Information technology—Common Biometric Exchange Formats Framework—Part 3: Patron format specifications	Under Development FDIS	ISO JTC1 SC37

ISO 19785-3 Not Yet Available. Track progress at:
http://www.iso.org/iso/iso_catalogue/catalogue_tc/catalogue_detail.htm?csnumber=43615

7.2.10.4 ISO 19794 Information Technology—Biometric Data Interchange Formats

ISO/IEC	CD	19794	Information technology—Biometric data interchange formats		ISO JTC1 SC37

ISO/IEC 19794 consists of the following parts, under the general title "Information Technology—Biometric data interchange formats:

Part 1: Reference Framework Model
Part 2: Finger Minutiae Data
Part 3: Finger Pattern Spectral Data
Part 4: Finger Image Data
Part 5: Face Image Data
Part 6: Iris Image Data
Part 7: Signature/Sign Behavioral Data
Part 8: Finger Pattern Skeletal Data
Part 9: Vascular Image Data
Part 10: Hand Geometry Silhouette Data Part 11: Signature/Sign Dynamic Data

ISO/IEC	CD	19794-1	Information technology—Biometric data interchange formats—Part 1: Framework	Published 2006	ISO JTC1 SC37

ISO 19794-1 Available from
http://www.iso.org/iso/iso_catalogue/catalogue_tc/catalogue_detail.htm?csnumber=38745

Standardized biometric data interchange formats are crucial to the interoperability of biometric components. ISO/IEC 19794-1:2006 describes general aspects of biometric data interchange formats and specifies requirements to be taken into account in standardizing specific formats. It classifies biometric data according to their processing level and establishes a naming concept for biometric data interchange formats on this basis.

ISO/IEC	CD	19794-2	Information technology—Biometric data interchange formats—Part 2: Finger minutiae data	Published 2005	ISO JTC1 SC37

ISO 19794-2 Available from
http://www.iso.org/iso/iso_catalogue/catalogue_tc/catalogue_detail.htm?csnumber=38746

ISO/IEC 19794-2:2005 specifies a concept and data formats for representation of fingerprints using the fundamental notion of minutiae. It is generic, in that it may be applied and used in a wide range of application areas where automated fingerprint recognition is involved. ISO/IEC 19794-2:2005 contains definitions of relevant terms, a description of how minutiae shall be determined, data formats for containing the data for both general use and for use with cards, and conformance information. Guidelines and values for matching and decision parameters are provided in an informative annex.

ISO/IEC 19794-2:2005 specifies the fundamental data elements used for minutiae-based representation of a fingerprint:

- Three data formats for interchange and storage of this data: a record-based format, and normal and compact formats for use on a smart card in a match-on-card application;
- Optional extended data formats for including additional data such as ridge counts and core and delta location.

ISO/IEC 19794-2:2005 provides for interchange of finger minutiae data between sensing, storage, and matching systems.

ISO/IEC	CD	19794-3	Information technology—Biometric data interchange formats—Part 3: Finger pattern spectral data	Published 2006	ISO JTC1 SC37

ISO 19794-3 Available from
http://www.iso.org/iso/iso_catalogue/catalogue_tc/catalogue_detail.htm?csnumber=38747

ISO/IEC 19794-3:2006, the finger pattern spectral data interchange format, specifies requirements for the representation of local or global spectral data derived from a fingerprint image. The format is designed to provide flexibility in the choice of spectral representation in that spectral components may be based on quantized cosinusoidal triplets, discrete Fourier transformations, or Gabor filters. The format also allows for a variable number of spectral components to be retained, which enables data representations in a form that is more compact than storage of the entire fingerprint image. ISO/IEC 19794-3:2006 provides example data records for each of the spectral representations.

ISO/IEC	CD	19794-4	Information technology—Biometric data interchange formats—Part 4: Finger image data	Published 2005	ISO JTC1 SC37

ISO 19794-4 Available from
http://www.iso.org/iso/iso_catalogue/catalogue_tc/catalogue_detail.htm?csnumber=38748

ISO/IEC 19794-4:2005 specifies a data record interchange format for storing, recording, and transmitting the information from one or more finger or palm image areas within an ISO/IEC 19785-1 CBEFF data structure. This can be used for the exchange and comparison of finger image data. It defines the content, format, and units of measurement for the exchange of finger image data that may be used in the verification or identification process of a subject.

The information consists of a variety of mandatory and optional items, including scanning parameters, compressed or uncompressed images, and vendor-specific information. This information is intended for interchange among organizations that rely on automated devices and systems for identification or verification purposes based on the information from finger image areas. Information compiled and formatted in accordance with ISO/IEC 19794-4:2005 can be recorded on machine-readable media or may be transmitted by data communication facilities.

ISO/IEC	FPD	19794-5	Information technology—Biometric data interchange formats—Part 5: Face image data	Published 2005	ISO JTC1 SC37

ISO 19794-5 Available from
http://www.iso.org/iso/iso_catalogue/catalogue_tc/catalogue_detail.htm?csnumber=38749

ISO/IEC 19794-5:2005 specifies scene, photographic, digitization, and format requirements for images of faces to be used in the context of both human verification and computer automated recognition.

The approach to specifying scene and photographic requirements in this format is to carefully describe constraints on how a photograph should appear rather than to dictate how the photograph should be taken. The format is designed to allow for the specification of visible information discernable by an observer pertaining to the face, such as gender, pose, and eye color.

The digital image format can be either ISO standard JPEG or JPEG2000. Finally, the "best practice" appendices provide guidance on photo capture for travel documents and face recognition performance versus digital compression.

ISO/IEC	FPD	19794-6	Information technology—Biometric data interchange formats—Part 6: Iris image data	Published 2005	ISO JTC1 SC37

ISO 19794-6 Available from
http://www.iso.org/iso/iso_catalogue/catalogue_tc/catalogue_detail.htm?csnumber=38750

ISO/IEC 19794-6:2005 specifies two alternative image interchange formats for biometric authentication systems that utilize iris recognition. The first is based on a rectilinear image storage format that may be a raw, uncompressed array of intensity values or a compressed format such as that specified by ISO/IEC 15444. The second format is based on a polar image specification that requires certain preprocessing and image segmentation steps, but produces a much more compact data structure that contains only iris information.

Data that comply with either one of the iris image formats specified in ISO/IEC 19794-6:2005 are intended to be embedded in a CBEFF-compliant structure in the CBEFF biometric data block as specified in ISO/IEC 19785-1.

ISO/IEC	FCD	19794-7	Information technology—Biometric data interchange formats—Part 7: Signature/sign time series data	Published 2007	ISO JTC1 SC37

ISO 19794-7 Available from
http://www.iso.org/iso/iso_catalogue/catalogue_tc/catalogue_detail.htm?csnumber=38751

ISO/IEC 19794-7:2007 specifies two data interchange formats for signature/sign behavioral data captured in the form of time series using devices such as digitizing tablets or advanced pen systems.

One data interchange format is for general use and the other one is a compact format for use with smart cards or other tokens. Both data interchange formats can be used for both acquired signature/sign samples (serving as a starting point for feature extraction) and for time-series features (to be compared directly by time-series based comparison algorithms).

Abstract syntax notation one (ASN.1) specifications of the data interchange formats and encoding instructions are provided in an informative annex.

ISO/IEC	FPD	19794-8	Information technology—Biometric data interchange formats—Part 8: Finger pattern skeletal data	Published 2006	ISO JTC1 SC37

ISO 19794-8 Available from
http://www.iso.org/iso/iso_catalogue/catalogue_tc/catalogue_detail.htm?csnumber=40715

ISO/IEC 19794-8:2006 specifies the interchange format for the exchange of pattern-based skeletal fingerprint recognition data. The data format is generic, in that it may be applied and used in a wide range of application areas where automated fingerprint recognition is involved.

The exchange format defined in ISO/IEC 19794-8:2006 describes all characteristics of a fingerprint in a small data record. Thus, it allows for the extraction of both spectral information (orientation, frequency, phase) and features (minutiae, core, ridge count). Transformations like translation and rotation can also be accommodated by the format defined herein.

ISO/IEC 19794-8:2006 supports the proliferation of low-cost commercial fingerprint sensors with limited coverage, dynamic range, or resolution. Thus, it defines a data record that can be used to store biometric information on a variety a storage media (including, but not limited to, portable devices and smart cards).

ISO/IEC	FDIS	19794-9	Information technology—Biometric data interchange formats—Part 9: Vascular image data	Under Development	ISO JTC1 SC37

ISO 19794-9 Available from
http://www.iso.org/iso/iso_catalogue/catalogue_tc/catalogue_detail.htm?csnumber=43637

ISO/IEC 19794-9:2007 defines the exchange of human vascular biometric image information. It defines a specific definition of attributes, a data record format for storing and transmitting vascular biometric images and certain attributes, a sample record, and conformance criteria.

ISO/IEC 19794-9:2007 is intended for applications requiring the exchange of raw or processed vascular biometric images. It is intended for applications not limited by the amount of storage required. It is a trade-off between the resources required for data storage or transmission and the potential for improved data quality/accuracy. Basically, it is to enable various algorithms to identify or verify the vascular biometric image data transferred from other image sources.

Currently available vascular biometric technologies that may utilize ISO/IEC 19794-9:2007 for image exchange are technologies that use the back of the hand, palm, and finger.

ISO/IEC	FCD	19794-10	Information technology—Biometric data interchange formats—Part 10: Hand geometry silhouette data	Published 2007	ISO JTC1 SC37

ISO 19794-10 Available from
http://www.iso.org/iso/iso_catalogue/catalogue_tc/catalogue_detail.htm?csnumber=43638

ISO/IEC 19794-10:2007 specifies a data record interchange format for storing, recording, and transmitting the information from one or more hand silhouettes within a CBEFF data structure. It defines the content, format, and units of measurement for the exchange of hand silhouette data that may be used in the verification or identification process of a subject.

The information consists of a variety of mandatory and optional items, including data capture parameters, standardized hand position, and vendor-specific information. This information is intended for interchange among organizations that rely on automated devices and systems for identification or verification purposes based on the information from hand geometry measurements.

ISO/IEC	WD	19794-11	Information technology—Biometric data interchange formats—Part 11: Signature/sign processed dynamic data	Under Development	ISO JTC1 SC37

ISO 19794-11 Not Yet Available. Track progress at:
http://www.iso.org/iso/iso_catalogue/catalogue_tc/catalogue_detail.htm?csnumber=43639

For the purposes of biometric verification and/or identification, this standard specifies a concept and data interchange format for signature/sign dynamic data captured in the form of time series using, for example, digitizing tablets or advanced pen devices. The data interchange format is "generic," in that it may be applied and used in a wide range of application areas where handwritten signs or signatures are involved.

No application-specific requirements or features are addressed in this standard. The standard contains definitions of relevant terms, a description of what data is captured, a data format for containing the data alongside examples of record contents and best practice in capture. It is advisable that stored and transmitted biometric data be time-stamped and that cryptographic techniques be used to protect their authenticity, integrity, and confidentiality; yet such provisions are beyond the scope of this standard.

ISO/IEC	NP	19794-12	Information technology—Biometric data interchange formats—Part 12: Face Identity Data	Under Development	ISO JTC1 SC37

ISO 19794-12 Not Yet Available. Track progress at:
http://www.iso.org/iso/iso_catalogue/catalogue_tc/catalogue_detail.htm?csnumber=43640

No documents are yet available for this work item.

7.2.10.5 ISO 24708 Information Technology—Biometrics—BioAPI Interworking Protocol

ISO/IEC	WD	24708	Information Technology—Biometrics—BioAPI Interworking Protocol (BIP)	withdrawn	ISO JTC1 SC37

This draft reached FCD status, but has subsequently been withdrawn.

7.2.10.6 ISO 24709 Information Technology—Conformance Testing for the Biometric Application Programming Interface (BioAPI)

ISO/IEC	FDIS	24709	Information technology—Conformance testing for the biometric application programming interface (BioAPI)	Under Development	ISO JTC / SC37

This series of standards is being developed in four parts (at the time of this writing).

ISO/IEC	FDIS	24709-1	Information technology—Conformance testing for the biometric application programming interface (BioAPI)—Part 1: Methods and procedures	Published 2007	ISO JTC1 SC37

ISO 24709-1 Available from
http://www.iso.org/iso/iso_catalogue/catalogue_tc/catalogue_detail.htm?csnumber=38819

ISO/IEC 24709-1:2007 specifies the concepts, framework, test methods, and criteria required to test conformity of biometric products claiming conformance to BioAPI (ISO/IEC 19784-1). Guidelines for specifying BioAPI conformance test suites, writing test assertions, and defining procedures to be followed during the conformance testing are provided. The conformance testing methodology is concerned with conformance testing of biometric products claiming conformance to BioAPI. Definitions of schemas of the assertion language are provided in normative annexes.

ISO/IEC	FDIS	24709-2	Information technology—Conformance testing for the biometric application programming interface (BioAPI)—Part 2: Test assertions for biometric service providers	Published 2007	ISO JTC1 SC37

ISO 24709-2 Available from
http://www.iso.org/iso/iso_catalogue/catalogue_tc/catalogue_detail.htm?csnumber=43352

ISO/IEC 24709-2:2007 defines a number of test assertions written in the assertion language specified in ISO/IEC 24709-1. These assertions enable a user of ISO/IEC 24709-2:2007 (such as a testing laboratory) to test the conformance to ISO/IEC 19784-1 (BioAPI 2.0) of any biometric service provider that claims to be a conforming implementation of that International Standard.

Each test assertion specified in ISO/IEC 24709-2:2007 exercises one or more features of an implementation under test. Assertions are placed into packages (one or more assertions per package), as required by the assertion language. These assertions allow for testing conformance of BSPs of all conformance subclasses, and are further organized according to conformance subclasses and claimed support of optional features.

ISO/IEC	NP	24709-3	Information technology—Conformance testing for the biometric application programming interface (BioAPI)—Part 3: Test assertions for BioAPI frameworks	Under Development NP	ISO JTC1 SC37

No documents or links are yet available.

ISO/IEC	NP	24709-4	Information technology—Conformance testing for the biometric application programming interface (BioAPI)—Part 4: Test assertions for biometric applications	Under Development	ISO JTC1 SC37

No documents or links are yet available.

7.2.10.7 ISO 24714 Cross-Jurisdictional and Societal Aspects of Implementation of Biometric Technologies

ISO/IEC	TR	24714	Cross-Jurisdictional and Societal Aspects of Implementation of Biometric Technologies	Under Development	ISO JTC1 SC37

This series of standards currently has two parts under development.

ISO/IEC	TR/ CD	TR 24714-1	Cross-Jurisdictional and Societal Aspects of Implementation of Biometric Technologies—Part 1: Guide to the Accessibility, Privacy, and Health and Safety Issues in the deployment of Biometric Systems for Commercial Application	Under Development	ISO JTC1 SC37

ISO 24714-1 Not Yet Available. Track progress at:
http://www.iso.org/iso/iso_catalogue/catalogue_tc/catalogue_detail.htm?csnumber=38824

ISO/IEC	WD	24714-2	Technical Report on Cross Jurisdictional and Societal Aspects of Implementations of Biometric Technologies—Part 2: Practical Application to Specific Contexts	Under Development	ISO JTC1 SC37

ISO 24714-2 Not Yet Available. Track progress at:
http://www.iso.org/iso/iso_catalogue/catalogue_tc/catalogue_detail.htm?csnumber=43607

7.2.10.8 ISO 24722 Information Technology—Multimodal Biometric Fusion

ISO/IEC	WD	TR 24722	Information technology—Multi-modal biometric fusion	Under Development	ISO JTC1 SC37

ISO 24722 Available from
http://www.iso.org/iso/iso_catalogue/catalogue_tc/catalogue_detail.htm?csnumber=38832

ISO/IEC TR 24722:2007 provides a description of and analysis of current practice on multimodal and other multibiometric fusion, including (as appropriate) reference to a more detailed description. It also discusses the need for, and possible routes to, standardization to support multibiometric systems.

7.3 Vehicle Identification

In the world of ITS, our principal interest lies in "automatic" vehicle identification (AVI)—that is, the use of ITS technology to identify a vehicle automatically, using wireless techniques. However, AVI identification schemes, for the good and sound reasons of consistency and migration, incorporate the manual and bar code identification schemes that have been in use by the automotive industry for many decades. These standards, though simple and largely old fashioned manual identification schemes and registries, form an integral part of the AVI systems that follow. The first section, therefore, briefly explains the principal "manual" identification standards for vehicles.

It is worth noting at this point that the well-known vehicle identification number (VIN) is not as global as it might appear. Japanese and Korean vehicles have always used a code that is similar but not the same. The Japanese/Korean code is shorter than the VIN. In the next section, however, you will see how they are both successfully accommodated in AVI standards.

It is worth commenting that these standards were developed before the introduction of bar code and obviously well before the introduction of RFID. The emphasis is therefore principally on human readability. Hence the letters O, Q, and I are generally not used. In today's world of electronic data, some of the codes in the following standards appear quaint and inefficient, and in today's world, they are far from optimal. However, the codes have been in operation for many years and so it would be more difficult to change their structure than to live with the inefficiencies.

These standards are developed by ISO/TC 22 (Motor Vehicles) and ISO/TC 204 (Intelligent Transport Systems).

The work program of TC 22 can be downloaded from:

http://www.iso.org/iso/iso_catalogue/catalogue_tc/catalogue_tc_browse.htm?commid=46706&development=true.

The work program of TC 204 can be downloaded from:

http://www.iso.org/iso/iso_catalogue/catalogue_tc/catalogue_tc_browse.htm?commid=54706&development=true.

With respect to container standards, these are developed by ISO/TC 104. The work program of TC 104 can be downloaded from:

http://www.iso.org/iso/iso_catalogue/catalogue_tc/catalogue_tc_browse.htm?commid=51156&development=true.

7.3.1 Manual Identification Standards Embedded in AVI/ERI Identification Standards

7.3.1.1 ISO 3779 Road Vehicles—Vehicle Identification Number (VIN)—Content and Structure

ISO	IS	3779	Road vehicles—Vehicle identification number (VIN)—Content and structure	Published Standard 1983	ISO TC22

ISO 3779 Available from
http://www.iso.org/iso/iso_catalogue/catalogue_tc/catalogue_detail.htm?csnumber=9305

The VIN is the most widely used vehicle identification code used by manufacturers, but it is not used globally. ISO 3779 was originally approved in February 1977 and was last revised in 1983. This International Standard was designed to uniquely identify motor vehicles, trailers, motorcycles, and mopeds, and it comprises three parts:

- The World Manufacturer Identifier (WMI). This part of the VIN identifies the maker of the vehicle in accordance with ISO 3780 (see Section 7.3.1.2 below). It is normally positioned as the first three positions of the VIN. Low volume manufacturers, constructing less than 500 vehicles per year, have slightly different rules (to accommodate a larger number of such manufacturers), and their third digit is always a 9, and the 12th, 13th, and 14th position of the VIN are used for the second part of the WMI.
- Vehicle Descriptor Section (VDS). The second part of the VIN (six characters which occupy positions 4 through 9 of the VIN) are for the manufacturers' use and normally identifies the vehicle model, or model range.
- Vehicle Identifier Section (VIS). The last eight characters of the VIN are used for the identification of a specific vehicle. The last four characters are numeric.

International Standard ISO 3779 also makes provision for identifying the year in which a vehicle is built, which is normally the 10th character. The character rotates A through Z and 0 through 9 (except the letters I, O, and Q, for obvious reasons with human readable characters). It should therefore be noted that this does not provide a truly unique year identification. However, the repeat cycle is long enough that all but museum vehicles are likely to have been scrapped before the recycle date reoccurs.

ISO also recommends that the 11th position of the VIN is used to represent the factory of manufacture, represented again by a single digit as capital letters A through Z and numbers 1 through 0 (except the letters I, O, and Q).

The VIN code is clearly based on human readable characters that predate barcode or RFID. It is therefore inefficient in its use as digital/binary representation.

There is room in ISO 3779 for regional variations.

Within Europe and North America, the VIN is mandatory for all new vehicles.

However, the specification for VIN in the United States is more rigid in that in North America some of the fields left for the manufacturer to define elsewhere are more tightly specified.

The U.S. VIN comprises six parts:

- The first three characters uniquely identify the manufacturer (the WMI of ISO 3780).
- The second part comprises five characters (VIN positions 4–8) and identifies the attributes of the vehicle.
- The third part is a one-character check digit.
- The fourth part consists of one character representing the model year (A–Z and 0–9).
- The fifth part is a character representing the factory of manufacture.
- The sixth part is a sequential production number (for manufacturers producing more than 500 vehicles per year). For other manufacturers, the sixth, seventh, and eighth position represent the sequential production number. This section confirms to the VIS in ISO 3779.

A comparison of the different implementations of VIN is shown in Table 7.1.

7.3.1.2 ISO 3780 Road Vehicles—World Manufacturer Identifier (WMI) Code

ISO	IS	3780	Road vehicles—World manufacturer identifier (WMI) code	Published Standard	ISO TC22

ISO 3780 Available from
http://www.iso.org/iso/iso_catalogue/catalogue_tc/catalogue_detail.htm?csnumber=9306

ISO 3780 provides WMIs and is used globally.

This International Standard specifies the content and structure of an identifier in order to establish, on a worldwide basis, the identification of road vehicle manufacturers. This identifier is the first section of the VIN described in ISO 3779.

Table 7.1 Different Implementations of VIN

Position	*1*	*2*	*3*	*4*	*5*	*6*	*7*	*8*	*9*	*10*	*11*	*12*	*13*	*14*	*15*	*16*	*17*
ISO 3779	WMI			VDS								VIS					
North American > 500 vehicles/year	Manufacturer ID			Vehicle Attributes					CHK Dgt	Model Yr	Factory	Sequential Number					
North American < 500 vehicles/year	Manufacturer ID			Vehicle Attributes					CHK Dgt	Model Yr	Factory	Manufacturer ID			Sequential Number		
Asia	Tbd																

This applies to motor vehicles, trailers, motorcycles, and mopeds as defined in ISO 3833.

This again is a coding scheme from the predigital age, with all of the inefficiencies that this brings.

The WMI comprises three characters. The first two characters are unique identifiers for a country and manufacturer and are registered. The third character is commonly used to identify a vehicle type (car, bus, or truck) or a division of the manufacturer, or both.

In Japan and Korea, there are two possibilities for the vehicle identification code.

- The international VIN code;
- ABCDE-1234567, where ABCDE identifies the manufacturer/model, and 1234567 identifies the sequential issued serial number for the manufacturer/model.

Within the AVI standards, the Japanese and Korean identity ABCDE-1234567 fits in to the relevant fields for the VIN and unassigned fields are padded with zeros.

7.3.1.3 ISO 4100 Road Vehicles—World Parts Manufacturer Identifier (WPMI) Code

ISO	IS	4100:1980	Road vehicles—World parts manufacturer identifier (WPMI) code	Published Standard	ISO TC22

ISO 4100 Available from
http://www.iso.org/iso/iso_catalogue/catalogue_tc/catalogue_detail.htm?csnumber=9841

This International Standard specifies the content and structure of an identifier in order to establish, on a worldwide basis, the identification of the manufacturers of parts for road vehicles.

This identifier may serve, at the manufacturer's option, as a substitute or an addition for various national or international systems which already exist.

The standard applies only to manufacturers of parts to be used on road vehicles (road vehicles as defined in ISO 3833).

7.3.1.4 ISO 8357 Road Vehicles—Instructions for the Implementation of WMI Codes for VIN Systems and for WPMI Codes

ISO	TR	8357	Road vehicles—Instructions for the implementation of the assignment of world manufacturer identifier (WMI) codes for vehicle identification number (VIN) systems and for world parts manufacturer identifier (WPMI) codes	Published TR 1996	ISO TC22

ISO 8357 Available from
http://www.iso.org/iso/iso_catalogue/catalogue_tc/catalogue_detail.htm?csnumber=15503

ISO 8357 provides instructions relating to the requesting and assignment of the WMI codes and WPMI codes according to ISO 3780 and ISO 4100.

7.3.2 Automatic Vehicle Identification

Having now understood the basis of the core vehicle identification schemes, we can turn our attention to the use of these identification schemes within an "automatic" vehicle identification context of ITSO.

Automatic vehicle identification lies on the cusp between an enabling technology and an application. As an application, an AVI identity can be used, for example, to provide unique identification for an access control system.

However, AVI is more widely used as a data component in tolling and road pricing systems and increasingly a broad range of ITS systems. It is important to understand that the AVI series of standards is technology neutral—that is, the air interface technology is not defined and the system works at OSI layer 7 (application layer).

There have been two generations of AVI standard/technical reports. The first, developed in the early to mid-1990s (CEN 12314/ISO/CEN 12414 and subsequently replaced by ISO 14814), dealt only with vehicle identification. The second generation of standards, developed in the late 1990s and now finally finished balloting, deals with both vehicle and (transport-related) equipment identification.

These families of AVI-related standards have a similar family grouping. For each family there is a reference architecture standard, a numbering and data structure standard, a system parameters standard, and often an interfaces standard.

There are three family groupings:

- Early AVI/AEI standards that related to vehicles and trailers (numbered 14xxx);
- Intermodal AVI/AEI standards (numbered 17xxx);
- Electronic registration identification (ERI) standards (numbered 2453x).

The early generations were published as technical reports and have subsequently been upgraded after 5-year review to full International Standards. Because of problems in the parallel CEN/ISO balloting system, this has taken longer than expected, but the process is now complete.

The intermodal AVI/AEI (1726x) series was also first passed as technical reports/specifications, also reviewed after 5 years, also suffered from problems with parallel balloting, and have recently been approved as full International Standards.

Unfortunately, because of the eccentricities of the ISO and CEN reference number allocation systems, the sequence within the families (reference architecture, numbering and data structure, system specification, and so forth) is not consistent. However, the numbers are allocated as a block of sequential numbers.

7.3.2.1 ISO 14814 Road Transport and Traffic Telematics—Automatic Vehicle and Equipment Identification—Reference Architecture and Terminology

CEN	ENv	12314	Automatic vehicle and equipment identification—Part 1: Reference architectures and terminology	Withdrawn	CEN TC278

This standard has been replaced by ISO 14814.

ISO	IS	14814	Road transport and traffic telematics—Automatic vehicle and equipment identification—Reference architecture and terminology	Published 2006	ISO TC204

ISO 14814 Available from
http://www.iso.org/iso/iso_catalogue/catalogue_tc/catalogue_detail.htm?csnumber=37046

ISO 14814:2005 establishes a common framework to achieve unambiguous identification in intelligent transport systems/road transport and traffic telematics (ITS/RTTT) automatic vehicle identification/automatic equipment identification (AVI/AEI) applications.

This scheme and reference architecture model is designed to be an enabling structure to allow interoperability between different commercial systems, and not prescriptive in determining any one system. It is not frequency, nor air interface protocol, specific. It provides maximum interoperability, has a high population capability, and provides the possibility of upwards migration to more capable systems.

ISO 14814:2005 provides a reference structure which enables an unambiguous identification and also identifies the data construct as an ITS/RTTT message. The construct also identifies which ITS/RTTT data structure is contained in the message.

The context of the AVI transaction has been identified as in Figure 7.2.

In Figure 7.2 the vertical arrows identify system interface points to which the standard assigns a Greek letter to identify them. The focus of the standard is on reference point Delta.

7.3.2.2 ISO 14815 Road Transport and Traffic Telematics—Automatic Vehicle and Equipment Identification—System Specifications

ISO	IS	14815	Road transport and traffic telematics—Automatic vehicle and equipment identification—System specifications	Published 2005	ISO TC204

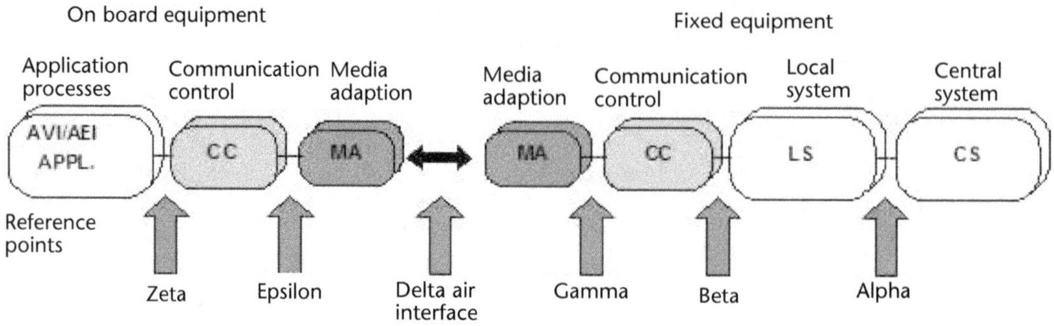

Figure 7.2 Overall conceptual reference architecture model showing the context of AVI. (*Source:* Microdesign (now Q-Free) ASA/Consultancy Services International Ltd. contribution to CEN TC278 WG12.)

ISO 14815 Available from
http://www.iso.org/iso/iso_catalogue/catalogue_tc/catalogue_detail.htm?csnumber=38666

ISO 14815:2005 defines a generic AVI/AEI system specification for nominal AVI/AEI to provide an enabling International Standard, which, while allowing the system specifier to determine the performance levels and operating conditions, provides a framework for nominal interoperability.

The standard provides classes for different operating characteristics to enable comparison between vendor offerings. It states that in order to obtain interoperability, it is a requirement that nominal AVI/AEI system equipment has the capability to operate with AVI/AEI system on-board equipment, and specifies how to achieve this.

It provides different classes of environmental and operating conditions within which such equipment has to function to accommodate different geographical location, traffic operating conditions, and so on, without onerously loading the cost of equipment that does not have to operate in the most demanding environment.

These generic system specifications provide a migration path to later generations of equipment and to equipment of greater capability.

7.3.2.3 ISO 14816 Road Transport and Traffic Telematics—Automatic Vehicle and Equipment Identification—Numbering and Data Structure

ISO	IS	14816	Road transport and traffic telematics—Automatic vehicle and equipment identification—Numbering and data structure	Published 2005	ISO TC204

ISO 14816 Available from
http://www.iso.org/iso/iso_catalogue/catalogue_tc/catalogue_detail.htm?csnumber=38667

ISO 14816:2005 establishes a common framework data structure for unambiguous identification in RTTT/ITS systems. It excludes any physical aspects such as interfaces. It is neither frequency- nor air interface protocol-specific.

Data elements that form part of transmission or storage protocols such as headers, frame markers, and checksums are thus excluded.

The specifications for protecting against changes, classifying and qualifying security aspects of the data structure elements are not included within ISO 14816:2005.

The standard specifies a data structure that enables upwards integration and expansion from the simplest low cost AVI/AEI system to more complex functions. The structure is designed to be flexible and enabling rather than prescriptive.

This standard has been designed to provide for the differing requirements of AVI and AEI by the use of separate application specific clauses. By retaining these differing requirements within one supervisory document, the interoperability is maximized, particularly in the case where both AVI and AEI are required at the same time in the road environment.

In order to support systems using both active and passive on-board equipments (OBEs), the basic data structures have been minimized. This enables any manufacturer/operator with an OBE with a user addressable memory of only 56 bits to be able to conform to a full core identification according to this standard.

The principles of data element structure and description determined in ISO/IEC 8824, ISO/IEC 8825-1, and ISO/IEC 8825-2 have been adopted to provide an interoperable architecture within a standard framework according to guidelines from ISO/TC 204 and CEN/TC 278.

ISO 14816:2005 defines data structures based on the ISO/IEC 8824-1 ASN.1 universal class types that may be directly imported to other application standards that would need only subsets of the full application class types. These universal class and application class types are uniquely defined as an ASN.1 module in Annex B of the standard. This module may be directly linked into an application data definition.

ISO 14816:2005 defines default encoding for simple AVI/AEI applications where no other relevant application standard exists.

Within this ISO 14816, data representation is defined using ASN.1. Its usage provides maximum interoperability and conformance to existing standards, and meets the specifically defined requirements for a generic standard model for RTTT in that it:

- Uses existing standard syntax notation and encoding rules;
- Is adaptable and expandable;
- Does not include unnecessary information for a specific system;
- Incurs a minimum of overhead in storage and transmission.

7.3.2.4 ISO TS/DIS 17261 Road Transport and Traffic Telematics—Automatic Vehicle and Equipment Identification—Intermodal

ISO	TS/ DIS	17261	Road transport and traffic telematics—Automatic vehicle and equipment identification—Intermodal goods transport: Architecture and terminology	TS Published 2005/Final ballot in process for IS	ISO TC204

ISO 17261 Available from
http://www.iso.org/iso/iso_catalogue/catalogue_tc/catalogue_detail.htm?csnumber=40206

The mission statement for this standard is to "to provide an 'enabling' Reference Architecture Model for Intermodal/Multimodal AEI. The Reference Architecture Model Standard is designed to accommodate, within the framework, a wide and diverse variety of ITS/RTTT applications from simple AVI/AEI to more complex transactions with a wide variety of uses, including the transfer of data relating to the manifest of loads and part loads and means of identification of loads and part loads, in an ITS/RTTT environment."

ISO TS 17261:2005 is a follow on to ISO 14814 and describes the conceptual and logical architecture for AVI/AEI and supporting services in an intermodal/multimodal environment. It presents a high level view of AEI intermodal and multimodal system architecture. ISO TS 17261:2005 describes the key subsystems, their associated interfaces and interactions, and how they fit into system-wide functions such as management, security, and information flow. This is represented by a series of architecture diagrams along the lines of Figure 7.3.

Application and physical architectures are considered and the framework for the family of intermodal AVI/AEI standards defined.

7.3.2.5 ISO TS/DIS 17262 Road Transport and Traffic Telematics—Automatic Vehicle and Equipment Identification—Intermodal Goods Transport Numbering and Data Structures

ISO	TS/ DIS	17262	Road transport and traffic telematics—Automatic vehicle and equipment identification—Intermodal Goods Transport Numbering and Data Structures	TS Published 2003/Final ballot in process for IS	ISO TC204

ISO 17262 Available from
http://www.iso.org/iso/iso_catalogue/catalogue_tc/catalogue_detail.htm?csnumber=29252

ISO 17262 is a follow on to ISO 14816, which extends to cover intermodal identifications and extends the use of ASN.1 within AVI/AEI numbering and data structures.

ISO/TS 17262:2003 defines generic numbering and data structures for unambiguous identification of equipment used for intermodal goods transport. These data are known as intermodal goods transport numbering and data structures.

ISO/TS 17262:2003 defines data independently of the data carrier. The modeling of data is based on ASN.1 as defined in ISO/IEC 8824. ISO/TS 17262:2003 excludes any physical aspects such as interfaces, dimensions, and so on. Data that form part of transmission or storage protocols (headers, frame markers, and checksums) are excluded.

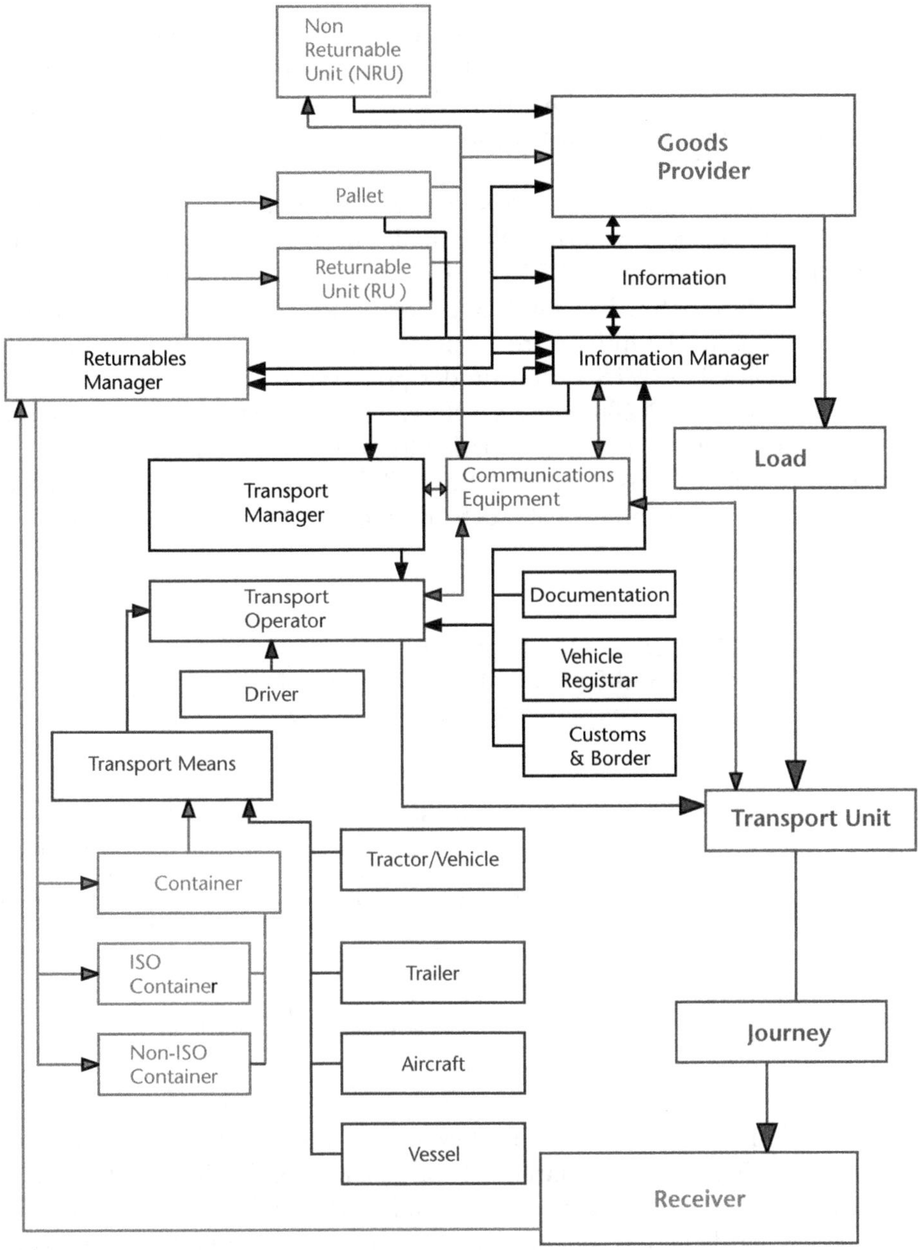

Figure 7.3 The distribution cycle from the view of the transport. (*Source:* Consultancy Services International Ltd. Contribution working document to CEN TC278.)

Data defined in ISO/TS 17262:2003 require a system for control and distribution of number series independent of the different AVI/AEI systems. This is required in order to avoid ambiguity and to provide the necessary level of security where appropriate. For this reason the registration authority defined in ENV ISO 14816 applies for ISO/TS 17262:2003.

ISO/TS 17262:2003 enables the use of optimized encoding schemes such as ASN.1 Basic Packed Encoding Rules (PER).

ISO/TS 17262:2003 provides interoperability, not only between simple AVI/AEI and more complex RTTT/TICS functions, but also with preexisting standards such as for containers (ISO 10374). Specifications for protecting against changes and classifying and qualifying security aspects of the data are out of the scope of ISO/TS 17262:2003.

ISO/TS 17262:2003 relates to AVI/AEI units, but not to smaller containers and units being transported. For smaller units (pallet loads, trays, parcels, and so forth), please refer to ISO/IEC SC31 standards, the ISO 18000 series. The numbering structure defined in ISO/TS 17262:2003 is designed to enable combinations with the data definitions from the ISO 18000 series. This combination will be covered in CIS/EN/TS 17264 (under preparation).

ISO/TS 17262:2003 provides the capability to carry application data, associated with the identification, to be carried as part of the AVI/AEI message. Within ISO/TS 17262:2003 this is provided as a "black box" facility. The definition of the structure and contents of such messages are outside the scope of ISO/TS 17262:2003.

7.3.2.6 ISO TS/DIS 17263 Road Transport and Traffic Telematics—Automatic Vehicle and Equipment Identification—Intermodal Goods Transport-System Parameters

ISO	TS/ DIS	17263	Road transport and traffic telematics—Automatic vehicle and equipment identification—Intermodal Goods Transport-System Parameters	TS Published 2003/Final ballot in process for IS	ISO TC204

ISO 17263 Available from
http://www.iso.org/iso/iso_catalogue/catalogue_tc/catalogue_detail.htm?csnumber=29253

ISO/TS 17263:2003 establishes an AEI system based on radio frequency technologies. This system is intended for general application in ITS. It allows for the transfer of the identification codes and further information about equipment and vehicles used in intermodal transport into such ITS and information systems related to intermodal transport processes. Within the intermodal context of the ITS sector, AEI systems have the specific objective of achieving an unambiguous identification of an ITU or related equipment or vehicle or item used in intermodal transport, and to make that identification automatically. Vehicles will be considered and handled under Intermodal aspects as "intermodal equipment." Therefore, a differentiation between AEI and AVI systems under the purpose of ISO/TS 17263:2003 is not required.

ISO/TS 17263:2003 is specifically aimed at DSRC-type air interfaces. The requirement and test methods may not apply for intermodal AEI systems using long-range communications such as cellular networks or satellite, or vicinity communication such as inductively coupled antennas. The interoperability across the air interface (reference point Delta; see Figure 7.2) is outside the scope of ISO/TS 17263:2003.

The aim of ISO/TS 17263:2003 is to define, describe, and specify the system parameters related to an intermodal AEI system to provide an enabling standard, which, while allowing the system specifier to determine the performance levels and operating conditions, provides a framework for interoperability.

Therefore ISO/TS 17263:2003 specifies:

- Parameters and requirements of the identification system itself;
- Performance criteria necessary to ensure consistent and reliable operation of AEI systems within international transport processing;
- Requirements of the performance and the position of the electronic devices (tag) when installed on intermodal equipment;
- Requirements for the installation of readers, and performance data related to these components.

The standard first defines the basic rules, purpose, and application categories. It then defines user requirements and technical and electronically features.

The standard goes on to define operational parameters and requirements to the system itself, "Basic requirements and performance criteria—System reliability and security."

It next provides definition for specific parameters and performance criteria for the reader, "General Application related performance criteria—Protection and safety—Specific operational parameters and performance criteria for TAG."

Further definition is provided with respect to data contents and storage capacity, mounting conditions, and lifetime and environmental requirements.

The final section provides a summary view of test requirements.

7.3.2.7 ISO TS/DIS 17264 Road Transport and Traffic Telematics—Automatic Vehicle and Equipment Identification—[AVI/AEI] Interfaces

ISO	TR/ DIS	17264	Road transport and traffic telematics—Automatic vehicle and equipment identification—[AVI/AEI] Interfaces	TS Published 2003/Final ballot in process for IS	ISO TC204

ISO 17264 Available from
http://www.iso.org/iso/iso_catalogue/catalogue_tc/catalogue_detail.htm?csnumber=43457

This European Standard/ISO Technical Specification has been evolving for a number of years and is still in voting process at the time of writing this book. It specifies an application interface for electronic fee collection (AVI) systems, which are based on dedicated short-range communication (DSRC), enabling interoperability between open AVI systems (i.e., between different AVI system operators) on an AVI-DSRC application interface level.

By using the layer 7 DSRC application layer standard, the deliverable makes clear specification requirements. However, ISO 15628 (DSRC Application Layer), while originally designed for European DSRC, is now supported by CALM and Japanese DSRC, so it is actually medium independent, preserving the nature of ISO AVI/AEI standards.

The deliverable provides specifications for the AVI transaction model, as well as AVI data elements (referred to as attributes) and functions, from which an AVI transaction can be built. The AVI transaction model provides a mechanism that allows handling of different versions of AVI transactions and associated contracts. A certain AVI transaction supports a certain set of AVI attributes and AVI functions as defined in the deliverable. It is not envisaged that the complete set of AVI attributes and functions is present in each piece of AVI equipment, be it OBE or RSE.

In order to achieve interoperability, operators have to agree on issues like:

- Which optional features are actually being implemented and used;
- Security policy (including encryption algorithms and key management, if applicable);
- Those needed between operators in order to regulate the handling of different AVI transactions.

This deliverable has the following structure: In the first four clauses, the scope, normative references, definitions of terms and abbreviations are accounted for. Next, in clause 5, AVI transaction requirements are defined, which are independent of any communication media. In a normative Annex A, the AVI application interface architecture is described in terms of its relation to the DSRC communication architecture, based on EN 12834/ISO 15628. An informative Annex D provides informative examples of AVI transactions using the specified AVI attributes and functions.

It was anticipated that the standard would finally be approved in late 2007.

7.3.3 Electronic Registration Identification

A rapidly emerging need has been identified with administrations to improve the unique identification of vehicles for a variety of services. Situations are already occurring where manufacturers intend to fit lifetime tags to vehicles. Various governments are considering the needs and benefits of electronic registration identification (ERI) as a legal proof of vehicle identity with potential mandatory uses. There is commercial and economic justification both with respect to tags and infrastructure that a standard enable an interoperable solution.

ERI is a means of uniquely identifying road vehicles. The application of ERI will offer significant benefits over existing techniques for vehicle identification. It will be a suitable tool for the future management and administration of traffic and transport, including applications in free-flow, multilane traffic conditions with the capability to support mobile transactions. ERI addresses the need of authorities and other road users for a trusted electronic identification, including roaming vehicles.

The unique vehicle identifier is held in a secure environment within an electronic registration tag (ERT) fitted to a vehicle. The identifier used to identify a vehicle is called the vehicle identifier, or vehicle ID. The preferred vehicle identifier is the VIN, assigned to the vehicle by its manufacturer in accordance with ISO 3779, or a variant of this vehicle identifier.

The ERT may contain vehicle data in addition to the unique identifier, as required by authorities or their agents for ERI applications (e.g., vehicle registration details). An ERT is the core component for simple to complex applications of ERI, ranging from a simple read-only device, with more complex applications requiring one or more communications systems.

The ERT may be accessed by an electronic registration reader (ERR), either to read, or read/write data, from or to an ERT. Optionally, the ERT may communicate with other on-board vehicle equipment. The potential range of ERI applications, simple to complex, will require interoperability to exist between an ERT and an ERR by application.

The various parts of ISO 24534 provide the overall framework for ERI and specification of requirements for "fully featured" ERI. An associated International Standard in this family of ERI standards, ISO 24535, provides a subset of these requirements and provides a "basic ERI" functionality.

7.3.3.1 ISO TS/DIS 24534-1 Road Transport and Traffic Telematics—Automatic Vehicle and Equipment Identification—Electronic Registration Identification (ERI) for Vehicles—Part 1: Architecture

ISO	TS/ DIS	24534-1	Road transport and traffic telematics—Automatic vehicle and equipment identification—Electronic Registration Identification (ERI) for vehicles—Part 1: Architecture	TS Approved/ DIS under ballot	ISO TC204

ISO TS 24534-1 Available from
http://www.iso.org/iso/iso_catalogue/catalogue_tc/catalogue_detail.htm?csnumber=41696

Part 1: Architecture of ISO/TS 24534 provides the generic requirements for electronic registration that is based on an identifier assigned to a vehicle (e.g., for recognition by national authorities), suitable to be used for:

- Electronic identification of local and foreign vehicles by national authorities;
- Vehicle manufacturing, in-life maintenance; and end-of-life identification (vehicle lifecycle management);
- Adaptation of vehicle data (e.g., for international resales);
- Safety-related purposes;
- Crime reduction;
- Commercial services.

It adheres to privacy and data protection regulations.

The standard defines the general context of the ERI standards 24534 and 24535.

ISO 24534-1 provides an overview of the ERI system concept, in terms of the on-board vehicle components and the external off-vehicle components required for an operational system. The detailed requirements are defined in the Parts 2, 3, 4 and 5 of ISO 24534 and for the more limited, relevant provisions of ISO 24535.

The context is described within Figure 7.4.

7.3.3.2 ISO TS/DIS 24534-2 Road Transport and Traffic Telematics—Automatic Vehicle and Equipment Identification—Electronic Registration Identification (ERI) for Vehicles—Part 2: Operational Requirements

ISO	TS/ DIS	24534-2	Road transport and traffic telematics—Automatic vehicle and equipment identification—Electronic Registration Identification (ERI) for vehicles—Part 2: Operational requirements	TS Approved/ DIS under ballot	ISO TC204

ISO TS 24534-2 Available from
http://www.iso.org/iso/iso_catalogue/catalogue_tc/catalogue_detail.htm?csnumber=41697

This part of ISO/TS 24534 defines the operational requirements for the remaining parts of ISO/TS 24534 and the more limited but relevant provisions of ISO/TS 24535.

While the definition of the organizational framework required to implement, operate, and maintain an ERI system is outside the scope of this part of ISO/TS 24534, a list of potential stakeholders in the public and private sector has been included.

7.3.3.3 ISO TS/DIS 24534-3 Road Transport and Traffic Telematics—Automatic Vehicle and Equipment Identification—Electronic Registration Identification (ERI) for Vehicles—Part 3: Vehicle Data

ISO	TS/ DIS	24534-3	Road transport and traffic telematics—Automatic vehicle and equipment identification—Electronic Registration Identification (ERI) for vehicles—Part 3: Vehicle data	TS Approved/ DIS under ballot	ISO TC204

ISO TS 24534-3 Available from
http://www.iso.org/iso/iso_catalogue/catalogue_tc/catalogue_detail.htm?csnumber=41698

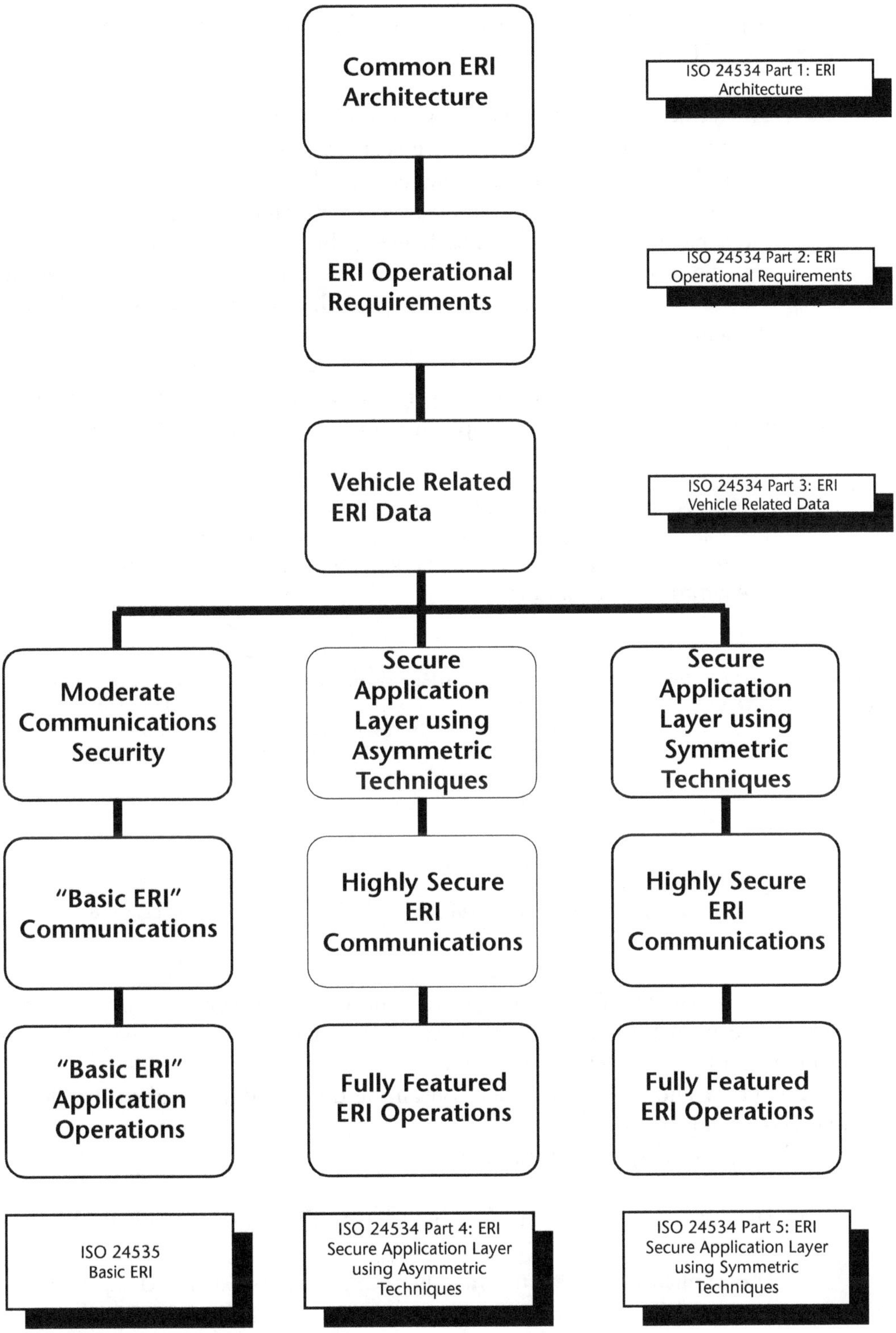

Figure 7.4 The architecture context for ERI. (*Source:* Consultancy Services International Ltd. Working draft contribution to CEN TC278.)

This part of ISO/TS 24534 defines the vehicle identification data. This data is called the ERI data and includes:

- The vehicle identifier, and possible additional vehicle-related information (as typically included in a vehicle registration certificate);
- All additional vehicle data elements are defined as optional. It is left to local legislation and/or the discretion of a registration authority to use or not to use a particular data element. If used, the value is assumed to be the one registered by the registration authority in accordance with local legislation. This part of ISO/TS 24534 only provides the syntax for all these data elements.

NOTE: The secure application layer interfaces for the exchange of ERI data with an ERI reader or writer are specified in Parts 4 and 5 of ISO/TS 24534.

7.3.3.4 ISO TS/DIS 24534-4 Road Transport and Traffic Telematics—Automatic Vehicle and Equipment Identification—Electronic Registration Identification (ERI) for Vehicles—Part 4: Secure Communications Using Asymmetric Techniques

ISO	TS/ DIS	24534-4	Road transport and traffic telematics—Automatic vehicle and equipment identification—Electronic Registration Identification (ERI) for vehicles—Part 4: Secure communications using asymmetric techniques	TS Approved/ DIS under ballot	ISO TC204

ISO TS 24534-4 Available from
http://www.iso.org/iso/iso_catalogue/catalogue_tc/catalogue_detail.htm?csnumber=41699

This part of ISO/TS 24534 specifies the interfaces for a secure exchange of data between an ERT and an ERI reader or ERI writer in or outside the vehicle using asymmetric encryption techniques.

NOTE 1: The on-board device containing the ERI data is called the electronic registration tag (ERT).

This technical specification includes:

- The application layer interface between an ERT and an on-board ERI reader or writer;
- The application layer interface between the on-board ERI equipment and external ERI readers and writers;
- Security issues related to the communication with the ERT.

NOTE 2: The vehicle identifiers and possible additional vehicle data (as typically contained in vehicle registration certificates) are defined in ISO/TS 24534-3.

NOTE 3: The secure application layer interfaces for the exchange of ERI data with an ERI reader or writer are specified in ISO/TS 24534-4 and a future ISO/TS 24534-5.

7.3.3.5 ISO TS 24534-5 Road Transport and Traffic Telematics—Automatic Vehicle and Equipment Identification—Electronic Registration Identification (ERI) for Vehicles—Part 5: Secure Communications Using Symmetric Techniques

ISO	TS	24534-5	Automatic vehicle and equipment identification—Electronic Registration Identification (ERI) for vehicles—Part 5: Secure communications using symmetrical techniques	TS Approved	ISO TC204

ISO TS 24534-5 Available from
http://www.iso.org/iso/iso_catalogue/catalogue_tc/catalogue_detail.htm?csnumber=46201

This part of ISO/TS 24534 specifies the interfaces for a secure exchange of data between an ERT and an ERI reader or ERI writer in or outside the vehicle using symmetric encryption techniques.

7.3.3.6 ISO FDIS 24535 Intelligent Transport Systems—Automatic Vehicle Identification—Basic Electronic Registration (Basic ERI)

ISO	IS	24535	Intelligent transport systems—Automatic vehicle identification—Basic electronic registration (Basic ERI)	Published 2007	ISO TC204

ISO 24535 Available from
http://www.iso.org/iso/iso_catalogue/catalogue_tc/catalogue_detail.htm?csnumber=39589

This International Standard supports simple systems for basic electronic registration identification (basic ERI) for use in intelligent road transport applications.

This International Standard basic ERI defines the:

- Specification of a unique vehicle identifier (using an International Standard, or nonstandard, data concept);
- Basic ERI functional capabilities, selectable for different basic ERI applications;
- Minimum data interoperability requirements between basic ERTs and ERRs.

This deliverable allows, but does not require, vehicle-related data storage in addition to the unique vehicle identifier.

The International Standard is consistent with the ERI architecture defined in ISO 24534 Part 1 and data concepts defined in ISO 24534 Part 3, but is not necessarily interoperable with the more capable fully featured ERI communication systems defined in ISO 24534 Part 4 or Part 5.

ISO 24535 defines a basic ERI system with security adequate for information that is currently available manually (such as license plate and/or VIN), but this International Standard does not purport to provide the high levels of security required for some administrative requirements, and those requiring high security are advised to follow ISO 24534.

Although not part of this International Standard, the employed technologies supporting ISO 24535 may enable additional nonstandard security measures to be added for specific applications.

NOTE: The Basic ERI system implies the use of a basic ERT. Interrogators may or may not support both ISO 24534 Part 4 or part 5 and ISO 24535, this being a commercial and not a Standardization decision.

It is not the purpose of this International Standard to define basic ERI applications.

This International Standard requires the use of an air interface communication that complies with an international or regional standard whose protocols are publicly defined and available in that standard, but this International Standard does not define such an air interface, nor specify which standard air interface is used for any particular implementation.

This International Standard limits its scope to:

- The requirement that a standard air interface with publicly available protocols is used;
- The use of data concepts consistent with ISO 24534 Part 3;
- The ability to additionally use private data concepts.

NOTE: It is important to understand that Parts 1–3 of 24534 and 24535 are common. The primary differences between 24534 Parts 4 and 5 and 24535 are that the form of security is specifically determined in either 24534-4 or 24534-5 (symmetric and asymmetric).

ISO 24535 recognizes that these levels of security are either too onerous for a particular application requirement or defined and controlled.

ISO 24535 permits the user to specify if security measures are used and to determine what security measures are used. Unless there is no security employed, this, of course, does not enable the interoperability of a system compliant with 24534 Parts 4 or 5. However, for closed systems, it provides the quality of a standard compliant system and the interoperability of data once read. Many ERI systems are locally specific and require security but do not require international interoperability. ISO 24535 provides a standard that satisfies these needs.

7.3.4 Cargo Shipment/Goods Item Identification

As more than 20% of vehicles on the road are commercial, one of the important sectors for ITS service provision is to improve the efficiency and security of the freight and fleet logistics chain. This is also seen as an area of strategic importance because the business case benefits are most easy to quantify. Since the 9/11 terrorist attack in the United States, freight security has acquired a higher significance. ISO/TC 204 (ITS) has a working group (WG7) which focuses on the needs for ITS standardization in this area. The WG7 standards are described in Part 3 of this book; however, one of these standards, ISO 24533, has a particular relevance to cargo shipment/goods item identification, and so it is described in this chapter.

Since the invention of the intermodal shipping container, or "box" as it is known in the sector, the whole concept of the movement of goods has changed. It is said that the intermodal shipping container is the biggest single factor in the development of the global economy. However, that success is entirely dependent on boxes being of standard design and quality, and the whole basis of this movement system is, and always has been dependent on International Standards, developed by ISO/TC 104 (Intermodal Containers), a technical committee that is older than the ITS technical committee by nearly two decades.

Some of the standards of TC 104, therefore, have direct relevance to ITS. While the published list is summarized for reference, description is limited to those which have a direct bearing on ISO/TC 204.

Similarly, as a significant proportion of movement of intermodal containers is across the seas, there are some standards from one of the oldest ISO committees (TC 8, Marine) that impact this area. In recent years, TC 8, has shown interest in the use of electronic monitoring of containers.

The work areas of ISO/TC 204, TC 104, and TC 8 potentially overlap in some areas, and efforts are made by these TCs and by ISO Central Secretariat to coordinate these areas through liaisons.

7.3.4.1 ISO 24533 Intelligent Transport Systems—Data Dictionary and Message Set for Tracking of Freight and Its Intermodal Transfer

ISO	CD	24533	Data Dictionary and Message Set for Tracking of Freight and its Intermodal Transfer	Under Development	ISO TC204

ISO 24533 Not Yet Available

This standard is essentially developed, but is being rigorously tested by U.S. Homeland Security and the U.S. DoT, which have requested that finalization of the standard be delayed until the experience resulting from the physical tests can be incorporated.

The seamless exchange of accurate, complete, and timely data at transportation hand-offs has always been important for efficiency and accountability. There is now a growing understanding of needs for security of transport information, as well as for transfer of information related to security against terrorism, theft, and traditional contraband. It is imperative for standards development organizations to address and facilitate these needs.

This International Standard specifies the data concepts applicable to the movement of freight and its intermodal transfer. It also addresses the business processes depicting the roles and responsibilities of the various participants in the international supply chain. This first version of the International Standard focuses on a single thread of the overall end-to-end supply chain consisting of a road-air-road combination. These data concepts include data elements, data frames (groups of data elements), and messages that comprise information exchanges at road transport interfaces along the chain of participants responsible for the delivery of goods from the point of origin through to the final recipient.

7.3.4.2 ISO 668 Freight Containers—Classification, Dimensions and Ratings

ISO	IS	668	Freight containers—Classification, dimensions and ratings	Published 1995	ISO TC104

ISO 668 Available from
http://www.iso.org/iso/iso_catalogue/catalogue_tc/catalogue_detail.htm?csnumber=24007

This classifies Series 1 freight containers based on external dimensions and specifies the associated ratings and, where appropriate, the minimum internal and door opening dimensions for certain types of containers.

7.3.4.3 ISO 830 Freight Containers—Vocabulary

ISO	IS	830	Freight containers—Vocabulary	Published 1999	ISO TC104

ISO 830 Available from
http://www.iso.org/iso/iso_catalogue/catalogue_tc/catalogue_detail.htm?csnumber=1238

7.3.4.4 ISO 1496 Freight Containers—Specification and Testing

ISO	IS	1496	Freight containers—Specification and Testing	Published 1995	ISO TC104

This standard is in multiple parts, only some of which have direct relevance for ITS services that use ISO containers.

ISO	IS	1496-1	Freight containers—Series 1 freight containers Specification and testing Part 1: General cargo containers for general purposes	Published 1990	ISO TC104

ISO 1496-1 Available from
http://www.iso.org/iso/iso_catalogue/catalogue_tc/catalogue_detail.htm?csnumber=6076

This part of ISO 1496 specifies the basic specifications and testing requirements for ISO Series 1 freight containers of the totally enclosed general purpose types and certain specific purpose types (closed, vented, ventilated, or open top) which are suitable for international exchange and for conveyance by road, rail, and sea, including interchange between these forms of transport.

The container types covered by this part of ISO 1496 are given in table 1 of the standard.

This part of ISO 1496 does not cover ventilation arrangements, either vented or ventilated.

The marking requirements for these containers are given in ISO 6346.

NOTE: There are several amendments to this part of the standard.

ISO	IS	1496-2	Freight containers—Series 1 freight containers—Specification and testing—Part 2: Thermal containers	Published 1996	ISO TC104

ISO 1496-2 Available from
http://www.iso.org/iso/iso_catalogue/catalogue_tc/catalogue_detail.htm?csnumber=20449

This specifies the basic specifications and testing requirements for ISO Series 1 thermal containers, which are suitable for international exchange and for conveyance of goods by road, rail, and sea, including interchange between these forms of transport.

ISO	IS	1496-3	Freight containers—Series 1 freight containers—Specification and testing—Part 3: Tank containers for liquids, gases and pressurized dry bulk	Published 1995	ISO TC104

ISO 1496-3 Available from
http://www.iso.org/iso/iso_catalogue/catalogue_tc/catalogue_detail.htm?csnumber=3047

This cancels and replaces the third edition (1991). It specifies the basic specifications and testing requirements for ISO Series 1 tank containers suitable for the carriage of gases, liquids, and solid substances (dry bulk) which may be loaded or unloaded as liquids by gravity or pressure discharge, for international exchange and for conveyance by road, rail, and sea, including interchange between these forms of transport. The requirements are minimum requirements.

ISO	IS	1496-4	Freight containers—Specification and Testing—Part 4: Nonpressurized containers for dry bulk	Published 1991	ISO TC104

ISO 1496-4 Available from
http://www.iso.org/iso/iso_catalogue/catalogue_tc/catalogue_detail.htm?csnumber=6082

This specifies the basic specifications and testing requirements for containers of the dry bulk nonpressurized type, which are suitable for international exchange and for conveyance by road, rail, and sea. Annexes F and G are for information only.

ISO	IS	1496-5	Freight containers—Specification and Testing—Part 5: Platform and platform-based containers	Published 1991	ISO TC104

ISO 1496-5 Available from
http://www.iso.org/iso/iso_catalogue/catalogue_tc/catalogue_detail.htm?csnumber=6084

This specifies the basic specifications and testing requirements for containers designated 1AA, 1A, 1AX, 1BB, 1B, 1BX, 1CC, 1C, and 1CX, which are suitable for international exchange and for conveyance by road, rail, and sea.

7.3.4.5 ISO 3874 Freight Containers—Handling and Securing

ISO	IS	3874	Freight containers—Handling and securing	Published 1991	ISO TC104

ISO 3874 Available from
http://www.iso.org/iso/iso_catalogue/catalogue_tc/catalogue_detail.htm?csnumber=26365

NOTE: There are several amendments to this standard.

7.3.4.6 ISO 6346 Freight Containers—Coding, Identification, and Marking

ISO	IS	6346	Freight containers—Coding, identification, and marking	Published 1995	ISO TC104

ISO 6346 Available from
http://www.iso.org/iso/iso_catalogue/catalogue_tc/catalogue_detail.htm?csnumber=20453

This standard provides a system for general application for the identification and presentation of information about freight containers. It specifies a identification system with mandatory marks for visual interpretation and optional features for automatic identification and electronic data interchange, as well as a coding system for data on container size and type. It replaces the second edition, which has been technically revised.

7.3.4.7 ISO 9711 Freight Containers—Information Related to Containers on Board Vessels—Part 1: Bay Plan System

ISO	IS	9711-1	Freight containers—Information related to containers on board vessels—Part 1: Bay plan system	Published Standard 1990	ISO TC104

ISO 9711 Available from
http://www.iso.org/iso/iso_catalogue/catalogue_tc/catalogue_detail.htm?csnumber=17568

NOTE: Although this standard is referred to as Part 1, there are no other published parts to this standard.

A second part, ISO-9711-2, "Information related to containers on board vessels—Part 2: Telex data," which specified a uniform data information system for transmitting all necessary data and loading plans in a short period of time to the next loading/discharging point using a telex or tele-copy (tele fax) communication system, was approved, but was withdrawn in 2006. (It still remains in some national body lists of standards.)

7.3.4.8 ISO 9897 Freight Containers—Container Equipment Data Exchange (CEDEX)—General Communication Codes

ISO	IS	9897	Freight containers—Container equipment data exchange (CEDEX)—General communication codes	Published Standard 1997	ISO TC104

ISO 9897 Available from
http://www.iso.org/iso/iso_catalogue/catalogue_tc/catalogue_detail.htm?csnumber=20455

7.3.4.9 ISO 10368 Freight Thermal Containers—Remote Condition Monitoring

ISO	IS	10368	Freight thermal containers—Remote condition monitoring	Published 2006	ISO TC104

ISO 10368 Available from
http://www.iso.org/iso/iso_catalogue/catalogue_tc/catalogue_detail.htm?csnumber=36595

ISO 10368:2006 establishes the information and interfaces required to permit complying central monitoring and control systems employed by one carrier or terminal to interface and communicate with complying remote communication devices of differing manufacture and configuration used by other carriers and terminals.

The data-logging formats and message protocols outlined in ISO 10368:2006 apply to all currently available data rate transmission techniques. These formats and protocols also apply to all future techniques designed to be an ISO International Standard compatible system.

7.3.4.10 ISO 10374 Freight Containers—Automatic Identification

ISO	IS	10374	Freight containers—Automatic identification	Published 1991 Standard Under Revision	ISO TC104

ISO 10374 Available from
http://www.iso.org/iso/iso_catalogue/catalogue_tc/catalogue_detail.htm?csnumber=20455

This is one of the oldest RFID standards, its original version being developed in the late 1980s. Since that time, while it is used around the world, it has failed to become pervasively used, and the technology has moved on. The standard is now therefore under a significant review and revision process.

This International Standard establishes:

- A container identification system which allows the transfer of information from a freight container to an automatic processing system by electronic means;
- A data coding system for container identification and permanent related information which resides within an electronic device called a tag installed on a freight container;
- A data coding system for the electronic transfer of both container identification and permanent related information from an electronic device installed on a freight container to automatic data processing systems;
- The description of the data to be included in the tag for transmission to the sensing equipment;
- Performance criteria necessary to ensure consistent and reliable operation of the AEI system within the international transportation community, requirements for the physical location of the electronic device on freight containers, and security features to inhibit malicious or unintentional alteration of the information content of the electronic device when installed on a freight container.

The standard specifies all necessary user requirements in order to permit international use of the tag without modification or adjustment.

This International Standard applies to freight containers as defined in IS0 668.

The use of AEI systems and the equipping of containers for automatic identification are not mandatory.

The purpose of this International Standard is to optimize the efficiency of equipment control systems.

For this reason, any AEI system used for identifying containers shall conform to and be compatible with this International Standard.

The standard defined the requirements and proposed a technical solution as an informative annex, inviting other technical submissions.

In 1995, no other technical solutions being proffered and accepted, amendment 1 to the standard upgraded the only provided technical solution to be a normative part of the standard.

7.3.4.11 ISO15069 Series 1 Freight Containers—Handling and Securing—Rationale for ISO 3874 Annex A

ISO		15069	Freight containers—Series 1 freight containers—Handling and securing—Rationale for ISO 3874 Annex A	Published 1997	ISO TC104

ISO 15069 Available from
http://www.iso.org/iso/iso_catalogue/catalogue_tc/catalogue_detail.htm?csnumber=26366

7.3.4.12 ISO 17712 Freight Containers—Mechanical Seals

ISO	CD	ISO/CD 17712	Freight containers—Mechanical seals	Under Development	ISO TC104

ISO PAS 17712 Available from
http://www.iso.org/iso/iso_catalogue/catalogue_tc/catalogue_detail.htm?csnumber=42791

The developing committee was unable to get enough consensus to develop a standard, so they have published a PAS.

ISO/PAS 17712:2006 establishes uniform procedures for the classification, acceptance, and withdrawal of acceptance of mechanical freight container seals. It provides a single source of information on mechanical seals which are acceptable for securing freight containers in international commerce.

7.3.4.13 Supply Chain Applications of RFID—Freight Containers

ISO		17363	Supply chain applications of RFID—Freight containers	Published 2007	ISO TC104

ISO 17363 Available from
http://www.iso.org/iso/iso_catalogue/catalogue_tc/catalogue_detail.htm?csnumber=38331

ISO 17363:2007 defines the usage of read/write RFID cargo shipment-specific tags on freight containers for supply chain management purposes (shipment tags). It defines the air interface communications, a common set of required data structures, and a commonly organized set of optional data requirements (through common syntax and semantics).

It contains recommendations about a containerized cargo supply chain RFID system, based on shipment tags; specific recommendations about mandatory nonreprogrammable information on the shipment tag; and specific recommendations about optional, reprogrammable information on the shipment tag.

Identified within ISO 17363:2007 are the air interface and communication parameters for active RFID communications using ISO/IEC 18000-7.

ISO 17363:2007 is applicable to freight containers, as defined in ISO 668, and to freight containers that are not defined by other ISO standards. It complements ISO 10374 for permanent container license plate tags.

It fully describes cargo shipment-specific tags. It does not address smart container technologies affixed to, or inside, freight containers (e.g., sensors) for supply chain management purposes.

7.3.4.14 ISO 18185 Freight Containers—Electronic Seals—Part 3: Environmental Characteristics

ISO	IS	18185	Freight containers—Electronic seals	Published Standard	ISO TC104

This standard is in five parts.

ISO	IS	18185-1	Freight containers—Electronic seals—Part 1: Communication protocol	Published Standard	ISO TC104

ISO 18185-1 Available from
http://www.iso.org/iso/iso_catalogue/catalogue_tc/catalogue_detail.htm?csnumber=41125

ISO 18185-1:2007 provides a system for the identification and presentation of information about freight container electronic seals. The identification system provides an unambiguous and unique identification of the container seal, its status and related information.

The presentation of this information is provided through a radio-communications interface providing seal identification and a method for determining whether a freight container's seal has been opened.

ISO 18185-1:2007 specifies a read-only, nonreusable freight container seal identification system, with an associated system for verifying the accuracy of use. It has:

- A seal status identification system;
- A battery status indicator;
- A unique seal identifier including the identification of the manufacturer;
- Seal (tag) type.

ISO 18185-1:2007 is used in conjunction with the other parts of ISO 18185.

It applies to all electronic seals used on freight containers covered by ISO 668, ISO 1496-1 to ISO 1496-5, and ISO 8323. Wherever appropriate and practicable, it also applies to freight containers other than those covered by these International Standards.

ISO	IS	18185-2	Freight containers—Electronic seals—Part 2: Application requirements	Published Standard	ISO TC104

ISO 18185-2 Available from
http://www.iso.org/iso/iso_catalogue/catalogue_tc/catalogue_detail.htm?csnumber=40807

ISO 18185-2:2007 specifies a freight container seal identification system, with an associated system for verifying the accuracy of use. It has:

- A seal status identification system;
- A battery status indicator;
- A unique seal identifier including the identification of the manufacturer;
- A seal (tag) type.

ISO 18185-2:2007 is used in conjunction with the other parts of ISO 18185.

ISO	IS	18185-3	Freight containers—Electronic seals—Part 3: Environmental characteristics	Published Standard	ISO TC104

ISO 18185-3 Available from
http://www.iso.org/iso/iso_catalogue/catalogue_tc/catalogue_detail.htm?csnumber=40808

ISO 18185-3:2006 specifies the minimum environmental characteristics for electronic seals.

ISO 18185-3:2006 describes the environmental requirements for the ISO 18185 series, for ISO 10374 (Freight containers—RF automatic identification) and for ISO 17363 (Supply chain applications of RFID—Freight containers), since it is expected that the implementation of these International Standards will face the same environmental conditions. However, each of these International Standards has its own unique requirements other than environmental conditions.

ISO	IS	18185-4	Freight containers—Electronic seals—Part 4: Data protection	Published Standard	ISO TC104

ISO 18185- 4 Available from
http://www.iso.org/iso/iso_catalogue/catalogue_tc/catalogue_detail.htm?csnumber=40809

ISO 18185-4:2007 specifies requirements for the data protection, device authentication, and conformance capabilities of electronic seals for communication to and from a seal and its associated reader. These capabilities include the accessibility, confidentiality, data integrity, authentication, and nonrepudiation of stored data.

ISO	IS	18185-5	Freight containers—Electronic seals—Part 5: Physical layer	Published Standard	ISO TC104

ISO 18185-5 Available from
http://www.iso.org/iso/iso_catalogue/catalogue_tc/catalogue_detail.htm?csnumber=40810

ISO 18185-5:2007 specifies the air interface between electronic container seals and reader/interrogators of those seals. It is to be used in conjunction with the other parts of ISO 18185.

ISO 18185-5:2007 describes the physical layer for supply chain applications of RFID for freight containers in accordance with the ISO 18185 series and ISO 17363, since it is expected that the implementation of these standards will face the same international conditions. However, each of these standards has its own unique requirements other than the physical layer. It is expected that RFID freight container identification (as specified in ISO 10374 and ISO 17363) and electronic seals (as specified in the ISO 18185 series) will be able to use the same infrastructure, while recognizing that that there may be requirements for different frequencies for passive devices as opposed to the active devices identified in ISO 18185-5:2007.

7.3.4.15 ISO 23389 Freight Containers—Read Write Radio Frequency Identification

ISO	IS	23389	Freight containers—Read write radio frequency identification		ISO TC104

ISO 23389 Not Available

This work item was approved but has been withdrawn in favor of ISO 17363:2000 (see Section 7.4.2).

7.4 Radio Frequency Identification

Radio frequency identification (RFID) is a technology that can be used as a component to support the provision of ITS services. Indeed, one of the first commercial applications of ITS—road tolling—has largely been based on the use of RFID technology. RFID provides a simple, low-cost means to identify an object wirelessly. Because an RFID transponder (commonly called a "tag") can carry a more complex data load than either a bar code or is practicably achievable using optical character recognition, it can carry key information in addition to a simple identification, and can effect a higher level of security. Some RFID devices can also interact with sensors—for example, to identify if a trailer refrigeration unit has been operating outside defined limits, and then pass that information via the wireless RFID link to a host when interrogated. Many of the data concepts described in Section 7.3 can and are effectively transmitted to their host using RFID technology. Within the freight and logistics chain, RFID is playing an increasingly important role in tracking and tracing containers, trailers, pallets, and items. In electronic ticketing, RFID and its sister technologies for smart cards (which are in effect short-range RFID devices) are playing an increasingly important role in the delivery of cost-efficient public transport. RFID is therefore a complementary component technology for ITS service provision.

Wireless communications, as we have stated several times during this book, is a key component of many ITS systems. While this book is not the appropriate place to detail RFID technology in any detail, it is important to explain and understand that RFID is a specific instance of a limited wireless technology, as well as to explain its relevance in ITS systems and its differences from the more capable in-vehicle equipment required for the provision of most of the ITS services described in this book. This means that some short explanation of the technical principles of RFID is useful, particularly if the appropriate RFID standards are to be used in an ITS context.

For most of this book, ITS wireless communications imply the use of some form of conventional two-way wireless system that enables a full bidirectional transmit and receive (Tx/Rx) capability. Therefore, whether we are using a dedicated 5.9-GHz or 63-GHz link, using cellular telephony or mobile wireless broadband, we can effect a normal full, usually duplex, interchange between two parties. Either party can, in many cases, initiate the communication session. The impact of this on the technology is that both the sender and receiver require Tx/Rx capability. This has impact on size, power requirements, and cost (although cost can be mitigated to some extent by volume).

RFID technology enables a limited (usually but not always simplex) communication enabling a tag to provide its identity, and for its memory to be searched and often updated in a limited transaction. Usually the interrogation of an RFID tag is initiated by an interrogator, and the whole process is managed by the interrogator. There are some exceptions to this (such as the continuously scrolling tag used in

ISO 10374, or a particular class of tags known as "active tags" (of which more later).

The principal characteristic of all RFID transponders (tags) is that they are designed to work with a very low power requirement with respect to the tag, and they are normally stand-alone entities. This means that an RFID transponder can be applied to any object and requires no external power supply or connection with the electronic capabilities of the object to which it is affixed. As we have said, it is *possible* for some classes of tag to be attached to sensors, or be wired to receive data updates in some way and store them in its memory, but this is an additional feature, rather than the essential nature of the RFID tag.

Some RFID devices use small low-power batteries to increase their effective range, but many obtain the power that they require in order to transmit their information from the carrier signal from the interrogator. These so-called passive devices have a particularly low cost and an infinite life, however, they are limited in effective range because of the weakness of the signals and relative insensitivity of their low-cost technologies. Battery-assisted passive devices increase the read range, and this is particularly important in order to achieve a useful reading range at higher frequencies. So-called active tags have a full Tx/Rx functionality and can operate at faster speeds and over much greater ranges, and indeed can in some cases operate their own mesh networks between tags. However, they remain stand-alone devices operating limited transactions in order to conserve battery power. Whereas true passive tags have a life limited only by the durability of the product and read/write limits of the technology, both battery-assisted passive tags and active tags are limited primarily by the useful lifetime of the battery.

RFID systems can work at a range of frequencies, of which the most common are 125 to 135 kHz, 13.56 MHz, 433 MHz, 860 to 930 MHz, 2.45 MHz, and 5.8 GHz.

Turning now to the standards that provide consistent and interoperable methodology for these technologies, we find most of the RFID techniques standardized by ISO/IEC SC31, although the first six items are from ISO/TC 122, Packaging.

7.4.1 ISO 17358 Supply Chain Application for RFID—Application Requirements

ISO/IEC	IS	17358	Supply chain application for RFID—Application requirements	Deleted	ISO JTC1 SC31

ISO 17358 is not available

This work item was deleted.

7.4.2 ISO 17363 Supply Chain Applications of RFID—Freight Containers

ISO/IEC	IS	17363	Supply chain applications of RFID—Freight containers	Published	ISO JTC1 SC31

ISO 17363 Available from
http://www.iso.org/iso/iso_catalogue/catalogue_tc/catalogue_detail.htm?csnumber=38331&commid=52040

ISO 17363:2007 defines the usage of read/write RFID cargo shipment-specific tags on freight containers for supply chain management purposes (shipment tags). It defines the air interface communications, a common set of required data structures, and a commonly organized set of optional data requirements (through common syntax and semantics).

It contains recommendations about a containerized cargo supply chain RFID system, based on shipment tags; specific recommendations about mandatory nonreprogrammable information on the shipment tag; and specific recommendations about optional, reprogrammable information on the shipment tag.

Identified within ISO 17363:2007 are the air interface and communication parameters for active RFID communications using ISO/IEC 18000-7.

ISO 17363:2007 is applicable to freight containers as defined in ISO 668 and to freight containers that are not defined by other ISO standards. It complements ISO 10374 for permanent container license plate tags. It fully describes cargo shipment-specific tags. It does not address smart container technologies affixed to, or inside, freight containers (e.g., sensors) for supply chain management purposes.

7.4.3 ISO 17364 Supply Chain Application for RFID—Transport Units

ISO/IEC	FDIS	17364	Supply chain application for RFID—Transport units	Under Development/ FDIS	ISO JTC1 SC31

ISO 17364 Not Yet Available. Track progress at:
http://www.iso.org/iso/iso_catalogue/catalogue_tc/catalogue_detail.htm?csnumber=38333

7.4.4 ISO 17365 Supply Chain Application for RFID—Returnable Transport Items

ISO/IEC	FDIS	17365	Supply chain application for RFID—Returnable transport items	Under Development/ FDIS	ISO JTC1 SC31

ISO 17365 Not Yet Available. Track progress at:
http://www.iso.org/iso/iso_catalogue/catalogue_tc/catalogue_detail.htm?csnumber=38335

7.4.5 ISO 17366 Supply Chain Application for RFID—Product Packaging

ISO/IEC	PRF	17366	Supply chain application for RFID—Product packaging	Under Development/ PRF	ISO JTC1 SC31

ISO 17366 Not Yet Available. Track progress at:
http://www.iso.org/iso/iso_catalogue/catalogue_tc/catalogue_detail.htm?csnumber=38336

7.4.6 ISO 17367 Supply Chain Application for RFID—Product Tagging

ISO/IEC	PRF	17367	Supply chain application for RFID—Product tagging	Under Development/ PRF	ISO JTC1 SC31

ISO 17367 Not Yet Available. Track progress at:
http://www.iso.org/iso/iso_catalogue/catalogue_tc/catalogue_detail.htm?csnumber=38337

7.4.7 ISO 18000-1 Radio Frequency Identification for Item Management—Part 1: Reference Architecture and Definition of Parameters to Be Standardized

ISO/IEC	IS	18000-1	Information technology—Radio frequency identification for item management—Part 1: Reference architecture and definition of parameters to be standardized	Published/ Under Revision	ISO JTC1 SC31

ISO 18000-1 Available from
http://www.iso.org/iso/iso_catalogue/catalogue_tc/catalogue_detail.htm?csnumber=34112

ISO/IEC 18000 has been developed by ISO/IEC SC31 WG4, Radio frequency identification for item identification and management, in order to provide parameter definitions for communications protocols within a common framework for internationally useable frequencies for RFID, and, where possible, to determine the use of the same protocols for ALL frequencies such that the problems of migrating from one to another are diminished; to minimize software and implementation costs; and to enable system management and control and information exchange to be common as far as is possible.

The adoption of a common format and common tables of characteristics also make it simpler to make comparison as to which technology is most likely to be suitable for a particular application. Unfortunately, SC31 has allowed a greater than desirable interpretation of 18000-1 in the subsequent standards: however, the basis for comparison remains in most places.

Informative annexes to this part of ISO/IEC 18000 provide contact information with respect to the radio regulations within which such systems have to operate, as well as some informational views of system architectures within which RFID for item management is likely to be used (ISO 18000-1, Annexes A and C).

There are no specific patents applicable to this part of ISO/IEC 18000. Known patents relating to other parts of ISO/IEC 18000 that may be applicable to one or more parts of ISO/IEC 18000 are provided in an annex to this part of ISO/IEC 18000.

The scope of the standard ISO/IEC 18000-1 explains that this standard is part of ISO/IEC 18000, which describes the generic architecture concepts in which item identification may commonly be required within the logistics and supply chain and defines the parameters that need to be determined in any standardized air interface definition in the subsequent parts of ISO/IEC 18000. The subsequent ISO/IEC 18000 parts of ISO/IEC 18000 provide the specific values for definition of the air interface parameters for a particular frequency/type of air interface from which compliance (or noncompliance) with this part of ISO/IEC 18000 can be established. ISO/IEC 18000-1 also provides a description of example conceptual architectures in which these air interfaces are often utilized.

ISO/IEC 18000-1 limits its scope to transactions and data exchanges across the air interface at reference point Delta (see Figure 7.5). The means of generating and managing such transactions, other than a requirement to achieve the transactional performance determined within this part of ISO/IEC 18000, are outside the scope of this part of ISO/IEC 18000, as is the definition or specification of any supporting hardware, firmware, software, or associated equipments.

The standardization of other reference points is outside the scope of this part of ISO/IEC 18000.

This part of ISO/IEC 18000 is an enabling standard which supports and promotes several RFID implementations without making conclusions about the relative technical merits of any available option for any possible application.

Part 1 of ISO/IEC 18000 also provides reference information with respect to patents that have been declared to the developers as being pertinent to ISO/IEC 18000, and provides reference addresses with respect to regulations under which the ISO/IEC 18000 standards must operate.

7.4.8 ISO 18000-2 Radio Frequency Identification for Item Management—Part 2: Parameters for Air Interface Communications Below 135 kHz

ISO/IEC	IS	18000-2	Information technology—Radio frequency identification for item management—Part 2: Parameters for air interface communications below 135 kHz	Published 2004/Under Revision	ISO JTC1 SC31

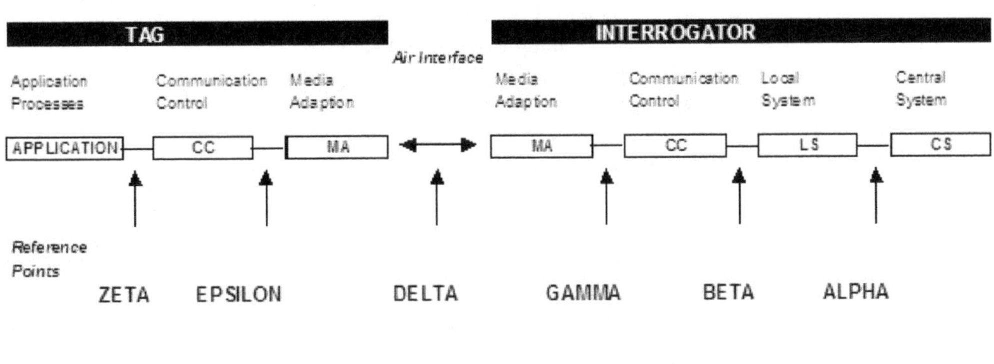

Figure 7.5 RFID reference communications architecture. (*Source*: CSI Library.)

ISO 18000-2 Available from
http://www.iso.org/iso/iso_catalogue/catalogue_tc/catalogue_detail.htm?csnumber=34113

This part of the International Standard ISO/IEC 18000 series relates to systems operating at frequencies less than 135 kHz.

This standard has been developed in accordance with the requirements determined in ISO 18000-1: "Definition of Physical Layer, Anti collision System and Protocols to be International Standardized."

This document specifies:

- The physical layer that shall be used for communication between the interrogator and the tag—the interrogator shall be capable of communicating with tags of both type A (FDX) and type B (HDX);
- The protocol and the commands;
- The method to detect and communicate with one tag among several tags (anti-collision).

Tag Types

This International Standard specifies two types of tags: type A (FDX) and type B (HDX).

These two types differ only by their physical layer. Both types support the same anti-collision and protocol. FDX tags are permanently powered by the interrogator, including during the tag-to-interrogator transmission. They operate at 125 kHz. HDX tags are powered by the interrogator except during the tag-to-interrogator transmission. They operate at 134.2 kHz. An alternative operating frequency is described in the informative Annex B. An optional anticollision mechanism is described in the informative Annex D.

7.4.9 ISO 18000-3 Radio Frequency Identification for Item Management—Part 3: Parameters for Air Interface Communications at 13.56 MHz

ISO/IEC	IS	18000-3	Information technology—Radio frequency identification for item management—Part 3: Parameters for air interface communications at 13.56 MHz	Published 2004/Under Revision	ISO JTC1 SC31

ISO 18000-3 Available from
http://www.iso.org/iso/iso_catalogue/catalogue_tc/catalogue_detail.htm?csnumber=34114

This part of ISO/IEC 18000 has been prepared in accordance with the requirements determined in ISO/IEC 18000-1.

This part of ISO/IEC 18000 has two modes of operation, intended to address different applications. The detailed technical differences between the modes are shown in the parameter tables.

This part of ISO/IEC 18000 relates solely to systems operating at 13.56 MHz.

This part of ISO/IEC 18000 is to provide physical layer, collision management system, and protocol values for RFID systems for item Identification operating at 13.56 MHz

The two modes are not interoperable. The two modes, while not interoperable, are noninterfering.

7.4.10 ISO 18000-4 Radio Frequency Identification for Item Management—Part 4: Parameters for Air Interface Communications at 2.45 GHz

ISO/IEC	IS	18000-4	Information technology—Radio frequency identification for item management—Part 4: Parameters for air interface communications at 2.45 GHz	Published 2004/Under Revision	ISO JTC1 SC31

ISO 18000-4 Available from
http://www.iso.org/iso/iso_catalogue/catalogue_tc/catalogue_detail.htm?csnumber=34115

This part of ISO/IEC 18000 is one of a series of standards and technical reports developed by ISO/IEC JTC 1/SC31 WG4 for the identification of items (Item Management) using RFID technology.

This part of ISO/IEC 18000 contains two modes. The first is a passive tag operating as an interrogator talks first, while the second is a battery assisted tag operating as a tag talks first.

The detailed technical differences between the modes are shown in the parameter tables.

NOTE: There is no 18000-5, which was originally designed to set parameters at 5.8 GHz, because there was an inadequate number of countries prepared to develop the standard.

7.4.11 ISO 18000-6 Radio Frequency Identification for Item Management—Part 6: Parameters for Air Interface Communications at 860 MHz to 960 MHz

ISO/IEC	IS	18000-6	Information technology—Radio frequency identification for item management—Part 6: Parameters for air interface communications at 860 MHz to 960 MHz	Published 2004/Under Revision to include Type C/CD	ISO JTC1 SC31

ISO 18000-6 Available from
http://www.iso.org/iso/iso_catalogue/catalogue_tc/catalogue_detail.htm?csnumber=34117

This International Standard specifies the physical and logical requirements for a passive-backscatter, interrogator-talks-first (ITF) RFID system operating in the 860 to 960-MHz frequency range. The system comprises interrogators (also known as readers) and tags (also known as labels).

The revisions are mainly to type C, and also to make provisions for battery assisted tags and tags linked to sensors. At the time of publication of this book, they are part way through the revision ballot process.

This part of ISO/IEC 18000 describes a passive backscatter RFID system that supports the following system capabilities:

- Identification and communication with multiple tags in the field;
- Selection of a subgroup of tags for identification or with which to communicate;
- Reading from and writing to or rewriting data many times to individual tags;
- User-controlled permanently lockable memory;
- Data integrity protection;
- Interrogator-to-tag communications link with error detection;
- Tag-to-interrogator communications link with error detection;
- Support for both passive backscatter tags with or without batteries.

This part of ISO/IEC 18000 contains one mode with three types. All three types are ITF. Type A uses pulse-interval encoding (PIE) in the forward link and an adaptive ALOHA collision-arbitration algorithm. Type B uses Manchester in the forward link and an adaptive binary-tree collision-arbitration algorithm. Type C uses PIE in the forward link and a random slotted collision-arbitration algorithm. The detailed technical differences between the three types are shown in the associated parameter tables. This part of ISO/IEC 18000 provides specifications for:

- Physical interactions (the signaling layer of the communication link) between interrogators and tags;
- Interrogator and tag operating procedures and commands;
- The collision arbitration scheme used to identify a specific tag in a multiple-tag environment.

7.4.12 ISO 18000-7 Radio Frequency Identification for Item Management—Part 7: Parameters for Active Air Interface Communications at 433 MHz

ISO/IEC	IS	18000-7	Information technology—Radio frequency identification for item management—Part 7: Parameters for active air interface communications at 433 MHz	Published/ Under Revision	ISO JTC1 SC31

ISO 18000-7 Available from
http://www.iso.org/iso/iso_catalogue/catalogue_tc/catalogue_detail.htm?csnumber=37978

This standard is intended to address RFID devices operating in the 433-MHz frequency band.

ISO 18000-7 defines the air interface for RFID devices operating as an active RF tag in the 433-MHz band used in item management applications.

7.4.13 ISO TR 18001 Information Technology—Radio Frequency Identification for Item Management—Application Requirements Profiles

ISO/IEC	TR	18001	Information technology—Radio frequency identification for item management—Application requirements profiles	Published TR 2004	ISO JTC1 SC31

ISO TR 18001 Available from
http://www.iso.org/iso/iso_catalogue/catalogue_tc/catalogue_detail.htm?csnumber=40733

ISO/IEC TR 18001:2004 provides:

- The result of three surveys identifying the applications for RFID in an item management environment, and the resultant classification of these applications based on various operational parameters, including operating range and memory size;
- An explanation of some of the issues associated with the parameters of distance and number of tags within an RFID interrogator's field of view;
- A means by which classification of RF tags may be accomplished based on the application requirements defined in the survey results;
- Recommendations for areas of standardization to the parent committee (ISO/IEC JTC 1/SC31 WG4) based on the results of these surveys.

7.4.14 ISO 15961 Radio Frequency Identification (RFID) for Item Management—Data Protocol: Application Interface

ISO/IEC	IS	15961	Information technology—Automatic identification and data capture—Radio frequency identification (RFID) for item management—Data protocol: application interface	Published 2004	ISO JTC1 SC31

ISO 15961 Available from
http://www.iso.org/iso/iso_catalogue/catalogue_tc/catalogue_detail.htm?csnumber=30528

The technology of RFID is based on noncontact electronic communication across an air interface. The structure of the bits stored on the memory of the RFID tag is invisible and accessible between the RF tag and the interrogator only by the use of an air interface protocol, as specified in the appropriate part of ISO/IEC 18000. The transfer of data between the application and the interrogator in open systems requires data to be presented in a consistent manner on any RF tag that is part of that open system. Application commands from the application and responses from the interrogator also require being processed in a standard way. This is not only to allow equipment to be interoperable, but in the special case of data carriers, for the data to be encoded on the RF tag in one systems implementation and for it to be read at a later time in a completely different and unknown systems implementation. The data bits stored on each RF tag must be formatted in such a way as to be reliably read at the point of use if the RF tag is to fulfill its basic objective. The integrity of this is achieved through the use of a data protocol as specified in this International Standard and ISO/IEC 15962.

Manufacturers of RFID equipment (interrogators, RF tags, and so forth) and the users of RFID technology require a publicly available data protocol for RFID for item management. This part of this International Standard and ISO/IEC 15962 specify this data protocol, which is independent of any of the air interface standards defined in ISO/IEC 18000. As such, the data protocol is a consistent component in the RFID system that may independently evolve to include additional air interface protocols.

The transfer of data to and from the application, supported by appropriate application commands is the subject of this part of this International Standard. The companion International Standard, ISO/IEC 15962, specifies the overall process and the methodologies developed to format the application data into a structure for storage on the RF tag.

The data protocol used to exchange information in an RFID system for item management is specified in this multipart International Standard and in ISO/IEC 15962. All the International Standards are required for a complete understanding of the data protocol in its entirety; but each focuses on one particular interface:

- ISO/IEC 15961 Part 1 addresses the information interface, which deals with item-related data, with the application system. This part of the standard is primarily of interest to those implementing software supporting the features of the application interface.
- ISO/IEC 15961 Part 2 addresses the registration of data constructs that define the application parameters of this International Standard. This part of the standard is primarily of interest to user organizations.
- ISO/IEC 15961 Part 3 addresses the technical definition of the data constructs. This part of the standard is primarily of interest to those implementing software supporting the features of the application interface.
- ISO/IEC 15961 Part 3 addresses the information interface, which deals with sensor-related data, with the application system. This part of the standard is primarily of interest to those implementing software supporting the features associated with sensor data of the application interface.
- ISO/IEC 15962 deals with the processing of item-related data and its presentation to the RFID tag, and the initial processing of data captured from the RFID tag. This International Standard is primarily of interest to those

implementing software supporting the item-related data features of the application interface.

- ISO/IEC 24753 deals with the processing of sensor-related data and its presentation to the RFID tag and the initial processing of sensor-related data captured from the RFID tag. ISO/IEC 24753 is primarily of interest to those implementing software supporting the sensor-related features of the application interface.

ISO/IEC	CD	15961-1	Information technology—Automatic identification and data capture—Radio frequency identification (RFID) for item management—Data protocol: Part 1: application interface	Under Development/ CD	ISO JTC1 SC31

ISO 15961-1 Not Yet Available. Track progress at:
http://www.iso.org/iso/iso_catalogue/catalogue_tc/catalogue_detail.htm?csnumber=43458

This part of this International Standard focuses on the interface between the application and the data protocol processor and includes the specification and definition of application commands and responses. It allows data and commands to be specified in a standardized way, independent of the particular air interface of ISO/IEC 18000.

This International Standard:

- Provides guidelines on how data shall be presented as objects;
- Defines the structure of object identifiers, based on ISO/IEC 9834-1;
- Specifies the commands that are supported for transferring data between the application and the RFID tag;
- Specifies the responses that are supported for transferring data between the RFID tag and the application;
- Does not specify any transfer syntax with ISO/IEC 15962, but provides sufficient information to enable different transfer mechanisms to be developed.

NOTE: In ISO/IEC 15961:2004 a transfer encoding, based on the basic encoding rules of ISO/IEC 8825-1, was specified as the transfer syntax with ISI/IEC 15962. Since publication of the original edition, it has been found that this transfer syntax can be a constraint on systems designs that integrate the application interface and ISO/IEC 15962 within one device or software configuration. In addition, other mechanisms have been found to provide similar functionality. Aspects of the original transfer encoding have been retained in this part of this International Standard for backward compatibility.

It is expected that this part of this International Standard will be used as a reference to develop software appropriate for particular applications, or for particular RFID equipment.

NOTE: Conventionally in International Standards, long numbers are separated by a space character as a thousands separator. This convention has not been followed in this International Standard, because the arcs of an object identifier are defined by a space separator (according to ISO/IEC 8824 and ISO/IEC 8825). As the correct representation of these arcs is vital to this International Standard, all numeric values have no space separators except to denote a node between two arcs of an object identifier.

ISO/IEC	CD	15961-2	Information technology—Automatic identification and data capture—Radio frequency identification (RFID) for item management—Data protocol: Part 2: Registration of RFID Data Constructs	Under Development/ CD	ISO JTC1 SC31

ISO 15961-2 Net Yet Available. Track Progress at:
http://www.iso.org/iso/iso_catalogue/catalogue_tc/catalogue_detail.htm?csnumber=43631

This part of ISO/IEC 15961:

- Specifies the procedural requirements to maintain specific RFID data constructs associated with managing open and closed applications that utilize RFID systems that are conforming to the data protocol defined in other parts of ISO/IEC 15961 and ISO/IEC 15962, and the air interface protocols of ISO/IEC 18000;
- Outlines the obligations of the registration authority and the application administrators, with respect to:
  - The allocation of AFIs to particular applications defined by the application administrator;
  - The allocation of data formats to particular applications defined by the application administrator;
  - The registration of root-OIDs, compliant with ISO/IEC 9834-1, to any unique item identifiers used in applications defined by the application administrator;
  - The registration of root-OIDs, compliant with ISO/IEC 9834-1, to any other data used in applications defined by the application administrator.

ISO/IEC	CD	15961-3	Information technology—Automatic identification and data capture—Radio frequency identification (RFID) for item management—Data protocol: Part 3	Under Development/ CD	ISO JTC1 SC31

ISO 15961-3 Not Yet Available. Track progress at:
http://www.iso.org/iso/iso_catalogue/catalogue_tc/catalogue_detail.htm?csnumber=43632

This part of ISO/IEC 15961 specifies rules and code structures associated with the data constructs for RFID for item management. In particular, it:

- Defines the application family identifier (AFI), including the range of code values that are available to use for RFID for item management;
- Defines the data format, including the range of code values that are available to use for RFID for item management;
- Describes the object identifier structure used for RFID for item management;
- Specifies the function of the object identifier for the unique item identifier (UII);
- Specifies the function of the object identifier for other item attendant data.

7.4.15 ISO 15962 Radio Frequency Identification (RFID) for Item Management—Data Protocol: Data Encoding Rules and Logical Memory Functions

ISO/IEC	IS	15962	Information technology—Automatic identification and data capture—Radio frequency identification (RFID) for item management—Data protocol: data encoding rules and logical memory functions	Published 2004	ISO JTC1 SC31

ISO 15962 Available from
http://www.iso.org/iso/iso_catalogue/catalogue_tc/catalogue_detail.htm?csnumber=30529

The data protocol used to exchange information in an RFID system for item management is specified in ISO/IEC 15961:2004 and in ISO/IEC 15962:2004. Both are required for a complete understanding of the data protocol in its entirety, but each focuses on one particular interface, as follows:

- ISO/IEC 15961:2004 addresses the interface with the application system.
- ISO/IEC 15962:2004 deals with the processing of data and its presentation to the RF tag, and the initial processing of data captured from the RF tag.
- ISO/IEC 15962:2004 focuses on encoding the transfer syntax, as defined in ISO/IEC 15961:2004, according to the application commands defined in that International Standard. The encodation is in a logical memory as a software analog of the physical memory of the RF tag being addressed by the interrogator.
- ISO/IEC 15962:2004:
  - Defines the encoded structure of object identifiers;
  - Specifies the data compaction rules that apply to the encoded data;
  - Specifies a precursor for encoding syntax features efficiently;
  - Specifies formatting rules for the data (e.g., depending on whether a directory is used or not);

- Defines how application commands (e.g., to lock data) are transferred to the tag driver;
- Defines other communication to the application.

7.4.16 ISO 15963 Radio Frequency Identification (RFID) for Item Management—Unique Identification for RF Tags

ISO/IEC	IS	15963	Information technology—Automatic identification and data capture—Radio frequency identification (RFID) for item management—Unique identification for RF tags	Published 2004	ISO JTC1 SC31

ISO 15963 Available from
http://www.iso.org/iso/iso_catalogue/catalogue_tc/catalogue_detail.htm?csnumber=30530

ISO/IEC 15963:2004 describes numbering systems that are available for the identification of RF tags.

A unique ID is required as part of the write operation to RFID tags. The unique ID guarantees that the information written to a tag is unambiguously written to the correct data carrier (tag). A unique ID is also required in many read situations where the contents of the tag are tied to a specific item and that item needs to be unambiguously identified.

The unique ID may also be used

- For the traceability of the integrated circuit itself for quality control in their manufacturing process;
- For the traceability of the RF tag during its manufacturing process and along its lifetime;
- For the completion of the reading in a multi-antenna configuration;
- By the anti-collision mechanism, to inventory multiple tags in the reader's field of view;
- For the traceability of the item to which the RF tag is attached.

7.4.17 ISO 18046 RFID Tag and Interrogator Performance Test Methods

ISO/IEC	IS	18046	Information technology—Automatic identification and data capture techniques—Radio frequency identification device performance test methods	Published 2004	ISO JTC1 SC31

ISO 18046 Available from
http://www.iso.org/iso/iso_catalogue/catalogue_tc/catalogue_detail.htm?csnumber=43174

RFID technology has broad applicability to the automatic identification and data capture (AIDC) industry in item management. As a wireless communication technique based on RF technology, the applications cover multiple levels of the industrial, commercial, and retail supply chains. These may include:

- Freight containers;
- Returnable transport items (RTIs);
- Transport units;
- Product packaging;
- Product tagging.

7.4.18 ISO 18047 Information Technology—Automatic Identification and Data Capture—RFID Device Conformance Test Methods

ISO/IEC	18047	Information technology—Automatic identification and data capture—RFID device conformance test methods	Published	ISO JTC1 SC31

This technical report is in multiple parts. The part number matches the part number for its equivalent frequency in ISO 18000 (see Sections 7.4.8 through 7.4.12) for which it provides conformance test methods.

There is, therefore, no part 1 (ISO 18000-1 does not specify a mode). Also, there is no part 5, because ISO 18000-5 failed to make progress and was deleted. Part 6 as 18000-6 does not include tests for 18000-6C, which is currently a major revision process of ISO 18000-6.

ISO/IEC 18000 defines the air interfaces for RFID devices used in item management applications.

The purpose of ISO/IEC TR 18047 is to provide test methods for conformance with the various parts of ISO/IEC 18000.

Each part of ISO/IEC TR 18047 contains all measurements required to be made on a product in order to establish whether it conforms with the corresponding part of ISO/IEC 18000.

It should be noted that measurement of tag and interrogator performance is covered by ISO/IEC TR 18046.

ISO/IEC	TR	18047-2	Information technology—Automatic identification and data capture—RFID device conformance test methods—Part 2: Test methods for air interface communication below 135 kHz	Published TR 2006	ISO JTC1 SC31

ISO 18047-2 Available from
http://www.iso.org/iso/iso_catalogue/catalogue_tc/catalogue_detail.htm?csnumber=41917

ISO/IEC 18000-2 defines the air interface for these devices operating in frequencies below 135 kHz.

ISO/IEC TR 18047-2:2006 defines test methods for determining the conformance of RFID devices (tags and interrogators) for item management with the specifications given in ISO/IEC 18000-2, but it does not apply to the testing of conformity with regulatory or similar requirements.

For ISO/IEC TR 18047-2:2006, each interrogator needs to be assessed with tags of both type A (FDX) and type B (HDX), while each tag needs to be assessed either with type A (FDX) or type B (HDX).

The test methods require only that the mandatory functions, and any optional functions which are implemented, be verified. This may, in appropriate circumstances, be supplemented by further, application-specific functionality criteria that are not available in the general case.

The interrogator and tag conformance parameters in ISO/IEC TR 18047-2:2006 are the following:

- Mode-specific conformance parameters including nominal values and tolerances;
- Parameters that apply directly affecting system functionality and interoperability.

The following are not included in ISO/IEC TR 18047-2:2006:

- Parameters that are already included in regulatory test requirements;
- High-level data encoding conformance test parameters (these are specified in ISO/IEC 15962);
- Unless otherwise specified, the tests in this part of ISO/IEC TR 18047-2:2006 are to be applied exclusively to RFID tags and interrogators defined in ISO/IEC 18000-2.

ISO/IEC TR 18047-2:2006 also describes all necessary conformance tests.

ISO/IEC	TR	18047-3	Information technology—Automatic identification and data capture—RFID device conformance test methods—Part 3: Test methods for air interface communications at 13.56 MHz	Published TR 2006	ISO JTC1 SC31

ISO 18047-3 Available From
http://www.iso.org/iso/iso_catalogue/catalogue_tc/catalogue_detail.htm?csnumber=39490

ISO/IEC TR 18047-3:2004 includes the following interrogator and tag conformance parameters:

- Mode-specific conformance parameters including nominal values and tolerances;

- Parameters that apply directly affecting system functionality and interoperability.

ISO/IEC TR 18047-3:2004 does not include the following:

- Parameters that are already included in regulatory test requirements;
- High-level data encoding conformance test parameters (these are specified in ISO/IEC 15962).

ISO/IEC	TR	18047-4	Information technology—Automatic identification and data capture—RFID device conformance test methods—Part 4: Test methods for air interface communications at 2.45 GHz	Published TR 2004	ISO JTC1 SC31

ISO 18047-4 Available From
http://www.iso.org/iso/iso_catalogue/catalogue_tc/catalogue_detail.htm?csnumber=39493

The interrogator and tag conformance parameters in ISO/IEC TR 18047-4:2004 are:

- Mode-specific conformance parameters including nominal values and tolerances;
- Parameters that apply directly affecting system functionality and inter-operability.

The following are not included in ISO/IEC TR 18047-4:2004:

- Parameters that are already included in regulatory test requirements;
- High-level data encoding conformance test parameters (these are specified in ISO/IEC 15962).

Unless otherwise specified, the tests in ISO/IEC TR 18047-4:2004 apply exclusively to RFID tags and interrogators defined in ISO/IEC 18000-4 mode 2.

ISO/IEC	IS	18047-6	Information technology—Automatic identification and data capture—RFID device conformance test methods—Part 6: Test methods for air interface communication at 860–960 MHz	Published TR 2006	ISO JTC1 SC31

ISO 18047-6 Available from
http://www.iso.org/iso/iso_catalogue/catalogue_tc/catalogue_detail.htm?csnumber=41916

NOTE: ISO 18047-6 does not yet include tests for the 18000-6C mode currently under development.

NOTE: The measurement of tag and interrogator performance is covered by ISO/IEC TR 18046.

ISO/IEC TR 18047-6:2006 defines test methods for determining the conformance of RFID devices (tags and interrogators) for item management with the specifications given in ISO/IEC 18000-6, but it does not apply to the testing of conformity with regulatory or similar requirements.

The test methods require only that the mandatory functions, and any optional functions which are implemented, be verified. This may, in appropriate circumstances, be supplemented by further, application-specific functionality criteria that are not available in the general case.

For ISO/IEC TR 18047-6:2006, each interrogator needs to be assessed for operation with both types A and B, while each tag is only required to support at least one of the types A or B.

The interrogator and tag conformance parameters in ISO/IEC TR 18047-6:2006 are as follows:

- Type-specific conformance parameters including nominal values and tolerances;
- Parameters that apply directly affecting system functionality and interoperability.

The following are not included in ISO/IEC TR 18047-6:2006:

- Parameters that are already included in regulatory test requirements;
- High-level data encoding conformance test parameters (these are specified in ISO/IEC 15962).

Unless otherwise specified, the tests in ISO/IEC TR 18047-6:2006 are to be applied exclusively to RFID tags and interrogators defined in ISO/IEC 18000-6.

ISO/IEC TR 18047-6:2006 also describes all necessary conformance tests.

ISO/IEC	IS	18047-7	Information technology—Automatic identification and data capture—RFID device conformance test methods—Part 7: Test methods for air interface communication at 433 MHz	Published TR 2005	ISO JTC1 SC31

ISO 18047-7 Available from
http://www.iso.org/iso/iso_catalogue/catalogue_tc/catalogue_detail.htm?csnumber=39496

ISO/IEC TR 18047-7:2005 defines test methods for determining the conformance of RFID devices (tags and interrogators) for item management with the specifications given in ISO/IEC 18000-7, but it does not apply to the testing of conformity with regulatory or similar requirements.

The test methods require only that the mandatory functions, and any optional functions which are implemented, be verified. This may, in appropriate circumstances, be supplemented by further, application-specific functionality criteria that are not available in the general case.

The interrogator and tag conformance parameters in ISO/IEC TR 18047-7:2005 are as follows:

- Mode-specific conformance parameters including nominal values and tolerances;
- Parameters that apply directly affecting system functionality and inter-operability.

The following are not included in ISO/IEC TR 18047-7:2005:

- parameters that are already included in regulatory test requirements;
- High-level data encoding conformance test parameters (these are specified in ISO/IEC 15962).

Unless otherwise specified, the tests in ISO/IEC TR 18047-7:2005 apply exclusively to RFID tags and interrogators defined in ISO/IEC 18000-7.

7.4.19 ARIB T92 Specified Low Power Radio Station 433 MHz-Band Data Transmission Equipment for International Logistics

ARIB	STD-T92	Specified Low Power Radio Station 433 MHz-Band Data Transmission Equipment for International Logistics	2007.3.14	Ver.1.0 546

ARIB T92 Available from
http://www.arib.or.jp/english/html/overview/st_j.html

Users at 433 MHz in Japan should be aware of this standard.

7.5 Track and Trace

Generic "track and trace" standards are based on identification, generally developed by the same SDO working groups, and so they are described in this section.

7.5.1 ISO 15459 Information Technology—Unique Identification of Transport Units

ISO/IEC	IS	15459	Information technology—Unique identification of transport units	Published	ISO JTC1 SC31

Unique identification can occur at many different levels in the supply chain, at the transport unit, at the item level, and elsewhere. Such distinct entities are often handled by several parties: the sender, the receiver, one or more carriers, customs authorities, and so on. Each of these parties must be able to identify and trace the item so that reference can be made to associated information such as address, order number, contents of the item, weight, sender, and batch or lot number.

The information is often held on computer systems, and may be exchanged between parties involved via electronic data interchange (EDI) and extensible markup language (XML) messages.

There are considerable benefits if the identity of the item is represented in bar code format, or other automatic identification and data capture (AIDC) media and attached to or made a constituent part of that which is being uniquely identified so that:

- It can be read electronically, thus minimizing errors;
- One identifier can be used by all parties;
- Each party can use the identifier to look up its computer files to find the data associated with the item;
- The identifier is unique within the class and cannot appear on any other item of the class during the lifetime of the item.

All AIDC technologies have the potential to encode a unique identifier. It is expected that application standards for items, using various automatic identification technologies, will be developed based upon the unique identifier as a prime key. These application standards may be made available from the issuing agency.

ISO/IEC	IS	15459-1	Information technology—Unique identification of transport units—Part 1: Technical Standard	Published	ISO JTC1 SC31

ISO 15459-1 Available from
http://www.iso.org/iso/iso_catalogue/catalogue_tc/catalogue_detail.htm?csnumber=43347

The unique identifier for transport units defined in ISO/IEC 15459-1:2006 and represented in a bar code label, two-dimensional symbol, radio-frequency identification tag, or other AIDC media attached to the item meets these needs.

ISO/IEC 15459-1:2006 specifies a unique, nonsignificant, string of characters for the identification of transport units. The character string is intended to be represented in a bar code label or other AIDC media attached to the item to meet item management needs. To address management needs different classes of items are recognized in the various parts of ISO/IEC 15459, which allows different requirements to be met by the unique identifiers associated with each class. The rules for the unique identifier for transport units, to identify physical logistical transfers, with the identity relevant for the duration of one or more items in the load being held or transported as part of that load, are defined and supported by an example.

ISO/IEC	IS	15459-2	Information technology—Unique identification of units in the Supply Chain—Part 2: Registration procedures	Published	ISO JTC1 SC31

ISO 15459-2 Available from
http://www.iso.org/iso/iso_catalogue/catalogue_tc/catalogue_detail.htm?csnumber=43348

ISO/IEC 15459-2:2006 specifies the procedural requirements to maintain a nonsignificant, unique identifier for item management applications, and outlines the obligations of the registration authority and issuing agencies.

ISO/IEC 15459-2:2006 excludes those items where ISO has designated maintenance agencies or registration authorities to provide identification schemes. It does not apply to:

- Freight containers, because their unique coding is specified in ISO 6346, Freight containers—Coding, identification and marking;
- Vehicles, because their unique identification is specified in ISO 3779, Road vehicles—Vehicle identification number (VIN)—Content and structure;
- Car radios, because their unique identification is specified in ISO 10486, Passenger cars—Car radio identification number (CRIN).

The exclusion also applies to ISO 2108, Information and documentation—International standard book number (ISBN), and ISO 3297, Information and documentation—International standard serial number (ISSN).

ISO/IEC	IS	15459-3	Information technology—Unique identifiers—Part 3: Common rules for unique identifiers	Published	ISO JTC1 SC31

ISO 15459-3 Available from
http://www.iso.org/iso/iso_catalogue/catalogue_tc/catalogue_detail.htm?csnumber=43349

ISO/IEC 15459-3:2006 specifies the common rules that apply for unique identifiers for item management that are required to ensure full compatibility across classes of unique identifiers.

ISO/IEC	IS	15459-4	Information technology—Unique identifiers—Part 4: Unique identifiers for supply chain management	Published	ISO JTC1 SC31

ISO 15459-4 Available from
http://www.iso.org/iso/iso_catalogue/catalogue_tc/catalogue_detail.htm?csnumber=43455

The unique identifier for supply chain management defined in ISO/IEC 15459-4:2006 and represented in a bar code label, two-dimensional symbol, radio-frequency identification tag, or other AIDC media attached to the item meets these needs.

All AIDC technologies have the potential to encode a unique identifier. It is expected that application standards for items, using various automatic identification technologies, will be developed based upon the unique identifier as a prime key. These application standards may be made available from the issuing agency.

ISO/IEC 15459-4:2006 specifies a unique, nonsignificant string of characters for the unique identifier for supply chain management. The character string is intended to be represented in a bar code label or other AIDC media attached to the item to meet supply chain management needs. To address management needs different classes of items are recognized in the various parts of ISO/IEC 15459, which allows different requirements to be met by the unique identifiers associated with each class. The rules are defined for the unique identifiers for supply chain management to identify the unique occurrence of an item, understood to mean the layers zero and one as will be defined in two future International Standards: ISO 17367 (Section 7.4.6) and ISO 17366 (Section 7.4.5), respectively.

ISO/IEC	CD	15459-5	Information technology—Unique identification of transport units—Part 5: Unique identification of returnable transport items (RTIs)	Under Development	ISO JTC1 SC31

ISO 15459-5 Available from
http://www.iso.org/iso/iso_catalogue/catalogue_tc/catalogue_detail.htm?csnumber=43346

ISO/IEC 15459-5:2007 specifies a unique, nonsignificant string of characters for the unique identification of returnable transport items (RTIs). The character string is intended to be represented in an RFID transponder, bar code label, or other AIDC media attached to the item to meet supply chain management needs. To address management needs, different classes of RTI are recognized in the various parts of ISO/IEC 15459, which allows different requirements to be met by the unique identifiers associated with each class. The rules for the unique identifier for RTIs, to identify the unique occurrence of an item, with the identity being relevant for the complete life cycle of the item, are defined and supported by an example.

ISO/IEC	IS	15459-6	Information technology—Unique identification of transport units—Part 6: Unique identification for product groupings in material lifecycle management	Published	ISO JTC1 SC31

ISO 15459-6 Available from
http://www.iso.org/iso/iso_catalogue/catalogue_tc/catalogue_detail.htm?csnumber=44149

ISO/IEC 15459-6:2007 specifies a unique, nonsignificant string of characters for the unique identifier of product groupings. The character string is intended to be represented in linear bar code and two-dimensional symbols, RFID transponder, or other AIDC media attached to the product and/or material to meet the management needs in a batch or lot unit. To address management needs, different classes of item are recognized in the various parts of ISO/IEC 15459. This allows different requirements to be met by the unique identifiers of each class.

The unique identifier for product grouping enables a product grouping defined by a batch or lot number to be uniquely identified from all other lots and batches compliant with ISO/IEC 15459-6:2007. Encoding this unique identifier in a data carrier enables information about the quality of product and end-of-life processing to be clearly identified.

The rules for the unique identifier for product grouping, to identify the unique occurrence of that quality, are defined and supported by an example.

7.5.2 ISO 24710 Information Technology AIDC Techniques—RFID for Item Management—ISO/IEC 18000 Air Interface Communications—Elementary Tag License Plate Functionality for ISO/IEC 18000 Air Interface Definitions

ISO/IEC	IS	24710	Information technology—Unique identification of transport units—Part 6: Unique identification for product groupings in material lifecycle management	Published	ISO JTC1 SC31

ISO TR 24710 Available from
http://www.iso.org/iso/iso_catalogue/catalogue_tc/catalogue_detail.htm?csnumber=38820

ISO/IEC 18000 defines the operation of RFID air interfaces for item identification and management.

ISO/IEC 18000 has been designed to encompass a full range of data capture and carrier functionality. Both read and write operations are enabled, and the interfaces can efficiently support both simple and complex data transactions.

This approach facilitates user implementation by providing consistency between differing types of RFID data transactions. Equally it provides architecture to guide future RFID development while maintaining the backward compatibility necessary to sustain market confidence.

Recent developments in the design and management of distributed databases holding item level information have focused attention on "identification data element" operation of RFID systems. In this application, the RFID tag carries only sufficient data to permit reference to attribute information held elsewhere. Typically this data does not change during the validity of the license and is of relatively low bit count.

ISO/IEC TR 24710:2005 has been prepared to assist users intending to implement ISO/IEC 18000 RFID air interface standards, with particular focus on so-called elementary tags (i.e., tags possessing limited memory—typically but not exclusively 256 bits or less—and lacking write capability, but not excluding WORM devices).

The annexes to ISO/IEC TR 24710:2005 describe the implementation of ISO/IEC 18000-2, -3, -4, -6, and -7 in such an application.

Users are strongly advised to refer to ISO/IEC 15961 and ISO/IEC 15962 for a full exposition of the management issues relating to data strings used for identification data element purposes.

Bodies external to ISO also specify identification data element length and structure for particular applications.

ISO/IEC TR 24710:2005 defines for each of ISO/IEC 18000-2, -3, -4, -6, and -7, and, where relevant, for each mode within each part, a transaction that achieves an elementary tag identification data element for item identification and management.

The transaction uses the existing air interface protocols defined in the corresponding parts of ISO/IEC 18000 or a subset thereof.

7.5.3 ISO 24720 Automatic Identification and Data Capture Techniques—Guidelines for Direct Part Marking (DPM)

ISO/IEC	TR/ CD	24720	Information technology—Automatic identification and data capture techniques—Guidelines for Direct Part Marking (DPM)	TR under development	ISO JTC1 SC31

ISO 24720 Available from
http://www.iso.org/iso/iso_catalogue/catalogue_tc/catalogue_detail.htm?csnumber=38830

Industries worldwide rely heavily on the use of various marking methods for parts identification. Because many of these methods were originally designed to apply human-readable marks, they frequently are not appropriate for applying high-density machine-readable symbols.

With the widespread implementation of machine-readable marking, the parts identification industry began to refine existing marking methods. Manual metal stamp and embossing techniques were replaced by dot peen machines. Photo stencils and thermal printing materials were developed to replace direct impact electro-chemical marking stencil materials. Desktop publishing systems were developed for the production of stencils. Ink jet machines were built to replace rubber stamps. Laser marking systems were designed to replace electric-arc etch and hot stamp processes.

For the purposes of this technical report, direct part marking (DPM) is considered a generic term referring to methods of applying a permanent mark directly onto a surface of an item. There are two techniques for applying a permanent mark described in this report, intrusive and nonintrusive.

Intrusive (or subtractive) marking methods alter the surface of a part and are considered controlled defects.

This technical report addresses dot peen and direct laser marking in depth, and briefly describes other technologies.

Nonintrusive, also known as additive markings, are produced as part of the manufacturing process or by adding a layer of media to the surface of a part. This report addresses ink jet marking in depth and other technologies only briefly.

This ISO/IEC technical report establishes uniform recommendations for applying permanent machine-readable marks to items—including components, parts, and products—using the DPM methods outlined in the report.

This document addresses marking method selection, marking surface preparation, marking location and protective coatings. The document also addresses the marking of human readable characters in conjunction with the two dimensional symbols and provides guidance to application standard development groups for specifying the symbol quality grade for direct marked symbols. This report does not specify the information to be encoded.

7.5.4 ISO 24729 Radio Frequency Identification for Item Management—Implementation Guidelines

ISO/IEC	TR/ CD	24729	Information technology—Radio frequency identification for item management—Implementation guidelines	Under development/ DTR	ISO JTC1 SC31

ISO/IEC 24729, "Information technology—Radio frequency identification for item management—Implementation guidelines," is in three parts:

Part 1: RFID-enabled labels and packaging;
Part 2: Recyclability of RF tags;
Part 3: RFID interrogator/antenna installation.

ISO/IEC	TR/ CD	24729-1	Information technology—Radio frequency identification for item management—Implementation guidelines—Part 1: RFID-enabled labels	DTR under development	ISO JTC1 SC31

ISO 24729-1 Not Yet Available. Track progress at:
http://www.iso.org/iso/iso_catalogue/catalogue_tc/catalogue_detail.htm?csnumber=41881

ISO/IEC	TR/ CD	24729-2	Information technology—Radio frequency identification for item management—Implementation guidelines—Part 2: Recyclability of RF tags	TR under development to IS	ISO JTC1 SC31

ISO 24729-2 Not Yet Available. Track progress at:
http://www.iso.org/iso/iso_catalogue/catalogue_tc/catalogue_detail.htm?csnumber=41882

ISO/IEC	TR/ NP	24729-3	Information technology—Radio frequency identification for item management—Implementation guidelines—Part 3: Implementation and operation of UHF RFID Interrogator systems in logistics applications	DTR under development	ISO JTC1 SSC31

ISO 24729-3 Not Yet Available. Track Progress at:
http://www.iso.org/iso/iso_catalogue/catalogue_tc/catalogue_detail.htm?csnumber=41883

7.5.5 ISO 24730 Real-Time Locating Systems (RTLS)

ISO/IEC	IS	24730	Information technology—Real-time locating systems (RTLS)	Published	ISO JTC1 SC31

ISO/IEC 24730 defines two air interface protocols and a single API for real-time locating systems (RTLS) for use in asset management and is intended to allow for compatibility and to encourage interoperability of products for the growing RTLS market.

Real-time locating systems are wireless systems with the ability to locate the position of an item anywhere in a defined space (local/campus, wide area/regional, global) at a point in time that is, or is close to, real time. Position is derived by measurements of the physical properties of the radio link.

Conceptually there are four classifications of RTLS:

- Locating an asset via satellite (requires line of sight)—accuracy to 10m;
- Locating an asset in a controlled area (e.g., warehouse, campus, airport) (area of interest is instrumented)—accuracy to 3m;
- Locating an asset in a more confined area (area of interest is instrumented)—accuracy to tens of centimeters;
- Locating an asset over a terrestrial area using a terrestrial mounted receiver over a wide area (e.g., cell phone towers)—accuracy to 200m.

There are two other methods of locating an object (which are really RFID rather than RTLS):

- Locating an asset by virtue of the fact that the asset has passed point A at a certain time and has not passed point B;
- Locating an asset by virtue of providing a homing beacon whereby a person with a handheld can find an asset.

The method of location is through identification and location, generally through multilateration. The different types are:

- Time of flight ranging systems;
- Amplitude triangulation;
- Time difference of arrival (TDOA);
- Cellular triangulation;
- Satellite multilateration;
- Angle of arrival.

ISO/IEC	IS	24730-1	Information technology—Real-time locating systems (RTLS)—Part 1: Application program interface (API)	Published 2006	ISO JTC1 SC31

ISO 24730-1 Available from
http://www.iso.org/iso/iso_catalogue/catalogue_tc/catalogue_detail.htm?csnumber=38840

ISO/IEC 24730-1:2006, the RTLS API, establishes a technical standard for RTLS. To be fully compliant with this standard, RTLS must comply with ISO/IEC 24730-1:2006 and at least one air interface protocol defined in ISO/IEC 24730. RTLS are wireless systems with the ability to locate the position of an item anywhere in a defined space (local/campus, wide area/regional, global) at a point in time that is, or is close to, real time. Position is derived by measurements of the physical properties of the radio link.

ISO/IEC 24730-1:2006 defines an API needed for utilizing an RTLS.

An API is a boundary across which application software uses facilities of programming languages to invoke services. These facilities may include procedures or operations, shared data objects, and resolution of identifiers. A wide range of services may be required at an API to support applications. Different methods may be appropriate for documenting API specifications for different types of services.

The information flow across the API boundary is defined by the syntax and semantics of a particular programming language, such that the user of that language may access the services provided by the application platform on the other side of the boundary. This implies the specification of a mapping of the functions being made available by the application platform into the syntax and semantics of the programming language. An API specification documents a service and/or service access method that is available at an interface between the application and an application platform.

This API describes the RTLS service and its access methods, to enable client applications to interface with the RTLS. This RTLS service is the minimum service that must be provided by an RTLS to be API compatible with this standard.

ISO/IEC 24730-1:2006 enables software applications to utilize an RTLS infrastructure to locate assets with RTLS transmitters attached to them. It defines a boundary across which application software uses facilities of programming languages to collect information contained in RTLS tag blinks received by the RTLS infrastructure.

ISO/IEC	IS	24730-2	Information technology—Real-time locating systems (RTLS)—Part 2: 2.4 GHz air interface protocol	Published 2006	ISO JTC1 SC31

ISO 24730-2 Available from
http://www.iso.org/iso/iso_catalogue/catalogue_tc/catalogue_detail.htm?csnumber=40508

ISO/IEC 24730-2:2006, the 2.4-GHz air interface protocol, establishes a technical standard for real-time locating systems that operate at an internationally available 2.4-GHz frequency band and that are intended to provide approximate location with frequent updates (e.g., several times a minute). In order to be compliant with this standard, compliance with ISO/IEC 24730-2:2006 and ISO/IEC 24730-1 is required.

ISO/IEC 24730-2:2006 specifies the air interface for a system that locates an asset in a controlled area (e.g., warehouse, campus, airport) (area of interest is instrumented)—accuracy to 3m.

ISO/IEC 24730-2:2006 defines the air interface protocol needed for the creation of an RTLS system. There are many types of location algorithms that could be used. An example location algorithm is given in Annex A of the deliverable.

ISO/IEC 24730-2:2006 defines a networked location system that provides X-Y coordinates and data telemetry. The system utilizes RTLS transmitters that autonomously generate a direct-sequence spread spectrum radio frequency beacon. These devices may be field programmable and support an optional exciter mode that allows modification of the rate of location update and location of the RTLS device.

ISO/IEC 24730 also defines these modes, but does not define the means by which they are accomplished.

ISO/IEC		24730-3	Information technology—Real-time locating systems (RTLS)—Part 3: 433 MHz air interface protocol	Deleted	ISO JTC1 SC31

This project was deleted in October 2006.

ISO/IEC	WD	24730-4	Information technology—Automatic identification and data capture techniques—Real Time Locating Systems (RTLS)—Part 4: Global Locating Systems (GLS)	Under Development	ISO JTC1 SC31

A letter ballot was proposed for this work item, but to date it has achieved no progress or status.

7.5.6 ISO 24753 Item Management—Application Protocol: Encoding and Processing Rules for Sensors and Batteries

ISO/IEC	WD	24753	Automatic identification and data capture techniques—Radio frequency identification (RFID) for item management—Application protocol: encoding and processing rules for sensors and batteries	Under Development	ISO JTC1 SC31

ISO 24753 Not Yet Available. Track progress at:
http://www.iso.org/iso/iso_catalogue/catalogue_tc/catalogue_detail.htm?csnumber=41523

7.5.7 ISO 24791 Radio-Frequency Identification (RFID) for Item Management—System Management Protocol

ISO/IEC		24791	Information technology—Automatic Identification and Data Capture Techniques—Radio-Frequency Identification (RFID) for Item Management—System Management Protocol	Under Development	ISO JTC1 SC31
ISO/IEC	CD	24791-1	Radio-Frequency Identification (RFID) for Item Management—System Management Protocol—Part 1: Architecture	Under Development/ CD	ISO JTC1 SC31

ISO 24791-1 Not Yet Available. Track Progress at:
http://www.iso.org/iso/iso_catalogue/catalogue_tc/catalogue_detail.htm?csnumber=46137

ISO/IEC	NP	24791-2	Radio-Frequency Identification (RFID) for Item Management—System Management Protocol—Part 2: Data management	Under Development/ NP	ISO JTC1 SC31

ISO 24791-2 Not Yet Available. Track Progress at:
http://www.iso.org/iso/iso_catalogue/catalogue_tc/catalogue_detail.htm?csnumber=46138

ISO/IEC	NP	24791-3	Radio-Frequency Identification (RFID) for Item Management—System Management Protocol—Part 3: Device management	Under Development/ NP	ISO JTC1 SC31

ISO 24791-2 Not Yet Available. Track Progress at:
http://www.iso.org/iso/iso_catalogue/catalogue_tc/catalogue_detail.htm?csnumber=46139

7.6 Other SC31 Automatic Identification Standards That May Be Useful for ITS Service Provision

7.6.1 ISO 9798 Information Technology—Security Techniques—Entity Authentication

ISO/IEC	IS	9798	Information technology—Security techniques—Entity authentication—Part 1: General	Published	ISO JTC1 SC31

This standard is in six parts:

Part 1: General;
Part 2: Mechanisms using symmetric encipherment algorithms;
Part 3: Mechanisms using digital signature techniques;
Part 4: Mechanisms using a cryptographic check functions;
Part 5: Mechanisms using zero knowledge techniques;
Part 6: Mechanisms using manual data transfer.

ISO/IEC	IS	9798-1	Information technology—Security techniques—Entity authentication—Part 1: General	Published 1999	ISO JTC1 SC31

ISO 9798-1 Available from
http://www.iso.org/iso/iso_catalogue/catalogue_tc/catalogue_detail.htm?csnumber=27743

ISO/IEC	IS	9798-2	Information technology—Security techniques—Entity authentication—Part 2: Mechanisms using symmetric encipherment algorithms	Published 1999	ISO JTC1 SC31

ISO 9798-2 Available from
http://www.iso.org/iso/iso_catalogue/catalogue_tc/catalogue_detail.htm?csnumber=30653

ISO/IEC	IS	9798-3	Information technology—Security techniques—Entity authentication—Part 3: Mechanisms using digital signature techniques	Published 1998	ISO JTC1 SC31

ISO 9798-3 Available from
http://www.iso.org/iso/iso_catalogue/catalogue_tc/catalogue_detail.htm?csnumber=29062

ISO/IEC	IS	9798-4	Information technology—Security techniques—Entity authentication—Part 4: Mechanisms using a cryptographic check function	Published 1999	ISO JTC1 SC31

ISO 9798-4 Available from
http://www.iso.org/iso/iso_catalogue/catalogue_tc/catalogue_detail.htm?csnumber=31488

ISO/IEC	IS	9798-5	Information technology—Security techniques—Entity authentication—Part 5: Mechanisms using zero-knowledge techniques	Published 2004	ISO JTC1 SC31

ISO 9798-5 Available from
http://www.iso.org/iso/iso_catalogue/catalogue_tc/catalogue_detail.htm?csnumber=39720

ISO/IEC	IS	9798-6	Information technology—Security techniques—Entity authentication—Part 6: Mechanisms using manual data transfer	Published 2005	ISO JTC1 SC31

ISO 9798-6 Available from
http://www.iso.org/iso/iso_catalogue/catalogue_tc/catalogue_detail.htm?csnumber=39721

7.6.2 ISO 19762 Automatic Identification and Data Capture (AIDC) Techniques—Harmonized Vocabulary

This standard is in three parts.

ISO/IEC	IS	19762	Automatic identification and data capture (AIDC) techniques—Harmonized vocabulary	Published 2005	ISO JTC1 SC31

ISO/IEC	IS	19762-1	Automatic identification and data capture (AIDC) techniques—Harmonized vocabulary—Part 1: General terms relating to AIDC	Published 2005	ISO JTC1 SC31

ISO 19762-1 Available from
http://www.iso.org/iso/iso_catalogue/catalogue_tc/catalogue_detail.htm?csnumber=37432

ISO/IEC 19762-1:2005 provides general terms and definitions in the area of automatic identification and data capture techniques on which are based further specialized sections in various technical fields, as well as the essential terms which should be used by nonspecialist users in communication with specialists in automatic identification and data capture techniques.

ISO/IEC	IS	19762-2	Automatic identification and data capture (AIDC) techniques—Harmonized vocabulary—Part 2: Optically readable media (ORM)	Published 2005	ISO JTC1 SC31

ISO 19762-2 Available from
http://www.iso.org/iso/iso_catalogue/catalogue_tc/catalogue_detail.htm?csnumber=37433

ISO/IEC 19762-2:2005 provides terms and definitions unique to optically readable media (ORM) in the field of automatic identification and data capture techniques. This glossary of terms enables the communication between nonspecialist users and specialists in optically readable media through a common understanding of basic and advanced concepts.

ISO/IEC	IS	19762-3	Automatic identification and data capture (AIDC) techniques—Harmonized vocabulary—Part 3: Radio frequency identification (RFID)	Published 2005	ISO JTC1 SC31

ISO 19762-3 Available from
http://www.iso.org/iso/iso_catalogue/catalogue_tc/catalogue_detail.htm?csnumber=37434

ISO/IEC 19762-3:2005 provides terms and definitions unique to RFID in the field of automatic identification and data capture techniques. This glossary of terms enables the communication between nonspecialist users and specialists in RFID through a common understanding of basic and advanced concepts.

PART III
ITS Services to Stakeholders

CHAPTER 8

ITS Service Groups and Domains

In this chapter we move on to the standards that have been and are being developed to directly support ITS service provision.

Since the end objective of intelligent transport systems is ITS service provision (i.e., doing something for the customers), the reader may expect this to be the largest chapter in the book. While one day it is possible that this might be the case, it is certainly not the case today, and before delving into the details of the service group standards themselves, it is worth considering why.

Standards enable markets. They enable level playing field competition and for the marketplace to stand the best chance to succeed. Standards provide a basis for quality of service, and for conformance and performance measurement and comparison. Where services are provided by multiple service providers, whether in an environment of open competition or using a chain of different service providers in order to get the service established and delivered, conformance to standards enables the interoperability required.

But it is the instantiation of systems that delivers service(s), not the standards themselves. As we saw in Chapter 2 of this book, there is a considerable number of ITS services that happen within a vehicle, without communication or interoperation with other actors in the system (e.g., lane departure warning, front and rear obstacle detection, electronic traction stability, air bags, antilock brakes, and so on).

Here, standards are not required to control how the service is provided. Standards may be required so that the performance and quality of service achieve certain performance levels, or can be compared one system to another. But whether BMW uses the same or a different technology than Nissan to make a system work really does not matter. Here standards to ensure the service is provided in the same way are more likely to impede progress by requiring adherence to a method that will become increasingly out of date, than to assist the market; and in any event, there is no benefit for the service provider to spend a large amount of time developing a standard. So do not expect to see many hundreds of standards that control product design.

Where a service can only be provided by the involvement of multiple actors, whether because of the communication requirements, the involvement of multiple actors to provide the service (as in car-to-car and infrastructure-to/from-car systems) or where multiple vendors are required to enable the market (such as in large scale electronic fee collection or electronic ticketing systems), expect to see many standards. We have already covered the communication, data transfer and identification technology standards in previous chapters. Those standards are essential to enable the service provision in the first place, but are not often visible at the point of service provision.

Finally, there is the state of development of the ITS market sector. As we have noted earlier in the book, most ITS service provision to date has happened within a vehicle without two-way communication with other parties. For those services that require two-way/multiparty involvement, the service itself if not possible until the communications infrastructure is deployed. Hence, the emphasis of work to date has been to reach that situation, and these standards have been described in the previous chapters. After these networks (CALM, VII, cellular, or other) are deployed, only then can the requirements for standards to support service instantiation be assessed and developed.

In this chapter the reader will see that the standards provision to date is patchy. Existing standards include: (1) those that ensure or measure performance of equipment in vehicles; (2) those for public transport and in particular interoperable fare collection and consistent display of information; (3) those for traveler information and some for traffic management; and (4) those concerned with the national/international operation of electronic toll collection/road pricing. The reader will see some standards supporting emergency services and some in the freight and hazardous materials sectors. But the reader will also see many identified service domains where, to date, there are few standards, some national standards, or no standards at all.

The principal committees for the development of international ITS standards at the "application" level are ISO/TC 204 Intelligent Transport Systems and CEN/TC 278 in Europe. We have seen in Chapter 4 how ISO/TC 204 has developed and is still developing standards throughout the OSI protocol stack, but the lower layer standards are generally developed in close cooperation with ETSI or IEEE standards; and in Chapter 5 how ISO/TC 204 has developed a number of architecture and support tool standards and deliverables. The remainder of its work lies at the application level, and this is, to date, less developed because the emphasis to a large part has been to establish the basic architectures and communications. There are also standards from ISO/TC 22 Road Vehicles concerning in-vehicle equipment, and some SAE standards.

The entire work program of ISO/TC 204 can be downloaded from:

http://www.iso.org/iso/iso_catalogue/catalogue_tc/catalogue_tc_browse.htm?commid=54706&development=true.

A list of its published deliverables at the time of this book's publication appears in an appendix at the end of the book. The up-to-date list of open and closed ballots can be found at:

http://www.tiaonline.org/standards/secretariats_tags/iso_tc204/index.cfm.

Within Europe ISO/TC 204's sister committee is CEN/TC 278. Whereas ISO/TC 204 has been developing standards since the end of 2003, CEN/TC 278 is its senior by 2 years.

The list of standards published by CEN/TC 278 can be downloaded from:

http://www.nen.nl/cen278/.

The current program of work of CEN/TC 278 can be obtained from:

http://www.nen.nl/cen278/.

CEN/TC 278 and ISO/TC 204 work in close harmony, often with joint working groups. Since the establishment of TC 204, CEN/TC 278 has led the completion of the work underway before TC 204's inception, and subsequently has focused on European-specific issues for CEN standards (in addition to the joint work on International Standards).

8.1 ITS Service Groups

As we have seen in Sections 2.1 through 2.3, and Section 5.1.1, the standard ISO 14813-1 describes the service groups and service domains of ITS as it is perceived to be at this point in time. It provides a sort of very high level abstract architecture of functions.

The remainder of ITS standards not yet described (i.e., excluding the communications and identification standards, and generic standards that can be used to support ITS service provision, which have already been described) exist to directly support specific ITS application services, or groups of ITS application services.

The remaining sections of Chapter 8 are therefore organized into the service groups and service domains as described in ISO 14813-1. Of course, there are some standards that support multiple service types, and, consistent with the rest of this book, these are cited where relevant, and the description is provided in the first instance of a standard being listed, with any repeat instances simply listing a reference to the section in which the standard is summarized.

It will become readily apparent that, while development of the basic core standards for communications and identification are well advanced, in many of the service groups and domains, there remains an absence of standards, or only standards under development.

This may seem strange, but is indeed logical. Application standards are developed as the applications come into (commercial) existence. Before that time they are the subject of research and development, followed by demonstration, and then revision with the benefit of experience.

Until we reach the point of actual tests and trials, it is unlikely that anything other than basic architecture types of standards will be developed. As the technology nears commercial deployment, the standards will start to be developed. Thus, in the following sections of this chapter, areas like electronic fee collection (principally road tolling and road pricing), which are already in commercial deployment, or areas like public transport, have a good number of developed and developing standards; domains associated with advanced collision avoidance, despite its high media profile, have few standards developed yet; and areas such as policing/enforcing traffic regulations and safety enhancements for vulnerable road users have no standards yet at all. This is not because these domains are not considered important, but because services can only be implemented once there is a more widespread infrastructure in place. Standards for these domains are only likely to be developed when this seems near term.

This actually produces an interesting and unplanned use for this book, because it helps to identify the gaps in the provision of standards, and standards developers may hopefully be able to identify gaps that need to be filled as a result of seeing the gaps within this chapter.

8.2 Traveler Information Service Groups

Service groups within the Traveler Information domain address the provision of both static and dynamic information about the transport network and services for users prior to and during the trip, and provide tools for transport professionals to collect, archive, and manage information for future trip planning activities.

NOTE: The systems used for on-trip information may sometimes also be used for pretrip information, and vice versa, so readers are advised to check both sections for standards necessary to their needs.

8.2.1 General

8.2.1.1 ISO 17267 Navigation System API

ISO	WD	17267	Navigation System API	CD ballot in process Ended Nov. 2007	ISO TC204

ISO 17267 Not Yet Available. Track progress at:
http://www.iso.org/iso/iso_catalogue/catalogue_tc/catalogue_detail.htm?csnumber=44700

The impetus for this standard was the recognition by ITS- related industry that standardization with respect to data access for map databases used by navigation applications is needed. As the vehicle navigation industry has grown, so has incompatibility between navigation systems and map databases. Both a standardized physical storage format (PSF) and a standardized navigation API can facilitate the interoperability between navigation systems and map databases.

The purpose of this standard is to define and structure the model for data access for vehicle navigation and traveler information systems. This standard is not restricted to physical media and will be independent of any underlying physical storage format. While this API is primarily targeted at self-contained in-vehicle systems, it is expected to be usable by other applications that use map data results in essentially the same way. For example, it may be usable by client/server or distributed navigation systems and location-based services without further specialization.

This standard will define an API for navigation and other location-based services targeted at transportation and mobile applications. It has taken some time to develop, but is currently under NP/CD ballots. Together with ISO 20452 (see Sections 5.1.2 and 8.2.1.2), it replaces ISO 14826, which is withdrawn.

8.2.1.2 ISO 20452 Requirements and Logical Data Model for a PSF and an API and Logical Data Organization for a PSF used in ITS Database Technology

ISO	TS	20452	Requirements and Logical Data Model for a PSF and an API and Logical Data Organization for a PSF used in ITS Database Technology	Published	ISO TC204

ISO 20452 Available from
http://www.iso.org/iso/iso_catalogue/catalogue_tc/catalogue_detail.htm?csnumber=39447

ISO/TS 20452:2007 describes the functional requirements and logical data model for PSF and API and the logical data organization for PSF that were completed under ISO/NP 14826. It does not specify a physical data organization.

See Section 5.1.12.

8.2.1.3 SAE J1746 ISP-Vehicle Location Referencing Standard

SAE	J1746_199912	ISP-Vehicle Location Referencing Standard	Published Dec. 1999	SAE

SAE J1746 Available from
http://www.sae.org/technical/standards/J1746_199912

This SAE standard is intended to be used for the communication of spatial data references between central sites and mobile vehicles on roads. References can be communicated from central sites to vehicles or from vehicles to central sites. The document may also be used where appropriate by other ITS applications requiring location references between data sets.

8.2.1.4 SAE J2353 Data Dictionary for Advanced Traveler Information Systems (ATIS)

SAE	J2353_199910	Data Dictionary for Advanced Traveler Information Systems (ATIS)	Published Oct. 1999	SAE

SAE J2353 Available from http://www.sae.org/technical/standards/J2353_199910

This SAE recommended practice provides a set of core data elements needed by information service providers for Advanced Traveler Information Systems (ATIS). The data dictionary herein provides the foundation for ATIS message sets for all stages of travel (pretrip and en route), all types of travelers (drivers, passengers), all categories of information, and all platforms for delivery of information (in-vehicle, portable devices, kiosks). The elements of this document are the basis for the SAE ATIS Message Set Standard J2354 and are entered into the SAE data registry for ITS-wide coordination.

8.2.1.5 SAE J2354 Message Sets for Advanced Traveler Information System (ATIS)

SAE	J2354_200402	Message Sets for Advanced Traveler Information System (ATIS)	Published Feb. 2004

SAEJ2354 Available from
http://www.sae.org/technical/standards/J2354_200402

This SAE standard describes standardized medium-independent messages needed by information service providers for ATIS. The messages contained herein address all stages of travel (pretrip and en-route), all types of travelers (drivers, passengers), all categories of information, and all platforms for delivery of information (in-vehicle, portable devices, kiosks).

8.2.1.6 SAE J2365 Calculation of the Time to Complete In-Vehicle Navigation and Route Guidance Tasks

SAE	J2365_200205	Calculation of the Time to Complete In-Vehicle Navigation and Route Guidance Tasks	Published May 2002	SAE

SAE J2365 Available from
http://www.sae.org/technical/standards/J2365_200205

This SAE recommended practice applies to both original equipment manufacturer (OEM) and aftermarket route-guidance and navigation system functions for passenger vehicles. This recommended practice provides a method for calculating the time required to complete navigation system-related tasks. These estimates may be used as an aid to assess the safety and usability of alternative navigation and route guidance system interfaces to assist in their design. This document does not consider voice-activated controls, voice output from the navigation system, communication between the driver and others, or passenger operation.

8.2.1.7 SAE J2369 Standards for ATIS Message Sets Delivered over Reduced Bandwidth Media

SAE	J2369_200003	Standards for ATIS Message Sets Delivered over Reduced Bandwidth Media	Published Mar. 2000	SAE

SAE J2369 Available from
http://www.sae.org/technical/standards/J2369_200003

This SAE standard outlines the U.S. standard for sending ATIS messages over various bandwidth reduced media (BRM). It specifies the "over the air" format and method for such data messages at a bit level.

The methodology allows a cooperative sharing of the media's total bandwidth with other non-ATIS data services. It allows for further expansion to other ATIS messages such as transit schedules and weather use. It uses a common directory of messages to divide out those messages which are the subject of this specification from others.

It is built upon the concept of "waveform neutrality" in that it can be hosted by a wide variety of media often found in the delivery of wireless data services (although it is also useful for wire line applications). This message set is incorporated into each medium (host medium) by a formatting process specific to each media. The capacity and services available over the host media can vary considerably in terms of capacity, other services present, one-way or two-way use, suitability for use, and a variety of commercial aspects not considered here. This document is divided into five primary sections. The first provides a conceptual background of various elements required to understand the message encoding process itself (Sections 4, 5, 6). The second section contains precise definitions of the messages and data elements in the messages (Section 7, 8, 9). The third section provides implementation advice and comments for both receiver and encoder manufacturers (Section 10, 11). The fourth section contains tables of constants used in various message encodings of the standard (Section 12). A fifth section, contained in the appendices, is an illustrative example of the message sets implemented with real data. Finally, a in set of appendices, SAE J2369-1 provides information regarding how to implement this document on a variety of host media.

8.2.1.8 SAE J2374 Location Referencing Message Specification

SAE	J2374_199907	Location Referencing Message Specification	Published July 1999	SAE

SAE J2374 Available from
http://www.sae.org/technical/standards/J2374_199907

The location referencing message specification (LRMS) is intended to provide a practical approach to standardization for location referencing within a mixed data set environment (i.e., where more than one kind of spatial data set exists, and where spatial references between these data sets must be made).

Although some ITS applications in local areas may be satisfied by having one common data set—for which location references may be implemented in any number of ways—many ITS applications will have broad interoperability requirements within the nation or a region. For example, a vehicle driven from California to Florida in the United States should be able to receive and understand spatial references for traffic information or routing instructions throughout the trip.

Similarly, information sent from a vehicle to a central site should be understood in any city regardless of the kinds of data sets in use, whether they are public or private, or how locations are referenced internally to particular data sets.

The LRMS can be applied to ITS systems involving mobile vehicles on roads, rails, and waterways. It can also be applied to location references to and from central sites to nonmobile sites such as kiosks, other central sites, or pedestrians. The broadest scope of the LRMS is therefore intermodal spatial data set interoperability at the national level and across all of ITS. Given the great variety of ITS systems, it is expected that individual LRMS profiles will generate location referencing standards for subsets of ITS applications, such as ISP-vehicle-ISP, or center-center.

8.2.1.9 SAE J2540 Messages for Handling Strings and Look-Up Tables in ATIS Standards

SAE	J2540_200207	Messages for Handling Strings and Look-Up Tables in ATIS Standards	Published July 2002	SAE

SAE J2540 Available from
http://www.sae.org/technical/standards/J2540_200207

This SAE standard provides a table of textual messages meeting the requirements for expressing International Traveler Information Systems (ITIS) phrases commonly used in the ITS industry. The tables provided herein follow the rules of SAE J2540 and therefore allow a local representation in various different languages and media expressions to allow true international use of these phrases.

The phrases are predominantly intended for use in the description of traffic-related events of interest to travelers and other traffic practitioners. Other phrases exit for other specific specialty areas of ITS, and all such phrases follow a set of encoding and decoding rules outlined in SAE J2540 to ensure that the use of these phrases in messages remain interoperable between disparate types of user equipment.

Implementers are cautioned to obtain the most recent set of tables by means of the ITS data registry, a process which involves the SAE and other standards setting organizations and which is intended to maintain and enhance the level of harmonization among ITS standards set by each of the organizations.

This document defines the normative index values to be used to provide phrases needed by ITS practitioners. This document provides nonnormative textual phrases which *may* be used by implementers to ensure intelligible results.

This document follows the formats and rules established in SAE J2540 in the expressions, manipulations, and use of such tables. It should be pointed out that within the rules established by this document a variety of final tables are all considered "compliant" with the document, and may vary as fits the needs of implementers.

8.2.1.10 SAE J2539 Comparison of GATS Messages to SAE ATIS Standards

SAE	J2539_200202	Comparison of GATS Messages to SAE ATIS Standards	Published Feb. 2002	SAE

J2539 Available from
http://www.sae.org/technical/standards/J2539_200202

This SAE information report provides a comparative summary between the various messages found in the SAE ATIS standards work (notably SAE J2313, J2353, J2354, J2369, and J2374) and that found in the Global Automotive Telematics Standard (GATS). GATS is a message set meant to be deployed on mobile phone systems based on the Global System for Mobile Communication (GSM) phone system which is being deployed in European markets and which the SAE may need to harmonize with as part of the world standards activities of TC 204. This document provides an overview of the various types of supported messages and how they compare with U.S. terms and messages. Some selected features of the GATS work are recommended for assimilation into the next revision of ATIS standards. No attempt at determining a U.S. policy in this regard is provided. This document seeks to provide the reader familiar with SAE ATIS with a high level overview of technical knowledge of the GATS approach in similar areas.

8.2.1.11 SAE J2540 ITIS Phrase Lists (International Traveler Information Systems)

SAE	J2540/2	ITIS Phrase Lists (International Traveler Information Systems)	Published Nov. 2006	SAE

J2540/2 Available from
http://www.sae.org/technical/standards/J2540/2_200611

This SAE standard provides a table of textual messages meeting the requirements for expressing ITIS phrases commonly used in the ITS industry. The tables provided in this deliverable follow the rules of SAE J2540 and therefore allow a local representation in various different languages, media expressions, and so forth, to allow true international use of these phrases.

The phrases are predominantly intended for use in the description of traffic-related events of interest to travelers and other traffic practitioners. Other phrases exist for other specific specialty areas of ITS, and all such phrases follow a set of encoding and decoding rules outlined in SAE J2540 to ensure that the use of these phrases in messages remain interoperable between disparate types of user equipment. Implementers are cautioned to obtain the most recent set of tables by means of the ITS data registry, a process which involves the SAE and other standards setting organizations and which is intended to maintain and enhance the level of harmonization among ITS standards set by each of the organizations.

This document defines the normative index values to be used to provide phrases needed by ITS practitioners. This document provides nonnormative textual phrases which *may* be used by implementers to ensure intelligible results. This document follows the formats and rules established in SAE J2540 in the expressions, manipulations, and use of such tables. It should be pointed out that within the rules established by this document, a variety of final tables are all considered "compliant" with the document, and may vary as fits the needs of implementers.

8.2.1.12 SAE J2630 Converting ATIS Message Standards from ASN.1 to XML

SAE	J2630_200312	Converting ATIS Message Standards From ASN.1 to XML	Published Dec. 2003	SAE

SAE J2630 Available from
http://www.sae.org/technical/standards/J2630_200312

See Section 5.2.2.3.

8.2.1.13 ITSO Specification v2.1.2 Interoperable Contactless Smart Customer Media, Public Transport Ticketing

ITSO	Converting ATIS Message Standards From ASN.1 to XML	Published Dec. 2003	SAE

ITSO Specifications available for download from
http://www.itso.org.uk/default.asp?ContentID=49

Part	*Title*	*Nomenclature*	*Version Release*
0	*Concept & Content*	ITSO TS 1000-0	Jun-07
1	*General Reference Customer Media Data and Customer Media*	ITSO TS 1000-1	Jun-07
2	*Architecture*	ITSO TS 1000-2	Jun-07
3	*Terminals*	ITSO TS 1000-3	Jun-07
4	*HOPS Customer Media Data*	ITSO TS 1000-4	Jun-07
5	*Record Definitions*	ITSO TS 1000-5	Jun-07
6	*Message Data ITSO Security*	ITSO TS 1000-6	Aug-07
7	*Subsystem*	ITSO TS 1000-7	Jun-07
8	ISAM Detailed Operation (available on request by non-disclosure agreement)		
9	*Communications Customer Media*	ITSO TS 1000-9	Jun-07
10	*Definitions*	10	Aug-07

Corrigendum

Title; Part and Version of the ITSO Spec Affected	*Released*
CORRIGENDUM 2; ITSO TS 1000-6 V2.1.1	*Aug-07*
CORRIGENDUM 1; ITSO TS 1000-6 V2.1.2	*Aug-07*
CORRIGENDUM 2; ITSO TS 1000-10 V2.1.2	*Aug-07*

ITSO is not a recognized standards development organization, but proposes specifications for

The current version of the ITSO specification is v2.1.2.

The purpose of the specification is to provide a platform and tool-box for the implementation of interoperable contactless smart customer media, public transport ticketing, and related services in the United Kingdom in a manner which offers end-to-end lossless data transmission and security. It has been kept as open as possible within the constraints of evolving national, European, and International Standards in order to maximize competition in the supply of systems and components to the commercial benefit of the industry as a whole. In general, it promotes open standards but it does not disallow proprietary solutions where they are offered on reasonable, nondiscriminatory terms and contribute towards the ultimate objective of interoperability.

The specification defines the key technical items and interfaces that are required to deliver interoperability. To this end, the end-to-end security system and shell layout are defined in detail, while other components (e.g., terminals, "back-office" databases) are described only in terms of their interfaces.

8.2.2 Pretrip Information

NOTE: The systems used for pretrip information may sometimes also be used for on-trip information, so readers are advised to check both sections for standards necessary to their needs.

This service group encompasses travel information received at home, work, hotels, major public locations (such as shopping centers), and on portable terminals prior to travel.

Pretrip information services may target road facilities, public transport, freight and intermodal users, and nonmotorized travel.

Depending on the service provided, pretrip information includes current information on traffic conditions, schedule adherence and location of public transport vehicles relative to the user's location, road and weather information, prevailing traffic regulations, and tolls.

8.2.2.1 ISO 14822-1 Traffic and Travel Information—General Specifications For Medium-Range Preinformation Via Dedicated Short-Range Communication—Part 1: Downlink

ISO/CEN TS/CD 14822-1	Traffic and Travel Information—General specifications for medium-range pre-information via dedicated short-range communication—Part 1: Downlink	Published 2005	CEN TC278

ISO 14822-1 Available from
http://www.iso.org/iso/iso_catalogue/catalogue_tc/catalogue_detail.htm?csnumber=40628

ISO 14822-1:2005 addresses the passive DSRC issues associated with medium range pre-information (MRPI) as applied to traffic and travel information (TTI) issued from an information service provider to a suitably equipped moving vehicle.

The application identification (AID) number for all MRPI application entities is defined as No. 8 in accordance with ISO 15628. See Sections 4.7.7.2 and 10.6.1.6.

NOTE: There is only a part 1 of this standard. A part 2, Uplink, was planned, but has been subsequently deleted in 2006.

This European standard specifies general purpose information components to be used in standards for information exchange and information models supporting various health-specific business requirements. The components defined in this standard are the most commonly needed basic building blocks for such standardization, but these components may require further specialization and be complemented by other objects required for specific purposes not met by these generally useful components.

Such standardization using these general purpose information components could be performed both on a European (CEN) level or be done nationally or for specific user communities regionally as well as internationally. This European standard provides an informative overview of this series of standards and includes rules for using the components defined in the other parts and on conformance claims.

8.2.2.2 CIS/EN/TS 14822-2 Traffic and Travel Information—Medium-Range Pre-Information Via DSRC—General Specification—Part 2: Uplink

ISO/CEN	TS	14822-2	Traffic and Travel Information—General specifications for medium-range pre-information via dedicated short-range communication—Part 2: Uplink	Deleted but being restored	ISO TC204/ CEN TC278

ISO 14822-2 is not currently available

This work item was automatically deleted by CEN BT because of lack of progress, but it is being prepared by the WG as a technical specification. Once completed, the work item will be reinstated and the item balloted.

8.2.2.3 Managing Demand Through Travel Information Services

FHWA	U.S. Department of Transportation Federal Highway Administration	1 Managing Demand Through Travel Information Services

Downloadable FOC 1 Managing Demand Through Travel Information Services can be downloaded at:
http://ops.fhwa.dot.gov/publications/manag_demand_tis/travelinfo.htm

This document is not a standard but is a very useful reference document for those planning or developing standards.

The introduction (source http://ops.fhwa.dot.gov/publications/manag_demand_tis/travelinfo.htm), "Note from the Director Office of Transportation Management, Office of Operations Federal Highway Administration," states:

> Information can influence transportation decisions in some very clear ways—time choices, mode choices, path (i.e., route) choices, and location choices. With better ways to learn how emergencies and other incidents affect travel times, workers can start for work and return home at times more responsive to actual traffic patterns than a fixed schedule. Better information can also influence the choice of the path to work or other destinations, specifically for private vehicles. As the effects of congestion become clearer to people, they are more able to consider alternate paths to their destination.

In this light, the Federal Highway Administration presents this brochure entitled *Managing Demand Through Travel Information Services*. It highlights the opportunities and benefits for using traveler information services to manage demand during periods of congestion, including congestion during commute periods, special events, and emergencies. The brochure aims to provide ideas for the use of traveler information in states, regions, and communities. The brochure presents the diversity of traveler information systems employed around the country and overseas and how agencies are using traveler information as a demand management tool. The results are summarized in an easy-to-read format. In the following pages you will find:

- A description of the many dimensions of traveler information;
- The types of travel situations where traveler information can make a difference;
- Examples of state-of-the-art traveler information systems;
- The benefits that traveler information offers;
- Lessons learned from past successes and failures;
- Future directions for traveler information systems;
- Resources for more information.

We believe that travel information systems can have a profound effect on managing the demand for transportation facilities and services. When applied as a demand management tool, it can help to improve the performance of the transportation system by facilitating the movement of people and the delivery of freight. With the help of our partners throughout the transportation community, the Federal Highway Administration looks forward to working with organizations, agencies, and interest groups to further advance the ideas and practices presented in this brochure.

8.2.3 On-Trip Information

NOTE: The systems used for on-trip information may sometimes also be used for pretrip information, so readers are advised to check both sections for standards necessary to their needs.

This service group encompasses information provided to travelers in the vehicle (either provided for a mass audience or tailored to the specific vehicle or traveler location) or along the travel route (provided as advisory information). Such information may include real-time travel information, including estimated time to a destination based on current conditions, as well as information on work zones, incidents, weather, tolls, parking availability, and other information of use to the traveler.

8.2.3.1 Traffic and Travel Information (TTI)—TTI Messages Via Traffic Message Coding

CEN	ENV	12313-1	Traffic and Travel Information (TTI)—TTI Messages via traffic message coding—Part 1: Coding protocol for Radio Data System—Traffic Message Channel (RDS-TMC) using ALERT-C	Replaced by IS/EN 14819-1	CEN TC278

This standard has been replaced by EN/ISO 14819-1.

CEN	ENV	ENV 12313-2: 1997	Traffic and Travel Information (TTI)—TTI Messages via traffic message coding—Part 2: Event and information codes for Radio Data System—Traffic Message Channel (RDS-TMC)	Replaced by IS/EN 14819-2

This standard has been replaced by EN/ISO 14819-2.

8.2.3.2 ENV 12313 Traffic and Travel Information (TTI)—TTI Messages Via Traffic Message Coding

CEN	ENV	12313	Traffic and Traveler Information (TTI)—TTI Messages Via Traffic Message Coding—Part 1: Coding Protocol for Radio Data System—Traffic Message Channel (RDS-TMC) (RDS-TMC Using ALERT C) Version 3.0	Withdrawn Superseded by ISO parts of 14819	CEN TC278

Parts 1–3 of this standard have been replaced by ISO/CEN 14819, and the relevant CEN ENV parts are withdrawn and superseded.

CEN	ENV	12313-1	Traffic and Traveler Information (TTI)—TTI Messages Via Traffic Message Coding—Part 1: Coding Protocol for Radio Data System—Traffic Message Channel (RDS-TMC) (RDS-TMC Using ALERT C) Version 3.0	Withdrawn Superseded by ISO 14819-1	CEN TC278

This standard was withdrawn in 1998, and superseded by IS/EN 14819-1:2003.

CEN	ENV	12313-2	Traffic and Travel Information (TTI)—TTI Messages via traffic message coding—Part 2: Event and Information codes for Radio Data System. Traffic Message Channel (TMC)	Withdrawn Superseded by ISO 14819-2	CEN TC278

This standard was withdrawn in 1998, and superseded by IS/EN 14819-2:2003.

CEN	ENV	12313-3	Traffic and Travel Information (TTI)—TTI Messages via traffic message coding—Part 3: Traffic and Traveler Information (TTI)—TTI Messages via traffic message coding—Part 3: Location Referencing for Alert C	Withdrawn Superseded by ISO 14819-3	CEN TC278

This standard was withdrawn in 1998, and superseded by IS/EN 14819-3:2003.

CEN	ENV	12313-4	Traffic and Travel Information (TTI)—TTI Messages via traffic message coding—Part 4: Coding protocol for Radio Data System—Traffic Message Channel (RDS-TMC)—RDS-TMC using ALERT-Plus with ALERT-C	CEN TC278

CEN ENV 12313-4 Available from any European National Standards organization
http://www.cen.eu/esearch/extendedsearch.aspx

The ALERT-Plus function is an extension of the ALERT-C function. While ALERT-C covers event-orientated information to be conveyed by the Radio Data System (RDS) medium, ALERT-Plus deals with status-orientated information to be conveyed by the same medium.

The ALERT-Plus function informs motorists about the changes affecting the status of traffic at predefined locations. Different status types are defined, such as level of service on road sections or areas, travel times on road sections, car park occupancy, and status of public transport traffic. A location can be a road section, a pole, a car park, referenced in the location table. Location referencing is dealt with separately by CEN/TC 278 SWG7.3 and is not included in this document. A part of the document is valid for other data transmission media such as AM or SWIFT broadcast data systems, the digital cellular radio system (GSM), and the digital audio broadcasting (DAB). At least the user messages content is medium independent.

This document deals with the coding protocol for the multiplex broadcasting of event orientated and status-orientated information, but it essentially describes the content of status-orientated messages. The event-orientated messages are described in the ENV 12313-1, ALERT-C Protocol. The presentation section describes the way in which messages are coded for broadcasting. The main coding principle for status-orientated messages is to gather together the information which relates to several locations in the same message, and to transmit a so-called collection number instead of directly transmitting location numbers like in the ALERT-C protocol.

8.2.3.3 ENV 12315 Traffic and Travel Information (TTI)—TTI Messages Via Dedicated Short-Range Communication

CEN	ENV	12315	Traffic and Travel Information (TTI)—TTI Messages via Dedicated Short-Range Communication	Published	CEN TC278

This ENV is in two parts.

CEN	ENV	12315-1	Traffic and Travel Information (TTI)—TTI Messages via Dedicated Short-Range Communication—Part 1: Data specification—Downlink (roadside to vehicle)	Published	CEN TC278

CEN ENV 12315-1 Available from any European National Standards organization
http://www.cen.eu/esearch/extendedsearch.aspx

CEN ENV 12315-1 describes common reference structure for messages regarding traffic information for drivers given by short-range equipment. Part 1 describes the roadside-to-vehicle downlink protocols.

CEN	ENV	12315-2	Traffic and Travel Information (TTI)—TTI Messages via Dedicated Short-Range Communication—Part 2: Data specification—Uplink (vehicle to roadside)	Published	CEN TC278

CEN ENV 12315-2 Available from any European National Standards organization
http://www.cen.eu/esearch/extendedsearch.aspx

CEN ENV 12315-2 describes common reference structure for messages regarding traffic information for drivers given by short-range equipment. Part 2 describes the vehicle-to-roadside uplink protocols.

8.2.3.4 ISO 14819 Traffic and Travel Information (TTI)—TTI Messages Via Traffic Message Coding

CEN/ISO	IS/EN	14819	Traffic and Travel Information (TTI)—TTI Messages via traffic message coding		CEN TC278

ISO 14819 consists of the following parts:

Part 1: Coding protocol for Radio Data System—Traffic Message Channel (RDS-TMC) using ALERT-C;
Part 2: Event and information codes for Radio Data System—Traffic Message Channel (RDS-TMC);
Part 3: Location referencing for ALERT-C;
Part 4: Reference model tutorial [deleted];
Part 5: Location referencing for ALERT-Plus [deleted];
Part 6: Encryption and conditional access for the Radio Data System—Traffic Message Channel ALERT C coding.

Traffic and traveler information may be disseminated through a number of services or means of communication, covering static displays, interactive terminals, and in-vehicle equipment. For all such services, the data to be disseminated and the message structure involved in the various interfaces require clear definition and standard formats in order to allow competitive products to operate with any received data.

CEN/ISO	IS/EN	14819-1	Traffic and Travel Information (TTI)—TTI Messages via traffic message coding—Part 1: Coding protocol for Radio Data System—Traffic Message Channel (RDS-TMC) using ALERT-C	Published	CEN TC278

ISO 14819-1 Available from
http://www.iso.org/iso/iso_catalogue/catalogue_tc/catalogue_detail.htm?csnumber=36922

ISO 14819-1:2003 specifies the coding protocol for Radio Data System—Traffic Message Channel (RDS-TMC)—RDS-TMC using the ALERT-C protocol that is designed to provide mostly event-orientated road driver information messages.

Many "hooks" have been left for future development, and indeed, a few status-orientated road driver information messages were included. This protocol is designed to be closely linked to the ALERT-Plus protocol, which is specifically designed for status-orientated road driver information; both protocols may be available in the same RDS transmission.

The ALERT-Plus protocol is specified in ENV 12313-4.

This standard focuses on the data specification for TTI messages, their network layer and their service layer, to be conveyed by the RDS-TMC feature, specified in IEC 62106:2000. Other standards are being developed by CEN/TC 278 WG4 to cover TTI messages that may be conveyed by other carriers.

The ALERT-C protocol defined in this specification supports a digital, silent broadcasting service for motorists, providing information about many kinds of traffic events. This includes roadwork, weather and traffic incident information relating to major national and international routes, regional routes, and local or urban roads.

CEN/ISO	IS/EN	IS/EN 14819-2	Traffic and Travel Information (TTI)—TTI Messages via traffic message coding—Part 2: Event and information codes for Radio Data System—Traffic Message Channel (RDS-TMC)	Published	CEN TC278

ISO 14819-2 Available from
http://www.iso.org/iso/iso_catalogue/catalogue_tc/catalogue_detail.htm?csnumber=29536

EN ISO 14819-2 is the second part of the EN 12313/EN ISO 14819 series of standards, covering the so-called ALERT-C protocol encoded for transmission into the RDS-TMC feature. Therefore, this standard must uniquely and solely be considered together with EN ISO 14819-1 for a complete understanding.

EN ISO 14819-1 fully describes the ALERT-C protocol concept and relationship with the RDS standard, IEC 62106:2000.

In this version of EN ISO 14819-2, the content and the structure of the "events list" have not been altered, but recent work from the FORCE/ECORTIS Projects regarding translations and a number of improved formatting ideas suggested by the EPISODE Project, have been introduced. Additionally mention is made of suggested events list subsets.

CEN/ISO	ENV/ ISO	14819-3	Traffic and Travel Information (TTI)—TTI Messages via traffic message coding—Part 3: Location referencing for ALERT-C	Published	CEN TC278

ISO 14819-3 Available from
http://www.iso.org/iso/iso_catalogue/catalogue_tc/catalogue_detail.htm?csnumber=35984

ISO 14819-3:2004 primarily addresses the needs of RDS-TMC ALERT-C messages, which are already being implemented. However, the modular approach used here is intended to facilitate future extension of the location referencing rules to other traffic and travel messaging systems.

The location referencing rules defined in ISO 14819-3:2004 address the specific requirements of traffic message channel (TMC) systems, which use abbreviated coding formats to provide TTI messages over mobile bearers (e.g., GSM, DAB) or via exchange protocols like DATEX. In particular, the rules address the RDS-TMC, a means of providing digitally coded traffic and travel information to travelers using a silent data channel (RDS) on FM radio stations, based on the ALERT-C protocol.

ISO	WD.TR	14819-4	Transport information and control systems—Reference model architecture(s) for the TICS sector—Part 4: Reference model tutorial	Deleted	CEN TC278

ISO 14819-4 has been deleted

CEN/ISO		14819-5	Traffic and Travel Information (TTI)—TTI Messages via traffic message coding—Part 5: Location referencing for ALERT-Plus	Deleted	CEN TC278

ISO 14819-5 has been deleted

ISO	IS	14819-6	Traffic and Traveler Information (TTI)—TTI Messages via Traffic Message Coding—Part 6: Encryption and conditional access for the Radio Data System—Traffic Message Channel ALERT-C coding	Published 2006	CEN TC278 / ISO TC204

ISO 14819-6 Available from
http://www.iso.org/iso/iso_catalogue/catalogue_tc/catalogue_detail.htm?csnumber=37267

ISO 14819-6:2006 establishes a method of encrypting certain elements of the ALERT-C coded data carried in the RDS-TMC type 8A data group, such that without application by a terminal or receiver of an appropriate keys, the information conveyed is virtually worthless.

Before a terminal is able to decrypt the data, the terminal requires two keys. The first is given in confidence by the service provider to terminal manufacturers with whom they have a commercial relationship; the second is broadcast in the Encryption Administration Group, which is also a type 8A group. This specification explains the purpose of the two keys and how often and when the transmitted key may be changed.

Before an individual terminal may present decrypted messages to the end user, it must have been activated to do so. Activation requires that a PIN code be entered. The PIN code controls access rights to each service and subscription period, allowing both lifetime and term business models to coexist.

8.2.3.5 ISO 14821 Traffic and Travel Information (TTI)—TTI Messages Via Cellular Networks (GATS)

CEN		14821	Traffic and Travel Information (TTI)—TTI messages via cellular networks		CEN TC278

These standards are commonly known as Global Automotive Telematics Standard (GATS).

In the field of traffic and traveler information (TTI), the innovation rate is high, with many research and development projects under way in many countries, and there is a need to establish prospective standards which allow manufacturers to introduce competitive products to the market with the knowledge that they can accommodate the future issues of the standard(s) without fundamental change to equipment.

TTI may be disseminated through a number of services or means of communication, covering static displays, portable terminals, and in-vehicle equipment.

For all such services, the data to be disseminated, and the message structure involved in the various interfaces, require clear definition and standards formats in order to allow competitive products to operate with any received data.

CEN/TS 14821 consists of eight parts; one part describing the framework and seven parts providing detailed specifications of all components, protocols, and services that are within the scope of CEN/TS 14821.

It is envisaged that future research and development will lead to improvements on the services listed above as well as to the creation of new services. Nevertheless, this standard provides the framework for seamless integration of new features and services into the existing architecture.

CEN	TS	14821-1	Traffic and Travel Information (TTI)—TTI messages via cellular networks—Part 1: General specifications	Published	CEN TC278

CEN TS 14821-1 Available from any European National Standards organization
http://www.cen.eu/esearch/extendedsearch.aspx

CEN/TC 14821-1 describes the specific interfaces and functionality of TT services based on the use of cellular networks.

CEN	TS	14821-2	Traffic and Travel Information (TTI)—TTI messages via cellular networks—Part 2: Numbering and ADP message header	Published	CEN TC278

CEN TS 14821-2 Available from any European National Standards organization
http://www.cen.eu/esearch/extendedsearch.aspx

CEN 14821-2 is a technical specification and states the specific interfaces and functionality of TT services based on the use of cellular networks.

Device manufacturers are enabled to develop terminal equipment compatible to services based on this standard. This will allow for interoperability of different terminal equipment and service providers, which allows competition between service providers and terminal manufacturers.

Furthermore, it sets the scene for international availability of these services.

It is envisaged that future research and development will lead to improvements on the services listed above as well as to the creation of new services. Nevertheless, this standard provides the framework for seamless integration of new features and services into the existing architecture.

CEN	TS	14821-3	Traffic and Travel Information (TTI)—TTI messages via cellular networks—Part 3: Basic information elements	Published	CEN TC278

CEN TS 14821-3 Available from any European National Standards organization
http://www.cen.eu/esearch/extendedsearch.aspx

CEN/TC 14821-3 states the specific interfaces and functionality of TT services based on the use of cellular networks.

This technical specification defines the specific interfaces and functionality of TT services based on the use of cellular networks. Device manufacturers are enabled to develop terminal equipment compatible to services based on this standard. This will allow for interoperability of different terminal equipment and service providers, which allows competition between service providers and terminal manufacturers. Furthermore, it sets the scene for international availability of these services.

CEN	TS	14821-4	Traffic and Travel Information (TTI)—TTI messages via cellular networks—Part 4: Service-independent protocols	Published	CEN TC278

CEN TS 14821-4 Available from any European National Standards organization http://www.cen.eu/esearch/extendedsearch.aspx

This states the specific interfaces and functionality of TT services based on the use of cellular networks.

This technical specification defines the specific interfaces and functionality of TT services based on the use of cellular networks. Device manufacturers are enabled to develop terminal equipment compatible to services based on this standard. This will allow for interoperability of different terminal equipment and service providers, which allows competition between service providers and terminal manufacturers. Furthermore, it sets the scene for international availability of these services.

It is envisaged that future research and development will lead to improvements on the services listed above as well as to the creation of new services. Nevertheless, this standard provides the framework for seamless integration of new features and services into the existing architecture.

CEN	TS	14821-5	Traffic and Travel Information (TTI)—TTI messages via cellular networks—Part 5: Internal services	Published	CEN TC278

CEN TS 14821-5 Available from any European National Standards organization http://www.cen.eu/esearch/extendedsearch.aspx

CEN/TS 14821-5 summary states the specific interfaces and functionality of TT services based on the use of cellular networks.

This technical specification defines the specific interfaces and functionality of TT services based on the use of cellular networks. Device manufacturers are enabled to develop terminal equipment compatible to services based on this standard. This will allow for interoperability of different terminal equipment and service providers, which allows competition between service providers and terminal manufacturers. Furthermore, it sets the scene for international availability of these services.

CEN	TS	14821-6	Traffic and Travel Information (TTI)—TTI messages via cellular networks—Part 6: External services	Published	CEN TC278

CEN TS 14821-6 Available from any European National Standards organization
http://www.cen.eu/esearch/extendedsearch.aspx

This states the specific interfaces and functionality of TT services based on the use of cellular networks.

This technical specification defines the specific interfaces and functionality of TT services based on the use of cellular networks. Device manufacturers are enabled to develop terminal equipment compatible to services based on this standard. This will allow for interoperability of different terminal equipment and service providers, which allows competition between service providers and terminal manufacturers. Furthermore, it sets the scene for international availability of these services.

CEN	TS	14821-7	Traffic and Travel Information (TTI)—TTI messages via cellular networks—Part 7: Performance requirements for onboard positioning	Published	CEN TC278

CEN TS 14821-7 Available from any European National Standards organization
http://www.cen.eu/esearch/extendedsearch.aspx

This technical specification defines the specific interfaces and functionality of TT services based on the use of cellular networks. Device manufacturers are enabled to develop terminal equipment compatible to services based on this standard. This will allow for interoperability of different terminal equipment and service providers, which allows competition between service providers and terminal manufacturers.

Furthermore, it sets the scene for international availability of these services.

CEN	TS	14821-8	Traffic and Travel Information (TTI)—TTI messages via cellular networks—Part 8: GSM-specific parameters	Published	CEN TC278

CEN TS 14821-8 Available from any European National Standards organization
http://www.cen.eu/esearch/extendedsearch.aspx

This technical specification defines the specific interfaces and functionality of TT services based on the use of cellular networks. Device manufacturers are enabled to develop terminal equipment compatible to services based on this standard. This will allow for interoperability of different terminal equipment and service providers, which allows competition between service providers and terminal manufacturers.

Furthermore, it sets the scene for international availability of these services.

8.2.3.6 ISO 14822 Traffic and Travel Information (TTI)—Medium-Range Pre-Information Via DSRC

ISO/CEN	TS	14822	Traffic and Travel Information—Medium-range pre-information via DSRC—General specification	Published	ISO TC204/ CEN TC278

ISO/CEN	TS	14822-1	Traffic and Travel Information—Medium-range pre-information via DSRC—General specification—Part 1: downlink	Published	ISO TC204/ CEN TC278

See Section 8.2.2.1.

ISO/CEN	TS	14822-2	Traffic and Travel Information—Medium-range pre-information via DSRC—General specification—Part 2: uplink	Deleted	ISO TC204/ CEN TC278

See Section 8.2.2.2.

8.2.3.7 CEN/TS 14823 Messages Via Media-Independent Stationary Dissemination Systems—Graphic Data Dictionary for Pre-Trip and In-Trip Information Dissemination System

ISO/CEN	TS/ DIS	14823	Traffic and Travel Information—Messages via media-independent stationary dissemination systems—Graphic data dictionary for pre-trip and in-trip information dissemination system	Published	ISO TC204/ CEN TC278

ISO TS 12823 Available from
http://www.iso.org/iso/iso_catalogue/catalogue_tc/catalogue_detail.htm?csnumber=43456

This final draft International Standard presents a graphic data dictionary (GDD) which has been developed with the intent of creating a common basis for transmitting graphic information data that can be, irrespective of language or regional differences, decoded and understood by the users who obtain TTI (pretrip and in-trip information) service through TTI system operators such as traffic management centers (TMCs), traffic information centers (TICs), and value-added service providers (VASPs) which add value(s) to the TTI. Adopting unified graphic data is expected to improve the understandability of the graphic information by the user and thereby increase the convenience of TTI systems.

The coding system can be used to form messages to be handled by respective media systems, graphic messages on on-board units, and media system information on TTI dissemination systems (VMS, PC, PAT), including graphic data. These types of information are required by travelers for their pre-trip planning as well as their in-trip plan modification based on information obtained through media systems.

The purpose of GDD is, in order to facilitate the data exchange between media, to catalog graphic images like traffic signs and pictograms specified and used uniquely in each country and to assign them a certain code.

8.2.3.8 ISO 14825 Intelligent Transport Systems—Geographic Data Files (GDF)—Overall Data Specification

CEN/ISO	ENV/ ISO	14825	Intelligent transport systems—Geographic Data Files (GDF)—Overall data specification	Published	CEN TC278

See Section 5.5.1.

8.2.3.9 ISO 15075 Transport Information and Control Systems—In-Vehicle Navigation Systems—Communications Message Set Requirements

CEN/ISO	ENV/ ISO	15075	Transport information and control systems—In-vehicle navigation systems—Communications message set requirements	Published 2003	CEN TC278

CEN ENV 15075 Available from
http://www.cen.eu/esearch/extendedsearch.aspx

8.2.3.10 ISO TR 17384 Requirements for Interactive Centrally Determined Route Guidance

CEN/ISO	ENV/ ISO	17384	Requirements for Interactive Centrally Determined Route Guidance	Published 2003	CEN TC278

CEN ENV 17364 Available from
http://www.cen.eu/esearch/extendedsearch.aspx

8.2.3.11 ISO PAS 17684 Transport Information and Control Systems—In-Vehicle Navigation Systems—ITS Message Set Translator to ASN.1 Format Definitions

ISO	ISO	17684	Transport information and control systems—In-vehicle navigation systems—ITS message set translator to ASN.1 format definitions	Published 2003	ISO TC204

ISO PAS 17684 Available from
http://www.iso.org/iso/iso_catalogue/catalogue_tc/catalogue_detail.htm?csnumber=38507

A Publicly Available Specification (PAS) is usually developed within an ISO technical committee where there is inadequate consensus to develop a TR, TS, or standard. It provides an option for a specification, but has no status as a standard or agreed technical specification.

ISO/PAS 17684:2003 specifies a method that can be used to define navigation message sets in tabular form with a subsequent translation into a corresponding ASN.1 description. An intermediate language called descriptor normal form (DNF), which is a subset of ASN.1, is specified and used as an intermediate description between a tabular form and its ASN.1 description. A tabular-form message-set description language called message set tabular form (MSTF) is included as an example of a tabular form definition.

8.2.3.12 ISO 18234 Traffic and Travel Information (TTI)—TTI Via Transport Protocol Expert Group (TPEG) Data-Streams

CEN/ISO	TS	18234	Traffic and Travel Information (TTI)—TTI via Transport Protocol Expert Group (TPEG) data-streams	Published	CEN TC278

During the development of the TPEG technology a number of versions have been documented and various trials implemented using various versions of the specifications. At the time of the publication of this technical specification, all parts are fully interworkable and no specific dependencies exist. This technical specification has the technical version number TPEG-INV/102 (Annex A)

TPEG technology uses a byte-oriented stream format, which may be carried on almost any digital bearer with an appropriate adaptation layer. TPEG messages are delivered from service providers to end-users and are used to transfer application data from the database of a service provider to an end-user's equipment. TPEG has initially been planned and designed to meet a particular brief, from the EBU's mBroadcast Management Committee, namely, "to develop a new protocol for Traffic and Travel Information, for use in the multimedia broadcasting environment to develop applications, service and transport features which will enable travel-related messages to be coded, decoded, filtered and understood both by humans (visually and/or audibly) and by agent systems." This brief is also endorsed by the EBU TTI Broadcast Strategy Team, which recognizes the vital importance of a bearer independent TTI protocol for broadcast applications.

TPEG technology has been designed to be usable for a wide range of applications that require the efficient transmission of point-to-multipoint data over potentially unreliable broadcast channels. It is also suitable for point-to-point and multicast applications and may easily be encapsulated in Internet Protocol.

CEN/ISO TS	18234-1	Traffic and Travel Information (TTI)—TTI via Transport Protocol Expert Group (TPEG) data-streams—Part 1: Introduction, Numbering and Versions	Published	CEN TC278

ISO TS 12834-1 Available from
http://www.iso.org/iso/iso_catalogue/catalogue_tc/catalogue_detail.htm?csnumber=44061

This technical specification, CEN ISO/TS 18234-1, provides an index to the other parts of CEN ISO/TS 18234 and describes the AID for TPEG (section 5). This document has been prepared by CEN/TC 278 Road Transport and Traffic Telematics in cooperation with ISO/TC 204, "Intelligent Transport Systems."

It provides an introduction and index to the initial set of TPEG applications and specifications. It allows the indexing of new applications as they are added to the TPEG applications family, by defining their application identification (AID).

CEN/ISO TS	18234-2	Traffic and Travel Information (TTI)—TTI via Transport Protocol Expert Group (TPEG) data-streams—Part 2: Syntax, Semantics and Framing Structure (SSF)		CEN TC278

ISO TS 18234-2 Available from
http://www.iso.org/iso/iso_catalogue/catalogue_tc/catalogue_detail.htm?csnumber=44062

This technical specification establishes the method of referencing used within a TPEG data stream to allow a service provider to signal availability of the same service on another bearer channel or similar service data from another service.

CEN/ISO TS	18234-3	Traffic and Travel Information (TTI)—TTI via Transport Protocol Expert Group (TPEG) data-streams—Part 3: Service and Network Information (SNI) Application		CEN TC278

ISO 18234-3 Available from
http://www.iso.org/iso/iso_catalogue/catalogue_tc/catalogue_detail.htm?csnumber=44063

This technical specification establishes the method of delivering service and network information within a TPEG service. The TPEG-SNI application is designed to allow the efficient and language-independent delivery of information about the availability of the same service on another bearer channel or similar service data from another service provider, directly from service provider to end users.

The term "application" is used in TPEG specifications to describe specific applications, which are at the highest layer of the ISO/OSI protocol stack (ISO/IEC 7498-1). Each TPEG application (e.g., TPEG-RTM) is assigned a unique number that is called the application identification (AID). An AID is defined whenever a new application is developed. The AID is used within the TPEG-service and network information application (this document) to indicate how to process TPEG content and allows routing of data to an appropriate application decoder. AID = 0000 is assigned to the TPEG-SNI application, described in this specification.

A number of tables of information are described, which provide comprehensive options for describing services, their timing, content, and geographical coverage. In all TPEG streams it is mandatory to deliver to so-called GST. Additionally it is possible to signal linkage of content between different bearers and services.

CEN/ISO TS	18234-4	Traffic and Travel Information (TTI)—TTI via Transport Protocol Expert Group (TPEG) data-streams—Part 4: Road Traffic Message (RTM) Application	Published	CEN TC278

ISO 18234-4 Available from
http://www.iso.org/iso/iso_catalogue/catalogue_tc/catalogue_detail.htm?csnumber=44064

This document establishes the method of delivering road traffic messages (RTM) within a TPEG service. The TPEG-RTM application is designed to allow the efficient and language-independent delivery of road information directly from service provider to end users. The information provided relates to events and some status information on the road network and on associated infrastructure affecting a road journey.

For example, limited information about abnormal operation of links in the network may be included, such as ferries and lifting bridges.

The term "application" is used in TPEG specifications to describe specific applications, such as in this case the RTM application, which comprises three information containers: the message management container, the application event container, and the TPEG-location container. The first two containers are fully described herein, and the TPEG-location container is described in CEN ISO/TS 18234-6.

Each TPEG application (e.g., TPEG-RTM) is assigned a unique number that is called the application identification. An AID is defined whenever a new application is developed. The AID is used within the TPEG-service and network information application (CEN ISO/TS 18234-3) to indicate how to process TPEG content and allows routing of data to an appropriate application decoder. AID = 0001 is assigned to the TPEG-road traffic message application, described in this specification.

A hierarchical methodology has been developed to allow the creation of messages from a set of TPEG-RTM tables, which are essentially word oriented and cover most needs. Many of the TTI descriptive words, in the TPEG-RTM tables, were obtained from the DATEX dictionary (ENV 13106), which embodies European TTI knowledge of the last 10 years or more, including a deconstruct of the phrase-oriented RDS-TMC events list (EN ISO 14819-2). These TPEG-RTM tables (essentially word-oriented data object dictionaries) comprise a wide ranging ability to describe a TTI event and some status information, introducing new precision in a number of areas such as "vehicle types," "positional information on the carriageway," and "diversion routing advice."

NOTE: Explicit backwards compatibility with the RDS-TMC events list (EN ISO 14819-2) could not be achieved since some update classes, such as "29 Reference to Audio Broadcasts" and "30 Service Messages," fall outside the TPEG-RTM remit.

CEN/ISO TS	18234-5	Traffic and Travel Information (TTI)—TTI via Transport Protocol Expert Group (TPEG) data-streams—Part 5: Public Transport Information (PTI) Application	Published	CEN TC278

ISO 18234-5 Available from
http://www.iso.org/iso/iso_catalogue/catalogue_tc/catalogue_detail.htm?csnumber=37265

This technical specification describes the public transport information (PTI) application, which is intended to cover all modes of public (i.e., collective) transport as well as interurban and intraurban travel. The application is designed to allow the efficient and language-independent delivery of public transport information directly from service provider to end users.

However, it is important to note that TPEG-PTI is not limited to describing single services, because it also allows the more general description of route, service, and area-wide problems.

Public (or collective) transport information is usually consumed in one of four principle ways, as follows:

- Leader board information as used at stations or terminals;
- A report on the state of a network;
- The description of an individual service;
- As a news flash report.

The elements needed to provide information for any one of the four end-user presentation modes are largely the same. The end-user focus of TPEG applications makes it useful to be able to mimic presentations, to which end users are accustomed, for example, a railway station indicator board. TPEG-PTI messages can therefore group data elements to present one of the following end-user presentation modes:

- Incident message report;
- Station/terminal information;
- Route information;
- Individual service information.

It is important to bear in mind that these end-user presentation modes are merely presentational aides; they have little to do with the content in the individual data elements. They do, however, indicate how data elements must be grouped if a presentation in any of these views is intended. Unlike the TPEG-RTM application (CEN ISO/TS 18234-4), TPEG-PTI benefits from the nodal structure of public transport, making use of its discrete start, end, and stopping points as well as being limited to fixed, be it real or virtual, routes.

CEN/ISO TS	18234-6	Traffic and Travel Information (TTI)—TTI via Transport Protocol Expert Group (TPEG) data-streams—Part 6: Location Referencing for applications (TPEG-Loc)	Published	CEN TC278

ISO 18234-6 Available from
http://www.iso.org/iso/iso_catalogue/catalogue_tc/catalogue_detail.htm?csnumber=37266

This technical specification establishes the method of location referencing used by TPEG applications such as TPEG-RTM or TPEG-PTI.

TPEG applications are specified to contain all the information required by a client TPEG decoder (i.e., both location referencing and event information) to present all the information intended for the end user when it was originated by the service provider.

The term "application" is used in TPEG specifications to describe specific applications, which are at the highest layer of the ISO/OSI protocol stack (ISO/IEC 7498-1). Each TPEG application (e.g., TPEG-RTM) is assigned a unique number that is called the application identification. In this respect TPEG-Loc is not an application, but it is an *essential* constituent part of an application.

Location referencing requires a service provider to give an impression or image to the human end user of where an event has taken place. This cannot be done easily because the human end user may or may not be familiar with the location. TPEG-Loc has the added challenge of attempting to be as language independent as possible. This is achieved by the use of TPEG-Loc tables (essentially word-oriented data object dictionaries).

TPEG-Loc also provides location data in a machine-readable form that allows a "thick" client such as a navigation system to map-match, on-the-fly, to locate the event being described onto a digital map display.

NOTE: Explicit backwards compatibility with the RDS-TMC location referencing method (EN ISO 14819-3) has NOT been attempted because RDS-TMC locations are finite in number and must be predetermined. TPEG technology does not suffer from this restriction.

CEN/ISO TS	18234-7	Traffic and Travel Information (TTI)—TTI via Transport Protocol Expert Group (TPEG) data-streams—Part 7: Parking Information (PKI) application (TPEG-PKI)	Under Development	CEN TC278

ISO 18234-7 is Not Yet Available. Track progress at:
http://www.iso.org/iso/iso_catalogue/catalogue_tc/catalogue_detail.htm?csnumber=44611

This work item was approved in February 2007. No summary or description of content is available.

CEN/ISO	18234-8	Traffic and Travel Information (TTI)—TTI via Transport Protocol Expert Group (TPEG) data-streams—Part 8: Congestion and Travel Time information (CTT) application (TPEG-CTT)	Under Development	CEN TC278

ISO 18234-8 is Not Yet Available. Track progress at:
http://www.iso.org/iso/iso_catalogue/catalogue_tc/catalogue_detail.htm?csnumber=44612

This work item was approved in February 2007. No summary or description of content is available.

CEN/ISO	18234-9	Traffic and Travel Information (TTI)—TTI via Transport Protocol Expert Group (TPEG) data-streams—Part 9: Traffic Event Compact (TEC) application (TPEG-TEC)	Under Development	CEN TC278

ISO 18234-9 is Not Yet Available. Track progress at:
http://www.iso.org/iso/iso_catalogue/catalogue_tc/catalogue_detail.htm?csnumber=44613

This work item was approved in February 2007. No summary or description of content is available.

CEN/ISO	prCIS/ EN/TS	18234-10	Traffic and Travel Information (TTI)—TTI via Transport Protocol Expert Group (TPEG) data-streams—Part 10: Weather information (WEA) application (TPEG-WEA)	Under Development	CEN TTC278

ISO Available from 18234-10 is Not Yet Available. Track progress at: http://www.iso.org/iso/iso_catalogue/catalogue_tc/catalogue_detail.htm?csnumber=44614

This work item was approved in February 2007. No summary or description of content is available.

8.2.3.13 ISO 24530 Traffic and Travel Information (TTI)—TTI Via Transport Protocol Expert Group (TPEG) Extensible Markup Language (XML)

CEN/ISO	prCIS/ EN/TS	24530	Traffic and Travel Information (TTI)—TTI via Transport Protocol Expert Group (TPEG) Extensible Markup Language (XML)	Published	CEN TC278

This series of standards and technical specifications originated with the Transport Protocol Experts Group, or TPEG for short (pronounced T-Peg), which was founded in 1997 by the European Broadcasting Union (EBU). This group of experts led by the EBU has brought together experts from across the TTI sector and the broadcasting sector who are largely the means of achieving the service provision. TPEG itself is an evolution of RDS-TMC (see Sections 8.2.3.1, 8.2.3.3, and 8.2.3.5), whereas RDS-TMS is machine readable, TPEG data is also understandable to humans. TPEG specifications for transmission are language independent and therefore understandable to the client, whichever country he or she happens to be in at the time.

TPEG binary data format is designed for transmission over DAB and DMB; tpegML is an XML implementation designed for use in editing systems and delivery via the Internet and DVB.

The subject area covers the following areas:

- RTM: road traffic message;
- PTI: public transport information;
- Loc: location referencing, used in conjunction with applications;
- PKI: parking Information;
- CTT: congestion and travel time;
- TEC: traffic event compact;
- WEA: weather information for travelers;
- RTM: radio traffic messages;
- PTI: public transport messaging.

This family of standards is being developed in seven parts. Parts 5–7 were approved only towards the end of 2006. However, a new work item proposal was submitted to ISO/TC 204 in July 2007 to amalgamate 14823, 17267, and 24530 into a single new work item "Navigation System Application Program Interface (API)." A work item number has not yet been allocated. See Figure 8.1.

CEN/ISO TS	24530-1	Traffic and Travel Information (TTI)—TTI via Transport Protocol Expert Group (TPEG) Extensible Markup Language (XML)—Part 1: Introduction, common data types and tpegML	Published	CEN TC278

ISO TS 24530-1 Available from
http://www.iso.org/iso/iso_catalogue/catalogue_tc/catalogue_detail.htm?csnumber=37268

ISO/TS 24530-1:2006 establishes the top-level containers for TPEG messages in XML and the common data types that are used by tpegML applications (e.g., tpeg-ptiML). Inherently, tpegML is designed to map the TPEG binary (ISO/TS 18234 series); however, additional tags are provided to create a message and message set structure to facilitate Internet file delivery.

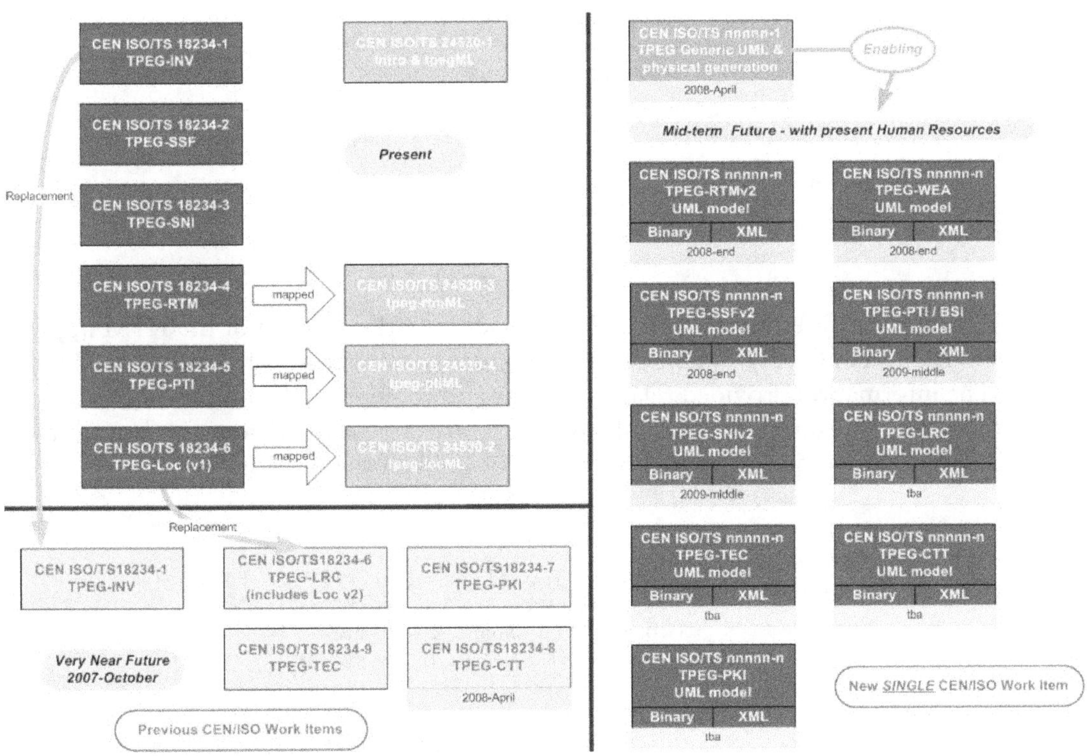

Figure 8.1 Navigation Systems Application Program Interface–Evolution. (*Source:* ISO/TC 204 WG10.)

CEN/ISO TS	24530-2	Traffic and Travel Information (TTI)—TTI via Transport Protocol Expert Group (TPEG) Extensible Markup Language (XML)—Part 2: tpeg-locML	Published	CEN TC278

ISO TS 24530-2 Available from
http://www.iso.org/iso/iso_catalogue/catalogue_tc/catalogue_detail.htm?csnumber=37500

ISO/TS 24530-2:2006 establishes the XML encoding of the method of location referencing used by TPEG applications.

TPEG applications contain the information required by a client TPEG decoder (i.e., both location referencing and event information), to present all the information intended for the end user when it was originated by the service provider.

Location referencing requires a service provider to give an impression or image, to the human end user, of where an event has taken place. This cannot be done easily because the human end user may or may not be familiar with the location. TPEG-Loc has the added challenge of attempting to be as language independent as possible. This is achieved by the use of TPEG-Loc tables (essentially word-oriented data object dictionaries).

CEN/ISO TS	24530-3	Traffic and Travel Information (TTI)—TTI via Transport Protocol Expert Group (TPEG) Extensible Markup Language (XML)—Part 3: tpeg-rtmML	Published	CEN TC278

ISO TS 24530-3 Available from
http://www.iso.org/iso/iso_catalogue/catalogue_tc/catalogue_detail.htm?csnumber=37501

This document establishes the XML encoding of the method of the RTM application. The TPEG-RTM application is intended to convey information to road users. The information provided relates to event and some status information on the road network and on associated infrastructure affecting a road journey (e.g., limited information about abnormal operation of links in the network).

CEN/ISO TS	24530-4	Traffic and Travel Information (TTI)—TTI via Transport Protocol expert Group (TPEG) Extensible Markup Language (XML)—Part 4: tpeg-ptiML	Published 2006	CEN TC278

ISO TS 24530-4 Available from
http://www.iso.org/iso/iso_catalogue/catalogue_tc/catalogue_detail.htm?csnumber=37502

ISO/TS 24530-4:2006 establishes the XML encoding of the method of the public transport information application.

The public transport information application is intended to cover all modes of public (i.e., collective) transport as well as interurban and intraurban travel. The application itself is designed to allow the efficient and language-independent transmission of public transport information either directly to an end user, be it the public or another service provider, such as broadcasters, service operators, or other information disseminating points, or to centers for onward transmission.

CEN/ISO	TS	24530-5	Traffic and Travel Information (TTI)—TTI via Transport Protocol expert Group (TPEG) Extensible Markup Language (XML)—Part 5: tpeg-pkiML	In development	CEN TC278

ISO TS 24530-5 Available. Track progress at:
http://www.iso.org/iso/iso_catalogue/catalogue_tc/catalogue_detail.htm?csnumber=38413

Work on this was terminated in favor of "Navigation System Application Program Interface (API)." See Section 8.2.3.14.

CEN/ISO	prCIS/ EN/ TS	24530-6	Traffic and Travel Information (TTI)—TTI via Transport Protocol expert Group (TPEG) Extensible Markup Language (XML)—Part 6: tpeg-cttML	Under Development	CEN TC278

ISO 24530-6 Not Yet Available

Work on this was terminated in favor of "Navigation System Application Program Interface (API)." See Section 8.2.3.14.

CEN/ISO	prCIS/ EN/ TS	24530-7	Traffic and Travel Information (TTI)—TTI via Transport Protocol expert Group (TPEG) Extensible Markup Language (XML)—Part 7: tpeg-weaML	Under Development	CEN TC278

ISO 24530-7 Not Yet Available

Work on this was terminated in favor of "Navigation System Application Program Interface (API)." See Section 8.2.3.14.

8.2.3.14 ISO ***** Traffic and Travel Information (TTI)—TTI Via Transport Protocol Expert Group (TPEG)—Navigation System Application Program Interface (API)

CEN/ISO	PWI	Work Item Reference Not yet allocated	Traffic and Travel Information (TTI)—TTI via Transport Protocol expert Group (TPEG)—Navigation System Application Program Interface (API)	Under Development	ISO TC204/ CEN TC278

ISO ***** (number not yet allocated) Not Yet Available

This standard will define an API for navigation and other location-based services targeted at transportation and mobile applications.

This standard will build on the previous work of TC 204/WG3. It is anticipated that development of this API will proceed more quickly than the WG3 API work to date because of this reuse of previous work, and because the higher-level interface will contain fewer functions and require the specification of far fewer details. See also Figure 8.1.

8.2.4 Route Guidance and Navigation Pretrip

This service group is categorized as a planning service undertaken pretrip and provides information on community and/or individual user optimum route options for specified destinations. Best route options may be calculated which take into account network and public transport information and they may incorporate multimodal options such as park-and-ride.

This service also includes the provision of route guidance to pedestrians, cyclists, and motorcyclists.

8.2.4.1 ISO 14826 Physical Storage for Database Technology

ISO/IEC	NP	14826	Physical Storage for Database Technology	Withdrawn	ISO TC204

ISO 14826 has been withdrawn in favor of ISO 17267 and ISO 20452

8.2.4.2 ISO 17267 Navigation System API Standard (API)

ISO/IEC	NP	17267	Navigation System API Standard (API)	Under Development/ CD ballot closed Oct. 2007	ISO TC204

ISO 17267 CD Not Yet Available

See Section 8.2.1.1.

8.2.4.3 ISO 24099 Data Structure for Map Data Provision and Update in ITS Applications

ISO/IEC	NP	24099	Data Structure for Map Data Provision and Update in ITS Applications	Under Development	ISO TC204

ISO 24099 Not Yet Available. Track progress at:
http://www.iso.org/iso/iso_catalogue/catalogue_tc/catalogue_detail.htm?csnumber=42016

This project will define data structures for provision and update of map-related data from data centers to navigation systems used in ITS applications.

8.2.5 Route Guidance and Navigation On-Trip

See also Section 8.2.4 with respect to systems that are used pretrip.

This service group covers services categorized as those that are used on-trip (i.e., dynamic services undertaken while the trip is in progress). Similar to its pretrip counterpart, it provides information on community and/or individual user optimum route options for specified destinations. Best route options may be calculated which take into account network and public transport information and they may incorporate multimodal options such as park-and-ride. Different to its pretrip counterpart, it provides the services during the trip and consequently can also provide additional services such as congestion avoidance route guidance.

This service also includes the provision of route guidance to pedestrians, cyclists, and motorcyclists.

8.2.5.1 SAE J2266 Navigation and Route Guidance Function Accessibility While Driving

SAE	J2266_200410	Navigation and Route Guidance Function Accessibility While Driving	Published Aug. 2004	SAE

SAE J2266 Available from
http://www.sae.org/technical/standards/J2266_200410

The location referencing message specification (LRMS) standardizes location referencing for ITS applications that require the communication of spatial data references between databases. ITS databases may reside in central sites, vehicles, or devices on or off roads or other transportation links. The LRMS is applicable to both homogeneous (same database) and mixed database environments that may be implemented on wireless or landline networks. While developed for ITS applications, the LRMS may be used for non-ITS applications as well within the field of geographic information processing.

8.2.5.2 J2678 Navigation and Route Guidance Function Accessibility While Driving Rationale

SAE	J2678_200408	Navigation and Route Guidance Function Accessibility While Driving Rationale	Published Aug. 2004	SAE

SAE J2678 Available from
http://www.sae.org/technical/standards/J2678_200408

This document provides the rationale used by the Navigation Function Accessibility Subcommittee (the Subcommittee) for the development and content of a SAE J2364 Recommended Practice: "*Navigation and Route Guidance Function Accessibility While Driving*." It provides both the reasoning for the overall recommended practice as well as each of its elements.

8.2.6 Trip Planning Support

This service group covers the use of ITS systems to provide data regarding traffic flows and travel demand for trip planning purposes, including collection, archiving, and retrieval of system data. Examples of such system data include:

- Current traffic flow data from traffic management systems;
- Current utilization levels from public transport information systems;
- Origin and destination data from route guidance systems or vehicle probes;
- Route choice data from route guidance systems or vehicle probes;
- Travel demand data from pretrip information systems.

NOTE: See also the previous parts of Section 8.2.

This service group is more likely to be serviced by commercial, regional, or state products. At this stage, no standards have been developed. However, the following freely downloadable document may be of use to those involved in service provision this sector.

8.2.6.1 U.S. Federal Transit Administration Trip Planning State of the Practice

USDoT Federal Transit Administration	Trip Planning State of the Practice

FTA Trip Planning State of the Practice Available from
http://www.itsdocs.fhwa.dot.gov/JPODOCS/REPTS_TE/13685.html

Automated trip planning is one of the easiest ways for transit users to identify their best choice of routes using the Internet. Trip planners use an input form to obtain information on desired trip characteristics, then automatically generate an itinerary for the user. This report identifies opportunities for FTA to facilitate the development of trip planners. The report:

- Summarizes the current state of the practice in Web-based single and multiagency transit trip planning;
- Identifies single and multiagency trip planner development issues;
- Groups transit agencies by likely capability and interest in developing trip planners;
- Recommends appropriate federal assistance for each group and research to overcome barriers.

The main text of the report summarizes the current status of trip planner deployment, expectations, benefits, development issues, standards, costs, and staffing. It also groups transit agencies by market segments. Appendices provide detail on methodology, related literature, and features of existing trip planners.

8.2.7 Travel Services Information

This service group contains activities that support travelers either in a pretrip or on-trip context. This information provides "yellow pages" type functions that can be allocated to different services based on the nature of information and/or the customer of the information.

Examples of the type of functions about which information could be provided under various services include hospitals, hotels, fuel stations, restaurants, truck stops, reservation services, and vehicle maintenance facilities.

8.2.7.1 ISO 14819 TTI Messages Via Traffic Message Coding—Conditional Access

ISO	DIS	14819	TTI Messages via Traffic Message Coding—Conditional access		ISO TC204

See Section 8.2.3.4.

8.2.7.2 ISO 14822 TTI Messages Via DSRC Beacons—Medium Range Pre-Information

ISO		14822-1	TTI Messages via DSRC Beacons—Medium Range Pre-Information—Part 1 Downlink	Published	ISO TC204

ISO 14822-1 Available from
http://www.iso.org/iso/iso_catalogue/catalogue_tc/catalogue_detail.htm?csnumber=40628

See Section 8.2.2.1.

ISO	TS	14822-2	TTI Messages via DSRC Beacons—Medium Range Pre-Information : Part 2 Uplink	Deleted	ISO TC204

See Section 8.2.2.2.

8.2.7.3 ISO TTI Messages Via Stationary Dissemination Systems

ISO	TS/CD 14823	TTI Messages via Stationary Dissemination Systems	Published TS under Development	ISO TC204

ISO 14823 Available from
http://www.iso.org/iso/iso_catalogue/catalogue_tc/catalogue_detail.htm?csnumber=43456

See Section 8.2.3.7.

8.2.8 Probe Data

A probe data system is a system that consists of a group of vehicles that collect and transmit various types of data using medium and wide area radio communication, and center functions for statistical processing of the received data to acquire information concerning traffic, road, and environment is called a probe vehicle system.

Probe data is the data sent from onboard systems in the vehicle to the centers and other external systems. The speed and other basic data elements in the probe data are called probe data elements and a compilation of multiple data elements is called a probe message.

Probe messages always contain position and time stamps. This area of standardization provides a data dictionary for the probe vehicle system, as well as standardization of the instructions for the probe data reporting management that is sent from the center side to the group of vehicles when collecting probe data.

ISO	CD	22837	Vehicle Probe Data for Wide Area Communications	Under Development	ISO TC204

22837 Available from
http://www.iso.org/iso/iso_catalogue/catalogue_tc/catalogue_detail.htm?csnumber=45418

This ISO International Standard specifies the core and initial sets of probe data elements and example probe messages, and a framework for defining and extending these probe data elements and probe messages. It facilitates the development and operation of probe vehicle systems by providing a standard set of probe data elements and probe messages along with the basic data framework to extend the set.

The ISO International Standard provides a reference architecture for probe vehicle systems and for probe data, a basic data framework for probe data elements and probe messages, the definition of core data elements, the definition of an initial set of additional probe data elements, and the definition of example probe messages.

Figure 8.2 depicts the scope described earlier.

Future Work

To completely define probe processing, the standardization of probe data elements and probe messages is not sufficient. Standards are also required for processed probe data (the output of probe processing) and downlink elements and messages (to convey these results to vehicles and other users). This International Standard prescribes an initial set of probe data elements that are important for transmission from vehicles to land-side processing centers. The standardization of other probe data issues will be addressed in future work.

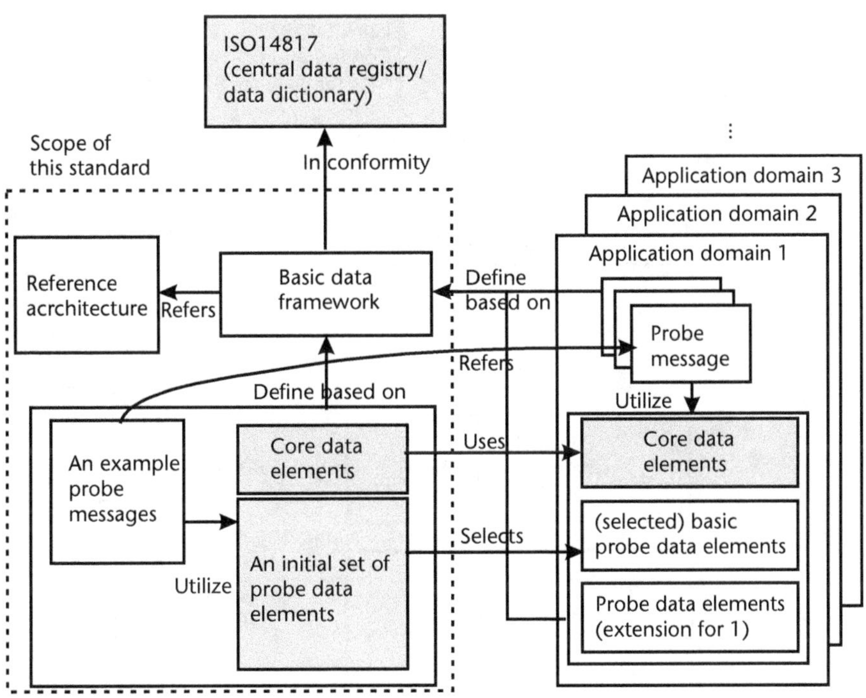

Figure 8.2 Scope of ISO 22837. (*Source:* ISO 22837.)

8.2.8.1 ISO 24100 Basic Principles for Personal Data Protection in Probe Vehicle Information Services

ISO	NP	24100	Basic Principles for Personal Data Protection in Probe Vehicle Information Services	Under Development	ISO TC204

ISO 24100 Not Yet Available. Track progress at:
http://www.iso.org/iso/iso_catalogue/catalogue_tc/catalogue_detail.htm?csnumber=42017

Figure 8.3 shows the scope of this standard.

See Section 4.6.2.18.

8.2.8.2 ISO 25114 Probe Data Reporting Management

ISO	PWI	25114	Probe Data Reporting Management	Under Development	ISO TC204

ISO 25114 Not Yet Available. Track progress at:
www.calm.hu

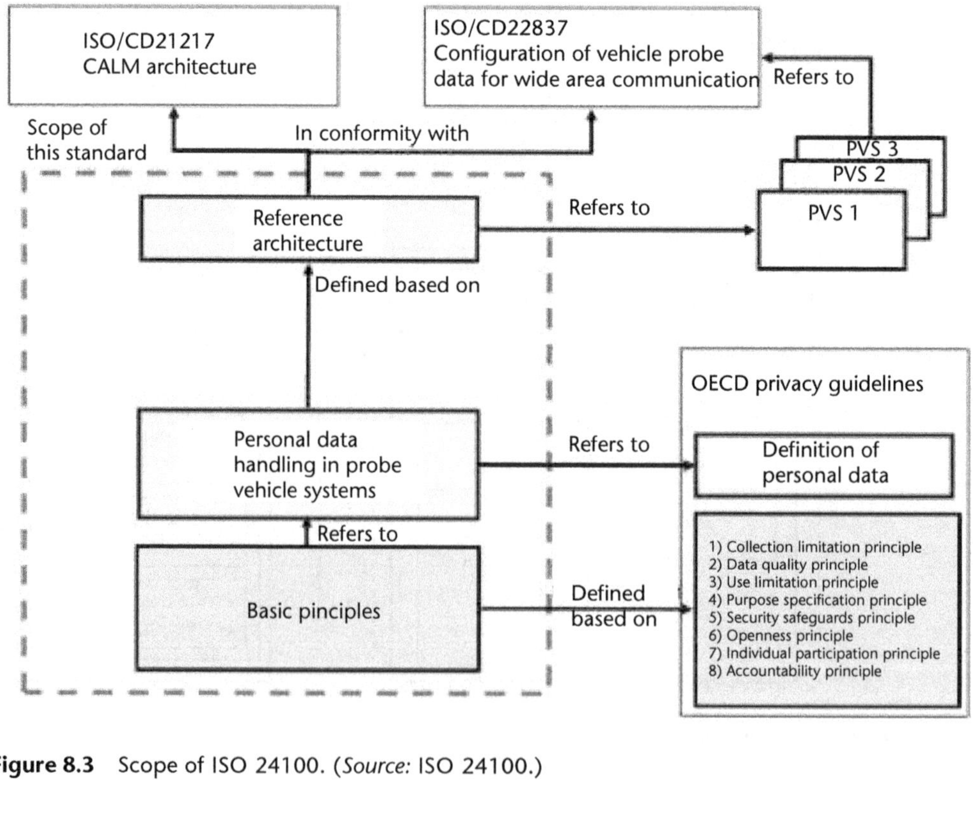

Figure 8.3 Scope of ISO 24100. (*Source:* ISO 24100.)

Probe vehicle systems are being investigated and deployed throughout the world. It is expected that the number of practical systems will grow steadily over the next few years. Since communications airtime will always be a scarce and expensive commodity, efficient probe data reporting systems must rely on techniques to use airtime efficiently and economically. One way to accomplish this is to have the probe processing center guide the economic collection of necessary probe information by sending reporting management instructions to probe vehicles; this is called probe data reporting management (PDRM).

As probe vehicle systems have to collect and manage probe data from a variety of vehicles from different vehicle manufacturers, the standardization of these reporting management instructions is essential. To do this, a common framework for PDRM is also required.

The scope of the standard will therefore be to provide:

- A common framework for defining PDRM instructions to facilitate the specification and design of probe vehicle systems;
- Concrete definition of PDRM instructions.

This standard will specify:

- Reference architecture for probe vehicle systems and probe data which incorporates PDRM, based on the reference architecture for ISO 22837;
- Basic data framework for PDRM instructions defines specifically:
    - Necessary conditions for probe data reporting management instructions;
    - Notations of these instructions (in XML);
- Rules for using PDRM instructions;
- The definition of PDRM instructions, as described earlier.

Different types of data reporting management messages are identified and defined, such as:

- Start/stop all probe reporting;
- Start/stop probe reporting of specific probe data elements;
- Generic scheme for conveying criteria for reporting specific probe data elements.

PDRM instructions may be structured in terms of a time period in which they are valid (duration), geographic region, and roadway heading in which the instruction applies.

See Figure 8.4.

8.2.8.3 ISO 29284 Event Based Probe Vehicle Data

ISO	PWI	29284	Event based probe vehicle data	Under Development	ISO TC204

ISO 29284 is Not Yet Available

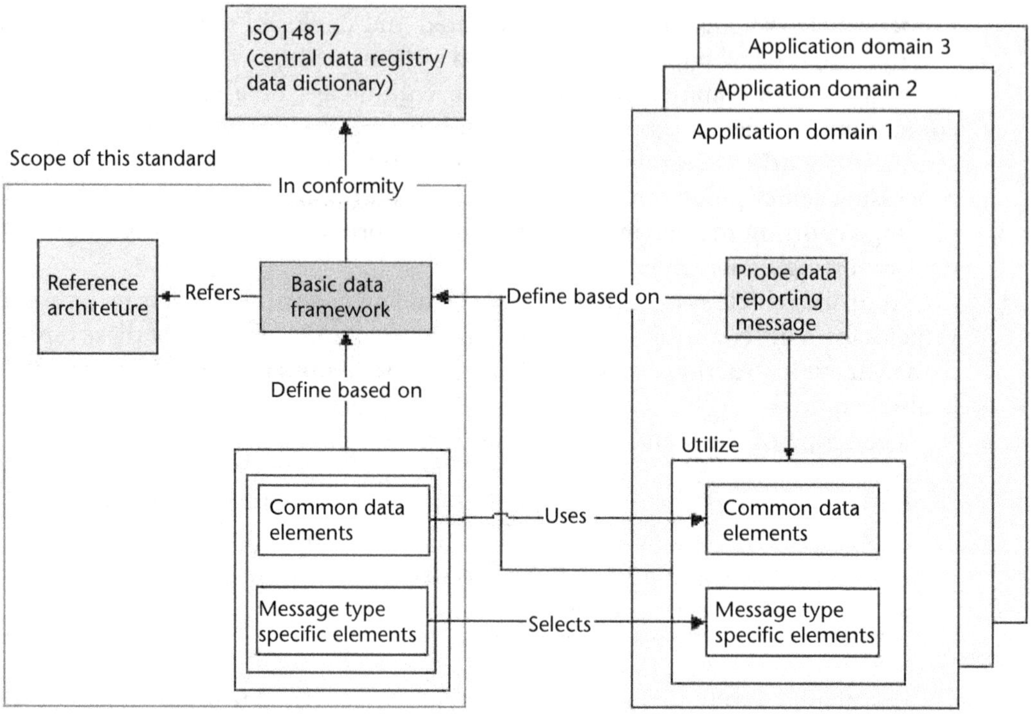

Figure 8.4 Scope of the PRDM standard. (*Source:* ISO 24100.)

As probe vehicle systems have to collect and manage probe data from a variety of vehicles from different vehicle manufacturers, the standardization of these event based messages is essential. To do this, a common framework for event based probe vehicle message reporting is also required.

The purpose of this project is to develop: (1) a reference architecture for event based probe data reporting within an architecture which encompasses both this function and standard probe data reporting defined in CD 22837; (2) the basic data framework for defining event based probe data messages; and (3) the concrete definition of these messages.

8.3 Traffic Management and Operations Service Groups

The service groups under the Traffic Management and Operations domain specifically address maintaining the movement of people, goods, and vehicles throughout the transportation network, and include both automated monitoring and control activities as well as decision-making processes (both automated and manual) that address real-time incidents and other disturbances on the transportation network, as well as managing travel demand as needed to maintain overall mobility. These service groups include the activities that have become known as Intelligent Highway and Cooperative Driving.

8.3.1 Traffic Management and Control

8.3.1.1 ISO 14827 TICS Data Interfaces Between Centres

ISO	IS	14827	TICS Data Interfaces between Centres	Published	ISO TC204

There are two parts to this International Standard.

ISO	IS	14827-1	TICS Data Interfaces between Centres: Part 1 message definition requirements	Published 2005	ISO TC204

ISO 14827-1 Available from
http://www.iso.org/iso/iso_catalogue/catalogue_tc/catalogue_detail.htm?csnumber=41361

A portion of the submittal dealt with the specification of messages. As this portion of the document could apply to various protocols, it was placed in ISO 14827-1, Message definition requirements. The remainder of the original submittal formed the basis of the application layer protocol and was placed in this part of ISO 14827.

Thus, this part defines only one way to implement the messages that are specified in the format defined by ISO 14827-1. This resulting International Standard supports existing and foreseen data exchange needs using modern design concepts.

Due to the flexibility required by the rapidly developing transport information and control systems (TICS) environment, this part of ISO 14827 uses a very generic structure. Thus, although it was initially intended to be an International Standard for TICS, it is flexible enough to be used for virtually any data exchange.

ISO 14827-1 explains how to define end-application messages that are to be exchanged between centers for TICS. This definition has been designed to be relatively generic to the selected protocol (e.g., DATEX-ASN, CORBA, and so forth). This part of ISO 14827 provides the specification of the Data Exchange protocol in ASN.1 (DATEX-ASN) used to exchange data between central systems. DATEX-ASN was the first protocol standardized because:

- The development of DATEX-Net could be leveraged.
- There was sufficient market interest to perform the required technical work.

ISO 14827-1:2005 defines the format that should be used to document those end-application messages that are to be exchanged between/among central systems. The format is protocol independent to the extent practical. For example, this one format can be used to define data exchanges that may apply to DATEX-ASN, CORBA, or other application protocols.

ISO	IS	14827-2	TICS Data Interfaces between Centers: Part 2 DATEX.ASN	Published	ISO TC204

ISO 14827-2 Available from
http://www.iso.org/iso/iso_catalogue/catalogue_tc/catalogue_detail.htm?csnumber=41362

Each end-application message is defined as either a "subscription" or a "publication" according to the format as specified in ISO 14827-1. DATEX-ASN defines how these end-application messages are packaged to form a complete data packet and also defines the rules and procedures for exchanging these data packets. Systems using DATEX-ASN are free to implement additional end-application functionalities according to the user requirements.

8.3.1.2 EN 12352 Traffic Control Equipment—Warning and Safety Light Devices

CEN	EN	12352	Traffic control equipment—Warning and safety light devices	Published 2005	CEN/TC 226

CEN ENV 12352 Available from any European National Standards organization
http://www.cen.eu/esearch/extendedsearch.aspx

This European standard specifies the requirements for individual electrically operated light devices, called warning lights, emitting a continuous or regular intermittent light of a single color which, by their color and position alone, are used to warn, inform, or guide road users. It specifies the requirements for visual, structural, and operational performances and the relevant test methods to be used. These devices rely upon existing furniture to provide the mounting.

This European standard is not applicable to lighting devices which convey messages by additional means (e.g., variable message signs), which convey a mandatory instruction (e.g., traffic signals), or which are covered by vehicle lighting regulations.

This European standard does not consider horizontal loads because it is the mounting to which they are fixed (which is not covered by this European standard) which has to resist applied horizontal loads.

8.3.1.3 EN 12368 Traffic Control Equipment—Signal Heads

CEN	EN	12368	Traffic control equipment—Signal heads	Published 2006	CEN/ TC226

CEN ENV 12368 Available from any European National Standards organization
http://www.cen.eu/esearch/extendedsearch.aspx

This standard may be required for reference for developers of ITS standards, but it is periphery to ITS. The content is therefore not detailed. In summary, it specifies: constructional requirements; signal head; mountings: poles, poles with bracket, and catenaries; deflection; environmental, electromagnetic compatibility (EMC) and electrical requirements; optical requirements; constructional and environmental test methods; optical test methods; and tolerances.

8.3.1.4 EN 12899 Fixed, Vertical Road Traffic Signs

CEN	EN	12899	Fixed, vertical road traffic signs	Published 2001	CEN/TC 226

This series of standards may be required for reference for developers of ITS standards, but they are periphery to ITS. The content of each part is therefore not detailed.

CEN	EN	12899-1: 2001	Fixed, vertical road traffic signs—Part 1: Fixed signs	Published 2001	CEN/TC 226

CEN ENV 12899-1 Available from any European National Standards organization
http://www.cen.eu/esearch/extendedsearch.aspx

This part of EN 12899:2001 specifies requirements for new fixed signs: nonretroreflective and retroreflective fixed signs; nonretroreflective and retroreflective fixed signs when they are illuminated at night by external lighting luminaires; and transilluminated signs. The main intended use of fixed signs is for the instruction and guidance of road users on public and private land. It defines performance limits and a range of performance classes for both sign assemblies without vertical supports and sign assemblies complete with vertical supports, as well as sign faces and supporting substrates, sign fixings and supports, and external lighting luminaires.

Colorimetric and retroreflective properties as well as the luminance are specified. The retroreflective properties are with respect to materials based on the use of glass bead technology only. Structural requirements for signs and sign supports include performance under static and dynamic loading. It also defines performance levels to be maintained after natural weathering exposure.

NOTE: Where tests for extremely low temperatures are required, they should be in accordance with the customer's requirements.

This standard does not require the replacement of existing signs. Products and requirements not covered by this standard include: (1) sign gantries, cantilevers, and sign foundations; (2) signs constructed from light emitting diodes (LED) or fiber optics; (3) variable message signs; (4) transilluminated retroreflective signs; (5) passive safety performance requirements of sign support structures against vehicle impact; and (6) signs used for temporary purposes.

CEN	prEN 12899-2	Fixed, vertical road traffic signs—Part 2: Transilluminated traffic bollards (TTB)	Under Approval	CEN/TC 226

CEN ENV 12899-2 Available from any European National Standards organization
http://www.cen.eu/esearch/extendedsearch.aspx

This document specifies requirements for new transilluminated traffic bollards (TTBs) (including fixing and installation), which may incorporate traffic signs of type 1 TTBs or type 2 TTBs. It covers performance requirements and test methods. Colorimetric and retroreflective properties as well as luminance of electrically illuminated portions are specified taking into account CIE recommendations. Structural properties include requirements under static and dynamic loading. Provision is made for safety in use.

Devices of similar function, but without transillumination or less than 600 mm in height, are not included.

CEN	prEN 12899-3	Fixed, vertical road traffic signs—Part 3: Delineator posts and retroreflectors	Approved	CEN/TC 226

CEN ENV 12899-3 Available from any European National Standards organization
http://www.cen.eu/esearch/extendedsearch.aspx

Part 3 of EN 12899 specifies requirements for new delineator posts and for new retroreflectors as separate products or combined together to be used in traffic circulation areas. It covers performance requirements and test methods. Colorimetric and retroreflective properties are specified taking into account CIE recommendations. Structural requirements include performance under static and dynamic loading. Provision is made for safety in use, including vehicle impact. To define durability, this standard also includes performance levels to be maintained after natural weathering exposure. No requirements are given for the use of colors, dimensions, and tolerances of delineator posts and retroreflectors.

CEN	prEN 12899-4	Fixed, vertical road traffic signs—Part 4: Factory Production Control	Under Approval	CEN/TC 226

CEN ENV 12899-4 Available from any European National Standards organization
http://www.cen.eu/esearch/extendedsearch.aspx

This standard specifies procedures to be followed during manufacture to ensure consistent compliance with the product standard prEN 12899.

CEN	prEN	12899-5	Fixed, vertical road traffic signs—Part 5: Initial type testing	Under Approval	CEN/TC 226

CEN ENV 12899-5 Available from any European National Standards organization
http://www.cen.eu/esearch/extendedsearch.aspx

Part 5 of EN 12899 describes the requirements for initial type testing (ITT) of Parts 1, 2, and 3.

8.3.1.5 EN 12966 Road Vertical Signs—Variable Message Traffic Signs

CEN	EN	12966	Road vertical signs—Variable message traffic signs	Published 2005	CEN/TC 226

This series of standards is also periphery to ITS but is most likely to be referenced for delivery of ITS traveler information using VMS.

This series of standards have been prepared under a mandate given to CEN by the European Commission and the European Free Trade Association, and it supports essential requirements of EU Directive(s).

The standards consist of the following parts under the general title, Road vertical signs:

- Part 1: Variable message traffic signs—Part 1: Product standard;
- Part 2: (this part) Variable message traffic signs—Part 2: Initial type testing;
- Part 3: Variable message traffic signs—Part 3: Factory production control.

As this is a periphery to ITS, only a brief summary of each part is provided.

CEN	EN	12966-1	Road vertical signs—Variable message traffic signs—Part 1: Product standard	Published 2005	CEN/TC 226

CEN ENV 12966-1 Available from any European National Standards organization
http://www.cen.eu/esearch/extendedsearch.aspx

This document specifies requirements and test methods for new variable message signs (VMS). VMS comprise two types: continuous and discontinuous signs. Continuous signs are those that are similar to fixed signs, the only difference being that by some electromechanical means they can show various messages (e.g., rotating prism signs, roller blinds). Discontinuous signs create messages using individual elements that can be in one of two states (or more) and can thereby create various messages on the same sign face (e.g., fiber optic signs, LED signs).

CEN	EN	12966-2	Road vertical signs—Variable message traffic signs—Part 2: Initial Type Testing	Published 2005	CEN/TC 226

CEN ENV 12966-2 Available from any European National Standards organization http://www.cen.eu/esearch/extendedsearch.aspx

This document specifies the requirements for Initial Type Testing (ITT) and is designed to be read in conjunction with Part 1.

CEN	EN	12966-3	Road vertical signs—Variable message traffic signs—Part 3: Factory production control	Published 2005	CEN/TC 226

CEN ENV 12966-3 Available from any European National Standards organization http://www.cen.eu/esearch/extendedsearch.aspx

The standard provides definitions for:

- System requirements;
- Common requirements;
- Product testing and evaluation.

8.3.1.6 ISO 14827 TICS Data Interfaces Between Centres

IS	IS	14827	TICS Data Interfaces between Centres	Published	ISO TC204

See Section 8.3.1.1.

8.3.1.7 ISO 15784 Transport Information and Communication System—Data Exchange Involving Roadside Module Communication

ISO	NP	15784	Parts 1-3—Transport Information and Communication System—Data Exchange involving roadside module communication		ISO TC204

The functional requirements for communication between a traffic management center and roadside modules used for traffic management are varied because there are many kinds of roadside modules for traffic management throughout the world, such as signal controller, dynamic message sign, and vehicle detectors. In the development of standards for data exchanges between a traffic management center and roadside modules used for traffic management, ISO/TC 204 WG9 agreed that the concept of a single standard for all countries and devices may not be appropriate, but a set of standards for different types of roadside module may be more appropriate.

As a result, ISO/TC 204 WG9 adopted the philosophy of producing profile documents to specify how data should be exchanged.

To make this standard, they refer to the existing standards about profiles, specifically NTCIP 8003 which is the U.S. standard for a profile framework, and ISO/IEC TR10000, which is a technical report on "Framework and Taxonomy of International Standardized Profiles."

The purpose of a profile is to specify the use of one or more base standards to provide requested function. Because there are multiple functional requirements to data exchange between a center and the roadside, this standard must be defined in multipart profiles.

This standard defines only the application profile. End application data is defined in the data registry. Each country should define lower layer profiles based on the internationally standardized protocols because each country has its own circumstance on communication infrastructure.

ISO	DIS	15784-1	Transport Information and Control Systems—Data exchange Involving Roadside Module Communication—Part 1: General principles and documentation framework of application profiles	Under Development DIS	ISO TC204

ISO 15784-1 Not Yet Available. Track progress at:
http://www.iso.org/iso/iso_catalogue/catalogue_tc/catalogue_detail.htm?csnumber=39586

Part 1 of this standard has been prepared in order to explain the principles of profile and documentation rules of the application profiles for traffic management. The following parts consist of the multiple application profiles to use for data exchange between traffic management center and roadside modules used for traffic management.

ISO	WD	15784-2	Data Exchange involving Roadside Modules—Part 2: Application Profile—TMP	Under Development	ISO TC204

ISO 15784-2 Not Yet Available. Track progress at:
http://www.iso.org/iso/iso_catalogue/catalogue_tc/catalogue_detail.htm?csnumber=39587

The message structures defined by the application layer are defined using ASN.1. Basically, ISO 14827-2 requires support of basic encoding rules (BER) and allows support of additional encoding rules through a negotiation scheme. The octet encoding rules is one of the options.

ISO	DIS	15784-3	Transport Information and Control Systems—Data exchange Involving Roadside Module Communication—Part 3: AP DATEX	Under Development DIS	ISO TC204

ISO 15784-3 Not Yet Available. Track progress at:
http://www.iso.org/iso/iso_catalogue/catalogue_tc/catalogue_detail.htm?csnumber=39588

The purpose of this part is to define an application profile referring ISO 14827 and other base standards. The application profile specified in this standard is used to exchange data and messages in the following cases:

- Between a traffic management center and roadside modules for traffic management;
- Between roadside modules used for traffic management.

The scope of this standard does not include the communication between roadside modules and in-vehicle units, in-vehicle communication, in-cabinet communication, or motion video transmission from a camera or recorded media.

This standard deals with the interface described in Figure 8.5.

The definition of application profiles and their functions and responsibilities are defined within clause 5.5.3 of ISO/IEC TR 10000-2:1998. This profile references

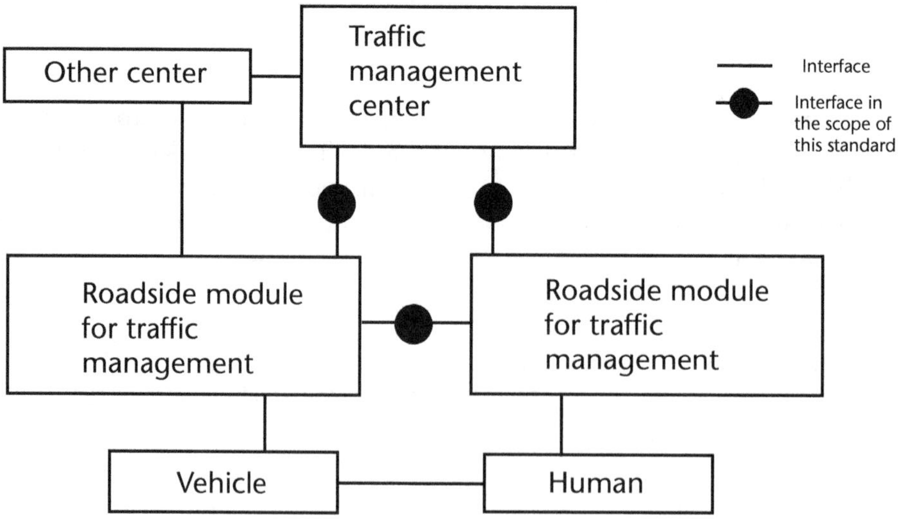

Figure 8.5 Example of AP scenario. (*Source:* ISO 15784-3 Scope.)

Table 8.1 ISO Layers, Base Standards, and Taxonomy

ISO Layer	*Base Standard*	*Taxonomy*
Application layer	ISO 14827-2 DATEX-ASN	AP-DATEX
Presentation layer	ISO 8825-1 basic encoding rules and others	Application profile
Session layer	<NULL>	

ISO standards for the application and presentation layers of the OSI reference model. The session layer is defined as being NULL because the selected application layer handles this functionality (see Table 8.1).

8.3.1.8 ENV DATEX Traffic and Travel Data Dictionary

CEN	ENV	13106	DATEX traffic and travel data dictionary	Published Revised	ISO TC204

CEN 13106 Available from any European National Standards organization
http://www.cen.eu/esearch/extendedsearch.aspx

This provides a data dictionary for DATEX. This was updated and revised in 2006–2007.

8.3.1.9 EN WI 00278074 Road Traffic Data—Elaboration, Storage, Distribution

CEN			Road traffic data—Elaboration, storage, distribution—Exchange formats (low level)	Published	CEN TC278

CEN WI 00278074 Available from any European National Standards organization
http://www.cen.eu/esearch/extendedsearch.aspx

Documents are not yet available.

8.3.1.10 ENV 13777 DATEX Specifications for Data Exchange Between Traffic and Travel Information Centers

CEN	ENV	13777	DATEX specifications for data exchange between traffic and travel information centers	Withdrawn	ISO TC204

ENV 13777 was withdrawn in July 2007.

8.3.1.11 ISO/EN Road Vehicles—Traffic Information and Control Systems—Ergonomic Aspects of In-Vehicle Visual Presentation of Information

ISO/CEN	IS/EN	15008	Road vehicles—Traffic information and control systems—Ergonomic aspects of in-vehicle visual presentation of information	Published 2003	ISO TC22

ISO 15008 Available from
http://www.iso.org/iso/iso_catalogue/catalogue_tc/catalogue_detail.htm?csnumber=34886

ISO 15008:2003 gives minimum specifications for the image quality and legibility of displays containing dynamic (changeable) visual information presented to the driver of a road vehicle by an onboard transport information and control system used while the vehicle is in motion. These specifications are intended to be independent of display technologies, while test methods and measurements for assessing compliance with them have been included where necessary.

ISO 15008:2003 is applicable to mainly perceptual, and some basic cognitive, components of the visual information: these include character legibility and color recognition. It is not applicable to other factors affecting performance and comfort such as coding, format, and dialog characteristics, nor to displays using the following: superimposed information on the external field (e.g., head-up displays), pictorial images (e.g., closed-circuit TV for reversing), maps and topographic representations (e.g., those for setting navigation systems), and static information (e.g., control labels, telltales).

8.3.1.12 CEN WI 00278213 Traffic Management Systems—Detection on Motorways for Traffic Information and Traffic Management Applications

CEN	EN	Not Yet Allocated	Traffic management systems—Detection on motorways for traffic information and traffic management applications	Under Development	CEN TC278

CEN ENV ***** Not Yet Available. Once completed it will be available from any European National Standards organization
http://www.cen.eu/esearch/extendedsearch.aspx

This is a recent work item and no information is available at the time of this writing. A reference number has not yet been allocated.

8.3.1.13 CEN WI 00278213 Traffic Management Systems—Detection on Motorways for Traffic Information and Traffic Management Applications

UMTC	Urban Traffic Management & Control

UMTC Specifications are available at
www.utmc.uk.com

These specifications are not International Standards, but they may be regarded as the U.K. Department for Transport recommendations for urban traffic management and control systems, and therefore have almost the status of a standard in the U.K., and are referred to in many countries. Until 2006 this was an official U.K. Department for Transport Web site (http://www.utmc.gov.uk/), transferred under U.K. government outsourcing arrangements in 2006 to its present organization and Web site, which is: http://www.utmc.uk.com.

TS0003:2005	The UTMC Framework Technical Specification: TS003

The UTMC Framework Technical Specification TS003:2005 is available at
http://www.utmc.uk.com/technical/pdf/ts003.pdf

The UTMC Framework Technical Specification TS003:2005 consolidates and supersedes TS001:1997 Part 1--Infrastructure and TS002:1997—Communications.

TS003:2005 takes into account feedback from the consultation process and ongoing international initiatives in this field. It presents the core technical standards recommended for use by U.K. traffic managers in their systems. Nonnormative annexes provide guidance on how to use the specification.

TS003 should be used in conjunction with the other main repository of UTMC technical recommendations, namely the UTMC Objects Registry, TS004.003.

TS0004:2005	UTMC Objects Registry TS004

The UTMC Objects Registry TS004:2005 is available at TS004.003:2007:
Main text and key annexes including MIB definitions
http://www.utmc.uk.com/technical/pdf/ts004.003apr07.pdf

Annex D to TS004.003:2007: Data object definitions
http://www.utmc.uk.com/technical/xls/ts004.003annexdapr07.xls

Annex F to TS004.003:2007: IDL script definitions
http://www.utmc.uk.com/technical/xls/ts004.003annexfapr07.xls

[*Archive: TS004—UTMC Objects Registry*] TS004.002:2005
http://www.utmc.uk.com/technical/pdf/ts004dec.pdf

Annex D to TS004.002:2005 http://www.utmc.uk.com/technical/xls/ts004dec-annexd02.xls

Annex F to TS004.002:2005 http://www.utmc.uk.com/technical/xls/ts004dec-annexf02.xls

TS004.001:2005 http://www.utmc.uk.com/technical/pdf/ts004.pdf

Annex D to TS004.001:2005 http://www.utmc.uk.com/technical/xls/ts004-d3.xls

Annex F to TS004.001:2005 http://www.utmc.uk.com/technical/xls/ts004-f.xls

The Objects Registry TS004.003:2007 UTMC Objects Registry supersedes TS004.002:2005 UTMC Objects Registry and TS001:1997 Part 2: Data Dictionary.

It presents the data standards recommended for use by U.K. traffic managers in their systems. Details are provided in normative annexes.

8.3.1.14 SAE J2540 Messages for Handling Strings and Look-Up Tables in ATIS Standards

SAE	J2540/200207	Messages for Handling Strings and Look-Up Tables in Atis Standards	Published July 2002	SAE

SAE J2540 Available from
http://www.sae.org/technical/standards/J2540/1_200207

This SAE standard defines methods and messages to efficiently translate sequences of text and other types of data into and out of indexed values and look-up tables for effective transmission. This document defines: methods and data elements for handling indexes and strings in ATIS applications and message sets; message sets to support the delivery and translations of tables used in such strings; tables of nationally standardized strings for use in ATIS message descriptions; and examples of each in illustrative portions. While developed for ATIS use, the methods defined in this document are useful for any textual strings in any telematics applications found both in intelligent vehicles and elsewhere.

8.3.1.15 SAE J2540 RDS Phrase Lists

SAE	J2540/1_200207	RDS Phrase Lists	Published July 2002	SAE

SAE J2540/1 Available from
http://www.sae.org/technical/standards/J2540/1_200207

This SAE standard provides a table of textual messages meeting the requirements for expressing radio data systems (RDS) phrases commonly used in the ITS industry.

They can be used both over the RDS subcarrier transmission media as part of a 37-bit long Group 8a message, as well as being used to provide a common content list of phrases used in a wide number of other media and applications. This document *shall* define the normative index values to be used, extending the CEN established list to provide phrases needed by U.S. practitioners.

This standard provides nonnormative textual phrases which *may* be used by implementers to ensure intelligible results. This document follows the formats and rules established in SAE J2540 in the expressions, manipulations, and use of such tables. It should be pointed out that within the rules established by this document, a variety of final table are all considered "compliant" with the document, and may vary as fits the needs of implementers.

8.3.1.16 ITE Traffic Management Data Dictionary (TMDD) and Message Sets for External Traffic Management Center Communications (MS/ETMCC)

ITE	Traffic Management Data Dictionary (TMDD) and Message Sets for External Traffic Management Center Communications (MS/ETMCC)	Published July 2002	SAE

ITE Available from http://www.ite.org/tmdd/

This U.S. development from the Institute of Traffic Engineers (ITE), working cooperatively with the Federal Highway Administration (FHWA) and the American Association of State Highway and Transportation Officials (AASHTO), is leading a U.S. national effort to develop a standardized traffic management data dictionary (TMDD). A companion effort in this cooperative program is developing message sets for external traffic management center communications (MS/ETMCC), which occur between TMCs and other external ITS centers. These message sets are based on the data elements as defined in the TMDD.

There are a number of documents available from the Web link above.

8.3.2 Transport-Related Incident Management

This service group provides the capability for detecting and responding to various incidents in the transport network which specifically involve conditions initiated in the network itself rather than purely from external sources (e.g., natural disasters, terrorist attacks). Examples of incident management activities include the following:

- Response to on-site incident information;
- Detection and confirmation of incident presence;
- Service patrols in circulation in the road network to spot and provide on-site assistance to stalled vehicles or accidents to be removed from the traveled way;

- Dispatch of police, emergency, and maintenance vehicles to address injuries, remove the incident, and restore normal operation in a timely fashion;
- Monitoring of hazardous materials movements along transportation networks in conjunction with hazardous materials and incident notification service.

8.3.2.1 CEN 15722 [formerly 24977] eCall Minimum Set of Data

CEN	NP	15722	eCall Minimum Set of data	Under TS Ballot Will then be progressed to full EN	CEN TC278

CEN 15772 Will be Available from any European National Standards organization
http://www.cen.eu/esearch/extendedsearch.aspx

NOTE: The communications media and means of transferring the eCall MSD are not defined in this technical specification.

This technical specification defines the standard data concepts that comprise the minimum set of data to be transferred from a vehicle to a public safety answering point (PSAP) in the event of a crash or emergency via an eCall communication session.

NOTE 1: The communications protocols and methods for the transmission of the eCall message are not specified in this standard.

NOTE 2: Additional data concepts may also be transferred, and any such data concepts should be registered using a data registry as defined in CEN ISO/TS 24978 (Intelligent transport systems, Emergency and safety messages, Data registry).

8.3.2.2 CEN WI 00278220 eCall Operating Requirements

CEN	WD WI	00278220	eCall operating requirements	Under Development

CEN WI 00278220 Will be Available from any European National Standards organization
http://www.cen.eu/esearch/extendedsearch.aspx

The objective of implementing the in-vehicle emergency call system (eCall) is to automate the notification of a traffic accident, wherever in the European Union, with the same technical standards and the same quality of service objectives by using the mobile telecommunication network (e.g., GSM) and the European preassigned emergency destination address (e.g., 112), and to provide a means of manually triggering an incident.

This European standard defines the general operating requirements for in-vehicle emergency call services in order to transfer an emergency message from a vehicle to a PSAP in the event of a crash or emergency via an eCall communication session and to establish a voice channel between the PSAP and the in-vehicle equipment.

NOTE: The communications protocols and methods for the transmission of the eCall message are not specified in this standard.

NOTE: This European standard determines the operating requirements for an eCall service. An important part of the eCall service is a minimum set of data. The operating requirements for the MSD are determined in this European standard, but the form and data content of the MSD is not defined herein. A common European MSD is determined in CEN 15722.

Additional data concepts may also be transferred, and it is recommended that any such data concepts be registered using a data registry as defined in ISO 24978 (Intelligent transport systems—Emergency and safety messages—Data registry) to ensure that they can be understood by the recipient.

8.3.2.3 ISO 24978 Emergency and Safety Message Data Registry

ISO	CD	24978	Emergency and Safety Message Data Registry	CD/Under DIS Ballot	ISO TC204

ISO CD 24978 Not Yet Available. Track progress at:
http://www.iso.org/iso/iso_catalogue/catalogue_tc/catalogue_detail.htm?csnumber=42727

This International Standard provides the framework for the standardized operation and quality of service for one or more freely available data registries for ITS safety messages and data concepts.

The definitions in this International Standard are consistent with ISO 14817 (ITS Data Registries) and ISO 11179 (General principles for data registries).

The scope is therefore to provide a standardized set of protocols, parameters, and a method of management of an updateable data registry and to provide application layers for ITS safety messages via any available wireless media.

8.3.2.4 IEEE Std 1512 Common Incident Management Message Sets for Use by Emergency Management Centers

IEEE	Std	1512	Common Incident Management Message Sets for Use by Emergency Management Centers	Published 2006

IEEE 1512 Available from
http://ieeexplore.ieee.org/xpl/freeabs_all.jsp?tp=&isnumber=35111&arnumber=1673309&punumber=11078

This standard addresses the exchange of vital data about transportation-related incidents among emergency management centers through common incident management message sets. Message sets specified are consistent with the National Intelligent Transportation Systems Architecture and are described using ASN.1 syntax. This standard comprises the base standard of a family of incident management standards; specific incident management message sets for traffic, public safety, and HAZMAT centers may be found in forthcoming companion volumes which build upon and augment this base standard.

8.3.2.5 IEEE Std 1512-1 Common Traffic Incident Management Message Sets for Use by Emergency Management Centers

IEEE	Std	1512-1	Common Traffic Incident Management Message Sets for Use by Emergency Management Centers	Published 2006

IEEE 1512-1 Available from
http://ieeexplore.ieee.org/xpl/freeabs_all.jsp?tp=&isnumber=4014404&arnumber=4014405&punumber=4014403

The purpose of this standard is to take the real-world, available-information situation confronting the incident commander and the several involved agencies, and then support the exchange of that information among agencies, to make further information available to the incident commander and involved agencies, to best support them in managing the event. That calls for messages to communicate whatever information is available, often quite partial, among the several involved agencies, to accomplish four decision-support functions, corresponding to the five kinds of information exchange listed earlier in this clause:

- Function 1: To disseminate information about current and future traffic flows in the transportation grid, including, in particular, impacts of an incident and particular route information, to assist in the real-time interagency management of a transportation-related incident or event.
- Function 2: To support the management of traffic to assist in the real-time interagency management of a transportation-related incident or event, including establishing reverse links and the use of priority/preemption technologies.
- Function 3: To support the management of TMC assets and other assets, as necessary to manage traffic, to assist in the real-time interagency management of a transportation-related incident or event.

- Function 4: To support the cleanup, repair, and replacement of damaged infrastructure, treating that operation as part of another incident, or as a separate incident.

8.3.2.6 IEEE Std 1512-2 Public Safety Traffic Incident Management Message Sets for Use by Emergency Management Centers

IEEE	Std	1512-2	Public Safety Traffic Incident Management Message Sets for Use by Emergency Management Centers	Published 2006

IEEE 1512-2 Available from
http://ieeexplore.ieee.org/xpl/freeabs_all.jsp?tp=&isnumber=31521&arnumber=1470008&punumber=9914

8.3.2.7 IEEE Std 1512-3 Hazardous Material Incident Management Message Sets for Use by Emergency Management Centers

IEEE	Std	1512-3	Hazardous Material Incident Management Message Sets for Use by Emergency Management Centers	Published 2006

IEEE 1512-3 Available from
http://ieeexplore.ieee.org/xpl/freeabs_all.jsp?tp=&isnumber=34722&arnumber=1658443&punumber=11015

This standard is a companion volume for IEEE Std 1512-2006, referred to here as the "base standard." This standard specifies messages, data frames, and data elements for communicating general and cargo information to other responders in support of real-time interagency transportation-related incident management. It addresses the unique disciplines associated with communications dealing with the control and confinement of hazardous materials during and after an incident.

The base standard specifies messages, data frames, and data elements for the basic underlying communication involved in real-time interagency transportation-related incident management. Refer to the base standard for specification of the scope of the combination of the base standard and companion volumes such as this one. Together, the base standard and companion volumes shall be referred to as the IEEE 1512 Family of Standards.

Clause 5 and Clause 6 of this standard present the specified messages, data frames, and data elements in ASN.1 format and in XML formats.

8.3.3 Demand Management

This service group covers the development and implementation of management and control strategies designed to influence the demand for travel.

These strategies influence the overall level of demand for travel at different times of the day and the relative demand for different modes of transport, through the management of pricing structures, area access control, or zone entry regulations. Demand management functions may include:

- Access control;
- Air quality based zone pricing;
- Congestion pricing;
- High occupancy vehicle facility management;
- Parking pricing;
- Public transport fares management.

NOTE: For standards to achieve demand management by fee collection, see Section 8.8.

8.3.3.1 Access Control

Vehicular access control systems are most frequently proprietary. However, increasingly they use the identification standards described in Section 7.4.

8.3.3.2 Air Quality Access Control

Air quality access control generally takes the form of banning classes of vehicles with poor pollution emissions at times of poor air quality. Another form of air quality access control is to allow vehicles on the road only every other day. This is normally achieved using odd and even numbered vehicle license plates or similar.

ITS can only assist where there is mandatory electronic identification of vehicles. In the long term this is likely to use the identification standards described in Section 7.4.

For such systems to work using ITS, it will, of course, be necessary for all vehicles to be equipped. The ISO 24534 series of standards described in Section 7.3 would seem to be the most relevant.

However, while there are many investigations, tests, and trials, to date there are no national instantiations.

Brazil, Mexico, and Pakistan have made decisions to proceed and it is likely that their systems will use the low-cost passive technologies defined in ISO 18000-6 (see Section 7.4.11), although this is not yet certain, and some Pakistani reports state that their system is infrared. However, much of the available data is from the press and may be inaccurate.

8.3.3.3 Congestion Pricing

Congestion management systems are most likely to use CEN DSRC or CALM technologies (see Chapter 3 and Sections 4.3 and 4.6) outside the United States (where there is no use of standards for this activity and most systems are proprietary 915-MHz systems). However, (currently) the worlds largest congestion pricing system (London) uses optical number plate recognition technology (which requires no International Standards except perhaps for performance/conformance), and some systems around the world are using infrared. Italy has its own national DSRC system that is not compatible with any International Standards.

The fastest growing tolling system in the United States uses ISO 18000-6B technology (see Section 7.4.11).

Fee collection system standards for EFC are to be found in Section 8.8.

No other specific standards for congestion pricing are known at the time of writing.

8.3.3.4 Parking Pricing

Parking systems are likely to use CEN DSRC, CALM, or U.S. VII technologies where these are already installed, or may use one trip low-cost RFID technology such as ISO 18000-6C or ISO 18000-3 Mode 1.

There are no parking standards produced by the major SDOs at the time of writing this book.

8.3.3.5 Public Transport Fares Management

See Section 8.6.

8.3.4 Transport Infrastructure Maintenance Management

This service group covers the application of ITS technologies to the management of road network maintenance, plus the maintenance of the communication and computer infrastructures used to support travelers using the road network. Included in this service group are:

- Highway maintenance management;
- Location and maintenance of utilities used by ITS equipment;
- The use of probe data to guide timing and location of planned road works and road closures;
- Highway sign maintenance management;
- Coordination with weather and environmental conditions monitoring service groups with respect to collection of condition information related to weather and road conditions that are needed as input to infrastructure maintenance management.

Example services include:

- Roadway construction and maintenance management;
- Winter maintenance;
- Pavement management;
- Automated road management;
- Work zone safety management.

The major SDOs have not published, nor have under development in their work programs, any standards in this service group. Any "standards" that exist are likely to be to specify national and/or local requirements.

8.3.5 Policing/Enforcing Traffic Regulations

This service covers the application of ITS technologies to the enforcement of traffic laws and regulations. Examples include the following:

- Access control;
- High occupancy vehicle facility usage;
- Parking regulation enforcement;
- Speed limit enforcement;
- Signal enforcement (e.g., red light violation);
- Emissions monitoring.

Although there is considerable interest in some countries to use communications with vehicles to enforce traffic regulations (especially speed), as there is not yet any deployment of enabling technology, it is premature to expect any standards. It is probable that these will first appear in areas such as around schools or other recognized high-risk locations in order to gain public acceptance.

National and local regulations and local performance requirements for such things as speed camera equipment, HOV enforcement, and signal enforcement exist, but there are, as yet, no International Standards.

Also, emissions monitoring is generally the purview of national regulations and there are no International Standards that we are aware of.

8.4 Vehicle Service Groups

The service groups in the Vehicle Services domain focus on specific services that improve the operational safety of vehicles. There are both services which use external information, as well as services which use in-vehicle information only.

8.4.1 Vehicle Data Systems

Safety readiness services include the use of monitoring and warning systems for both private car driver and vehicle. Examples include the following:

- Critical component monitoring;
- Driver alertness monitoring;
- Engine temperature;
- Oil pressure;
- Road condition monitoring;
- Weather and visibility monitoring.

NOTE: The lead on standardization for the onboard issues will be agreed between TC 204 and TC 22.

Example services include:

- Vehicle internal systems monitoring;
- Vehicle external conditions monitoring.

8.4.1.1 ISO 9141 Road Vehicles—Diagnostic Systems—Requirements for Interchange of Digital Information

ISO	IS	11898	Road vehicles—Diagnostic systems—Requirements for interchange of digital information	Published	TC22

ISO 9141 Available from
http://www.iso.org/iso/iso_catalogue/catalogue_tc/catalogue_detail.htm?csnumber=16737

This standard specifies the requirements for setting up the interchange of digital information between onboard ECUs of road vehicles and suitable diagnostic testers. This communication is established in order to facilitate inspection, test diagnosis, and adjustment of vehicles, systems, and ECUs. It does not apply when system-specific diagnostic test equipment is used.

8.4.1.2 ISO 11898 Road Vehicles—Controller Area Network (CAN)

ISO	IS	11898	Road vehicles—Controller area network (CAN)	Published	TC22
ISO	IS	11898-1	Road vehicles—Controller area network (CAN)—Part 1: Data link layer and physical signaling	Published 2003	TC22

ISO 11898-1 Available from
http://www.iso.org/iso/iso_catalogue/catalogue_tc/catalogue_detail.htm?csnumber=33422

ISO 11898-1:2003 specifies the data link layer (DLL) and physical signaling of the controller area network (CAN)—a serial communication protocol that supports distributed real-time control and multiplexing for use within road vehicles. While describing the general architecture of CAN in terms of hierarchical layers according to the ISO reference model for OSI established in ISO/IEC 7498-1, it provides the characteristics for setting up an interchange of digital information between modules implementing the CAN DLL—itself specified according to ISO/IEC 8802-2 and ISO/IEC 8802-3—with detailed specification of the logical link control sublayer and medium access control sublayer.

ISO	IS	11898-2	Road vehicles—Controller area network (CAN)—Part 2: High-speed medium access unit	Published 2003	TC22

ISO 11898-2 Available from
http://www.iso.org/iso/iso_catalogue/catalogue_tc/catalogue_detail.htm?csnumber=33423

ISO 11898-2:2003 specifies the high-speed (transmission rates of up to 1 Mbps) medium access unit (MAU), and some medium dependent interface (MDI) features (according to ISO 8802-3), which comprise the physical layer of CAN—a serial communication protocol that supports distributed real-time control and multiplexing for use within road vehicles.

ISO	IS	11898-3	Road vehicles—Controller area network (CAN)—Part 3: Low-speed, fault-tolerant, medium-dependent interface	Published 2006	TC22

ISO 11898-3 Available from
http://www.iso.org/iso/iso_catalogue/catalogue_tc/catalogue_detail.htm?csnumber=36055

ISO 11898-3:2006 specifies characteristics of setting up an interchange of digital information between electronic control units of road vehicles equipped with CAN at transmission rates above 40 Kbps up to 125 Kbps.

CAN is a serial communication protocol which supports distributed control and multiplexing.

ISO	IS	ISO 11898-4	Road vehicles—Controller area network (CAN)—Part 4: Time-triggered communication	Published 2004	TC22

ISO 11898-4 Available from
http://www.iso.org/iso/iso_catalogue/catalogue_tc/catalogue_detail.htm?csnumber=36306

ISO 11898-4:2004 specifies time-triggered communication in CAN—a serial communication protocol that supports distributed real-time control and multiplexing for use within road vehicles. It is applicable to setting up a time-triggered interchange of digital information between electronic control units of road vehicles equipped with CAN, and specifies the frame synchronization entity that coordinates the operation of both logical link and media access controls in accordance with ISO 11898-1, to provide the time-triggered communication schedule.

ISO	IS	ISO 11898-5	Road vehicles—Controller area network (CAN)—Part 5: High-speed medium access unit with low-power mode	Published 2004	TC22

ISO 11898-5 Available from
http://www.iso.org/iso/iso_catalogue/catalogue_tc/catalogue_detail.htm?csnumber=41284

ISO 11898-5:2007 specifies the CAN physical layer for transmission rates up to 1 Mbps for use within road vehicles. It describes the medium access unit functions as well as some medium-dependent interface features according to ISO 8802-2.

ISO 11898-5:2007 represents an extension of ISO 11898-2 dealing with new functionality for systems requiring low-power consumption features while there is no active bus communication.

Physical layer implementations according to ISO 11898-5:2007 are compliant with all parameters of ISO 11898-2, but are defined differently within ISO 11898-5:2007. Implementations according to ISO 11898-5:2007 and ISO 11898-2 are interoperable and can be used at the same time within one network.

8.4.1.3 ISO 11992 Road Vehicles—Interchange of Digital Information on Electrical Connections Between Towing and Towed Vehicles

ISO	IS	ISO 11992	Road vehicles—Interchange of digital information on electrical connections between towing and towed vehicles	Published	ISO TC22

This standard is in four parts.

ISO	IS	ISO 11992-1	Road vehicles—Interchange of digital information on electrical connections between towing and towed vehicles—Part 1: Physical and data-link layers	Published 2003	ISO TC22

ISO 11992-1 Available from
http://www.iso.org/iso/iso_catalogue/catalogue_tc/catalogue_detail.htm?csnumber=33467

ISO 11992-1:2003 specifies the interchange of digital information between road vehicles with a maximum authorized total mass greater than 3,500 kg, and towed vehicles, including communication between towed vehicles in terms of parameters and requirements of the physical and data link layer of the electrical connection used to connect the electrical and electronic systems. It also includes conformance tests of the physical layer.

ISO	IS	11992-2	Road vehicles—Interchange of digital information on electrical connections between towing and towed vehicles—Part 2: Application layer for brakes and running gear	Published 2003	ISO TC22

ISO 11992-2 Available from
http://www.iso.org/iso/iso_catalogue/catalogue_tc/catalogue_detail.htm?csnumber=33469

ISO 11992-2:2003 specifies the parameters and messages for electronically controlled braking systems, including antilock braking systems (ABS) and for running gear equipment (i.e., systems for steering, suspension, and tires), to ensure the interchange of digital information between road vehicles with a maximum authorized total mass greater than 3,500 kg, and their towed vehicles, including communication between towed vehicles. The objective of the data structure is to optimize the use of the interface, while preserving a sufficient reserve capacity for future expansion.

ISO	IS	11992-3	Road vehicles—Interchange of digital information on electrical connections between towing and towed vehicles—Part 3: Application layer for equipment other than brakes and running gear	Published 2003	ISO TC22

ISO 11992-3 Available from
http://www.iso.org/iso/iso_catalogue/catalogue_tc/catalogue_detail.htm?csnumber=33470

ISO 11992-3:2002 specifies the parameters and messages for electronically controlled braking systems other than systems for braking and running gear (i.e., steering, suspension, and tires), to ensure the interchange of digital information between road vehicles with a maximum authorized total mass greater than 3,500 kg and their towed vehicles, including communication between towed vehicles.

ISO	IS	11992-4	Road vehicles—Interchange of digital information on electrical connections between towing and towed vehicles—Part 4: Diagnostics	Published 2005	ISO TC22

ISO 11992-4 Available from
http://www.iso.org/iso/iso_catalogue/catalogue_tc/catalogue_detail.htm?csnumber=36252

ISO 11992-4:2005 specifies the data communication for diagnostic purposes on a serial data link between a road vehicle and its towed vehicle(s). It is applicable to road vehicles of a maximum authorized total mass greater than 3,500 kg.

8.4.1.4 ISO 15031 Road Vehicles—Communication Between Vehicle and External Equipment for Emissions-Related Diagnostics

ISO	IS	15031	Road vehicles—Communication between vehicle and external equipment for emissions-related diagnostics	Published	TC22

This standard is in seven parts. By itself it is not directly an ITS standard, but is referred to by several standards that are ITS related or are directly used in ITS. ISO 15031 parts are therefore provided for reference, together with hyperlinks; short summaries are provided where appropriate, but the contents are not detailed.

ISO	IS	15031-1	Road vehicles—Communication between vehicle and external equipment for emissions-related diagnostics—Part 1: General information	Published 2001	TC22

ISO 15031-1 Available from
http://www.iso.org/iso/iso_catalogue/catalogue_tc/catalogue_detail.htm?csnumber=29105

ISO	IS	15031-2	Road vehicles—Communication between vehicle and external equipment for emissions-related diagnostics—Part 2: Terms, definitions, abbreviations and acronyms	Published 2004	TC22

ISO 15031-2 Available from
http://www.iso.org/iso/iso_catalogue/catalogue_tc/catalogue_detail.htm?csnumber=38939

ISO/TR 15031-2 is a guide to terms, definitions, abbreviations, and acronyms used in emissions-related diagnostics, with respect to the communication between road vehicles and external equipment used in that field. It also specifies a procedure for constructing new terms. As it gives recommended usage of diagnostic terms applicable to electrical/electronic systems, it also makes reference to related mechanical terms, definitions, abbreviations, and acronyms.

ISO	IS	15031-3	Road vehicles—Communication between vehicle and external equipment for emissions-related diagnostics—Part 3: Diagnostic connector and related electrical circuits, specification and use	Published 2004	TC22

ISO 15031-3 Available from
http://www.iso.org/iso/iso_catalogue/catalogue_tc/catalogue_detail.htm?csnumber=29021

ISO 15031-3:2004 specifies a minimum set of requirements for a diagnostic connector used in communication between a vehicle and external test equipment for emissions-related diagnostics. Its aim is to promote the use of a common diagnostic connector throughout the motor vehicle industry.

The diagnostic connection consists of two mating connectors: the vehicle connector and the external test equipment connector. Applicable to all types of road vehicles, the connector specified has no positive locking feature and is intended for short-term diagnostic connection only.

ISO	IS	15031-4	Road vehicles—Communication between vehicle and external equipment for emissions-related diagnostics—Part 4: External test equipment	Published 2005	TC22

ISO 15031-4 Available from
http://www.iso.org/iso/iso_catalogue/catalogue_tc/catalogue_detail.htm?csnumber=40087

ISO 15031-4:2005 specifies a means of establishing communications between an OBD-equipped vehicle and external test equipment, and a set of diagnostic services to be provided by the external test equipment in order to exercise the services defined in ISO 15031-5, conformance criteria for the external test equipment.

ISO	IS	15031-5	Road vehicles—Communication between vehicle and external equipment for emissions-related diagnostics—Part 5: Emissions-related diagnostic services	Published 2006	TC22

ISO 15031-5 Available from
http://www.iso.org/iso/iso_catalogue/catalogue_tc/catalogue_detail.htm?csnumber=29108

ISO 15031-5:2006 specifies diagnostic services and functionally addressed request/ response messages required to be supported by motor vehicles and external test equipment for diagnostic purposes which pertain to motor vehicle emission-related data. Any external test equipment meeting the requirements of ISO 15031-4 uses these messages to retrieve emissions-related information from the vehicle.

Each section of ISO 15031-5:2006, which specifies additional detail to existing sections of ISO 9141-2, ISO 14230-4, SAE J1850, and ISO 15765-4, supersedes those specifications.

ISO	IS	15031-6	Road vehicles—Communication between vehicle and external equipment for emissions-related diagnostics—Part 6: Diagnostic trouble code definitions	Published 2005	TC22

ISO 15031-6 Available from
http://www.iso.org/iso/iso_catalogue/catalogue_tc/catalogue_detail.htm?csnumber=29109

ISO 15031-6:2005 provides recommended uniformity for alphanumeric trouble codes. It further provides guidance for uniform messages associated with these codes.

ISO 15031-6:2005 specifies several sections addressing format, structure, messages, and a few examples, and it is applicable to electrical/electronic systems diagnostics of motor vehicles.

ISO	IS	15031-7	Road vehicles—Communication between vehicle and external equipment for emissions-related diagnostics—Part 7: Data link security	Published 2001	TC22

ISO 15031-7 Available from
http://www.iso.org/iso/iso_catalogue/catalogue_tc/catalogue_detail.htm?csnumber=29110

8.4.1.5 ISO 15764 Road Vehicles—Extended Data Link Security

ISO	IS	15764	Road vehicles—Extended data link security	Published 2004	TC22

ISO 15674 Available from
http://www.iso.org/iso/iso_catalogue/catalogue_tc/catalogue_detail.htm?csnumber=28775

ISO 15764:2004 describes an extension of data link protocols for enhancing the security of data transfers between ECUs connected by a communication network used in road vehicles. It is based on cryptographic methods that include encryption, digital signatures, and message authentication codes. It provides a description of services to establish ECUs as trusted parties with respect to one another and to protect against specific threats. It is applicable to all data links between pairs of ECUs capable of storing and processing secret data so that unauthorized third parties are denied access to it. Parameters are provided to enable the level of security in the data link to be selected.

8.4.1.6 ISO 15765 Road Vehicles—Diagnostics on Controller Area Networks (CAN)

ISO	IS	15765	Road vehicles—Diagnostics on Controller Area Networks (CAN)	Published	TC22

This standard is in four parts.

ISO	IS	15765-1	Road vehicles—Diagnostics on Controller Area Networks (CAN)—Part 1: General information	Published 2004	TC22

ISO 15765-1 Available from
http://www.iso.org/iso/iso_catalogue/catalogue_tc/catalogue_detail.htm?csnumber=33615

ISO 15765-1:2004 gives an overview of the structure and the partitioning of ISO 15765, and it shows the relation between the different parts. It also defines the diagnostic network architecture.

The terminology defined in ISO 15765-1:2004 is common for all diagnostic networks and is used throughout all parts of ISO 15765.

ISO	IS	15765-2	Road vehicles—Diagnostics on Controller Area Networks (CAN)—Part 2: Network layer services	Published 2004	TC22

ISO 15765-2 Available from
http://www.iso.org/iso/iso_catalogue/catalogue_tc/catalogue_detail.htm?csnumber=33616

ISO 15765-2:2004 specifies a network protocol tailored to meet the requirements of CAN-based vehicle network systems on controller area networks as specified in ISO 11898.

It has been defined in accordance with the diagnostic services established in ISO 14229-1 and ISO 15031-5, but is not limited to use with them, and is also compatible with most other communication needs for in-vehicle networks. The protocol specifies an unconfirmed communication.

ISO	IS	15765-3	Road vehicles—Diagnostics on Controller Area Networks (CAN)—Part 3: Implementation of unified diagnostic services (UDS on CAN)	Published 2004	TC22

ISO 15865-3 Available from
http://www.iso.org/iso/iso_catalogue/catalogue_tc/catalogue_detail.htm?csnumber=33618

ISO 15765-3:2004 specifies the implementation of a common set of unified diagnostic services (UDS), in accordance with ISO 14229-1, on CANs as specified in ISO 11898.

It gives the diagnostic services and server memory programming requirements for all in-vehicle servers connected to a CAN network and external test equipment. It does not specify any requirement for the in-vehicle CAN bus architecture.

ISO	IS	15765-4	Road vehicles—Diagnostics on Controller Area Networks (CAN)—Part 4: Requirements for emissions-related systems	Published 2005	TC22

ISO 15765-4 Available from
http://www.iso.org/iso/iso_catalogue/catalogue_tc/catalogue_detail.htm?csnumber=33619

ISO 15765-4:2005 specifies requirements for the emissions-related systems of legislated OBD-compliant controller area networks—such communications networks consisting of a road vehicle equipped with a single or multiple emissions-related ECUs and external test equipment.

It is based on the specifications of ISO 15765-2, ISO 11898-1, and ISO 11898-2, while placing restrictions on those standards for legislated-OBD purposes. It does not specify in-vehicle CAN bus architecture. Legislated OBD-compliant vehicles are to comply with external test equipment requirements.

8.4.1.7 ISO 16845 Road Vehicles—Controller Area Network (CAN)—Conformance Test Plan

ISO	IS	16845	Road vehicles—Controller area network (CAN)—Conformance test plan	Published 2004	TC22

ISO 16845 Available from
http://www.iso.org/iso/iso_catalogue/catalogue_tc/catalogue_detail.htm?csnumber=30478

ISO 16845:2004 provides the methodology and abstract test suite necessary for checking the conformance of any CAN implementation of the CAN specified in ISO 11898-1.

8.4.1.8 ISO/EN 15005 Road Vehicles—Ergonomic Aspects of In-Vehicle Presentation of Traffic Information and Control Systems—Dialogue Management Principles and Compliance Procedures

CEN/ISO	EN/ ISO	15005	Road vehicles—Ergonomic aspects of in-vehicle presentation of traffic information and control systems—Dialogue management principles and compliance procedures	Published	CEN TC278 ISO TC22

ISO 15005 Available from
http://www.iso.org/iso/iso_catalogue/catalogue_tc/catalogue_detail.htm?csnumber=34085

This International Standard presents ergonomic principles for the design of the dialogues that take place between the driver of a road vehicle and the vehicle's transport information and control systems (TICS) while the vehicle is in motion. It also specifies compliance verification conditions for the requirements related to these principles.

This International Standard is applicable to TICSs consisting of either single or multiple devices, which can be either independent or interconnected. It is not applicable to TICSs without dialogues, TICS failures or malfunctions, or controls or displays used for non-TICS functions.

8.4.1.9 ISO 15006 Road Vehicles—Ergonomic Aspects of Transport Information and Control Systems—Specification and Compliance Procedures for In-Vehicle Auditory Presentations

CEN/ISO	EN/ ISO	15006	Road vehicles—Ergonomic aspects of transport information and control systems—Specification and compliance procedures for in-vehicle auditory presentations	Published 2003	CEN TC278/ ISO TC22

ISO 15006 Available from
http://www.iso.org/iso/iso_catalogue/catalogue_tc/catalogue_detail.htm?csnumber=34885

ISO 15006:2004 establishes ergonomic specifications for the presentation of auditory information related to TICS through speech or sounds. It is applicable only to the use of auditory displays when the vehicle is in motion. It presents a set of requirements and recommendations for in-vehicle auditory messages from TICS, and it provides message characteristics and functional factors for maximizing message intelligibility and utility while helping to prevent auditory or mental overload.

8.4.1.10 ISO/EN 15007 Road Vehicles—Man Machine Interfaces—Visual Demand Measurement Method

CEN/ISO	EN/ ISO	15007	Road vehicles—Man Machine Interfaces—Visual demand measurement method	Published	CEN TC278/ ISO TC22

This standard is in two parts.

ISO	EN/ ISO	15007-1	Road vehicles—Man Machine Interfaces—Visual demand measurement method—Part 1: Requirements	Published	CEN TC278 ISO TC204

ISO 15007-1 Available from
http://www.iso.org/iso/iso_catalogue/catalogue_tc/catalogue_detail.htm?csnumber=26194

This part of ISO 15007 defines key terms and parameters applied in the analysis of driver visual behavior. It can be applied in environments from real-world trials to laboratory-based driving simulator studies. Minimum requirements for reporting the results of TICS evaluations are provided.

The procedures described in this part of ISO 15007 could also apply to more general assessments of driver visual behavior without the introduction of TICS-specific systems. The parameters and definitions described are intended to assist development of a common source of reference for driver visual behavior data.

Due to the limitation of visual behavior measurement techniques (e.g., related to the effects of accommodation and adaptation of the eyes), this part of ISO 15007 does not apply to the evaluation of head-up displays. Further guidance, including the specification of analysis methodologies and results presentation for visual behavior analysis, is available in other ISO publications. Data collated and analyzed in this way allow comparisons to be performed across different TICS applications and experimental scenarios.

ISO	EN/ ISO	15007-2	Road vehicles—Measurement of driver visual behavior with respect to transport information and control systems—Part 2: Equipment and procedures	Published 2002	CEN TC278 ISO TC204

ISO 15007-2 Available from
http://www.iso.org/iso/iso_catalogue/catalogue_tc/catalogue_detail.htm?csnumber=32149

This technical specification gives guidelines on equipment and procedures for analyzing driver visual behavior, intended to enable assessors of TICS to plan evaluation trials, specify (and install) data capture equipment, and analyze, interpret, and report visual-behavior metrics (standards of measurement).

It is applicable to both road trials and simulated driving environments. It is not applicable to the assessment of head-up displays.

8.4.1.11 ISO 15008 Road Vehicles—Traffic Information and Control Systems—Ergonomic Aspects of In-Vehicle Visual Presentation of Information

ISO	IS	15008	Road vehicles—Traffic information and control systems—Ergonomic aspects of in-vehicle visual presentation of information	Published 2003	ISO TC22

ISO 15008 Available from
http://www.iso.org/iso/iso_catalogue/catalogue_tc/catalogue_detail.htm?csnumber=34886

See Section 8.3.1.11.

8.4.1.12 ISO/EN 16951 Road Vehicles—Ergonomic Aspects of Transport Information and Control Systems—Procedure for Determining Priority of On-Board Messages Presented to Drivers

CEN/ISO	EN/ ISO	16951	Road vehicles—Ergonomic aspects of transport information and control systems—Procedure for determining priority of on-board messages presented to drivers	Published 2004	CEN TC278 ISO TC22

ISO 16951 Available from
http://www.iso.org/iso/iso_catalogue/catalogue_tc/catalogue_detail.htm?csnumber=29024

ISO/TS 16951:2004 provides formal procedures and two alternative methods for determining the priority of onboard messages presented to drivers of road vehicles by TICS and other systems.

It is applicable to the whole range of TICS in-vehicle messages, including traveler information, navigation, travel and traffic advisories, yellow pages information, warnings, systems status, emergency calling system information, and electronic toll/fee collection, as well as to messages from non-TICS sources such as telephone, warnings, and telltales.

8.4.1.13 ISO/EN 17287 Road Vehicles—Ergonomic Aspects of Transport Information and Control Systems—Procedure for Assessing Suitability for Use When Driving

CEN/ISO	EN/ ISO	17287	Road vehicles—Ergonomic aspects of transport information and control systems—Procedure for assessing suitability for use when driving	Published	CEN TC278 ISO TC22

ISO 17287 Available from
http://www.iso.org/iso/iso_catalogue/catalogue_tc/catalogue_detail.htm?csnumber=30597

ISO 17287:2002 specifies a procedure for assessing whether specific TICS, or a combination of TICS with other in-vehicle systems, are suitable for use by drivers while driving. It addresses user-oriented TICS description and context of use, TICS task description and analysis, assessment process, and documentation.

The TICS description and context of use includes consideration of improper use, reasonably foreseeable misuse, and TICS failure. The TICS description, analysis, and assessment include a process for identifying and addressing suitability issues.

ISO 17287:2002 does not recommend specific variables for assessing suitability nor does it define criteria for establishing the suitability of use of a TICS table while driving.

8.4.1.14 SAE J2361 Bluetooth Wireless Protocol for Automotive Applications

SAE	J2561_200112	Bluetooth Wireless Protocol for Automotive Applications	Published Dec. 2001	SAE

SAE J2561 Available from
http://www.sae.org/technical/standards/J2561_200112

See Section 4.2.6.42.

8.4.1.15 SAE J2497 Power Line Carrier Communications for Commercial Vehicles

SAE	J2497_200210	Power Line Carrier Communications for Commercial Vehicles	Published Oct. 2002	SAE

SAE J2497 Available from
http://www.sae.org/technical/standards/J2497_200210

This SAE recommended practice defines a method for implementing a bidirectional, serial communications link over the vehicle power supply line among modules containing microcomputers. This document defines those parameters of the serial link that relate primarily to hardware and software compatibility such as interface requirements, system protocol, and message format that pertain to power line communications (PLC) between tractors and trailers. This document defines a method of activating the trailer ABS indicator lamp that is located in the tractor.

8.4.1.16 SAE J1939 Recommended Practice for a Serial Control and Communications Vehicle Network

SAE	J1939_200311	Recommended Practice for a Serial Control and Communications Vehicle Network	Published Nov 2003	SAE

See Section 8.4.5.24.

8.4.1.17 SAE J1698 Vehicle Event Data Interface

SAE	J1698	Vehicle Event Data Interface	Published	SAE
SAE	J1698/1_200502	Vehicle Event Data Interface—Output Data Definition	Published Feb. 2005	SAE

SAE J1698/1 Available from
http://www.sae.org/technical/standards/J1698/1_200503

This document is part of the J1698 document family and provides the definitions for event-related data items.

SAE	J1698/ 2_200405	Vehicle Event Data Interface-Vehicular Data Extraction	Published May 2004	SAE

SAE J1698/2 Available from
http://www.sae.org/technical/standards/J1698/2_200405

This recommended practice is intended to define a common method for determining how to extract event data from a motor vehicle, including the event data set needed to output the event record of data elements defined in SAE J 1698. It is intended for use by those developing tools for the purpose of event data set extraction.

This recommended practices aims to utilize existing industry standards to define a common physical interface and the protocols necessary to event data set extraction. To accomplish this, the SAE J 1962 Diagnostics Connector has been designed and the primary physical interface and associated industry standard diagnostic protocols have been designated for communications.

8.4.1.18 SAE J2640 General Automotive Embedded Software Design Requirements

SAE	J2640_200604	General Automotive Embedded Software Design Requirements	Published April 2006	SAE

SAE J2640 Available from
http://www.sae.org/technical/standards/J2640_200604

The scope of this recommended practice encompasses the following objectives: concentrate on the general best practices for vehicular embedded software design; establish programming language-independent best practices; establish hardware/ software interface best practices; establish preemptive system best practices; and provide verification criteria to evaluate product compliance with this best practice.

8.4.1.19 SAE J2748 VHDL-AMS Statistical Analysis Packages

SAE	J2748_200610	VHDL-AMS Statistical Analysis Packages	Published Oct. 2006	SAE

SAE J2748 Available from
http://www.sae.org/technical/standards/J2748_200610

This document specifies the interface and the behavior of the VHDL-AMS packages for use in modeling statistical behavior. These packages are useful in defining the statistical variation of parameters of electrical, electronic, and mechatronic components and subsystems. These can then be used with simulation tools to analyze the performance and reliability of systems composed of these components and subsystems. Providing a standard definition of the package interfaces and their behavior is intended to facilitate the exchange of models between component and system manufacturers and the use of different CAE simulation tools.

8.4.1.20 SAE USCAR 30 Performance Specification for Automotive Universal Serial Bus (USB) Connection System

SAE	USCAR30	Performance Specification for Automotive Universal Serial Bus (USB) Connection System	Published Oct. 2006	SAE

SAE USCAR30 Available from
http://www.sae.org/technical/standards/USCAR30

8.4.1.21 SAE J2735 Dedicated Short Range Communications (DSRC) Message Set Dictionary

SAE	J2735_200612	Dedicated Short Range Communications (DSRC) Message Set Dictionary	Published Dec. 2006

SAE J2735 Available from
http://www.sae.org/technical/standards/J2735_200612

This SAE recommended practice is intended as a guide towards standard practice and is subject to change to keep pace with experience and technical advances. It specifies standard message sets, data frames, and data elements for use by applications intended to utilize the 5.9-GHz dedicated short range communications for wireless access in vehicular environments (DSRC/WAVE, referenced in this document simply as "DSRC") communication systems. The scope is limited to specifying initial representative message structure and providing sufficient background information to allow readers to properly interpret the DSRC standards and message definitions from the point of view of an application developer.

8.4.2 Assistance to the Driver

Vehicle data system services include the use of sensors and control systems to detect potential for collisions and either prompt the driver to take action or automatically initiate action (see Section 8.4.5).

Longitudinal collision avoidance includes the application of obstacle detection systems. Lateral collision avoidance is the use of systems (such as sensors and control systems) to monitor the potential hazards involved in lane keeping, lane changing, entering and leaving high-speed roads, and overtaking.

Example services include:

- Longitudinal collision avoidance;
- Lateral collision avoidance;
- Intersection collision avoidance.

These systems are sometimes called "driver support systems" and they provide technology for vehicles directly linked to the driver. The purpose is to reduce driver workload, improve convenience, and create awareness of dangers, as well as to avoid accidents and decrease damage by the use of advanced technologies.

8.4.2.1 ISO 22840 Extended Range Backing Aids Systems

ISO	NP	22840	Extended Range Backing Aid Systems	Under Development	ISO TC204

ISO 22840 Not Yet Available. Track progress at:
http://www.iso.org/iso/iso_catalogue/catalogue_tc/catalogue_detail.htm?csnumber=36521

Extended range backing aids (ERBA) are detection devices with noncontact sensors which assist the driver during low to mid-speed backing maneuvering. These systems detect and warn the driver of objects in the pathway of the vehicle. In comparison to low-speed only devices whose main purpose is assisting in parking maneuvers (e.g., ISO 17386 Maneuvering Aids for Low Speed Operation), the purpose of ERBA is to assist in higher speed backing maneuvers associated with traversing longer distances.

This International Standard (draft) for ERBA addresses light-duty vehicles (e.g., passenger cars, pick-up trucks, light vans, and sport utility vehicles; motorcycles excluded) equipped with such ERBA systems. This standard establishes minimum functionality requirements that the driver can expect of the system, such as the detection of and information on the presence of relevant obstacles within a defined detection range. This standard also sets minimum requirements for failure indication as well as performance test procedures. It includes rules for the general information strategy but does not restrict the kind of information or display system.

8.4.2.2 ISO 17361 Lane Departure Warning

ISO	IS	17361	Intelligent transport systems—Lane departure warning systems—Performance requirements and test procedures	Published 2007	ISO TC204

ISO 17361 Available at
http://www.iso.org/iso/iso_catalogue/catalogue_tc/catalogue_detail.htm?csnumber=41105

Lane departure warning systems (LDWS) are based on fundamental traffic rules. The main focus of LDWS is to help the driver keep the vehicle in the lane on highways and highway-like roads. Accordingly, a warning is issued to alert the driver in case of lane departure caused, for example, by inattention. LDWS are not intended to issue warnings with respect to collisions with other vehicles or control vehicle motions.

ISO 17361:2007 specifies the definition of the system, classification, functions, human-machine interface (HMI), and test methods for lane departure warning systems. These are in-vehicle systems that can warn the driver of a lane departure on highways and highway-like roads. The subject system, which may utilize optical, electromagnetic, GPS, or other sensor technologies, issues a warning consistent with the visible lane markings.

8.4.2.3 ISO 17387 Lane Change Decision Aids Systems

ISO	DIS	17387	Lane Change Decision Aids Systems	Under Development	ISO TC204

ISO 17387 Not Yet Available. Track progress at:
http://www.iso.org/iso/iso_catalogue/catalogue_tc/catalogue_detail.htm?csnumber=43654

This standardization draft specifies system requirements and test methods for lane change decision aid systems (LCDAS). LCDAS are fundamentally intended to warn the driver of the subject vehicle against potential collisions with vehicles to the side and/or to the rear of the subject vehicle and moving in the same direction as the subject vehicle during lane change maneuvers. This standardization addresses LCDAS for use on forward moving cars, vans, and straight trucks in highway situations. This standardization does not address LCDAS for use on motorcycles or articulated vehicles such as tractor/trailer combinations and articulated buses.

An LCDAS is intended to supplement the interior and exterior rear-view mirrors, but not eliminate the need for such mirrors. The system is intended to detect vehicles to the rear and sides of the subject vehicle. When the subject vehicle driver indicates the desire to make a lane change, the system will evaluate the situation and warn the driver if a lane change is not recommended. LCDAS are not meant to encourage aggressive driving.

8.4.2.4 ISO 22178 Low Speed Following Systems

ISO	DIS	22178	Intelligent Transport Systems—Low Speed Following Systems—Performance requirements and test procedures	Under Development DIS	ISO TC204

ISO 22178 Not Yet Available. Track progress at:
http://www.iso.org/iso/iso_catalogue/catalogue_tc/catalogue_detail.htm?csnumber=40752

The main system function of low speed following is to control vehicle speed adaptively to a forward vehicle by using information about:

- Ranging to forward vehicles;
- The motion of the subject (LSF equipped) vehicle;
- Driver commands (see Figure 8.6).

Based upon the information acquired, the controller (identified as "LSF control strategy" in Figure 8.6) sends commands to actuators for carrying out its longitudinal control strategy and also sends status information to the driver.

The goal of LSF is a partial automation of the longitudinal vehicle control to reduce the driver's workload.

This International Standard may be used as a system level standard by other standards, which extend the LSF to a more detailed standard (e.g., for specific detection and ranging sensor concepts or higher level of functionality. So, issues like specific requirements for the detection and ranging sensor function and performance or communication links for cooperative solutions will not be considered here.

This International Standard contains the basic control strategy, minimum functionality requirements, basic driver interface elements, minimum requirements for diagnostics and reaction to failure, and performance test procedures for low speed following systems.

8.4.2.5 ISO 15662 Adaptive Cruise Control

ISO	IS	15622	Adaptive Cruise Control Systems	Published	ISO TC204

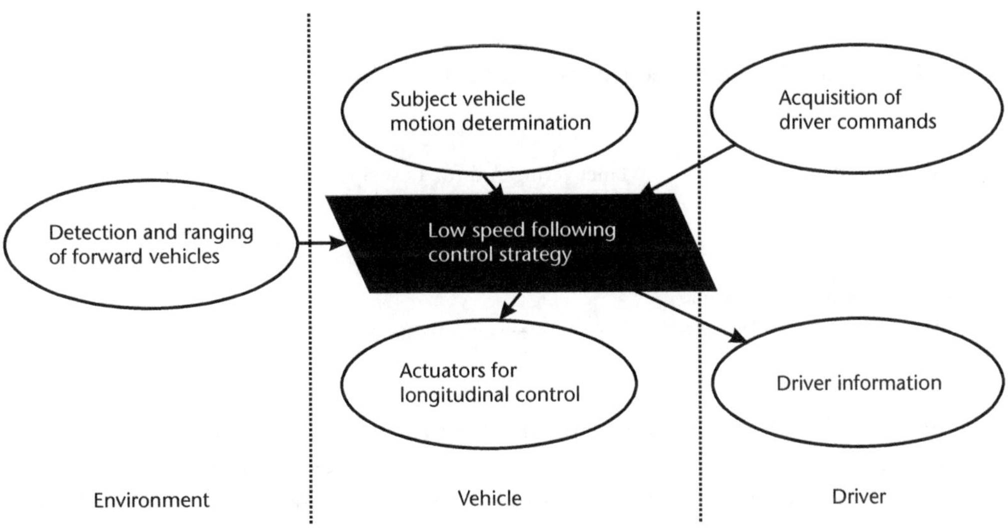

Figure 8.6 Functional LSF elements. (*Source:* TC 204 WG 14 document circulated for comment.)

ISO 15622 Available from
http://www.iso.org/iso/iso_catalogue/catalogue_tc/catalogue_detail.htm?csnumber=27834

ISO 15622:2002 specifies the basic control strategy, minimum functionality requirements, basic driver interface elements, minimum requirements for diagnostics and reaction to failure, and performance test procedures for adaptive cruise control (ACC) systems. ACC is fundamentally intended to provide longitudinal control of equipped vehicles while traveling on highways under free-flowing traffic conditions. ACC may be augmented with other capabilities, such as forward obstacle warning.

This standard is in the process of revision by the working group.

8.4.2.6 ISO 15623 Forward Vehicle Collision Warning System

ISO	IS	15623	Transport information and control systems—Forward vehicle collision warning systems—Performance requirements and test procedures	Published 2002	ISO TC204

ISO 15623 Available from
http://www.iso.org/iso/iso_catalogue/catalogue_tc/catalogue_detail.htm?csnumber=27835

ISO 15623:2002 specifies performance requirements and test procedures for systems capable of warning the driver of short intervehicle distance and closing speed which may cause a rear-end collision with other vehicles, including motorcycles, ahead of the subject vehicle while it is operating at ordinary speed.

ISO 15623:2002 is applicable to operations on roads with curve radii over 125m as well as higher radius curves.

8.4.2.7 ISO 17386 Maneuvering Aid for Low Speed Operation

ISO	IS	17386	Maneuvering Aid for Low Speed Operation	Published 2004	ISO TC204

ISO 17386 Available from
http://www.iso.org/iso/iso_catalogue/catalogue_tc/catalogue_detail.htm?csnumber=30634

ISO 17386:2004 for maneuvering aids for low speed operation (MALSO) addresses light-duty vehicles (e.g., passenger cars, pick-up trucks, light vans and sport utility vehicles; motorcycles excluded) equipped with such MALSO systems. It specifies minimum functionality requirements which the driver can generally expect of the device (i.e., detection of and information on the presence of relevant obstacles within a defined (short) detection range). It defines minimum requirements for failure indication as well as performance test procedures; it includes rules for the general information strategy but does not restrict the kind of information or display system.

MALSO systems use object-detection devices (sensors) for ranging in order to provide the driver with information based on the distance to obstacles. The sensing technology is not addressed. The current test objects are defined based on systems using ultrasonic sensors, which reflect the most commonly used available technology.

Visibility enhancement systems like video camera aids without distance ranging and warning and reversing aids and obstacle-detection devices on heavy commercial vehicles are not covered by ISO 17386:2004.

This standard is currently being updated and revised by the working group.

8.4.2.8 ISO 22179 Full Speed Range Adaptive Cruise Control Systems—Performance Requirements and Test Procedures

ISO	DIS	22179	Intelligent Transport Systems—Full Speed Range Adaptive Cruise Control Systems—Performance requirements and test procedures	Under Development DIS	ISO TC204

ISO 22179 Not Yet Available. Track progress at:
http://www.iso.org/iso/search.htm?qt=22179&sort=rel&type=simple&published=true

The main system function of full speed range adaptive cruise control (FSRA) is to control vehicle speed adaptively to a forward vehicle by using information about:

- Distance to forward vehicles;
- The motion of the subject (FSRA equipped) vehicle;
- Driver commands (see Figure 8.7).

Based upon the information acquired, the controller (identified as "FSRA control strategy" in Figure 8.7) sends commands to actuators that carry out its longitudinal control strategy, and sends status information to the driver.

The goal of FSRA is partial automation of longitudinal vehicle control to reduce drivers' workload.

This International Standard may be used as a system level standard by other standards, which extend FSRA to a more detailed standard (e.g., for specific detection and ranging sensor concepts or higher levels of functionality). Issues like specific requirements for the detection and ranging sensor function and performance or communication links for cooperative solutions will not be considered here.

This International Standard contains the basic control strategy, minimum functionality requirements, basic driver interface elements, minimum requirements for

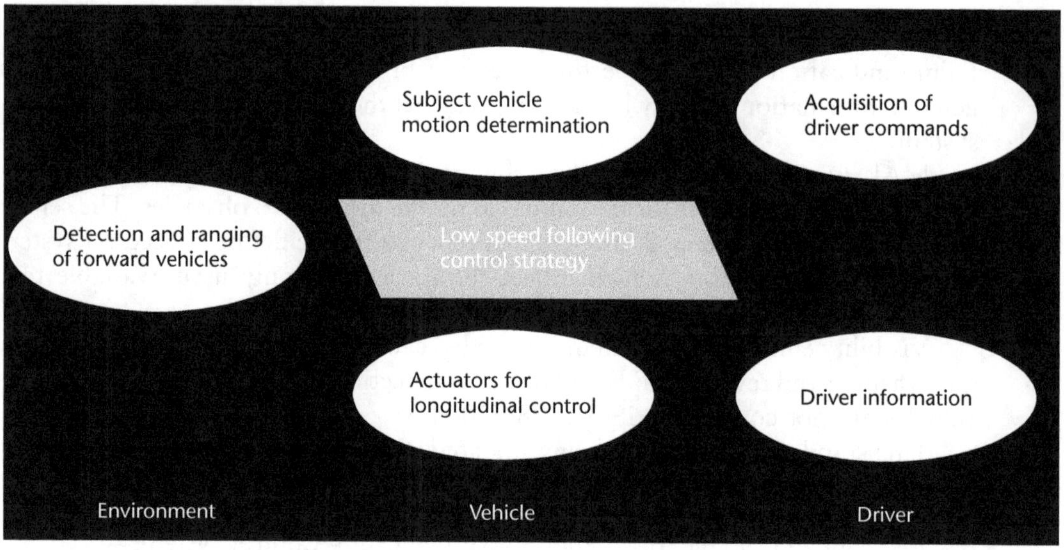

Figure 8.7 Functional FSRA elements. (*Source*: TC 204 WG 14 document circulated for comment.)

diagnostics and reaction to failure, and performance test procedures for FSRA systems. FSRA is fundamentally intended to provide longitudinal control of equipped vehicles while traveling on highways (roads where nonmotorized vehicles and pedestrians are prohibited) under free-flowing and congested traffic conditions. FSRA provides support within the speed domain of standstill up to the designed maximum speed of the system. The system will attempt to stop behind an already tracked vehicle within its limited deceleration capabilities and will be able to start again after the driver has input a request to the system to resume the journey from standstill. The system is not required to react to stationary or slow moving objects (in accordance with ISO 15622 ACC).

8.4.2.9 ISO 26684 Intersection Signal Information and Violation Warning Systems (ISIVWS)

ISO	PWI	26684	Intelligent Transport Systems—Intersection Signal Information and Violation Warning Systems (ISIVWS)	Under Development PWI	ISO TC204

ISO 26684 Not Yet Available

This will determine systems based on roadside and vehicle cooperation that display current traffic light information on onboard equipment and activates a warning system using onboard equipment when the driver is about to ignore a red light.

Studies of the central features, such as basic functions, standardization items, and information contents.

8.4.2.10 ISO 11067 Curve Speed Warning Systems (CSWS)

ISO	PWI	11067	Intelligent Transport Systems—Curve Speed Warning Systems (CSWS)	Under Development PWI	ISO TC204

ISO 11067 Not Yet Available

This is a recent work item. Curve speed warning systems use a combination of GPS and digital maps to assess threat levels for a driver approaching a curve too quickly.

No further details are yet available.

8.4.2.11 ISO 11270 Lane Keeping Assist Schemes

ISO	PWI	11270	Intelligent Transport Systems—Lane Keeping Assist Schemes	Under Development PWI	ISO TC204

ISO 11270 Not Yet Available

Vision-based lane recognition systems are relatively mature and have already been introduced to market for lane departure warning. This recent work item is to specify vehicle positioning with a high accuracy using enhanced lane-level digital mapping and inertial system (INS) for enhanced positioning performance and availability.

8.4.2.12 SAE J2399 Adaptive Cruise Control (ACC) Operating Characteristics and User Interface

SAE	J2399_200312	Adaptive Cruise Control (ACC) Operating Characteristics and User Interface	Published	SAE

SAE J2399 Available from
http://www.sae.org/technical/standards/J2399_200312

Adaptive cruise control (ACC) is an enhancement of conventional cruise control systems which allows the ACC equipped vehicle to follow a forward vehicle at a preselected time gap by controlling the engine, power train, and/or service brakes.

This SAE standard focuses on specifying the minimum requirements for ACC system operating characteristics and elements of the user interface. This document applies to original equipment and aftermarket ACC systems for passenger vehicles (including motorcycles). This document does not apply to commercial vehicles. Furthermore, this document does not address future variations on ACC, such as "stop and go" ACC, which can bring the equipped vehicle to a stop and then reaccelerate. Future revisions of this document should consider enhanced versions of ACC, as well as the integration of ACC with forward collision warning (FCW).

8.4.3 Cooperative Driving

This service group is the application of ITS functionality to completely automate the driving process, creating a "hands-off" driving environment, or it may be a partially automated operation that supports drivers. For public transport, specific tools can be used to permit vehicles to be aligned properly at stops, assuring access either to same-height platforms or disabled travelers (e.g., wheelchair lifts, bus lowering mechanisms). Examples include the following:

- Automatic lane keeping;
- Automatic parking operation;
- Vehicle platooning;
- Very low speed cruise control (inching);

Example services include:

- Automated highway operation;
- Automated low speed maneuvering;
- Precision docking for public transport vehicles;
- Automated cruise control.

8.4.3.1 ISO 22839 Rear-End Collision Mitigation Braking Systems

ISO	NP	22839	Rear-End Collision Mitigation Braking Systems	Under Development	ISO TC204

ISO 22839 Not Yet Available

This is a current work item under development. Rear-end collision mitigation systems predict rear-end collisions and assist brake operation to reduce both impact on occupants and vehicle damage. These systems predict the probability of a collision based on driving conditions, distance to the vehicle ahead, and relative speeds. They use visual and audio warnings to prompt the driver to take preventative action. Some systems may also initiate braking to reduce the vehicle's speed, and may also include pretension seatbelts.

Millimeter-wave radar is used in some systems to detect forward vehicles within a range of 100m, and then calculate the distance between the vehicles, the relative vehicle speeds, and the anticipated vehicle path to determine the likelihood of a collision.

No documents are yet available.

8.4.3.2 Automated Vehicle Operation

Although automated vehicle operation will only work with the benefit of International Standards, it is still very much in the research stage, and, unlike many advisory systems will only successfully operate where a high percentage of vehicles are equipped. While the research demonstrations are interesting to watch and fascinating to participate in, it is premature to expect standards at this point in time. Watch the work program of ISO/TC 204 WG 14 to see this domain evolve.

This section needs to be read in conjunction with Section 8.4.2.

8.4.3.3 Collision Avoidance

The collision avoidance service group includes the use of sensors and control systems to detect potential for collisions and either prompt the driver to take action or automatically initiate avoiding action. Longitudinal collision avoidance includes the application of obstacle detection systems. Lateral collision avoidance is the use of systems (such as sensors and control systems) to monitor the potential hazards involved in lane keeping, lane changing, entering and leaving high-speed roads, and overtaking. Example services include:

- Longitudinal collision avoidance;
- Lateral collision avoidance;
- Intersection collision avoidance.

For forward and rear obstacle warning, lane departure warning, and decision aid systems see Section 8.4.2. Other standards are shown next.

SAE	2400_200308	Human Factors in Forward Collision Warning Systems: Operating Characteristics and User Interface Requirements	Published Aug. 2003

SAE J2400 Available from
http://www.sae.org/technical/standards/J2400_200308

Forward collision warning (FCW) systems are onboard systems intended to provide alerts to assist drivers in avoiding striking the rear-end of another moving or stationary motorized vehicle. This SAE information report describes elements for a FCW operator interface, as well as requirements and test methods for systems capable of warning drivers of rear-end collisions. This information report applies to original equipment and aftermarket FCW systems for passenger vehicles including cars, light trucks, and vans. This report does not apply to heavy trucks. Furthermore, this document does not address integration issues associated with adaptive cruise control (ACC), and consequently, aspects of the document could be inappropriate for an ACC system integrated with a FCW system.

8.4.3.4 Safety Readiness

There are no standards yet under development for this domain.

8.4.3.5 Precrash Restraint Deployment

This service group uses ITS functionality to determine the velocity, mass, and direction of vehicles and objects involved in a potential collision, as well as the number, location, and major physical characteristics of occupants. A system's use of this data to determine a response strategy may include the following elements:

- Arming and deploying air bags;
- Deploying lateral protection systems;
- Deploying rollbars;
- Tightening seatbelts.

NOTE: The lead on standardization for the onboard issues will be agreed between TC 204 and TC 22.

Example services include precrash restraint deployment.

SAE	J2431_199710	The Effects of Front-Mounted Accessories on Air Bag Sensors and Crashworthiness	Published Oct. 1997

SAE J2431 Available from
http://www.sae.org/technical/standards/J2431_199710

The crushability of the front bumper and supporting structure are key elements in the system, so alterations to that area become logical concerns. The accessory manufacturers, dealers, and installers are concerned because the very core of their business could be at risk. The unknowns can range from fear of setting off an airbag while working on the vehicle to liability issues in an injury accident situation. In some cases, the installation of the product is contrary to recommendations from the vehicle manufacturer and may void the warranty.

SAE	J1855_200110	Deployment of Electrically Activated Automotive Air Bags for Automobile Reclamation	Published Oct. 2001

SAE J1855 Available from
http://www.sae.org/technical/standards/J1855_200110

This SAE recommended practice describes the method for safe deployment of air bag modules in vehicles equipped with electrically actuated air bag systems for the purpose of disposal. It is intended to provide a procedure which does not require significant technical expertise, is easy to operate, and is readily available to be used by automobile dismantlers or vehicle shredders to deploy air bag modules prior to automobile reclamation.

SAE	J2238_200110	Airbag Inflator Ballistic Tank Test Procedure Gas Generators Used In Inflatable Restraint Systems	Published Oct. 2001

SAE J2238 Available from
http://www.sae.org/technical/standards/J2238_200110

This SAE recommended practice establishes a ballistic tank test procedure for evaluating inflator assemblies used in inflatable restraint systems. It is intended to be a general procedure for repetitive testing and suggests only general guidelines for the safe conduct of tests and data correlation. Uniform test requirements, test procedures, and data recording requirements are specified. The intent of the document is to provide a procedure employing a ballistic tank test method for determining the ability of an inflator to meet requirements for deploying inflatable restraint systems. A ballistic tank test is described which will yield repeatable and comparable results for evaluating a given inflator configuration's pressure output versus time. Use of the ballistic tank test for comparison of various inflator configurations may be of limited value due to differences in gas temperature and mass flow effects on airbag performance as it relates to occupant protection.

SAE	J1980_200112	Guidelines for Evaluating Out-of-Position Vehicle Occupant Interactions with Deploying Frontal Airbags	Published Dec. 2001

SAE J1980 Available from
http://www.sae.org/technical/standards/J1980_200112

An airbag generates a considerable amount of kinetic energy during its inflation process. As a result, substantial forces can be developed between the deploying airbag and the out-of-position occupant. Accident data and laboratory test results have indicated a potential for head, neck, chest, abdominal, and leg injuries from these forces. This suggests that mitigating such forces should be considered in the design of airbag restraint systems.

This document outlines a comprehensive set of test guidelines that can be used for investigating the interactions that occur between the deploying airbag and the occupant who is near the module at the time of deployment. Static and dynamic tests to investigate driver and passenger systems are given. Static tests may be used to sort designs on a comparative basis. Designs that make it through the static sorting procedure may be subjected to the appropriate dynamic tests. On a specific vehicle model, engineering judgment based upon prior experience in airbag testing may make it unnecessary to conduct the tests identified by the document or may indicate that different tests should be conducted.

SAE	J2189_200112	Guidelines for Evaluating Child Restraint System Interactions with Deploying Airbags	Published Dec. 2001

SAE J2189 Available from
http://www.sae.org/technical/standards/J2189_200112

This SAE information report prescribes dummies, procedures, and configurations that can be used for investigating the interactions that might occur between a deploying airbag and a child restrained by a child restraint system (CRS). During the inflation process, airbags generate a considerable amount of kinetic energy which can result in substantial forces being applied to a child who is restrained in a CRS in the front seat of a vehicle.

SAE	J1538_200112	Glossary of Automotive Inflatable Restraint Systems	Published Dec. 2001

SAE J1538 Available from
http://www.sae.org/technical/standards/J1538_200112

The terms included in the glossary are general in nature and usually apply to more than one manufacturer's system. All terms in section 3 of the glossary apply to automotive restraint systems in general which are initiated by an electric or mechanical stimulus upon receipt of a signal from a sensor. All terms in section 4 of the glossary apply primarily to the initiator used in automotive restraint systems. In some cases, the same terms appear in both sections to allow for the general definition and for definitions applying specifically to initiators. These terms are intended to reflect existing designs and the glossary will be updated as information on other types of systems becomes available. Appendix A of the document is included to identify terminology that is no longer in common use or specifically applicable to inflatable restraint systems, but was published in the April 1988 version of SAE J1538.

SAE	J1630_200112	Driver or Passenger Airbag Module Deployment Test Procedure	Published Dec. 2001

SAE J1630 Available from
http://www.sae.org/technical/standards/J1630_200112

This SAE recommended practice describes a method to be used for the static deployment of airbag module assemblies. The results obtained from the deployment tests will be used to verify compliance with design requirements and/or specifications.

The intent of this procedure is to describe recommended test methods to evaluate performance characteristics of driver airbag modules, passenger airbag modules, or other airbag modules; for example, side airbags (door or seat mounted), roof rail airbags, and knee bolster airbags. Performance limits or acceptance criteria are not established as they are typically defined based on specific vehicle design requirements. It is intended to be a general procedure for repetitive testing and suggests only general guidelines for the safe conduct of tests and reliable data correlation.

8.4.4 In-Vehicle Advice and Control

These standards are principally developed by ISO/TC 22 Road Vehicles and are more concerned with internal vehicle control than ITS systems or systems that will be directly used or affected by ITS service provision. However, many may be connected to ITS systems and are therefore provided for reference. Test standards regarding radio emissions may be particularly important.

8.4.4.1 ISO 2575 Road Vehicles—Symbols for Controls, Indicators and Tell-Tales

ISO	IS/CD/ ISO 2575 WD	Road vehicles—Symbols for controls, indicators and tell-tales—Amendment 4	Published Standard (Amnd 3) AMD 4 2004 FDIS: 2007-10-07 AMD5 Under development	ISO TC22

ISO 2575 Available from
http://www.iso.org/iso/iso_catalogue/catalogue_tc/catalogue_detail.htm?csnumber=39704

ISO 2575:2004 establishes symbols (i.e., conventional signs) for use on controls, indicators, and tell-tales applying to passenger cars, light and heavy commercial vehicles, and buses, to ensure identification and to facilitate use. It also indicates the colors of possible optical tell-tales, which inform the driver of either correct operation or malfunctioning of the related devices.

8.4.4.2 ISO 3833 Road Vehicles—Types—Terms and Definitions

ISO	IS	ISO 3833	Road vehicles—Types—Terms and definitions	Published Standard	ISO TC22

ISO 2575 Available from
http://www.iso.org/iso/iso_catalogue/catalogue_tc/catalogue_detail.htm?csnumber=9389

Although this standard is not directly ITS related, it is referred to in many ITS and ITS related standards. ISO 3833 defines terms relating to some types of road vehicles designated according to certain design and technical characteristics. It applies to all vehicles designated for road circulation, except for agricultural tractors, which are only incidentally used for the carriage of persons or goods by road. This second edition cancels and replaces the first edition (1976).

8.4.4.3 ISO 4138 Passenger Cars—Steady-State Circular Driving Behavior—Open-Loop Test Procedure

ISO	IS	ISO 4138	Passenger cars—Steady-state circular driving behavior—Open-loop test procedure	Published 2004	ISO TC22

ISO 4138 Available from
http://www.iso.org/iso/iso_catalogue/catalogue_tc/catalogue_detail.htm?csnumber=30084

ISO 4138:2004 specifies open-loop test methods for determining the steady-state circular driving behavior of passenger cars as defined in ISO 3833 and light trucks.

8.4.4.4 ISO 4513 Road Vehicles—Visibility—Method for Establishment of Eyellipses for Driver's Eye Location

ISO	WD	ISO/WD 4513	Road vehicles—Visibility—Method for establishment of eyellipses for driver's eye location	Published/ revised 2003 (Revision of ISO 4513:2003) CD: 2008-02-21	ISO TC22

ISO 4513 Available from
http://www.iso.org/iso/iso_catalogue/catalogue_tc/catalogue_detail.htm?csnumber=36126

ISO 4513:2003 specifies a method for establishing an eyellipse for locating driver's eyes inside a road vehicle for the purpose of measuring the driver's field of view. Elliptical (eyellipse) models in both two and three dimensions are used to represent 95th and 99th percentiles of driver eye locations. Its procedures, which differ depending on the type of vehicle considered, are applicable to passenger cars (and light trucks), and to buses and heavy vehicles, as defined in ISO 3833. The statistical representation of the driver's eye locations it provides can be used as a design tool for passenger cars (V-points can be used in lieu of the complete eyellipse to standardize the driver's field of view for regulation purposes).

8.4.4.5 ISO 7401 Road Vehicles—Lateral Transient Response Test Methods—Open-Loop Test Methods

ISO	IS	ISO 7401	Road vehicles—Lateral transient response test methods—Open-loop test methods	Published 2003	ISO TC22

ISO 7401 Available from
http://www.iso.org/iso/iso_catalogue/catalogue_tc/catalogue_detail.htm?csnumber=36998

ISO 7401:2003 specifies open-loop test methods for determining the transient response behavior of road vehicles. It is applicable to passenger cars, as defined in ISO 3833, and to light trucks.

8.4.4.6 ISO 7639 Road Vehicles—Diagnostic Systems—Graphical Symbols

ISO	IS	7639	Road vehicles—Diagnostic systems—Graphical symbols	Published Standard 1985	ISO TC22

ISO 7639 Available from
http://www.iso.org/iso/iso_catalogue/catalogue_tc/catalogue_detail.htm?csnumber=14454

This specifies graphical symbols for diagnostic testers: marking of controls, indicators and tell-tales; use for indications on screens, and similar variable indicating systems; and marking of connections and other input and output openings. It tabulates the terms and functions in English and French together with the graphical symbols of concern.

8.4.4.7 ISO 9141 Road Vehicles—Diagnostic Systems—Requirements for Interchange of Digital Information

ISO	IS	9141	Road vehicles—Diagnostic systems—Requirements for interchange of digital information	Published 1989	ISO TC22

This family of standards is in several parts but does not follow the normal ISO numbering (i.e., starting at xxx-1.)

ISO 9141 Available from
http://www.iso.org/iso/iso_catalogue/catalogue_tc/catalogue_detail.htm?csnumber=16737

This specifies the requirements for setting up the interchange of digital information between onboard ECUs of road vehicles and suitable diagnostic testers. This communication is established in order to facilitate inspection, test diagnosis, and adjustment of vehicles, systems, and ECUs. It does not apply when system-specific diagnostic test equipment is used.

ISO	IS	9141-2	Road vehicles—Diagnostic systems—Part 2: CARB requirements for interchange of digital information	Published 1994 Amd 1996	ISO TC22

ISO 9141-2 Available from
http://www.iso.org/iso/iso_catalogue/catalogue_tc/catalogue_detail.htm?csnumber=16738
9141-2 Amd 1 Available from
http://www.iso.org/iso/iso_catalogue/catalogue_tc/catalogue_detail.htm?csnumber=24607

This is limited to vehicles with nominal 12V supply voltage. It describes a subset of ISO 9141:1989. It specifies the requirements for setting up the interchange of digital information between onboard emission-related electronic control units of road vehicles and the SAE OBD II scan tool as specified in SAE J1978. This communication is established to facilitate compliance with California Code of Regulation, Title 13, 1968.1.

The amendment contains minor amendments in three subclauses.

ISO	IS	9141-3	Road vehicles—Diagnostic systems—Part 3: Verification of the communication between vehicle and OBD II scan tool	Published 1998	ISO TC22

ISO 9141-3 Available from
http://www.iso.org/iso/iso_catalogue/catalogue_tc/catalogue_detail.htm?csnumber=28621

The contents are as implied in the title.

8.4.4.8 ISO 10305 Road Vehicles—Calibration of Electromagnetic Field Strength Measuring Devices

ISO	TR	10305	Road vehicles—Calibration of electromagnetic field strength measuring devices	Published TR ISO TC22
ISO	TR	10305-1	Road vehicles—Calibration of electromagnetic field strength measuring devices—Part 1: Devices for measurement of electromagnetic fields at frequencies > 0 Hz	Published TR ISO TC22 2003

ISO 10305-1 Available from
http://www.iso.org/iso/iso_catalogue/catalogue_tc/catalogue_detail.htm?csnumber=36575

ISO/TR 10305-1:2003 specifies techniques for calibrating field strength measuring devices used in automotive testing for the measurement of magnetic fields at frequencies greater than 0 Hz, for both EMC and human protection applications. It has been prepared by German experts using devices including capacitor or coil arrangements, TEM cells, and antenna arrangements in absorber-lined chambers. In the automotive field, these field strength measuring devices are used for measurements specified in the various parts of ISO 11451 and ISO 11452.

ISO	TR	10305-2	Road vehicles—Calibration of electromagnetic field strength measuring devices—Part 2: IEEE standard for calibration of electromagnetic field sensors and probes, excluding antennas, from 9 kHz to 40 GHz	Published TR ISO TC22 2003

ISO 10305-2 Available from
http://www.iso.org/iso/iso_catalogue/catalogue_tc/catalogue_detail.htm?csnumber=37229

ISO/TR 10305-2:2003 specifies techniques for calibrating electromagnetic field sensors and probes, excluding antennas, used in automotive testing for the measurement of magnetic fields at frequencies from 9 kHz to 40 GHz. In the automotive field, these field strength measuring devices are used for measurements specified in the various parts of ISO 11451 and ISO 11452.

8.4.4.9 ISO 11451 Road Vehicles—Vehicle Test Methods for Electrical Disturbances from Narrowband Radiated Electromagnetic Energy

ISO	IS	11451	Road vehicles—Vehicle test methods for electrical disturbances from narrowband radiated electromagnetic energy	Published 1994 Subsequently revised	ISO TC22

This series of standards is in four parts.

ISO	IS	11451-1	Road vehicles—Vehicle test methods for electrical disturbances from narrowband radiated electromagnetic energy—Part 1: General principles and terminology	Published 1994 Revised 2005	ISO TC22

ISO 11451-1 Available from
http://www.iso.org/iso/iso_catalogue/catalogue_tc/catalogue_detail.htm?csnumber=37998

ISO 11451-1:2005 specifies general conditions, defines terms, gives practical guidelines, and establishes the basic principles of the vehicle tests used in the other parts of ISO 11451 for determining the immunity of passenger cars and commercial vehicles to electrical disturbances from narrowband radiated electromagnetic energy, regardless of the vehicle propulsion system (e.g., spark-ignition engine, diesel engine, electric motor).

The electromagnetic disturbances considered are limited to continuous narrowband electromagnetic fields. A wide frequency range (0.01 to 18,000 MHz) is allowed for the immunity testing in this and the other parts of ISO 11451.

ISO	IS	11451-2	Road vehicles—Vehicle test methods for electrical disturbances from narrowband radiated electromagnetic energy—Part 2: Off-vehicle radiation sources	Published 1994 Revised 2005	ISO TC22

ISO 11451-2 Available from
http://www.iso.org/iso/iso_catalogue/catalogue_tc/catalogue_detail.htm?csnumber=37999

ISO 11451-2:2005 specifies a vehicle test method for determining the immunity of passenger cars and commercial vehicles to electrical disturbances from off-vehicle radiation sources, regardless of the vehicle propulsion system (e.g., spark ignition engine, diesel engine, electric motor). It can also be readily applied to other types of vehicles. The electromagnetic disturbances considered are limited to narrowband electromagnetic fields.

ISO	IS	11451-3	Road vehicles—Vehicle test methods for electrical disturbances from narrowband radiated electromagnetic energy—Part 3: On-board transmitter simulation	Published 1994 Revised 2007	ISO TC22

ISO 11451-3 Available from
http://www.iso.org/iso/iso_catalogue/catalogue_tc/catalogue_detail.htm?csnumber=37439

ISO 11451-3:2007 specifies methods for testing the immunity of passenger cars and commercial vehicles to electromagnetic disturbances from onboard transmitters connected to an external antenna and portable transmitters with integral antennas, regardless of the vehicle propulsion system (e.g., spark ignition engine, diesel engine, electric motor).

ISO	IS	11451-4	Road vehicles—Vehicle test methods for electrical disturbances from narrowband radiated electromagnetic energy—Part 4: Bulk current injection (BCI)	Published 1995 Revised 2006	ISO TC22

ISO 11451-4 Available from
http://www.iso.org/iso/iso_catalogue/catalogue_tc/catalogue_detail.htm?csnumber=38000

ISO 11451-4:2006 specifies bulk current injection (BCI) test methods and procedures for testing the electromagnetic immunity of electronic components for passenger cars and commercial vehicles regardless of the propulsion system (e.g., spark-ignition engine, diesel engine, electric motor). The electromagnetic disturbance considered in ISO 11451-4:2006 is limited to continuous narrowband electromagnetic fields.

8.4.4.10 ISO 11452 Road Vehicles—Component Test Methods for Electrical Disturbances from Narrowband Radiated Electromagnetic Energy

ISO	IS	11452	Road vehicles—Component test methods for electrical disturbances from narrowband radiated electromagnetic energy	Published 1994	ISO TC22

This group of standards is in eight parts.

ISO	IS	11452-1	Road vehicles—Component test methods for electrical disturbances from narrowband radiated electromagnetic energy—Part 1: General principles and terminology	Published 2001 Revised 2004	ISO TC22

ISO 11452-1 Available from
http://www.iso.org/iso/iso_catalogue/catalogue_tc/catalogue_detail.htm?csnumber=38001

ISO 11452-1:2005 specifies general conditions, defines terms, gives practical guidelines, and establishes the basic principles of the component tests used in the other parts of ISO 11452 for determining the immunity of electronic components of passenger cars and commercial vehicles to electrical disturbances from narrowband radiated electromagnetic energy, regardless of the vehicle propulsion system (e.g., spark-ignition engine, diesel engine, electric motor). The electromagnetic disturbances considered are limited to continuous narrowband electromagnetic fields. A wide frequency range (0.01 to 18,000 MHz) is allowed for the immunity testing of the components in this and the other parts of ISO 11452.

ISO	IS	11452-2	Road vehicles—Component test methods for electrical disturbances from narrowband radiated electromagnetic energy—Part 2: Absorber-lined shielded enclosure	Published 1995 Revised 2004	ISO TC22

ISO 11452-2 Available from
http://www.iso.org/iso/iso_catalogue/catalogue_tc/catalogue_detail.htm?csnumber=36570

ISO 11452-2:2004 specifies an absorber-lined shielded enclosure method for testing the immunity (off-vehicle radiation source) of electronic components for passenger cars and commercial vehicles regardless of the propulsion system (e.g., spark-ignition engine, diesel engine, electric motor). The device under test (DUT), together with the wiring harness (prototype or standard test harness), is subjected to an electromagnetic disturbance generated inside an absorber-lined shielded enclosure, with peripheral devices either inside or outside the enclosure. It is applicable only to disturbances from continuous narrowband electromagnetic fields. See ISO 11452-1 for general test conditions.

ISO	IS	11452-3	Road vehicles—Component test methods for electrical disturbances from narrowband radiated electromagnetic energy—Part 3: Transverse electromagnetic mode (TEM) cell	Published 1995 Revised 2001	ISO TC22

ISO 11452-3 Available from
http://www.iso.org/iso/iso_catalogue/catalogue_tc/catalogue_detail.htm?csnumber=24569

ISO	IS	11452-4	Road vehicles—Component test methods for electrical disturbances from narrowband radiated electromagnetic energy—Part 4: Bulk current injection (BCI)	Published 2001 Revised 2005	ISO TC22

ISO 11452-4 Available from
http://www.iso.org/iso/iso_catalogue/catalogue_tc/catalogue_detail.htm?csnumber=37414

ISO 11452-4:2005 specifies bulk current injection (BCI) test methods for determining the immunity of electronic components of passenger cars and commercial vehicles regardless of the propulsion system (e.g., spark-ignition engine, diesel engine, electric motor). The electromagnetic disturbances considered in ISO 11452-4 are limited to continuous narrowband electromagnetic fields. See ISO 11452-1 for general test conditions.

ISO	IS	11452-5	Road vehicles—Component test methods for electrical disturbances from narrowband radiated electromagnetic energy—Part 5: Stripline	Published 1995 Revised 2002	ISO TC22

ISO 11452-5 Available from
http://www.iso.org/iso/iso_catalogue/catalogue_tc/catalogue_detail.htm?csnumber=24571

This part of ISO 11452 specifies stripline tests for determining the immunity of electronic components for passenger cars and commercial vehicles to electrical disturbances from narrowband electromagnetic energy, regardless of the vehicle propulsion system (e.g., spark-ignition engine, diesel engine, electric motor). As the performance of these tests depends on the exposure of the equipment harness to a disturbance field, the applicability of this part of ISO 11452 is limited to equipment harnesses, which have a maximum diameter of one-third the stripline height or less.

The electromagnetic disturbances considered are limited to continuous narrowband electromagnetic fields.

ISO	IS	11452-6	Road vehicles—Component test methods for electrical disturbances from narrowband radiated electromagnetic energy—Part 6: Parallel plate antenna	Deleted 2002	ISO TC22

ISO 11452-6 is no longer available

ISO	IS	11452-7	Road vehicles—Component test methods for electrical disturbances from narrowband radiated electromagnetic energy—Part 7: Direct radio frequency (RF) power injection	Published 1995 Revised 2003	ISO TC22

ISO 11452-7 Available from
http://www.iso.org/iso/iso_catalogue/catalogue_tc/catalogue_detail.htm?csnumber=38302

ISO 11452-7:2003 specifies a direct RF power injection test for determining the immunity of electronic components of passenger cars and commercial vehicles to electrical disturbances from narrowband electromagnetic energy, regardless of the propulsion system (e.g., spark-ignition engine, diesel engine, electric motor). The test method, which involves providing differential mode excitation to the DUT, is applicable to all DUT leads except RF ground. Applicable over the frequency range 0.25 to 500 MHz, the method can be used to predict the compatibility in the vehicle environment with respect to radiated and conducted RF energy, including conducted transient RF energy, and is especially useful as a means of isolating the susceptible circuits within a DUT and evaluating potential solutions.

ISO	IS	11452-8	Road vehicles—Component test methods for electrical disturbances from narrowband radiated electromagnetic energy—Part 8: Immunity to magnetic fields	Published 2007	ISO TC22

ISO 11452-8 Available from
http://www.iso.org/iso/iso_catalogue/catalogue_tc/catalogue_detail.htm?csnumber=40831

ISO 11452-8:2007 specifies tests for the electromagnetic immunity of electronic components for passenger cars and commercial vehicles, regardless of the propulsion system (e.g., spark-ignition engine, diesel engine, electric motor), to magnetic fields generated by power transmission lines and generating stations and some powerful electrical equipment, such as motors. To perform this test, the DUT is exposed to a magnetic disturbance field.

8.4.4.11 ISO 12155 Commercial Vehicles—Obstacle Detection Device During Reversing—Requirements and Tests

ISO	TR	12155	Commercial vehicles—Obstacle detection device during reversing—Requirements and tests	Published TR 1994	ISO TC22

ISO 12155 Available from
http://www.iso.org/iso/iso_catalogue/catalogue_tc/catalogue_detail.htm?csnumber=20639

This specifies requirements and tests for detection devices which indicate to the driver of a commercial road vehicle, when he is reversing, the presence of objects which are within the monitoring range of the device. Applies to detection devices with noncontact sensors which can be fitted on commercial vehicles to improve safety during maneuvering at a speed of up to 5 km/h, but they do not relieve the driver of his responsibility when reversing.

8.4.4.12 ISO 12364 Two-Wheeled Motorcycles—Antilock Braking Systems (ABS)—Tests and Measurement Methods

ISO	IS	12364	Two-wheeled motorcycles—Antilock braking systems (ABS)—Tests and measurement methods	Published 2001	ISO TC22

ISO 12364 Available from
http://www.iso.org/iso/iso_catalogue/catalogue_tc/catalogue_detail.htm?csnumber=1680

This International Standard specifies tests and measurement methods for solo motorcycles (defined in ISO 3833) equipped with one or more antilock, and one or more independent, braking systems. It sets out procedures for the following types of straight-line braking tests: utilization-of-adhesion test; wheel-lock check test; tests with ABS failed; and complementary tests that could assist in the assessment and development of braking systems.

8.4.4.13 ISO 12366 Two-Wheeled Mopeds—Antilock Braking Systems (ABS)—Tests and Measurement Methods

ISO	IS	12366	Two-wheeled mopeds—Antilock braking systems (ABS)—Tests and measurement methods	Published 2001	ISO TC22

ISO 12366 Available from
http://www.iso.org/iso/iso_catalogue/catalogue_tc/catalogue_detail.htm?csnumber=1194

8.4.4.14 ISO 14229 Road Vehicles—Diagnostic Systems—Diagnostic Services Specification

ISO	IS	14229	Road vehicles—Diagnostic systems—Diagnostic services specification	Published 1998	ISO TC22

There is currently only one part to this standard. No other parts have achieved any status in the ISO system.

ISO	IS	14229-1	Road vehicles—Unified diagnostic services (UDS)—Part 1: Specification and requirements	Published 2006	ISO TC22

ISO 14229-1 Available from
http://www.iso.org/iso/iso_catalogue/catalogue_tc/catalogue_detail.htm?csnumber=45293

ISO 14229-1:2006 specifies data link independent requirements of diagnostic services, which allow a diagnostic tester (client) to control diagnostic functions in an on-vehicle ECU (server) such as an electronic fuel injection, automatic gear box, or antilock braking system connected on a serial data link embedded in a road vehicle. It specifies generic services, which allow the diagnostic tester (client) to stop or to resume nondiagnostic message transmission on the data link. ISO 14229:2006 does not apply to nondiagnostic message transmission, or use of the communication data link between two ECU. It does not specify any implementation requirements.

The vehicle diagnostic architecture of ISO 14229:2006 applies to a single tester (client) that may be temporarily or permanently connected to the on-vehicle diagnostic data link and several on-vehicle ECU (servers) connected directly or indirectly.

8.4.4.15 ISO 14230 Road Vehicles—Diagnostic Systems—Keyword Protocol 2000

ISO	IS	14230	Road vehicles—Diagnostic systems—Keyword protocol 2000	Published	ISO TC22

ISO	IS	14230-1	Road vehicles—Diagnostic systems—Keyword protocol 2000—Part 1: Physical layer	Published 1999	ISO TC22

ISO 14230-1 Available from
http://www.iso.org/iso/iso_catalogue/catalogue_tc/catalogue_detail.htm?csnumber=23919

This part of SSF 14230 describes the physical layer on which the diagnostic services will be implemented. It is based on the physical layer described in ISO 9141-2, but expanded to allow for vehicles with either 12V or 24V supply.

ISO	IS	14230-2	Road vehicles—Diagnostic systems—Keyword Protocol 2000—Part 2: Data link layer	Published 1999	ISO TC22

ISO 14230-2 Available from
http://www.iso.org/iso/iso_catalogue/catalogue_tc/catalogue_detail.htm?csnumber=23920

This part of ISO 14230 specifies common requirements of diagnostic services which allow a tester to control diagnostic functions in an on-vehicle ECU (e.g., electronic fuel injection, automatic gearbox, antilock braking system, and so forth) connected on a serial data link embedded in a road vehicle.

Communication between ECUs are not part of this part of ISO 14230.

ISO	IS	14230-3	Road vehicles—Diagnostic systems—Keyword protocol 2000—Part 3: Application layer	Published 1999	ISO TC22

ISO 14230-3 Available from
http://www.iso.org/iso/iso_catalogue/catalogue_tc/catalogue_detail.htm?csnumber=23921

ISO	IS	14230-4	Road vehicles—Diagnostic systems—Keyword Protocol 2000—Part 4: Requirements for emission-related systems	Published 2000	ISO TC22

ISO 14230-4 Available from
http://www.iso.org/iso/iso_catalogue/catalogue_tc/catalogue_detail.htm?csnumber=28826

8.4.4.16 ISO 20119 Road Vehicles—Test Method for the Quantification of On-Centre Handling—Determination of Dispersion Metrics for Straight-Line Driving

ISO	TS	20119	Road vehicles—Test method for the quantification of on-centre handling—Determination of dispersion metrics for straight-line driving	Published TS 2002	ISO TC22

ISO TS 20119 Available from
http://www.iso.org/iso/iso_catalogue/catalogue_tc/catalogue_detail.htm?csnumber=34081

ISO/TS 20119:2002 specifies a test schedule that addresses certain aspects of the on-center handling characteristics of a vehicle—on-center handling being used to describe the steering "feel" and precision of the vehicle during nominally straight-line driving and in negotiating large-radius bends at high speeds but low lateral accelerations. It is applicable to passenger cars in accordance with ISO 3833, and to light trucks.

8.4.4.17 ISO 22239 Road Vehicles—Child Seat Presence and Orientation Detection System

ISO	CD	22239	Road vehicles—Child seat presence and orientation detection system (ISOcare)	Committee Draft DIS: 2007-12	ISO TC22
ISO	CD	22239-1	Road vehicles—Child seat presence and orientation detection system (ISOcare)—Part 1: Specifications and test methods	Committee Draft DIS: 2007-12	ISO TC22

ISO 22239-1 Not Yet Available. Track progress at:
http://www.iso.org/iso/iso_catalogue/catalogue_tc/catalogue_detail.htm?csnumber=40791

ISO	CD	22239-2	Road vehicles—Child seat presence and orientation detection system (ISOcare)—Part 2: Resonator specification	Committee Draft DIS: 2007-12	TC22

ISO 22239-2 Not Yet Available. Track progress at:
http://www.iso.org/iso/iso_catalogue/catalogue_tc/catalogue_detail.htm?csnumber=44905

8.4.4.18 ISO 22240 Road Vehicles—Vehicles Safety Information Model (VSIM)

ISO	TS	22240	Road vehicles—Vehicles safety information model (VSIM)	DIS FDIS: 2007	TC22

ISO CTS 22240 Not Yet Available. Track progress at:
http://www.iso.org/iso/iso_catalogue/catalogue_tc/catalogue_detail.htm?csnumber=40792

8.4.4.19 ISO 27957 Road Vehicles—Temperature Measurement in Anthropomorphic Test Devices—Definition of the Temperature Sensor Locations

ISO	WD	27957	Road vehicles—Temperature measurement in anthropomorphic test devices—Definition of the temperature sensor locations	Working Draft	TC22

ISO 27957 Not Yet Available. Track progress at:
http://www.iso.org/iso/iso_catalogue/catalogue_tc/catalogue_detail.htm?csnumber=36564

8.4.5 Vehicle Dynamics and Road Holding

These standards are principally developed by ISO/TC 22 Road Vehicles and are more concerned with internal vehicle control than ITS systems or systems that will be directly used or affected by ITS service provision. However, there are in addition some SAE standards that may be useful to or used for ITS service provision.

8.4.5.1 ISO 12021 Road Vehicles—Sensitivity to Lateral Wind

ISO	IS	7975	Passenger cars—Braking in a turn—Open-loop test method	Published 1996 Revised 2006	TC22

ISO 7975 Available from
http://www.iso.org/iso/iso_catalogue/catalogue_tc/catalogue_detail.htm?csnumber=37217

ISO 7975:2006 specifies an open-loop test procedure to examine the effect of braking on course holding and directional behavior of a vehicle. Specifically, the procedure determines how the steady-state circular response of a vehicle is altered by a braking action only. ISO 7975:2006 applies to passenger cars as defined in ISO 3833 and to light trucks.

The open-loop maneuver specified in this test procedure is not representative of real driving conditions but is useful to obtain measures of vehicle braking behavior resulting from control inputs under closely controlled test conditions.

8.4.5.2 ISO 8349 Road Vehicles—Measurement of Road Surface Friction

ISO	IS	8349	Road vehicles—Measurement of road surface friction	Published 2002	TC22

ISO 8349 Available from
http://www.iso.org/iso/iso_catalogue/catalogue_tc/catalogue_detail.htm?csnumber=25944

This test method covers the measurement of skid resistance of paved surfaces with a specified full-scale automotive tire.

This test method utilizes a measurement representing the steady-state friction force on a locked test wheel as it is dragged over a wetted pavement surface under constant load and at a constant speed while its major plane is parallel to its direction of motion and perpendicular to the pavement.

8.4.5.3 ISO Road Vehicles—Transient Open-Loop Response Test Method with One Period of Sinusoidal Input

ISO	TR	8725	Road vehicles—Transient open-loop response test method with one period of sinusoidal input	Published TR 1988	TC22

ISO TR 8725 Available from
http://www.iso.org/iso/iso_catalogue/catalogue_tc/catalogue_detail.htm?csnumber=16128

This technical report specifies a method for determining transient response behavior at approximately constant speed. It is not fully representative of real driving conditions but similar to lane change maneuvers in real traffic. It applies to passenger cars as defined in ISO 3833. In a simplified form this test method is also specified in ISO 7401, together with alternative and complementary procedures.

8.4.5.4 ISO 8726 Road Vehicles—Transient Open-Loop Response Test Method with Pseudo-Random Steering Input

ISO	TR	8726	Road vehicles—Transient open-loop response test method with pseudo-random steering input	Published TR 1988	TC22

ISO TR 8726 Available from
http://www.iso.org/iso/iso_catalogue/catalogue_tc/catalogue_detail.htm?csnumber=16129

This technical report specifies a method for determining transient response behavior at approximately constant speed. The quasi-open-loop maneuver used in this method is not representative of real driving conditions but is useful in obtaining measures of vehicle transient behavior in terms that will be enable the response to any deterministic input to be calculated. It applies to passenger cars as defined in ISO 3833. In a simplified form this test method is also specified in ISO 7401, together with alternative and complementary procedures.

8.4.5.5 ISO 8855 Road Vehicles—Vehicle Dynamics and Road-Holding Ability—Vocabulary

ISO	IS	8855	Road vehicles—Vehicle dynamics and road-holding ability—Vocabulary	Published 1991 Add 1992	TC22

ISO 8855 Available from
http://www.iso.org/iso/iso_catalogue/catalogue_tc/catalogue_detail.htm?csnumber=16332
Add 1992 http://www.iso.org/iso/iso_catalogue/catalogue_tc/catalogue_detail.htm?csnumber=16333

This specifies 129 terms and definitions in English and French. It contains the alphabetical indexes in English and French. Annexes A and B are for information only.

8.4.5.6 ISO 9816 Passenger Cars—Power-Off Reactions of a Vehicle in a Turn—Open-Loop Test Method

ISO	IS	9816	Passenger cars—Power-off reactions of a vehicle in a turn—Open-loop test method	Published Standard 1993	TC22

ISO 9816 Available from
http://www.iso.org/iso/iso_catalogue/catalogue_tc/catalogue_detail.htm?csnumber=42037

ISO 9816:2006 specifies open-loop test methods to determine the reactions of a vehicle in a turn to a sudden drop in motive power resulting from release of the accelerator pedal. It applies to passenger cars as defined in ISO 3833.

The open-loop maneuver specified in this test method is not representative of real driving conditions, but is useful to obtain measures of a vehicle's power-off behavior resulting from specific types of control inputs under closely controlled test conditions.

8.4.5.7 ISO 12021 Road Vehicles—Sensitivity to Lateral Wind

There is only one part to this 1996 standard.

ISO	IS	12021-1	Road vehicles—Sensitivity to lateral wind—Part 1: Open-loop test method using wind generator input	Published 1996	TC22

ISO 12021-1 Available from
http://www.iso.org/iso/iso_catalogue/catalogue_tc/catalogue_detail.htm?csnumber=1063

This standard gives an open-loop test method to determine the sensitivity to lateral wind of a vehicle by means of a wind generator. It applies to passenger cars as defined in ISO 3833, passenger car-trailer combinations, and light trucks.

8.4.5.8 ISO 13674 Road Vehicles—Test Method for the Quantification of On-Centre Handling

ISO	IS	13674	Road vehicles—Test method for the quantification of on-centre handling		TC22
ISO	IS	13674-1	Road vehicles—Test method for the quantification of on-centre handling—Part 1: Weave test	Published 2003	TC22

ISO 13674-1 Available from
http://www.iso.org/iso/iso_catalogue/catalogue_tc/catalogue_detail.htm?csnumber=33402

ISO 13674-1:2003 specifies a test schedule that addresses a particular aspect of the on-center handling characteristics of a vehicle: the weave test. It is applicable to passenger cars in accordance with ISO 3833, and to light trucks.

ISO	IS	13674-2	Road vehicles—Test method for the quantification of on-centre handling—Part 2: Transition test	Published 2006	TC22

ISO 13674-2 Available from
http://www.iso.org/iso/iso_catalogue/catalogue_tc/catalogue_detail.htm?csnumber=41146

ISO 13674-2:2006 specifies a test schedule that addresses the transition test, a particular aspect of the on-center handling characteristics of a vehicle. It is applicable to passenger cars in accordance with ISO 3833, and to light trucks.

8.4.5.9 ISO 15037 Road Vehicles—Vehicle Dynamics Test Methods

ISO	IS	15037	Road vehicles—Vehicle dynamics test methods	Published Standard	TC22
ISO	IS	15037-1	Road vehicles—Vehicle dynamics test methods—Part 1: General conditions for passenger cars	Published Standard 1998	TC22

ISO 15037-1 Available from
http://www.iso.org/iso/iso_catalogue/catalogue_tc/catalogue_detail.htm?csnumber=38887

ISO 15037-1:2006 specifies the general conditions that apply when vehicle dynamics properties are determined according to ISO test methods.

ISO 15037-1:2006 is applicable to passenger cars as defined in ISO 3833, and to light trucks.

ISO	IS	ISO 15037-2	Road vehicles—Vehicle dynamics test methods—Part 2: General conditions for heavy vehicles and buses	Published Standard 2002	TC22

ISO 15037-2 Available from
http://www.iso.org/iso/iso_catalogue/catalogue_tc/catalogue_detail.htm?csnumber=28824

ISO 15037-2:2002 specifies the general conditions that apply when vehicle dynamics properties are determined according to ISO test methods carried out on heavy vehicles. These are commercial vehicles, combinations, buses, and articulated buses, as defined in ISO 3833 for trucks and trailers above 3.5-ton and buses above 5-ton maximum weight, and in United Nations Economic Commission for Europe (UNECE) and EC vehicle classification, categories M3, N2, N3, O3, and O4.

8.4.5.10 ISO 16234 Heavy Commercial Vehicles and Buses—Straight-Ahead Braking on Surfaces with Split Coefficient of Friction—Open-Loop Test Method

ISO	IS	16234	Heavy commercial vehicles and buses—Straight-ahead braking on surfaces with split coefficient of friction—Open-loop test method	Published 2006	TC22

ISO DIS 16234 Available from
http://www.iso.org/iso/iso_catalogue/catalogue_tc/catalogue_detail.htm?csnumber=37788

ISO 16234:2006 describes an open-loop test method for determining vehicle reactions during a straight-line braking maneuver on a surface having a split coefficient of friction.

It applies to heavy vehicles—that is, commercial vehicles, commercial vehicle combinations, buses, and articulated buses as defined in ISO 3833 (trucks and trailers with maximum weight above 3.5 ton and buses and articulated buses with maximum weight above 5 ton, according to ECE and EC vehicle classification, categories M3, N2, N3, O3, and O4).

The method is limited to vehicles in which at least the first unit is equipped with an antilock braking system. It is valid for braking with service-brake systems only or in combination with retarders and/or engine brakes.

8.4.5.11 ISO Heavy Commercial Vehicles and Buses—Steady-State Rollover Threshold—Tilt-Table Test Method

ISO	IS	16333	Heavy commercial vehicles and buses—Steady-state rollover threshold—Tilt-table test method	Published 2004	TC22

ISO 16333 Available from
http://www.iso.org/iso/iso_catalogue/catalogue_tc/catalogue_detail.htm?csnumber=37857

ISO 16333:2004 specifies a tilt-table test method for estimating the steady-state rollover threshold of a heavy commercial vehicle or bus (i.e., the maximum lateral acceleration that the test vehicle could sustain in steady-state turning without rolling over). It is applicable to complete roll units/combinations of roll-coupled vehicle units (e.g., single-unit vehicles, tractor-semitrailer combinations, articulated buses, full trailers, B-train combinations) of commercial vehicles, commercial vehicle combinations, buses, or articulated buses as defined in ISO 3833, and under categories M3, N2, N3, O3 and O4 of ECE and EC vehicle regulations (trucks and trailers with maximum weights above 35 ton and buses and articulated buses with maximum weights above 5 ton).

8.4.5.12 ISO 17288 Passenger Cars—Free-Steer Behaviour

ISO	IS	17288	Passenger cars—Free-steer behaviour	Published	TC22

ISO	IS	17288-1	Passenger cars—Free-steer behaviour—Part 1: Steering-release open-loop test method	Published 2002	TC22

ISO 17288-1 Available from
http://www.iso.org/iso/iso_catalogue/catalogue_tc/catalogue_detail.htm?csnumber=33526

ISO 17288-1 specifies an open-loop test method for determining the free control stability of a passenger car as defined in ISO 3833, by measurement of the transient behavior following steering release, starting from a steady-state cornering status.

ISO	IS	17288-2	Passenger cars—Free-steer behaviour—Part 2: Steering-pulse open-loop test method	Published 2004	TC22

ISO 17288-2 Available from
http://www.iso.org/iso/iso_catalogue/catalogue_tc/catalogue_detail.htm?csnumber=30598

ISO 17288-2:2004 specifies a procedure for determining the free control stability of a passenger car as defined in ISO 3833, by measurement of the transient behavior following steering pulse input, starting from a straight-ahead, steady-state status.

8.4.5.13 ISO 20119 Road Vehicles—Test Method for the Quantification of On-Centre Handling—Determination of Dispersion Metrics for Straight-Line Driving

ISO	IS	20119	Road vehicles—Test method for the quantification of on-centre handling—Determination of dispersion metrics for straight-line driving	Published 2002	TC22

ISO 20119 Available from
http://www.iso.org/iso/iso_catalogue/catalogue_tc/catalogue_detail.htm?csnumber=34081

ISO/TS 20119:2002 specifies a test schedule that addresses certain aspects of the on-center handling characteristics of a vehicle—on-center handling being used to describe the steering "feel" and precision of the vehicle during nominally straight-line driving and in negotiating large-radius bends at high speeds but low lateral accelerations.

It is applicable to passenger cars in accordance with ISO 3833, and to light trucks.

8.4.5.14 ISO 21994 Passenger Cars—Stopping Distance at Straight-Line Braking with ABS—Open-Loop Test Method

ISO	IS	21994	Passenger cars—Stopping distance at straight-line braking with ABS—Open-loop test method	Published	TC22

ISO 21994 Available from
http://www.iso.org/iso/iso_catalogue/catalogue_tc/catalogue_detail.htm?csnumber=40588

ISO 21944:2007 specifies an open-loop test method to determine the stopping distance of a vehicle during a straight-line braking maneuver, with ABS fully engaged. It applies to passenger cars as defined in ISO 3833, and to light trucks.

ISO 21944:2007 specifies a reference method and is especially designed to ensure high repeatability.

8.4.5.15 SAE J1113 Immunity to Radiated Electromagnetic Fields

SAE	J1113	Immunity to Radiated Electromagnetic Fields	Published 2004	SAE
SAE	J1113/ 4_200408	Immunity to Radiated Electromagnetic Fields-Bulk Current Injection (BCI) Method	Published 2004	SAE

SAE J1113/4 Available from
http://www.sae.org/technical/standards/J1113/4_200408

This SAE recommended practice defines a method for evaluating the immunity of automotive electrical/electronic devices to radiated electromagnetic fields coupled to the vehicle wiring harness. The method, called bulk current injection (BCI), uses a current probe to inject RF current from 1 to 400 MHz into the wiring harness of automotive devices. BCI is one of a number of test methods that can be used to simulate the electromagnetic field.

SAE	J1113/13	Electromagnetic Compatibility Measurement Procedure for Vehicle Components, Part 13—Immunity to Electrostatic Discharge	Published Revised Nov. 2004	SAE

SAE J1113/13 Available from
http://www.sae.org/technical/standards/J1113/13_200411

This SAE standard specifies the test methods and procedures necessary to evaluate electrical components intended for automotive use to the threat of electrostatic discharges (ESDs). It describes test procedures for evaluating electrical components on the bench in the powered mode and for the packaging and handling nonpowered mode.

A procedure for calibrating the simulator that is used for electrostatic discharges is given in Appendix A. Functional performance status classifications for immunity to ESD are given in Appendix B. Sensitivity classifications for ESD sensitive devices are given in Appendix B.

8.4.5.16 SAE J1213 Glossaries Relating to Vehicles

SAE	J1213	Glossaries relating to vehicles	Published	SAE

SAE	J1213/1	Glossary of Vehicle Networks for Multiplexing and Data Communications	Published SAE	

SAE J1213/1 Available from
http://www.sae.org/technical/standards/J1213/1_199709

This document covers the general terms and corresponding definitions that support the design, development, implementation, testing, and application of vehicle networks.

The terminology also covers some terms and concepts of distributed embedded systems, network hardware, network software, physical layers, protocols, and other related areas.

SAE	J1213/2	Glossary of Reliability Terminology Associated with Automotive Electronics	Published	SAE

SAE J1213/2 Available from
http://www.sae.org/technical/standards/J1213/2_198810

This glossary has been compiled to assist, by serving as a reference, in the communication between the automotive electronics engineer and the reliability engineer. The compilation of terms, acronyms, and symbols was drawn from usage and which should be familiar to those working in automotive electronics reliability. Terms are included which are used to describe how items, materials, and systems are evaluated for reliability, how they fail, how failures are modeled, and how failures are prevented. Terms are also included from the disciplines of designing for reliability, testing, and failure analysis as well as the general disciplines of quality and reliability engineering. This glossary is intended to augment SAE J1213, Glossary of Automotive Electronic Terms.

8.4.5.17 SAE J1455 Joint SAE/TMC Recommended Environmental Practices For Electronic Equipment Design (Heavy-Duty Trucks)

SAE	J1455	Joint SAE/TMC Recommended Environmental Practices For Electronic Equipment Design (Heavy-Duty Trucks)	Published Currently under revision (Work in Progress)	SAE

SAE J1455 Available from
http://www.sae.org/technical/standards/J1455_200606

The climatic, dynamic, and electrical environments from natural and vehicle-induced sources that influence the performance and reliability of vehicle and tractor/trailer electronic components are included in this SAE recommended practice. Test methods that can be used to simulate these environmental conditions are also included. This information is applicable to diesel power trucks in Classes 6, 7, and 8.

This guideline is intended to aid the designer of automotive electronic systems and components by providing material that may be used to develop environmental design goals.

8.4.5.18 SAE J1698 Joint SAE/TMC Electronic Data Interchange Between Microcomputer Systems in Heavy-Duty Vehicle Applications

SAE	J1698	Joint SAE/TMC Electronic Data Interchange Between Microcomputer Systems In Heavy-Duty Vehicle Applications	Published	SAE

SAE J1698 Available from
http://www.sae.org/technical/standards/J1587_200202

This SAE recommended practice defines a document for the format of messages and data that is of general value to modules on the data communications link. Included are field descriptions, size, scale, internal data representation, and position within a message. This document also describes guidelines for the frequency of and circumstances in which messages are transmitted. In order to promote compatibility among all aspects of electronic data used in heavy-duty applications, it is the intention of the SAE Truck and Bus Low Speed Communications Network Subcommittee (formerly Data Format Subcommittee) (in conjunction with other industry groups) to develop recommended message formats for:

- *Vehicle and component information:* This includes all information that pertains to the operation of the vehicle and its components (such as performance, maintenance, and diagnostic data).
- *Routing and scheduling information:* Information related to the planned or actual route of the vehicle. It includes current vehicle location (e.g., geographical coordinates) and estimated time of arrival.
- *Driver information:* Information related to driver activity. It includes driver identification, logs (e.g., DOT), driver expenses, performance, status, and payroll data.
- *Freight information:* Provides data associated with cargo being shipped, picked up, or delivered. It includes freight status, overage, shortage and damage reporting, billing and invoice information, and customer and consignee data.

This document represents the recommended formats for basic vehicle and component identification and performance data. This document is intended as a guide towards standard practice and is subject to change to keep pace with experience and technical advances.

8.4.5.19 SAE J1698 Vehicle Event Data Interface

SAE	J1698	Vehicle Event Data Interface-Vehicular Data Extraction	Published May 2004	SAE

SAE J1698 Available from
http://www.sae.org/technical/standards/J1698_200502

This recommended practice aims to establish a common format for displaying and presenting crash-related data recorded and stored within certain electronic components currently installed in many light-duty vehicles. This recommended practice pertains only to the post-download format of such data and is not intended to standardize the format of the data stored within any onboard storage unit, or to standardize the method of data recording, storing, or extraction.

SAE	J1698/ 1_200405	Vehicle Event Data Interface—Output Data Definition	Published Mar. 2005	SAE

SAE J1698/1 Available from
http://www.sae.org/technical/standards/J1698/1_200503

This document is part of the J1698 document family and provides the definitions for event-related data items.

SAE	J1698/ 2_200405	Vehicle Event Data Interface-Vehicular Data Extraction	Published May 2004	SAE

SAE J1698/2 Available from
http://www.sae.org/technical/standards/J1698/2_200405

This recommended practice is intended to define a common method for determining how to extract event data from a motor vehicle, including the event data set needed to output the event record of data elements defined in SAE J 1698. It is intended for use by those developing tools for the purpose of event data set extraction.

This recommended practice aims to utilize existing industry standards to define a common physical interface and the protocols necessary for event data set extraction. To accomplish this, the SAE J 1962 Diagnostics Connector has been designed as the primary physical interface and associated industry standard diagnostic protocols have been designated for communications.

8.4.5.20 SAE J1708 Serial Data Communications Between Microcomputer Systems in Heavy-Duty Vehicle Applications

SAE	J1708_200408	Serial Data Communications Between Microcomputer Systems in Heavy-Duty Vehicle Applications	Published Aug. 2004 SAE	SAE

J1708 Available from
http://www.sae.org/technical/standards/J1708_200408

This SAE recommended practice document defines a recommended practice for implementing a bidirectional, serial communication link among modules containing microcomputers. This document defines those parameters of the serial link that relate primarily to hardware and basic software compatibility such as interface requirements, system protocol, and message format. The actual data to be transmitted by particular modules, which is an important aspect of communications compatibility, is not specified in this document. These and other details of communications link implementation and use should be specified in the separate application documents referenced in Section 2 of the document.

8.4.5.21 SAE J1760 Data Security Services

SAE	J1760_200112	Data Security Services	Published Dec. 2001	SAE

SAE J1760 Available from
http://www.sae.org/technical/standards/J1760_200112

The scope of this SAE recommended practice is to require the use of the same security services as defined by the International Standard ISO/CD 15764, modified by the class of security as determined by the resource provider and referenced in Table 1, Extended Data Link Security References.

8.4.5.22 SAE J1843 Accelerator Pedal Position Sensor for Use with Electronic Controls in Medium- and Heavy-Duty Vehicle Applications

SAE	J1843	Accelerator Pedal Position Sensor for Use with Electronic Controls in Medium- and Heavy-Duty Vehicle Applications	Published Currently under revision (Work in Progress)	SAE

SAE J1843 Available from
http://www.sae.org/technical/standards/J1843_199304

The purpose of this SAE recommended practice is to provide a common electrical and mechanical interface specification that can be used to design electronic accelerator pedal position sensors and electronic control systems for use in medium- and heavy-duty vehicle applications.

8.4.5.23 SAE J1930 Electrical/Electronic Systems Diagnostic Terms, Definitions, Abbreviations, and Acronyms

SAE	J1930	Electrical/Electronic Systems Diagnostic Terms, Definitions, Abbreviations, and Acronyms	Published	SAE

SAE J1930 Available from
http://www.sae.org/technical/standards/J1930_200204

This SAE recommended practice is applicable to all light-duty gasoline and diesel passenger vehicles and trucks, and to heavy-duty gasoline vehicles. Specific applications of this document include diagnostic, service and repair manuals, bulletins and updates, training manuals, repair data bases, under-hood emission labels, and emission certification applications.

This document focuses on diagnostic terms applicable to electrical/electronic systems, and therefore also contains related mechanical terms, definitions, abbreviations, and acronyms.

8.4.5.24 SAE J1939 Recommended Practice for a Serial Control and Communications Vehicle Network

SAE	J1939_200501	Recommended Practice for a Serial Control and Communications Vehicle Network	Published Jan. 2005	SAE

SAE J 1930series Available from links below And from
http://www.sae.org/technical/standards/J1939_200501

The SAE J1939 documents are intended for light, medium, and heavy-duty vehicles used on or off road as well as appropriate stationary applications which use vehicle derived components (e.g., generator sets). Vehicles of interest include, but are not limited to, on- and off-highway trucks and their trailers, construction equipment, and agricultural equipment and implements.

The purpose of these documents is to provide an open interconnect system for electronic systems. It is the intention of these documents to allow ECUs to communicate with each other by providing a standard architecture.

The J1939 standards family is maintained to ensure CAN device interoperability. J1939, the top-level document in the collection, is the master control for definitions common to many applications and industries, while companion documents explain component rationalization and product standardization for a particular application or industry. Communication systems designed according to J1939 standards are EMI/RFI tolerant, free of connection wires, easy to install, and feature log, record, remote access, and self-diagnosis capabilities.

These standards can be downloaded via the Web but are not free of charge.

Core J1939 Standards J1939 Recommended Practice for a Serial Control and Communications Vehicle Network Revised Jan 05
http://www.sae.org/technical/standards/J1939_200501

J1939-01 Recommended Practice for Control And Communications Network for On-Highway Equipment
http://www.sae.org/technical/standards/J1939/1_200009 J1939-11

Physical Layer—250k bits/s, Twisted Shielded Pair
http://www.sae.org/technical/standards/J1939/11_200609

J1939-13 Off-Board Diagnostic Connector Revised Mar 04
http://www.sae.org/technical/standards/J1939/13_200403

J1939-15 Reduced Physical Layer, 250K bits/sec, Un-Shielded Twisted Pair (UTP)
http://www.sae.org/technical/standards/J1939/15_200311

J1939-21 Data Link Layer
http://www.sae.org/technical/standards/J1939/21_200612

J1939-31 Network Layer Revised Apr 04
http://www.sae.org/technical/standards/J1939/31_200404

J1939-71 Vehicle Application Layer Revised Jun 06
http://www.sae.org/technical/standards/J1939/71_200611 J1939-73

Application Layer—Diagnostics Revised Sep 06
http://www.sae.org/technical/standards/J1939/73_200609

J1939-74 Application—Configurable Messaging Revised Sep 04
http://www.sae.org/technical/standards/J1939/74_200611

J1939-75 Application Layer—Generator Sets and Industrial
http://www.sae.org/technical/standards/J1939/75_200706

J1939-81 Network Management
http://www.sae.org/technical/standards/J1939/81_200305

Also included: Related Technical Papers

931809 J1939 High Speed Serial Communications, the Next Generation Network for Heavy Duty Vehicles
http://www.sae.org/technical/papers/931809

940361 CAN Controller Architecture Optimized for SAE J1939 Applications
http://www.sae.org/technical/papers/940361

950043 Development Tools for SAE J1939 Networks on Tractors
http://www.sae.org/technical/papers/950043

972757 J1939 Serial Vehicle Network Explanation and Tutorial
http://www.sae.org/technical/papers/972757

972758 System Level Testing Using the J1939 Datalink Adapter
http://www.sae.org/technical/papers/972758

972759 Dynamic Address Configuration in SAE J1939
http://www.sae.org/technical/papers/972759

1999-01-2840 Large-Scale Application of J-1939 CAN
http://www.sae.org/technical/papers/1999-01-2840

Also included: Parameter Tool

CS1939 J1939 Companion Spreadsheet
http://www.sae.org/servlets/otherProduct?PROD_TYP=SFTWR&PROD_CD=CS1939_032006

8.4.5.25 SAE J1979 E/E Diagnostic Test Modes

SAE	J1979	E/E Diagnostic Test Modes	Published	SAE

SAE J1979 Available from
http://www.sae.org/technical/standards/J1979_200705

This document supersedes SAE J1979 Apr 2002, and is technically equivalent to ISO 15031-5:2006, with the addition of new capabilities required by revised regulations from the California Air Resources Board (see Section 1.2 of the document). This document is intended to satisfy the data reporting requirements of onboard diagnostic (OBD) regulations in the United States and Europe, and any other region that may adopt similar requirements in the future. This document specifies:

- Message formats for request and response messages;
- Timing requirements between request messages from external test equipment and response messages from vehicles, and between those messages and subsequent request messages;
- Behavior of both the vehicle and external test equipment if data is not available;
- A set of diagnostic services, with corresponding content of request and response messages, to satisfy OBD regulations.

This document includes capabilities required to satisfy OBD requirements for multiple regions, model years, engine types, and vehicle types. Those regulations are not yet final for some regions, and are expected to change in the future.

8.4.5.26 SAE J2403 Medium/Heavy-Duty E/E Systems Diagnosis Nomenclature

SAE	J 2403	Medium/Heavy-Duty E/E Systems Diagnosis Nomenclature	Published Revised Aug. 2004	SAE

SAE J2403 Available from
http://www.sae.org/technical/standards/J2403_200708

This SAE recommended practice is applicable to all E/E systems on MD and HD vehicles. The terms defined are largely focused on compression-ignited and spark-ignited engines. Specific applications of this document include diagnostic, service, and repair manuals, bulletins and updates, training manuals, repair data bases, under hood emission labels, and emission certification applications. This document focuses on diagnostic terms, definitions, abbreviations, and acronyms applicable to E/E systems. It also covers mechanical systems which require definition.

8.4.5.27 SAE J2496 Transport Area Network Cabling

SAE	J2496	Transport Area Network Cabling	Published	SAE

SAE J2496 Available from
http://www.sae.org/technical/standards/J2496_200702

This series of SAE recommended practices was developed to provide an open architecture system for onboard electronic systems. It is the intention of these documents to allow electronic devices to communicate with each other by providing a standard architecture. This particular document describes the network interface and cabling which defines the requirements needed for communicating between devices that are on different segments of the SAE J2496 Transport Area Network. While these recommended practices may be used in retrofitting older vehicles, the primary intent is for implementation in new bus procurements.

8.4.5.28 SAE J2497 Power Line Carrier Communications for Commercial Vehicles

SAE	J2497	Power Line Carrier Communications for Commercial Vehicles	Published	SAE

SAE J2497 Available from
http://www.sae.org/technical/standards/J2497_200706

See Section 8.4.1.15.

8.4.5.29 SAE J2178 Class B Data Communication Network Messages

SAE	J2178/ 3_200407	Class B Data Communication Network Messages—Part 3—Frame IDs for Single-Byte Forms of Headers	Published July 2004	SAE

SAE J2178/3 Available from
http://www.sae.org/technical/standards/J2178/3_200407

This SAE recommended practice defines the information contained in the header and data fields of nondiagnostic messages for automotive serial communications based on SAE J1850 Class B networks. This document describes and specifies the header fields, data fields, field sizes, scaling, representations, and data positions used within messages. The general structure of a SAE J1850 message frame without in-frame response is shown in Figure 1 of that standard. The structure of a SAE J1850 message with in-frame response is shown in Figure 2 of that standard. Figures 1 and 2 also show the scope of frame fields defined by this document for nondiagnostic messages.

Refer to SAE J1979 for specifications of emissions related diagnostic message header and data fields.

Refer to SAE J2190 for the definition of other diagnostic data fields.

8.4.5.30 SAE J2178 Class B Data Communication Network Messages

SAE	J2178	Class B Data Communication Network Messages	Published July 2004	SAE
SAE	J2178/ 2_200407	Class B Data Communication Network Messages—Part 2: Data Parameter Definitions	Published July 2004	SAE

SAE J2178/2 Available from
http://www.sae.org/technical/standards/J2178/2_200407

This SAE recommended practice defines the information contained in the header and data fields of nondiagnostic messages for automotive serial communications based on SAE J1850 Class B networks. This document describes and specifies the header fields, data fields, field sizes, scaling, representations, and data positions used within messages. The general structure of a SAE J1850 message frame without in-frame response is shown in Figure 1 of that standard. The structure of a SAE J1850 message with in-frame response is shown in Figure 2 of that standard. Figures 1 and 2 of the standard also show the scope of frame fields defined by this document for nondiagnostic messages.

Refer to SAE J1979 for specifications of emissions related diagnostic message header and data fields.

Refer to SAE J2190 for the definition of other diagnostic data fields.

8.4.5.31 SAE J2178 Class B Data Communication Network Messages—Message Definitions for Three Byte Headers

SAE	J2178/ 4_200407	Class B Data Communication Network Messages—Message Definitions for Three Byte Headers	Published July 2004	SAE

SAE J2178/4 Available from
http://www.sae.org/technical/standards/J2178/4_200407

This SAE recommended practice defines the information contained in the header and data fields of nondiagnostic messages for automotive serial communications based on SAE J1850 Class B networks. This document describes and specifies the header fields, data fields, field sizes, scaling, representations, and data positions used within messages. The general structure of a SAE J1850 message frame without in-frame response is shown in Figure 1 of that standard. The structure of a SAE J1850 message with in-frame response is shown in Figure 2 of that standard.

Figures 1 and 2 of the standard also show the scope of frame fields defined by this document for nondiagnostic messages.

Refer to SAE J1979 for specifications of emissions related diagnostic message header and data fields.

Refer to SAE J2190 for the definition of other diagnostic data fields.

8.4.5.32 SAE J2186 E/E Data Link Security

SAE	J2186_200506	E/E Data Link Security	Published June 2005	SAE

SAE J2186 Available from
http://www.sae.org/technical/standards/J2186_200506

This SAE recommended practice establishes a uniform practice for protecting vehicle components from "unauthorized" access through a vehicle data link connector (DLC). The document defines a security system for motor vehicle and tool manufacturers. It will provide flexibility to tailor systems to the security needs of the vehicle manufacturer.

The vehicle modules addressed are those that are capable of having solid-state memory contents accessed or altered through the data link connector. Improper memory content alteration could potentially damage the electronics or other vehicle modules; could risk the vehicle compliance to government requirements; or could risk the vehicle manufacturer's security interests. This document does not imply that other security measures are not required or possible.

8.4.5.33 SAE J2366 ITS Data Bus

SAE	J2366	Its Data Bus	Published Nov. 2001	SAE
SAE	J2366/ 1_200111	Its Data Bus—Idb-C Physical Layer	Published Nov. 2001	SAE

SAE J2366/1 Available from
http://www.sae.org/technical/standards/J2366/1_200111

This SAE recommended practice details the physical layer of the ITS data bus on CAN (IDB-C), which is generally intended for in-vehicle use. It has been developed by the ITS Data Bus (IDB) Physical Layer Subcommittee of the IDB Committee. The objectives of the subcommittee are to develop information reports, recommended practices, and standards concerned with the requirements, design, and usage of devices that communicate electronic signals and control information among ITS related components within the vehicle environment.

SAE	J2366/ 2_200111	Link Layer	Published Nov. 2001	SAE

SAE J2366/2 Available from
http://www.sae.org/technical/standards/J2366/2_200111

This SAE recommended practice details the link layer of the ITS data bus, which is generally intended for in-vehicle use. The IDB is a nonproprietary token passing bus, designed to allow disparate consumer, vehicle, and commercial electronic components to communicate and share information across a standard, open data bus. This first version of the IDB is called IDB-C.

This recommended practice describes the link layer of the IDB-C, as shown in Section 1 of the recommended practice. Below the link layer of the IDB-C is the CAN 2.0B link layer. The physical layer of the IDB-C incorporates the physical layer as specified by SAE J2366-1 and J2590.

SAE	J2366/ 4_200203	Its Data Bus—Thin Transport Layer	Published Mar. 2002	SAE

SAE J2366/4 Available from
http://www.sae.org/technical/standards/J2366/4_200203

This SAE recommended practice details the thin transport layer of the ITS data bus, which is generally intended for in-vehicle use. The thin transport layer sits between SAE J2366-2 and J2366-7. It provides the handling of such activities as the packetizing of long messages and message reassembly. Design of the messages and headers has stressed economy, in terms of bits within a CAN 2.0B frame.

The IDB is a nonproprietary virtual token passing bus, designed to allow disparate consumer, vehicle, and commercial electronic components to communicate and share information across a standard, open data bus.

SAE	J2366/ 7_200204	Its Data Bus—Application Message Layer	Published Apr. 2002	SAE

SAE J2366/7 Available from
http://www.sae.org/technical/standards/J2366/7_200204

NOTE: SAE J2366-7LX—ITS Data Bus Application Message Layer Lexicon is now combined (as an appendix) with SAE J2366-7 to provide for a quicker revision process in the future.

This SAE surface vehicle recommended practice defines an application message layer, which may be used as part of a complete protocol stack with the other protocol layers in the SAE J2366 family. The application message layer provides application message support for devices that are interconnected via a bus or network. Design of the messages and headers has stressed flexibility, expandability, economy (in terms of octets on the bus), and reusability.

8.4.5.34 SAE J2395 ITS In-Vehicle Message Priority

SAE	J2395_200202	ITS In-Vehicle Message Priority	Published Feb. 2002	SAE

SAE J2395 Available from
http://www.sae.org/technical/standards/J2395_200202

This SAE recommended practice applies to both OEM and aftermarket ITS message-generating systems for passenger vehicles and heavy trucks. The recommended practice describes the method for prioritizing ITS in-vehicle messages and/or displayed information based on a defined set of criteria.

Each criterion has a fixed number of levels that are used to rate/rank a given message or information item to determine its prioritization value. The prioritization value is used to determine the priority in which simultaneous, or overlapping, in-vehicle messages are presented to the driver.

8.4.5.35 SAE J2556 Radiated Emissions (RE) Narrowband Data Analysis—Power Spectral Density (PSD)

SAE	J2556_200503	Radiated Emissions (RE) Narrowband Data Analysis—Power Spectral Density (PSD)	Published Mar. 2005	SAE

SAE J2556 Available from
http://www.sae.org/technical/standards/J2556_200503

This SAE information report defines a procedure for indicating the severity of narrowband emissions from an electronic system-component.

8.4.5.36 SAE J2590 Pmode for In-Vehicle Networks

SAE	J2590_200112	Pmode for In-Vehicle Networks	Published Dec. 2001	SAE

SAE J2590 Available from
http://www.sae.org/technical/standards/J2590_200112

This SAE recommended practice describes the power mode requirements for in-vehicle networks that conform to the Automotive Multimedia Interface Collaboration (AMI-C) specifications. These networks include, but are not limited to, the IDB-C (SAE J2366), IDB-1394, and MOST. This version of the document covers primarily IDB-C and may be revised when the PMODE requirements for the other networks are more fully developed by AMI-C.

8.4.5.37 SAE USCAR17 Performance Specification for Automotive RF Connector Systems

SAE	USCAR17-2	Performance Specification for Automotive RF Connector Systems	Published Nov. 2004/2	SAE

SAE USCAR 17-2 Available from
http://www.sae.org/technical/standards/USCAR17-2

Procedures included within this specification are intended to cover performance testing at all phases of development, production, and field analysis of electrical terminals, connectors, and components for coaxial cable connection systems (hereafter referred to as RF connectors) intended for road vehicle applications.

The intent of this specification is to qualify RF connectors that operate at frequencies greater than 200 MHz. This does not exclude lower frequency applications; the acceptance criteria may not apply, however (i.e., the AM/FM acceptance criteria is specified on Page 3 of USCAR print 999-U-001-1-A01).

This specification does not apply to single conductor wire or twisted pair connection systems.

8.5 Freight Transport Service Groups

The service groups addressed in the Freight Transport domain specifically address activities that facilitate both commercial vehicle operations and multimodal logistics, including interjurisdictional coordination.

8.5.1 Architecture

CEN	TR 00278094	Freight and Fleet Management Systems—Reference architecture and terminology—Part 1: High level architecture and terms	Deleted

This work item has been deleted, and the working group is dormant/closed.

8.5.2 Commercial Vehicle Preclearance

The commercial vehicle preclearance service group provides services that enable commercial vehicles, including trucks and buses, to have credentials and other documents, safety status, and weights checked automatically at normal road speeds. A principal objective of this service is to effect preclearances with minimal disruption to the vehicle journey and the traffic flow.

Examples services include:

- Weigh-in-motion;
- Nonstop preclearance;
- Vehicle safety records monitoring.

There are no International Standards for this activity at this point in time; however, the U.S. program International Border Electronic Clearance is worth referencing. It also provides links to more local implementations in the United States. The Web site can be found at:

http://www.calccit.org/itsdecision/serv_and_tech/Freight_operations/freight.html#border.

The International Border Clearance (IBC) program was initiated by the U.S. DOT to facilitate faster, safer vehicle inspections at border crossings by reducing the need for manual inspections and evaluations of trucks entering the United States. Without some form of automation, inspections become necessarily rushed at peak travel times, making it easier for customs and safety violations to go undetected.

Before a commercial vehicle arrives at a border, trading firms (typically through customs brokers) electronically file the appropriate declarations to U.S. Customs through the North American Trade Automation Prototype (NATAP) in Washington, D.C. In current tests, the Freight and Trade Processing System (FTPS) receives this data and makes it available to state motor vehicle enforcement agencies that perform credentials and safety screening of the carrier, vehicle, and driver. Each agency returns its prearrival screening results to the border via the FTPS.

When a commercial vehicle arrives at a border crossing, an onboard DSRC transponder transmits its unique identifier to a roadside reader. This reader communicates with the local trade processing system, which forwards the unique identifier to NATAP. NATAP then retrieves the appropriate prearrival screening results and returns the results to the customs inspector. The inspector may use that information to determine if further inspections are necessary.

8.5.3 ISO 24533 Commercial Vehicle Administrative Processes

This service group is complementary to the above service group. It enables haulers and shippers to purchase annual and ad hoc credentials, using communications and computer technologies.

Example services include:

- Automated credential filing;
- Automated commercial vehicle administration;
- Automated border crossings.

ISO	CD	24533	Intelligent Transport Systems—Data dictionary and message set to facilitate the movement of freight and its intermodal transfer—Road transport information exchanges	Under Development DIS Ballot	ISO TC204

ISO 24533 Not Yet Available. Track progress at:
http://www.iso.org/iso/iso_catalogue/catalogue_tc/catalogue_detail.htm?csnumber=37505

See Section 7.3.4.1.

8.5.4 ISO 26683 Freight Conveyance Content Identification and Communication Architecture—Application Profile

ISO	CD	26683	Freight conveyance content identification and communication architecture—Application profile	Under Development	ISO TC204

ISO 26683 Not Yet Available

This International Standard defines the aggregated data structure for communications between cargo elements and infrastructure in the multimodal environment. Cargo elements include: vehicle, container, pallet, carton, and items.

This International Standard is based on current standards for onboard equipment, RF-tags, and other identifiers.

This International Standard is also going to define the integrated architecture and relevant application and security profile needed to track and trace cargo.

This International Standard covers the road environment and cooperative transfers to and from other transport modes.

See also Section 8.5.5.

8.5.5 Automated Roadside Safety Inspection

The automated roadside safety inspection service group covers the use of ITS functionality to enable roadside access to safety performance records of haulers, vehicles, and drivers. This will enhance existing systems of spot checks by providing inspectors with easy access to current data relevant to the inspection.

Example services include: remote access to commercial vehicle safety data.

As yet, there are no International Standards in this domain; however, some U.S. ANSI standards and technical specifications exist.

ANSI	TS	284	Commercial Vehicle Safety Reports	Published 1998	ANSI

ANSI TS 284 Available from
www.disa.org www.ansi.org

This involves an electronic data interchange (EDI) transaction set to allow authorized parties to electronically request and send reports on information related to the safe operation of commercial road vehicles, such as inspection reports, safety and compliance review reports, and hazardous material incident reports.

ANSI	TS	285	Commercial Vehicle Safety and Credentials Information Exchange	Published 2006	ANSI

ANSI TS 285 Available from
www.disa.org www.ansi.org

This involves an EDI transaction set to permit enforcement officials, government administrators, and other authorized parties to retrieve information electronically on the safety performance, regulatory compliance, and credentials status of commercial motor vehicles, carriers, and drivers.

ANSI	TS	286	Commercial Vehicle Credentials	Published 1997	ANSI

ANSI TS 286 Available from
www.disa.org www.ansi.org

This involves an EDI transaction set that can be used by owners, leasers, and drivers of commercial motor vehicles to apply electronically for credentials necessary to legally operate those vehicles, and by authorizing jurisdictions to electronically transmit credential data to applicants and other authorized entities.

8.5.6 Commercial Vehicle On-Board Safety Monitoring

This service group covers the use of onboard monitoring systems to oversee the safety status of commercial vehicles, commercial vehicle drivers, and cargo during the entire course of the trip. This may include sensing and collecting data on the following:

- Brakes;
- Driver alertness;
- Driving time;
- Lights;
- Shifted cargo;
- Tires.

Warning may be provided to both the driver and/or remote monitoring facilities. Example services include:

- Commercial vehicle internal systems monitoring;
- Commercial vehicle driver alertness monitoring.

There are currently no standards published or known to be under development for the service group.

8.5.7 Freight Transport Fleet Management

At a multimodal level, commercial fleet management includes logistics and freight management systems. It also covers the use of automatic vehicle location (AVL) to achieve automatic freight carrier location/container location and the use of vehicle-to-control center communications to provide vehicle location and other status information to the fleet operators. This facilitates the use of dynamic dispatching systems to improve the efficiency of the fleet management process. These services are implemented in conjunction with the traffic management service group services. This service group includes:

- Pretrip information;
- Intermodal terminal conditions.

Example services include:

- Commercial vehicle fleet tracking;
- Commercial vehicle fleet dispatching;
- Freight container tracking.

At this point in time these capabilities are provided as proprietary commercial products and few relevant International Standards exist.

See Sections 7.3.2.4 through 7.3.2.6, 7.3.4.1, 8.2.1.8, and 8.5.4.

8.5.7.1 ISO 9897 Freight Containers—Container Equipment Data Exchange (CEDEX)—General Communication Codes

ISO	IS	9897	Freight containers—Container equipment data exchange (CEDEX)—General communication codes	Published 1997	TC104

ISO 9897 Available from
http://www.iso.org/iso/iso_catalogue/catalogue_tc/catalogue_detail.htm?csnumber=20455

See Section 7.3.4.8.

8.5.7.2 ISO 9711 Freight Containers—Information Related to Containers On Board Vessels—Part 1: Bay Plan System

ISO	IS	9711-1	Freight containers—Information related to containers on board vessels—Part 1: Bay plan system	Published Standard 1990	TC104

ISO 9711 Available from
http://www.iso.org/iso/iso_catalogue/catalogue_tc/catalogue_detail.htm?csnumber=17568

See Section 7.3.4.7

8.5.8 Intermodal Information Management

The ITS reference model architecture for transport information and control systems (ISO TR 14813 Part 2: Core TICS reference architecture) identifies a commercial vehicle functional domain including:

> . . .transactions to maintain the TICS information about a shipment from the time of the order by the consignor to the reception of goods by the consignee. The key TICS transactions are to provide registers of service providers and to enable the goods to be tracked throughout intermodal journeys.

This service group covers the exchange of information about transport of goods across modes. This includes knowledge of where the units transporting the goods are located, plus their condition and status, as well as similar information about the vehicle transporting the unit. It is also possible to locate subunits and provide customers with information about the progress with the movement of the goods.

Example services include:

- Vehicle and container arrival information exchange (users are fleet and intermodal carriers and nodes);
- Customer freight information access (users are customers and shippers).

See Sections 7.3.2.4 through 7.3.2.6, 7.3.4.1, 8.2.1.8, 8.5.4, and 8.5.7.

8.5.9 Management and Control of Intermodal Centers

This service group includes services that manage the operation of the intermodal center, including those concerned with parking, operation and maintenance of buildings and equipment, the operation of the internal infrastructure, and the interfaces to the various modes in the external infrastructure. This differs from the previous service group (intermodal information management), in that it provides management and control capabilities based on the information collected and received. This includes managing the operation of the modal interchanges, and managing the personnel concerned with the transportation of the goods.

Example services include:

- Intermodal center facility management;
- Intermodal vehicle and container control.

NOTE: This service group includes services which cross modal boundaries. These services are therefore being (or will be) developed in conjunction with other standardization committees or by other standardization committees in conjunction with ISO/TC 204.

No International Standards exist as yet.

8.5.10 Management of Dangerous Freight

This service group includes services that manage the operation of transport fleets concerned with the movement of dangerous goods, including the monitoring of its status and condition, and its movement along the infrastructure of the transport modes that will be used. Also, the activities include exchange of information with organizations responsible for the actual transportation of the dangerous goods. Example services include:

- Dangerous goods movement data sharing;
- Dangerous goods movement data registry;
- Dangerous goods movement fleet coordination;
- Dangerous goods movement police/safety coordination.

8.5.10.1 ISO 17687 General Fleet Management and Commercial Freight Operations—Data Dictionary and Message Sets for Electronic Identification and Monitoring of Hazardous Materials/Dangerous Goods Transportation

ISO	IS	17687	Transport Information and Control Systems (TICS)—General fleet management and commercial freight operations—Data dictionary and message sets for electronic identification and monitoring of hazardous materials/ dangerous goods transportation	Published 2007	ISO TC204

ISO 17687 Available from
http://www.iso.org/iso/iso_catalogue/catalogue_tc/catalogue_detail.htm?csnumber=33155

ISO 17687:2007 supports the application of automated identification, monitoring, and exchange of emergency response information regarding dangerous goods carried onboard road transport vehicles. Such information may include the identification, quantity, and current condition (such as pressure and temperature) of such goods, as well as any relevant emergency response information. When equipped with appropriate electronics and communications capabilities, vehicles carrying dangerous goods may respond to queries regarding their status or self-initiate a message.

The information defined here, electronically carried onboard the road transport vehicle, may be transferred to interested roadside systems by whatever communications means are appropriate to that roadside system.

The primary intent of ISO 17687:2007 is not trade, economic, or commercial, but to help save lives by facilitating emergency response. ISO 17687:2007 supports local on-site needs in the same manner as conventional visual placards do but with an optional, complementary, enhanced, and more versatile electronic version.

8.6 Public Transport Service Groups

The service groups within this domain describe activities that result in more timely and efficient operation of public transport services and provision of operational information to the operator and passenger.

8.6.1 Public Transport Management

This service group covers the application of ITS functionality to the operation, planning, and management of public transport operations. It includes the provision of real-time information on vehicle location and status, enabling the identification of departures from schedules and dynamic rescheduling. This also includes the monitoring of public transport vehicle status such as passenger loadings, engine management system functions, and tire pressures. This service group also includes the application of public transport scheduling and planning systems for assuring reliable connections with minimum time between different modes (e.g., bus and rail services). This is often called *transfer connection protection*, and can be addressed through one or more of the example services below. These services are implemented in conjunction with the traffic management service group services. Example services include:

- Public transport vehicle internal systems monitoring;
- Public transport vehicle fleet tracking;
- Public transport scheduling services;
- Public transport service dispatch;
- Public transport service planning.

8.6.1.1 ISO 24014 Public Transport—Interoperable Fare Management System [IFMS] Architecture

ISO	NP	24014	Public transport—Interoperable fare management system (IFMS) architecture	Published 2007	CEN TC278

ISO/CEN	DIS	24014-1	Interoperable fare management system—Part 1: Architecture	Published 2007	CEN TC278

ISO 24014-1 Available from
http://www.iso.org/iso/iso_catalogue/catalogue_tc/catalogue_detail.htm?csnumber=41985

ISO 24014-1:2007 provides the basis for the development of multioperator/multiservice interoperable public surface (including subways) transport fare management systems (IFMSs) on a national and international level.

ISO 24014-1:2007 is applicable to bodies in public transport and related services which agree that their systems need to interoperate.

While this standard does not imply that existing interoperable fare management systems need to be changed, it applies, so far as it is practically possible, to extensions of these.

The standard covers the definition of a conceptual framework, which is independent of organizational and physical implementation. Any reference within ISO 24014-1:2007 to organizational or physical implementation is purely informative.

ISO/CEN	DIS	24014-2	Public transport—Interoperable fare management system—Part 2: Recommended business practices for set of rules	Under Development	CEN TC278

ISO 24012-2 Not Yet Available. Track progress at:
http://www.iso.org/iso/iso_catalogue/catalogue_tc/catalogue_detail.htm?csnumber=50127

8.6.1.2 ISO 22951 Data Dictionary and Message Sets for Pre-Emption and Prioritization Signal Systems for Emergency and Public Transport Vehicles (PRESTO)

ISO	IS	22951	Data Dictionary and Message Sets for Pre-emption and Prioritization Signal Systems for Emergency and Public Transport Vehicles (PRESTO)	ISO TC204

ISO 22951 Available from
http://www.iso.org/iso/iso_catalogue/catalogue_tc/catalogue_detail.htm?csnumber=46009

The scope of standardization includes message sets and a data dictionary related to the communications as follows:

- Between a roadside communication unit and each in-vehicle unit,
- Between roadside communication unit and roadside units, in-vehicle units, and roadside units.

The standardization targets only information related to priority signal control and does not deal with information provision such as that of the situations at scenes. Since it is necessary to handle public transport vehicles in accordance with the conditions of individual cities and regions, the section in the messages and the data dictionary that are concerned with priority signal control for the vehicles are treated as an option. Furthermore, the standardization shall not depend on the type of communication medium used.

8.6.1.3 ISO 28701 Public Transport—Identifications of Fixed Objects in Public Transport (IFOPT)

ISO	WD/ TS	28701	Public Transport—Identifications of Fixed Objects in Public Transport (IFOPT)	Under Development	ISO TC204

ISO 28701 Not Yet Available. Track progress at:
http://www.iso.org/iso/iso_catalogue/catalogue_tc/catalogue_detail.htm?csnumber=44887

This work item failed ballot for lack of participating countries. Its future is therefore uncertain.

8.6.1.4 CEN 12896 Public Transport—Reference Data Model

CEN	ENV	12896	Public transport—Reference data model	Published 1997. Revised and republished 2006	CEN TC278

Text Downloadable FOC www.transmodel.org	Transmodel Documentation

See Section 5.2.10.

8.6.2 Demand Response and Shared Transport

This service group covers the provision of on-demand transport services to individual travelers. This will provide demand responsive transport services to the user, while enabling transport operators to dispatch and schedule vehicles.

Typically, travelers may request service by specifying destination and any special needs such as pram conveyancing, wheelchair lifts, or other special services for the disabled. Vehicles, which cover a corridor or area, are then dispatched to the traveler by a dispatching system. The public transport fleet deployed on this service group may include buses, vans, and taxis.

This service group addresses the needs of commuters by providing a viable shared transport alternative to the single occupancy private car and also addresses the needs of specific groups such as the elderly and disabled. Example services include:

- Paratransit fleet dispatch;
- Dynamic ridesharing.

8.6.3 Public Transport Information

8.6.3.1 ISO 17685 Standards Numbering System for Public Transport Stops (SNSPTS)

ISO	NP	17685	Intelligent Transport Systems—Standards Numbering System for Public Transport Stops (SNSPTS)	Under Development	ISO TC204

ISO 17685 Not Yet Available

This work item is under development.

8.6.3.2 CEN ***** Public Transport—Public Interactive Information Terminals—Traveler Interface

CEN			Public transport—Public interactive information terminals—Traveler interface	Under Development	CEN TC278

CEN ***** (Number not yet allocated) This standard will be available from any European National Standards organization
http://www.cen.eu/esearch/extendedsearch.aspx

No documents are yet available.

8.6.3.3 ENV 13998 Public Transport—Noninteractive Dynamic Passenger Information on Ground

CEN	ENV	13998	Public transport—Noninteractive dynamic passenger information on ground	Published 2001	CEN TC278

CEN ENV 13998 Available from any European National Standards organization
http://www.cen.eu/esearch/extendedsearch.aspx

The passenger information variable message sign in this standard is defined as the noninteractive dynamic man-machine interface between a vehicle scheduling and control system, usually including an automatic vehicle monitoring (AVM) system, and the passenger at a stop point or other location. This interface is basically visual. It may also incorporate an audio communicator to give audio information.

8.6.4 Pubic Service Vehicle Environment

8.6.4.1 ISO/CEN 24014 Public Transport—Interoperable Fare Management System

CEN/ISO	24014	Public transport—Interoperable fare management system	Under Development

See Section 8.6.1.1.

8.6.4.2 CEN 12896 Public Transport—Reference Data Model

CEN	ENV	12896	Public transport—Reference data model	Published 2006	CEN TC278

Text Downloadable FOC Transmodel Documentation
www.transmodel.org

See Section 5.2.10.

8.6.4.3 ENV 13149 Public Transport—Road Vehicle Scheduling and Control Systems

CEN	ENV	13149	Public transport—Road vehicle scheduling and control systems	Published	CEN TC278

These documents specify the choice and the general application rules of an onboard data transmission bus between the different equipment for service operations and monitoring of the fleet. This applies to equipment installed onboard buses, trolley buses, and tramways only as part of a bus fleet operation. It excludes tramways when they are operated as part of a train, subway, or metro operation. This equipment includes operations aid systems, automatic passenger information systems, and fare collection systems. Equipment directly related to the functioning of the vehicle (e.g., driver dashboard, engine management, brake systems, door opening systems) are excluded from the scope of this document and are dealt with in other standardization bodies.

For the described application, two bus systems are standardized. Parts 1 to 3 describe the WORLDFIP bus system and Parts 4 to 6 describe the CAN open bus system. There is no ranking between the two bus systems. This document covers the link between equipments inside a single vehicle. Although it could be applied to multiple vehicles, this application is not explicitly covered by this document.

CEN	ENV	13149-1	Public transport—Road vehicle scheduling and control systems—Part 1: WORLDFIP definition and application rules for onboard data transmission	Published 1999	CEN TC278

CEN ENV 13149-1 Available from any European National Standards organization
http://www.cen.eu/esearch/extendedsearch.aspx

The equipment includes operations aid systems, automatic passenger information systems, fare collection systems, and so on. The equipment directly related to the functioning of the vehicle (driver dashboard, engine management, brake systems, door opening systems) are excluded from the scope of the present document and are dealt with in other standardization bodies.

Two alternative transmission buses will be accepted.

This document refers to the so-called WORLDFIP bus. A second set of documents will be published for the second solution (so-called CAN).

CEN	EN	13149-2	Public transport—Road vehicle scheduling and control systems—Part 2: WORLDFIP cabling specifications	Published 2000	CEN TC278

CEN 13149-2 Available from any European National Standards organization
http://www.cen.eu/esearch/extendedsearch.aspx

This document defines the cabling specifications for an on-board data transmission bus between the different parts of equipment for service operations and monitoring of the fleet.

Two alternative transmission buses will be accepted. This document refers to one of them (known as WORLFIP as specified in EN 50170).

CEN	TS	13149-3	Public transport—Road vehicle scheduling and control systems—Part 3: WORLDFIP message content	Published	CEN TC278

CEN TS 13149-3 Available from any European National Standards organization
http://www.cen.eu/esearch/extendedsearch.aspx

This European standard is Part 3 of ENV 13149, which gives rules for onboard data transmission systems. Part 3, together with Parts 1 and 2, describe a complete solution independent from Parts 4 through 6.

This document uses terms which are already used in other standards (e.g., ENV 12896:1997 Public transport—Reference data model) when applicable.

Part 3 of this European standard specifies in detail the application profiles for a simple network.

CEN	EN	13149-4	Public transport—Road vehicle scheduling and control systems—Part 4: General application rules for CAN open transmission busses	Published	CEN TC278

CEN 13149-4 Available from any European National Standards organization
http://www.cen.eu/esearch/extendedsearch.aspx

The equipment directly related to the safety-related functioning of the vehicle (propulsion management, brake systems, door opening systems) is excluded from the scope of the present document and are dealt with in other standardization bodies.

Part 4 of this document specifies the CAN open-based network. This specification describes the general architecture in terms of hierarchical layers according to the ISO reference model for OSI specified in ISO 7498.

CEN	EN	13149-5	Public transport—Road vehicle scheduling and control systems—Part 5: CAN open cabling specifications	Published	CEN TC278

CEN 13149-5 Available from any European National Standards organization
http://www.cen.eu/esearch/extendedsearch.aspx

This document covers the link between equipment inside a single vehicle. Although it could be applied to multiple vehicles, this application is not explicitly covered by this document. Part 4 of this document specifies the CAN open-based network. This specification describes the general architecture in terms of hierarchical layers according to the ISO reference model for OSI specified in ISO 7498. Part 5 of this document specifies in detail the connectors and the connector pin assignment and the cabling.

CEN	TS	13149-6	Public transport—Road vehicle scheduling and control systems—Part 6: CAN message content	Published	CEN TC278

CEN TS 13149-6 Available from any European National Standards organization
http://www.cen.eu/esearch/extendedsearch.aspx

This technical specification is Part 6 of EN 13149, which gives rules for onboard data transmission systems.

8.6.4.4 CEN ***** Public Transport—Road Vehicles—AVMS Onboard Equipment—Environmental and Electrical Conditions and Limits

CEN			Public Transport—Road vehicles—AVMS onboard equipment—Environmental and electrical conditions and limits	On Hold	CEN TC278

CEN ***** (Number not yet allocated) Will be available from any European National Standards organization
http://www.cen.eu/esearch/extendedsearch.aspx

Documents are not yet available.

8.6.4.5 ENV 12694 Public Transport—Road Vehicles—Dimensional Requirements for Variable Electronic External Signs

CEN	ENV	12694	Public transport—Road vehicles—Dimensional requirements for variable electronic external signs	Published 1997	CEN TC278

CEN ENV 12694 Available from any European National Standards organization
http://www.cen.eu/esearch/extendedsearch.aspx

This refers to variable external signs (mechanical signs without electronic control excluded) when installed in public transport vehicles such as buses, trams, trolley-buses (undergrounds and railway vehicles are excluded), and specifies location, dimensions, display characteristics, contents of information, and wiring.

8.6.4.6 CEN 15504 Public Transport—Road Vehicles—Visible Variable Passenger Information Devices Inside the Vehicle

CEN	TS	15504	Public transport—Road vehicles—Visible variable passenger information devices inside the vehicle	Published 2007	CEN TC278

CEN TS 15504 Available from any European National Standards organization
http://www.cen.eu/esearch/extendedsearch.aspx

This standard applies to different IVMS systems mounted in public transport vehicles, like, for example, buses, tramways, trolley-buses, and it specifies the installation location, dimensions, characteristics of the sign system, information contents, and cabling. At present there are several technologies for these kinds of IVMS (e.g., LCD, LED, VFD).

8.6.4.7 CEN 13093 Public Transport—Road Vehicles—Driver's Console Mechanical Interface Requirements—Minimum Display and Keypad Parameters

CEN	ENV	13093	Public transport—Road vehicles—Driver's console mechanical interface requirements—Minimum display and keypad parameters	Published 1998	CEN TC278

CEN 13093 Available from any European National Standards organization
http://www.cen.eu/esearch/extendedsearch.aspx

The driver console is a device fitted on road, urban, interurban, and rural public transport vehicles as an interface between the driver and all onboard equipment for AVMS. The driver console must respect a set of given conditions in order to be compatible with the ergonomic design of the driver environment and functional ergonomics for driver use.

8.6.4.8 CEN 12796 Public Transport—Road Vehicles—Validators

CEN	ENV	12796	Public transport—Road vehicles—Validators	Published 1997	CEN TC278

CEN 12796 Available from any European National Standards organization
http://www.cen.eu/esearch/extendedsearch.aspx

This prestandard refers to the validators, excluding ticket vending machines, installed on-board public transport road vehicles (buses, trams, light rail). It specifies: location, dimensions, display and keypad characteristics, and functions and data processing of this equipment.

8.6.4.9 CEN TS 15531 Public Transport—Service Interface for Real-Time Information Relating to Public Transport Operations

CEN	TS	15531	Public transport—Service interface for real-time information relating to public transport operations		CEN TC278
CEN	TS	15531-1	Public transport—Service interface for real-time information relating to public transport operations—Part 1: Context and framework	Published	CEN TC278

CEN TS 15531-1 Available from any European National Standards organization
http://www.cen.eu/esearch/extendedsearch.aspx

Real-time information may be exchanged between a number of different organizations, or between different systems belonging to the same organization. Key interfaces include the following:

- Between public transport vehicle control centers, generally, for fleet and network management;
- Between a control center and an information provision system, generally, to provide operational information for presentation.

CEN	TS	15531-2	Public transport—Service interface for real-time information relating to public transport operations—Part 2: Communications infrastructure	Published	CEN TC278

CEN TS 15531-2 Available from any European National Standards organization
http://www.cen.eu/esearch/extendedsearch.aspx

Service interface for real-time information (SIRI) uses a consistent set of general communication protocols to exchange information between client and server. The same pattern of message exchange may be used to implement different specific functional interfaces as sets of concrete message content types. Two well-known specific patterns of client server interaction are used for data exchange in SIRI: request/response and publish/subscribe.

CEN	TS	15531-3	Public transport—Service interface for real-time information relating to public transport operations—Part 3: Functional service interfaces	Published	CEN TC278

CEN TS 15531-3 Available from any European National Standards organization
http://www.cen.eu/esearch/extendedsearch.aspx

There are many potential ways for passenger transport operations centers to interact. The approach taken by SIRI is for an open-ended set of standard data structures, carried over a communications channel constructed using one of a small number of specific options. Part 2 of this technical specification specifies the communications channel. This part specifies a number of functional modules.

CEN	TS	15531-4	Public transport—Service interface for real-time information relating to public transport operations—Part 4: Real-time status monitoring information of facilities	Published	CEN TC278

CEN TS 15531-4 Not Yet Available. Will be available later from any European National Standards organization
http://www.cen.eu/esearch/extendedsearch.aspx

SIRI is a specification for an interface that allows systems running computer applications to exchange information about the planned, current, or projected performance of the public transport operations.

CEN	TS	15531-5	Public transport—Service interface for real-time information relating to public transport operations—Part 5: Traffic incident monitoring service	Published	CEN TC278

CEN TS 15531-5 Not Yet Available. Will be available later from any European National Standards organization
http://www.cen.eu/esearch/extendedsearch.aspx

SIRI is a specification for an interface that allows systems running computer applications to exchange information about the planned, current, or projected performance of the public transport operations.

The current version of SIRI identifies a number of known functional gaps. Partly for this reason, SIRI was developed with an extensible architecture, so that additional services and capabilities can be added as necessary. This proposal covers one such service.

The scope of this work item is to develop a structured incident model for describing disruptions to services, in terms that relate directly to the entities of other SIRI services. Incidents can then be directly linked to stops, lines, and journeys in two ways: as the cause of disruption or as the result of service problems.

8.6.4.10 CEN WI 00278207 Public Transport—Identification of Fixed Objects in Public Transport (IFOPT)

CEN	NP		Public transport—Identification of Fixed Objects in Public Transport (IFOPT)	Under Development	CEN TC278

CEN ***** (Number not yet allocated) Not Yet Available. Will be available later from any European National Standards organization
http://www.cen.eu/esearch/extendedsearch.aspx

8.7 Emergency Service Groups

The service groups in this domain describe activities that permit emergency services to be more quickly initiated and expedited throughout the transportation network

8.7.1 Transport-Related Emergency Notification and Personal Security

This service group applies ITS functionality to provide both driver/personal security services and automatic incident notification for private car drivers and goods vehicle drivers. This may include:

- Automatic collision notification;
- User initiated distress calls;
- Third-party emergency notification.

NOTE: The lead on standardization for the onboard issues will be agreed between TC 204 and TC 22.

Example services include: automated emergency call and mayday dispatch.

8.7.1.1 ISO 24978 Emergency and Safety Message Data Registry

ISO	CD	24978	Emergency and Safety Message Data Registry	CD/Under DIS Ballot	ISO TC204

ISO CD 24978 Not Yet Available. Track progress at:
http://www.iso.org/iso/iso_catalogue/catalogue_tc/catalogue_detail.htm?csnumber=42727

See Section 8.3.2.3.

8.7.1.2 ISO 25109 Emergency Services Architecture

ISO		25109	Emergency Services Architecture	Issued as Internal TR only Not Published. Superseded by ISO 26682	ISO TC204

ISO 25109 Not publicly available. Superseded by ISO 26682

8.7.1.3 ISO 26682 Crash and Emergency Notification Reference Architecture

ISO	NP	26682	Crash and Emergency Notification Reference Architecture	Under Development	ISO TC204

ISO 26682 Not Yet Available. Track progress at:
http://www.iso.org/iso/iso_catalogue/catalogue_tc/catalogue_detail.htm?csnumber=43651

There are no drafts available at this point in time.

8.7.1.4 CEN 15722 [formerly 24977] eCall Minimum Set of Data

CEN	NP	15722	eCall Minimum Set of data	Under CD Ballot	CEN TC278

CEN 15772 Available from any European National Standards organization
http://www.cen.eu/esearch/extendedsearch.aspx

See Section 8.3.2.1.

8.7.1.5 CEN WI 00278220 eCall Operating Requirements

CEN	WD	WI 00278220	eCall operating requirements	Under Development	

CEN WI 00278220 will be available from any European National Standards organization
http://www.cen.eu/esearch/extendedsearch.aspx

See Section 8.3.2.2.

8.7.1.6 SAE J2313_199909 On-Board Land Vehicle Mayday Reporting Interface

SAE	J2313_199909	On-Board Land Vehicle Mayday Reporting Interface	Published Sep. 1999	SAE

SAE J2313 Available from
http://www.sae.org/technical/standards/J2313_199909

This SAE standard describes the interface between an on-vehicle mayday detection reporting system and the off-vehicle response center that will manage the response to the vehicle's call for assistance. The automatic detection and reporting by either the intelligent vehicle itself or by the intelligent roadway of a vehicle that is disabled or involved in an accident (referred to as automatic mayday) is one of the key services identified by the ITS America program plan road map. This effort has been identified, as STD SAE J2313, in recent DOT-FHWA efforts to advance needed national standards.

8.7.2 After-Theft Vehicle Recovery

This service group applies ITS functionality to immobilize or recover stolen vehicles.
Example services include:

- User initiated distress calls;
- Automated theft warning;
- Automated vehicle intrusion and stolen vehicle monitoring;
- Stolen vehicle tracking;
- Remote vehicle immobilization.

8.7.2.1 CEN TS 15213

These technical specifications were developed to define an architecture within the CEN/TC 278 guidelines through which a level of interoperability can be achieved between systems operating centers (SOC) and law enforcement agencies (LEA), both nationally and internationally.

The documents will provide minimum standards of information and assurance to users regarding the functionality of systems, so as to enable the recovery of vehicles, detection of offenders, and reduction in crime.

These technical specifications should be read in conjunction with CEN/TS 15213-1, Road transport and traffic telematics—After-theft systems for the recovery of stolen vehicles—Reference architecture and terminology, which provides the preliminary framework for ATSVR concepts.

CEN		15213	After-theft systems for the recovery of stolen vehicles		CEN TC278
CEN	TS	15213-1	After-theft systems for the recovery of stolen vehicles—Part 1: Reference architecture and terminology	Published	CEN TC278

CEN TS 15213-1 Available from any European National Standards organization
http://www.cen.eu/esearch/extendedsearch.aspx

This preliminary document aims to provide the preliminary framework of ATSVR concepts and definitions for the purpose of following one. It will:

- Define the concepts and global architecture models for ATSVR and the appropriate terminology;
- Identify the various elements that may comprise an ATSVR.

Specifically, the format of the technical specification is as follows:

- ATSVR conceptual architecture model;
- General;
- Functions within the process;
- Short and long-range concepts;
- ATSVR operations with short-range concept;
- ATSVR operations with long-range concept.

CEN	TS	15213-2	After-theft systems for the recovery of stolen vehicles—Part 2: Common status message elements	Published	CEN TC278

CEN 15213-2 Available from any European National Standards organization
http://www.cen.eu/esearch/extendedsearch.aspx

This document specifies the basic structure of the message elements, or items of information, that are put together to form the common message sets used in exchanging information in an ATSVR.

This part of CEN/TS 15213 aims to identify the main elements and illustrate the data concepts and way forward.

CEN	TS	15213-3	After-theft systems for the recovery of stolen vehicles—Part 3: Interface and system requirements in terms of short-range communication system	Published	CEN TC278

CEN 15213-3 Available from any European National Standards organization
http://www.cen.eu/esearch/extendedsearch.aspx

This technical specification focuses on short-range (SR) interface/systems requirements. SR systems use an interface that allows detection equipment to operate some ATSVR functions in the direct line of sight of vehicles.

This part of CEN/TS 15213 describes the structure, bit arrangements, number representation, and coding of message elements that are typically transmitted as data. There is no requirement to make the messages as short or as effective as possible. Emphasis is placed on making them as clear and unambiguous as possible.

CEN	TS	15213-4	After-theft systems for the recovery of stolen vehicles—Part 4: Interface and system requirements in terms of long-range communication system	Published	CEN TC278

CEN 15213-4 Available from any European National Standards organization
http://www.cen.eu/esearch/extendedsearch.aspx

This technical specification specifies the characteristics required to operate the long-range ATSVR architecture.

This technical specification permits existing proprietary systems to operate using these interface specifications at ATSVR application level.

CEN	TS	15813-5	After-theft systems for the recovery of stolen vehicles—Part 5: Messaging interface	Published	CEN TC278

CEN 15813-5 Available from any European National Standards organization
http://www.cen.eu/esearch/extendedsearch.aspx

This technical specification specifies guidelines for cooperation and the procedures to be followed between the LEA and ATSVR SOCs in response to alarm signals by ATSVR systems. For purposes of optimum mutual communication, this technical specification also includes suggestions and a format for the electronic exchange of information.

CEN	TS	15813-6	After-theft systems for the recovery of stolen vehicles—Part 6: Test Procedures	Under Development	CEN TC278

CEN TC 15213-6 Not yet available
CEN 15813-6 Will be Available from any European National Standards organization
http://www.cen.eu/esearch/extendedsearch.aspx

The current scope of the work item under development is the test criteria for ATSVR and their control and use with electronic and electromechanical inhibitor control equipment utilizing both conventional switched outputs and/or soft-coded outputs of setting and unsetting devices, detectors, warning devices, and ancillary equipment, for fitting to vehicles operating on 12/24V negative earth electrical systems.

The requirements and tests specified in this standard enable reasonable assessment of components performance with regard to safety, reliability, functionality, security, and documentation. To provide reproducible test methods and to avoid the proliferation of technically similar test methods, the test procedures have been chosen, where possible, from internationally accepted standards. For specific guidance on these tests, reference is made to the appropriate document. In the context of the test procedures the term "specimen(s)" shall refer to the component or components of the ATSVR under test. To identify the tests that are to be applied to each type of component, reference shall be made to Table 23 in Annex A1.

8.7.3 Emergency Vehicle Management

The emergency vehicle management service group includes the application of fleet management, route guidance, and traffic signal priority techniques to the management of emergency vehicles such as fire, police, and ambulance. These services are implemented in conjunction with the traffic management service group services described in Section 6.2.1.

Example services include:

- Emergency vehicle fleet tracking;
- Emergency vehicle traffic management coordination.

ISO	DIS	22951	Data Dictionary and Message Sets for Preemption and Prioritization Signal System for Emergency and Public Transport Vehicles (PRESTO)	In Development DIS	ISO TC204

ISO 22951 Not Yet Available. Track progress at:
http://www.iso.org/iso/iso_catalogue/catalogue_tc/catalogue_detail.htm?csnumber=46009

See Section 8.6.1.2.

8.7.4 Hazardous Materials and Incident Notification

This service group covers the use of ITS functionality to provide authorities with data on the nature, location, and condition of hazardous goods cargoes. This facilitates the enforcement of routing instructions and the effective response to any incident involving the load. Data to be provided may include:

- Routing data:
  - Route guidance;
  - Route enforcement;
  - Incident data:
- Issuing post-incident instructions to driver;
- Location of vehicle;
- Nature of incident;
- Nature of cargo.

Example services include:

- Hazmat vehicle tracking;
- Automated hazmat emergency call/mayday notification;
- Hazmat preclearance services.

For HAZMAT related standards, see Sections 8.3.2.4 and 8.5.10.1.

This area is largely controlled by regional and national regulations. For example, U.S. regulations can be obtained from:

http://ecfr.gpoaccess.gov/cgi/t/text/text-idx?sid=585c275ee19254ba07625d8c92fe925f&c=ecfr&tpl=/ecfrbrowse/Title49/49cfrv2_02.tpl.

In Europe, many HAZMAT requirements are to be found in several "End of Life" recycling directives.

8.8 Transport-Related Electronic Payment Service Groups

This domain addresses activities that permit revenues for transportation services and facilities to be collected through cashless and nonstop payment.

8.8.1 Transport-Related Electronic Financial Transactions

This service group includes the use of electronic, or cashless, payment systems for transportation services, along with the implementation of automated systems to collect road user fees based on overall use of transportation services (e.g., distance based), not just use of particular facilities.

Example services include:

- Electronic transit fare payment;
- Electronic toll collection;

- Electronic parking payment;
- Electronic services payment (e.g., traveler information, reservations);
- Electronic distance-based road user fee payment services.

8.8.1.1 ISO/CEN 14904 Electronic Fee Collection (EFC)—Interface Specification for Clearing Between Operators

ISO/CEN	ENv/ ISO	14904	Electronic Fee Collection (EFC)—Interface specification for clearing between operators	Published	CEN TC278/ ISO TC204

ISO/CEN 14904 Available from
http://www.iso.org/iso/iso_catalogue/catalogue_tc/catalogue_detail.htm?csnumber=35986

ISO/TS 14904:2002 specifies the interfaces for clearing between operators and gives a framework of the common message structure and data elements to be used on the interfaces. Its objective is to make the transfer of payment and electronic fee collection (EFC) related data possible both between different payment systems and between different operators such as collection agents, clearing operators, or providers of public and private transport services.

ISO/TS 14904:2002 supports:

- Different payment modes (e.g., pre-payment, post-payment);
- A wide variety of transport and transport related services (tolling, parking, ferry/bridge/tunnel, public transport, payment for route guidance);
- Operator services (coordination between collectors of money and charge points);
- Security and privacy.

It is not within the scope of ISO/TS 14904:2002 to define administrative procedures and organizational structures. The specification of a higher (e.g., pan-European) level interoperable payment system is outside the scope of ISO/TS 14904:2002.

Not described within ISO/TS 14904:2002 are indirect (external) participants such as authorities enacting general or special legislation concerning the payment system and other national regulations.

The models presented in this standard are generic. Simple systems (closed systems) can be designed by selecting subsets of the interface framework described.

8.8.1.2 ISO/CEN 14906 EFC—Application Interface Definition for DSRC

ISO/EN		14906	Road transport and traffic telematics—EFC—Application interface definition for DSRC	Published 2004	CEN TC278/ ISO TC204

ISO/CEN 14906 Available from
http://www.iso.org/iso/iso_catalogue/catalogue_tc/catalogue_detail.htm?csnumber=35987

ISO 14906:2004 specifies the application interface in the context of EFC systems using the dedicated short-range communication (CEN-DSRC). The EFC application interface is the EFC application process interface to the CEN-DSRC application layer.

It should be noted that the base standard IS/EN 14906:2004 is subject to a near standing review. The next edition of IS/EN 14906 will incorporate advancements made since its publication, such as, for example, the definition of additional Euro classes (i.e., Euro-4 and Euro-5). Hence, such amendments have not been made in this standard as it would jeopardize the consistency with the base standard and violate the ISP concept.

8.8.1.3 ISO/CEN 14907 EFC—Test Procedures User and Fixed Equipment

ISO/EN	14907	Road transport and traffic telematics—EFC—Test procedures user and fixed equipment		CEN TC278/ ISO TC204

This standard is in two parts.

ISO/EN	14907-1	Road transport and traffic telematics—EFC—Test procedures user and fixed equipment—Part 1: Description of test procedures	Published 2004	CEN TC278/ ISO TC204

ISO/CEN 14907 Available from
http://www.iso.org/iso/iso_catalogue/catalogue_tc/catalogue_detail.htm?csnumber=39390

ISO/TS 14907-1:2005 specifies the test procedures of EFC roadside equipment and onboard equipment (OBE) with regard to the conformance to standards and requirements for type approval and acceptance testing which is specifically within the realm of EFC application.

The scope of ISO/TS 14907-1:2005 is restricted to systems operating within the radio emission, EMC regulations, traffic, and other regulations of the countries in which they are operated.

ISO/TS 14907-1:2005 identifies a set of suitable parameters and provides test procedures to enable the proof of a complete EFC system, as well as components of an EFC system (e.g., OBE related to the defined requirements of an application). The defined parameter and tests are assigned to the following parameter groups:

- Functionality;
- Quality;
- Referenced pretests.

ISO/EN	TS	14907-2	Road transport and traffic telematics—EFC—Test Procedures user and fixed equipment—Part 2: Conformance test specification for onboard units	Published 2005	CEN TC278/ ISO TC204

ISO/CEN 14907-2 Available from
http://www.iso.org/iso/iso_catalogue/catalogue_tc/catalogue_detail.htm?csnumber=40634

ISO TS 14907-2:2006 describes tests that verify OBU conformance of implementations of functions and data structures, as defined in the implementation conformance statement based on ISO 14906, for EFC applications.

8.8.1.4 ISO/CEN 17573 EFC—System Architecture for Vehicle Related Transport Services

ISO/EN	EN/IS	17573	Road transport and traffic telematics—EFC—System architecture for vehicle related transport services	Published 2002 Under revision	CEN TC278/ ISO TC204

ISO TS 17573 Available from
http://www.iso.org/iso/iso_catalogue/catalogue_tc/catalogue_detail.htm?csnumber=30757

ISO/TS 17573:2003 specifies a system architecture for EFC systems concerning vehicle related transport services such as the use of toll roads, zone access, parking, and route guidance.

ISO/TS 17573:2003 does not cover person related transport services such as public transport. However, some of the clauses in ISO/TS 17573:2003 may also be applicable for fare collection.

8.8.1.5 ISO/CEN 17574 EFC—Security Services Framework—Guidelines for EFC Security Protection Profiles

ISO/IEC	TS	17574	Road transport and traffic telematics—Electronic Fee Collection (EFC)—Guidelines for EFC security protection profiles	Approved ENV/TS Due for reaffirmment/ replacement/ withdrawal 2007	ISO TC204

ISO TS 17574 Available from
http://www.iso.org/iso/iso_catalogue/catalogue_tc/catalogue_detail.htm?csnumber=30759

The scope of the deliverable is to provide guidelines for the preparation and evaluation of security requirements specifications, referred to as protection profiles (PP) in ISO/IEC 15408 Evaluation criteria for IT/ICT security (see Section 5.3.1) and ISO/IEC PDTR 15446 Guide for the production of protection profiles and security target (see Section 5.3.2). A protection profile is a set of security requirements for a category of products or systems which meet specific needs. A typical example would be a PP for OBEs to be used in an EFC system, and in this case the PP would be an implementation-independent set of security requirements for the OBEs meeting the operators and users needs for security.

The document uses an OBE with an integrated circuit(s) card (ICC) as an example describing both the structure of the PP as well as the proposed content.

8.8.1.6 ISO/CEN 17575 EFC—Application Interface Definition for EFC Based on Global Navigation Satellite Systems and Cellular Network (GNSS/CN)

ISO	PDTS 17575	Road transport and traffic telematics—Electronic fee collection—Application interface definition for electronic fee collection (EFC) based on Global Navigation Satellite Systems and Cellular Network (GNSS/CN)	CEN TC278/ ISO TC204

ISO/CEN 17575 Not Yet Available

These technical specifications provide a basis for technical interoperability between contexts. The key assumption is that the OBE installed in a vehicle is able to adapt to the rules of a context when the vehicle approaches the part of the road network subject to fee in this context.

For this the OBE must be capable of continuously monitoring the position of the vehicle. This capability is assumed to be supported by GNSS. The OBE then has to compare the vehicle position with coordinates stored in its data memory. It has to react in a specific way when finding a match between these coordinates and the vehicle position. A set of coordinates, together with further attributes, telling the OBE how to match and react, is called a geographic object.

To adapt to a context, the OBE has to know where its geographic objects are and what rules apply to them. These technical specifications support the use of the CN communication link for the OBE to exchange data with various central equipments (CE). The OBE may download from the CE all data required to properly adapt to a context. With this update capability the OBE is able to comply with new or updated contexts, even if their implementation or update is dated later than the OBE was installed in the vehicle.

To support this type of technical interoperability, the assumed EFC system architecture distinguishes three functional levels, with the OBE exchanging at each level data with specific elements of CE using transactions via the CN communication link. These technical specifications provide the required details to implement these transactions.

For most of the development to date (up to version 13) ISO WD TS17575 has existed as a single document. It is in the process of being divided into four parts.

ISO	PDTS 17575-1	Road transport and traffic telematics—Electronic fee collection—Application interface definition for electronic fee collection (EFC) based on Global Navigation Satellite Systems and Cellular Network (GNSS/CN)—Part 1: Charging	CEN TC278/ ISO TC204

ISO/CEN 17575-1 Not Yet Available

This technical specification defines the transaction structure and the possible transaction contents for the information exchange over a cellular network (CN) interface between an OBE and a CE for the EFC application based on a GNSS. The application is intended to automatically collect fees for the use of road infrastructures including, among others, motorway fees, zone based fees in urban areas, congestion tolling, tolls for special road infrastructures like bridges and tunnels, and parking fees.

This technical specification defines the services provided by the application layer to the EFC application for the execution of transactions, and for each transaction the service data units.

ISO	PDTS 17575-2	Road transport and traffic telematics—Electronic fee collection—Application interface definition for electronic fee collection (EFC) based on Global Navigation Satellite Systems and Cellular Network (GNSS/CN)—Part 2: Communication	CEN TC278/ ISO TC204

ISO/CEN 17575-2 Not Yet Available

This part specifies the communication architecture supporting the communication between an EFC OBU and one or more CEs. It defines how EFC applications software modules or EFC support library modules may access the communication port, defines how to set up a communication channel, defines how to select the required security characteristics, and defines how to transfer data packets from and to the CE/OBE and close the communication channel.

It supports the usage of OBEs supporting different sets of available communication media which will be transparent for the EFC application software and its support library functions.

This part will also define how to set up the communication server of an EFC CE in order to allow communication with OBEs communication according to this standard.

Figure 8.8 illustrates the generic communication architecture. The broken line box indicates the scope of this part.

ISO	PDTS 17575-3	Road transport and traffic telematics—Electronic fee collection—Application interface definition for electronic fee collection (EFC) based on Global Navigation Satellite Systems and Cellular Network (GNSS/CN)—Part 3: Update	CEN TC278/ ISO TC204

ISO/CEN 17575-3 Not Yet Available

This technical specification defines the transactions for the information exchange over a cellular network between OBE and CE for EFC applications based on a GNSS. The transactions are used to determine the toll for the use of a vehicle in a toll domain (i.e., motorways, other roads, urban areas, bridges, and tunnels).

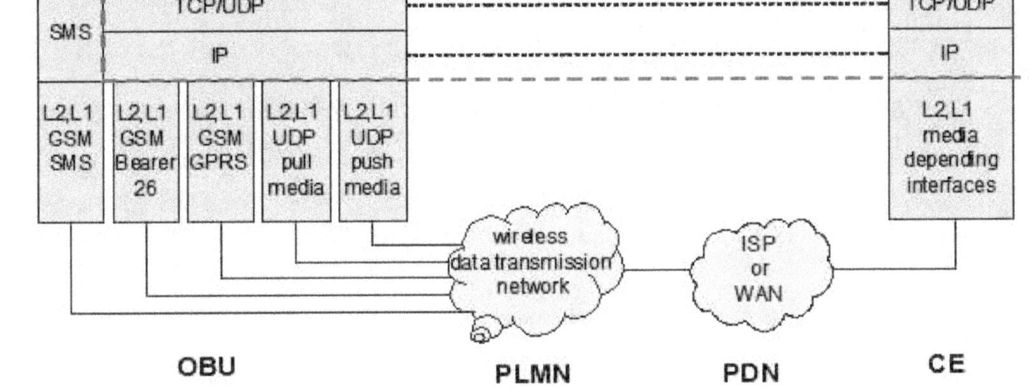

Figure 8.8 ISO 17575 communication architecture and scope. (*Source:* CSI Library.)

ISO	PDTS	17575-4	Road transport and traffic telematics—Electronic fee collection—Application interface definition for electronic fee collection (EFC) based on Global Navigation Satellite Systems and Cellular Network (GNSS/CN)—Part 4: Roaming		CEN TC278/ ISO TC204

ISO/CEN 17575-4 Not Yet Available

This part specifies the communication architecture supporting the communication between an EFC OBU and one or more CEs. It defines how EFC applications software modules or EFC support library modules may access the communication port, defines how to set up a communication channel, defines how to select the required security characteristics, and defines how to transfer data packets from and to the CE/OBE and close the communication channel.

It supports the usage of OBEs supporting different sets of available communication media which will be transparent for the EFC application SW and its support library functions.

This specification will also define how to set up the communication server of an EFC CE in order to allow communication with OBEs communication according to this standard.

8.8.1.7 ISO/CEN WI 00278192 EFC—Information Flows Between Operators of EFC Systems

CEN/ISO	WD	Not yet Allocated	Electronic Fee Collection (EFC)—Information flows between operators of EFC systems	Under Development	CEN TC278/ ISO TC204

ISO/CEN XXXXX Not Yet Available

No documents are yet available.

8.8.1.8 EN 15509 EFC—Interoperable Application Profile for Dedicated Short Range Communication

CEN	PrEN	15509	Road transport and traffic telematics—Electronic fee collection—Interoperable application profile for dedicated short range communication	Under Development	CEN TC278

CEN 15509 Available from any European National Standards organization
http://www.cen.eu/esearch/extendedsearch.aspx

See Section 4.6.1.13.

8.8.1.9 CEN WI 00278216 Electronic Fee Collection (EFC)—Conformity Evaluation of On-Board Unit And Roadside Equipment to EN 15509—Part 1: Test Suite Structure and Test Purposes

CEN	WI	Not yet allocated	Electronic fee collection (EFC)—Conformity evaluation of on-board unit and roadside equipment to EN 15509—Part 1: Test Suite Structure and Test Purposes	Under Development	CEN TC278

CEN ***** Number not yet allocated. Not Yet Available

No documents are yet available.

8.8.1.10 CEN WI 00278119 Dedicated Short Range Communication—Physical Integration with the Vehicle of On-Board Units (OBU) for Electronic Fee Collection (EFC)

CEN	WI	Deleted	Ensuring the correct function of ETC equipment installed behind metallized windshield	Deleted	CEN TC278

This work item has been deleted in favor of work item 0278215. See Section 8.8.1.11.

8.8.1.11 CEN WI 00278215 Ensuring the Correct Function of ETC Equipment Installed Behind Metallized Windshield

CEN	WI	Not yet allocated	Ensuring the correct function of ETC equipment installed behind metallized windshield	Under Development	CEN TC278

CEN ***** Number not yet allocated. Not Yet Available

European car manufacturers are introducing metallized windscreens at a wide scale throughout Europe.

These windscreens have interesting thermal qualities but risk compromising the correct functioning of OBUs for electronic toll collection (ETC) communicating with the roadside.

As it is essential to come to Europe-wide solutions, an ERTICO committee was created in February 1998 to work toward consensus on a common solution. Overall, 25 organizations have actively followed the work of this committee by attending the meetings and/or by otherwise contributing. The committee proposed the use of a nonmetallized window as a Europe-wide solution. This solution consists of providing a zone in the windscreen that is free of metal coating. The ERTICO committee produced a document, "Ensuring the correct functioning of ETC equipment installed behind metallized windscreens." During the 37th TC 278 meeting, TC 278 adopted the resolution 037/05/2006 to publish this document as a technical report.

8.8.1.12 CEN WI 00278217 Electronic Fee Collection (EFC)—Conformity Evaluation of On-Board Unit and Roadside Equipment to EN 15509—Part 2: Abstract Test Suite

CEN	WI	Electronic fee collection (EFC)—Conformity evaluation of on-board unit and roadside equipment to EN 15509—Part 2: Abstract test suite	Under Development	CEN TC278

CEN ***** Number not yet allocated. Not Yet Available

Work item currently under development.

8.8.2 Integration of Transport-Related Electronic Payment Services

This service group includes development of integrated payment systems and mechanisms between jurisdictions as well as between modes.

Example services include:

- Integration of multijurisdictional electronic payment systems;
- Integration of regional multimodal payment systems.

8.8.2.1 ISO/CEN 14904 EFC—Interface Specification for Clearing Between Operators

ISO/EN	EN/IS 14904	RTTT—EFC—Interface specification for clearing between operators	Published 2002	CEN TC278/ ISO TC204

ISO/CEN 14904 Available from
http://www.iso.org/iso/iso_catalogue/catalogue_tc/catalogue_detail.htm?csnumber=35986

See Section 8.8.1.1.

8.8.2.2 ISO/CEN 25110 EFC—Interface Definition for On-Board Account Using ICC

ISO	NP	25110	RTTT—EFC—Interface definition for on-board account using ICC	Under Development	CEN TC278/ ISO TC204

ISO/CEN 25110 Not Yet Available. Track progress at:
http://www.iso.org/iso/iso_catalogue/catalogue_tc/catalogue_detail.htm?csnumber=42750

Documents are not yet available.

8.9 Road Transport-Related Personal Safety

The service groups in this domain describe activities that protect the personal safety of pedestrians and individuals using road transportation facilities.

8.9.1 Public Travel Security

The public travel security service group includes the surveillance and monitoring systems for public transport facilities, car parks, and onboard public transport vehicles. Systems may be automatic, sending a distress call when specified conditions are encountered, or manually initiated. This also covers the use of security systems designed to protect public transport vehicle operators.

Example services include:

- Silent alarm;
- Emergency call/mayday alert for public transport;
- Intrusion detection;
- Public transport surveillance.

Even accounting for a rise in terrorist incidents in the past decade, the risk of a fatal accident on any form of transport is low. Even on the most dangerous types of transport, there are less than 120 fatalities for every billion kilometers traveled, according to UK DfT. The relative risk varies depending on how people choose to travel; these figures show that public transport is safer than personal motorized forms of transport such as cars, and that pedal bicycles and particularly motorbikes have the highest risk of fatalities.

Achievement will be particularly challenging in parts of the transport industry that are essentially open in nature, such as the railways. Such security is, at the present time, handled more by national government programs and regulations than by International Standards.

For example, in the United Kingdom, government policy across all forms of transport focuses on four key themes:

- The continuous development of domestic security programs;
- Driving up standards of transport security within Europe and worldwide;
- Investment in research and development to ensure that the staff responsible for security benefit from the most appropriate recruitment and training procedures and are supported by the best available technology;
- Compliance monitoring with enforcement action to ensure that the industries meet the required standards.

In United States, the ANSI Homeland Security Standards Panel (HSSP) has as its mission to identify existing consensus standards, or, if none exist, assist the Department of Homeland Security (DHS) and those sectors requesting assistance to accelerate development and adoption of consensus standards critical to homeland security. The ANSI-HSSP promotes a positive, cooperative partnership between the public and private sectors in order to meet the needs of the nation in this critical area.

To date the ANSI-HSSP has ongoing work examining standardization in the following workshop areas:

- Credentialing/access control for disaster management;
- Transit security;
- Emergency communications.

Formal standards have not yet been produced, but final deliverables have been produced for the following workshops:

- Lessons learned from Hurricane Katrina;
- Perimeter security;
- Enterprise power security and continuity;
- Training program standardization for first response to WMD events;
- Private sector emergency preparedness and business continuity;
- Biometrics;
- Biological and chemical threat agents;
- Citizen preparedness.

The Panel recognizes efforts underway on the following subjects and stands ready to provide any additional support that is needed:

- Risk analysis and management;
- Cyber security.

Further information is available from:

http://www.ansi.org/standards_activities/standards_boards_panels/hssp/overview.aspx?menuid=3.

ISO standards activity at this point of time is limited to the following new work items from ISO/TC 204 WG16 (Wide Area Communications).

8.9.1.1 ISO ***** Lawful Interception in ITS and CALM

ISO	PWI	Lawful Interception in ITS and CALM	Under Development PWI	ISO TC204 WG16

ISO ***** Number not yet allocated. Not Yet Available

The scope of this new project is to:

Outline the core requirements for LI in ITS networks, using CALM as an example technology, identifying the role of ISO/TC 204 WG16 and its relationship with other standards bodies. The technical content of the report will highlight data in scope of LI and the characteristics of the Point of Interception to be met by ITS (and CALM in particular). Summary reports of regional legislation may be given in annexes.

Purpose and justification: To assist deployment of network based ITS, it is required in many regions and jurisdictions to support lawful interception.

ISO	PWI	Data retention for law enforcement in ITS and CALM	Under Development PWI	ISO TC204 WG16

ISO ***** Number not yet allocated. Not Yet Available

The scope of this new project is to:

Outline the core requirements for data retention (DR) in network based ITS, using CALM as an example technology, identifying the role of ISO/TC 204 WG16 and its relationship with other standards bodies. The technical content of the report will highlight data in scope of DR and the characteristics of the Point of Retention to be met by ITS (and CALM in particular). Summary reports of regional legislation may be given in annexes.

Purpose and justification: To assist deployment of network based ITS, it is required in many regions and jurisdictions to support data retention for examination by both law enforcement and civilian agencies.

8.9.2 Safety Enhancements for Vulnerable Road Users

This service group covers the application of ITS functionality to the enhancement of safety levels for vulnerable road user groups (particularly the elderly, the disabled, and road maintenance workers). These groups include:

- Motorcyclists;
- Pedal cyclists;
- Pedestrians.

Safety enhancement measures may include measures such as:

- Smart pedestrian crossings (e.g., automatic warning of pedestrians for drivers, prolonging crossing times for elderly users, and changing pedestrian prioritization);
- Vehicle speed warning systems;
- Vehicle presence detection;
- Automatic advice to drivers by vulnerable road users (e.g., presence of road maintenance workers).

Example services include:

- Nonmotorized vehicle and pedestrian monitoring systems;
- Systems to monitor specialized vehicles.

Most standards achieved to date in these areas could not be described as ITS standards, but they relate to things like security belt standards, crash helmet standards, and crash test standards. In Europe the European Transport Safety Council (ETSC) (http://www.etsc.be/home.php) has projects in key safety areas such as drunk driving, enforcement, seat belts, and so on, but does not write standards. In particular, it has a project called VOICE (for vulnerable road user organizations in cooperation across Europe; at http://www.etsc.be/Voice.php), which is a Europe-wide campaign to protect vulnerable road users.

The principle aim of this ETSC activity is to raise awareness of the needs of vulnerable road users among EU policy makers so that they more readily accept responsibility for the implementation of the measures necessary for the protection of cyclists and pedestrians.

The Global Road Safety Partnership has similar types of initiatives for vulnerable road users, but again no standards. Find our more at http://www.grsproadsafety.org/?pageid=18&newsid=104.

This is another area where, to date, national regulations apply in preference to International Standards.

8.9.3 Safety Enhancements for Disabled Road Users

This service group covers the application of ITS functionality to the enhancement of safety levels for road user groups (particularly pedestrians with physical impairment).

Safety enhancement measures may include measures such as:

- Smart pedestrian crossings (e.g., prolonging crossing times for elderly and disabled users);
- Vehicle presence detection (either to the vehicle or from the vehicle to the pedestrian);
- Automatic advice to drivers by disabled road users (e.g., presence of wheel chair).

Example services include:

- Intersection monitoring of specialized conveyances (e.g., wheelchairs, carts);
- Driver warnings for specialized conveyances.

Disabled road users are a class of "vulnerable road users," so see Section 8.9.2 for activity group links. Similarly, it is an area covered, to date, by national regulations in preference to International Standards.

8.9.4 Safety Provisions for Pedestrians Using Intelligent Junctions Links

This service group covers the application of ITS technologies to the provision of monitoring and warning systems at junctions (including modal, multimodal, or intermodal), both signal controlled and priority, to enhance pedestrian safety. Warnings may include:

- Clarification of right-of-way rules;
- Onboard echo of warning signs;
- Presence of oncoming vehicles;
- Warning of imminent signal phase change.

Example services include:

- Signal display advance warning;
- Oncoming vehicle advance warning (for nonsignalized junction);
- In-vehicle signage and warning systems.

Standards in this domain are still sometime in the future.

8.10 Weather and Environmental Conditions Monitoring Service Groups

The service groups in this domain describe activities that monitor weather and environmental conditions that have an impact upon the transport network and its users.

NOTE: Information obtained by these services will be provided to users by using traveler information services (see Sections 2.3.2 and 8.2).

8.10.1 Weather Monitoring

This service group contains activities that result in monitoring of weather conditions, including fog, ice, snow, wind, rain, and heat; along with prediction of specific conditions as they impact the condition of the pavement and the overall traveled way, including icing and visibility.

Example services include:

- Road weather information monitoring;
- Road weather prediction.

To date, there are no International Standards for weather monitoring. Efforts to date tend to be national, with perhaps the most progress being in the United States. U.S. Road Weather Information Systems (RWIS) have become a critical component of many agencies' maintenance efforts. Transportation agencies in the United States are now considering ways to integrate RWIS with their other ITS so that weather information can support a broader range of ITS services. Standards are being identified to help ensure the fidelity of weather data being collected as agencies begin to share weather data with a wider audience.

Three categories of standards are being considered for RWIS applications:

- Siting standards;
- Calibration standards;
- Communication standards.

However, these "standards" are guidelines, recommended procedures, protocols, and other practices that formalize some of the processes involved in deploying and maintaining RWIS sensors rather than formal standards. These deliverables are still being developed and are not mandated by the U.S. DOT but their use is encouraged. The following summaries are from various DoT and state published information.

There are two categories of communication standards: (1) communication protocols, used to exchange data between RWIS devices and other ITS devices, and (2) display and message set standards, used to communicate weather and road conditions to end users. Communication protocols for RWIS are being developed under the National Transportation Communications for ITS Protocol (NTCIP) standards development effort. These are open (nonproprietary), industry-based standards that make it possible for RWIS and other ITS devices from multiple vendors to exchange information—both with each other and with a central system—through a common communications interface. Further information can be obtained from various Web sites:

- http://www4.trb.org/trb/onlinepubs.nsf/web/shrp_publications;
- http://www.ofcm.gov/siting/text/a-cover.htm;

- http://www.itsmichigan.org/pdf/RWIS_brochure.pdf;
- http://www.aurora-program.org;
- http://www4.trb.org/trb/crp.nsf/NCHRP+projects;
- http://www.its-standards.net;
- http://www.aurora-program.org/matrix.cfm.

8.10.2 Environmental Conditions Monitoring

This service group contains activities that result in monitoring of conditions such as flooding (due to high tides), land movement (earthquakes, mud slides), and pollution levels. Also, the service group may include services that can predict specific conditions that are likely to occur based on current and historical trends.

Example services include:

- Water level/tidal monitoring and prediction;
- Seismic monitoring;
- Pollution monitoring;
- Avalanche, mud slide, fallen rock monitoring.

See Section 8.10.1 for U.S. initiatives in this domain (where both weather and environmental issues are intertwined). Tidal and water level monitoring is more often covered by proprietary and local systems than by systems in accordance to standards. Seismic and pollution monitoring standards tend to be national standards for performance measurement of equipment.

There appear to be no standards at all, existing or under development, with respect to avalanche, mud slide, and rockfall monitoring.

ISO	TS	15624	Transport information and control systems—Traffic Impediment Warning System (TIWS)—System requirements	Published	34

ISO TS 15624 Available from
http://www.iso.org/iso/iso_catalogue/catalogue_tc/catalogue_detail.htm?csnumber=27833

This technical specification characterizes a system that identifies obstacles in turns ahead of the vehicle through roadside sensors and that informs the driver using roadside message boards.

It will not be progressed to a full standard because infrastructures will vary form country to country.

8.11 Disaster Response Management and Coordination Service Groups

The service groups in this domain describe ITS activities that manage resources from multiple jurisdictions in their response to natural disasters, civil disturbances, or terrorism.

8.11.1 Disaster Data Management

This service group includes services that collect data about a disaster from the appropriate agencies.

Example services include:

- Disaster and emergency data collection;
- Disaster and emergency data sharing.

The World Association for Disaster and Emergency Medicine has been requested by the U.N. World Health Organization (WHO) to assist with the development of a standardized set of applied tools for data collection in the field. They have not yet, at the time of publication of this book, produced these deliverables, but have active working groups working on these issues. More information can be found at: http://wadem.medicine.wisc.edu/ and http://wadem.medicine.wisc.edu/TFC.htm.

See also Section 8.11.2.

8.11.2 Disaster Response Management

This service group includes services that manage the use of the road network to minimize the impact of a disaster on its use.

Example services include:

- Disaster response planning for the transport network;
- Disaster response implementation.

There are no ISO or CEN standards under development for this domain. The most International Standards in this domain have been or are being developed by ETSI EMTEL.

ETSI EMTEL is currently defining the user requirements for the four main areas of emergency communications:

- Communication from citizens to authorities/organizations (emergency calls);
- Communication between authorities/organizations (public safety communications);
- Communication from authorities/organizations to citizens (warning systems);
- Communication among citizens during emergencies.

These user requirements will influence the development of the standards adopted by the ETSI members. While the initiating objective for these deliverables is not ITS, but general response to disaster and emergency situations (in a European context), there is a clear potential use of much of their work in an ITS context.

EMTEL has published two special reports, and five other documents are being currently drafted, as described below.

8.11.2.1 Requirements for Communication of Citizens with Authorities/Organizations in Case of Distress (Emergency Call Handling)

ETSI	SR	002 180	Requirements for communication of citizens with authorities/organizations in case of distress (emergency call handling)	Published	ETSI/ EMTEL

ETSI SR 002 180 Available from
http://webapp.etsi.org/exchangefolder/sr_002180v010101p.pdf

The special report provides an overview of the requirements for communication from citizens to authorities and organizations in all types of emergencies. It collects operational and organizational requirements as a basis for a common 112 service, including caller location information (E112).

8.11.2.2 Emergency Communications; Collection of European Regulatory Principles

ETSI	SR	002 299	Emergency Communications; Collection of European Regulatory Principles	Published	ETSI/ EMTEL

ETSI SR 002 299 Available from
http://webapp.etsi.org/exchangefolder/sr_002299v010101p.pdf

The need of ensuring a high level of support from communications systems in emergency conditions and a permanent increase of the effectiveness of the responsible bodies in such situations require interoperability and consistency. In fact, if in a single network an emergency situation is already a delicate case to consider, when two or more telecommunications networks are interconnected and the corresponding operators have different and some times opposite interests, the solution for each problem may be more complex and difficult to find.

In this context, standardization may be the only solution to promote freely and widely accepted solutions fulfilling the needs of the population, particularly those identified in regulatory documents.

The goal of the present document is to facilitate a stronger standardization in this area by bringing together the most important, easily identified rules in EU.

8.11.2.3 ETSI TR 102 181 Requirements for Communications Between Authorities/Organizations During Emergencies

ETSI	TR	102 181	Requirements for communications between authorities/organizations during emergencies	Published	ETSI/ EMTEL

ETSI SR 002 181 Available from
http://webapp.etsi.org/exchangefolder/ts_102181v010101p.pdf

The document addresses the requirements for communications between the authorized representatives who can be involved in the responses and actions when handling an emergency.

8.11.2.4 ETSI TR 102 182 Requirements for Communications from Authorities/Organizations to Citizen During Emergencies

ETSI	TR	102 182	Requirements for communications from authorities/organizations to citizen during emergencies	Published	ETSI/EMTEL

ETSI TR 102 182 Available from
http://webapp.etsi.org/exchangefolder/ts_102182v010201p.pdf

The document gives an overview of the requirements for communication from authorities/organizations to citizens in all types of emergencies. It collects operational and organizational requirements as a basis for a common notification service, including targeting of the area to be notified. Although many of the requirements relate to national public policies and regulation, there are a number of service and technical aspects which are better dealt with on the European level to ensure harmonized access and services over Europe and service effectiveness through increased user awareness by using standardized solutions.

The document also collects already established requirements for notification and gives guidance on how to find the standardization work published or ongoing. The document identifies the areas needing particular attention from the experts and refers to identified documents in preparation in SDOs.

The document provides a collection of technical requirements and recommendations.

The document is applicable to ETSI technical bodies for the defining of services and specifying technical solutions.

8.11.2.5 ETSI TR 102 410 Emergency Communications (EMTEL): Basis of Requirements for Communications Between Individuals and Between Individuals and Authorities While Emergencies Are in Progress

ETSI	TR	102 410	Emergency Communications (EMTEL); Basis of requirements for communications between individuals and between individuals and authorities while emergencies are in progress	Published	ETSI/EMTEL

ETSI TR 102 410 Available from
http://webapp.etsi.org/exchangefolder/tr_102410v010101p.pdf

The document addresses the requirements for communication facilities among individuals and to authorities/organizations, nongovernmental organizations (NGO), and media while emergencies are in progress, not including alerting communication. The proposals given here complement the facilities specified in TR 102 180, TS 102 181, and TS 102 182.

Although many suggested requirements collected from network operators, service providers (e.g., emergency response organizations), and users relate to national public policies and regulation, there are a number of service and technical aspects which are better dealt with on the European level to ensure harmonized access and services. Standardized solutions may foster increased user awareness and better utilization of the services.

The present document identifies potential areas needing particular attention from the experts and refers to identified documents in preparation in SDOs.

The scope of the present document covers aspects from small scale to large scale incidents.

The present document outlines the basis for technical, network operational, network organizational, and regulatory requirements which could be applied to existing as well as future networks and infrastructure.

8.11.2.6 ETSI TR 102 444 Analysis of the Short Message Service (SMS) and Cell Broadcast Service (CBS) for Emergency Messaging Applications

ETSI	TR	102 444	Analysis of the Short Message Service (SMS) and Cell Broadcast Service (CBS) for Emergency Messaging Applications	Published	ETSI/ EMTEL

ETSI TR 102 444 Available from
http://webapp.etsi.org/exchangefolder/tr_102444v010101p.pdf

The document covers the analysis of SMS and CBS for emergency applications using mobile radio systems, although TS 123 040 makes provision for short messages to be originated or terminated at a suitably equipped fixed network termination.

8.11.2.7 ETSI TR 102 445 Emergency Communications (EMTEL); Overview of Emergency Communications Network Resilience and Preparedness

ETSI	TR	102 445	Emergency Communications (EMTEL); Overview of Emergency Communications—Network Resilience and Preparedness	Published	ETSI/ EMTEL

ETSI TR 102 445 Available from
http://webapp.etsi.org/exchangefolder/tr_102445v010101p.pdf

The document presents resilience concepts and considers their application within technical systems enabling emergency communications and also considers network preparedness and requirements for specialized systems and capabilities.

8.11.3 Coordination with Emergency Agencies

This service group includes services that coordinate the use of the road network by vehicles belonging to emergency agencies. This may coordinate with the traffic control services activities reflected in Sections 8.3 and 2.6.

Example services include: disaster response coordination.

See Section 8.11.2 for EMTEL deliverables in this domain.

8.12 National Security Service Groups

The service groups in this domain describe activities that directly protect or mitigate physical or operational harm to persons and facilities due to natural disasters, civil disturbances, or terror attacks.

8.12.1 Monitoring of Suspicious Vehicles

This service group includes remote monitoring of vehicles for explosives or HAZMAT detection, and operational control of such vehicles, permitting shut down of vehicle operations if it is currently occupied by terrorists or known to be equipped (e.g., rigged with explosives) to cause destruction.

Example services include:

- Vehicle HAZMAT and explosives monitoring;
- Vehicle disablement;
- Road traffic management;
- Identification of suspicious vehicles.

For HAZMAT related standards, see Sections 8.3.2.4 and 8.5.10.1.

This area is largely controlled by regional and national regulations. For example, U.S. regulations can be obtained from:

http://ecfr.gpoaccess.gov/cgi/t/text/text-idx?sid=585c275ee19254ba07625d8c92fe925f&c=ecfr&tpl=/ecfrbrowse/Title49/49cfrv2_02.tpl

In Europe, many HAZMAT requirements are to be found in several "End of Life" recycling Directives.

8.12.2 Utility and Pipeline Monitoring

This service group includes systems designed to address stoppage of flow or detection of foreign or hazardous matter within the utility or pipeline facility and emergency notification to relevant agencies.

NOTE: Although utilities and pipelines are not directly related to ITS, the common practice of siting on or near roadways means that an incident against such utilities or pipelines is likely to disrupt the transport system and ITS may be involved.

Example services include:

- Pipeline and utility HAZMAT/explosives monitoring;
- Emergency notification to key agencies.

For relevant standards, see Sections 8.11 and 8.12.

PART IV

Standard National/Regional Architecture

CHAPTER 9

National and Regional Architecture

This is a book about ITS standards. It is not a book about elaborating ITS architecture. There is considerable confusion, however, both about what is available and what are standards, as well as the desirability of a "standard ITS architecture," so it is appropriate to allocate a small part of this book to clarify these issues and provide pointers for further study.

Perhaps the most important and fundamental issue with any business architecture—no less so for ITS—is that it requires an investment to create, and the return on that investment is that it makes the evolution of ITS systems and networks of systems a practical and predictable activity. Without a systems architecture to guide this evolution, it is doomed to frequent failures, cost and time overruns, and reduced safety. The reduction in risk and savings in cost and time penalties should provide the return on the investment made to develop the architecture.

The corollary of this is that the architecture itself can never be standardized because it is always evolving. The closest that can be achieved is to standardize a reference architecture, which is a very high level abstract model of what real architecture and their implementation will generally follow.

This is also covered in more detailed abstraction processes that are beyond the scope of a book on standards. Suffice to say that metamodels have been defined to describe what models should contain; and there are even meta-metamodels to describe metamodels. ITS architecture deals with the metamodel and model levels. Some countries, such as the United States, have developed such metamodel architectures and use these architectures as a framework for the application of policy. This chapter provides examples of such national and regional architectures.

The International Standards relating to ITS architecture are summarized and referenced in Chapter 5. Specifically, Sections 5.1.2 through 5.1.4 describe technical reports that assist in the development and elaboration in ITS architecture.

Section 5.2.7 describes a (not yet finalized work item) regarding how to use process-oriented methodology to elaborate architecture. Section 6.2.12 describes the unified modeling language (UML) (ISO 19501), which is the other principal method to elaborate architecture; Section 5.2.3 provides a summary of the deliverable, "Using UML for defining and documenting ITS Interfaces" (ISO 24097); and Section 5.2.4 provides a summary of the deliverable, "Using UML in ITS Standards" (ISO 24529).

Throughout this book you will see references to using these techniques, particularly UML, in other standards.

But note, all of the standards and deliverables just listed provide tools to assist the development of ITS architecture. Nowhere will you will find a "standard" ITS architecture. The closest that you will find to an ITS architecture is the Standard ISO 14813-1 (Section 5.1.1) "ITS service domains, service groups and services."

This standard has been recently revised and upgraded from a technical report to a full International Standard. This standard, as its name implies, attempts to provide a common understanding of what service domains and service groups comprise the ITS sector as we understand it today, and some of the services that we are likely to find in those domains. The actual services being defined in the standards are described in Chapter 8. But this International Standard stops short of defining an architecture—the relationship between the services and the actors in the system. This is deliberate.

In the early 1990s there was much initial enthusiasm for a single standard overarching ITS architecture, but it was quickly concluded that this was both impracticable and undesirable. Further, the ITS architecture for any region or country would develop and evolve rapidly over time, and "fossilizing" the architecture into a standard would be completely counterproductive. Certainly, requiring compliance to something that was by definition out of date would be undesirable.

So the standards committees have concentrated their efforts into providing the tools to help countries and regions develop architectures, and to assist standards developers to describe and define the architecture behind their standards.

Apart from the problem of the architecture evolving faster than it could pass through the standardization system approval process, why else is it inappropriate to have an agreed international standardized ITS architecture?

The principal reason is that the situation in each country or region differs greatly. For example, the architecture in a federal country of 50 states like United States, but with a single Department of Transportation managing overarching national budgets for transport, is fundamentally different from that of, say, Europe, where although there are 27 countries in the EU, there is no central Department of Transportation, but rather each country is responsible for its own transport system. Compare this again with countries like Japan or Korea, where there is a recognized formal role of an institutional architecture in a way that is fundamentally different from anything to be found in Europe or the United States. Then again, the highly centralized, centrally planned, and regionally managed state of China has fundamentally different relationships between the actors in its ambitious plans for ITS, which are controlled by the countries 5-year planning system.

These are fundamental differences of approach and relationships and this is a consequence of the political governmental structures in each of the countries mentioned.

It should be understood that this difference is more than the fact that some ITS services are present or planned in some countries and not in others; it affects the whole fundamental nature of the system architecture.

9.1 The Role of National/Regional Architecture

In general, it is only practicable for nations to develop and elaborate their architectures to fit their own national needs. Here again there will be much difference according to the political and institutional structures in the country.

But there are situations where regional ITS architectures are required. For example, in the European Union, although each country controls its own transportation network, there is a "single market" and more or less open and free movement of people and goods between and through countries. A common architecture, at least at a very high abstract level, is therefore desired by many; and for detailed ITS service provision in areas such as toll collection and international rail transport, more detailed common architectures are needed.

In Europe, there have been a series of EU projects over the years, starting with a project called KAREN and more recently (2004) with a framework architecture called FRAME; high level architecture has been developed as projects (not standards).

Now there is clamor once more for further work to be undertaken on ITS architecture. Once again the clamor was originally for a standard, but is now accepted to be for further project work. A successful ITS Architecture Workshop held by CEN/TC 278 in Madrid in June 2007 attracted more than 50 delegates from 16 countries and featured keynote presentations from two Directorates of the European Commission.

Its presentations and debates will be available until mid-2009 at: www.2007CEN-ITS-ArchitectureWorkshop.org.

The role of standardization is to provide tools to help create and maintain these national and regional architectures.

There is an opportunity to write a whole book on the different approaches and work on architectures around the world. This book is not the appropriate place for anything other than a short summary of initiatives under development and in use during the same timeframe as the standards described in this book.

The remainder of this chapter is limited to providing links to where you can find more about individual initiatives, as well as very short overview summaries of the initiatives to help you to decide which ones you wish to investigate further.

9.2 Japan—National ITS Architecture

Details of the Japanese ITS architecture can be obtained from: http://www.its-jp.org/english/arch_e/index.htm.

To promote the application of information technologies on roads, traffic, and vehicles, the five related government bodies (National Police Agency, Ministry of International Trade and Industry, Ministry of Transport, Ministry of Posts and Telecommunications, and Ministry of Construction) jointly finalized a "Comprehensive Plan for ITS in Japan" in July 1996, which is based on the "Basic Guidelines for the Promotion of an Advanced Information and Telecommunications Society" (determined by the Advanced Information and Telecommunications Society Promotion Headquarters in February 1996).

This architecture was developed and issued a decade before the writing of this book, and although it was updated in 1999, it probably does not fully reflect the current state of ITS in Japan. However, it is true to say that of all countries the Japanese ITS architecture has been used to study and shape strategic development of ITS worldwide more than any other country. Developed in 1996 and following very shortly after version 1 of the U.S. ITS architecture, the basis of the reference model uses process-oriented methodology.

The Japanese ITS architecture demonstrated a long-term vision of basic ideas on ITS development as it was in 1996, and identified implementation and user services in which ITS was to be promoted systematically and efficiently from the users' view point, and promoted in Japan based on the "comprehensive plan" in cooperation with industrial and academic sectors.

At the same time, the five government bodies recognized the need to create a grand design which was more detailed than the long-term vision in order to efficiently realize an integrated, highly expandable ITS system to respond to changes in social needs and future technology development.

In August 1999, the five government bodies organized a draft copy entitled "System Architecture for ITS." Subsequently, the draft was released so as to collect opinions from a broad range of the industrial and academic sectors and to actively address information overseas. Recently, "System Architecture for ITS" has been composed.

Japan continues to promote ITS-related projects through extensive cooperation among the five government bodies, with the industrial and academic sectors, as well as with other nations targeting an early and efficient realization of diverse ITS services based on the comprehensive plan and system architecture.

9.3 The United States—National ITS Architecture

Details of the U.S. national ITS architecture can be obtained from: http://www.its.dot.gov/arch/index.htm and http://www.its.dot.gov/arch/arch_howto.htm.

The United States was the first nation to develop a formal ITS architecture, which was published in 1994.

The architecture is based on process-oriented methodology, and is extremely detailed. It is maintained regularly, and the current version is Version 6. Because it is a POM-based architecture, the maintenance and revision is a complex process and has been maintained courtesy of a reasonably large budget to do so.

The motivation for this update process is that U.S. DOT has built compliance to the national ITS architecture into a requirement to obtain national funding for state transportation projects. Implementation by states varies, and it is not possible to say it is consistent throughout the country; however, some states, such as Minnesota (http://www.dot.state.mn.us/guidestar/about.html) and Michigan (www.itsmichigan.org/) have very detailed architectures of their own.

The national ITS architecture Web site has been updated with new features that enhance the architecture definition. The principal recent changes include increasing consistency with the Vehicle Infrastructure Integration (VII) initiative, added support for additional DSRC applications, and updates addressing CVISN, the Clarus initiative, the Border Information Flow Architecture, transit, and incident management standards.

The currently available version of the national ITS architecture addresses 33 user services (and is aligned with the latest version of ISO 14813-1); it is dated

May 2007. There are several options for obtaining the documentation, depending upon what you are looking for, and your level of familiarity with the architecture.

The World Wide Web version of the architecture is available at: http://www.iteris.com/itsarch/.

The WWW version allows the more technically minded to actually navigate the data flows of the architecture. This is a fully interactive system designers' tool, and it provides access to the text of the national architecture documentation. There are three ways to navigate through the national ITS architecture:

- The hypertext view provides immediate, interconnected access to all the elements of the architecture definition.
- The document view presents the complete set of architecture documentation as it exists in printed format.
- The database view provides developers direct access to the source material for the national ITS Architecture in several connected relational database.

The architecture can be obtained on the CD-ROM (Version 5.0) by e-mailing ITSPUBS@igate.fhwa.dot.gov.

This free CD-ROM contains viewable and printable versions of all documents, including the recently developed security document, plus the hyperlinked "browseable version" of the architecture. Adobe Acrobat Reader software is included for convenience.

You can also view the documents that make up the national ITS architecture using the "ITS Electronic Document Library," available at: http://www.its.dot.gov/library.htm.

For a good introduction, you can also read the Vision Statement, http://www.itsdocs.fhwa.dot.gov//JPODOCS/REPTS_PR//5389.pdf, and the Executive Summary, http://www.itsdocs.fhwa.dot.gov//JPODOCS/REPTS_PR//5388.pdf.

9.4 European Union—ITS Architecture Framework

Details of the European FRAME ITS architecture can be obtained from: http://www.frame-online.net/FrameForum.htm and http://www.frame-online.net/eits-fa2.htm.

This pan-European high-level architecture is also based on process-oriented methodology. The current version was completed in 2004 and is currently maintained by voluntary effort.

Printable versions of the documentation can be found in the LIBRARY: http://www.frame-online.net/library.htm.

The functional architecture can be navigated using the BROWSING TOOL: http://www.frame-online.net/BrowsingTool/welcome_v3.htm.

You can create your own architecture using the SELECTION TOOL: http://www.frame-online.net/selectiontool.htm.

9.5 Other National ITS Architecture

9.5.1 Australian National ITS Architecture

Details of the Australian ITS architecture can be obtained from: www.its-australia.com.au/KMXServer3/Portals/0/NationalRefArchFull.pdf.

As an integral part of the development process for the national strategy ITS, Australia engaged PPK Environment and Infrastructure (in conjunction with CSIRO and Intelligent Transport Specialists) to develop a National Reference Architecture for Intelligent Transport Systems (ITS). This is the third and final report to be produced by the consultant team for the current project and presents the final reference architecture following completion of consultations with stakeholders.

Development of the national reference architecture is an important component in the development of the overall national ITS strategy. The final version was issued in 1999. The Australian ITS architecture has been developed using UML techniques. The reference architecture has been based on ISO 14813. This document describes an object oriented system architecture based on the ITS/TICS fundamental services.

ITSs actually depend upon a large-scale transport information and control system (TICS). Therefore, ITS architecture is about the architecture of an information and control system. From here on it refers to the reference architecture as the TICS reference architecture.

A reference architecture is the first of all architectures. It is a concise generic framework that guides the development of more concrete system architectures. It is large enough that distinct concepts are not merged out of necessity, and small enough that it does not become unwieldy.

The objective in defining a TICS reference architecture is to provide a concise reference point which is both educational and a framework for the standards process. The reference architecture will be used in the future by particular industry groups to develop specific, logical, and physical architectures in a cohesive manner.

The logical architecture elaborates the conceptual behavior, and in so doing it provides more detail about the modularity. There is no firm demarcation between a reference architecture and a logical architecture.

The physical architecture is reached when the actual distribution of the system modules is defined, thus leading to important implications for communications. It is not until a physical architecture is defined that the major expectations for architecture identified in the first part of this report can be attained.

In establishing a national TICS reference architecture in Australia it is hoped that a degree of consensus and awareness can be achieved among stakeholders in the Australian ITS industry. This will allow for more efficient continued development of a national ITS system architecture by identifying those issues which must be addressed in the Australian context and those issues where considerable experience can be obtained directly from overseas developments.

9.5.2 French ACTIF National ITS Architecture

Details of the French ACTIF ITS architecture can be obtained from: http://www.its-actif.org/gb/links.asp.

Although ACTIF originated from FRAME, it uses and object oriented tool for its development. The tool is not UML but is quite similar in many ways.

The ACTIF model is structured into nine functional domains which model, through a common vocabulary and grammar, the functions and information exchanged between functions of transport systems. This model was the subject of an important consultation and stems from the experience of partners of different sectors. It allows one to gain in efficiency in the modeling and analysis of transport systems.

Nine fields of activity have been selected:

- DF1: Provide electronic means of payment;
- DF2: Manage emergency and safety services;
- DF3: Manage traffic and journeys;
- DF4: Operate public transport;
- DF5: Provide advanced drivers' assistance systems;
- DF6: Inform on journeys;
- DF7: Manage law enforcement;
- DF8: Operate the freight and fleet;
- DF9: Manage shared data.

9.5.3 Korean National ITS Architecture

Korea completed a national architecture in 1999. It was based on an adaptation of the U.S. architecture, adapted to the Korean situation.

Unfortunately, it was only ever available in Korean and is not available on the Internet. However, we are advised that it will be made available and that the links will be as follows:

MOCT: www.moct.go.kr;
MOCT ITS Web page: http://www.moct.go.kr/mct_hpg/Html/10/its.jsp.

9.5.4 Italian National ITS Architecture

Details of the Italian ITS architecture can be obtained from: http://www.its-artist.rupa.it/inglese/home.htm.

The ARTIST project was officially launched on March 11, 2003, in Rome. ARTIST, which used the European ITS framework architecture as a basis, took just 16 months to produce.

The Italian ITS architecture, ARTIST, is designed to be fully integrated within the European context. It is compatible with the European ITS framework architecture and is closely coordinated with French national architecture ACTIF.

The ARTIST project has introduced some new elements, and in particular the organizational architecture. While other architectures have focused mainly on functional aspects, ARTIST has also dealt specifically with the organizational and business aspects of ITS.

Another feature of ARTIST is the attention to multimodal aspects of ITS systems. ARTIST is designed to provide an "open," complete, and innovative tool within the Italian context.

9.5.5 Czech Republic TEAM National ITS Architecture

Details of the Czech ITS architecture can be obtained from: http://www.its-portal.cz.

Initiated by the Czech Ministry of Transport, the Czech TEAM ITS is evolving a national architecture.

The main technical advantages are seen to be the achievement of interoperability between ITS applications, and the coordination and synchronization of information systems. It is also hoped to reduce the costs of ITS deployments. From the transport policy point of view, ITS is seen as a way of achieving maximum use of the transport infrastructure.

The architecture will consist of a reference architecture, a functional architecture, and also information, physical, communications, and organization architectures. The architecture is intended to cover all modes (including road, rail, and waterborne transport).

The architecture is created through *process analysis*. (Processes are defined by chaining system components through information links. The chains of functions are mapped onto physical subsystems or modules.)

9.5.6 Netherlands National ITS Architecture

Details of the Netherlands ITS architecture can be obtained from: http://www.avb-bureau.nl/.

An ITS reference model called the KOEPEL Architecture was developed in 1999–2000. This model was the initial phase for developing a national ITS architecture, but so far no initiatives have been taken to develop it further.

An interurban traffic management architecture, Architectuur voor VerkeersBeheersing (AVB) was developed by Rijkswaterstaat (Netherlands Ministry of Transport) during the period of 1997–2000.

STIS is a national architecture for waterborne transport (developed from the European RIS).

AVB is a *system-owner* architecture, covering functionality, responsibility, and technicality. The ITS reference model covers the principal processes in logistics and transport and the way these transport systems can migrate to ITS.

9.5.7 Austrian National ITS Architecture

Austria has a strong interest in developing an ITS architecture, but it is placing all of its efforts towards an pan-European ITS architecture.

TTS-A is multimodal; this is seen as necessary to meet strategic transport goals: rail +20%, airspace +30%, motorway capacity +10%, shift of private to → public transport +10%.

General aims of ITS are to increase safety, efficiency, quality of life, and applicability (i.e., usability or general conditions required for ITS to achieve these objectives).

A high profile is given to intermodality, for example, regarding public transport and river traffic (VIA DONAU is closely involved in TTS-A development).

The aim regarding intermodality is to achieve integration, not only a bilateral approach.

It is considered important to give more emphasis to safety related ITS applications.

FRAME has been used as reference for the basic architecture principles and structure.

9.5.8 Norwegian ARKTRANS National ITS Architecture

Details of the Norwegian ITS architecture can be obtained from: http://www.sintef.no/units/informatics/projects/arktrans/arktransweb/English/english-content.htm.

ARKTRANS is a system framework architecture for intelligent transport systems. ARKTRANS establishes a common view upon the transport domain for all transport modes (road, sea, rail, and air).

ARKTRANS provides a template for intelligent transport systems. Standard functionality and common information elements are defined, as well as the interfaces that arrange for interoperability between ITSs. However, ARKTRANS does not dictate how the functionality is to be realized.

When ITS solutions are to be established, ARKTRANS can be used as a template. Functionality can be omitted or added—system components or services from different providers are combined by means of the interfaces specified by ARKTRANS, and they can operate together.

A system framework architecture like ARKTRANS will support the establishment of new and improved services for the transport domain. Information exchange and better coordination between actors can be supported in a way that arranges for improved efficiency and safety as well as solutions that are favorable to the environment

9.5.9 Finnish National ITS Architecture

Details of the Finnish ITS architecture can be obtained from: http://www.traficon.fi/tetra9.

The Finnish National ITS Architecture TelemArk consists of four main components:

- The architecture description, including service process descriptions (conceptual architecture) and logical architecture. Service processes are the main content that are used in system or subarchitecture design. Physical architecture is not described by TelemArk.
- The development plan. This document shows step-by-step how TelemArk can and should be utilized. It also states what is not included in the architecture but is still needed to make the telematic systems interoperable. For example, many interoperability problems arise from organizational barriers and lack of cooperation.

- Multimodal Traffic Data Library, KALKATI. The library contains lists of data objects, data models, and XML schemas that are needed for traffic information exchange. The KALKATI library is available on the Web (http://www.kalkati.net).
- Web pages for monitoring of ITS standardization work. The pages are available on the Web: http://www.traficon.fi/tetra9.

Also a new freight ITS architecture (TARKKI) has been constructed using the TelemArk methodology. Whereas TelemArk describes service processes for travelers and drivers, freight ITS architecture describes supply chain information processes.

In addition to these main components, the essential sixth element in the architecture is the "help desk" function, which was initiated after completion of TelemArk. The Finnish architecture components and relations between them are illustrated

The European framework architecture (KAREN) and the Finnish architecture (TelemArk) differ from each other, although their method of use and basic idea is similar. Since the basic idea is similar, the conclusion is that the architectures can be utilized together, supplementing each other.

9.5.10 Canadian National ITS Architecture

Details of the Canadian ITS architecture can be obtained from: http://www.its-sti.gc.ca/Architecture/english/static/content.htm.

The ITS architecture for Canada, like the U.S. architecture, is based on process oriented methodology. This provides a common framework for planning, defining, and integrating intelligent transportation systems. It reflects the contributions of a broad cross-section of the ITS community (transportation practitioners, systems engineers, system developers, technology specialists, and consultants). The architecture defines:

- The functions that are required for ITS (e.g., gather traffic information or request a route);
- The physical entities or subsystems where these functions reside (e.g., the roadside or the vehicle);
- The information flows that connect these functions and physical subsystems together into an integrated system.

A comprehensive glossary of architecture terms is on the menu and is also linked with the architecture content.

A summary document is available that presents the key architecture concepts, found at: http://www.its-sti.gc.ca/Architecture/english/static/keycon.htm.

The hypertext view provides immediate, interconnected access to all the elements of the architecture definition.: http://www.its-sti.gc.ca/Architecture/english/static/homepg.htm.

The document view presents the complete set of architecture documentation as it exists in printed format: http://www.its-sti.gc.ca/Architecture/english/archdoc.pdf.

9.5.11 Romanian National ITS Architecture

Details of the Romanian NARITS ITS architecture project, which is currently under development, can be obtained from: http://www.its-romania.ro/eng/index.php?option=com_content&task=view&id=8&Itemid=18.

The specific link to the NARITS page is: www.its-romania.ro, but at the time of writing, it is password protected.

9.5.12 Hungarian National ITS Architecture

Work has started on developing a Hungarian national ITS architecture. No links or descriptions are yet available.

9.5.13 Slovenian CONNECT National ITS Architecture

Work began on a Slovenian ITS architecture in 2005. It is called CONNECT. CONNECT intends to elaborate and focus on the harmonization of national system architectures (ITS architectures) in order to support interoperability of ITS applications and systems planned and deployed on the CONNECT road network. For these activities it has been agreed that ITS architecture in CONNECT will be in line with the recommendations of previous and ongoing European projects, initiatives like KAREN and FRAME.

The results, or Web site links, were not available at the time of this writing.

9.5.14 Spanish National ITS Architecture

Spain is developing an ITS architecture. At the time of writing, nothing is available, but when it is available it will probably be obtainable from: http://www.fomento.es.

9.5.15 Swiss National ITS Architecture

A Swiss ITS architecture is under development and is believed to be at "a stage of functional analysis." There is no Web site yet available, but the project leader has offered that those interested may contact him directly at: franz.muehlethaler@ptvswiss.ch.

PART V
Strategies to Use Standards in ITS

CHAPTER 10

Planning, Development, Deployment, and Operations

This chapter will not attempt to allocate nor prescribe how or where the comprehensive lists and descriptions of standards in the chapters above should be used. Rather, a few other standards which may well be useful, but did not appropriately fit in with any of the other chapter subject groupings are detailed here. But the main purpose of this chapter is to make some brief strategic considerations for the role of ITS standards in planning, development, deployment, and operations.

10.1 ITS System Requirements, Analysis, Design, and Delivery

ISO	TS	28682	A joint APEC-ISO study of progress to develop and deploy ITS Standards	Published	ISO TC204

ISO 28682 Available from
http://www.iso.org/iso/iso_catalogue/catalogue_tc/catalogue_detail.htm?csnumber=44878

The scope of this technical report is to:

- Provide a survey of the current status and plan of ITS standards and their deployment;
- Identify common problems which members are facing related to international standardization activities at the ISO with a focus on ISO technical committee (TC) 204 (ITS);
- Provide collective opinions to improve ITS standardization activities and their implementations.

It makes a number of recommendations regarding the improvement of the take up and use of ISO ITS standards.

The study does not, of course, reproduce the highly specific and detailed recommendations here. But it is worthy to note that where standards have been developed, their take up is different in different countries. Some countries mandate the use of certain, and in some cases all, International Standards; in most cases, however, they are advisory. Some countries give preference to International Standards in contracts bidding for public sector contracts, but this is not universal. The United States gives preference to national standards over International Standards, and has generally developed its national standards first, and offered them to the rest of the world to adopt as International Standards, rather than fully share in the development of International Standards based consensus. This "take ours if you want access to our markets" attitude is a regrettable situation for a major country that espouses free trade.

Many other countries have a different view: they take International Standards and then adapt them to their national domestic needs and produce national standards. This has the same effect of creating trading barriers. China for example, adopts almost all International ITS Standards, uses them as a basis (hence claiming compliance), and then adapts them to national standards which usually differ in some or many respects to the International Standards (therefore creating invisible trade barriers).

The EU mandates some CEN/ISO or ETSI standards, but in most cases only implies that it is preferable to use standards.

The report highlights some of these issues and concludes that the situation is far from satisfactory, and makes recommendations to improve the situation.

10.2 Systems Engineering for ITS

ISO	NP	25104	Training requirements for ITS System Architecture	Under Development	ISO TC204

ISO 25104 Available from
http://www.iso.org/iso/iso_catalogue/catalogue_tc/catalogue_detail.htm?csnumber=42745

This technical report, developed by ISO/TC 204 WG1, discusses the development for generic education and training requirements for the teaching of ITS architecture, as well as the acquisition of skills to interpret and develop ITS architectures.

The document provides suggestions to those planning education and/or training courses associated with ITS system architecture as to the subjects that should be studied.

This is an example of a standards WG taking the initiative where it sees potential problems in the marketplace. It is not a typical standardization deliverable, but has been well received.

10.3 Business Case and Benefit-Cost Analysis for ITS

ISO	NP	25103	Business justification for ITS Architecture	Under Development	ISO TC204

ISO 25103 Not Yet Available. Track Progress at:
http://www.iso.org/iso/iso_catalogue/catalogue_tc/catalogue_detail.htm?csnumber=42744

This deliverable is still under development with a new editor. Again, it is an initiative by the architecture WG, which that group hopes will also be taken up by other WGs for their domains.

One of the problems in getting resources allocated to develop standards is the cost justification of the time and travel costs of the experts involved. This to some extent is the fault of the expert groups, because in a world where everything needs to be cost justified, there are few examples of a business case being presented to potential sponsors of the work regarding the benefits that they will gain from their investment. These issues were briefly discussed in Section 10.1.

If resources are to continue to be made available for standards development, at least the generic business case, as opposed to just the technical requirements, need to be argued.

As Convenor of ISO/TC 204 WG1, I was once shocked at a working group meeting in Australia when Tony Kursius, a Senior Executive of Queensland Roads, came to our meeting to explain why he was not sending experts to the group. He explained. At the strategic level, using a project for a new bypass as an example, that we would do the equivalent of going into great detail about the foundations, the road surfacing, the signage and linage and so forth, and how the road was to be built, but we would forget to put the case as to why the bypass was needed, and its quantified benefits. If we didn't start by doing that, we would never get the bypass project approved, and the technical details were then irrelevant. Of course, as an accountant once said, it was so obvious: we were just taking the business case justification for granted. I am ever grateful to Tony for this lesson.

Business case justification, not just for the time and costs for the expert to participate, but the business case for having the standard—the benefits that it will bring—should therefore either be written as a separate technical report, or should be argued in the case for the development of an ITS standard. Then we may see more resources allocated to the tasks.

However, as stated in Chapter 9, the fundamental return on investment for the development of ITS architecture is the reduced risk and cost and time reductions during the life of the systems covered by the architecture as they are changed many times to meet evolving business and community requirements. In other words, the investment in ITS architecture reduces the cost of ownership taken over the full life of the assets involved

10.4 Use Case Requirements Specification for ITS

ISO	TR	25102	ITS Use Case pro forma Template	Approved TR awaiting publication	ISO TC204

ISO 25102 Available from
http://www.iso.org/iso/iso_catalogue/catalogue_tc/catalogue_detail.htm?csnumber=42743

See Section 5.1.8 for a more descriptive summary of this deliverable, which is an attempt to provide a consistent approach for the development of use cases in the ITS sector.

Use cases are now a popular way to describe the functional requirements of a system. Use cases describe the interaction between a primary actor (the initiator of an interaction) and the system itself, described as a sequence of simple steps. Each use case is a complete series of events, from the point of view of the actor.

According to Bittner and Spence: "Use cases, stated simply, allow description of sequences of events that, taken together, lead to a system doing something useful" [Bittner and Spence, *Use Case Modeling*].

A use case should avoid as far as possible the use of technical jargon, and should use the language of the end user or domain expert. A use case describes how an actor will interact with the system to achieve a specific objective. This is sometimes described from a single view, or may be described using multiple views of different actors.

An important part of a fully developed use case is that it not only describes the steps for normal operation but also includes the sequences of actions to be expected when abnormal or exceptional behavior occurs.

Sections 5.1.1 through 5.1.4, which provide architecture elaborations and training, describe use cases to explain their system elaborations.

10.5 Integrated Test, Evaluation, and Acceptance of ITS

Systems are becoming more and more complex with the overarching needs being focused around integration and interoperability. As a result their test, evaluation, and acceptance have evolved into a multifaceted through-life process.

Integrated test, evaluation, and acceptance (ITEA) planning is a cohesive process that combines integration and interoperability to confirm that all the clients needs have been met.

ITEA seeks to monitor a project throughout its life cycle, developing specific plans for:

- Through-life management in aspects such as:
  - Project, mission, and objectives;
  - Stakeholder responsibilities;
  - Strategies;
  - Methods, tools, and techniques;
  - Whole life costs and resources;
  - Evaluation and success criteria;
- Integrated testing, evaluation, and acceptance.

ITEA also deals with aspects such as time, cost, and performance, as well as gaining client acceptance of the ITEA.

To date, there are no current work items to develop ITEA standards for the ITS sector, or for specific ITS service provision. This will clearly be a requirement for the future if consistency and quality of service are to be attained and maintained.

10.6 ITS Software Development and Integration

10.6.1 Software Development Standards

ITS software development is and will largely be a commercial implementation, and so it is quality of service, performance and conformance, and maintenance standards that are of greatest import.

The software industry has many standards to measure performance and conformance in this respect. The lead standards are those of ISO, but other SDO and industry association standards and recommended practices should also be consulted.

The following standards are generally considered the most appropriate, of which the ISO Quality Software system has been built upon many years of experience and takes into account the requirements of manufacturing companies running various quality standards on a daily basis.

10.6.1.1 ISO 9001 Quality Management Systems—Requirements

ISO	IS	9001	Quality management systems—Requirements	Published 2000	ISO

ISO 9001 Available from
http://www.iso.org/iso/catalogue_detail?csnumber=21823

ISO 9001:2000 specifies requirements for a quality management system where an organization needs to demonstrate its ability to consistently provide product that meets customer and applicable regulatory requirements, and aims to enhance customer satisfaction through the effective application of the system, including processes for continual improvement of the system and the assurance of conformity to customer and applicable regulatory requirements.

All requirements of this International Standard are generic and are intended to be applicable to all organizations, regardless of type, size, and product provided.

Where any requirement(s) of this International Standard cannot be applied due to the nature of an organization and its product, this can be considered for exclusion.

Where exclusions are made, claims of conformity to this International Standard are not acceptable unless these exclusions are limited to requirements within clause 7, and such exclusions do not affect the organization's ability, or responsibility, to provide product that meets customer and applicable regulatory requirements.

10.6.1.2 ISO 90003 Software Engineering—Guidelines for the Application of ISO 9001:2000 to Computer Software

ISO	IS	90003	Software engineering—Guidelines for the application of ISO 9001:2000 to computer software	Published 2004	ISO

ISO 90003 Available from
http://www.iso.org/iso/iso_catalogue/catalogue_tc/catalogue_detail.htm?csnumber=35867

ISO/IEC 90003:2004 provides guidance for organizations in the application of ISO 9001:2000 to the acquisition, supply, development, operation, and maintenance of computer software and related support services. ISO/IEC 90003:2004 does not add to or otherwise change the requirements of ISO 9001:2000.

The guidelines provided in ISO/IEC 90003:2004 are not intended to be used as assessment criteria in quality management system registration/certification.

The application of ISO/IEC 90003:2004 is appropriate to software that is:

- Part of a commercial contract with another organization;
- A product available for a market sector;
- Used to support the processes of an organization;
- Embedded in a hardware product;
- Related to software services.

Some organizations may be involved in all the above activities; others may specialize in one area. Whatever the situation, the organization's quality management system should cover all aspects of the business (both software related and otherwise).

ISO/IEC 90003:2004 identifies the issues that should be addressed and is independent of the technology, life cycle models, development processes, sequence of activities, and organizational structure used by an organization. Additional guidance and frequent references to the ISO/IEC JTC 1/SC 7 software engineering standards are provided to assist in the application of ISO 9001:2000, in particular ISO/IEC 12207, ISO/IEC TR 9126, ISO/IEC 14598, ISO/IEC 15939, and ISO/IEC TR 15504.

10.6.1.3 ISO 12207 Information Technology—Software Life-Cycle Processes

ISO	IS	12207	Information technology—Software life cycle processes	Published	ISO

ISO 12207 Available from
http://www.iso.org/iso/iso_catalogue/catalogue_tc/catalogue_detail.htm?csnumber=21208

This establishes a system for software life cycle processes with well-defined terminology. It contains processes, activities, and tasks that are to be applied during the acquisition of a system that contains software, a stand-alone software product and software services.

10.6.1.4 ISO 14764 Software Engineering—Software Life Cycle Processes—Maintenance

ISO	IS	14764	Software Engineering—Software Life Cycle Processes—Maintenance	Published	ISO

ISO 14764 Available from
http://www.iso.org/iso/iso_catalogue/catalogue_tc/catalogue_detail.htm?csnumber=39064

ISO/IEC 14764:2006 describes in greater detail management of the maintenance process described in ISO/IEC 12207, including amendments. It also establishes definitions for the various types of maintenance. ISO/IEC 14764:2006 provides guidance that applies to planning, execution and control, review and evaluation, and closure of the maintenance process. The scope of ISO/IEC 14764:2006 includes maintenance for multiple software products with the same maintenance resources. "Maintenance" in ISO/IEC 14764:2006 means software maintenance unless otherwise stated.

ISO/IEC 14764:2006 provides the framework within which generic and specific software maintenance plans may be executed, evaluated, and tailored to the maintenance scope and magnitude of given software products. It provides the framework, precise terminology, and processes to allow the consistent application of technology (tools, techniques, and methods) to software maintenance.

ISO/IEC 14764:2006 provides guidance for the maintenance of software. The basis for the maintenance process and its activities comes from the definitions of ISO/IEC 12207. It defines the activities and tasks of software maintenance, and provides maintenance planning requirements. It does not address the operation of software and the operational functions (e.g., backup, recovery, and system administration) which are normally performed by those who operate the software.

ISO/IEC 14764:2006 is written primarily for maintainers of software and additionally for those responsible for development and quality assurance. It may also be used by acquirers and users of systems containing software that may provide inputs to the maintenance plan.

10.6.1.5 ISO 15271 Information Technology—Guide for ISO/IEC 12207 (Software Life Cycle Processes)

ISO	IS	15271	Information technology—Guide for ISO/IEC 12207 (Software Life Cycle Processes)	Published	ISO

ISO 15271 Available from
http://www.iso.org/iso/iso_catalogue/catalogue_tc/catalogue_detail.htm?csnumber=27148

10.6.1.6 ISO 15628 Systems Engineering—System Life Cycle Processes

ISO	IS	15288	Systems engineering—System life cycle processes	Published	ISO

ISO 15288 Available from
http://www.iso.org/iso/iso_catalogue/catalogue_tc/catalogue_detail.htm?csnumber=27166

10.6.1.7 ISO 15289 Systems and Software Engineering—Content of Systems and Software Life-Cycle Process Information Products (Documentation)

ISO	IS	15289	Systems and software engineering—Content of systems and software life cycle process information products (Documentation)	Published	ISO

ISO 15289 Available from
http://www.iso.org/iso/iso_catalogue/catalogue_tc/catalogue_detail.htm?csnumber=43790

ISO/IEC 15289:2006 was developed to assist users of systems and software life cycle processes to manage information items (documents). It is based on the life cycle processes specified in ISO/IEC 15288 or ISO/IEC 12207:1995/AMD 1:2002/ AMD 2. Information items are essential to preserving what transpired when using system life cycle processes, and may be identified as deliverable documents.

ISO/IEC 15289:2006 identifies the purpose and content of all identified systems and software life cycle information items as required for the various life cycle processes. The information item contents are defined according to generic document types (which may be referred to as information item types) and the specific purpose of the document.

ISO/IEC 15289:2006 may be applied to any of the activities and tasks of a project, system or software product, or service life cycle. It is not limited by the size, complexity, or criticality of the project. It may be applied to all forms of information items, information item content, and document delivery media. Information items may be combined or subdivided as needed for project or organizational purposes. The nomenclature for information items, document titles, and contents is informative.

10.6.1.8 ISO15504 Information Technology—Process Assessment

ISO	IS	15504	Information technology—Process assessment	Published	ISO

This standard is in five parts.

ISO	IS	15504-1	Information technology—Process assessment—Part 1: Concepts and vocabulary	Published	ISO

ISO 15504-1 Available from
http://www.iso.org/iso/iso_catalogue/catalogue_tc/catalogue_detail.htm?csnumber=38932

This part of ISO/IEC 15504:2004 provides overall information on the concepts of process assessment and its use in the two contexts of process improvement and process capability determination. It describes how the parts of the suite fit together, and provides guidance for their selection and use. It explains the requirements contained within ISO/IEC 15504, and their applicability to performing assessments.

Readers of this guide should familiarize themselves with the terminology and structure of the document suite, and then reference the appropriate parts of the suite for the context in which they propose to conduct an assessment. A more detailed description of the use of ISO/IEC 15504 is given in clause 4.

ISO	IS	15504-2	Information technology—Process assessment—Part 2: Performing an assessment	Published	ISO

ISO 15504-2 Available from
http://www.iso.org/iso/iso_catalogue/catalogue_tc/catalogue_detail.htm?csnumber=37458

ISO/IEC 15504-2:2003 defines the requirements for performing process assessment as a basis for use in process improvement and capability determination.

Process assessment is based on a two-dimensional model containing a process dimension and a capability dimension. The process dimension is provided by an external process reference model, which defines a set of processes characterized by statements of process purpose and process outcomes. The capability dimension consists of a measurement framework comprising six process capability levels and their associated process attributes.

ISO	IS	15504-3	Information technology—Process assessment—Part 3: Guidance on performing an assessment	Published	ISO

ISO 15504-3 Available from
http://www.iso.org/iso/iso_catalogue/catalogue_tc/catalogue_detail.htm?csnumber=37454

ISO/IEC 15504 (all parts) provides a framework for the assessment of processes. This framework can be used by organizations involved in planning, managing, monitoring, controlling, and improving the acquisition, supply, development, operation, evolution, and support of products and services.

ISO/IEC 15504-3:2004 provides guidance on meeting the minimum set of requirements for performing an assessment contained in ISO/IEC 15504-2.

It provides an overview of process assessment and interprets the requirements through the provision of guidance on:

- Performing an assessment;
- The measurement framework for process capability;
- Process reference models and process assessment models;
- Selecting and using assessment tools;
- Competency of assessors;
- Verification of conformity.

ISO/IEC 15504-3:2004 also provides an exemplar documented assessment process that conforms to the requirements of 4.2 in ISO/IEC 15504-2.

ISO	IS	15504-4	Information technology—Process assessment—Part 4: Guidance on use for process improvement and process capability determination	Published	ISO

ISO 15504-4 Available from
http://www.iso.org/iso/iso_catalogue/catalogue_tc/catalogue_detail.htm?csnumber=37462

ISO/IEC 15504-4 provides a framework for the assessment of processes. This framework can be used by organizations involved in planning, managing, monitoring, controlling, and improving the acquisition, supply, development, operation, evolution, and support of products and services.

ISO/IEC 15504-4:2004 provides guidance on how to utilize a conformant process assessment within a process improvement program or for process capability determination.

ISO	IS	15504-5	Information technology—Process Assessment—Part 5: An exemplar Process Assessment Model	Published	ISO

ISO 15504-5 Available from
http://www.iso.org/iso/iso_catalogue/catalogue_tc/catalogue_detail.htm?csnumber=38934

ISO/IEC 15504-5:2006 describes an exemplar process assessment model (PAM), conformant with the ISO/IEC 15504-2 requirements for PAM. It provides guidance by way of example on the nature and structure of PAMs and on the variety and function of different indicators of process performance and capability. It also provides guidance, through example, on the requirements for conformance of PAMs and on the approaches for demonstration of conformance.

ISO/IEC 15504 provides a framework for the assessment of process capability. This framework can be used by organizations involved in planning, managing, monitoring, controlling, and improving the acquisition, supply, development, operation, evolution, and support of products and services. It is also intended for use by assessors in the performance of process assessment and by organizations involved in the development of process reference models, process assessment models, or process assessment processes.

10.6.1.9 ISO 15940 Information Technology—Software Engineering Environment Services

ISO	IS	15940	Information Technology—Software Engineering Environment Services	Published	ISO

ISO 15940 Available from
http://www.iso.org/iso/iso_catalogue/catalogue_tc/catalogue_detail.htm?csnumber=29573

ISO/IEC 15940:2006 defines the software engineering environment (SEE) services conceptually in a reference model that can be adapted to any SEEs to automate one or more software engineering activities. It describes services that support the process definitions as in ISO/IEC 12207 so that the set of SEE services are compatible with ISO/IEC 12207.

ISO/IEC 15940:2006 can be used either as a general reference, or to define an automated software process.

10.6.1.10 ISO 16085 Systems and Software Engineering—Life Cycle Processes—Risk Management

ISO	IS	16085	Systems and software engineering—Life cycle processes—Risk management	Published	ISO

ISO 16085 Available from
http://www.iso.org/iso/iso_catalogue/catalogue_tc/catalogue_detail.htm?csnumber=40723

ISO/IEC 16085:2006 defines a process for the management of risk in the life cycle. It can be added to the existing set of system and software life cycle processes defined by ISO/IEC 15288 and ISO/IEC 12207, or it can be used independently.

ISO/IEC 16085:2006 can be applied equally to systems and software.

Risk management is a key discipline for making effective decisions and communicating the results within organizations. The purpose of risk management is to identify potential managerial and technical problems before they occur so that actions can be taken that reduce or eliminate the probability and/or impact of these problems should they occur. It is a critical tool for continuously determining the feasibility of project plans, for improving the search for and identification of potential problems that can affect life cycle activities, and the quality and performance of products, and for improving the active management of projects.

10.6.1.11 ISO 16236 Software Engineering—Guide for the Application of ISO/IEC 12207 to Project Management

ISO	IS	16326	Software engineering—Guide for the application of ISO/IEC 12207 to project management	Published	ISO

ISO 16326 Available from
http://www.iso.org/iso/iso_catalogue/catalogue_tc/catalogue_detail.htm?csnumber=30930

10.6.1.12 ISO 17799 Information Technology—Security Techniques—Code of Practice for Information Security Management

ISO	IS	17799	Information technology—Security techniques—Code of practice for information security management	Published	ISO

ISO 17799 Available from
http://www.iso.org/iso/iso_catalogue/catalogue_tc/catalogue_detail.htm?csnumber=39612

ISO/IEC 17799:2005 establishes guidelines and general principles for initiating, implementing, maintaining, and improving information security management in an organization. The objectives outlined provide general guidance on the commonly accepted goals of information security management.

10.6.2 Software Integration Standards

Computer-aided software engineering (CASE) tool integration standards, developed and standardized from the early 1990s, have had a significant a positive impact on the development of software engineering environments. This domain addresses control and data integration, which provide the keys to tool interoperability.

"CASE" has becoming a term applied to any tools used to develop, maintain, and reengineer software. The successful use of CASE is predicated on establishing sound engineering fundamentals, including process management, project management, configuration management, documentation management, team communications, and project verification and validation.

The software industry has many standards to measure performance and conformance in this respect. The lead standards are those of ISO and IEEE, but other SDO and industry association standards and recommended practices should also be consulted.

ISO 12207 (see Section 10.6.1.3), ISO 15628 (see Section 10.6.1.6), and ISO 16236 (see Section 10.6.1.10) address many of the issues relating to software integration.

10.6.2.1 ISO 20000-1 Information Technology—Service Management—Part 1: Specification

ISO	IS	20000-1	Information technology—Service management—Part 1: Specification	Published	ISO

ISO 20000-1 Available from
http://www.iso.org/iso/catalogue_detail?csnumber=41332

ISO/IEC 20000-1:2005 defines the requirements for a service provider to deliver managed services. It is based on BS 15000-2, which has been superseded.

It may be used by businesses that are going out to tender for their services:

- To provide a consistent approach by all service providers in a supply chain;
- To benchmark IT service management;
- As the basis for an independent assessment;
- To demonstrate the ability to meet customer requirements;
- To improve services.

ISO/IEC 20000-1:2005 promotes the adoption of an integrated process approach to effectively deliver managed services to meet business and customer requirements. For an organization to function effectively, it has to identify and manage numerous linked activities. Coordinated integration and implementation of the service management processes provides ongoing control, greater efficiency, and opportunities for continual improvement.

10.6.2.2 ISO 20000-2 Information Technology—Service Management—Part 2: Code of Practice

ISO	IS	20000-2	Information technology—Service management—Part 2: Code of practice	Published	ISO

ISO 20000-2 Available from
http://www.iso.org/iso/iso_catalogue/catalogue_tc/catalogue_detail.htm?csnumber=41333

ISO/IEC 20000-2:2005 represents an industry consensus on guidance to auditors and offers assistance to service providers planning service improvements or to be audited against ISO/IEC 20000-1. ISO/IEC 20000-2:2005 is based on BS 15000-2, which has been superseded.

10.6.2.3 IEEE 11175 CASE Tool Interconnections

Reference models for tool-to-organization interconnections, tool-to-platform interconnections, and information transfer among tools are provided. The purpose is to establish agreements for information transfer among tools in the contexts of human organization, a computer system platform, and a software development application. To make the transfer of semantic information among tools easier, a semantic transfer language (STL) is also provided. Interconnections that must be considered when buying, building, testing, or using computing system tools for specifying behavioral descriptions or requirements of system and software products are described.

The standard is in three parts.

IEEE	1175.1	IEEE guide for CASE tool interconnections—classification and description	Published	IEEE

IEEE 11175.1 Available from
http://ieeexplore.ieee.org/Xplore/login.jsp?url=/iel5/8401/26478/01178882.pdf

This guide describes the scope of application and interrelationships for the members of the IEEE 1175 family of standards, and it points the reader to the appropriate standards for clarifying issues involved in effectively integrating computing system tools into a productive engineering environment.

IEEE	1175.2	IEEE Recommended Practice for CASE Tool Interconnection-Characterization of Interconnections	Published	IEEE

IEEE 11175.2 Available from
http://ieeexplore.ieee.org/xpl/standardstoc.jsp?isnumber=4068342&isYear=2007

Interconnections that need to be understood and evaluated when buying, building, testing, or using CASE tools are described in this recommended practice. This recommended practice is intended to help people interconnect tools by identifying and characterizing various contexts for tool interconnection. Each context serves to define a group of interconnections pertinent to various functional perspectives. Each group contains interconnections that have a common kind of endpoint in the environment. This recommended practice considers four contexts: (1) an organizational context for a tool, (2) the individual user context for a tool, (3) the platform context for a tool, and (4) a peer context for a tool. Within a context, subsets of interconnections are characterized by a collection of common features applicable to a given functional perspective. The purpose of this recommended practice is to establish sets of interconnection features with which each perspective on a CASE tool's interconnections can be characterized.

The IEEE approved IEEE 1175.2—Recommended Practice for CASE Tool Interconnection—Characterization of Interconnections, which identifies attributes that characterize contexts for CASE tool operation.

These attributes provide a checklist of interconnection concerns for use in selecting, adopting, and using CASE tools. IEEE 1175.2 also will help users organize and apply other industry standards in designing and realizing large, integrated, multivendor software engineering environments.

IEEE	IS	1175.3	IEEE Standard for CASE Tool Interconnections—Reference Model for Specifying Software Behavior	Published	IEEE

IEEE 11175.3 Available from
http://ieeexplore.ieee.org/xpl/standardstoc.jsp?isnumber=29690

CASE tools are used to describe the behavior of software using a variety of different design notations. These may be graphical or textual in nature, or they may be a combination. This standard provides a reference model of fundamental software concepts that form the building blocks for a number of these commonly used notations. This standard also includes a textual language, the semantic transfer language (STL), for representing software application behavior descriptions. A software behavior description consists of a collection of sentences that conform to the formal syntax of the STL and that are to be interpreted in terms of the software concepts defined in this standard. The STL syntax is designed to be computer-parsable, while remaining easy for users to read and write. This reference model and transfer syntax may be used for directly recording, storing, and analyzing a software behavior description, as well as for transferring elements of a software behavior description between CASE tools.

10.7 ITS Data Collection and Analysis

ITS data collection is an application specific task, with the exception of data required to establish a communications session or an application session on top of a communications session. Therefore, the data concepts specific to the application are/will be specified in any relevant application standard where appropriate, or may be specific to a commercial application.

Where data is to be of use communally, the relevant standard is ISO 14817 (Section 5.1.7) and ISO 24978 (Section 8.3.2.3), and both of these Standards are ITS adaptations of ISO 11179 (Section 6.2.2.24)—ISO 11179 being the base ISO standard for data registry management.

Data analysis tends to be proprietary algorithm based and embedded in commercial products rather than standards.

Data mining is a specific data analysis technique. There are no standards by leading SDOs that we have been able to identify, but the consortium, the Data Mining Group (DMG) have developed standards, or common practices, in this area.

Members of the DMG include: IBM, Microsoft, National Center for Data Mining, University of Illinois, and Oracle Corporation.

The DMG is an independent vendor-led group of volunteers which develops data mining standards, such as the predictive model markup language (PMML). DMG's XML related standards can be found at Source Forge (link below).

PMML Version 3.2 was approved in May 2007. The XML schemas defining Version 3.2 can be found at the PMML project home page on Source Forge: http://sourceforge.net/projects/pmml.

10.8 Simulation and Modeling for ITS

10.8.1 Simulation Standards

There are no specific simulation standards for ITS, although simulation techniques have been widely used to enable regulation changes to enable ITS.

Although there are provisions regarding simulation in ISO 9001, general simulation standards to date have largely been generated by IEEE in its series IEEE 1278.

10.8.1.1 IEEE 1278 Standard for Distributed Interactive Simulation

IEEE	IS	1278.1	IEEE Standard for Distributed Interactive Simulation—Application Protocols	Published	IEEE

IEEE 1278.1 Available from
http://ieeexplore.ieee.org/search/freesearchresult.jsp?history=yes&queryText=%281278.1%29

Data messages, known as protocol data units (PDUs), that are exchanged on a network between simulation applications are defined here. These PDUs are for interactions that take place within specified domains called protocol families, which include entity information interaction, warfare, logistics, simulation management, distributed emission regeneration, and radio communications.

IEEE	IS	1278.1a	IEEE Supplement to Standard for Distributed Interactive Simulation—Application Protocols	Published	IEEE

IEEE 1278.1a Available from
http://ieeexplore.ieee.org/search/freesearchresult.jsp?history=yes&queryText=%281278.1%29

Data messages, known as PDUs, that are exchanged on a network between simulation applications are defined here. These PDUs are for interactions that take place within specified domains called protocol families, which include entity information/interaction, warfare, logistics, simulation management, distributed emission regeneration, radio communications, entity management, minefield, synthetic environment, simulation management with reliability, live entity information/interaction, and nonreal time.

IEEE	IS	1278.2	IEEE Supplement to Standard for Distributed Interactive Simulation—Communication Services and Profiles	Published	IEEE

IEEE 1278.2 Available from
http://ieeexplore.ieee.org/search/freesearchresult.jsp?history=yes&queryText=%28%281278.2%29%29

Communication services to support information exchange between simulation applications participating in the distributed interactive simulation (DIS) environment are defined here. These communication services describe a connectionless information transfer that supports real-time, as well as nonreal-time, exchange. Several communication profiles specifying communication services are provided.

IEEE	IS	1278.3	IEEE Recommended Practice for Distributed Interactive Simulation—Exercise Management and Feedback	Published	IEEE

IEEE 1278.3 Available from
http://ieeexplore.ieee.org/search/freesearchresult.jsp?history=yes&queryText=%28%281278.3%29%29

Guidelines are established for exercise management and feedback in DIS exercises. Guidance is provided to sponsors, providers, and supporters of DIS-compliant systems and exercises, as well as to developers of DIS exercise management and feedback stations. The activities of the organizations involved in a DIS exercise and the top-level processes used to accomplish those activities are addressed here.

The functional requirements of the exercise management and feedback process are also addressed. This standard is one of a series of standards developed for DIS to assure interoperability between dissimilar simulations for currently installed and future simulations developed by different organizations.

IEEE	IS	1278.4	IEEE Trial-Use Recommended Practice for Distributed Interactive Simulation—Verification, Validation, and Accreditation	Published	IEEE

IEEE 1278.4 Available from
http://ieeexplore.ieee.org/search/freesearchresult.jsp?history=yes&queryText=%28%281278.3%29%29

Guidelines are established for the verification, validation, and accreditation (VV&A) of DIS exercises. "How-to" procedures for planning and conducting DIS exercise VV&A are provided. Intended for use in conjunction with IEEE Std 1278.3-1996, this recommended practice presents data flow and connectivity for all proposed verification and validation activities and provides rationale and justification for each step. VV&A guidance is provided to exercise users/sponsors and developers.

10.8.2 Modeling for ITS

The principal modeling standard used in ITS today is the unified modeling language (UML) (see Sections 6.2.12.1 and 1.9), although process oriented architecture aficionados would also claim that ISO 26999 (Section 5.2.7) also described a modeling technique of equal status, and certainly one that has been used to describe more of the ITS system architectures than UML. However, UML is the modeling technique now favored by software engineers today.

Nevertheless, there are other modeling notations and tools in widespread use, and these approaches should not be discounted.

PART VI
The Process of Standards Development

CHAPTER 11

International Standards Development Organizations for ITS

The purpose of this chapter is to review the processes used by standards development organizations (SDOs) to develop and approve standards, and to provide their methods and compare their processes. The sections of this chapter represent a précis of the published "directives" of each of the standards organizations.

Full copies of the directives for each of the SDOs are openly available for download from their respective Web sites. This section, therefore, is not a full authoritative guide for standards developers (please download from the relevant SDO Web site and use the directives for this purpose) but rather provides a summary about the way that each of the SDOs process documents to become standards, and enables the reader to compare the processes.

The relevant SDO is acknowledged as the source of the information within this chapter. No additions have been made, but only the more significant processes are summarized and précis used.

The procedures that the different SDOs use to develop a standard differ, and this affects the nature, in some cases the robustness, and in most cases the time it takes to develop standards and other deliverables.

It is relevant at this stage to ask: Why there is more than one SDO for ITS? What is the difference between standards and regulations?

Let us first answer with the second question, and explain the difference between regulations and standards.

Regulations exist mostly at a national level, and they exist to meet the political and practical requirements of a country, or in cases of federal countries, maybe the requirements of a state within a country. Some of these regulations are specific to the trading situation of a country, and in many cases its trading or safety agreements with other countries. International agreements like the World Trade Organization or the United Nations make requirements on countries for fair trade. There are two aspects of regulations that most significantly affect ITS. The first are regulations with respect to vehicle safety, and the second are radio regulations.

Vehicle safety regulations are, of course, essential, but the requirements have evolved from country to country, and even today the requirements in countries like the United States and United Kingdom are significantly different from those in, say, India and Indonesia. This makes it difficult for automotive manufacturers which are at present mostly either global or members of international combines or internationally related companies, and which wish to make as far as possible one model to sell around the world. That dream is never quite realized, but it is generally realized through committees such as ISO/TC 22 which develops International Standards for road vehicles; for regulators the best way to enforce a regulation is by reference to a standard specification. As trading blocks like the European Union, NAFTA, APEC, and ASEAN have developed, these standards are a good way to ensure fair access to each others' markets, and so compliance to specified standards are often written into international trading agreements.

With respect to radio regulations, the problems are more complex.

The radio spectrum is both a scarce resource and a national asset. In nearly all cases the radio spectrum is carved up and allocated or licensed out by the national radio regulator. As radio emissions do not respect national boundaries, this has always been a problem that had to be dealt with at an international level, particularly in places like Europe, with many national regulators within the range of many of the emitted signals. Hence, organizing bodies, like the European Conference of Postal and Telecommunications Administrations (Conference Européenne des administration des Postes et des Télécommunications, CEPT) have taken on a role not as radio regulator, but rather coordinator of radio regulations. On an international scale the International Telecommunication Union (ITU) provides a similar role. These bodies develop, usually by consensus, "recommendations," and national regulators are encouraged to apply the recommendations as regulations within their countries. Of course, this does not always happen and there are variations as well, perhaps because a country is using a military radar at a particular frequency or has a lower emissions safety limit at a particular frequency. However, in general it enables efficient band sharing across regions and the development of regionally or internationally agreed uses for particular application types at common frequency band allocations.

Figure 11.1 shows a global regulatory situation.

Wherever possible, and particularly in the case of CEPT, the best way to have enforceable regulations are to base them on internationally or regionally agreed standards. In Europe this caused the creation of the European Telecommunications Standards Institute (ETSI). Almost all CEPT recommendations are now related to ETSI standards, and this, of course, makes life easier for national regulators.

The development of the European Union, with its single market objectives, implies that common standards are required, and so harmonized European standards (HENs)—market enabling devices rather than protocol specifications—are now developed where European directives are probable. These are also developed by ETSI for radio matters. The involvement of the European Commission in radio spectrum policy has complicated matters in Europe so there is now also the ECC and ERO.

The complex radio regulation situation in Europe (Region 1) is discussed in Section 4.1. The radio regulation situation under the U.S. Federal Communications Commission (FCC) and the rest of Region 2 is discussed in Section 4.1.2. The radio regulation situation in other countries, primarily Region 3, is discussed in Section 4.1.3. Suffice to say here that radio spectrum is allocated nationally within a framework of international agreements.

In Europe, ETSI develops standards to support CEPT inspired regulations; in the United States a number of local SDOs write standards that support FCC regulations. The most significant American SDO, at least so far, is the Institute of

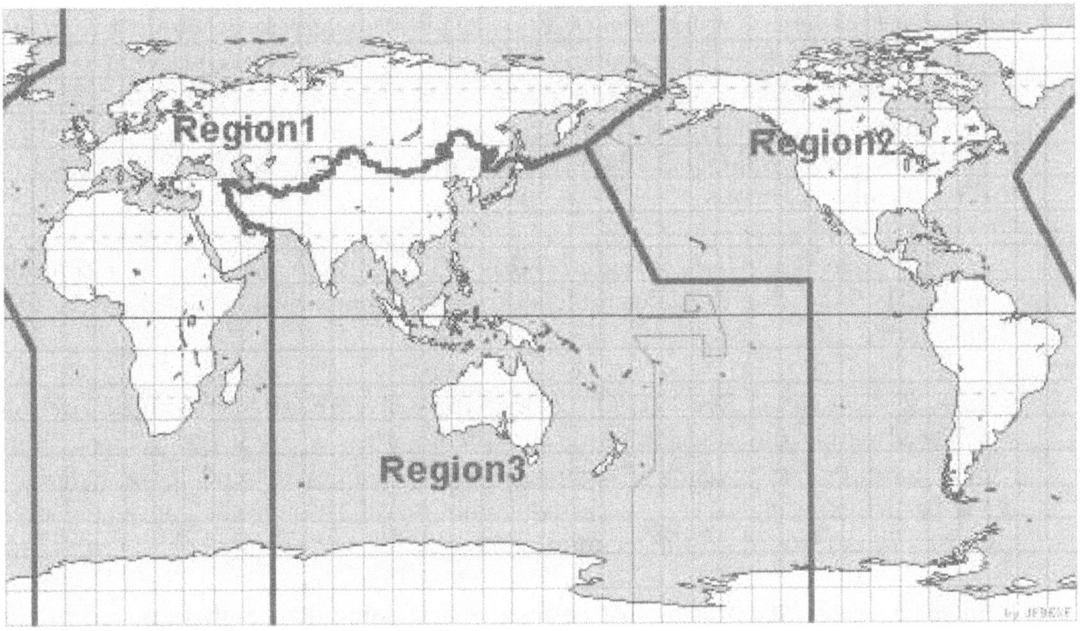

Figure 11.1 ITU regions. (*Source:* ITU.)

Electronic and Electrical Engineers (IEEE), which has developed the 802 series of standards, particularly 802.11p (WAVE) and the associated IEEE 1609 standards. See Sections 4.2.3, 4.6.7.1.1, and 4.6.7.1.2.

11.1 International Standards Organization

International Standards Organization (ISO) directives can be downloaded from: http://www.iso.org/iso/standards_development/processes_and_procedures/iso_iec_directives_and_iso_supplement.htm.

Although not the first SDO to become involved in ITS (that honor goes to CEN TC 278), ISO/TC 204 is currently the lead SDO with respect to ITS standardization, and is supported with in-vehicle aspects (such as human-machine interface) being led by TC 22, international intermodal shipping containers being led by TC 104, and on board marine vessels issues handled by TC 8.

ISO comprises member bodies from around the world. These members are national standards organizations, such as British Standards Institute (BSI), AFNOR in France, and ANSI in the United States).

ISO work is progressed through *technical committees* (TC).

The member bodies choose whether or not to participate in any ISO standards committee, and if they elect to be involved, choose to be a *participating* (P) member or an *observer* (O) member—these roles are as implied in their name.

Standards and deliverables are decided by ballot of the participating members.

There are a number of stages to developing an ISO standard.

Stage 0: Preliminary Work Item (PWI). This is an agreement within a technical committee to work on a particular subject, usually to undertake preliminary work to identify a standard or other ISO deliverable (such as a technical specification or technical report) and to identify its scope. (Typically, 18 months is allowed for this work.) Within ISO/TC 204 a PWI can only be approved if it is supported by three member countries.

Stage 1: New Work Item Proposal (NP). This is an agreement by a resolution of the TC to develop the deliverable. (Typically, 18 months is allowed for the first deliverable.)

Stage 2: Committee Draft (CD). Once the deliverable is stable and consensus achieved within the working group or subcommittee where it is developed, it its put in front of the technical committee, which agrees to send it for ballot by the participating members. Successive committee drafts may be considered until consensus is reached on the technical content.

Stage 3: Committee Stage. As soon as a first committee draft is agreed by the TC, it is registered by the ISO Central Secretariat. It is distributed for voting, by the participating members of the TC/SC. This is a 3-month ballot. At the end of the ballot a comment resolution is posted and held. The resultant document is known as a CD.

The document is then back in the hands of the working group or subcommittee to make any further technical changes that may have evolved. It is accepted that the document should be 90% complete. If more significant changes are made, the document is sent back for another CD ballot. Once agreed, it is then approved by resolution of the TC to go to draft International Standard (DIS) ballot.

Stage 4: Enquiry Stage. The DIS is circulated to all ISO member bodies by the ISO Central Secretariat for voting and comment within a period of 5 months. It is approved for submission to the next stage—as a final draft International Standard (FDIS)—if a two-thirds majority of the participating members of the TC/SC are in favor and not more than one-quarter of the total number of votes cast are negative. If the approval criteria are not met, the text is returned to the originating TC/SC for further study and a revised document will again be circulated for voting and comment as a draft International Standard.

Once approved, the document is referred to as a DIS.

If a DIS is approved without any negative votes, it may move straight to publication stage.

Stage 5: Approval Stage. The FDIS is circulated to all ISO member bodies by the ISO Central Secretariat for a final yes/no vote within a period of 2 months. If technical comments are received during this period, they are no longer considered at this stage, but registered for consideration during a future revision of the International Standard. The text is approved as an International Standard if a two-thirds majority of the participating members of the TC/SC are in favor and not more than one-quarter of the total number of votes cast are negative. If these approval criteria are not met, the standard is referred back to the originating TC/SC for reconsideration in the light of the technical reasons submitted in support of the negative votes received.

Stage 6: Publication Stage. Once a final draft International Standard has been approved, only minor editorial changes, if and where necessary, are introduced into the final text. The final text is sent to the ISO Central Secretariat which prepares the document for printing and publishing as an International Standard.

Technical Specifications and Technical Reports

Technical specifications and technical reports may be approved by ballot of the TC.

NOTE: Joint Technical Committee 1 (JTC1) documents have a slightly different balloting process involving Final Committee Draft (FCD) status. Neither ISO/TC 204 nor TC 22 are members of JTC1, so these procedures are not detailed here, but they can be obtained from: http://isotc.iso.org/livelink/livelink/fetch/2000/2122/327993/customview.html?func=ll&objId=327993.

As you can see, the process has several stages and therefore it takes time to achieve a full standard.

Excluding standards that are fast tracked from other SDOs, the minimum period of time before a standard is achieved is a little more than 2 years, but it is more typical to take 4 or 5 years, and some, not uncommonly, take longer.

The latest version of ISO directives for the development of standards can be obtained from: http://www.iso.org/iso/standards_development/processes_and_procedures/iso_iec_directives_and_iso_supplement.htm.

11.2 International Telecommunication Union

ITU directives can be obtained from: http://www.itu.int/ITU-T/.

The ITU is the United Nations specialized agency in the field of telecommunications. The ITU Telecommunication Standardization Sector (ITU-T) is a permanent organ of ITU. ITU-T is responsible for studying technical, operating, and tariff questions and issuing recommendations on them with a view to standardizing telecommunications on a worldwide basis.

The ITU does not develop standards (see the opening paragraphs of this chapter). It develops "recommendations." Originally these were only recommendations to regulators, but in recent years it has increased its scope to provide the equivalent of what others would call standards.

ITU recommendations are developed by study groups and are normally ratified at the World Telecommunication Standardization Assembly (WTSA), held every 4 years. However, there are faster procedures (which are described below).

The traditional approval process (TAP) is that study groups should apply the process described below for seeking the approval of all draft new and revised recommendations as soon as they have been developed to a mature state.

Cases where approval of new or revised recommendations should be deferred for consideration at a WTSA are:

- For recommendations of an administrative nature concerning ITU-T as a whole;
- Where the study group concerned considers it desirable that WTSA itself debate and resolve particularly difficult or delicate issues;
- Where attempts to gain agreement within the study groups have failed due to nontechnical issues such as differing views on policy.

Study groups are encouraged to establish an internal editing group to review the texts of new and revised recommendations for suitability in each of the official and working languages.

The text of the draft new or revised recommendation must be available to TSB in a final edited form in at least one of the official and working languages at the time that the director makes the announcement of the intended application of the approval procedure. A summary is provided that reflects the final edited form of the draft.

Approval may only be sought for a draft new or revised recommendation, within the study group's mandate as defined by the "Questions" allocated to it, in accordance with No. 192 of the Convention. Alternatively, or additionally, approval may be sought for amendment of an existing recommendation within the study group's responsibility and mandate (see Resolution 2).

In the interests of stability, once a new or revised recommendation has been approved, approval should not normally be sought within a reasonable period of time for any further amendment of the new text or the revised portion, respectively, unless the proposed amendment complements rather than changes the agreement reached in the previous approval process or a significant error or omission is discovered. As a guideline, in this context "a reasonable period of time" would be at least 2 years in most cases.

Any member state considering itself to be adversely affected by a recommendation approved in the course of a study period may refer their case to the Director, who shall submit it to the relevant study group for prompt attention.

Consultation of the member states encompasses the time period and procedures beginning with the announcement by the Director of TSB of the intention to apply the approval procedure. The Director shall request member states' opinions within this period on whether they assign authority to the study group that the draft new or revised recommendations should be considered for approval at the study group meeting.

Should any member states be of the opinion that consideration for approval shall not proceed, they should advise their reasons for disapproving and indicate the possible changes that would facilitate further consideration and approval of the draft new or revised recommendation.

If 70% or more of the replies from member states support consideration for approval at the study group meeting (or if there are no replies), the Director should advise the chairman that consideration of the approval may proceed.

If less than 70% of the replies received by the due date support consideration for approval at the study group meeting, the Director should advise the chairman that consideration of the approval may not proceed at that meeting.

Any comments received along with responses to the consultation shall be collected by TSB and submitted as a temporary document to the next meeting of the study group.

This process provides an interim recommendation, which is subsequently ratified at the next WTSA, or World Radio Conference (the ITU-R) equivalent.

As this only happens every 4 years, there is also an accelerated approval process (AAP), which is an ITU-T fast-track approval procedure for technical standards that claims that a standard can be approved in as little as 8 weeks. This, of course, almost never happens; however, since AAP was launched in January 2001, more than 1,000 standards have been approved with this process.

Under the AAP process, where a draft new or revised recommendation falls within the mandate of more than one study group, the chairman of the study group proposing the approval should consult and take into account the views of any other study group chairmen concerned before proceeding with the application of this approval procedure. There shall then be a "last call and additional review."

The last call encompasses the 4-week time period and procedures beginning with the Director's announcement of the intention to apply the alternative approval process.

If TSB has received a statement(s) indicating that the use of intellectual property, protected by one or more copyright(s) or patent(s), issued or pending, may be required in order to implement a draft recommendation, the Director shall post this information on the ITU-T Web site.

The Director of TSB shall advise the Directors of the other two bureaus that member states and sector members are being asked to comment on approval of a proposed new or revised recommendation.

During the last call, should any member state or sector member be of the opinion that the draft new or revised recommendation should not be approved, they should advise their reasons for disapproving and indicate the possible changes that would facilitate further consideration and approval of the draft new or revised recommendation. TSB will make the comments available to the membership of ITU-T.

If no comments other than comments indicating typographical error(s) are received by the end of the last call, the draft new or revised recommendation is considered as approved, and the typographical errors are corrected.

If comments, other than those indicating typographical errors, are received by the end of the last call, the study group chairman, in consultation with TSB, makes the judgment whether:

- A planned study group meeting is sufficiently close to consider the draft recommendation for approval, in which case the procedures regarding approval at a study group meeting are applied.
- To save time and/or because of the nature and maturity of the work, comment resolution should be initiated under the direction of the study group chairman. This will be accomplished by appropriate study group experts, via electronic correspondence or at meetings. Revised, edited draft text is prepared as appropriate and the procedures beginning in 4.4.3 are applied.

After comment resolution is completed, and the revised and edited draft text is made available, the study group chairman, in consultation with TSB, makes the judgment whether:

- A planned study group meeting is sufficiently close to consider the draft Recommendation for approval, in which case the procedures in 4.6 are applied.
- To save time and/or because of the nature and maturity of the work, an additional review should be initiated, in which case the procedures in 4.5 are applied.

The additional review encompasses a 3-week time period and will be announced by the Director. The text (including any revisions as a result of comment resolution) of the draft recommendation in a final edited form and comments from the last call must be made available to TSB at the time that the Director makes the announcement of the additional review. Reference shall be provided to the documentation where the text of the draft recommendation and last call comments to be considered may be found. If no comments other than comments indicating typographical error(s) are received by the end of the additional review, the recommendation is considered as approved, and the typographical errors are corrected by TSB.

If comments other than comments indicating typographical errors(s) are received by the end of the additional review, then the procedures regarding approval at a study group meeting are applied.

The Director shall explicitly announce the intention to approve the draft recommendation at least 3 weeks prior to the study group meeting. The Director shall include the specific intent of the proposal in summarized form.

The Director of TSB shall promptly notify the membership of the results (indicating approval or nonapproval) of the last call and additional review. The Director shall arrange for this information to also be included in the next available ITU Operational Bulletin. Within this same time period, the Director shall also ensure that any recommendation approved is available online, with an indication that the recommendation may not be in its final publication form.

See ITU T Recommendation A.11 concerning the publication of new and revised recommendations.

11.2.1 International Telecommunication Union—Telecommunications

The procedures for ITU-T are as described above.

11.2.2 International Telecommunication Union—Radiocommunications

The procedures for ITU-R are as described above, except that the World Radio Conference, which also meets every 4 years, may replace the WTSA.

11.3 International Electrotechnical Commission

International Electrotechnical Commission (IEC) procedures are generally the same as ISO. See Section 11.1 for the general process. For IEC-specific requirements see: http://www.iec.ch/tiss/iec/Directives-IECSup-Ed2.pdf.

11.4 Comité Européen de Normalisation

The European Committee for Standardization (CEN) has aligned its process to be in line with that of ISO, the only difference being that there is a 6-month comment period for drafts before they are submitted to CD ballot. Therefore, see Section 11.1 above.

CEN Directives can be obtained from: http://www.cen.eu/boss/production/production+processes+-+index/production+processes+-+index.asp.

11.5 European Telecommunications Standards Institute

ESTI standards development directives can be obtained from: http://portal.etsi.org/Directives/home.asp.

One of ETSI's advantages amongst SDOs is that it is able to develop and approve standards more rapidly.

The standards making process (SMP) is the process applied for the technical organization's production of standards and deliverables. The general conditions for the SMP are defined by the ETSI directives, in particular the technical working procedures.

The input to the process is an existing (as yet known or unknown) market need for standardization. The output is the broad application of the produced deliverables in the market place.

The SMP consists of five main elements (subprocesses) with their own distinct objectives, inputs, and outputs. Each subprocess is defined to the level of which operational tasks are performed, and where in the ETSI organization they are performed.

The whole ETSI organization is in one way or the other involved in either operation of the SMP or in direct or indirect support of it. The main technical activities are performed in the technical bodies of the technical organization. The main direct support to those activities is provided by the Standards Making Support (SMS) department of the ETSI Secretariat.

The process recognizes that the times when initiatives to standardization were taken only when products and services were already available are since long gone. This is particularly the case for telecommunications where standardization precedes or goes hand in hand with the design and development processes.

The process itself consists, in broad terms, of identifying needs for standardization in the subject areas defined by the ETSI Statutes and Rules of Procedure and defining (or creating) the suitable organization for such standardization within ETSI.

The creation of a new standardization area is manifested by the establishment of the new technical body or the amendment of the Terms of Reference of the Project Requirements Definition of an existing technical body.

The identification, definition, approval, and adoption of work items are the main elements of the conception phase (even if work items may have been envisaged already during the inception process).

The input is identified standardization needs in the area. These work items may either be entirely new, leading to new deliverables, or a new version of an existing deliverable (maintenance work item).

The output is a work item, adopted by the ETSI membership (normally by ballot within an extant TC).

A proposal for a work item may come from inside or outside the technical body. The technical body may approve the work item if at least four ETSI members volunteer to support the work. The adoption is formally done by the ETSI membership (the existence of new work items is made known via the ETSI Web site, and members who disagree with the item may within a 30-day period oppose its adoption into the ETSI Work Program.

The drafting usually takes place in a small team (rapporteur group) lead by a rapporteur. The work is largely done by "correspondence" (i.e., by exchange of documents via the ETSI DocBox server and LISTSERV email exploder facilities).

When the draft by the rapporteur group is considered ready, the draft deliverable is handed over to the working group (when it exists) for approval. The formal approval for further processing or, in the case of ETSI technical specifications or ETSI technical reports, approval and adoption can only be done by the technical body, either at a meeting or by correspondence.

Some drafting activities for a technical body are performed by specialist task forces (STF) located at the ETSI Secretariat.

While the drafting process is, in principle, the same for all ETSI deliverables, the process elements of the adoption process depend on the type of deliverable being processed.

ETSI Technical Specification and ETSI Technical Report

For ETSI technical specifications (TS) and ETSI technical reports (TR), the technical body approval and adoption take place at the same time (one combined decision). Publication is then the only element in the adoption process.

The publication process element consists of final editing of the Word for Windows version of the adopted TS or TR, archiving, and publication in PDF format. The published deliverable is then made available for distribution via the ETSI Web server and constitutes a part of the ETSI Documentation Service (EDS).

ETSI Standard and ETSI Guide

After technical body approval, the draft ETSI standard (ETSI ES) or ETSI guide (ETSI EG) is edited by the SMS technical officer and the editing and document management team (SMS/EDM) and made available to the ETSI membership (full members and associate members) for voting in accordance with the membership approval procedure (MAP) defined in the ETSI Rules of Procedure and Technical Working Procedures.

Voting with MAP is done via a Web browser interface (member voting application). The voting period is 60 days (in compliance with World Trade Organization timing requirement). The deliverable is adopted if at least 71% of the weighted member votes cast are in favor of the draft.

After adoption, the deliverable is finally edited, archived, and published in PDF format.

European Standard—EN (Telecommunications Series)

European standards produced by ETSI, ETSI EN (telecommunications series), are after the technical body approval entered into one of the two approval procedures stipulated by the ETSI Rules of Procedure:

- One-step approval procedure (OAP);
- Two-step approval procedure (TAP);

EN—One-Step Approval Procedure. This procedure is used when the draft is considered by the technical body to be mature, or is a new version of an ETSI EN. After editing, the draft is made available to the ETSI national standards organizations (NSOs) for a process where each NSO establishes the national position for the vote (i.e., performs national consultation in the territory of the NSO concerned) (the exact implementation may vary from one NSO to another).

Although the vote is formally cast by the Head of National Delegation to ETSI (who in the majority of cases is not a representative of the NSO, but may be—in the case where the NSO is a member of ETSI), the procedure including the voting is often referred to as "NSO voting."

The procedure is basically the same as that for the membership approval procedure, except that the period for NSO voting is longer (120 days). The deliverables are made available to the NSOs via file transfer via Internet and CD-ROM (to be phased out). The NSO sends the national position for the vote to ETSI via a Web-based electronic voting application.

The deliverable will be adopted if at least 71% of the weighted national votes cast are in favor of the draft.

EN—Two-Step Approval Procedure. This procedure, which is normally obligatory for so-called harmonized standards, involves the NSOs at two stages with, as necessary, resolution actions taken by the technical body responsible for the draft.

The first NSO involvement, called NSO public enquiry, lasts 120 days. The second, called NSO voting, is 60 days.

The comments received, if any, from the public enquiry are used by the technical body to decide on whether changes should be made to the draft before it is sent to the NSOs for their consultation and establishment of national position for the vote.

Combined Processes

In order to make the results of the work of the technical body available to the market at an early stage, some of the above processes may be combined in such a way that two deliverables with identical content are processed/published in parallel.

For example, if the intention is to publish the draft as an ETSI EN (telecommunications series), but only after application of the TAP, the editing of the ETSI EN (subprocess editing prior to public enquiry) also covers the publication of an ETSI TS with identical contents.

Parallel ETSI ES and ETSI TS processing is also possible, but the time gain in this case is less pronounced.

NOTE: Detailed rules for the approval procedures described above may be found in the Technical Working Procedures (TWP). TWP also defines the rules which apply to the previous regime deliverables—for example, European Telecommunication Standards (ETS)—maintained by ETSI.

The operations of ETSI and its bodies are governed by the ETSI Directives (containing ETSI Statutes, ETSI Rules of Procedure, ETSI Board Working Procedures, Powers and Functions of the Board, Terms of Reference of the Operational Coordination Group, and ETSI Technical Working Procedures).

11.6 Internet Engineering Task Force

The Internet Engineering Task Force (IETF) directives can be obtained from: http://www.ietf.org/IETF-Standards-Process.html.

Internet Standards

NOTE: Peculiarly, all IETF documents have to be provided in courier/ASCII.

The Internet is a loosely organized international collaboration of autonomous, interconnected networks that supports host-to-host communication through voluntary adherence to open protocols and procedures defined by Internet Standards. There are also many isolated interconnected networks, which are not connected to the global Internet but use the Internet Standards.

The Internet Standards process described in this document are concerned with all protocols, procedures, and conventions that are used in or by the Internet, whether or not they are part of the TCP/IP protocol suite. In the case of protocols developed and/or standardized by non-Internet organizations, however, the Internet Standards process normally applies to the application of the protocol or procedure in the Internet context, not to the specification of the protocol itself.

In general, an Internet Standard is a specification that is stable and well understood, is technically competent, has multiple, independent, and interoperable implementations with substantial operational experience, enjoys significant public support, and is recognizably useful in some or all parts of the Internet.

The process of creating an Internet Standard is straightforward: a specification (1) undergoes a period of development and several iterations of review by the Internet community and revision based upon experience, (2) is adopted as a standard by the appropriate body (see below), and (3) is published.

In practice, the process is more complicated, due to the following:

- The difficulty of creating specifications of high technical quality;
- The need to consider the interests of all of the affected parties;
- The importance of establishing widespread community consensus;

- The difficulty of evaluating the utility of a particular specification for the Internet community.

The goals of the Internet Standards process are:

- Technical excellence;
- Prior implementation and testing;
- Clear, concise, and easily understood documentation;
- Openness and fairness;
- Timeliness.

At each stage of the standardization process, a specification is repeatedly discussed and its merits debated in open meetings and/or public electronic mailing lists, and it is made available for review via worldwide online directories.

- These procedures are explicitly aimed at recognizing and adopting generally accepted practices. Thus, a candidate specification must be implemented and tested for correct operation and interoperability by multiple independent parties and utilized in increasingly demanding environments before it can be adopted as an Internet Standard.
- These procedures provide a great deal of flexibility to adapt to the wide variety of circumstances that occur in the standardization process. Experience has shown this flexibility to be vital in achieving the goals listed above.

The goal of technical competence, the requirement for prior implementation and testing, and the need to allow all interested parties to comment, all require significant time and effort. On the other hand, today's rapid development of networking technology demands timely development of standards. The Internet Standards process is intended to balance these conflicting goals. The process is believed to be as short and simple as possible without sacrificing technical excellence, thorough testing before adoption of a standard, or openness and fairness.

From its inception, the Internet has been, and is expected to remain, an evolving system whose participants regularly factor new requirements and technology into its design and implementation. Users of the Internet and providers of the equipment, software, and services that support it should anticipate and embrace this evolution as a major tenet of Internet philosophy.

Requests for Comments (RFCs)

Each distinct version of an Internet Standards–related specification is published as part of the Request for Comments (RFC) document series. This archival series is the official publication channel for Internet Standards documents and other publications of the Internet Engineering Steering Group (IESG), Internet Architecture Board (IAB), and the Internet community. RFCs can be obtained from a number of Internet hosts using anonymous FTP, gopher, World Wide Web, and other Internet document-retrieval systems.

The RFC series of documents on networking began in 1969 as part of the original ARPA wide-area networking (ARPANET) project. RFCs cover a wide range of topics in addition to Internet Standards, from early discussion of new research concepts to status memos about the Internet. RFC publication is the direct responsibility of the RFC Editor, under the general direction of the IAB.

The status of Internet protocol and service specifications is summarized periodically in an RFC entitled "Internet Official Protocol Standards." This RFC shows the level of maturity and other helpful information for each Internet protocol or service specification.

Not all specifications of protocols or services for the Internet should or will become Internet Standards or BCPs. Such nonstandards track specifications are not subject to the rules for Internet standardization. Nonstandards track specifications may be published directly as "experimental" or "informational" RFCs at the discretion of the RFC Editor in consultation with the IESG.

Internet-Drafts

During the development of a specification, draft versions of the document are made available for informal review and comment by placing them in the IETF's "Internet-Drafts" directory, which is replicated on a number of Internet hosts. This makes an evolving working document readily available to a wide audience, facilitating the process of review and revision.

An Internet-Draft that is published as an RFC, or that has remained unchanged in the Internet-Drafts directory for more than 6 months without being recommended by the IESG for publication as an RFC, is simply removed from the Internet-Drafts directory. At any time, an Internet-Draft may be replaced by a more recent version of the same specification, restarting the 6-month timeout period.

An Internet-Draft is NOT a means of "publishing" a specification; specifications are published through the RFC mechanism described in the previous section. Internet-Drafts have no formal status, and are subject to change or removal at any time.

RFCs are not standards and should not be used as reference documents.

Internet Standard Specifications

Specifications subject to the Internet Standards process fall into one of two categories: technical specification (TS) and applicability statement (AS).

Technical Specification. A technical specification is any description of a protocol, service, procedure, convention, or format. It may completely describe all of the relevant aspects of its subject, or it may leave one or more parameters or options unspecified. A TS may be completely self-contained, or it may incorporate material from other specifications by reference to other documents (which might or might not be Internet Standards).

A TS shall include a statement of its scope and the general intent for its use (domain of applicability). Thus, a TS that is inherently specific to a particular context shall contain a statement to that effect. However, a TS does not specify requirements for its use within the Internet; these requirements, which depend on the particular context in which the TS is incorporated by different system configurations, are defined by an applicability statement.

Applicability Statement. An applicability statement specifies how, and under what circumstances, one or more TSs may be applied to support a particular Internet capability. An AS may specify uses for TSs that are not Internet Standards, as discussed in Section 7.

An AS identifies the relevant TSs and the specific way in which they are to be combined, and may also specify particular values or ranges of TS parameters or subfunctions of a TS protocol that must be implemented. An AS also specifies the circumstances in which the use of a particular TS is required, recommended, or elective.

An AS may describe particular methods of using a TS in a restricted "domain of applicability," such as Internet routers, terminal servers, Internet systems that interface to ethernets, or datagram-based database servers.

The broadest type of AS is a comprehensive conformance specification, commonly called a "requirements document," for a particular class of Internet systems, such as Internet routers or Internet hosts.

Requirement Levels. An AS shall apply one of the following requirement levels:

- *Required:* Implementation of the referenced TS, as specified by the AS, is required to achieve minimal conformance.
- *Recommended:* Implementation of the referenced TS is not required for minimal conformance, but is desirable.
- *Elective:* Implementation of the referenced TS is optional within the domain of applicability of the AS.

Two additional requirement level designations are available:

- *Limited Use:* The TS is considered to be appropriate for use only in limited or unique circumstances.
- *Not Recommended:* A TS that is considered to be inappropriate for general use is labeled "Not Recommended."

The Internet Standards Track

Specifications that are intended to become Internet Standards evolve through a set of maturity levels known as the "standards track."

Internet specifications go through stages of development, testing, and acceptance. Within the Internet Standards process, these stages are formally labeled "maturity levels."

Proposed Standard. The entry-level maturity for the standards track is Proposed Standard. A specific action by the IESG is required to move a specification onto the standards track at the Proposed Standard level.

A Proposed Standard specification is generally stable, has resolved known design choices, is believed to be well understood, has received significant community review, and appears to enjoy enough community interest to be considered valuable. However, further experience might result in a change or even retraction of the specification before it advances.

Implementers should treat Proposed Standards as immature specifications. It is desirable to implement them in order to gain experience and to validate, test, and clarify the specification.

Draft Standard. A specification from which at least two independent and interoperable implementations from different code bases have been developed, and for which sufficient successful operational experience has been obtained, may be elevated to the Draft Standard level. For the purposes of this section, "interoperable" means to be functionally equivalent or interchangeable components of the system or process in which they are used. If patented or otherwise controlled technology is required for implementation, the separate implementations must also have resulted from separate exercise of the licensing process. Elevation to Draft Standard is a major advance in status, indicating a strong belief that the specification is mature and will be useful.

The requirement for at least two independent and interoperable implementations applies to all of the options and features of the specification. The working group chair is responsible for documenting the specific implementations which qualify the specification for Draft or Internet Standard status along with documentation about testing of the interoperation of these implementations.

Internet Standard. A specification for which significant implementation and successful operational experience has been obtained may be elevated to the Internet Standard level. An Internet Standard (which may simply be referred to as a standard) is characterized by a high degree of technical maturity and by a generally held belief that the specified protocol or service provides significant benefit to the Internet community.

A specification that reaches the status of standard is assigned a number in the STD series while retaining its RFC number.

Non-Standards Track Maturity Levels. Specifications that are not on the standards track are labeled with one of three "off-track" maturity levels: Experimental, Informational, or Historic. The documents bearing these labels are not Internet Standards in any sense.

Advancing in the Standards Track. A specification shall remain at the Proposed Standard level for at least 6 months.

A specification shall remain at the Draft Standard level for at least 4 months, or until at least one IETF meeting has occurred, whichever comes later.

11.7 World Wide Web Consortium

Directives for World Wide Web Consortium (W3C) specifications and guidelines, called "technical reports" can be obtained from: http://www.w3.org/2005/10/Process-20051014/tr.html#Reports.

The maturity level of a published technical report indicates its place in the development process. The maturity levels "Working Draft" and "Working Group

Note" represent the possible initial states of a technical report in the development process. The maturity levels "Recommendation," "Working Group Note," and "Rescinded Recommendation" represent the possible end states.

Working Draft (WD)
A Working Draft is a document that W3C has published for review by the community, including W3C members, the public, and other technical organizations. Some, but not all, Working Drafts are meant to advance to Recommendation; see the document status section of a Working Draft for the group's expectations.

Maturity Levels of the Recommendation Track
In addition to Working Drafts that are meant to advance to Recommendation, the other maturity levels of the Recommendation Track are as follows.

Candidate Recommendation (CR). A Candidate Recommendation is a document that W3C believes has been widely reviewed and satisfies the Working Group's technical requirements. W3C publishes a Candidate Recommendation to gather implementation experience.

Proposed Recommendation (PR). A Proposed Recommendation is a mature technical report that, after wide review for technical soundness and implementability, W3C has sent to the W3C Advisory Committee for final endorsement.

W3C Recommendation (REC). A W3C Recommendation is a specification or set of guidelines that, after extensive consensus-building, has received the endorsement of W3C members and the Director. W3C recommends the wide deployment of its Recommendations.

NOTE: W3C Recommendations are similar to the standards published by other organizations.

Reviews and Review Responsibilities
Experience shows that the following elements help build consensus around technical reports:

- Frequent publication (see the Working Group "Heartbeat" requirement);
- Early review, to find errors quickly and decrease the chances of diverging technologies;
- Wide review, including from other groups in and outside of W3C.

A document receives review from the moment it is first published. Starting with the First Public Working Draft until the start of a Last Call review, a Working Group should formally address any substantive review comment about a technical report and should do so in a timely manner.

Starting with a Last Call review up to the transition to Proposed Recommendation, a Working Group must formally address any substantive review comment about a technical report and should do so in a timely manner. When a Working Group requests to advance to Candidate Recommendation or beyond, the Director expects positive documentation that issues have been formally addressed (e.g., in an issues list that shows their disposition).

The Director must formally address any substantive issue raised by Advisory Committee representatives during Proposed Recommendation, Proposed Edited Recommendation, and Proposed Rescinded Recommendation review periods. The Working Group must communicate to the Director (usually through the Team Contact) any substantive issues raised during Proposed Recommendation, Proposed Edited Recommendation, and Proposed Rescinded Recommendation review periods by parties other than Advisory Committee representatives.

Reviewers should not send substantive technical reviews late on the Recommendation track. Reviewers should not expect that a Working Group will readily make substantive changes to a mature document. The more evidence a Working Group can show of wide review, the less weight substantive comments will carry when provided late on the Recommendation Track. Worthy ideas should be recorded even when not incorporated into a mature document.

The Working Group must be able to show evidence of having attempted to respond to and satisfy reviewers. Reviewers may register a Formal Objection any time they are dissatisfied with how a Working Group has handled an issue.

A Working Group should negotiate review schedules with other groups expected to review a document, including relevant liaisons.

There are two formal review periods with fixed durations when advancing to Recommendation: after a Last Call announcement and after a Call for Review of a Proposed Recommendation. Out of consideration for the Working Group, reviewers should send their comments early in a review period.

A Working Group should not start a new review before the scheduled end of an ongoing review (e.g., do not start a new Last Call review before the scheduled end of an ongoing Last Call review).

Ordinarily, reviewers should not raise substantive technical issues about a technical report after the end of a Last Call review period. However, this does occur, and as stated above, a Working Group's requirement to formally address those issues extends until the start of a Proposed Recommendation review period. However, to allow the Working Group to make progress on a technical report, the Working Group may decline to make substantive changes to address issues raised between the end of a Last Call review period and publication of a Recommendation. A reviewer may register a Formal Objection.

Advancing a Technical Report to Recommendation

W3C follows these steps when advancing a technical report to Recommendation:

1. Publication of the First Public Working Draft;
2. Last Call announcement;
3. Call for Implementations;
4. Call for Review of a Proposed Recommendation;
5. Publication as a Recommendation.

In general, Working Groups embark on this journey with the intent of publishing one or more Recommendations. However, W3C may end work on a technical report at any time, or may require a Working Group to conduct further work, possibly repeating one or more steps.

Candidate Recommendation
At this step, W3C believes the technical report is stable and appropriate for implementation. The technical report *may* still change based on implementation experience.

11.8 Object Management Group

Object Management Group (OMG) is an international, open membership, not-for-profit computer industry consortium.

OMG uses Task Forces to develop enterprise integration standards for a wide range of technologies, and an even wider range of industries. OMG's claim that its modeling standards "enable powerful visual design, execution and maintenance of software and other processes."

OMG's middleware standards and profiles are based on the Common Object Request Broker Architecture (CORBA) (see Section 5.2.6) and support a wide variety of industries. All of its specifications may be downloaded without charge from the OMG Web site: http://www.omg.org.

Any organization may join OMG and participate in its standards-setting process. OMG uses a one-organization-one-vote policy.

OMG membership comprises hundreds of organizations, with "half being software end-users in over two dozen vertical markets, and the other half representing virtually every large organization in the computer industry and many smaller ones."

11.9 United Nations Centre for Trade Facilitation and Electronic Business

United Nations Centre for Trade Facilitation and Electronic Business (UN/CEFACT) procedures can be obtained from: http://www.unece.org/cefact/cf_docs.htm.

UN/CEFACT has a mission "to improve the ability of business, trade and administrative organizations, from developed, developing and transitional economies, to exchange products and relevant services effectively—and so contribute to the growth of global commerce."

CEFACT comprises participation from United Nations member states, intergovernmental agencies, sector and industry associations recognized by the Economic and Social Council of the United Nations (ECOSOC), and the private sector.

The principal focus of UN/CEFACT is facilitating national and international transactions through the simplification and harmonization of processes, procedures and information flows, and in so doing to contribute to the growth of global commerce. One way in which UN/CEFACT fulfils this mission is by publishing standards, specifications, recommendations, and user guides (collectively referred to as "publications").

UN/CEFACT facilitates the development of e-business standards that have the objective to cross all international boundaries and help lower transaction costs, simplify data flow, and reduce bureaucracy.

A cornerstone of the UN/CEFACT approach is the Core Component Technical Specification (CCTS).

Publication Types

UN/CEFACT produces the following types of publications:

- *UN/CEFACT Technical Specifications:* Specifications established by consensus within the UN/CEFACT Forum to establish how one or more Business Standards and/or Recommendations shall be developed;
- *UN/CEFACT Business Standards:* Specifications established by consensus within the Forum that provide rules, guidelines, and/or principals related to activities in the context of trade facilitation or electronic business;
- *UNECE Recommendations:* Trade facilitation or electronic business standards that provide formal guidance to governments, the private sector, and the business community;
- *UN/CEFACT Implementation Guides:* Informative (in contrast to normative) documents and/or audio/video productions that provide guidance to publication implementers.

UN/CEFACT produces initial versions of publications by executing a process called the Open Development Process (ODP).

UN/CEFACT Activity Initiation

The first step is when a stakeholder expresses a need, called the UN/CEFACT Activity Request. A stakeholder is a person or organization that would like UN/CEFACT to do something for them. Stakeholders may initially express their need in any written form and consider it officially submitted once it has been delivered to any Forum Management Group (FMG) member or to TBG16. The need, once written and delivered, is called a CEFACT Activity Request (CAR).

When an FMG member receives a CAR, they will immediately forward it to TBG16 where initial processing occurs. TBG16 reviews the CAR, categorizes it as a Simple CAR or a Project CAR, and assigns the CAR to the appropriate permanent group.

Open Development Process Overview

ODP steps are as follows:

- ODP1: Project Proposal and Team Formation;
- ODP2: Requirements Gathering;
- ODP3: Internal Draft Development;
- ODP4: Internal Review;
- ODP5: Public Review;
- ODP6: Implementation Verification;

- ODP7: Publication;
- ODP8: Maintenance.

Typical artifacts produced by ODP1 include:

- Project proposal;
- Call for participation;
- Initial contributions.

Open Development Process 5: Public Review

The UNECE Secretariat provides links on the UNECE Web site to the Public Draft and related information. The FMG notifies Heads of Delegation and various e-mail distribution list subscribers that the Public Draft is available for review and provides them with review-process details. The project team processes comments and posts updated Public Drafts and comment logs to the PG Web site or the UNECE Web site (through the Secretariat). The comment/update/posting cycle continues until the PG approves a project team recommendation to conclude ODP5. While the criteria, evaluation, and ultimate decision to conclude ODP5 is left to the PG, the PG must ensure that the project team has met all comment processing requirements. The draft resulting from this step is called an Implementation Draft.

Open Development Process 6: Implementation Verification

The UNECE Secretariat provides links on the UNECE Web site to the implementation. The FMG notifies Heads of Delegation and various e-mail distribution list subscribers that the Implementation Draft is available for implementation verification and provides them with details regarding the process for submitting comments. The project team processes comments and posts updated Implementation Drafts and comment logs to the PG Web site or UNECE Web site (through the Secretariat). The comment/update/posting cycle continues until at least two independent implementations have been confirmed and the PG approves a project team recommendation to conclude ODP6. While the criteria, evaluation, and ultimate decision to conclude ODP6 is left to the PG, the PG must ensure that the project team has met all comment processing requirements. The draft resulting from this step is called a Final Draft.

Open Development Process 7: Publication

The UNECE Secretariat updates the Final Draft to meet UNECE requirements and publishes the result (called a publication) to the UNECE Web site.

11.10 Institute of Electrical and Electronics Engineers

Institute of Electrical and Electronics Engineers (IEEE) directives and documents for standards development can be downloaded from: http://standards.ieee.org/resources/index.html#guides.

A project first requires a sponsor.

Sponsor

The sponsor is the organization that assumes responsibility for a particular standards idea within the IEEE. There has to be some part of the IEEE that takes responsibility for the technical content of the document, and this oversight is provided by the sponsor. The sponsor is responsible for determining the scope and nature of the technical content.

Project Authorization Request

The Project Authorization Request (PAR) is the official document that authorizes work on a standards project in the IEEE. PARs are approved by the IEEE Standards Board based on a review and recommendation from the New Standards Committee (NesCom), one of several Board committees.

Officially, a standards project does not exist until a PAR is approved. However, often the members of a potential working group will have gathered to work on the PAR and to gain the support of their potential sponsor.

The PAR is a small, structured, and highly detailed document that essentially states the reason why the project exists and what it intends to do.

Officially, a standard cannot take longer than 4 years to complete. After this, the sponsor has to request an extension of time to finalize the project.

Types of Balloting

The IEEE now has three types of standards balloting. The first is the traditional balloting process by individuals only. The second allows a mixed balloting group made of nonindividuals and individuals. The third allows a ballot group made up of nonindividuals only (nonindividuals can be entities like corporations, organizations, and so on).

In all of these types of balloting groups, each entity (individual or not) has one vote. This reflects the membership options in the IEEE Standards Association. Entities can name a representative and an alternate to cover personnel issues. In mixed balloting, a person can vote for himself or herself and also represent an organization.

Working Group

Once the PAR is approved, there is then an officially sanctioned working group preparing an IEEE document. Keep in mind that the working group is independent from the balloting group. The latter is formed by special rules and usually is not determined until right before the ballot starts.

IEEE working groups are open to anyone to participate—participants do not have to be IEEE-SA members.

However, it is usual to have a procedure for determining voting rights in the working group. This can be as simple as "anyone can vote anytime," to more elaborate rules that allow someone to vote after attending a certain number of meetings (and the right of voting is contingent on continued attendance). However, none of these voting rules would preclude an individual's right to participate and comment at any meeting.

CHAPTER 12

National and Other Standards Development Organizations (SDO) for ITS

12.1 National Standards

National standards relevant to ITS are provided in Annex H of the e-book version of this publication.

12.2 Other Organizations Relevant to ITS Standards

12.2.1 Asia-Pacific Economic Cooperation

The Asia-Pacific Economic Cooperation (APEC) works in three broad areas to meet the Bogor Goals of free and open trade and investment in the Asia-Pacific region, by 2010 for developed economies and by 2020 for developing economies.

Known as APEC's Three Pillars, APEC focuses on three key areas:

- Trade and investment liberalization;
- Business facilitation;
- Economic and technical cooperation.

The outcomes of these three areas enable APEC Member Economies to strengthen their economies by pooling resources within the region and achieving efficiencies. Tangible benefits are also delivered to consumers in the APEC region through increased training and employment opportunities, greater choices in the marketplace, cheaper goods and services, and improved access to international markets.

Trade and Investment Liberalization

Trade and investment liberalization reduces and eventually eliminates tariff and nontariff barriers to trade and investment. Protectionism is expensive because it raises prices for goods and services. Thus, trade and investment liberalization focuses on opening markets to increase trade and investment among economies, resulting in economic growth for APEC Member Economies and increased standard of living for all.

Business Facilitation

Business facilitation focuses on reducing the costs of business transactions, improving access to trade information, and aligning policy and business strategies to facilitate growth and free and open trade. Essentially, business facilitation helps importers and exporters in the Asia-Pacific region to meet and conduct business more efficiently, thus reducing costs of production and leading to increased trade, cheaper goods and services, and more employment opportunities due to an expanded economy.

ITS and APEC

ITS is perceived by APEC as a tool to implement its policies. As many APEC countries are developing countries, APEC sees ITS as being of strategic importance to these economies.

APEC meetings often hold sessions on ITS related subjects such as secure trading and movement of goods.

APEC funded the development of ISO TR 28682 "A joint APEC-ISO study of progress to develop and deploy ITS Standards" (see Section 10.1).

PART VII
Conclusions

CHAPTER 13

Summary

Well, if you have made it all the way through from beginning to end, I am impressed. But the realistic likelihood is that you have not. You have, in all probability, skipped to the end for a status report.

This book is a reference work, and like a dictionary or encyclopedia, it is primarily designed for quick and easy reference, rather than an enthralling end-to-end read.

Like a dictionary or encyclopedia, an end-to-end read, or even a detailed read of the parts and chapters of this book of most interest to you, would, I hope, be an educational experience and a good way to learn about interoperability, functionality, and the usefulness of standards in the ITS sector. But the reality is that most of you will quite rightly use this work for reference, and particularly with the "soft" copy installed in your computer, it will enable you to quickly find details for relevant standards, and indeed, identify where there are relevant standards (or lack thereof) without the need to ever read the book cover to cover.

This conclusion offers my insights, which may hopefully be of use to you.

In the development of ITS standardization, a number, indeed a growing number, of people and teams have foreseen the need for standardization in the ITS sector if the aspirations for the sector are to stand a chance of being achieved. Many have worked hard and devotedly, and continue to do, so in order to start this process. No one can say how many standards will be needed or when the process will end. But the many pages of this book show that the journey is well underway.

> "The journey of a thousand miles begins with one step."
> —Lao-Tzu, *The Way of Lao-Tzu*, Chinese philosopher (604 B.C.–531 B.C.)

The very simplicity of this Chinese aphorism belies the complexity of the journey ahead and emphasizes the importance of beginnings. Every research scientist will tell you that before proceeding to a consideration of any proposition, it is first necessary to check the premises upon which the proposition rests. Every civil engineer knows that the foundations of a building must be strong and proportioned if the structure is to be viable.

The successful America author, Lisa Alther, on being questioned by some aspiring students about whether or not these students should embark on careers as authors, replied, "If you have something of value to say, and feel an impulse to write, then the most helpful thing I can tell you is to start writing." With very little adaptation, her advice seems to transfer well to the world of standards development for ITS. Those who believe in ITS standardization have something to say, and are beginning to make their mark to enable the achievement of the potential of ITS.

There is a more recent aphorism, which I believe originated in the United States, but I can neither recall nor find the author, which runs something like this: "The pioneering of any enterprise contains the soul of it." What "soul" of ITS have our pioneering efforts achieved?

No one has the power to call up in advance the total energy needed for the whole task—nor is that necessary. But everyone has the power to start. Everyone can take the first step, and use the strength he or she has, to drive forward the generation of results. By our efforts in research laboratories, and in the standards committees, have we started an evolution that will deliver our expectations for ITS and hopefully exceed them?

> In ITS we have embarked on a journey, which we now know will be long but full of promise.

Staying with the Chinese analogies, 紅軍長征 (in western script the "Cháng Zhēng"), which refers to the "Long March" of the First (or Central) Red Army from Yudu in Jiangxi province to Yan'an in Shaanxi province, started simply as a military necessity—a strategic retreat. But its journey took Mao Zedung to prominence and power, and it subsequently led to the success of the revolution, to inefficiency and pain through idealism, to the excesses of the cultural revolution, resulting in the insights of Mao's colleague and successor Deng Xiaoping and the now famous practical policy of 具有中国特色的社会主义, or "Socialism with Chinese Characteristics," which is leading to the transformation of a society that had been inward looking since 1423, and causing it to emerge as the great powerhouse economy of the world for the 21st century. That whole process has taken little more than half a century. Did Mao Zedung understand where it would lead when he coaxed his troops into the Long March? I doubt it. But he knew that it was a direction that had to be taken to move forward, and that it was necessary to move forward.

And we in ITS have set out on a journey, with our first steps, full of aspirations (and the trepidations that subsequently occur as we realized what journey we have embarked on). But in the first one to two decades of this journey, how effective have we been? What are the consequences of what we have achieved, and even if we cannot accurately envisage or predict the end point, where do the next steps lead us?

If you reread Chapter 2 of this book, with its lists of identified ITS service domains and services, it is clear that although we have achieved much and ITS is already saving lives and reducing injury and pollution, we still have a very long way to go.

But once we achieve the publication of standards and practical demonstrations of our second generation ITS communications technologies, we can rapidly embark on stage two of our journey.

Could we have reached this point earlier? If resources had been available, possibly we could have progressed standards more quickly, but as I pointed out in Section 1.1, ITS is a developing sector, and standards are being developed as the technology evolves. Indeed, the technology can *only* evolve to its potential if the technical developments are undertaken cooperatively, often in precompetitive phases, and the development of standards is an intrinsic part of this process.

If we succeed in producing widely accepted standards which result in workable systems for the next generations of ITS communications, standardization will have accelerated the speed of progress, not trailed behind it. That in itself is no mean achievement. We are not quite there yet, but we are nearly there, and no doubt revisions will have to be made as we learn from experience. But if we succeed, it will be real progress.

- The standardization process in ITS is leading the evolution of the technologies to provide ITS service provision, not trailing them.
- ITS is not an isolated universe.

This might sound to be a statement of the obvious. But if you squirrel experts in a closed group, lock them away in a meeting room, then, as fervent believers in their cause and objectives, the first thing that they are tempted to do is to begin with reinventing the wheel (I know, I have been there many times).

Instead, it is wise to think of ITS primarily as a specific application domain for communications and IT. And like all such developments, "convergence" is the name of the game. Yes, ITS may be carried by dedicated media, especially in the domains of safety and security, but in many, if not most cases, ITS data will need to be carried over backbone IT/ICT networks. So before we start to reinvent the wheel, we should always first see what is already generally available. I hope that Chapters 5, 6, and 7 will provide most of these generic references for you. The advantage here is that by using commonly accepted and used standards, you will minimize the risk for your project. And I cannot shake off my accountancy background far enough to believe anything other than that minimizing unnecessary risk to be highly desirable to help to ensure a successful project that works, and a project that stands the best chance to be cost efficient, and a project that is most likely to be timely. With new technology implementations there are plenty of risks, both technical and commercial, so why make life even more difficult and risky?

Yet, under the pressure of modern business life, too many standardization groups have spent *man years* of base R&D because they didn't have *man days* to first find out what was already available. Hopefully, this book will reduce much of that research to man minutes, or even further to man seconds (or person seconds to be politically correct), and will allow research to be quickly done in a meeting, without access to researchers or the Internet.

My chief technical editor for this project was most concerned by the length of this work—that is, the number of ITS standards, particularly in Chapter 8. For me, who was familiar with the work programs, it was not the size of Chapter 8 that surprised me during the research, but the number of generic standards in Chapter 6 and 7 that can be used or adapted. It is my personal opinion that many of us in the sector do not spend enough time becoming aware of these standards. Hopefully this book will help make this a more achievable task.

- ITS standards development should start with thorough investigation of what generic IT/ICT standards might be useful to provide the desired service, before inventing a totally custom solution for ITS.
- ITS is not an isolated universe—it is an IT/ICT application domain in the area of transportation.

I know that this statement is likely to make some of my friends, colleagues, and other dedicated professionals working in the sector to throw their arms up in horror. But at the most abstracted level, this is really the case.

Yes, much of ITS may involve wireless communications, but these are digital communications—an IT/ICT application over the air rather than copper wire or fiber optics. Yes, it is about vehicles and services to the transport sector. Sure, the ITS business is about such applications, but many of the routes to achieve these applications are generic. Yes, very specialized skills in radio, in engineering, in architecture, and in location are needed. But much of the time these skills are common to several sectors.

I believe that it was John Donne, the Jacobean metaphysical poet and preacher, who wrote, "No man is an island entire of itself," and that is the point that I am trying to make. Because although ITS is very special and a unique opportunity to those of us developing the sector, we should not believe that is an "island entire to itself." There is a lot of help out there from specialists who may not know much about your ITS application, but may have many of the IT/ICT skills that you require—indeed need. There are already many IT/ICT standards that you can use. Do not forget them. You do not have to sit in the mushroom shed and reinvent the world. Use their tools and you will probably, in most cases, make much faster progress and better standards.

What *is* special about ITS is how these unique combinations of skills can provide an improved travel experience, and, most importantly, do more to further reduce deaths and injuries in transport environments, especially road transport, than almost any other initiative, and can also significantly contribute to the achievement of reductions in the levels of pollution and emissions.

But the mission in writing this book is based on the belief—no, more than a belief, knowledge—that these objectives can only be achieved through the use of standards. *Why?* Because there is no other way to achieve the communication levels necessary or provide common mass-produced physical components. There is no other way to provide migration paths as the technology evolves. And there is no other way that equipment in the vehicle of one manufacturer can reliably and quickly communicate with the equipment in a vehicle of another manufacturer.

- ITS is in a unique position to save lives, reduce injuries, and reduce pollution and emissions. Standards play an important role in achieving these objectives.

Some ITS services rely mainly on a backbone infrastructure architecture (for example, most traffic control and prioritization systems), using wireless communications only for the "last link" where necessary. However, many ITS services will be vehicle↔vehicle or vehicle↔infrastructure. These links will work in some of the most difficult conditions imaginable. They will test communications theories and engineering to the limits, indeed beyond the limits of what is actually achievable today.

Picture, for example, an eight-lane highway (four lanes going in each direction) with vehicles traveling from 50 to 130 km/h. In any kilometer, at any moment there may be more than 1,000 vehicles present. This may be complicated by several bridges or underpasses, where for a short time vehicles crossing over or under the highway come into range (but at a different height) introducing potentially several hundred more vehicles. Of the vehicles on the highway, half will be in a relatively stable platoon network (i.e., vehicles traveling at roughly the same speed in the same direction), although "stable" is perhaps an exaggeration because the vehicles in the slow lane will be traveling at, say, 50 km/h and those in the fast lane at two and a half times their speed. And in many countries lane speed may vary regardless of lane (not a progression from the slowest lane to the fastest lane as is common in most European countries).

However, the other 500 vehicles will be traveling in the opposite direction, and will wirelessly be in contact for only fleeting moments as they progress towards each other at closing speeds in excess of 250 km/h. Even with directional antennas, radio is no respecter of the central crash barrier (where there is one present), nor do we necessarily want it to be, for one of the situations we want ITS to help us to avoid is the high-speed head-on collision.

These communications environments will test the limits of communications technologies. With 2G and 3G networks, we have made incredible progress in the last few years. We can now use GSM in crowded places at times of high demand (New Year celebrations, goals at football matches); we can use them in the metal cage of an elevator that a few years ago would have been regarded as an impenetrable "faraday cage;" we can use them on the metro and on the high-speed train. But we all know the "Network Busy" signal. We all know what happens when the train enters a tunnel. We all know the "blue screen of death" on operating systems.

Such things are not tolerable on safety-critical ITS systems, and they will have to be minimized to a higher degree than we can yet achieve with cellular phones and mobile wireless broadband.

Only military technology in high-speed aircraft and large-scale battlefield communications come anywhere near these communications traffic loadings in adverse conditions, but the demands for ITS will be beyond anything yet achieved even in the military scenarios.

Why, then, should we bother if it is all so difficult? Well, as I stated near the start of this book, 1.25 million road deaths and 30 to 35 million road injuries each and every year are compelling reasons.

There was a time when it was accepted that the only way to keep the factories of the industrial revolution of the eighteenth century and early nineteenth century functioning, was to send small boys up the chimneys to clean them. When the technology became available to use mechanical means to clean chimneys and burn furnaces more efficiently so that the chimneys clogged up less, it very quickly became unacceptable in a civilized society to allow such attrition of death, injury, and long-term health risks to young children. People quickly questioned, "How can we allow this to happen? It must stop, now." *How much more unacceptable is it in the twenty-first century to allow 1.25 million people to die and 35 million to be injured if the technology exists or can be developed to reduce this carnage?*

Another compelling reason for ITS lies in the reduction of emissions and pollution. Regardless of your position on global warming, that debate has highlighted the growth of pollution caused by transport. While individual vehicles may pollute far less than they did a decade or two ago, the growth in numbers of vehicles, and more particularly the strain on the infrastructure causing ever greater congestion, is a major problem. Vehicles emit most pollution in congested situations. ITS has already proven, even in its early stages, to have the potential to reduce congestion. Reducing congestion has a direct relationship to reducing emissions and pollution. I am not going to claim that ITS can save the planet, nor even whether the planet needs saving. But pollution kills. It damages the health of people and animals and damages plants and trees. ITS can make a significant contribution.

If then, there are compelling reasons for ITS service provision, then we have to make these systems interoperate, and the only way to do this is through standards.

- ITS is needed, but it will require robust communications in the most difficult environments; these ambitious goals will only be achieved with the support of robust standards.

Security is another problem. It is one which is not yet very well addressed outside of the electronic fee collection and possibly traffic control environments. We live in a world of hacking, especially in wireless and Internet environments. Secure freight movement issues and national security issues are raising these profiles, but there is much yet to be done, and work needs to be started in these areas.

- Robust security and encryption standards are urgently required for safe and secure ITS service provision.

In future years, while we concentrate much of our next phases of development on ITS Web service provision, which in my experience will be essential, it will be the convergence and use of IT/ICT generic Web services that will probably have most impact for ITS service provision. Between the provision of robust communications and the delivery of a specific ITS service, there are many common building blocks that are not end application specific: location, identification, communications sequences, and so on. By developing these intermediate building blocks in a generic and interoperable manner, and enabling their movement around the Internet, we can make them more secure and enable the construction and implementation of an ITS service a much faster and lower risk process. Unfortunately, separated from the end user service, it is often far harder to provide an evaluated business case for a generic Web service than an end application, which is more clearly linkable to the "bottom line," and getting the resources/funding for such Web service developments is not an easy task.

- Web service standards are needed for ITS in order to increase security and robustness, as well as to increase the speed and reduce the risk of ITS system development.

Despite my pleas to use generic standards wherever possible, we nevertheless have to recognize that ITS is a special case and the demands on its communications, and the nature of many of its services, are likely to remain unique for many years to come. This means that a considerable number of standards—far more than are currently under development—will still need to be developed, especially towards the application layer at or near that of service provision itself. Although in many cases such specifications will be purely commercial products, where the systems have to operate in many different countries, especially in Europe and South Asia, this will be particularly important.

- The work is not complete; many more standards will be required than are currently being developed, especially at the application layers.

The reader who has waded through several chapters of this book will have noted my mission to try to make SDOs cooperate and work together in this field, rather than to compete with each other. My involvement in and chairmanship of committees in multiple SDOs is not the best way to maximize consultancy income, but it is a good way to encourage cooperation and coworking between SDOs.

Despite some problems of bureaucracy, ISO and CEN cooperate well in the field of ITS, and, indeed, have done so for 15 years. CEN and ETSI cooperate well, largely by not treading in each other's areas. ITU is beginning to work in a joined up way, and through ITU-T TELEMOV, is working well to achieve cooperation and to act as a focal point for prestandardization cooperation of common investigative issues (such as software defined radio and safety messaging).

National SDOs deal quite well with national level issues, which are often special cases. However, there are instances of these special cases being used as impediments to free trading to protect their domestic market places. This is highly regrettable and counter-productive, and it includes major countries which claim to be the exponents of free trade, as well as major developing nations. In the author's opinion this is counter-productive, and in view of the life-saving and pollution-reduction potential of ITS, is immoral; particularly as these countries are also major polluters and have major problems with death and injury on their roads.

- Countries should not use national ITS standards as a barrier to trade.

Alas, good cooperation and coworking between international SDOs is not universal. For example, at an expert and working group level (in the context of ITS) cooperation is good between ISO and IEEE, but IEEE has a very restricted view at its institutional level which restricts common specifications in the standards of both organizations and use of common texts (unless it is done solely by reference to IEEE standards). These efforts to try to be seen to be more important in the international pecking order of SDOs are petty and counter-productive. Another example is SAE. SAE claims to be an international organization, but it makes no attempts at international collaboration with other SDOs when they write their allegedly "international" standards. This again is counter-productive as it makes their standards seem to be U.S. standards, and not international standards. The days when U.S. vehicle manufacturers were the dominant force in the automotive world is long gone. SAE did make some efforts for some years by its Secretariat of ISO/TC 204, but even then could not introduce coworking or exchange of deliverables at draft stage of its own standards with the ISO community. Indeed, in some cases, seeing the work being undertaken to obtain international consensus in the international community, it has raced to get its own standards in the same area approved first.

- SDOs that wish to be regarded as "international" should encourage coworking, joint development of standards, and common use of text and specification in deliverables.

So in conclusion, how are we doing? Do we know where we are going? Are we getting to the destination?

We set out on a mission some 10 years ago—others of us more than 20 years ago—to discover a new world for transportation, and in particular road transportation. We had some evangelists then, some of whom have died, some have retired, and some are still with us. And we have some new young blood, especially from Asian countries, who have more recently arrived on the scene with fresh new ideas and insights. Like all exploration, it has taken us into new and far more complex environments than we would ever have imagined. It has turned out to be much more complex than we expected, but the potential benefits have also increased. We are a long way off from being "on time and on budget" to those original expectations, and we are in a different place than where we expected to be, but we have achieved much, and it is now clear that we will achieve far more than those original expectations.

In standardization efforts we spent too much time learning to work together. We spent too much time trying to work de facto and proprietary IP-based solutions into de jure standards. Now we have begun to learn that pooling expertise before there are such competitive positions will in general help everyone reach their targets more quickly and grow the markets more quickly.

We have focused, in my opinion quite correctly, much of the emphasis on communication, identification, and location standards. These are essential building blocks for ITS system provision. There has also been good focus on fee collection and public transport, both "close to market" areas where the benefits of ITS standardization have been quickly realized.

We have grasped the political and technical nettle that there will be no one size fits all grand solution for ITS communications, but we have moved, through projects like the CALM initiative, to enable media provision to be separated from service provision, and to work on whatever media are available in the vicinity and supported by the vehicle. This can now be compared to the generic Internet architecture where the service provider has no idea what Internet linkage his customer is using, nor does it care—unless the user is attempting to download large videos—but we did not realize this when we set out to achieve multimedia support for ITS services. We set off simply in the knowledge that we would never get all parties—governments and road operators from around the world, radio regulators, automobile manufacturers, and equipment vendors—to agree on one communications solution, and the knowledge that it probably was not desirable in any case;

so how could we cope with this situation and come up with something workable? The fact that our architectures are now common to and normally based on an enhanced IPv6 backbone helps us to be confident of success and minimize risk.

But the limitations on our ability to use the radio spectrum mean that in some life threatening situations we need a faster and simpler means of communication as an integral part of the ITS communications architecture, and work on this is now well advanced.

- ITS must support multiple communications media, networked to support ITS service provision, where the service provider is separated from the medium supporting the service provision, and that will provide workable solutions today and a migration path as technology evolves.
- Fast communications solutions for time-critical situations must coexist with more elegant IPv6 based solutions.

There have been many frustrating years where huge amounts of resource seemed to produce only a trickle of completed standards. Now much of the labors of past years are coming to fruition, and, at least in the case of CEN and ISO, it is gratifying to see the work of whole work group areas coming to successful conclusion. The speed of completed standards becoming available has accelerated.

- After many years of effort, results are now coming through quite quickly.

So the mid-term report, in my humble opinion, and to steal the words of my science teacher in my school report of so many years ago, is as follows:

> [ITS has made] a promising start, but sometimes lacks focus. As a result progress could have been better. He must learn to work better with his classmates to achieve successful classwork, rather than insisting on going his own way and conducting his own experiments. If he focuses on the curriculum ahead, applies himself better, and continues to work hard, and together with his classmates, he could do well.

Summary

The summary of my observations and conclusions is therefore as follows:

- "The journey of a thousand miles begins with one step."
- In ITS we have embarked on a journey, which we now know will be long but full of promise.
- The standardization process in ITS is leading the evolution of the technologies to provide ITS service provision, not trailing them.
- ITS is not an isolated universe.
- ITS standards development should start with thorough investigation of what generic IT/ICT standards might be useful to provide the desired service, before inventing a totally custom solution for ITS.
- ITS is not an isolated universe; it is an IT/ICT application domain in the area of transportation.

- ITS is in a unique position to save lives, reduce injuries, and reduce pollution and emissions; standards play an important role in achieving these objectives.
- ITS is needed, but will require robust communications in the most difficult communications environments; these ambitious goals will only be achieved with the support of robust standards.
- Robust security and encryption standards are urgently required for safe and secure ITS service provision.
- Web service standards are needed for ITS in order to increase security and robustness, as well as to increase the speed and reduce the risk of ITS system development.
- The work is not complete; many more standards will be required than are currently being developed, especially at the application layers.
- Countries should not use national ITS standards as a barrier to trade.
- SDOs that wish to be regarded as "international" should encourage coworking, joint development of standards, and common use of text and specification in deliverables.
- ITS must support multiple communications media, networked to support ITS service provision, where the service provider is separated from the medium supporting the service provision, and that will provide workable solutions today and a migration path as technology evolves.
- Fast communications solutions for time-critical situations must coexist with more elegant IPv6 based solutions.
- After many years of effort, results are now coming through quite quickly.
- ITS has made "A promising start, but sometimes lacks focus. As a result progress could have been better. He must learn to work better with his classmates to achieve successful classwork, rather than insisting on going his own way and conducting his own experiments. If he focuses on the curriculum ahead, applies himself better, and continues to work hard, and together with his classmates, he could do well."

Finally, this book represents the status of ITS standardization as it stands at the time of this writing. We will need to revisit progress in a year or to so monitor progress and see what standards deliverables are then available.

About the Author

Although a qualified accountant (FCMA) by profession, Bob Williams found himself a director of a successful and rapidly growing automatic identification company in the early 1980s when the technology was in its infancy and has subsequently been a consultant in the automatic data capture sector for more than 25 years. He is well known and respected in the automatic data capture and Intelligent Transport System (ITS) sectors and has been involved in ITS Standards developments since 1991. He has been the editor of more than 20 International Standards, and has been involved in the organization and management of ITS Standards committees throughout this period. Mr. Williams has helped to make standards organizations and committees work together and work more efficiently to provide useful and timely standards to the sector. In 2003 he was awarded an honorary doctorate for his work "The Future of Automatic Product Identification."

Mr. Williams is currently the chairman of ITU-T ASPC TELEMOV, a cross standards-organization group dedicated to assisting standards organizations to better work together and providing a cross-organization reference point for users. He is the chairman of the newly formed ETSI committee TC-ITS and its older sister radio matters committee ETSI ERM TG37. He has been the head of the U.K. delegation to ISO TC204 since 1993 and is the convenor of its Working Group 1, System Architecture, and convenor of its Working SubGroup 16.1 CALM Media. He is also the convenor of CEN TC278 Working Group 13, System Architecture (since 1992), and the convenor of CEN TC278 Working Group 15, eSafety.

Mr. Williams is the author of *The Future of Automatic Product Identification* (PIRA, 2002), *Understanding Barcoding* (PIRA, 2003), and *Automatic Identification* (PIRA, 2004). This is his fourth technical work.

www.ingramcontent.com/pod-product-compliance
Lightning Source LLC
LaVergne TN
LVHW062241200926
848484LV00006B/1589